Advanced genetic analysis

Advanced genetic analysis

Genes, genomes, and networks in eukaryotes

Philip Meneely

Haverford College

With contributions from Matthew R. Willmann,
University of Pennsylvania

OXFORD
UNIVERSITY PRESS

Great Clarendon Street, Oxford OX2 6DP

Oxford University Press is a department of the University of Oxford.
It furthers the University's objective of excellence in research, scholarship,
and education by publishing worldwide in

Oxford New York

Auckland Cape Town Dar es Salaam Hong Kong Karachi
Kuala Lumpur Madrid Melbourne Mexico City Nairobi
New Delhi Shanghai Taipei Toronto

With offices in

Argentina Austria Brazil Chile Czech Republic France Greece
Guatemala Hungary Italy Japan Poland Portugal Singapore
South Korea Switzerland Thailand Turkey Ukraine Vietnam

Oxford is a registered trade mark of Oxford University Press
in the UK and in certain other countries

Published in the United States
by Oxford University Press Inc., New York

First published 2009

British Library Cataloguing in Publication Data
Data available

Library of Congress Cataloging in Publication Data
Data available

Typeset by Graphicraft Limited, Hong Kong
Printed in Italy by Legoprint S.p.A

ISBN 978-0-19-921982-7

1 3 5 7 9 10 8 6 4 2

Dedication

I have been fortunate to have excellent teachers and mentors.

Thanks, Bob.

Thanks, Bill.

BRIEF CONTENTS

DETAILED CONTENTS

Unit 3 Gene activity

PREFACE

The goal of this book is to illustrate how classical genetics and molecular biology provide important insights for understanding genomics and bioinformatics. In an era when every day brings another sequenced genome, it is reasonable to ask how the wealth of genomic information can be captured in a book. My approach is to acknowledge that the information itself might rapidly become outdated. However, this is also an age when we carry hand-held devices that allow us to search the World Wide Web while waiting in the grocery line. Up-to-date information is readily available and we don't rely on a book to provide it. So if you are planning to use this book as a reference to look up the most recent information on a gene or a biological process, my advice is to search the Web instead. It will be easier to browse for information and the data could be more current.

On the other hand, if your interest is in thinking about what the genomic data tells us, this book might be able to help. The emphasis in this book is on the experimental strategies and intellectual foundations that allow us to interpret genetic information and to plan approaches for the next experimental questions. I view genomics and bioinformatics through the lenses of classical genetics and molecular biology. I have consciously tried to integrate what we are learning from studying genomics with what has also been learned from genetics and molecular biology while studying individual genes. Each of these approaches has a starting point that acts as the key experimental tool in probing new questions: classical genetics has mutant alleles, molecular biology depends on cloned genes, and genomics relies on genome sequence information. But these tools aren't mutually exclusive in terms of the questions they can answer: the same biological questions can be probed from different experimental starting points. In fact, when taken together, classical genetics, molecular biology, and genomics all help to provide a more *complete* answer than any of them would provide alone. It is not accurate to think of these as separate approaches using different tools but rather as different parts of the same approach, namely **genetic analysis**. Thus, this book is not so much about *what* we have learned—that will continue to change—but rather about *how* we have learned it. In other words, the book is about analysis, strategy, and experimental logic.

Course suitability

I have taught a course on advanced genetic analysis for more than a decade to undergraduate students (juniors and seniors) at Haverford College. These students have taken introductory genetics and have a general familiarity with biochemistry and with molecular and cellular biology; the book is written with that level of student preparation in mind. The topics included in this course change every year, but the approach of describing a few examples in depth has been consistent; this approach has been followed in writing this book. The book includes many more topics than could be included in my course, and yet I am keenly aware of how many additional topics could be included. I have had to be selective in the topics, the organisms, and even the experimental approaches that I have presented.

I limited the book to genetic analysis in eukaryotes rather than attempting to include bacterial genetics, although many of the experimental strategies used in eukaryotic organisms are derived from similar approaches first developed in bacteria. I focus on five model eukaryotic organisms—the yeast *Saccharomyces cerevisiae*, the nematode worm *Caeneorhabditis elegans*, the fruit fly *Drosophila melanogaster*, the flowering plant *Arabidopsis thaliana*, and the mouse *Mus musculus*. No one is likely to question the inclusion of these five organisms. However, some may ask why another model organism has been excluded. I apologize to those whose preferred model organism has been omitted.

For clarity, I have relied on relatively few examples from these organisms and discussed them repeatedly throughout the book. For instance, the *patched* gene in *Drosophila* (and other animals) and genes affecting dauer larva formation in *C. elegans* are discussed in four or more different chapters. Sex determination in flies and worms and hormone signaling in *Arabidopsis* are also discussed in more than one chapter. These particular examples illustrate well the powerful approaches arising from contemporary genetic analysis.

But this is not to suggest that these examples are the most important ones to know, or even necessarily the best ones to illustrate a particular aspect of genetic analysis. Dozens, even hundreds, of other genes and biological processes could have been used as effectively. I hope and expect that those who are teaching with this book will have their own examples. In fact, I welcome hearing about other genes and biological processes that could have been covered instead. Perhaps it would be helpful to describe how these examples are employed in my own class. Although I spend many class periods on *patched*, dauer larva formation, and some of the other examples included in the book, none of my examination questions and none of the student papers requires any specific information about *patched* itself or dauer larva formation. Instead, the students are asked to apply the experimental strategies from these examples to the analysis of other genes and biological questions. I would be pleased to learn that a reader has come away with an appreciation of these biological examples because they illustrate important principles of genetic analysis. But I would be even more pleased to learn that the examples described here served as a springboard for analyzing other genes and biological questions.

Organization of the book

The book is structured as four units. The first unit, *Genes and genomes*, reviews the fundamental findings of genetics and molecular biology from an historical perspective in Chapter 1 and introduces the five model organisms and their genome databases in Chapter 2. No biological question in these model organisms can be investigated or understood without relying on the genome databases, and all readers are encouraged to explore these remarkable resources. These two chapters lay out the principles of genetics and molecular biology that are the foundation for the remaining chapters.

The second unit, *Genes and mutants*, explains the crucial tools of mutant alleles, cloned genes, and genome sequences. Chapter 3 discusses the principles of genetic screens and selections to find mutants that affect a biological process, while Chapter 4 describes how these mutants are mapped, classified into genes, and characterized. Chapter 5 presents an overview of how genes are cloned, i.e. how the mutant phenotypes from Chapters 3 and 4 are connected to a DNA sequence. Chapter 6 reverses the process of Chapters 3, 4, and 5, and describes how a mutant phenotype is found for genes that have been previously cloned. Chapter 7 introduces the concept of genome-wide mutant screens, i.e. the chapter describes how deletion mutations have been created for every gene in yeast and how RNAi has been used to find mutant phenotypes for every gene in other organisms. The important comparison between genetic screens for mutations in Chapter 3 and genome-wide mutant screens in Chapter 7 is made.

The third unit, *Gene activity*, assumes that mutant alleles, cloned genes, and genome sequences are available for further genetic analysis. Chapter 8 discusses molecular approaches to the analysis of gene expression. This includes both the analysis of the expression pattern of individual genes and the analysis of entire genomes using microarrays and expression profiles. Lessons learned from expression studies with individual genes are applied to examining microarray data. Chapter 9 returns to the theme of using mutant alleles to examine gene activity and discusses approaches such as temperature-sensitive periods and mosaic analysis. These are compared with what is learned from molecular analysis of gene expression.

The fourth unit, *Gene interactions*, discusses the identification and analysis of the many genes typically involved in a single biological process. Chapter 10 describes how one gene is used to find other genes with related functions. Both genetic interactions involving suppressor and synthetic enhancer mutations and physical interactions between proteins are included. Chapter 11 provides a detailed description of the use of epistasis and other genetic interactions to create a logical pathway of interacting components. Chapter 12 moves from the linear genetic interaction pathways derived from comparing two genes to the more complex networks and systems of interactions that occur within an organism. This chapter provides an overview of the field known as systems biology or network biology and examines it in light of the more traditional genetic analysis discussed in the preceding chapters.

Pedagogical features

Genetic analysis can be considered from a number of different angles including the experimental techniques adopted, the information yielded from different model organisms, or the biological principles these studies illuminate. This multilayered quality is mirrored by the pedagogical structure of the book.

The first layer is the main **narrative text**, which leads the reader through the field in what I hope will be a progressive and engaging way.

The second layer is the inclusion of **case studies**. Almost every chapter includes a case study (and some include more than one). The case studies describe in more detail how the genetic approaches discussed in the chapter were employed to study a particular gene or biological question. Many of the case studies are connected to one another or to another chapter. For example, the *patched* gene was found among the segment polarity mutants affecting *Drosophila* embryogenesis, as described in the case studies for Chapters 3 and 4; a description of how the *patched* gene was cloned from *Drosophila* and mammals is found in the case study for Chapter 5 and a gene disruption for *patched* in mice is described in the case study in Chapter 6. The case study in Chapter 8 describes how molecular analysis of gene expression was done in *Drosophila* embryogenesis (although not the *patched* gene itself), and a case study in Chapter 9 focuses on a mosaic analysis of *patched* in the *Drosophila* wing. Finally, genetic and physical interactions among *patched*, *hedgehog*, and *smoothened* are described in Chapter 11. Although the case studies are designed to be read in connection with the chapter itself, they can also be read as separate miniature histories of the genetic analysis of a gene or biological process.

The third and final layer of presentation comes in the form of **text boxes**. Almost all the chapters include text boxes that expand or speculate on a topic in the main text. Some of these text boxes arose directly from questions that students posed in class, and a few lay the groundwork for separate lectures in my advanced genetics or genomics courses. Although the text boxes are not crucial for understanding the main threads of the chapter, my opinion is that the topics in the text boxes are often as interesting as the topics in the main chapter.

Other learning features

The book includes a number of other learning features, to make it as reader-friendly as possible.

Frequent cross-references provide links to other sections and chapters.

Key literature articles are highlighted immediately after the section of text to which they relate.

Potentially unfamiliar terms are highlighted and defined in the margin of the page on which they appear.

A glossary is included at the end of the book, listing key terms used in the study of genetic analysis that may not be familiar from earlier courses.

One of the main goals of my teaching is to rouse students' imaginations to think creatively about new biological questions, including ones that I have not thought about. A similar goal lies behind this book. It is important to know the right answers, but it is even more important to be able to ask the right questions and know why an answer is right. I hope that this book will provide some stimulus to ask new questions. I believe that I sometimes learn as much from my students as they learn from me (which may be a commentary on my teaching ability). With that same philosophy of learning from each other, I encourage you to contact me with additional examples, ideas for new topics, corrections and updates, and other comments.

Philip Meneely
April 2008

Online resource centre

www.oxfordtextbooks.co.uk/orc/meneely

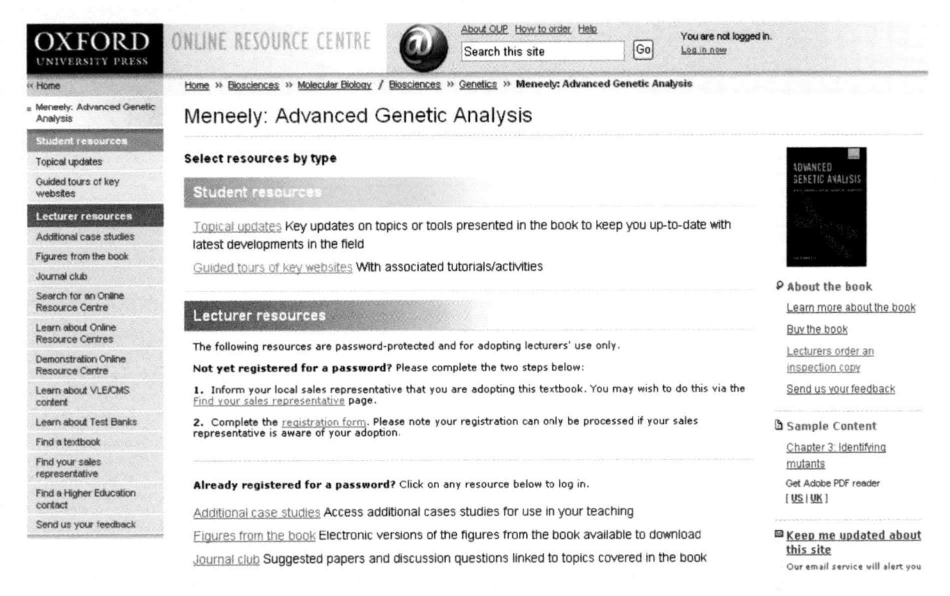

This book has been written with web usage in mind, so the reader is urged to consult the website associated with the book (www.oxfordtextbooks.co.uk/orc/meneely) for updates and additional resources as they are developed. Key resources on this site include the following

For registered adopters of the book:

- Figures from the book in electronic format, ready to download.
- Journal club: suggested papers and discussion questions linked to topics covered in the book.

Additional case studies for use in teaching.

For students:

- Guided tours of key websites, with associated tutorials/activities.
- Topical updates: key updates on topics or tools presented in the book to keep you up to date with the latest developments in the field.

Bookmark the site, or register for updates by visiting **www.oxfordtextbooks.co.uk/orc/meneely** and clicking on the 'Keep me updated about this site' link.

ACKNOWLEDGMENTS

I would like to recognize and express my appreciation for the editorial work of Jonathan Crowe at Oxford University Press, who saw this book from pre-conception to birth. I never intended to write a textbook, but having done so, I thank Jonathan for his encouragement, patience, and insights. He also played a significant role in the cover design, as we attempted to capture both classical genetics and genomics in one image. The editorial team at OUP patiently guided a novice writer through the long process of producing a textbook, and I thank them.

I also thank Eric Alani, Greg Prelich, and Paula Cohen for ideas that got the book started and comments and suggestions on early drafts of what became Chapters 10 and 11. I am grateful for their insights and assistance. My colleague Matthew Willmann made significant contributions that improved every chapter, in addition to teaching me some plant genetics. I hope that the lessons don't stop with the publication of the book.

The biology students at Haverford College have contributed immensely to this book. Their comments on my class greatly shaped the approach that I took in this book. Many of them in the last few years patiently read fragments of chapters, scratched their heads over poorly drawn figures, and helped me think carefully about better ways to describe a subject. The organization of Chapter 5 is entirely the result of a student question that I could not answer cogently when it was asked, and other student questions and comments appear in many forms throughout the book. I am fortunate to be able to teach such inquisitive and enthusiastic students each year.

Foremost among those to thank are the members of my family for their patience and support, not only for this book but for my love of genetics. My wife Deb accepted that most evenings and weekends involved me working at the computer in my office. My grown children, Alison and Andy, contributed ideas and comments on figures, editorial suggestions on pedagogical features, help with computer files and software, and general encouragement throughout. One particular comment inspired me from the beginning—'Come on, Dad. Yours can't be that much worse than some of the textbooks I had to buy.' Few children express such confidence in the abilities of their parents, and I trust that it has been earned.

FIGURE ACKNOWLEDGMENTS

Figure 2.1 (A) Fly: istock.com; Mouse: Photodisc; Worm: Nancy Nehring/istock.com (B) Cricket: Photodisc; Fly: istock.com; Butterfly: Photodisc.

Figure 2.2 Nancy Nehring/istock.com.

Figure 2.15 Copyright Cold Spring Harbor Press and used by permission. Also courtesy of Zeynep Altun of Wormatlas.

Figure 2.16 Used by permission of Cold Spring Harbor Press and Zeynep Altun, courtesy of Wormatlas.

Figure 2.22 Courtesy of Min Han, University of Colorado.

Figure 2.27 Courtesy of Peter Carlson and John Sedat, University of California, San Francisco.

Figure 2.28 Courtesy of Matthew Willmann, University of Pennsylvania.

Figure 2.42 Copyright Nature Publishing Group and used by permission; and Photodisc.

Case Study Figure C2.1 Copyright Nature Publishing Group and used by permission; and Nancy Nehring/istock.com.

Case Study Figure C2.2 Copyright Nature Publishing Group and used by permission.

Case Study Figure C2.3 Nancy Nehring/istock.com.

Figure 3.7 Courtesy of Maize Genetics and Genomics Database.com.

Case Study Figure C6.2 Copyright AAAS and used by permission.

Figure 7.11 White: R. Daniaud/istock.com and Purple: R. Sanchez/istock.com, © Stephen Vickers, Fotolia.com.

Figure 8.4 Courtesy of Andrea Morris, Haverford College.

Case Study Figure C8.2 Copyright Nature Publishing Group and used by permission.

Figure 8.6 (A) Copyright AAAS and used by permission. (B) Copyright AAAS and used by permission. (C) Courtesy of Harald Hutter, Simon Fraser University. (D) Copyright Elsevier Limited for Cell Press, and used by permission.

Figure 8.7 Courtesy of Charles Braben and Alison Woolard, Oxford University.

Figure 8.12 Copyright Elsevier Limited for Cell Press, and used by permission.

Figure 8.14 Copyright Nature Publishing Group and used by permission.

Figure 8.16 Copyright Elsevier Limited for Cell Press, and used by permission.

Figure 8.19 Copyright Nature Publishing Group and used by permission.

Case Study Figure 9.1 Used by permission of Cold Spring Harbor Press and courtesy of Zeynep Altun from Wormatlas.

Case Study Figure 9.2 Used by permission of Cold Spring Harbor Press and courtesy of Zeynep Altun from Wormatlas.

Figure 10.3 Copyright *Molecular Biology of the Cell* and used by permission

Figure 10.12 Courtesy of Min Han, University of Colorado.

Figure 10.15 Copyright Elsevier Limited for Cell Press and used by permission.

Case Study Figure 10.3 Copyright *Molecular Biology of the Cell* and used by permission.

Case Study Figure 10.5 Copyright *Molecular Biology of the Cell* and used by permission.

Case Study Figure 10.7 Copyright *Molecular Biology of the Cell* and used by permission.

Figure 11.9 Used by permission of Cold Spring Harbor Press and courtesy of Zeynep Altun from Wormatlas.

Figure 12.1 Copyright Elsevier Limited for Cell Press and used by permission.

Figure 12.19 Copyright Nature Publishing Group and used by permission.

Figure 12.20 Copyright Nature Publishing Group and used by permission.

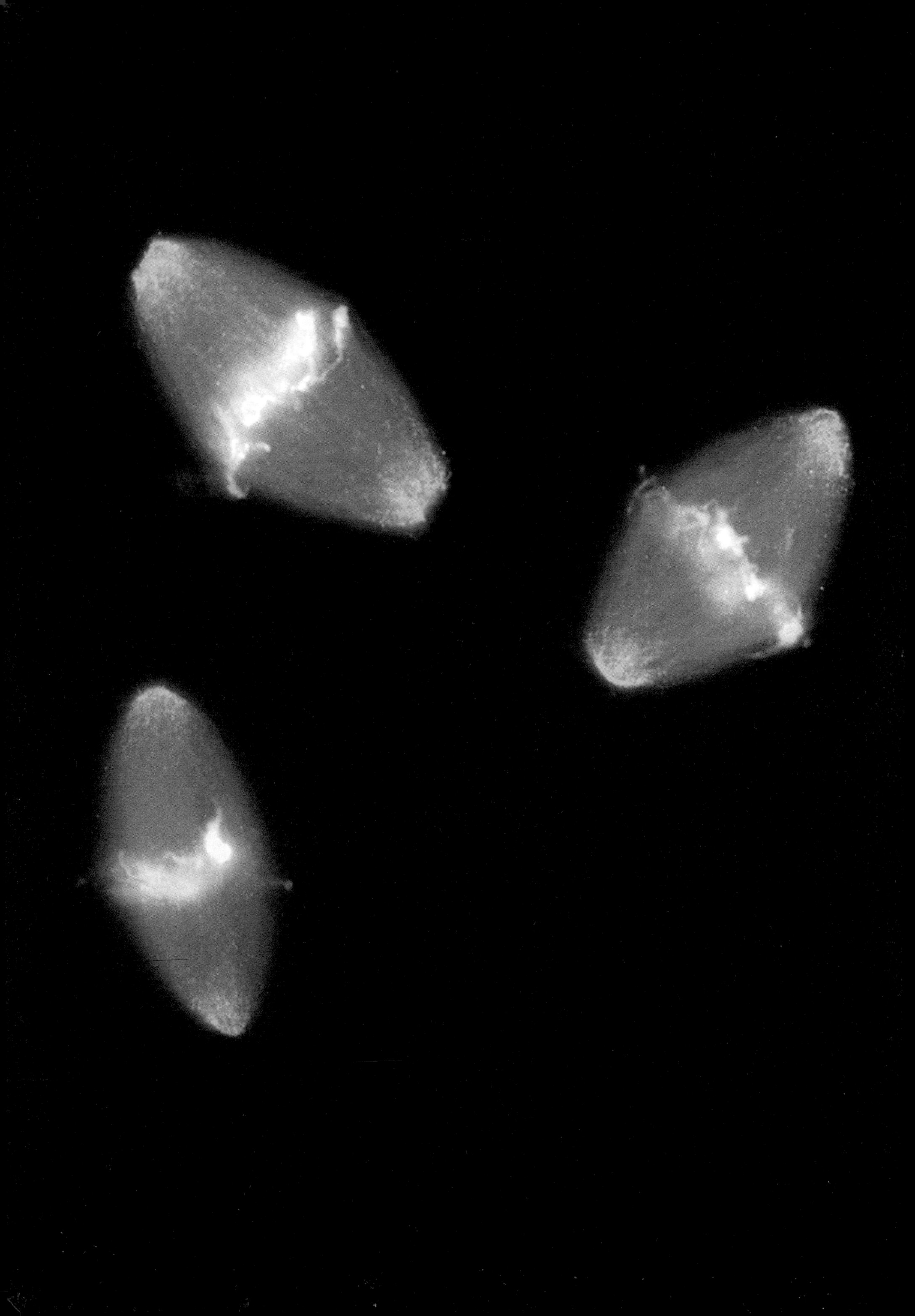

Unit 1
Genes and genomes

We review the properties of genes as learned from Mendelian and molecular genetics and put these in the context of genome analysis. We then describe five model eukaryotic organisms with sequenced genomes that will form the foundation for the remainder of the book.

CHAPTER MAP

▼ Unit 1 Genes and genomes

1 The logic of genetic analysis

The outline of the book and the connections between the chapters are illustrated in this flowchart. Chapters that emphasize experimental approaches based primarily on mutant phenotypes are shown in red, whereas those that rely on primarily on cloned genes are shown in blue. Chapters that closely integrate mutant phenotypes and cloned genes are shown in purple.

2 Model organisms and their genomes

▶ Unit 2 Genes and mutants

▶ Unit 3 Gene activity

▶ Unit 4 Gene interactions

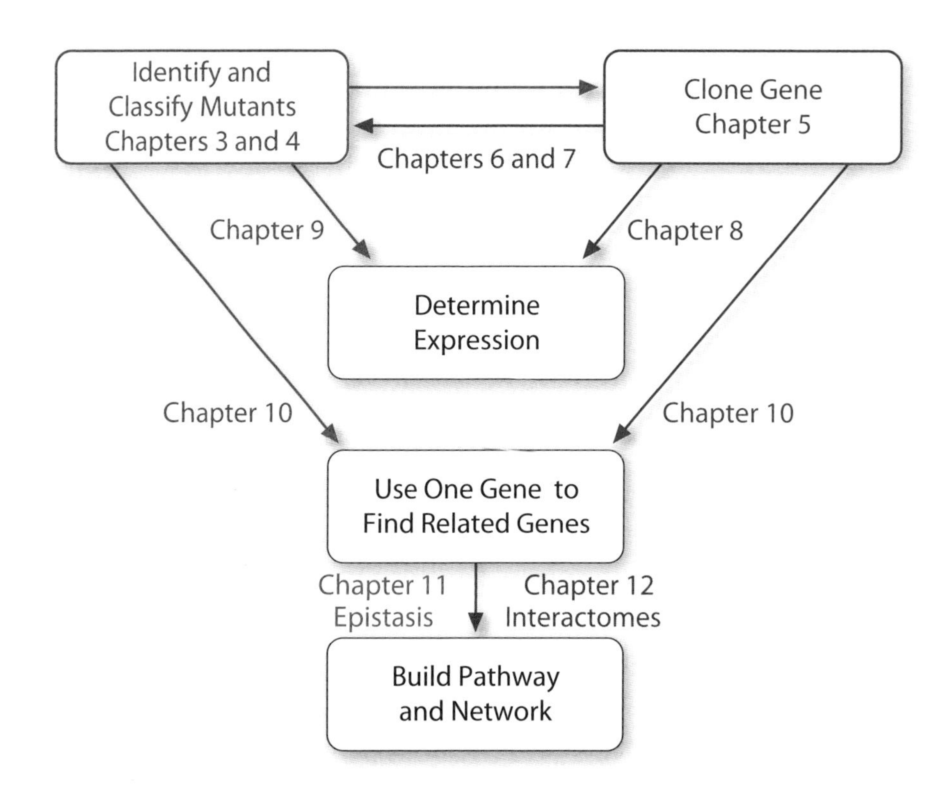

1

CHAPTER 1
The logic of genetic analysis

TOPIC SUMMARY

The fundamental findings from introductory genetics include the elucidation of properties of genes such as inheritance, allelism, linkage, and mutation. These properties are extended to gene activity through the activity of proteins and RNA. Ultimately, all of these properties are explained by the molecular properties of DNA complexed with proteins in the form of chromosomes. These key findings are summarized using a historical perspective. Genomic analysis is rooted in all of these earlier approaches and contributes many additional insights about the structure and inheritance of our genomes.

IN BRIEF

Genetics can be understood as a set of principles but it also provides a means to analyze complicated biological problems.

INTRODUCTION

Genetics has a long and rich history. Parents have always been interested in knowing how their children will be like them and different from them. Our earliest forebears domesticated plants and animals and selected for favorable traits among their offspring. References to hereditary traits are found in the ancient literature of nearly every culture. Modern geneticists, eager to take the next steps forward, stand at the leading edge of a long line of previous investigators, some known to us but most unknown.

Although genetics in the form of observations on inheritance is an old science, genetics as a means of experimental analysis really only began in the twentieth century. The outline of the story is well known. During the 1850s and 1860s in the town of Brno in what is now the Czech Republic, Gregor Mendel, an Augustinian monk with training in botany from the University of Vienna, grew vegetables in the abbey garden plot. Photographs of the abbey and his garden are shown in Figure 1.1. He engaged in a lengthy series of plant breeding experiments and kept detailed and meticulous records of his results. Mendel presented his results to the local natural history society and sent a paper of his results and interpretation to

Figure 1.1 The Abbey of St. Thomas in Brno. (A) The courtyard viewed from the street in front. Note the statue of Mendel to the left of center and the brick wall along the right side. (B) The statue of Gregor Mendel in the courtyard. (C) The garden plot where Mendel grew his peas. The plot is no more than 20 square meters in area. The abbey windows on the right overlooking the garden were the dining room. (D) The monks with Mendel at the abbey, with Mendel in the back row. (Photographs taken by the author in April 2003.)

the great botanist Naegeli who failed to realize their significance. Published in the proceedings of the local natural history society, Mendel's paper lay in obscurity for 35 years before being simultaneously and independently rediscovered and referenced in 1900 by Correns and de Vries who recognized its importance. Mendel himself went on to work on honeybees, meteorology, and the administration of the abbey, with no additional papers on peas or heredity. (One can imagine the relief of his brother monks who must have tired of eating all of those peas.)

In 1906, the British naturalist William Bateson first used the word 'genetics' to describe the then new science of heredity, and he became one of Mendel's earliest and most effective champions. Geneticists quickly emerged on both sides of the Atlantic Ocean. At first, the emphasis was on the laws of genetics themselves and showing that they applied to all organisms, including humans. Soon afterwards, genetics became a set of methods that could be used to analyze other experimental questions, including sex determination, color patterns, behavior, fecundity, human diseases, and many others.

A lengthy and often acrimonious dispute arose between those who regarded genetics in terms of discrete phenotypes and individual inheritance—as did Mendel and Bateson—and those who regarded genetics in terms of continuous phenotypes and inheritance in populations—led by Pearson and the biometricians. Many volumes have been written about the history of these two schools of thought and the seemingly larger-than-life men who were involved. One reason that the dispute between the Mendelians and the biometricians has been so thoroughly analyzed by historians is because of the talents and genius of the men on each side. Another reason is that genetics lies at the center of every other field of biology. The history of the ideas in genetics illuminates every other area of biology.

One aspect of the dispute that should be appreciated is that the schools were debating genetics as a **method of analysis**—in other words, genetics as a tool to explore other biological questions. The fundamental rules by which genes were inherited were laid out by Mendel. The analytical challenge was to explain the inheritance and behavior of **phenotypes** in terms of the inheritance and behavior of **genes**. The biometricians created much of the field of statistics in order to explain their experimental results. The Mendelians relied on mutations and chromosomes to explain theirs. With the benefit of hindsight, greatly helped along by the insights of other great geneticists and evolutionary biologists, we now can see how the explanations of both these schools were correct. We can also see how the explanations of each school were incomplete.

In this respect, nothing has changed when we use genetic analysis in the twenty-first century. Those of us who rely on genetics as an analytical approach to explain some complicated phenotype believe that we are right in interpretation—for now. We expect to see some of our favorite ideas elaborated or even overturned by new findings; in fact, we look forward to it because it means that we have encountered something novel and unexpected. We stand now with an avalanche of genomic data sweeping over us, and we look for the flakes of insight that will carry us forward. Like those who worked before us, we can be sure that we are both correct and incomplete in our understanding of biology.

1.1 The logic of genetic analysis: a historical overview

In order to discuss genetic analysis, it may help to define one of its most basic terms: ***What is a gene?***

The question is deceptively simple. Even those of us who have studied genes for most of our adult lives would struggle to present a full and comprehensive definition of what it is we study. We usually fall back on a definition that includes what a gene does rather than attempting to define its essential nature. When we describe a gene, the description may include a mutant phenotype, an expression pattern, a DNA sequence, an inheritance pattern, and more. The definition that is being used often depends on the property of the gene that is being used for the investigation. Even more intriguing is that the definition being offered now will probably have to be modified as more genes are discovered and analyzed.

So what is a gene? Let us approach this question from the perspectives of those who have defined genes so far.

As we take this historical tour of the fundamental principles of genetics, we will consider a highly selective summary of the findings and interpretations. We assume that much of this material will be familiar but that our perspective could be new. Thus, we will emphasize the key findings without much additional explanation or background.

Genes are the units of inheritance

or Mendel was right—up to a point

Mendel was right. Genes—which he did not name—are inherited as individual particles and an individual has two copies of each gene. Mendel recognized that these individual particle genes can have different forms, which we now call **alleles**. He recognized from his experiments that the genes affecting different phenotypes—such as tall vs. dwarf plants or round vs. wrinkled seeds—are inherited separately and independently of each other. In addition, in his experiments he found that different genes acted independently of each other in affecting the phenotype, i.e. the four phenotypic combinations of tall round, tall wrinkled, dwarf round, and dwarf wrinkled were found in predictable frequencies.

We can now recognize that the inheritance patterns of phenotypes that Mendel was observing are the result of the behavior of chromosomes in meiosis. This is shown diagrammatically in Figure 1.2. In a diploid eukaryote, there are two homologs of each chromosome—each homolog with its own allele—that separate from each other in the first meiotic division. This is known as Mendel's Rule of Segregation: the alleles *segregate* from each other in meiosis I (Figure 1.2A).

In addition, the assortment of each pair of homologous chromosomes is independent of the assortment of any other pair of homologous chromosomes, as shown in Figure 1.2B. In cytological terms, the orientation of one pair of chromosomes on the meiotic spindle is independent of the orientation of another pair of chromosomes. Thus genes on different chromosomes—as all of Mendel's genes appear to be—are inherited independently of each other. This is Mendel's Rule of Independent Assortment.

Of course, on these two points of independent inheritance and independent activity of different genes, Mendel was not entirely correct. We now know that many genes are linked together on each chromosome, and their inheritance is not independent of each other. Mendel did not encounter linkage for the genes he studied or, if he did encounter a pair of linked genes, he did not use them in reporting his results. In retrospect, working with unlinked genes was a fortunate circumstance. It is likely that the simplicity of Mendel's results was crucial in the rapid spread of genetics upon their rediscovery. More complicated results that included linkage might well have been more difficult to accept.

Likewise, phenotypes that arise from the interactions between genes were also not part of Mendel's results. In this, we know that his choice of phenotypes was fortuitous. Phenotypes arising from gene interactions lay at the core of the dispute between the Mendelians and the biometricians. If Mendel had not recognized that his phenotypes were two discrete alternatives for a gene, his results would certainly have been lost. Mendel did not work with complex phenotypes that arise from the action of multiple different genes and from the interaction between genotypes and the environment. Peas certainly have complex phenotypes—the number of pods

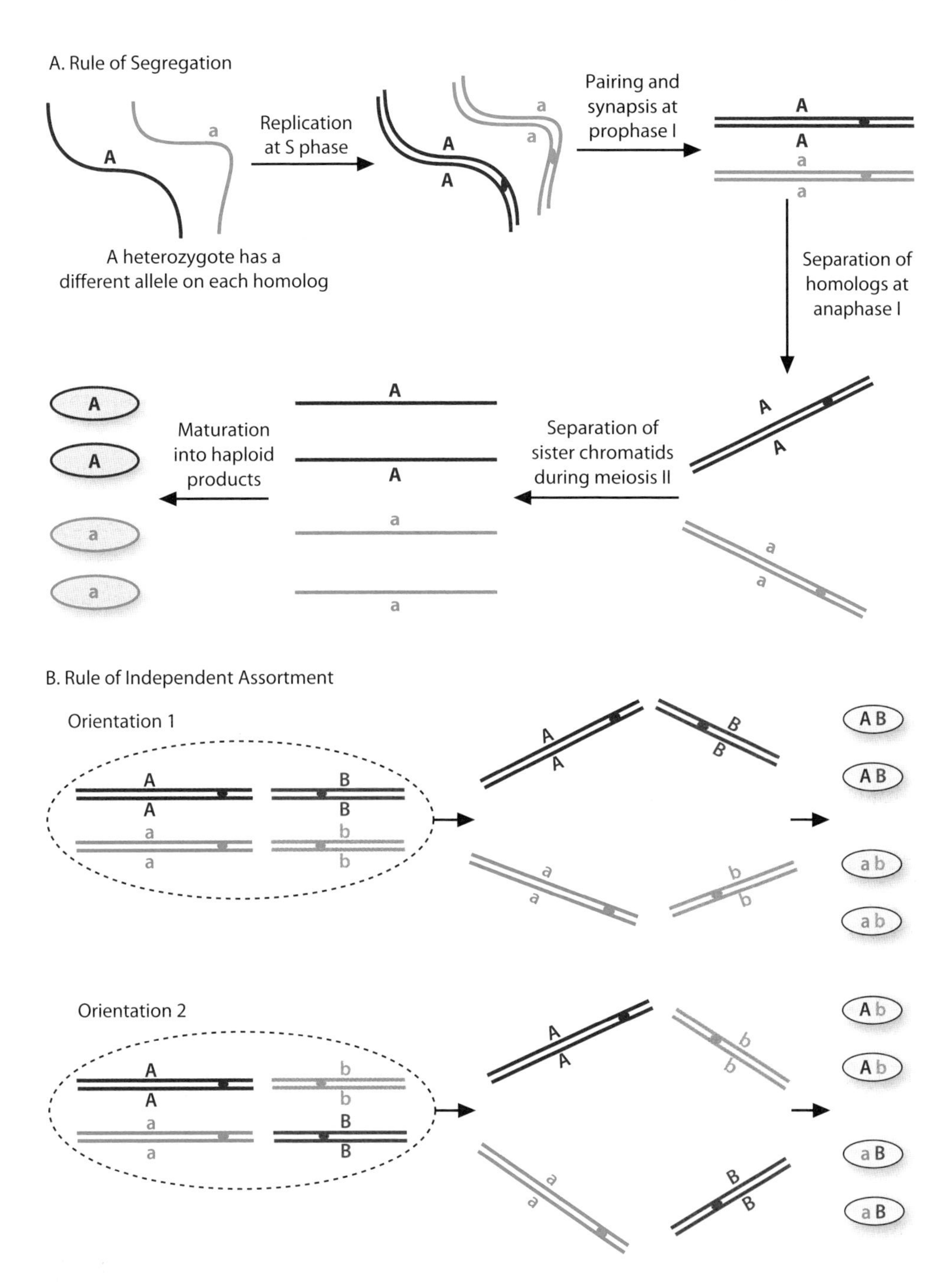

Figure 1.2 (A) Mendel's Rule of Segregation. Two homologous chromosomes are shown, with different alleles of a gene. With replication at the S phase, each chromosomes has two sister chromatids. The homologous chromosomes pair at prophase I of meiosis, and separate at anaphase I. During meiosis I, the sister chromatids cohere to each other but the homologous chromosomes separate. The sister chromatids separate during meiosis II, resulting in four haploid products, each with one sister chromatid. **(B) Mendel's Rule of Independent Assortment.** Two different pairs of non-homologous chromosomes are shown, one in shades of blue with two alleles of the *A* gene and the other in shades of red with two alleles of the *B* gene. When the homologs pair at meiosis I, either of the two orientations of the non-homologous chromosomes can occur with equal probability. With the chromosomes in orientation 1, the *A* and *B* alleles segregate together at meiosis I. With the chromosomes in orientation 2, the *A* and *b* alleles segregate together at meiosis I. Four types of haploid product (*AB*, *ab*, *Ab*, and *aB*) result with equal frequency.

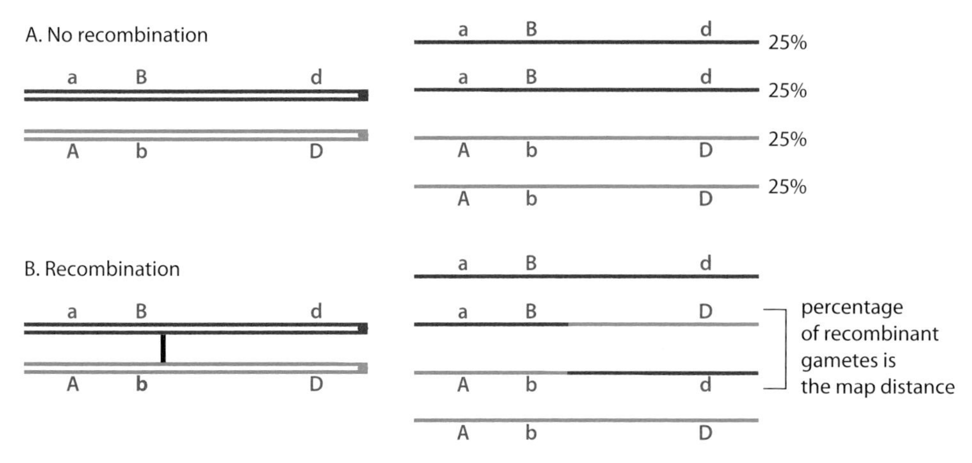

Figure 1.3 Linkage and map distances. A pair of homologous chromosomes from an individual that is heterozygous for the genes *A*, *B*, and *D* is shown. (A) If no recombination occurs between any pair of the genes, the alleles will be inherited together, with half of the haploid gametes having the genotype of each parent. (B) If recombination occurs in the interval between genes *B* and *D*, some of the gametes will have the parental arrangements *a B d* and *A b D* whereas others will have the recombinant arrangements *a B D* and *A b d*. The percentage of gametes that have the recombinant arrangement defines the genetic map distance between genes *B* and *D*. In normal meiosis, all homologous pairs of chromosomes have at least one recombination event; the genetic map is based on the probability that recombination occurred in a particular interval.

on a plant is an obvious one—and Mendel may well have seen these differences. If so, he did not attribute them to genetic differences. Mendel read Darwin—the abbey's copy of *The Origin of Species* with Mendel's handwritten notes in the margin is on display in the Mendel Museum in Brno—but the evolutionary synthesis explaining complex phenotypes in Mendelian terms lay many decades in the future.

Mendel was right. But areas that he left incomplete were the substance for another set of great geneticists.

Genes are found on chromosomes

or Morgan and the Fly Room were right—mostly

Thomas Hunt Morgan and his students Calvin Bridges, Alfred Sturtevant, H.J. Muller, and many others (including Morgan's wife Lillian, reputed to be among the most gifted observers) comprised the denizens of the different 'fly rooms' at Columbia University in New York and then at California Institute of Technology in Pasadena. The core group was a colorful cast of characters, each meriting an individual biography. This would have been an exciting time and place to be a biologist.

Morgan and the scientists in the Fly Room were right. As noted above, they recognized that Mendelian inheritance describes the behavior of chromosomes at meiosis. The gene particles whose inheritance Mendel described lie at fixed positions on chromosomes. By proving that genes lie on chromosomes, Morgan and his students were introduced to many other properties of genes, a brief list of which is given below. These are also summarized in Figures 1.3 and 1.4.

- Genes lie at positions on chromosomes, and genes on the same chromosome will tend to be inherited together (Figure 1.3A).
- Recombination or crossing over between the chromosomes of a homologous pair means that genes on the same chromosome are not *always* inherited together. In fact, the frequency with which genes are co-inherited, i.e. linked in their inheritance pattern, can be used to infer their relative position on the chromosome. A map of the position of the genes on a chromosome can be constructed using the frequency of crossing over between two genes (Figure 1.3B).
- Although a diploid individual only has two alleles for a gene, as Mendel observed, genes themselves can have more than two alleles; in fact, a gene can have a nearly limitless number of different alleles. The same gene can have mutant alleles that are dominant to wild-type (another concept from *Drosophila*) and others that are recessive to wild-type (Figure 1.4A). Most mutant alleles are recessive to wild-type and

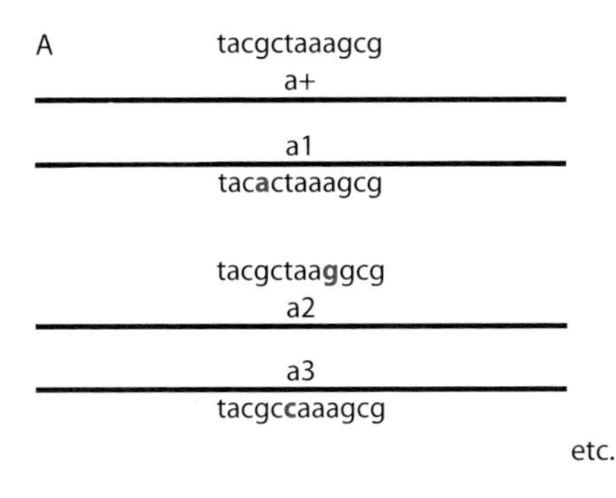

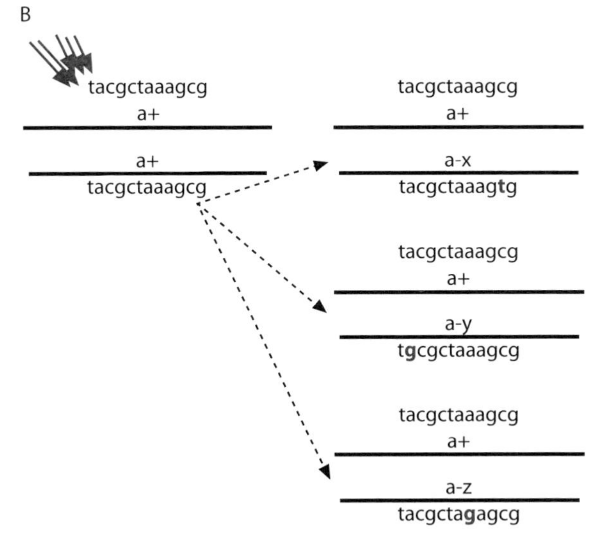

Figure 1.4 Multiple alleles for a gene. Although a diploid individual has only two alleles for each gene, many possible alleles can occur since any nucleotide pair in the DNA sequence can be changed to another pair, as shown in panel A, or can be deleted or duplicated. Once the mutation to a new allele has occurred, as shown in panel B, the new allele is itself stable.

arise from a loss or reduction in gene activity. The residual gene activity of mutant alleles can be inferred and different mutant alleles can be placed in a series, an idea that we explain more fully in Chapter 3.

- The alleles of a gene are not themselves fixed. Genes can be mutated to new and different allelic forms. Mutations occur spontaneously by natural processes but can also be induced by radiation and chemicals. These newly arisen mutations are inherited just like other alleles of the gene (Figure 1.4B).
- Contrary to what Mendel observed, the expression or phenotype of one gene can in fact affect the phenotype of another gene. We illustrate this in Figure 1.5. The genes themselves are inherited by predicted rules to give genotypes at the expected frequencies, but the phenotypes arising from those different genotypes could vary. Mutations in one gene could **suppress** those in another gene and make it more like wild-type, or they could **enhance** one another and make the phenotype more severely mutant, or one phenotype could mask the other in the process of epistasis. Discussion of these interactions between gene activities forms the core of Chapters 10 and 11.

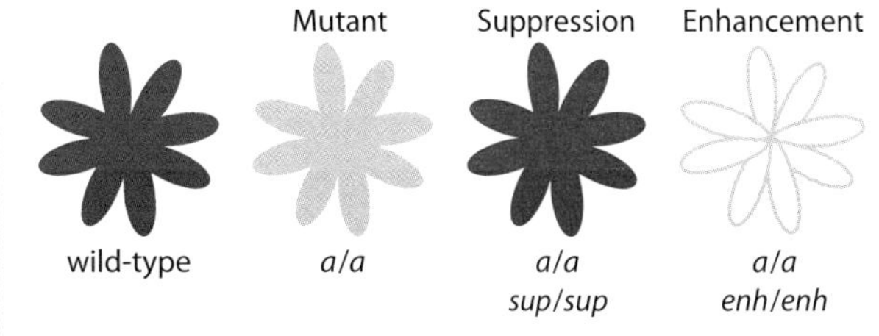

Figure 1.5 Phenotypes arising from gene interactions. A hypothetical flower is shown, whose wild-type color is purple. A mutation in the *a* gene results in a lighter purple color. An interacting mutation in a second gene can suppress the mutant phenotype, resulting in a color more similar to wild-type, or can enhance the mutant phenotype, resulting in a more mutant flower, in this case white.

This brief list only begins to describe what the early *Drosophila* geneticists learned. Many of the classical genetics concepts in the rest of this book, i.e. those aspects of genetic analysis that can be performed using mutant alleles, have their origins in the work on Morgan and his students in the Fly Room.

But their work was still incomplete. A gene can be mutated; it exists in different forms; it lies a defined position on a chromosome; and it interacts with the phenotype of another gene. But how does the gene carry out its activities? There was room for more geneticists, many more.

One gene–one protein

or Garrod, Beadle, Ephrussi, and Tatum were right—sort of

If we say that a gene gives rise to a phenotype—or more precisely, that a mutated version of the gene results in a mutant phenotype—we are describing the activity of the gene. Mendel worked with tall and dwarf plants, so the activity of the particular gene he was studying affected the height of the plant. But how exactly does it do this? What is the biochemical explanation for the phenotypes observed from different alleles?

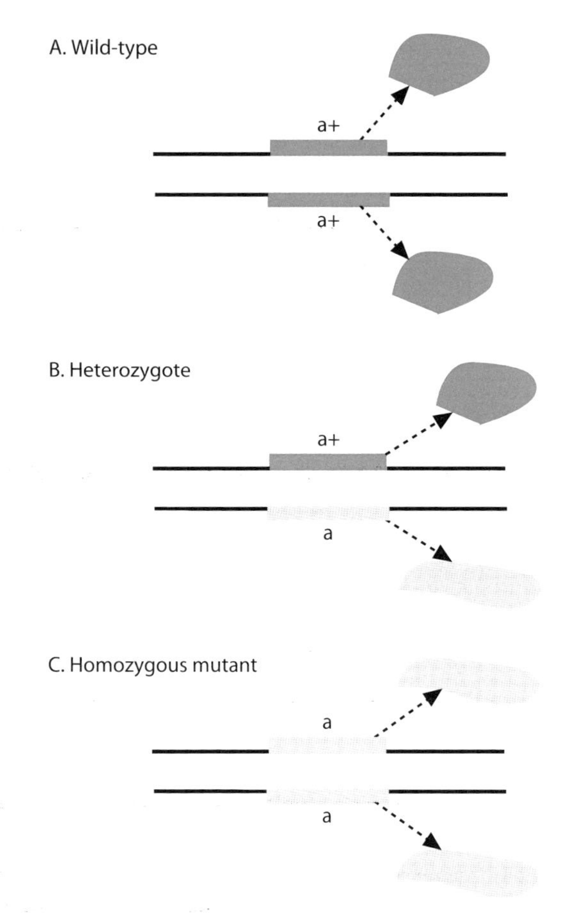

Figure 1.6 One gene–one protein. The *a+* allele of the gene makes a protein with normal function, and the mutant allele ***a*** produces an altered form of the protein. The heterozygote produces both the normal and the altered form.

Shortly after Mendel's work was rediscovered, the British physician Archibald Garrod described 'in-born errors of metabolism' in infants, including the biochemical bases for the rare genetic diseases phenylketonuria (PKU) and alkaptoneuria. Thirty years later, George Beadle and his coworkers Boris Ephrussi and Edward Tatum used pigmentation mutants in the *Drosophila* eye and metabolic mutants in the fungus *Neurospora crassa* to examine the biochemical basis for these metabolic defects. This work led to the conclusion that genes encode enzymes: one gene–one enzyme. This is illustrated in Figure 1.6.

Although much of the general conclusion is right, the 'one gene–one enzyme' hypothesis has been much modified by subsequent research. The first modification is that the term 'enzyme' should be replaced by 'protein': most genes encode the information to make a **protein**. The different alleles observed by classical genetics were different forms of the same protein, with changes that altered or eliminated its function by changing or truncating the amino acid sequence. This modification in 'one gene–one enzyme' was made very soon after the original findings, and has generally stood up well until the last few years. But now, even this generalization should be treated with caution, and our definition of a gene is challenged once again as the exceptions to this principle proliferate.

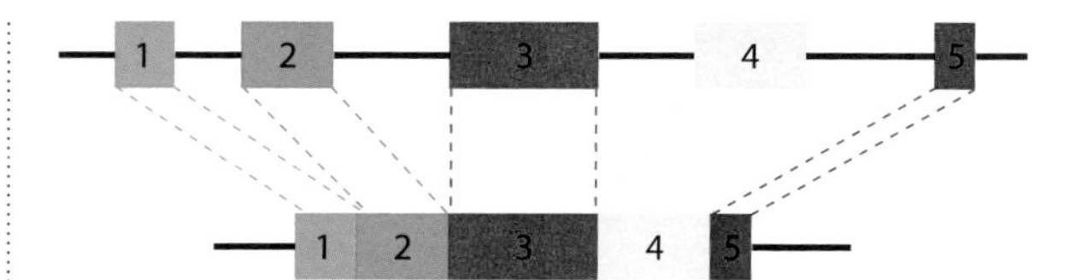

Figure 1.7 Colinearity of genes and proteins. Each of the five exons of a gene is shown, in different colors. The processing of the transcript to eliminate the introns and join the exons produces a protein that is colinear with the gene.

One very important exception began with the recognition of RNA splicing. When splicing itself was first described—that is, that the initial transcript from a gene in a multicellular organism is edited to the much smaller mRNA before translation (Figure 1.7)—many questions about the colinearity of genes and proteins were re-examined. The result was to re-affirm the colinear relationship between the DNA sequence and the amino acid sequence.

The coding sequence of the gene is interrupted by sequences that are transcribed and then removed from the transcript. These intervening sequences, the introns, could be much longer in sequence than the translated or expressed sequences, the exons. In fact, the number and length of introns has greatly increased with evolutionary complexity. In the yeast *Saccharomyces cerevisiae*, only about 5 percent of the genes have an intron, and those genes with introns typically have only one intron. In humans, genes without introns are a distinct minority, and most genes have many introns that are longer in sequence than the flanking exons. However, even with RNA splicing, the principle of 'one gene–one protein' still held. The exceptions to the rule had not yet been found.

Alternative splicing results in 'one gene–many proteins'

The mere presence of introns does not violate the 'one gene–one protein' principle but alternative patterns of

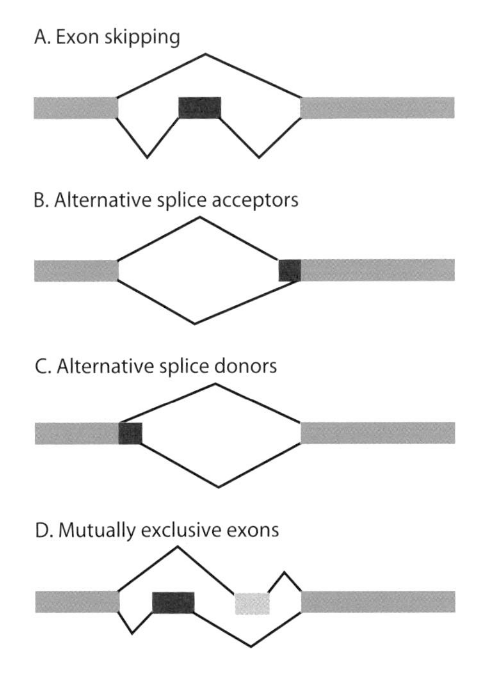

Figure 1.8 Alternative splicing of coding regions. Four different types of alternative splicing are shown, each of which changes the amino acid sequence of the proteins. The common exons are shown in green, and the alternative exons in purple or yellow. Splice patterns are shown by the lines connecting exons. In panel A, one transcript includes the additional exon and one transcript skips it. In panels B and C, alternative splice acceptors or splice donors produce proteins in which an exon is a different size. In panel D, there are two alternative exons and two common exons. One splicing pattern results in a transcript that includes the first (purple) alternative exon but not the second one. Another splicing pattern results in a transcript that includes only the second (yellow) alternative exon but not the first. Many other forms of alternative splicing that affect 5′ and 3′ untranslated regions are also known.

splicing the exons together definitely do make a violation. Some types of alternative splicing are shown in Figure 1.8. Alternative splicing of genes is the rule rather than the exception, at least for genes in mammals. While the exact percentage of genes that are alternatively spliced may never be accurately known, different splice variants have been found or postulated for as many as two-thirds of mammalian genes.

Alternative splicing could and often does occur in the non-translated parts of the transcript, so those splice variants may differ in stability or expression, but the protein products will have the same amino acid sequences, one gene–one protein, despite the splice variants. However, alternative splicing is also common for the protein-coding portions of the transcript. Exons may differ by a few codons at the upstream donor or the downstream acceptor sites, resulting in a few amino acid differences in the protein products at these locations. More profoundly, exons may be present in some splice variants and absent in others. This could result in a significant difference in the protein products, with some portions of the sequence in common between the proteins and others that are specific to one form or another. Moreover, some genes are spliced in many different ways, so that some genes can encode dozens or even hundreds of related but different proteins.

Therefore, the principle that one gene encodes the information for one protein has broken down. A more accurate but less memorable version would be 'one gene–one or more proteins (and possibly many more)'.

The difference in splice variants can also have physiological significance for fundamental biological processes. One well-studied example is sex determination in *Drosophila melanogaster* (Figure 1.9). Males and females are quite different and the differences are easily spotted—students who have never seen a fly before can learn to distinguish the sexes reliably in a few minutes. At the level of the fly, the differences are profound and pervasive—genitalia, germline, behavior, pigmentation, and so on. At the level of the gene, these biological differences arise from the choice of a splice acceptor site in the gene *doublesex* (*dsx*). The first three exons are common to males and females, but an alternative splice results

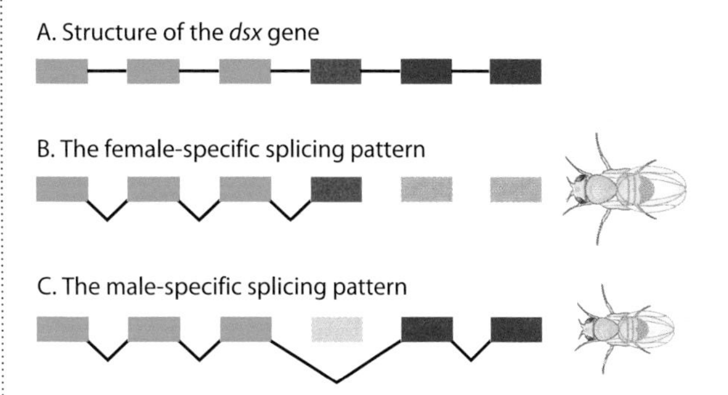

Figure 1.9 Alternative splicing and sex determination in *D.melanogaster*. The dsx gene encodes a transcription factor that regulates sexually dimorphic gene expression. The gene has six exons, as shown in panel A. The first three exons, shown in green, are not sex specific. In females, these three exons are spliced to a fourth exon, shown in red, to produce a female-specific transcript that lacks exons 5 and 6. In males, these three common exons are spliced to the fifth and sixth exons, shown in blue, skipping the fourth exon, to produce a male-specific transcript. Although only one male-specific transcript is shown here, a second one is produced by an alternative polyadenylation site at the 3′end. The drawing is not to scale; the introns are actually much longer than the exons.

in a protein in males with two exons that are absent in females, and a female version that has an exon that is absent in males. The amino acid sequence is quite different, and the proteins encoded by the same gene have different functions. (As will be described in Chapter 2, the sex-specific alternative splicing of *doublesex* is itself the result of alternative splices in the genes that regulate it.)

Alternative splicing is so common that it has important implications for genetics, genomics, and evolution. We have seen an example in which different splice variants result in biologically significant differences in protein function. Many more examples of functionally significant alternative splices could be described, to the extent that more genes are known to be alternatively spliced than are uniformly spliced. Nearly every biological process in mammals includes at least one gene that is alternatively spliced. From the point of view of genome annotation, this has important implications that will be discussed in Chapter 7. When an investigator is annotating a genome and identifying transcripts, alternative splice variants are difficult to predict accurately.

From an evolutionary perspective, alternative splicing provides an important source of genetic variation. It may be helpful to consider alternative splicing in the same way that we consider mutation—as a balance between opposing forces of precision and efficiency. Perhaps a reproducible splicing pattern that is completely precise is simply not possible for the existing spliceosome. If so, some alternative splices are simply the result of an occasionally imprecise process with no other physiological significance. On the other hand, the occasional imprecision may give rise to proteins with amino acid sequences that are slightly or even significantly different, which can then be subjected to different selective pressures.

Some genes have RNA rather than a protein as their functional product

If alternative splicing challenges our concept of a gene as the protein-coding unit, another class of genes completely demolishes the concept of one gene–one protein. It has been known for decades that some genes do not code for protein at all but instead have RNA itself as their ultimate and functional product. The genes that encode ribosomal RNA (rRNA) and transfer RNA (tRNA) molecules are long-standing examples. Our gene concept could handle such genes because they clearly appeared to be exceptions to the general principle. Furthermore, they were well-behaved exceptions. These exceptional genes are shown in Figure 1.10A. For example, rRNA and tRNA are stable molecules; the genes encoding them are generally collected together in their own locations in the genome, and transcription of these genes even involves different RNA polymerases—RNA polymerase I in the case of rRNA and RNA polymerase III in the case of tRNA genes. All these properties are distinctive for these genes and different from protein-coding genes. Investigators who studied these RNA genes knew the situation with these genes before beginning to work on them. In the world of genes, these types of RNA gene exist within their own culture—interesting and with its own set of defined rules.

But then even more genes were found that had RNA as the final functional product. The first few of these new RNA genes, such as the gene encoding telomerase RNA and the one for the X inactivation product Xist, were interesting curiosities—different from the main class of protein-coding genes and certainly worth studying as expansions of the gene concept. For example, each of these genes affects a very specific biological process and thus could easily be placed into context as 'one more exception' to the gene concept. Our definition of a gene as the information needed to make a protein was being stretched, but was not yet broken, by the discovery of these new RNA genes.

However, the last decade has shown that the one gene–one protein principle was woefully naive, and perhaps even misleading, about the definition of the gene. We now know that there are hundreds and possibly thousands of different genes in the genome of multicellular organisms whose functional product is an RNA. By some estimates, there may be nearly as many genes in the human genome whose functional product is an RNA as there are genes that encode proteins. These are the genes that encode the regulatory molecules known as microRNAs (miRNAs) (Figure 1.10B). These miRNAs, which are discussed in more detail in Chapter 7, are not like the well-behaved exceptions that we had expanded the gene definition to accommodate.

- First, the miRNA genes are found scattered at hundreds of sites throughout the genome; an investigator never knows where the next one will be found.

A. rRNA and tRNA genes

5′ 3′ 5′ 3′ 5′ 3′

Ribosomal RNA (rRNA) and transfer RNA (tRNA) genes:

- are clustered in the genome
- are transcribed by different polymerases
- produce stable RNA products

B. mRNA and microRNA genes

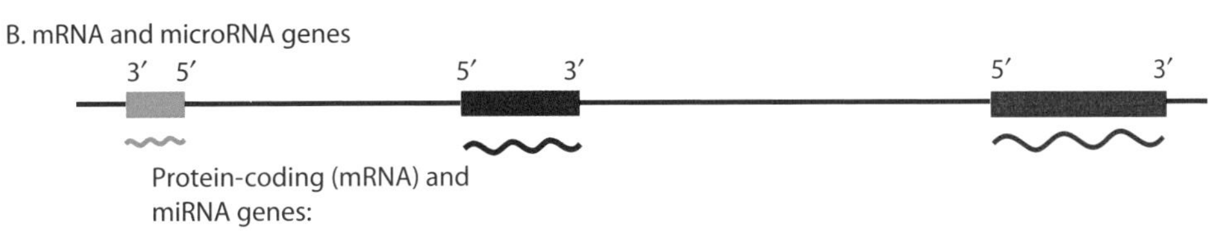

- are scattered throughout the genome
- are transcribed by RNA polymerase II
- produce highly unstable RNA products

Figure 1.10 Comparison of rRNA and tRNA genes with mRNA and miRNA genes. (A) Ribosomal RNA genes are clustered head to tail in the genome, are transcribed by RNA polymerase I, and produce stable RNA products. A similar pattern is seen for many tRNA genes, which are also found in clusters and are transcribed by RNA polymerase III. (B) Protein-coding genes and miRNA genes are not clustered in the genome and can be in either orientation with respect to each other. Transcription is carried out by RNA polymerase II. The genes vary widely in size, and the transcript is highly unstable.

Some of the genes are found in clusters, but others sit in the introns of perfectly standard protein-coding genes. Like protein-coding genes, but unlike rRNA and tRNA genes, these genes can be found in either orientation with respect to each other.

- Secondly, their RNA products are not large (like rRNA and Xist) or stable (like the other functional RNAs). The RNAs are very small, with a final product of 22 nucleotides, much smaller even than any mRNA. Furthermore, they are synthesized and degraded with a half-life comparable to mRNAs. From a general perspective, stable RNA molecules are likely to perform a structural role, whereas ones that are turned over are more likely to be involved in gene regulation or expression.
- Thirdly, these genes are not transcribed by their own RNA polymerase but instead use the same polymerase and cellular resources as a typical mRNA, namely RNA polymerase II.
- Fourthly, and most significantly for our gene definition, the miRNA genes have a wide-ranging and ever-expanding set of functions. A given miRNA can regulate the expression of a few genes, a few dozen genes, or a few hundred other genes; most of their targets have not yet been defined. It is no longer remarkable to find that a gene or a biological process involves a miRNA in its regulation. In fact, miRNAs are so widespread that it is an open question as to whether *any* biological process in a multicellular organism does not involve miRNA regulation.

With so many genes whose final product is RNA involved in so many processes, 'one gene–one protein' cannot even begin to define gene activity. More than a century after the gene concept, we are still making significant adjustments in our understanding of genes. These miRNA genes comprise an entirely new world sitting largely undetected in our very own genomes. Our basic ideas of gene regulation simply were not equipped to handle their existence. Fortunately, we have made rapid changes to our view of gene regulation, and now miRNA genes are included as significant members of the genome. So perhaps we need to try yet another modification to our gene concept—'one gene–one or more RNA or proteins.' Somehow that does not seem like a definition that is destined to resonate through the years as 'one gene–one enzyme' did. But that is where our gene definition is—for now.

Genes consist of DNA

or Watson, Crick, Franklin, Wilkins, and Avery–MacLeod–McCarty were right

At least one part of the gene definition is very unlikely to change with further experimentation. The properties of genes are explained by the properties of DNA. (Of course, some viruses use RNA rather than DNA as their genetic material, but these are well-behaved exceptions again so we do not need to modify our definition too much.) We cannot describe the findings of molecular biology in a few paragraphs, or even in a few chapters—we could not describe the details even in a few books. However, here are some of the fundamentals that we need to know for the rest of this book.

- A chromosome consists of one double-stranded DNA molecule extended from end to end (Figure 1.11). The shortest human chromosome is about 56 million base pairs (56 Mb) long, and the longest is about 260 Mb. The DNA in a chromosome is complexed with a range of different proteins to make chromatin. Some chromosomal proteins change during the cell cycle and throughout the life of the organism, and vary between chromosomes and between different regions of the same chromosome. Some of the chromosomal proteins, such as the histones, are common to the structure of all chromosomes at all times in the life of the organism. However, even these structural proteins are modified post-translationally in a variety of ways, including phosphorylation, methylation, and acetylation. These post-translational modifications *do* vary during the cell cycle and the life of the organism, and changes in the post-translational modifications are critically important in regulating gene expression.

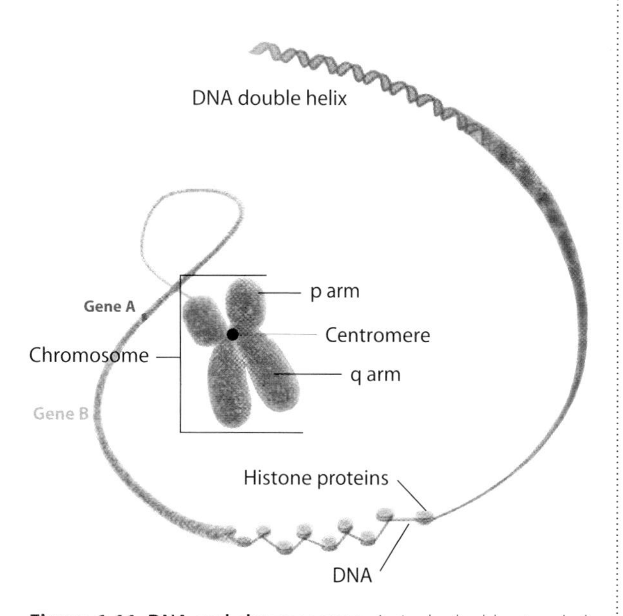

Figure 1.11 DNA and chromosomes. A single double-stranded DNA molecule extends the length of the chromosome. The genes, represented as gene **A** in red and gene **B** in green, occupy different sites on the DNA molecule. The DNA is complexed with histone proteins to make a nucleosome, and the nucleosomes are packaged to make chromatin.

- Genes occupy specific locations on the DNA molecule in chromatin, with the distance between them measured in base pairs or, more frequently, in kilobase pairs (kb), as shown in Figure 1.12. A gene could be on the same DNA strand as its neighboring genes, and thus oriented in the same direction, or it could be on the opposite DNA strand and thus oriented in the opposite direction. The region of the nematode genome shown in Figure 1.12 has genes in both orientations.
- Different alleles arise from changes in the DNA sequence of a gene. These changes can occur in any part of the gene. Alleles may differ by one or more base substitutions or by deletions and insertions.
- DNA replication begins at a region referred to as the origin of replication and is carried out by a large protein complex called DNA polymerase. There are many origins of replication per chromosome, not all of which are used in any given cell cycle.
- Regulation of gene expression occurs principally by the regulation of transcription initiation (we think), although many other regulatory mechanisms are known and are important for some genes. The general structure of a gene is shown in Figure 1.13. A protein-coding gene usually begins with an ATG (methionine) codon, designated +1. Transcription begins upstream of this start codon, often about 10–20 bases upstream, but much longer untranslated regions (UTRs) are known and are significant for some genes. Transcription is initiated by the binding of a large protein complex called RNA polymerase to an AT-rich region upstream of the start codon, often called a TATA box. Most genes have a TATA box and other conserved sequence features in the region 100–120 bases upstream of +1, although recent work with

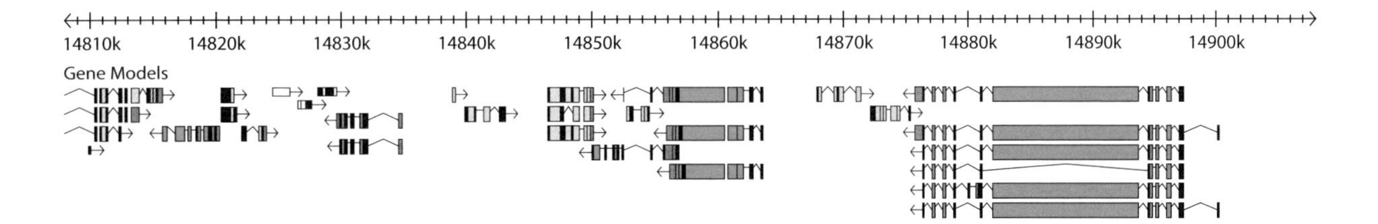

Figure 1.12 A region from chromosome I of *Caenorhabditis elegans*. A region of 100 kb is shown from the screen capture image of the *C. elegans* genome database Wormbase. At least eight different genes are indicated; the turquoise and pink colors of the exons indicate the direction of transcription of the genes. As shown by the arrow, genes in turquoise are transcribed from right to left, whereas those in pink are transcribed from left to right. The orientation is because the genes are encoded on different DNA strands. The right-most gene is the myosin heavy chain gene *unc-54*; note that there are six different splice variants. In fact, at least five of the genes are alternatively spliced. The number of genes in any region is widely variable, and other segments of similar length could have more or fewer genes, or even none.

the human genome has found many genes that do not have this canonical TATA box structure. The region with the TATA box and the associated upstream sequences is usually called the **core promoter** and, generally speaking, can be experimentally swapped between different genes without affecting transcription very much.

- Specific regulation of transcription, i.e. turning the transcription of a specific gene on or off at the proper time and in the proper tissues, is controlled by other DNA sequences in the regulatory region. This is also shown in Figure 1.13. The **regulatory region** has DNA sequences called enhancers that stimulate transcription, and sequences called silencers that reduce transcription. Enhancers and silencers are often 8–12 base sequences that serve as sequence-specific binding sites for proteins known as transcription factors. Transcription factors are among the proteins whose association with the DNA sequence and chromosome vary at different locations and at different times. The regulatory region with the enhancers and silencers is usually found upstream of the core promoter, and can be several kilobases or more upstream of the core promoter. Enhancers can also be found downstream of the gene or in its introns, although this is less common.

This is a very quick and superficial summary of many crucial findings in molecular biology and genetics, but it will suffice for now. Some of these points will be elaborated as needed in later chapters.

New features in the structure and organization of genomes are emerging routinely

Genomes from different species are being sequenced daily, and new information about the organization and evolution of genomes is routinely being revealed. However, some important features of the structure of genomes were recognized even before DNA sequencing was possible or routine.

For example, even 35 years ago, experiments showed that the genome was composed of different types of sequences at different frequencies. The relative frequencies of these different structural features were characteristic for the species. Figure 1.14 shows how hybridization experiments known as Cot curves were used to recognize that a substantial portion of the genome in a multicellular organism consisted of repeated sequences. Some of these repeated sequences were less than 1 kb long and

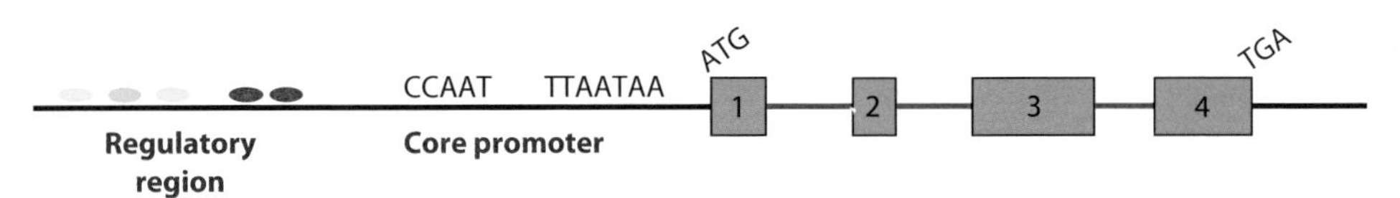

Figure 1.13 The structure of a typical protein-coding gene. The gene shown has four exons, with an ATG codon at the beginning of exon 1 and a TGA stop codon at the end of exon 4. The core promoter consists of approximately 120–200 bp upstream, with the TATA box at –25 and the CCAAT box at –80. The distances and sequences are widely variable, and not all genes have such regions in their promoters. The regulatory region is typically further upstream, and can be several kilobases or more away. The binding sites for different transcription factors are shown as colored ovals, but these are usually identified by functional assays rather than by sequence.

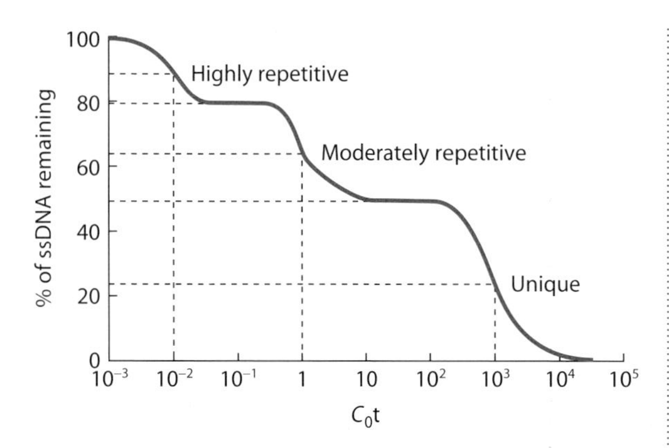

Figure 1.14 A Cot curve of the human genome. The initial concentration of DNA (C_0) with time on a log scale is plotted on the *x* axis; this gives rise to the name Cot curve. DNA is extracted, sheared into fragments of approximately 1 kb, and then denatured into single strands. The DNA is allowed to re-anneal and the percentage that remains single-stranded (ss) is plotted on the *y* axis. Three different populations of DNA sequences are usually seen in such experiments, as shown here. Some DNA re-anneals rapidly because there are multiple copies with similar sequences in the genome; these are referred to as highly repetitive DNA. The moderately repetitive or middle repetitive fraction takes longer to re-anneal because it is longer in sequence and/or present in fewer copies in the genome. The 'unique' fraction is not necessarily single-copy in the genome but rather sequences found at low frequency. Most protein-coding genes are found in the unique fraction.

were present thousands or even tens of thousands of times in the genome. These sequences are collectively known as **highly repetitive DNA**.

The exact functional role of highly repetitive DNA is not known but different sequence families are known. Some of these repeat families are unique to a species and others are found in related species. One such highly repetitive sequence in primates is the Alu element, a sequence about 300 bases long and present in about a million copies in primate genomes, comprising perhaps 10 percent of our total genome. The location and sequence changes in Alu elements have been an important tool in studying primate phylogeny. At least a dozen other highly repetitive sequence families are known in humans.

A second category of repeated DNA sequences are somewhat longer (1.5 kb long), and are present in hundreds of copies in the genome. Foremost among these middle repetitive sequences are **transposable elements**, which can be of many different types. Transposable elements are discrete mobile sequences. However, transposition is infrequent, and most elements have mutations that render them incapable of movement. Nonetheless, transposition occurs often enough for the location and number of these elements to differ between individuals of the same species. Repetitive sequences are a characteristic of heterochromatin, the intensely staining chromatin found at the centromere and telomeres of chromosomes, as well as at other locations.

Other sequences in the genome are present in one or a few copies, and many of the conventional protein-coding genes fall into this category. Along the length of a chromosome in a multicellular organism, the protein-coding portions are in a distinct minority. In humans, for example, less than 2 percent of the genome encodes proteins. Furthermore, these protein-coding regions are usually widely separated from each other, at least in most multicellular organisms. Although most of our experimental attention has focused on the protein-coding regions of the genome, these are relatively rare; the more common sequence in the genome is a member of a repeat that may or may not be transcribed and probably does not contribute to the overall protein capacity of the genome.

Gene families are extremely common in the genome

Even among the protein-coding genes, families of related sequences were recognized. Although gene (or protein) families were known before genome sequences were available, the extent to which genes exist in gene families was not widely appreciated until genomes were sequenced. A gene family can be defined as genes that are highly related in sequence and whose products carry out similar biochemical functions. Different members of the gene family are termed **paralogs** and frequently appear to have arisen by gene duplication and divergence as shown in Figure 1.15. Some family members known as **pseudo-genes** have been rendered non-functional by mutations. The vast majority of genes in humans have a related family member elsewhere in the genome, and the existence of a gene family is the rule rather than the exception.

One of the earliest recognized gene families in humans was the β-globin family of five genes and two pseudogenes (Figure 1.16). The predominant form of hemoglobin in adults has two α-globin chains (also encoded by a gene family) and two β-globin chains, as well as the heme molecule. Other members of the β-globin gene family are expressed at different times in human development (or at the same time but in different amounts).

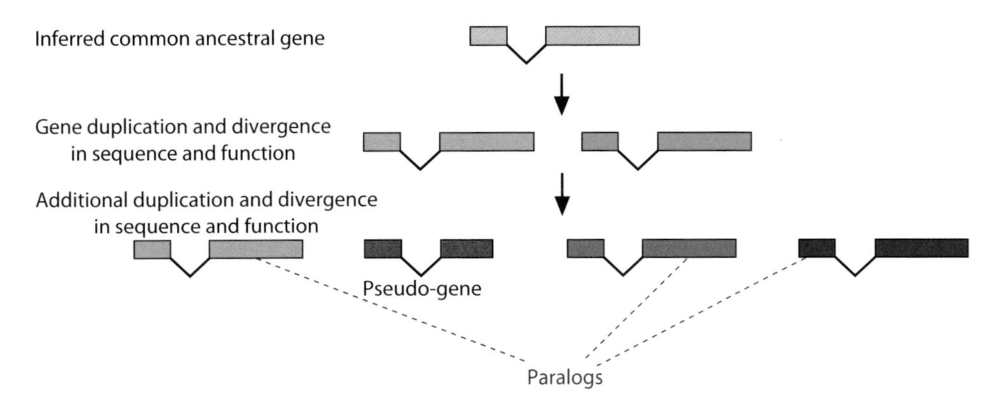

Figure 1.15 Gene families. The inferred ancestral gene is shown in light blue. Gene duplication results in two copies of the gene, which diverge somewhat from each other and from the ancestral gene in both sequence and function. An additional duplication produces further divergences. A pseudo-gene (in red) has arisen as a truncated version of the gene that is no longer functional. The related genes within a species are called paralogs. A gene family may be clustered together in the genome or dispersed.

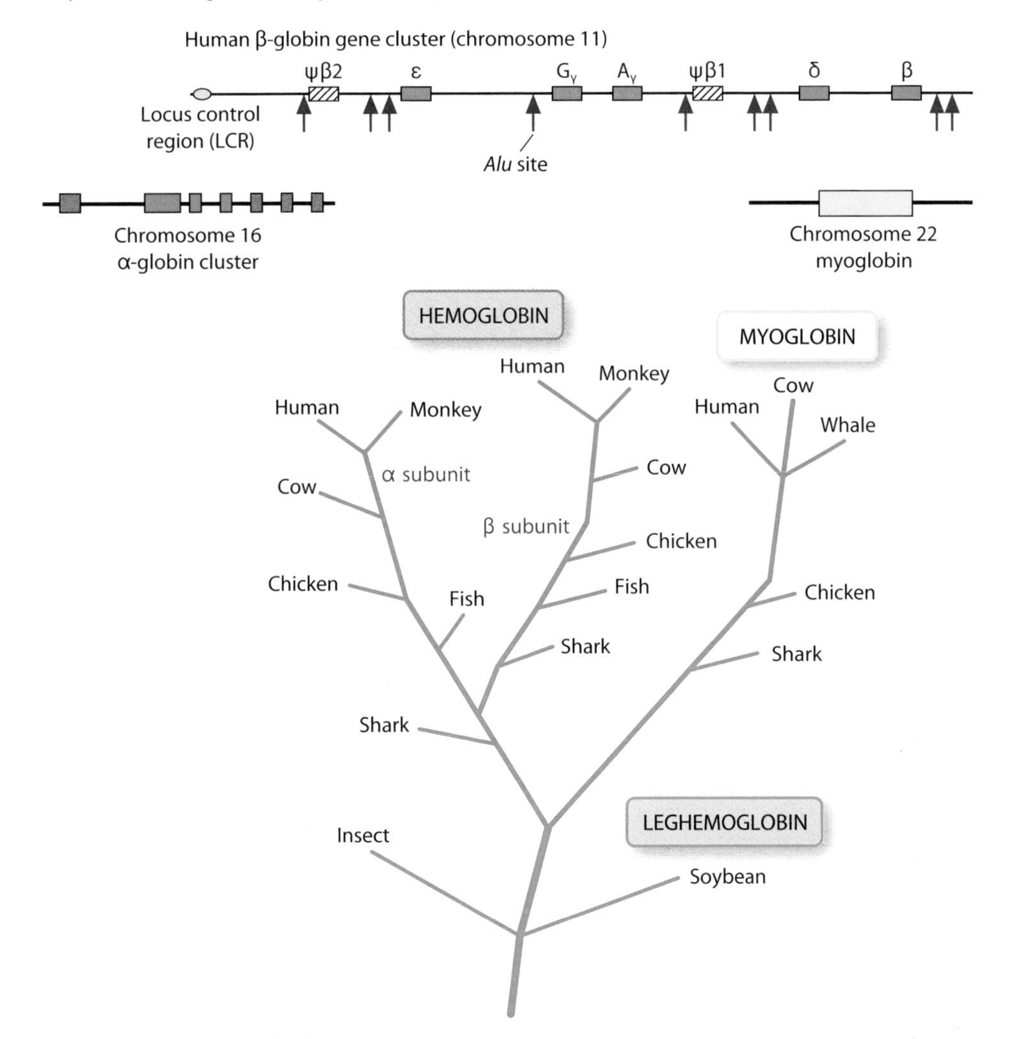

Figure 1.16 The β-globin gene family in humans. The β-globin gene family is found in a cluster of five genes and two pseudo-genes on chromosome 11, as shown at the top of the figure. The β-globin family is related to the α-globin gene family on chromosome 16 and the myoglobin gene on chromosome 16. The branch diagram shows the evolutionary relationship among the gene families.

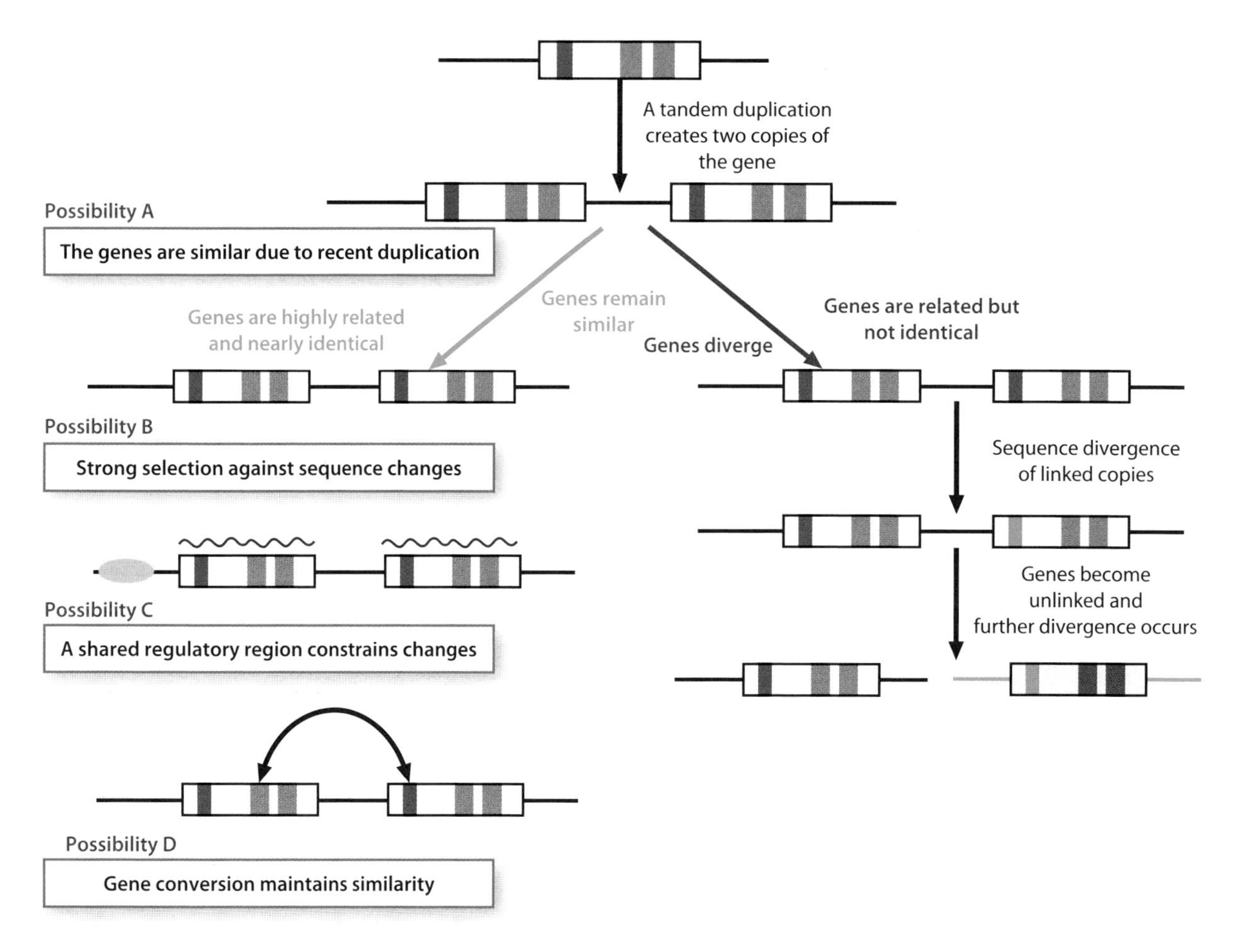

Figure 1.17 Sequence similarity in gene families. The gene shown at the top of the figure encodes a protein with two different domains, one shown in blue and two copies of the domain in orange. A local tandem duplication event results in two identical copies of the gene. Over time, the two copies of the gene will diverge in sequence and function, as illustrated on the right. The first divergence can be seen in the regions other than those encoding the domains, with further divergence occurring in the domain regions. Eventually the genes may become unlinked to each other so that even more divergence occurs. The two genes will encode proteins that are related but not identical. However, if the two genes remain similar in sequence, four different explanations are possible, as shown on the left. In possibility A, the duplication may be recent so that the two genes have not had time to diverge. If they have been able to diverge, three mechanisms could result in highly related or nearly identical sequences: strong selection against any sequence change (possibility B); a shared regulatory region upstream of the genes constrains changes (possibility C); or a gene conversion mechanism maintaining similarity by using one gene as the template to repair sequence changes in the other gene. These mechanisms can also work jointly.

These members include the embryonic ε-globin, the two different fetal γ-globins, and the minor adult form, δ-globin. The switch from the expression of one form to another during development occurs principally (if not entirely) by regulating transcription initiation, but the exact mechanism of the switch is not known.

The β-globin family consists of genes that are clustered together on chromosome 11. The clustering of the gene family is related to the regulation of their expression, with a locus control region located upstream of the ε-globin gene. Most other gene families are not clustered like this but are instead dispersed throughout the genome. Although there are exceptions, gene families that are clustered tend to be more similar to one another in sequence than gene families that are dispersed. There are several explanations for this, as shown in Figure 1.17. Many gene families arise from a tandem duplication of the region of a genome (Figure 1.17). Over time, the two genes are expected to diverge in sequence; eventually the genes may become unlinked so that the genes are recognizably related but not highly similar, as illustrated for the gene pair on the right of Figure 1.17. If the gene pair is linked and highly similar or identical, four explanations could apply. First, some clustered gene families are the result of relatively recent evolutionary events so the genes have not diverged yet in sequence or in location (possibility A in Figure 1.17). Secondly, strong selection against any changes can maintain very similar sequences (possibility B in Figure 1.17). Thirdly, as with the globin genes and some other gene families such as the *Hox* genes, clustered genes may share a regulatory mechanism that imposes selection against mutation in the genes (possibility C in Figure 1.17). Fourthly, gene conversion among the members of the family will keep the sequences similar, even for ancient families (possibility D in Figure 1.17). A gene conversion process likely explains the clustering and sequence similarity of the histone genes, which are evolutionarily ancient. Any or all of these mechanisms might be at work in a clustered gene family.

The sequence similarity of the members of a gene family can have consequences for genetic analysis. Some members of gene families, such as the two γ-globin genes in humans, are functionally equivalent to each other so that a mutation that eliminates the function of one gene may not have an overall effect on the phenotype of the organism. Other paralogs have different functions, which can often be recognized by mutations. Anemias arising from mutations in the β-globin gene family are among the best-studied genetic diseases and mutations in humans. The existence of mutant phenotypes from a gene indicates that other genes must have somewhat different functions and cannot completely functionally substitute for one another. For the β-globin gene family, the paralogs are not expressed at the same time during development and some functionally important differences in the amino acid sequences have occurred. We return to the concept of functional redundancy among gene family members in several other chapters, and deal with it extensively in Chapters 10 and 12.

The existence of a mutant phenotype from a recessive mutation in a gene in the gene family implies that no other gene in the family has exactly the same function. (If another gene had the same function, it would compensate and no mutant phenotype would be observed.) Dominant mutations, which alter the function, of a gene, can sometimes be found that affect one member of a gene family even when the genes within the family are functionally redundant. We discuss this concept in Chapter 4.

Family relationships are detected from aligning the nucleotide or amino acid sequences

The globin genes are similar, although not identical, over the entire length of their gene or amino acid sequences. Because of this similarity over the length of the proteins, it is possible to perform a **global alignment** in which the residue at each position in one paralog is compared with the corresponding position in the other paralogs. A global alignment of the globins is shown in Figure 1.18A. This type of gene family with global similarity is easy to identify because so many of the positions have the same amino acid.

A. A global alignment of human ε globin and β globin

```
        Score = 238 bits (607), Expect = 8e-62
        Identities = 111/147 (75%), Positives = 130/147 (88%), Gaps = 0/147 (0%)

Epsilon MVHFTAEEKAAVTSLWSKMNVEEAGGEALGRLLVVYPWTQRFFDSFGNLSSPSAILGNPK  60
        MVH T EEK+AVT+LW K+NV+E GGEALGRLLVVYPWTQRFF+SFG+LS+P A++GNPK
Beta    MVHLTPEEKSAVTALWGKVNVDEVGGEALGRLLVVYPWTQRFFESFGDLSTPDAVMGNPK  60

Epsilon VKAHGKKVLTSFGDAIKNMDNLKPAFAKLSELHCDKLHVDPENFKLLGNVMVIILATHFG  120
        VKAHGKKVL +F D + ++DNLK  FA LSELHCDKLHVDPENF+LLGNV+V +LA HFG
Beta    VKAHGKKVLGAFSDGLAHLDNLKGTFATLSELHCDKLHVDPENFRLLGNVLVCVLAHHFG  120

Epsilon KEETPEVQAAWQKLVSAVAIALAHKYH  147
        KEFTP VQAA+QK+V+ VA ALAHKYH
Beta    KEFTPPVQAAYQKVVAGVANALAHKYH  147
```

Figure 1.18 A global and a local alignment. (A) The global alignment of the ε- and β-globin proteins of humans. For comparison, the protein sequences are divided into blocks of 60 amino acids, with the ε-protein on top and the β-protein on the bottom; the middle sequence is the identity or similarity between the two proteins. The proteins are each 147 amino acids long, with 111 identical amino acids at the same positions, and 130 amino acids either identical or similar. The 17 amino acids that are considered to be different are shown in red. The proteins align over their entire length, and no gaps are needed to make the alignment. A '+' indicates a conserved substitution.

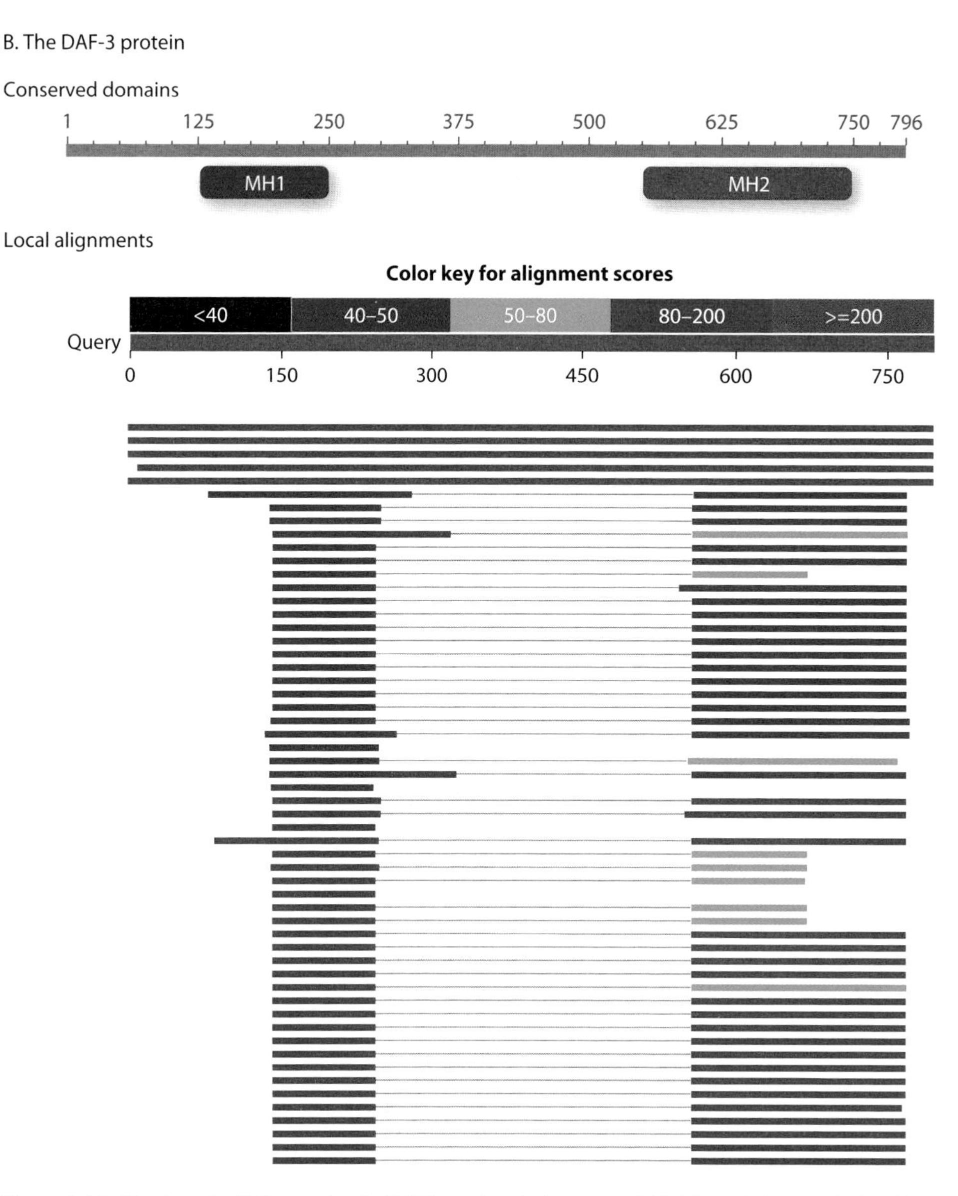

Figure 1.18 (Continued). (B) The results of a BLAST search with the protein DAF-3 from *C.elegans*, which will be discussed in more detail in later chapters. The protein has two conserved domains, as shown at the top of the results figure. BLAST lists the most similar proteins in order, each represented by a horizontal bar. The top four red bars that extend the length of the protein are the *daf-3* gene itself, represented many times in the database. Note that the alignments from other organisms are to the two conserved domains but not to the other part of the protein.

However, this is only one type of gene family. Another and more common type of gene family involves those that encode proteins which have similarities with each other in only some regions of the amino acid sequence rather than over the entire sequence (Figure 1.18B). A **local alignment** of the sequence, in which only a portion of the sequences are similar, is important for these proteins. Notice in Figure 1.18B that the alignment for the DAF-3 protein is only found in the two conserved domains; the other parts of the proteins do not have significant sequence similarity to each other. Proteins consist of different amino acids motifs, or blocks, or domains—the names have somewhat different meanings but all refer to the same fundamental concept. Figure 1.19 shows the domains in some common proteins. Proteins are often a composite of these different domains. Each domain mediates a specific biochemical function, and so the presence of shared domains indicates some underlying similarity of function. Proteins that share a domain did not necessarily arise from a gene

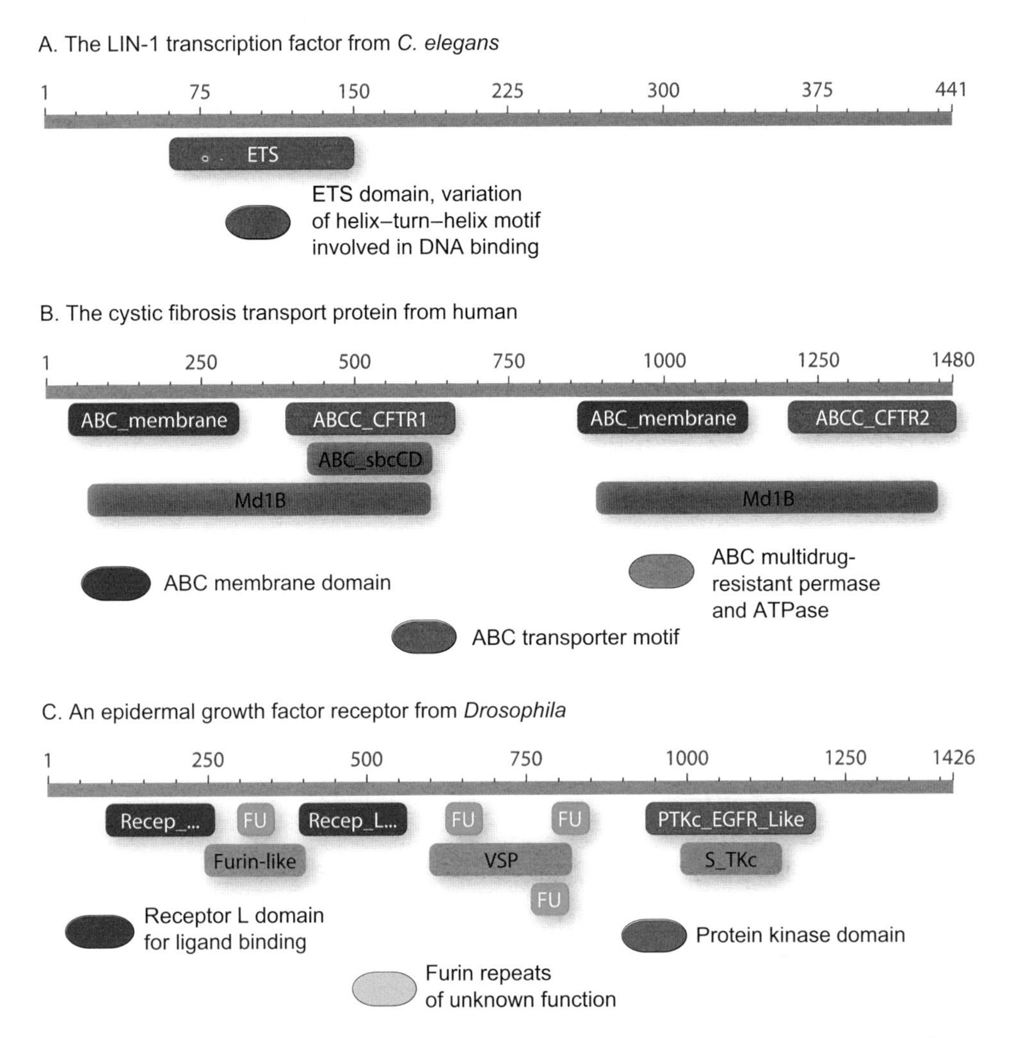

Figure 1.19 Domains and motifs in some proteins. Three proteins are shown: the LIN-1 transcription factor from *C. elegans*, the cystic fibrosis protein from human, and an epidermal growth factor (EGF) receptor from *D. melanogaster*. Notice the scales showing that the proteins are different lengths. Most proteins have at least one recognizable motif; some have several different motifs or domains. The furin repeats found in the EGF receptor gene from *Drosophila* are found in many different proteins but their exact function is not known. (These images and annotations are modified from those found at the conserved domain database at www.ncbi.nlm.nih.gov/.)

duplication event, or at least the origin of these domains cannot easily be traced back to a common ancestor by a series of gene duplications.

The limits of a gene (or protein) family begin to be strained when two proteins have only one domain in common. It may be helpful to compare gene families with a human family. Siblings share similarities and are clustered into a nuclear family. As family members diverge from one another in space and time, they are also less similar to one another in appearance. My aunt is identifiably related to me, but not so similar to me as my siblings, and my great-great-grandfather had only a few features in common with me. Related individuals may diverge so much from one another that the similarity is found in only one feature (e.g. a shared last name) and really does not give much insight into the behavior and function of the individuals. (Most human families probably also have non-functional members that are the familial equivalents of pseudo-genes.)

So it is with shared domains. Closely related protein family members are likely to be similar over their entire length and be clustered together in the genome. However, as the amount of evolutionary distance between two proteins increases, the amount of similarity over the

```
Human    MVHLTPEEKSAVTALWGKVNVDEVGGEALGRLLVVYPWTQRFFESFGDLS    50
Mouse    MVHLTAEEKSLVSGLWGKVNVDEVGGEALGRLLIVYPWTQRFFDSFGDLS    50
Dog      MVNFTAEEKTLINGLWSKVNVEEVGGEALGRLLVVYPWTQRFFDSFGNLS    50
         **::*.***: :..**.****:***********:*********:***:**

Human    TPDAVMGNPKVKAHGKKVLGAFSDGLAHLDNLKGTFATLSELHCDKLHVD   100
Mouse    TPDAVMSNAKVKAHGKKVLNSFSDGLKNLDNLKGTFAKLSELHCDKLHVD   100
Dog      SASAIMGNPRVKAHGKKVLTAFGESIKNLDNLKSALAKLSELHCDKLHVD   100
         :..*:*.*.:********* :*.:.: :*****.::*.************

Human    PENFRLLGNVLVCVLAHHFGKEFTPPVQAAYQKVVAGVANALAHKYH   147
Mouse    PENFKLLGNVLVCVLAHHFGKEFTPQVQAAYQKVVAGVANALAHKYH   147
Dog      PENFKLLGNVLVIVLASHFGNEFTAEMQAAWQKLVAGVATALSHKYH   147
         ****:******* *** ***:***. :***:**:*****.**:****
```

Figure 1.20 Orthologs of β-globins. A multiple sequence alignment of the β-globin proteins from humans, mice, and dogs is shown. The alignment was done by CLUSTALW, and the sequences are shown in 50 amino acid blocks. In CLUSTALW, identical amino acids are designated by an asterisk (*), those that are similar in both charge and hydrophobicity are designated by a colon (:), and those that are similar in either charge or hydrophobicity are designated by a period (.). The high degree of similarity of these sequences makes it easy to identify them as orthologs.

entire length decreases. Two proteins that have C2H2 zinc-finger domains are both DNA-binding proteins. However, the functions that they carry out once they are bound to DNA could be quite different if the similarity is restricted to the presence of these zinc fingers. This is because the presence of the zinc fingers does not indicate what the protein does (the effect it has) once it is bound to DNA—does it activate or repress transcription, regulate chromosome structure, affect DNA replication, or something else?

Genomes of widely divergent species share many of the same genes

Another important, albeit anticipated, finding from genome sequencing projects is that most genes have related genes in other organisms. Related genes in different organisms are known as **orthologs**. The mouse β-globin gene, the dog β-globin gene, and the human β-globin gene are orthologs. The presence of orthologs indicates that some common ancestral species possessed a copy of this gene, and so its evolutionary descendants will have the gene. Orthologs are genes related by evolutionary descent, as shown in Figure 1.20.

Most genes have an ortholog in another species, and some genes have orthologs that can be found in all, or nearly all, eukaryotes. For example, about half of the genes from budding yeast have a recognizable ortholog in humans. The presence of an ortholog gives the investigator clues about the function of the gene in the absence of direct tests, even if the orthologous region is limited to a particular domain or few domains. As the number of recognized functional domains has increased, more proteins can be assigned to some functional category or some family. Most proteins can in fact be assigned to more than one family since they contain more than one recognizable domain. The parallel to human families is again illustrative. None of us belongs to a single family; we have two parental families, four grandparental families, and so on. Each of these families has left a trace in our character, some of them more significant than others. With gene families, the sequence similarities may be so remote (albeit detectable) that they do not provide much insight into the functional role of the protein. Other sequence similarities can provide significant clues to their function.

It should not be assumed that every gene can be assigned to a gene family. As each genome is sequenced, some genes are found that have no recognized homologs and even no recognizable domains. These genes could be newly arisen in that particular species or may be evolving so rapidly that sequence similarity to their ancestral genes is obscured. It could be argued that genes with no homologs are the most interesting for understanding the unique characteristics of the species. It could equally well be argued that novel or pioneering genes are among the least informative in understanding the biology of an organism, and that important processes involve conserved proteins. Both statements have some supporting examples, and the reality is a mixture of the two points of view.

Genome sequences from related organisms have introduced a new twist to the comparative studies that characterize biology. Rather than comparative anatomy or comparative physiology, we now can talk about comparative genomics. Genetic changes in the genome are now beginning to be connected to the morphological differences within a species and between species. In general, species differ in morphology not because each genome has many novel genes but because the expression of the same genes is regulated differently—to happen in different cells or tissues or at different times—in the two species. This is illustrated in Figure 1.21. Evolution has altered the regulation of gene expression much more than it has the structure of the gene product. In the vocabulary used by François Jacob (1977) in an essay from three decades ago, evolution involves 'tinkering' with gene expression and employing their products to novel uses rather than re-designing or engineering new genes. Comparative genomics promises that geneticists will have biological questions to investigate for many years to come. However, this book describes what we know—for now.

KEY ARTICLE

Jacob, F. (1977). Evolution and tinkering. *Science*, **196**, 1161–6.

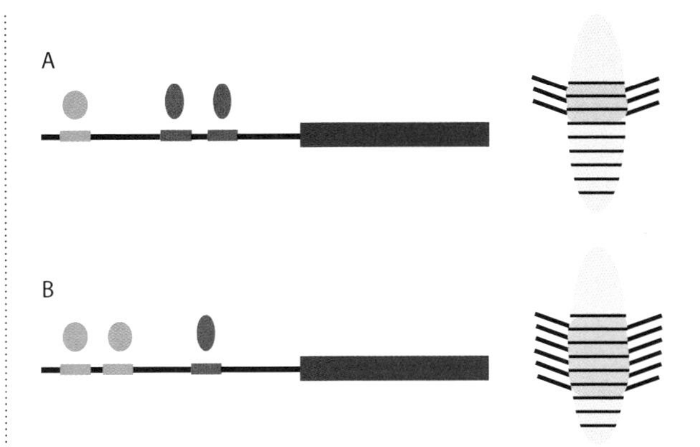

Figure 1.21 Evolutionary change by alterations in expression patterns. Two related imaginary species are depicted. Each is segmented but one has three appendages and the other has six. The expression pattern of the gene is shown in blue. The segments in which the gene is expressed correspond to the locations of the appendages. In the species shown in panel A, the gene is expressed under the control of three regulatory regions, two of the red type and one of the light-blue type. In the species shown in panel B, one of the red regulatory regions has been lost and one of the light-blue regions has been duplicated. This results in a change in the expression pattern of the gene, extending expression to additional segments, with additional appendages. The function of the gene has not changed, but because its expression pattern has changed, the morphology of the two species is different. This type of change is thought to underlie morphological change between closely related species; patterns of change like this are seen with *Hox* gene expression and the expression of the gene *Distalless* in insects.

1.2 Genetic analysis

Our attempt to define a gene has been a limited success at best. We are able to describe many of its essential properties—its inheritance, location, mutation, activity, and so on. We are able to define its essential nature—a DNA sequence embedded in a genome of other DNA sequences. It is more challenging and rewarding to attempt to connect these two concepts, and to explain the essential properties of a gene in terms of its essential nature. This is the substance of genetics and genetic analysis.

Genetic analysis attempts to use the properties of genes that we have encountered in this chapter as an approach to solving or dissecting a complicated biological problem. Let us imagine a biological problem and apply genetic analysis to it. Here is our biological question: What is the mechanism by which insects fly to a light bulb? This definitely qualifies as a complicated biological problem. Consider a few of the processes that we know have to occur in order for this familiar behavior to happen, as summarized in Figure 1.22.

- The insect has to have light-sensing genes.
- The input from the light-sensing genes has to be integrated in the brain with other sensory inputs so that directionality and orientation are maintained during flight.
- The signal has to be processed as an attractant rather than a repellent.
- Nerve impulses have to be transmitted to the appropriate muscles.
- The muscles have to contract and relax, causing the wings to beat.

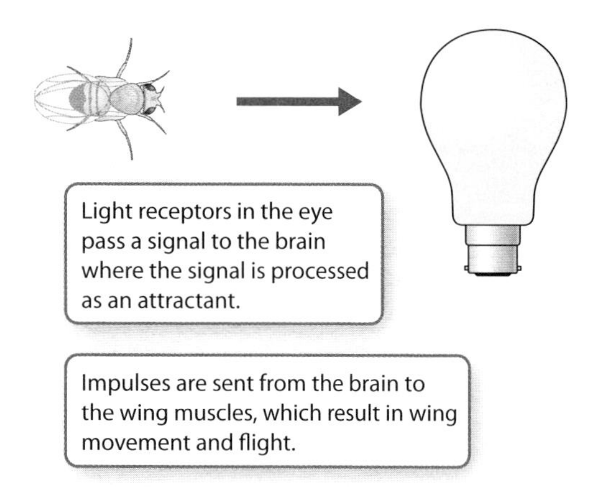

Figure 1.22 A complicated biological problem. As described in the text, genetic analysis provides one approach to the study of a complicated biological question, such as the attraction of an insect to a light bulb as shown here.

- The air currents have to be detected so that directional flight is possible.

As we reflect on this, it is remarkable that any insect ever makes its way to the light—except that anyone who has left a window open on a summer evening knows how reliably this occurs.

How does an investigator begin to understand the mechanism by which this occurs? The investigator wants to identify and characterize all of the component parts. As suggested in Figure 1.23 (experimental approach A), one could do this by grinding up or dissecting insect eyes, brains, muscles, and wings, and characterizing all of the RNA and proteins that are present during the response, as well as the changes in them. There would be thousands of molecules, many of them having little to do with the attractive response. Although we may want to know all these molecules eventually, this might not be the most productive way to begin the analysis of the insect response to light.

A **genetic approach** is different, as illustrated in experimental approach B (Figure 1.23). Fundamentally, genetic analysis begins by finding insects that cannot fly to the light normally, i.e. they carry a mutation affecting one of the many genes involving the light response. Then one has to determine the nature of the altered response—the mutant phenotype. From there, the investigator finds the gene, determines its product, attempts to infer its function from similarity to other proteins, and closely examines the mutant phenotype.

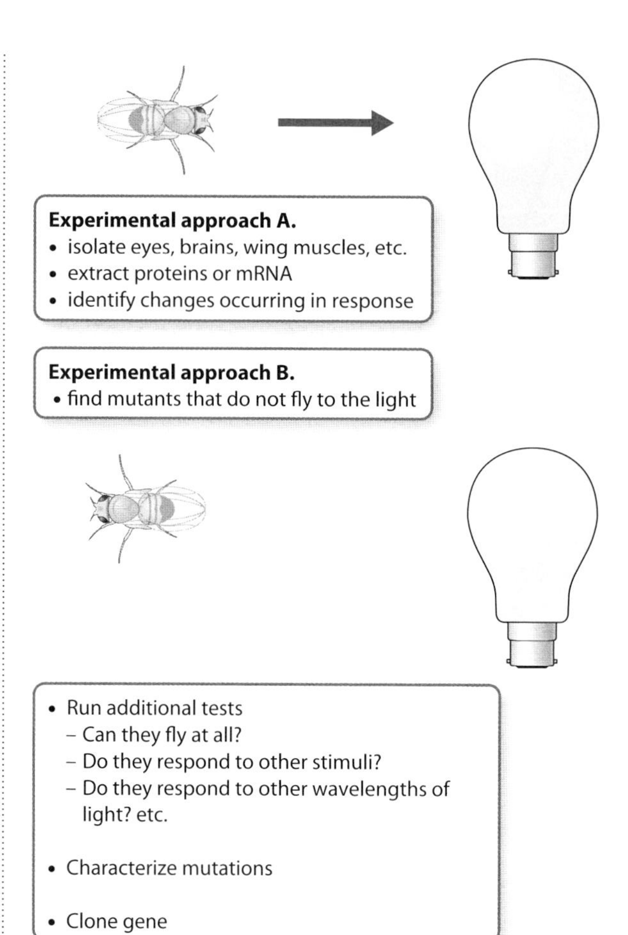

Figure 1.23 Two possible approaches to the complicated biological problem in Figure 1.22. Experimental approach A isolates proteins and RNA from the tissues thought to be involved in the response and looks for changes that occur during the response. Many changes are possible, including new transcription, phosphorylation of existing proteins, and so on. In experimental approach B, the investigator identifies a mutant that cannot fly to the light. Additional tests are run to determine what part of the response is not occurring properly. The mutations are characterized by procedures described in Chapters 3 and 4, and the corresponding genes are cloned by procedures described in Chapter 5.

The expression pattern and the activity of the gene are determined. In order to understand the light response thoroughly, other genes that also affect the response must be found and their interactions with each other have to be determined. If all of that is done carefully, it is possible that at least some aspect of the light response (e.g. the process by which the receptor cells in the eye signal to the brain) will be understood. One can imagine (or more accurately, dream) that the entire process could be understood this way.

The preceding paragraph reflects the overall outline of this book. First, we describe some of the organisms that we will be discussing. Then we go through the steps in the order outlined above—finding the mutation, characterizing the mutant phenotype, identifying the molecular nature of the affected gene, examining its activity, and discovering its interactions. With that introduction, let us turn on the lights, open the window, and see what comes in.

Chapter Capsule

Genetic analysis

Genes are defined by properties such as inheritance, phenotypes, map positions, interactions, and activity. These properties are integrated with the molecular nature of the gene's DNA sequence. Genetic analysis uses the established properties and nature of genes to analyze other complex biological problems.

2

CHAPTER 2

Model organisms and their genomes

TOPIC SUMMARY

The life-cycles and genetics of some of the best-studied model organisms are reviewed. The models include the yeast *Saccharomyces cerevisiae*, the nematode *Caenorhabditis elegans*, the fruit fly *Drosophila melanogaster*, the flowering plant *Arabidopsis thaliana*, and the mouse *Mus musculus*. We explore the advantages and potential shortcomings of each model organism, with an example of a biological question that has been investigated in each one. Furthermore, we provide an introductory tutorial to the genome database for each organism.

IN BRIEF

A few model organisms have been studied intensively to provide insights for many other eukaryotic species.

INTRODUCTION

Any lover of nature is amazed at the diversity of life on earth. More than 1.75 million species have been named but, since most species are microscopic or live in unexplored habitats, these named species are only a fraction of the number of species alive. In fact, in the fall of 2007 two new species of mammals living on a remote island in Indonesia were described—even among our own class, new species are still being found. Most of these species have not been studied in any detail either in nature or in the laboratory, so it is impossible to imagine that we could ever understand them all.

Although much about the behavior, development, natural history, and physiology of all of these species will be unknown to us, the basic *principles* of their genetics and inheritance are known. As described in Chapter 1, the fundamental principles that Mendel found for traits in garden peas have been extended to many other eukaryotes and, with some occasional adjustments, will undoubtedly apply to all others. We have confidence that this assertion is true because all species evolve from common ancestors. The point is made by a well-known statement made in 1954 by Jacques Monod, co-discoverer of the *lac* operon and mRNA, and one of the founding fathers of molecular biology: 'What is true for *E. coli* must be true for elephants.' What we learn from one organism provides the starting point for understanding others.

We assert this confidently but also humbly; the humility is needed because species are different and even a casual observer can tell many of them apart. *Escherichia coli* is a prokaryote, and an elephant is a eukaryote; we now know many fundamental aspects of genetics and biochemistry that are not true for both. Like many of the ideas described in Chapter 1, Monod's point is helpful but incomplete. Monod's statement—which he surely made hyperbolically—does a disservice both to *E. coli* and to elephants. (It is fortunate that we do not harbor elaborate populations of elephants in our gut, for example.) Each of these is a fantastically complex organism, worthy of study on its own merits without the need to justify the study by some broader human-centered explanation. As biologists, we would like to thoroughly investigate *E. coli* and elephants and all the species in between. But that is simply not possible.

Instead of spreading our research efforts among thousands of different species, geneticists have focused on a handful of **model organisms**. The choice to study these few species among the many possible ones was often made on the weakest of rationales. Yet having made the choice, geneticists have investigated these few model organisms thoroughly for decades; all or nearly all of the tools of genetic analysis have been applied to the study of these few model organisms. We have genetic maps, mutations, and cloned genes. We have gene expression patterns and interaction networks. We have a sequenced genome. We have in place all the tools that will be developed and used in the chapters to come.

From these model organisms, we are able to make intellectual extensions to all others. Our confidence that all other eukaryotes, including those we have not yet studied, will follow these same fundamental principles comes not from the brashness of knowing too little but from the humility of knowing how little we truly understand, even about these well-studied models. Perhaps we can re-state what Monod said with a perspective arising from more than 50 years of investigation. If it is true for *Saccharomyces cerevisiae*, *Caenorhabditis elegans*, *Drosophila melanogastor*, *Arabidopsis thaliana*, and *Mus musculus*, it is probably true for elephants as well. Among our models have been a yeast, a nematode, an insect, a flowering plant, and a rodent. We have asked these organisms to serve as stand-ins for all the diverse eukaryotes that we encounter, including humans. We have placed a lot of responsibility on these unassuming creatures, and they have responded by yielding a wealth of insights.

2.1 Model organisms: an overview

The importance of model organisms is underlined from the beginning of genetic analysis. Mendel drew his laws of inheritance from a single organism, the garden pea, *Pisum sativum*. His garden probably included other domesticated vegetables that he might have used instead, but he reported only on peas. An intriguing aspect of the story of the origins of genetics comes from Mendel's correspondence with the distinguished botanist Naegeli who encouraged Mendel to apply his principles to another plant, in particular hawkweeds. Naegeli's recommendation to study hawkweeds is a mistake dreaded by every research mentor, since his suggestion may have been among the worst possible choices for follow-up experiments. Hawkweeds are not a single species, and homozygous true-breeding lines cannot be established. Peas were a good model organism for understanding the principles of inheritance; hawkweeds were a terrible model that may have set genetics back by 35 years. But neither Mendel nor Naegeli knew which plant was the good model organism and which was the poor choice.

Like peas and hawkweeds, many different organisms have been used for genetic analysis. Some have been like peas and were used to establish the foundations of modern molecular genetics. Others, like hawkweeds, seemed like a good idea at the time but, for one reason or another, little research of fundamental importance has arisen from them. What makes the difference between an experimental organism that becomes a valued model and one that does not? Often the answer is complex. Sometimes it lies in some properties of the biology of the organism itself. Sometimes it lies in the personalities and relationships among its early investigators. Sometimes it lies in the willingness of funding sources to invest in a new model.

Let's look at that question a little more closely. ***What makes an experimental organism a good model?*** A brief comparison of the model organisms described in this chapter generates the following list.

- It is easy to maintain in the laboratory. Ideally, it should not be difficult or dangerous to work with or require any peculiar growth conditions. If it is capable of escaping from the laboratory, it should not present a significant health or environmental hazard. Many biologically interesting organisms fail on this criterion, simply because they are so difficult to grow in the laboratory in sufficient quantity for research.
- Homozygous true-breeding lines can be established.
- Mutations can be induced and recognized easily.
- It produces a large number of offspring in a relatively short period of time. Because a large number of offspring allows the investigator to screen for mutations easily, it is also helpful if the organism is comparatively small.
- Genes can be cloned by standard procedures, and cloned genes can be reintroduced into the organism.

'Standard procedures' are those that have been developed for other experimental organisms, although all procedures have to be altered for each research application. We discuss reintroducing genes later in this chapter. But, to give one example, protozoa with codon assignments differing from the universal genetic code present significant challenges for protein expression studies. In a more contemporary sense, it helps to have a substantial amount of genome sequence or an ongoing genome project.

- It attracts enough young investigators to sustain the analysis beyond the founding generation. This usually requires a complex combination of intriguing biological properties, compelling and persuasive personalities, and available funding and permanent employment.

Many experimental organisms have these qualities yet have not become widely used models for genetics. Thus, there is probably an ill-defined component that could be considered luck or happenstance—having the right combination of people at the same place, for example. If we were able to create a computer simulation of twentieth-century genetic research and run it many times, it seems likely that some of the model organisms that we use now would not become established and that others would emerge as new models. Although such an imagined simulation would change the short-term history of genetics, it would be unlikely to result in long-term changes. The model organisms have served us very well, but much of what we have learned would probably have been discovered eventually no matter what models were used. It is partly because the model organisms have served so well that contemporary genetics is able to study many new models.

Having defined some of the characteristics that make an experimental organism a good model, it is appropriate to ask another question. ***What do we want the model organisms to model?*** Figure 2.1 provides one view of some of ways to answer this question.

The first answer to our question is that we want the model organisms to tell us about the basic principles of molecular biology—in short, the organisms in Figure 2.1A have **universal properties**. By using *E. coli* to model elephants, Monod was saying that the basic processes in molecular biology are the same in all organisms —transcription, translation, replication, mutation, and so on. It is always dangerous to believe that we have now learned the fundamental processes of eukaryotic molecular biology, but the range of model organisms that are studied suggests that most of these fundamentals have been or will be uncovered from one or more of them.

The second answer to the question looks at these organisms from an evolutionary perspective. These organisms provide the molecular and genetic tools, as well as the overall intellectual framework, to consider experimental questions that are common to their phylogenetic group. The organisms in Figure 2.1B have **taxon-specific properties**. *D. melanogaster* serves as a model to study other insects, *A. thaliana* to study other flowering plants, mice to study other mammals including humans, and so on. This use of model organisms underlies the field of evolutionary developmental biology, nicknamed 'evo-devo'.

The third answer recognizes that these organisms are distinct species. *D. melanogaster* is different from every other insect, from every other dipteran, and from every other species of *Drosophila*, as shown in Figure 2.1C. As the genomes of closely related species are sequenced and annotated, for instance, with the release of 12 different genomes of *Drosophila* species and five different genomes of *Caenorhabditis* species, we can begin to think about organisms and their **species-specific properties**. That is, what makes one species different in morphology, behavior, habitat, and so on from its closely related species? What are the evolutionary processes that were involved in the divergence of *D. melanogaster* from the other *Drosophila* species?

Model organisms and human biology

Perhaps you are thinking that we have overlooked the most obvious answer to the question of why we study model organisms—we want to be able to apply the insight from them to human biology and disease. While it may be true that all of us want to believe that our research will have an important application to humans, the more realistic attitude is that findings with model organisms will provide insight into some of these other properties. In addition to serious ethical and moral issues, humans as the subjects for experimental research fail on nearly all of the properties listed above. Our major advantages are that mutations can be recognized

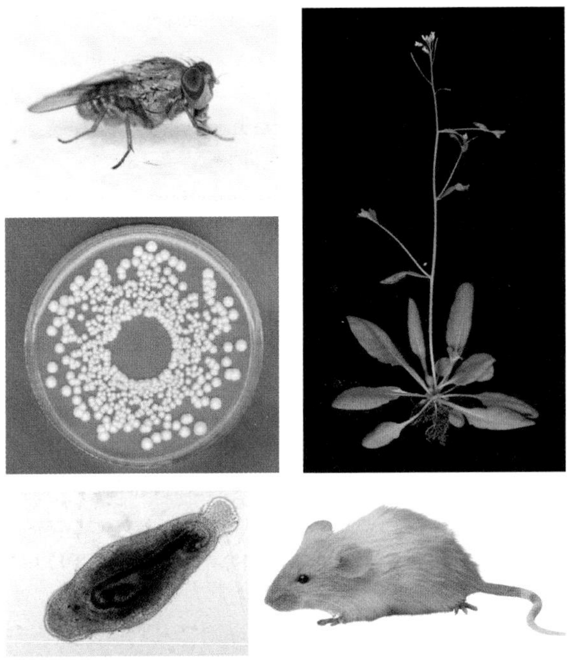

Figure 2.1 (A) The five model organisms and universal properties: the fruit fly *Drosophila melanogaster*, the budding yeast *Saccharomyces cerevisiae*, the flowering plant *Arabidopsis thaliana*, the nematode worm *Caenorhabditis elegans*, and the mammal *Mus musculus*. Properties shared among these organisms are probably universal among eukaryotes. (B) Taxon-specific properties: analysis in *D. melanogaster* can be used to understand and compare properties that are common among different insects, including crickets and butterflies. (C) Species-specific properties: comparison of closely related species, such as these three species of *Drosophila*, provides insights into characteristics that are specific to each one. The figures are not to scale, and the three different species of *Drosophila* are nearly the same size. The pigmentation patterns are different, as suggested here.

(although the genetic basis for phenotypic differences is often very difficult to ascertain) and that young investigators continue to move into the research field. All the other desirable characteristics from the list above for an experimental organism have presented challenges for human genetics, some of which we will describe throughout the book.

We are certainly unlikely to discover universal biological properties from humans, and even mammal-specific and primate-specific properties are probably best studied in other species. However, as an increasing number of different genomes are sequenced and annotated, it becomes possible to think about the species-specific properties of humans. Even before Linnaeus produced the standards that we still use, biological organisms were being classified and compared. Genomics has allowed comparative analysis to reach a new and exciting level of research. Biologists who were trained before the arrival of molecular biology may have had classes such as comparative anatomy, which examined the skeletons and muscle structure of different vertebrates, or comparative physiology, which examined respiration or excretion to determine both the common and the distinctive characteristics of each group. We are

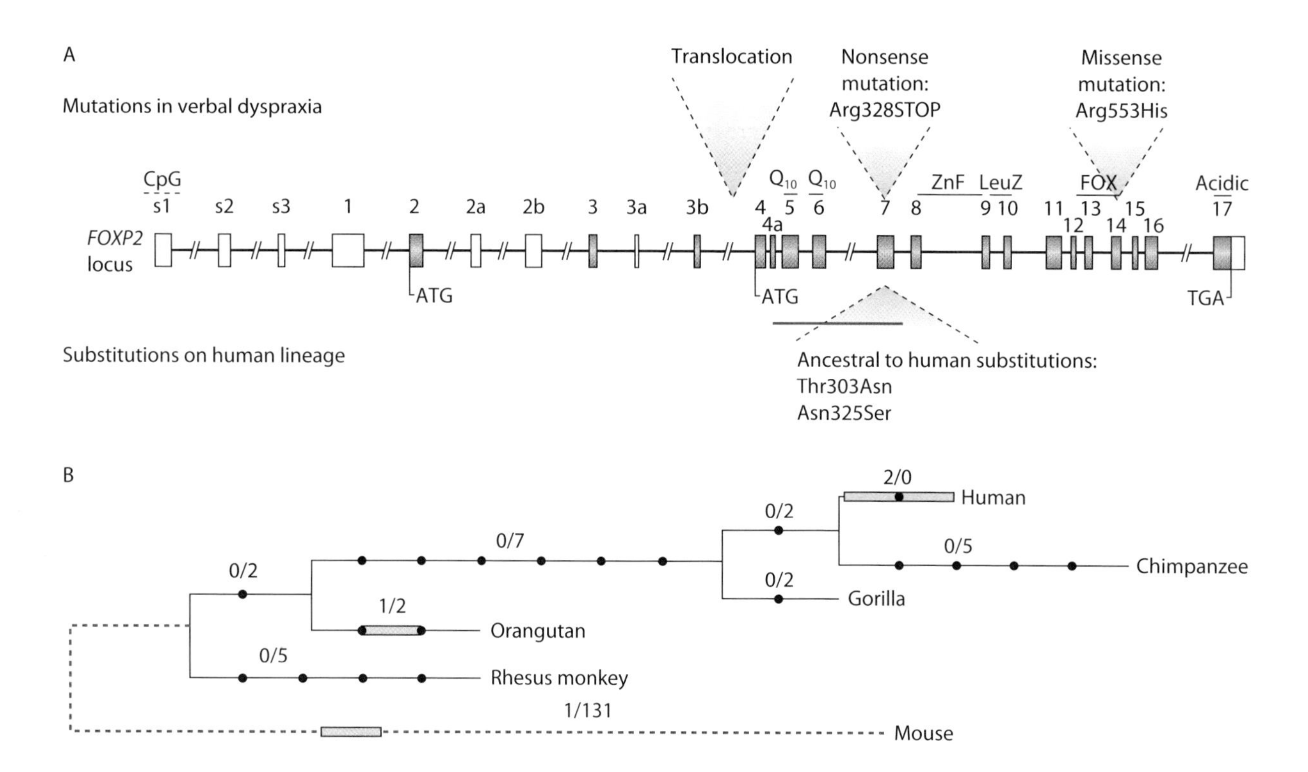

Figure 2.2 Human-specific properties. The FoxP2 transcription factor is expressed in areas of the brain associated with speech in humans, and mutations in this gene cause loss of speech in humans. These are shown in the triangles around the gene diagram. Base substitutions in exon 7 resulting in amino acid changes in the protein are found only in humans. Recent analysis of the *FoxP2* gene in Neanderthal specimens indicate that these substitutions occurred after the divergence of humans from chimpanzees but prior to the divergence of modern humans from Neanderthals.

currently in the age of **comparative genomics**, when each new genome is understood in light of the others that have been sequenced and analyzed. One can even imagine that in a few decades, if not a few years, comparative genomics will be the foundation of a new way to think about biology as a whole.

Our classification of what can be learned from different model organisms becomes most helpful when put into the context of the comparative analysis of genomes. We are beginning to identify genomic characteristics and genes that are common to *all* eukaryotes, as well as those that are common to multicellular organisms, animals, insects, mammals, and so on. As the genomes of other primates are studied, we can also begin to recognize a few genetic properties that are peculiar, if not unique, to humans.

One example comes from one of the most distinctive properties of humans—our ability to form and process speech—and from genetic analysis with mutations. Some people who are aphasic, with an inability to produce or comprehend language, have mutations in a transcription factor called FoxP2 (Figure 2.2). The *FoxP2* gene is also expressed in the regions of the brain associated with speech. It is found in other vertebrates, and its expression has been correlated with vocalizations in mice, songbirds, and bats. The FoxP2 protein in chimpanzees has two amino acid substitutions compared with humans, but its expression pattern is somewhat different than in humans. Some have speculated that these changes in amino acid sequence and/or expression pattern are important in the complexities of human speech. While it is certainly naive to believe that human speech is the product of changes in a single gene, the FoxP2 transcription factor does provide a starting point to think about the species-specific evolutionary changes that produced our ability to speak.

However, this thinking may need to include more than current living species. With DNA sequence analysis of specimens of *Homo neanderthalis*, we will soon have substantial genome sequence information from another species in the *Homo* genus. This will provide another important point of reference for thinking about *Homo sapiens*; in fact, the two human-specific changes in exon 7 of the FoxP2 transcription factor that parallel speech acquisition are found in the Neanderthal gene. Perhaps Neanderthals were capable of a form of speech; there is no fossil record, but there may be a genetic record.

Furthermore, the analysis of the genomes of individual people is beginning. This is both exciting and controversial in what it may tell us. For humans, we can go beyond the species-specific properties to begin to think about how to identify and understand **individual-specific properties**. The genetic answer to the question 'What is man?' may not be far off.

In the meantime, we describe the properties of five model organisms that are used in this book. Certainly other model organisms could be introduced and discussed, since these five are far from the only species that contribute to modern genetics. A summary of many of the key features of these model organisms appears in Table 2.1.

The most important information in Table 2.1 is undoubtedly the genome database for each organism. These websites provide the most accurate, comprehensive, and up-to-date information for these model organisms and for some of the most closely related species whose genomes have also been sequenced. The amount of information available through these databases is staggering, and is sometimes intimidating to the newcomer. We provide a very brief guide for each database with this chapter, and a more detailed tutorial online. Because updates to these databases are posted monthly, the tutorials with the chapter may become out of date in content or format if the structures of the websites change. The online database tutorials will reflect any significant changes to the websites. Geneticists working with one of these model organisms access the genome databases frequently, and we encourage the readers to explore them thoroughly.

2.2 The awesome power of yeast genetics

A story is told about three scientists, a mathematician, a biochemist, and a yeast geneticist, flying together in a small airplane. The plane developed engine trouble and a crash seemed imminent. Unfortunately, there were only two parachutes for the three scientists, so each was asked to explain why he should be rescued. The mathematician explained how mathematics was the queen of the sciences, and all other analysis depends on the foundations of mathematics. The biochemist spoke at length about the many medical and economic benefits that we have gained from understanding biochemistry. After a pause, the geneticist began, 'Let me tell you about the awesome power of yeast genetics.' The biochemist screamed, 'Not again!' and promptly jumped out of the plane.

While the story is unlikely to be true in its specifics, geneticists who work with the budding yeast *Saccharomyces cerevisiae* can take a justifiable pride in their ability to manipulate the organism in order to study nearly any process in a eukaryotic cell. Some of its properties are shown in Table 2.1. For fundamental intracellular properties that occur in all eukaryotes, including meiosis and mitosis, DNA replication, the cell cycle, protein trafficking, and many more, the comparison is nearly always made with yeast. Although industrial applications for yeast genetics date back centuries, the use of yeast as a laboratory research organism, apart from brewing, began in the 1930s and greatly increased during the 1950s.

Yeast cells can be grown in liquid culture or on agar plates with defined growth requirements (Figure 2.3).

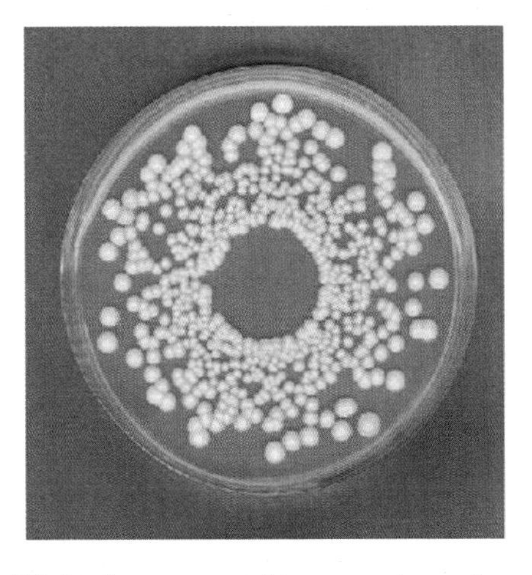

Figure 2.3 *Saccharomyces cerevisiae* as seen by most geneticists. Yeast cells grown as colonies on an agar plate with defined nutrients, which allows easy use of auxotrophic mutations.

Table 2.1 A comparison of selected model organisms

	Saccharomyces cerevisiae	*Caenorhabditis elegans*	*Drosophila melanogaster*	*Arabidopsis thaliana*	*Mus musculus*
Common names	'Yeast', budding yeast, baker's or brewer's yeast	'The worm'	Fruit fly	*Arabidopsis*	Mouse
Genome size	16 chromosomes, 6608 genes, 12 Mb	Six chromosomes, approximately 18 500 genes, 105 Mb	Four chromosomes, the fourth with very few genes approximately 13 750 genes 170 Mb	Five chromosomes, approximately 27 000 genes, 125 Mb	Approximately 30 000 genes, 3000 Mb or 3 billion bases
Attractiveness as a research organism	Industrial applications for brewing. Single-celled eukaryote that can be grown in large quantities. Can be grown as either haploids or diploids.	Defined cell lineages. Organs and systems including the nervous system have few cells with highly defined connections. Self-fertilizing hermaphrodites produce homozygous strains, with males arising to allow crosses.	Easily grown in the laboratory with easy-to-score morphological differences. More than a century of genetic research with vast number of mutant lines and rearrangements available.	Flowering plant. Short life-cycle. Small size allows many individuals to be grown in a small space. Naturally self-fertilizing, and produces many seeds. Small genome for a flowering plant. Easiest plant for making transgenics.	Mammals raised by hobbyists. Small size and fast life-cycle.
Method of making transgenics	Transformation with plasmids. Homologous recombination with artificial chromosomes.	Injection with dsDNA, maintained as an extrachromosomal multicopy array.	Injection with P transposable element, which integrates randomly in genome.	*A. tumefaciens*-mediated transformation by floral dip.	Embryonic stem cells with targeted gene insertion and disruptions.
Mutant phenotypes used for markers	Nutritional auxotrophs, drug resistance.	Morphological, body shape, and movement.	Morphological, including eye color, wing shape, body color, etc.	Morphological, including leaf hairs, flower and seed pod characters, etc., flowering time, hormone responses.	Coat colors.

Particular advantages	Ideal for studying nearly every intracellular property of eukaryotic cells. Particular contributions have come in cell cycle and cell division, meiosis, intracellular secretory mechanisms and protein trafficking, chromosome structure, signal transduction, eukaryotic transcription and replication, and many more. First eukaryotic genome to be sequenced.	Cellular simplicity contributed to embryology, programmed cell death (apoptosis), neurobiology and behavior, most signal transduction pathways. First multicellular genome to be sequenced. Discovery of microRNAs and RNA interference (RNAi) in animals.	Long history of genetic research; provides insights into nearly all topics in animal genetics. Polytene chromosomes are easily visible.	Ideal for studying the biology of flowering plants. Multicellularity arose independently in plants and animals; comparisons with the animal models provide a better understanding of the requirements for multicellularity and the make-up of the earliest common unicellular ancester. Key for delineating the mechanisms of RNA silencing and other epigenetic phenomena.	Compared with humans in nearly all areas of mammalian biology, including immunology, neurobiology, embryology, cancer biology, etc.
Limitations and peculiarities	Extracellular signaling is limited. Very few introns. Centromeres are unusually small. May lack microRNAs as a means of regulating gene expression.	No established tissue culture. Targeted integration of introduced DNA is difficult. Many genes are in polygenic transcripts with trans-spliced leader sequences	No recombination in males. Targeted gene disruptions are difficult.	No homologous recombination; successful targeted gene disruptions so infrequent that they are rarely attempted. Many seemingly functionally redundant genes.	
Genome database	The *Saccharomyces* Genome Database http://www.yeastgenome.org/	Wormbase http://www.wormbase.org/	Flybase http://flybase.bio.indiana.edu/	TAIR—The *Arabidopsis* Information Resource http://www.Arabidopsis.org/	Mouse Genome Informatics http://www.informatics.jax.org/

Since the nutritional requirements are defined, mutations that cannot synthesize some key molecule are easy to identify and use. These mutations are known as **auxotrophic.** For example, wild-type (or **prototrophic**) yeast can produce their own uracil. By growing cells in the absence of added uracil, i.e. starving them of uracil, it is possible to find mutations that have a defect in the uracil biosynthetic pathway. These mutants cannot grow unless uracil is added to the growth medium, so they are uracil auxotrophs. Auxotrophic *ura3* mutations are useful as genetic markers. As discussed more fully in Chapter 3, auxotrophic mutations provide a powerful selection scheme since only cells with the desired genotype (e.g. uracil prototrophs) will grow and divide.

Yeast forms colonies, each consisting of thousands of cells, under normal growth conditions on agar plates in the laboratory, and much of the analysis is done by examining colonies of genetically identical cells rather than individual cells. Thus, yeast can be treated much like *E. coli* or other bacteria. Growth rates depend on environmental conditions, of course, but a typical doubling time is less than 2 hours, so a few cells on an agar plate can produce easily recognized colonies in less than 48 hours when grown at 30°C. Tens of thousands of yeast cells can be grown on a standard petri dish.

Yeast is often grown as a haploid

Nearly all eukaryotic organisms have both a haploid and a diploid phase of their life cycle. Haploid cells have one copy of each chromosome in the genome, and diploids have two copies. Unlike the other organisms we will discuss, yeast is often grown and maintained as a haploid. By examining haploids, issues arising from dominance (as discussed in Chapter 3) are avoided, which makes recessive mutations easy to spot. The life cycle of budding yeast is shown in Figure 2.4.

Haploid yeast have one of two possible mating types, termed **a** and **α**. Haploid cells of one mating type can fuse or mate with that of the other mating type to form an

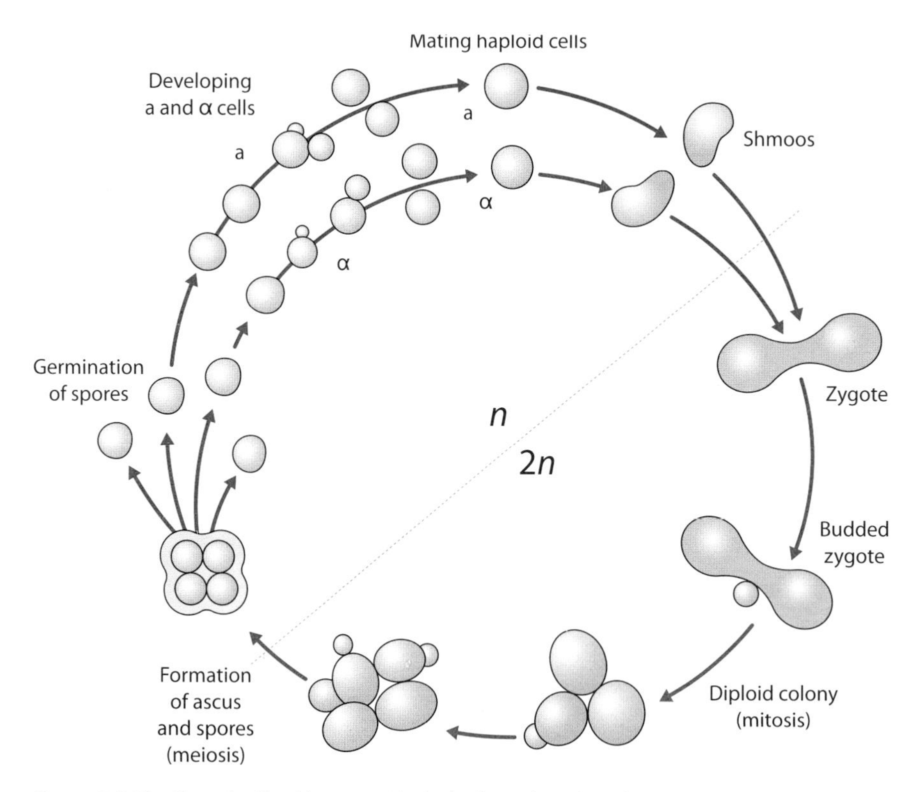

Figure 2.4 The life cycle of budding yeast. Haploid cells can be either of two mating types, a or α. These haploids can fuse into a diploid cell. The diploid cell reproduces by budding. Under starvation or other stimuli, the diploid cell enters meiosis and forms four haploid cells known as ascospores. These four spores, two of mating type a and two of mating type α, are found together in a sac called an ascus. Thus, all four products of a single meiotic division can be identified together. The ascospores can be germinated and grown as haploid cells, which also reproduce by budding.

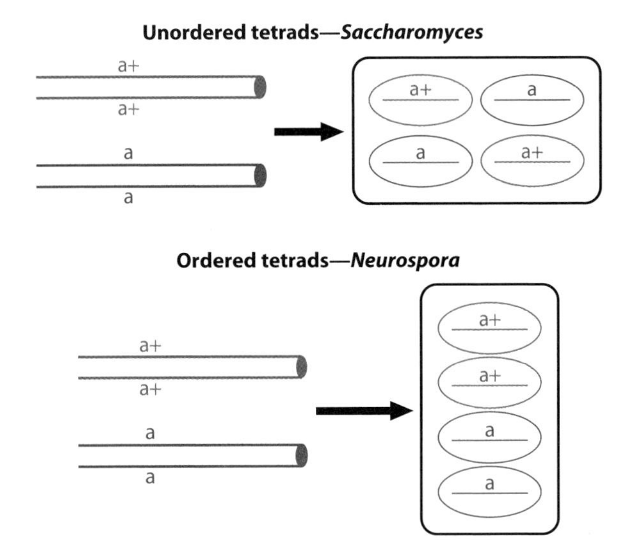

Figure 2.5 Unordered and ordered tetrads. In ascomycete fungi such as *Saccharomyces, Schizosaccharomyces, Neurospora*, and *Aspergillus*, the four products of a single meiotic division can be found together in an ascus. In *Saccharomyces* and *Schizosaccharomyces*, the four haploid products can be oriented in any formation in the ascus. In *Neurospora* and *Aspergillus*, the four ascospores are found in order of the sister chromatids at meiosis I. Many fungi with ordered tetrads, including *Neurospora*, have a mitotic division immediately following meiosis so that there are actually eight ascospores rather than four.

a/α diploid. A diploid strain divides by mitotic division and can be propagated indefinitely. Under conditions of starvation (or other stimuli), the diploid initiates meiosis, similar to cells in the germline of animals. Chromosome pairing, synapsis, and recombination occur to produce four haploid products. These four products, a **tetrad**, remain together in a structure called an **ascus**, so all of the products of one meiotic division can be recovered together (Figure 2.5). (*Saccharomyces* has unordered tetrads because the four haploid ascospores can be found in any orientation and ones arising from sister chromatids need not lie adjacent to each other. In some other fungi, such as *Neurospora crassa*, the haploid ascospores are located in the ascus in the order of the sister chromatids during meiosis, as ordered tetrads.) Tetrad analysis has allowed genetic mapping and analysis of many of the events of meiosis that occur in all eukaryotes, particularly gene conversion.

The processes of cell division, including the cell cycle, mitosis, chromosome segregation, cytokinesis, and so on, have been thoroughly studied by genetic analysis in yeast. One unusual but useful feature of yeast chromosomes deserves particular mention. Chromosomes are positioned on the spindle by the attachment of the centromere to the spindle fiber. In yeast, a single spindle fiber attaches at the centromere, which is an unusually small and well-defined sequence—126 bp in yeast compared with millions of base pairs of repetitive sequences in most other organisms, as shown in Figure 2.6. Despite the difference in the length and composition of the underlying DNA sequence, the protein complex that forms at the centromere is composed of the same proteins in yeast as in other organisms. The small size of the yeast centromere sequence has allowed the construction of some useful cloning vectors, as described below.

Transformation in yeast involves naturally occurring plasmids

DNA can be introduced into yeast cells by transformation using either electroporation or osmotic shock. The ease with which transformation occurs has been one of the key advantages for yeast genetics, since foreign DNA can readily be cloned into yeast cells. The form that is used to introduce this DNA depends on the experimental application. Many of the transformation vectors are engineered to be capable of growth in either *E. coli* or yeast, and have selectable markers appropriate for each host, as shown diagrammatically in Figure 2.7. Growth in bacteria provides a sufficient quantity of DNA to conduct the experiments in yeast. These transformation vectors are referred to as **shuttle vectors** because they can be moved back and forth between yeast and bacteria.

Yeast cells have a naturally occurring plasmid referred to as the 2 μm plasmid, which in its native form is about 6.3 kb length (Figure 2.8A). This plasmid replicates separately from the rest of the yeast genome and has its own origin of replication. The 2 μm plasmid can exist in up to 20 copies in the cell and segregates at random during mitosis; because of its high copy number, both daughter cells receive copies of the plasmid. As shown in Figure 2.8B, the 2 μm plasmid forms the backbone of YEp plasmids.

Genes cloned into YEp plasmids are present at a high copy number in the cell, which is useful for expressing a gene at high level. In addition to its value as a high-copy-number cloning vector, the 2 μm plasmid has another special importance for genetic analysis. The naturally occurring 2 μm plasmid encodes a site-specific recombination enzyme known as FLP; the target sites for FLP are

A. Sequence structure of the *Saccharomyces* centromere

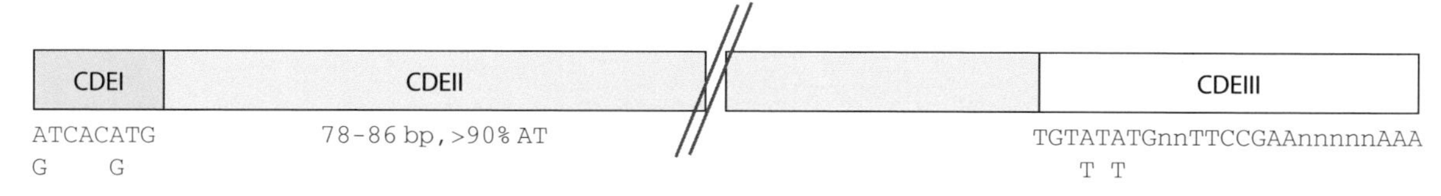

B. Spindle fiber attachment

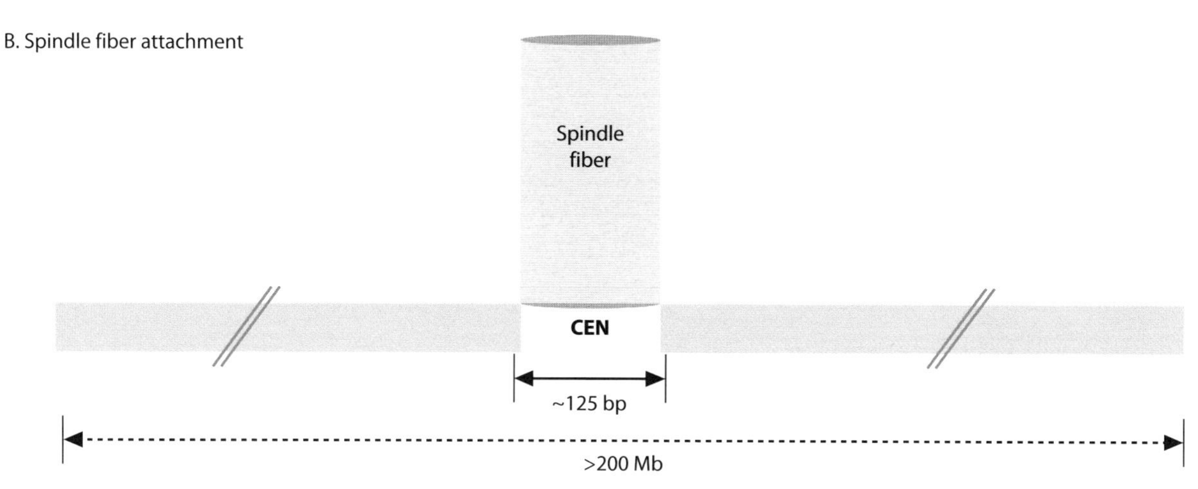

Figure 2.6 Centromere sequence and structure in *Saccharomyces*. (A) The sequence of a yeast centromere. There are three sequence blocks: CDEI, CDEII, and CDEIII. CDEI and CDEIII have highly conserved sequences among the centromeres on different chromosomes. CDEII is slightly variable in length and sequence between different chromosomes but is more than 90 percent AT in overall composition. The total length is about 125 bp, much smaller than the centromeres in other organisms. (B) The centromere in the context of a chromosome. A single spindle fiber attaches to the centromere in *Saccharomyces*, unlike other organisms, which have multiple spindle fibers per centromere.

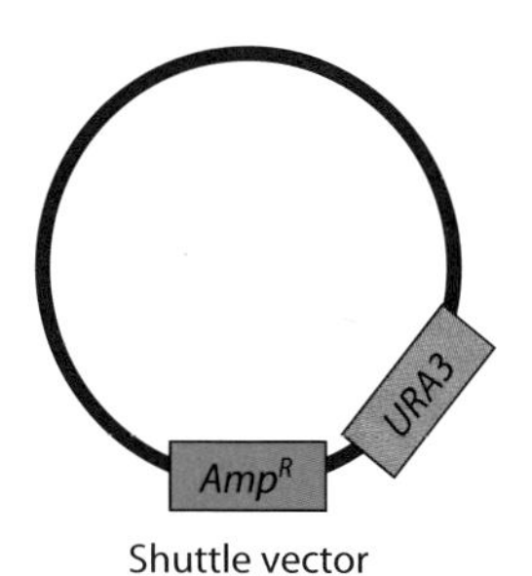

Figure 2.7 The general structure of yeast shuttle vectors. These plasmids can be grown in either *E. coli*, using a drug-resistance marker such as the ampicillin-resistance gene, or yeast using a nutritional marker such as the *URA3*+ genes. By using bacteria, large quantities of the plasmid can be obtained for transformation into yeast

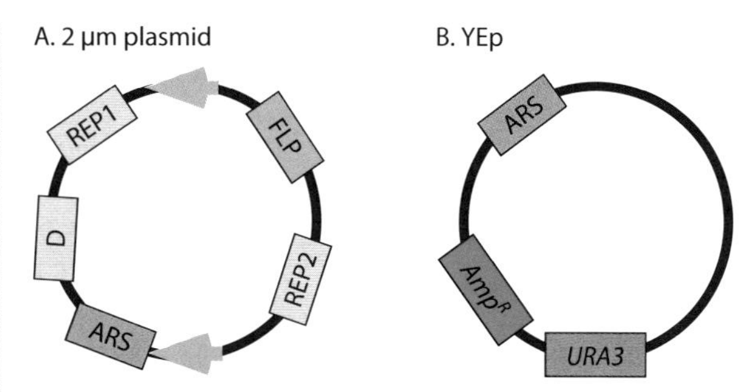

Figure 2.8 High-copy-number plasmids in yeast. Yeast has a naturally occurring plasmid referred to as the 2 μm plasmid, shown in panel A. The 2 μm plasmid has an origin of replication, shown in orange as ARS, and a site-specific recombination system known as FLP/FRT, shown in green. FLP is a site-specific recombination enzyme that catalyzes recombination at the FRT sequences, shown as green arrows. Use of FLP/FRT recombination is discussed in more detail in Chapter 6. The three genes in gray regulate FLP. Because of the site-specific recombination, the 2 μm plasmid exists in different configurations, which are 'flipped' images of each other. A yeast cell can have as many as 20 copies of the 2 μm plasmid. In panel B, the 2 μm plasmid has been modified to produce a YEp plasmid vector. A YEp plasmid can be grown in either bacteria or yeast, as in Figure 2.7, and genes cloned into it will be present at high copy number in the cell.

the FRT repeat sequences. The FLP/FRT recombination system will be discussed again in Chapter 6 as a means of targeting insertions to specific sites in the genome.

Some artificially engineered plasmids have other uses. Yeast has an extremely high rate of meiotic recombination, much higher than any of the other experimental organisms we will discuss. Thus, at detectable frequency, DNA introduced into yeast is integrated into the genome by a double crossover, as shown in Figure 2.9A. Because sequence homology between the introduced donor DNA and the recipient host DNA directs the integration, homologous recombination targets

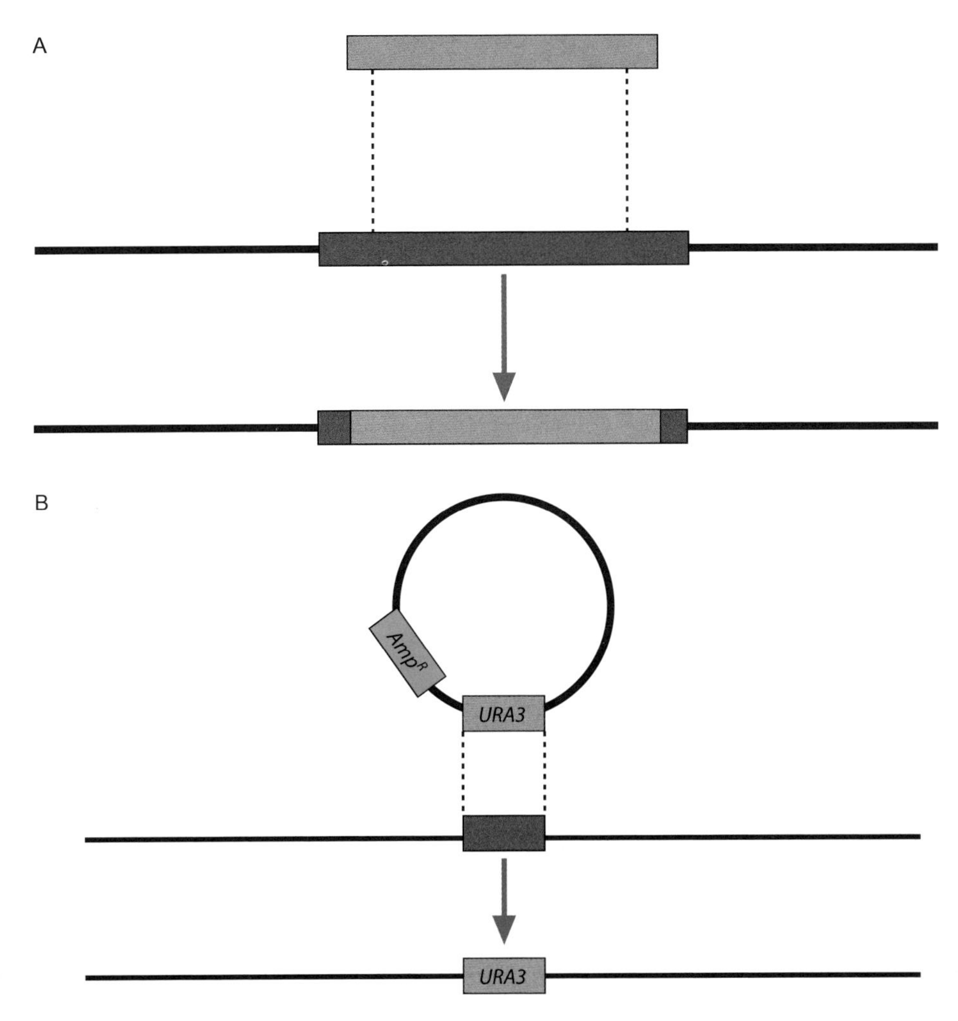

Figure 2.9 Yeast integrative plasmid (YIp). Yeast has a very high rate of recombination, and any DNA introduced into yeast cells can be integrated by homologous recombination. As shown in panel A, the introduced DNA replaces the chromosomal copy by a double crossover. The ability to target integration and replace chromosomal genes is an important tool in yeast genetics. In panel B, the *URA3* gene on the YIp vector replaces the chromosomal *ura3* gene.

the donor sequences to a specific site, and the donor sequences can replace the genomic copy at that site in the recipient. The yeast integrative plasmids known as the YIp plasmids (Figure 2.9B) are particularly useful for this purpose. They do not have an origin of replication and thus do not replicate themselves or accumulate to a high copy number. However, a gene present on a YIp can be integrated into the genome, replacing the homologous gene at the target site. The ability to target genes to specific sites in the genome using integrative plasmids is a significant strength of yeast genetics.

Because the yeast centromere consists of a short defined sequence, centromere-containing plasmids, shown diagrammatically in Figure 2.10A, have also been useful for genetic analysis, particularly for experiments that require a copy number and expression level of the gene that is similar to what occurs physiologically in yeast. The circular CEN plasmids have an origin of replication and a centromere, so they replicate autonomously but do not accumulate to high copy number; they also segregate normally (although with higher rates of loss) during meiosis. Yeast centromere plasmids are in effect small circular chromosomes, and were used in many of the early experiments on chromosome structure. Yeast centromere plasmids are used in the yeast two-hybrid assays described in Chapters 10 and 12.

An important modification to the centromere plasmid was to make a linear version, as a yeast artificial

A. Yeast CEN plasmid

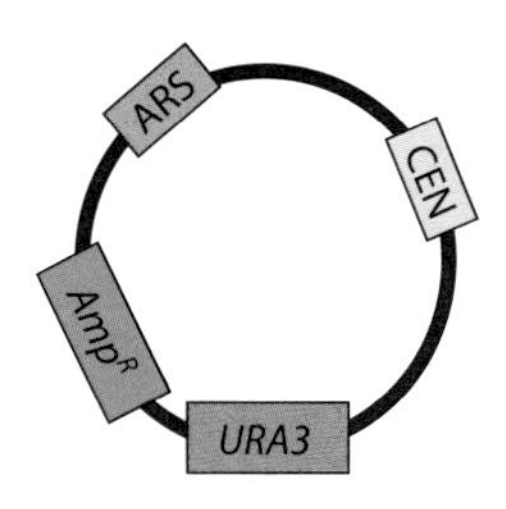

B. Yeast artificial chromosome (YAC)

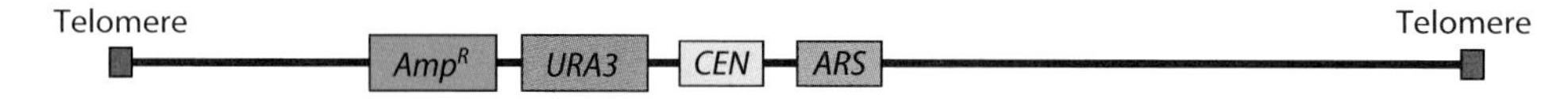

Figure 2.10 Yeast centromere plasmids and chromosomes. In panel A, a yeast centromere (CEN) plasmid includes the centromere, as shown in Figure 2.6, on the shuttle vector shown in Figure 2.8B. In the presence of the centromere, the plasmid does not accumulate to high copy number and segregates mitotically, as do the regular chromosomes. The centromere plasmid forms the basis for the yeast artificial chromosome (YAC), as shown in panel B. The linear chromosome includes telomere sequences at the termini, as well as the elements of the CEN plasmid. A YAC includes 200 kb or more of other DNA, which makes YACs widely used as cloning vectors.

chromosome (YAC), shown in Figure 2.10B. To make a circular plasmid into a linear chromosome required more than simply opening the circle; telomeres were added to each end to make the YAC stable mitotically and meiotically. In addition to the telomeres, YACs have an origin of replication and a centromere, and so they are stable chromosomes present at a single copy. YACs are an important experimental vector because they can consist of more than a million base pairs of additional DNA that can be derived from any source. YACs are used to clone large fragments of the genomes of multicellular organisms, including humans.

Protein trafficking is one example of a fundamental process in eukaryotic cells studied by genetic analysis in yeast

Yeast has contributed so much to our understanding of eukaryotic molecular biology that it is almost impossible to single out one example for further discussion. Many examples appear throughout the book including the cell cycle (in Chapter 9) and an extended description of the tubulin genes in Case Study 10.1 (Chapter 10). A very brief example will be used here to show how the awesome power of yeast genetics has been applied to understanding a key property of eukaryotic cells.

Eukaryotic cells are characterized by different cellular components, or organelles with different functions. In order to carry out these functions, different proteins must be appropriately localized to the correct organelle; the particular constellation of proteins is characteristic for the organelle. Proteins are made on ribosomes, which are associated with the endoplasmic reticulum (ER); the presence of ribosomes on the ER led electron microscopists to refer to it as the rough ER. From the rough ER, proteins are transported to other parts of the cell including the Golgi apparatus, the lysosome, and the plasma membrane, as summarized in Figure 2.11. The process of protein trafficking is

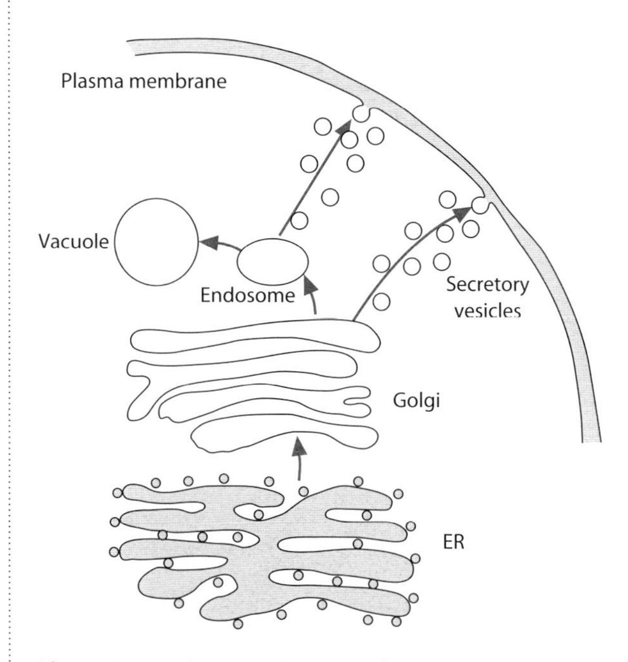

Figure 2.11 Schematic diagram of protein trafficking in eukaryotic cells. The yeast vacuole is the functional equivalent of the lysosome in other cells. Proteins from the rough ER move to the Golgi where they are sorted to further processing in the endosome or the vacuole or secreted to the plasma membrane.

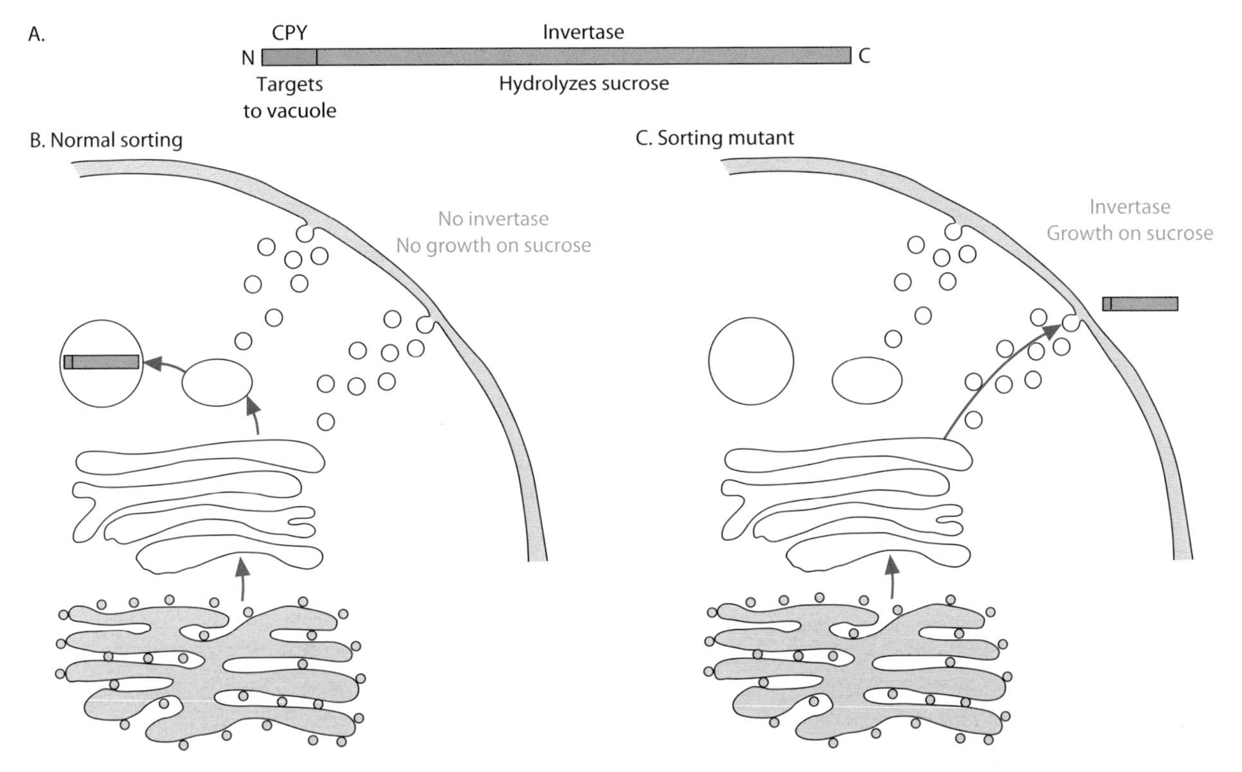

Figure 2.12 One method of identifying genes involved in protein trafficking. (A) A fusion protein between carboxypeptidase Y and invertase is shown. CPY is normally found in the vacuole while invertase is normally secreted from cells and breaks down sucrose. The N terminus of the CPY protein is fused to the N terminus of invertase. (B) CPY targets this fusion protein to the vacuole. As a result of this location, invertase is not secreted from the cell and the cell cannot grow on sucrose. (C) The scheme used to find protein sorting mutations. A mutation that mis-sorts the fusion protein will send it to the plasma membrane rather than the vacuole. As a result of the mutation in protein trafficking, the fusion protein is secreted from the cell and the cell can grow on sucrose.

fundamental to eukaryotic cells. Different proteins must have characteristics that target them to particular locations—signal sequences and post-translational modifications such as glycosylation, for example. However, there must also be 'traffic controller' proteins that recognize the molecular signals and sort the proteins to the proper compartment.

Many of the traffic controller proteins, or more properly the sorting and secretion proteins that are common to all eukaryotic cells, were identified by genetic analysis in *S. cerevisiae*. Different investigators, often using related but distinct approaches, have looked for yeast mutants in which proteins are mis-sorted to the wrong cellular compartment. For example, one powerful genetic selection shown diagrammatically in Figure 2.12 exploited the nature of sorting of a fusion protein made by attaching part of the N terminus of carboxypeptidase Y (CPY) to the enzyme invertase. The fusion protein is shown in Figure 2.12A.

CPY is normally found in the vacuole in yeast cells, the *Saccharomyces* equivalent of the lysosome. Its enzymatic role is unimportant for this mutant selection; the important feature is its localization. CPY is synthesized as a longer inactive precursor protein (preproCPY) in the lumen of the ER. Here, a signal sequence is cleaved from it to yield proCPY, which is then glycosylated and moved via transport vesicles to the Golgi. In the Golgi, the proCPY is sorted away from proteins that are destined for the cell surface and instead is transported to the vacuole.

The portion of the fusion protein that provided the genetic selection comes from invertase. Invertase is the enzyme that hydrolyzes sucrose to glucose and fructose, providing a carbon source that yeast cells can utilize. (An organism lacking invertase cannot grow on sucrose, because it cannot metabolize the sucrose to generate glucose or fructose.) Invertase is secreted from the cells, and so its localization is completely different from that of CPY. However, when the N terminus from preproCPY is attached to invertase, invertase is sorted according to the CPY localization and ends up in the vacuole, as shown in Figure 2.12B. Invertase is still enzymatically active in this fusion protein, but the cells cannot grow on sucrose because invertase is not being secreted from the cell and sucrose is not hydrolyzed. However, a mutation that *fails* to sort the CPY–invertase fusion protein to the vacuole but instead sends it to the cell surface will secrete invertase and will be able to use sucrose as a carbon source, as shown in Figure 2.12C.

Hundreds of such mutations that affect protein sorting have been found by yeast genetic selections and screens like this one. By using the procedures described in Chapters 3 and 4, these mutations have been assigned to different genes by complementation tests; more than 40 such vacuole protein sorting or *vps* genes are known in yeast, nearly all of which are also found in all other eukaryotes. In this genetic analysis, yeast served as a model for universal properties of protein trafficking.

The *Saccharomyces* Genome Database

Our brief introduction to yeast concludes with a quick tour of the genome database. The yeast genome database is called the *Saccharomyces* Genome Database (SGD). An annotated set of screen captures from the website is presented in Figure 2.13, and a more detailed tutorial on the use of the database is on the website associated with this book. The screen for the *TUB2* gene is shown; *TUB2* will be used for other examples in the book. Many of the headings are self-explanatory and encourage further exploration. The term 'gene ontology' is used for all the genome databases and is discussed in more detail in Chapter 3 in the discussion of gene names and in Chapter 8 in the discussion of interpreting microarray data. The Gene Ontology (GO) project is an attempt to classify proteins and genes by their general molecular functions and cellular location, among other features. The GO project uses a standard nomenclature with a defined vocabulary. Also available on this page are links to many other research resources, including mutant phenotypes, DNA sequences, map positions, interactions, and papers describing *TUB2*. *TUB2* is a particularly well-studied gene, but the information is clearly laid out and available to a patient and curious reader.

A

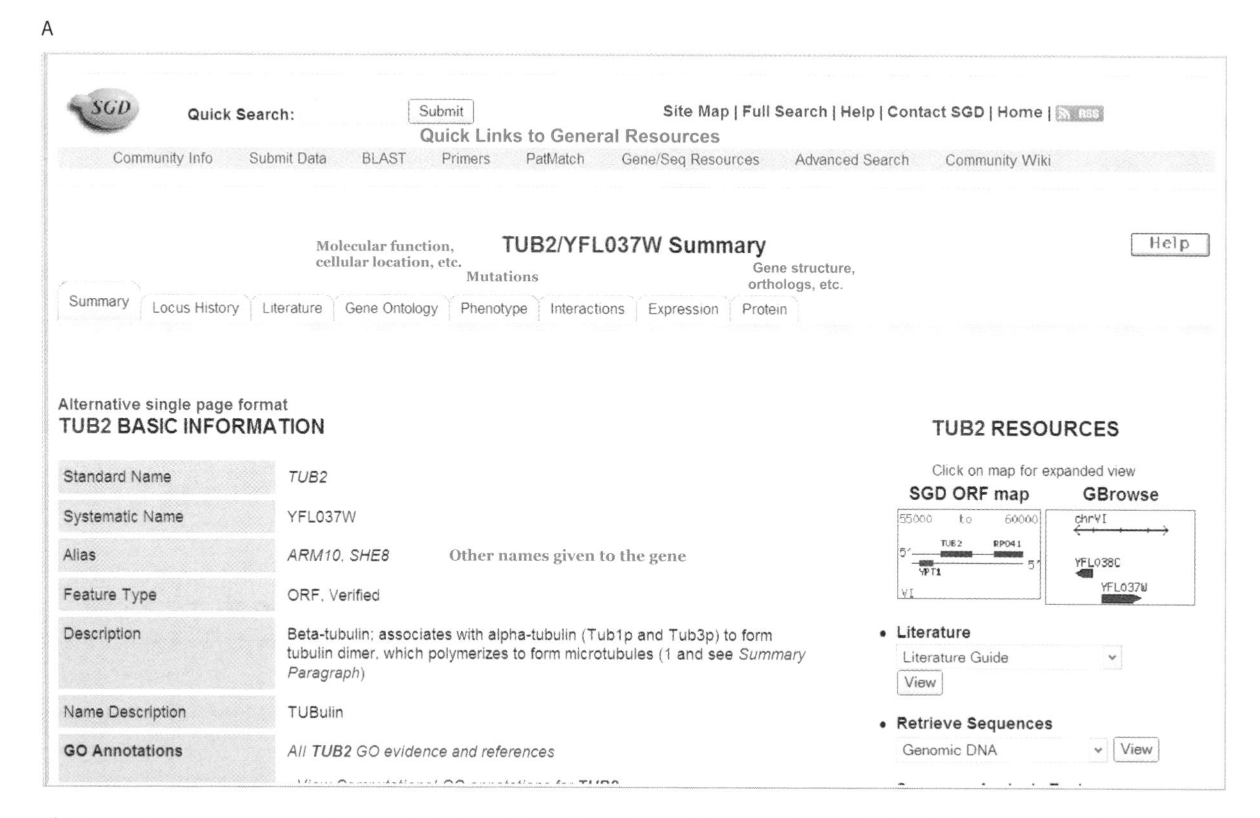

Figure 2.13 The Saccharomyces Genome Database for *TUB2*. *TUB2* is one of the genes that will be described in other parts of the book, most extensively in Case Study 10.1. Our annotations that do not appear on the web page itself are shown in green. (A) The top part of the page with the general layout. Note the Quick Links across the top of the page and the tabs to Locus History, Literature, Gene Ontology, and so on. *TUB2* encodes β-tubulin. The top of the page provides a quick summary of the gene.

B

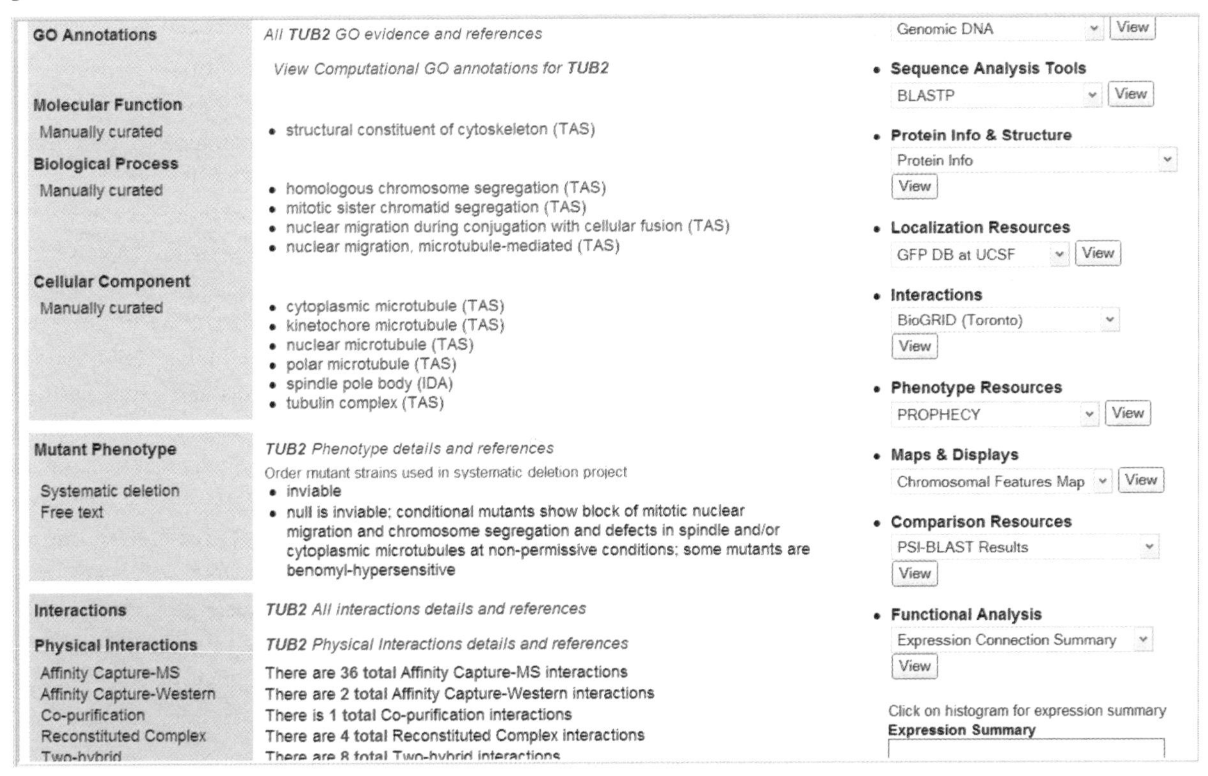

C

Sequence Information	ChrVI:56335 to 57708 | ORF Map | GBrowse
	chrVI 56k 57k YFL038C YFL037W Regulatory regions & binding sites FKH1-binding-site
	Genetic position: -59 cM
Last Update	Coordinates: 1996-07-31 | Sequence: 1996-07-31
Subfeature details	

	Relative Coordinates	Chromosomal Coordinates	Most Recent Updates Coordinates	Sequence
CDS	1..1374	56335..57708	1996-07-31	1996-07-31

ORF Genomic DNA | Get Sequence

External Links	All Associated Seq | Entrez Gene | Entrez RefSeq Protein | MIPS | UniProt/Swiss-Prot
Primary SGDID	S000001857

ADDITIONAL INFORMATION for *TUB2*

Community wiki	Domains/Motifs	Expression Connection	Function Junction	Gene/Sequence Resources
Global Gene Hunter	Locus History	Mapping Data	PDB Homologs	Protein Info
Researchers				

SUMMARY PARAGRAPH for *TUB2*

In *S. cerevisiae*, *TUB2* encodes the single essential beta-tubulin (1). *Tub2*p belongs to the tubulin superfamily, which includes alpha- and gamma-tubulin and the prokaryotic tubulin-like gene FtsZ (reviewed in 2, 3). Beta- and alpha-tubulin form tubulin heterodimers, which polymerize into microtubules. Microtubules are

Figure 2.13 (Continued). (B) Scrolling further down the *TUB2* page provides more information about the gene ontology (GO), with a defined and limited vocabulary that describes the functions of the gene and gene product. Note the resource links on the right-hand side of the page. (C) Scrolling further down the page shows the structure of the gene and provides links to the map position, the DNA sequence, and other resources.

D

SUMMARY PARAGRAPH for *TUB2*

In *S. cerevisiae*, *TUB2* encodes the single essential beta-tubulin (1). *Tub2p* belongs to the tubulin superfamily, which includes alpha- and gamma-tubulin and the prokaryotic tubulin-like gene FtsZ (reviewed in 2, 3). Beta- and alpha-tubulin form tubulin heterodimers, which polymerize into microtubules. Microtubules are conserved cytoskeletal elements that function in nuclear processes: chromosome segregation in mitosis and meiosis, spindle orientation, and nuclear migration during mitosis and mating (4; for reviews, see 5, 6). All microtubules in *S. cerevisiae* emanate from a microtubule organizing center called the spindle pole body (SPB), which is embedded in the nuclear envelope (for review, see 7). Microtubules extend from both faces of the SPB, generating two types of microtubules: nuclear and cytoplasmic microtubules (8; for review, see 7). The distribution and length of these two types of microtubules is regulated throughout the cell cycle (8; reviewed in 9).

TUB2 was cloned based on its strong homology with its counterparts in other eukaryotes (1). There is an abundance of *tub2* conditional mutants resulting from genetic screens for chromosome loss and sensitivity/resistance to anti-microtubule drugs (such as benomyl), suppressor analysis, and *in vitro* mutagenesis (10, 11, 12, 13). One benomyl-resistant allele of *TUB2*, *tub2*-150, actually requires benomyl for growth at high temperatures, suggesting that microtubules in this mutant are hyper-stable (12, 14). Most conditional *tub2* mutants are cold sensitive, presumably reflecting the intrinsic cold-sensitivity of the microtubule polymer. *Tub2p* interacts with numerous proteins involved in the regulation of microtubules, such as microtubule motors, SPB components, kinetochore components, tubulin biogenesis factors, and alpha-tubulin (encoded by *TUB1* and *TUB3*) (reviewed in 5, 6).

Tub2p is a GTP-binding protein (for review, see 15). *Tub2p* hydrolyzes its GTP following tubulin dimer addition to the microtubule end, whereas the GTP bound to Tub1p and Tub3p is non-hydrolyzable (15). The structure of tubulin has been crystallized in the polymerized state; Tub3p and Tub1p, rather than *Tub2p*, are believed to interact directly with the SPB (16).

Last updated: 2003-12-18

REFERENCES CITED ON THIS PAGE [View Complete Literature Guide for *TUB2*]

1) **Neff NF, *et al.* (1983)** Isolation of the beta-tubulin gene from yeast and demonstration of its essential function in vivo. *Cell* 33(1):211-9
SGD Curated Paper | PubMed | Access Full Text

2) **McKean PG, *et al.* (2001)** The extended tubulin superfamily. *J Cell Sci* 114(Pt 15):2723-33
SGD Curated Paper | PubMed | Access Full Text

3) **Nogales E, *et al.* (1998)** Tubulin and FtsZ form a distinct family of GTPases. *Nat Struct Biol* 5(6):451-8
SGD Curated Paper | PubMed

4) **Jacobs CW, *et al.* (1988)** Functions of microtubules in the Saccharomyces cerevisiae cell cycle. *J Cell Biol* 107(4):1409-26

Figure 2.13 (Continued). (D) Near the bottom of the page is a summary for the gene, with links to the literature and the key terms. References are shown at the bottom, providing ready access to the original literature on *TUB2*.

2.3 *Caenorhabditis elegans*: 'amenable to genetic analysis'

During the early years of nematode research, the phrase 'amenable to genetic analysis' appeared in the introduction of nearly every paper on *C. elegans*. But who was the first to use the phrase? It did not appear in the classic paper by Brenner that introduced *C. elegans* to the world of genetics in 1974, so where did it come from? In the days before computers and web searches, it was the type of argument that post-docs could only settle by spending hours in the library. Now, the worm has reached a status so that it needs no such phrase in the introduction. Two different Nobel Prizes in the past 10 years have been given to researchers who worked primarily on *C. elegans*.

C. elegans was employed as a model organism from the beginning of its use in the laboratory. Some of its properties are shown in Table 2.1. Worms fit our criteria for model organisms perfectly, in part because Brenner sorted through potential model eukaryotes before settling on this one. Although the wild worm lives in the soil, little is known about its ecological niche or natural history. It is a laboratory organism, as shown in Figure 2.14. In the laboratory, it lives on agar plates spread with *E. coli*, crawling along the surface eating bacteria. (It may say something about our choice of model organisms that some of them feed on each other. Worms eat *E. coli* and *Drosophila* eats yeast. And humans eat both *E. coli* and yeast.) The worms are about 1 mm long and are barely visible without a microscope; nearly all genetics experiments are done using a dissecting microscope, and worms are transferred manually from one plate to another.

Worms are morphologically simple, at least at the resolution of a dissecting microscope, with few obvious anatomical features. Most of the mutations used as genetic markers in *C. elegans* affect either the movement or the morphology of the worm; typical genetic markers include uncoordinated (*unc*) mutations that affect the motion of the worm on the surface of the agar plant and dumpy (*dpy*) mutations that result in a short fat worm.

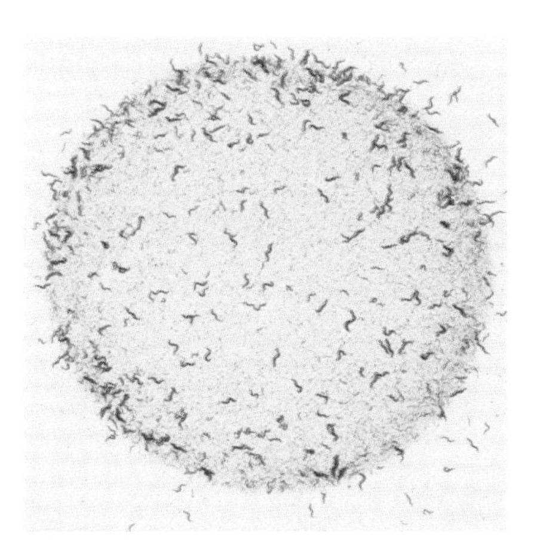

Figure 2.14 *C. elegans* in the laboratory. *C. elegans* crawls on the surface of agar plates, eating bacteria that have been spread as a lawn. Most manipulations are done using a wire pick while viewing the worm through a dissecting microscope.

C. elegans has two sexes, but no females

The life cycle of *C. elegans* consists of an egg stage, four larval (or juvenile) stages termed L1 through L4, and an adult stage (Figure 2.15). The stages of the post-hatching worm are separated by molts at which the worm secretes a new cuticle and sheds the old one. The entire life cycle at 20°C takes about 4 days from the time an egg is laid until that worm lays eggs. An alternative to the L3 stage, the dauer larva, is formed under stress conditions. Mutations that affect dauer larva formation are one of the major topics in later chapters of the book.

One of the distinctive features of *C. elegans* is that it has self-fertilizing hermaphrodites rather than true females (Figure 2.16A). A hermaphrodite, which is morphologically very similar to the females of related

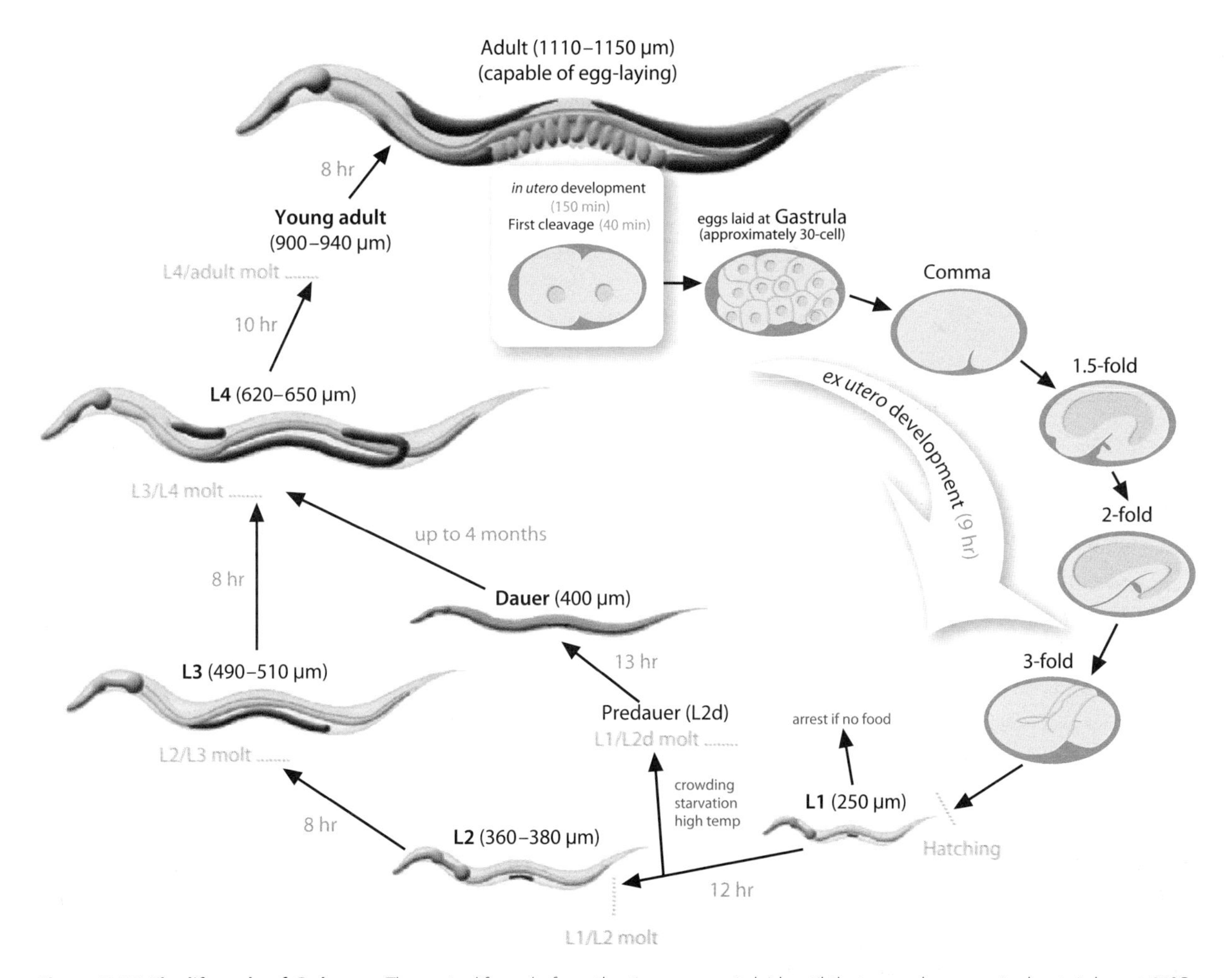

Figure 2.15 The life cycle of *C.elegans*. The entire life cycle from the time an egg is laid until that worm lays eggs is about 4 days at 20°C. The cycle consists of an egg, four larval stages (L1–L4) separated by molts when the cuticle is replaced, and finally an adult. The dauer larva is an alternative to the L3 stage, which forms under stress conditions. Dauer larva formation is described in more detail in Chapters 9–12.

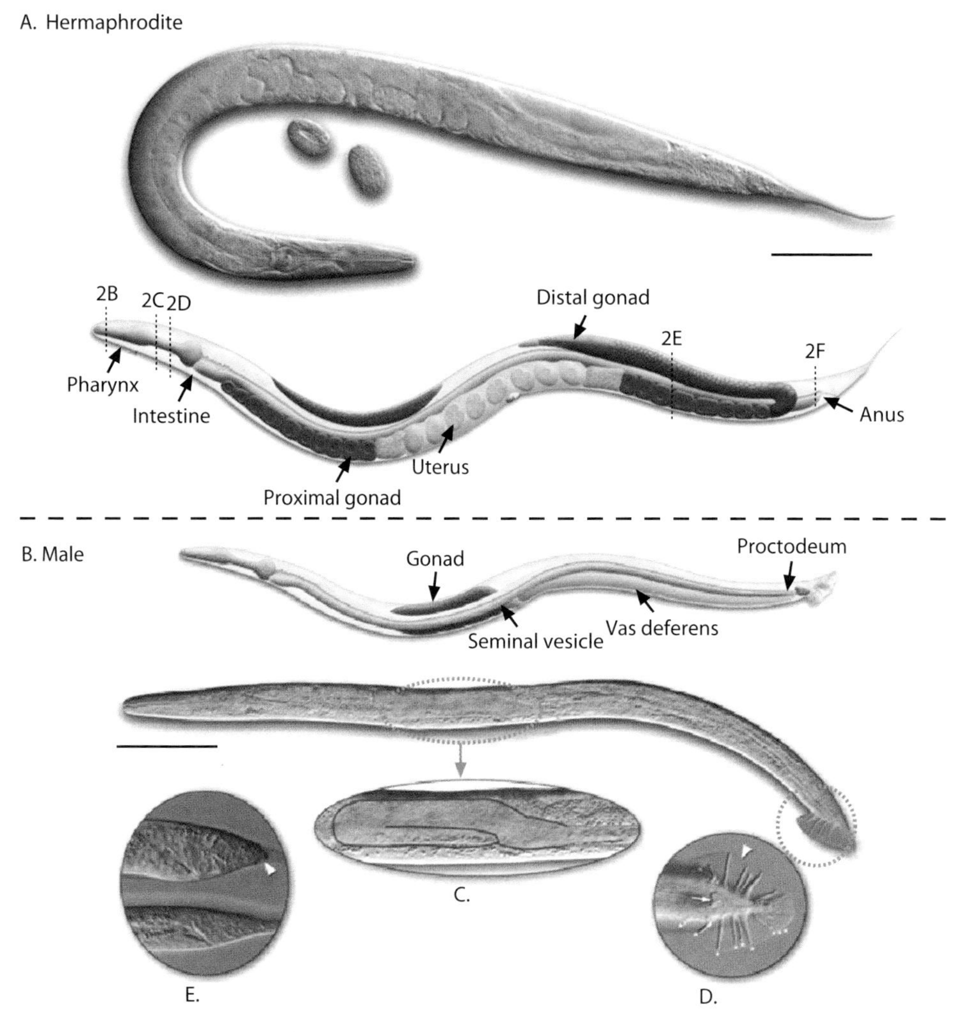

Figure 2.16 Sex differences in *C. elegans*. Worms have two sexes, hermaphrodite (panel A) and male (panel B). Hermaphrodites have two X chromosomes (XX), whereas males have one X chromosome and no other sex chromosome (X0). Hermaphrodites have a gonad with two arms centered on the uterus and a vulva in the middle of the animal. The male gonad has one arm that begins in midbody and extends to the posterior (C). Males have an elaborate fan-shaped tail that is used in mating (D), while the hermaphrodite tail is a simple spike (E). Scale bar, 0.1 mm.

nematode species, produces and stores about 300 sperm at the beginning of meiosis. Spermatogenesis is then shut off and oogenesis begins. The sperm will fertilize an ovum from the same germline internally if no sperm from males are present. The capacity for self-fertilization makes it easy to maintain homozygous strains, and even mutations with severe effects on movement and morphology (such as paralyzed mutants) can be maintained as homozygous hermaphrodites. The number of sperm produced by a hermaphrodite determines the number of self-progeny, so a wild-type hermaphrodite produces about 300 offspring.

In order to perform a genetic cross, males are mated to hermaphrodites. A male is shown in Figure 2.16B. Males arise spontaneously during hermaphrodite self-fertilization at a frequency of about 0.2 percent. Sex is determined by the number of X chromosomes. Two modes of reproduction, self-fertilization by hermaphrodites and cross-fertilization between males and hermaphrodites, are possible (Figure 2.17). Most of the offspring of a hermaphrodite are themselves hermaphrodites. Hermaphrodites have a pair of X chromosomes (abbreviated XX), whereas males have a single X chromosome and no Y chromosome (abbreviated X0). Thus, loss of an X chromosome during meiosis in the hermaphrodite produces a nullo-X gamete, which results in a male when fertilized by a gamete with the usual one X chromosome (Figure 2.17A). Mating

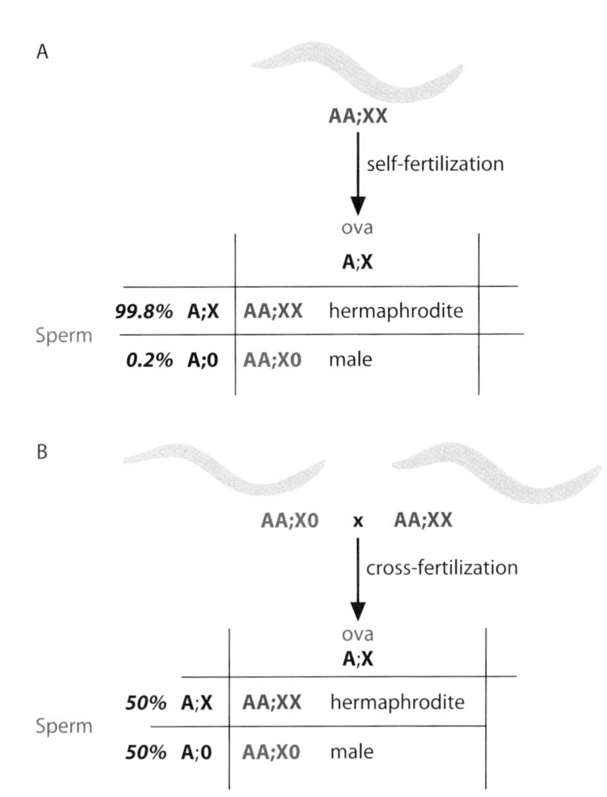

Figure 2.17 Modes of reproduction in *C. elegans*. *C. elegans* can reproduce by either self-fertilization in a hermaphrodite (panel A) or cross-fertilization between a male and a hermaphrodite (panel B). A haploid set of five autosomes is indicated by **A** and the X chromosome by **X**. A diploid hermaphrodite has two X chromosomes (**AA;XX**) whereas a male has only one X chromosome (**AA;X0**). As shown in panel A, a hermaphrodite produces both sperm and ova which can fertilize each other internally. Approximately 99.8 percent of the gametes have an X chromosome, so most of the offspring of self-fertilization are hermaphrodites. About 0.2 percent of the gametes lack an X chromosome, symbolized by **A;0**. When a gamete without an X chromosome fertilizes a gamete with an X chromosome, the offspring is **AA;X0**, which is male. As shown in panel B, a male can mate with a hermaphrodite for cross-fertilization. Since half the sperm produced by a male have an X chromosome (symbolized **A;X**) and half lack an X chromosome (symbolized **A;0**), half of the offspring of cross-fertilization are hermaphrodites and half are males.

of a male to a hermaphrodite produces 50 percent males and 50 percent hermaphrodites among the cross progeny (Figure 2.17B), and male strains can easily be created and maintained by mating with hermaphrodites. Although sperm from a male out-compete self-sperm from an hermaphrodite by an unknown mechanism during a mating, for all practical purposes the offspring of a male-by-hermaphrodite mating include both cross-progeny and self-progeny. Males and hermaphrodites are readily distinguished, with the elaborate fan-like tail in the male being the feature that is most easily recognized. The sexes also differ in size, movement, and behavior.

Nematodes have precisely defined cell lineages

One of the unusual features of *C. elegans*, like all nematodes, is that the number of cells is small and largely invariant between individuals. This feature was noted more than a century ago by classical embryologists such as Theodor Boveri, who was able to track the early cell divisions in the embryo of the parasitic nematode *Ascaris*. *C. elegans* is transparent, so all the internal structures and cells can be observed in live worms. All the cell lineages from fertilization to adulthood have been traced in *C. elegans*, and so the complete ancestry of every cell at every stage is known. The cell lineage of the early embryo is shown in Figure 2.18. The precision in the location and timing of the divisions is impressive; an observer can watch a cell divide, take a break for a few minutes, and come back in time to see the next cell division occur on schedule. The relatively few variable cell fates involve discrete 'decisions' between cells; if a cell adopts cell fate and lineage pattern A, its partner cell adopts cell fate and lineage pattern B, as illustrated in Figure 2.19 for the anchor cell and the ventral uterine cells. Because they are so precisely delimited, many of these cell signaling events have been thoroughly studied by genetic analysis.

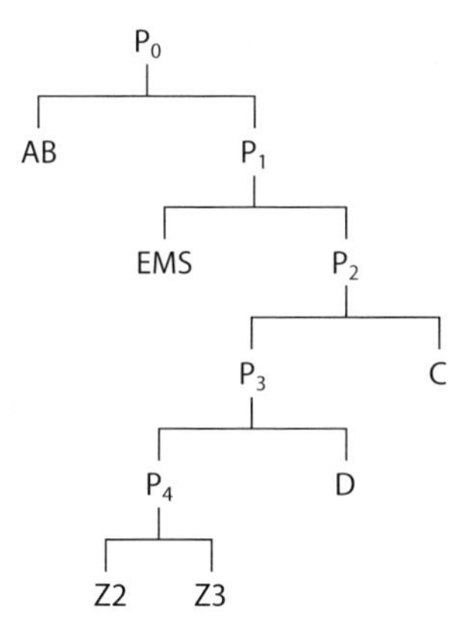

Figure 2.18 The early embryonic lineage of *C. elegans*. The lineage of every cell in *C. elegans* has been determined by observing cell divisions in living worms. This diagram shows the establishment of the early cell lineages in the embryo. The lineage diagrams are drawn with time on the vertical axis, and the earliest cells at the top. In this and many lineage diagrams, anterior cells are on the left and posterior cells on the right. Thus, the fertilized egg P_0 divides to produce two cells, the anterior cell named AB and the posterior cell named P_1, and so on.

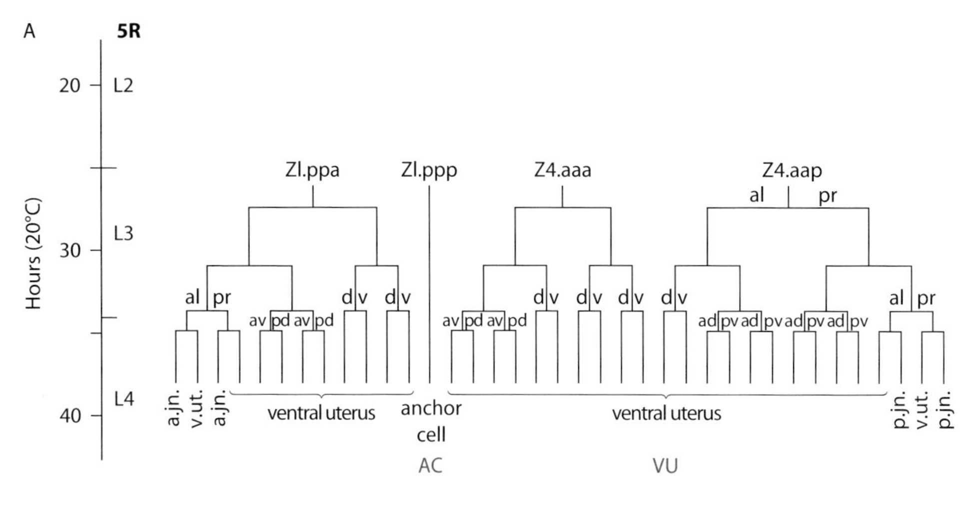

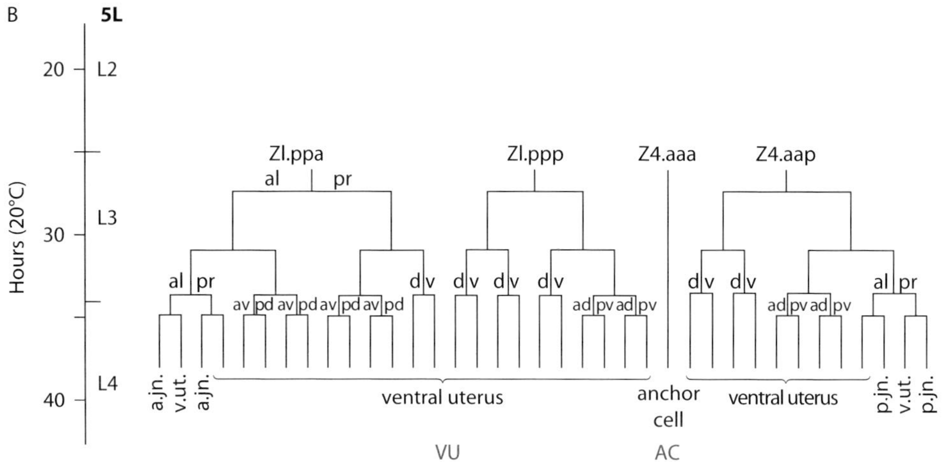

Figure 2.19 Cell decisions in *C. elegans*. Most cell lineages and fates are invariant between individuals in *C. elegans*. Even for those few variable cell lineages, the variation is strictly limited. This example shows the cell lineages of four cells—Z1.paa, Z1.ppp, Z4.aaa, and Z4.aap—which produce the anchor cell (AC) and the ventral uterus (VU). The AC can be derived from either Z1.ppp (at the top as pattern 5R) or from Z4.aaa (at the bottom as pattern 5L). Note that in each case, either Z1.ppp or Z4.aaa does not divide and forms the AC. The other cell undergoes a series of mitotic divisions and produces cells of the VU. In either pattern, the flanking cells Z1.paa and Z4.aap do not change their cell lineage. The regulation begins during the L3 stage with the cell divisions occurring during the L4 stage, as shown in the time axis at the left of the figure.

The precise cell *lineages* also result in precise cell *locations* since there is relatively little cell migration. Thus, the intercellular connections have also been described precisely and reconstructed by electron microscopy. For example, there are exactly 302 neurons in the adult hermaphrodite, and the complete wiring diagram with all the connections has been compiled from electron micrographs. The roles of individual cells have been determined by genetic analysis (i.e. finding a mutation in which the cell does not form) and by laser ablation. Typically, if a cell is destroyed by a laser beam, no other cell replaces it and the worm will lack the function of that cell. For those cells whose functions are replaced, the pool of cells that are capable of adjusting their fate and cell division pattern is known. This inability to replace missing cells has been a particularly useful quality for studying cell interactions and neurobiology. Laser ablation of a certain sensory neuron can result in failure to respond to a specific environment signal such as some class of chemical or a gentle touch.

It was noted during observations of the cell lineage pattern that some cells are produced and then die. These were among the first detailed descriptions of the programmed cell deaths that occur during metazoan development. Genetic, cellular, and molecular analysis of this cell death process in *C. elegans* formed the foundation for understanding the broader process of **apoptosis** in other animals. Apoptosis is a central process in immunology, neurobiology, developmental biology, and cancer biology, among other fields. Thus, what might have been thought to be a taxon-specific feature of nematode cell lineages in fact turned out to be a universal feature of animals. This pioneering analysis was made possible by the precise cell lineages in *C. elegans*, where mutant worms that did not undergo the programmed cell deaths could be found. Nearly all the genes identified as affecting cell death in *C. elegans*, termed the *ced* genes, have one or more orthologous genes in humans, often carrying out a very similar cellular function. In fact, some of the mammalian genes involved in apoptosis are capable of functionally replacing the equivalent gene in *C. elegans*, indicating that not only the function of the gene itself but also its ability to trigger responses from other genes is conserved during evolution.

Transformation of *C. elegans* is done by microinjection

Foreign DNA is introduced into *C. elegans* by one of the most inelegant methods, direct injection of linear DNA into the gonad using a needle, as illustrated in Figure 2.20. The injected DNA forms a complex linear array, usually a head-to-tail concatamer of 50–200 copies. Somewhat remarkably, this DNA array is maintained stably during both mitotic and meiotic divisions, although it can be lost at unpredictable and sometimes high frequencies.

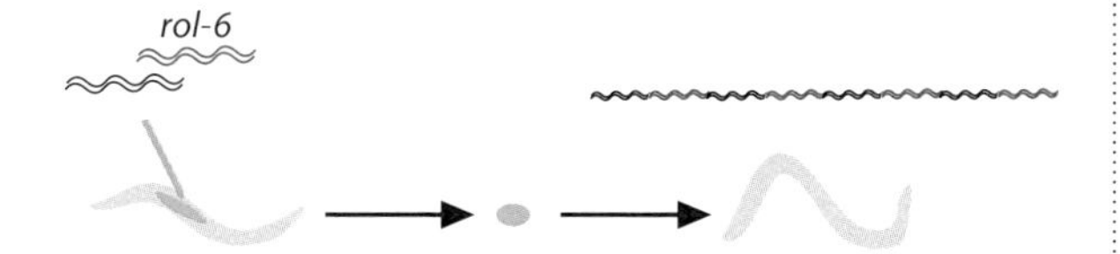

Figure 2.20 Transgenic worms and microinjection. Transgenic worms are produced by injecting DNA directly into the gonad. As shown in the figure, the DNA mix to be injected often includes a dominant mutant allele of the *rol-6* gene, which causes the worm to move in a corkscrew motion. The DNA is often, but not always, co-injected with a dye, shown here in green. The resulting egg grows into a worm with a rolling movement. The DNA has been made into a complex array of both the *rol-6* gene and the other DNA, with as many as 200 copies in a long linear array.

Maintenance of the injected extrachromosomal array is a function of an unusual property of nematode chromosomes. Unlike the situation in most other eukaryotes, nematode chromosomes do not have localized centromeres. Instead, microtubules from the mitotic spindle can attach anywhere along the length of the chromosome. In classical cytology, chromosomes like this without localized centromeres and spindle attachment are termed **holocentric**. The proteins associated with the kinetochore in other organisms are also found in *C. elegans*, but they localize along the length of the chromosome rather than being targeted to one specific region at the centromere.

Thus, when DNA is injected into the gonad of a *C. elegans* hermaphrodite in a transformation experiment, it essentially forms a complex repeating sequence array; complex repeat sequences are found at the centromeres of nearly every other eukaryote. Apparently, a complex repeat structure provides enough structure to function for spindle attachment, and the DNA array is transmitted during cell division. In order to determine that a particular offspring of an injected hermaphrodite has inherited the extrachromosomal array, some additional genetic marker is usually included in the injection mixture. One widely used marker is a dominant mutant allele of the collagen gene *rol-6*, which causes the worm to move in a corkscrew rolling motion; any rolling offspring have inherited the injected DNA.

Extrachromosomal arrays of injected DNA work very well for many types of transgenic experiments, as described in subsequent chapters. However, other experiments such as targeted gene replacements and disruptions require that the DNA integrate into the chromosome. *Stable* integration of the DNA is rare in *C. elegans*, although transformation using ballistic pellets has been more successful for this purpose than microinjection. One limitation of working with *C. elegans* continues to be that targeted gene replacements cannot be done efficiently.

Mutations and the cell lineage patterns are used together for genetic analysis in *C. elegans*

Many examples from genetic analysis in *C. elegans* are used throughout the book. In particular, we describe dauer larva formation in the case studies in

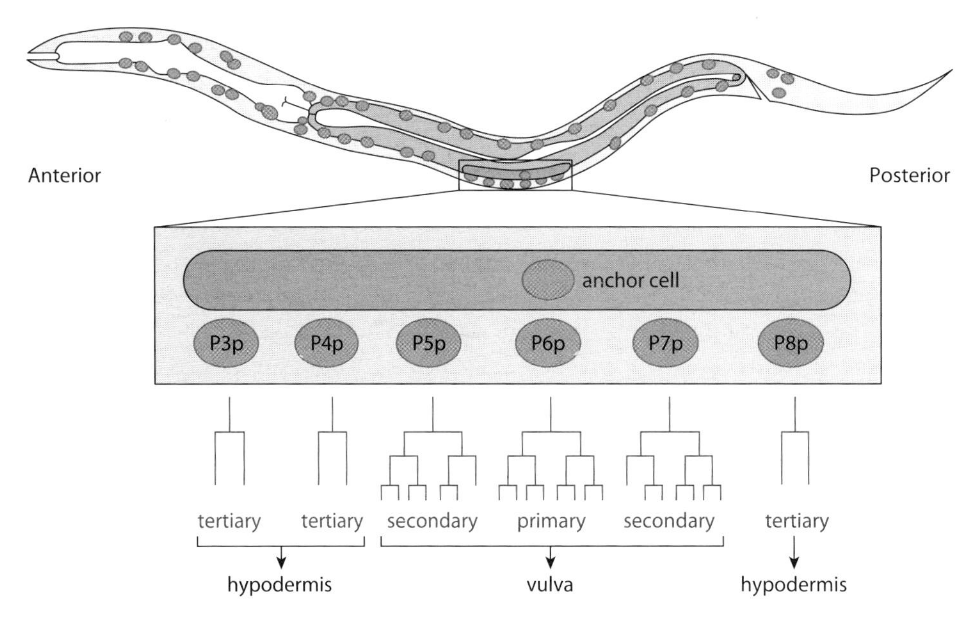

Figure 2.21 The cell lineages that produce the vulva in *C. elegans* hermaphrodites. As described in the text, the vulva and surrounding structures arise from the cell divisions of three cells, known as P5.p, P6.p, and P7.p. In wild-type worms, P6.p responds to a signal from the anchor cell, which lies directly above it in this diagram, and divides to produce eight cells. This is known as the primary lineage. The flanking cells, P5.p and P7.p, each produce seven cells by the secondary lineage. P3.p, P4.p, and P8.p each form part of the hypodermis in wild-type worms by the tertiary lineage. P3.p, P4.p, and P8.p can regulate their cell division pattern and produce the primary or secondary lineages in various mutants or with different experimental manipulations.

Chapters 9 and 12 and in the main text of Chapters 10 and 11; sex determination in worms is also used as an example in Chapters 10 and 11. In this section, we describe the development of the vulva to illustrate how the cell lineages and mutations were used together in a genetic analysis of a signal transduction pathway.

The vulva in the adult hermaphrodite has 22 cells in precisely defined locations arising from invariant cell lineages. Three cells, called P5.p, P6.p, and P7.p, undergo a series of cell divisions during the L4 stage to produce the vulva in the wild-type (Figure 2.21). Two different patterns of cell lineages occur: the primary lineage, which generates the eight cells produced by P6.p, and the secondary lineage, which generates the seven cells produced by P5.p and P7.p. The neighboring cells known as P3.p and P4.p (on the more anterior side) and P8.p (on the posterior side) do not participate in vulva formation in the wild-type worm, but may regulate their cell lineage pattern to produce vulva cells in mutant worms, so they are included in the figure; their cell division pattern is called the tertiary lineage. Collectively, the cells P3.p–P8.p are termed the vulva precursor cells (VPCs). The anchor cell, which does not form part of the vulva, lies adjacent to P6.p. When the anchor cell is destroyed using a laser, a vulva does not form and none of the cells adopt the cell lineage pattern that they do in the wild-type. This provides evidence that the anchor cell sends a signal to the VPCs to trigger vulva formation. In particular, other laser ablations have shown that the signal between the anchor cell and P6.p is critical for normal vulval development.

Many different mutations that affect the lineages of the vulva have been found among mutant worms that are defective in egg-laying. Broadly speaking, these mutants have one of two phenotypes (Figure 2.22). Some do not form a vulva at all, a phenotype known as Vulvaless (Vul, Figure 2.22A). In view of the role of the anchor cell in relation to P6.p, some of these mutants are easy to understand. In some, the anchor cell itself does not form or it lies in an unusual location. In others, P6.p and the other VPCs do not form normally. Thus, these do not form a vulva because one of the component parts, either the signaling anchor cell or the receiving VPC, is absent.

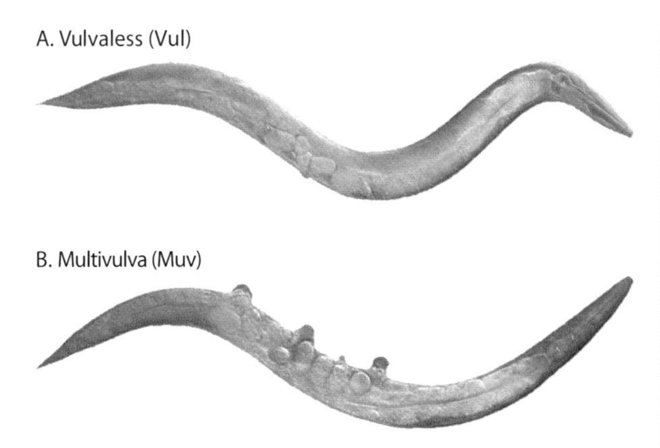

Figure 2.22 Mutant phenotypes from the vulva lineages. Two principal classes of mutations that affected the vulva lineages in Figure 2.20 were found. Panel A shows a mutant that lacks a vulva, a phenotype known as Vulvaless or Vul. Among others, genes that affect the receptor on the surface of the P6.p cell are in this category. Panel B shows a mutant in which more than one of the Pn.p cells has adopted a primary lineage, resulting in multiple partial vulvae, a phenotype known as Multivulva or Muv. Among others, genes that affect the ability of the P6.p cell to transmit a lateral inhibitory signal to the P5.p and P7.p cells are in this category.

Other Vulvaless mutants have both the anchor cell and the vulva precursor cells, but still do not form a vulva. Subsequent analysis, using procedures described in subsequent chapters, showed that, in some of these mutations, the defect lies in the anchor cell, which seems not to send a required signal. In other mutations, the defect lies in the VPCs, which behave as if they are not receiving the signal or are not passing the signal on within the cell (or 'transducing' the signal) properly.

Another class of mutant phenotype is shown by worms in which more than one of the VPCs follows the primary lineage that only P6.p does in wild-type. These mutants have multiple partial vulvae, a phenotype known as Muv (Figure 2.22B). At first sight, these mutants are not so easy to explain. Some mutations cause transduction of the intracellular signal in the absence of the signal from the anchor cell. The cells such as P4.p or P8.p, which normally do not receive the signal from the anchor cell and thus do not respond by producing vulva lineages, are now responding constitutively. Another group of mutations affects a different signal transduction process among the vulva precursor cells. Once P6.p receives the signal from the anchor cell, it signals its neighboring cells and inhibits them from responding to the anchor cell signal—a process known as **lateral inhibition**. The normal response of P5.p, for example, arises from a composite of the amount of the signal it receives from the anchor cell and the amount of the inhibitory signal from P6.p. If the inhibitory signal is removed by mutation, P5.p or P7.p can also follow the primary lineage. By combining their knowledge of the cell lineages with the mutant phenotypes, the investigators were able to recognize how the cells signal to each other during vulval formation.

These mutations identified two different cellular pathways involved in human cancers

The analysis of this complex biological question relied on knowing the cell lineages in worms and the ability to find mutations that affected these cells and their interactions. All the analysis was done in the absence of knowledge of the molecular identity of the signals and without knowing initially how many different extracellular signal molecules were involved. By having genes that affect the processes, i.e. mutants that failed in some parts of the differentiation but not in others, it was possible to learn the developmental and cellular logic of the signal transduction process. Once the genes were cloned, by processes similar to those described in Chapter 5, the molecular functions of these signal transduction pathways were revealed: the signal from the anchor cell is a molecule similar to epidermal growth factor, and the receptor on P6.p is a protein in the class of epidermal growth factor receptors. The Vulvaless mutations that failed in the intracellular signal transduction process were even more interesting. These genes defined the components of the Ras/Raf Signal transduction pathway, a highly conserved signal transduction pathway in animals. Interestingly, some of the individual genes and proteins, including Ras and Raf, had been identified in mammals as oncogenes since many tumors have mutations in one of these genes. The pathway was learned from genes, and the connection to cancer biology was made when the genes were cloned.

(To be historically accurate, at about the same time that the vulva lineages in worms were being analyzed, the response of *Drosophila* to blue light was being studied. This response also depends on the Ras/Raf pathway, and the two genetic analyses together connected the signal transduction pathway to the defects in the cancer cells.)

The lateral inhibition response depends on a different signal transduction pathway. The genes in this pathway were first described by mutations in *D. melanogaster*, including the gene *Notch* by which the pathway is now known. Again, the signal transduction pathway among these cells was inferred from the phenotypes of the mutant worms and by the interactions among the mutations, the analysis described in Chapter 10 and 11. Later, some human tumors and other genetic diseases were found to have mutations in the Notch pathway.

By the genetic analysis of a very specific phenotype, the ability of a worm to make a vulva and lay eggs, two signal transduction pathways that are used in all animals were uncovered. These are further examples of universal properties of eukaryotes being revealed by the model organism. The equivalent signal transduction pathways in humans govern entirely different developmental processes to vulva formation. The pathway is the same, but evolution has 'tinkered' with its use and applied it to different developmental processes.

The *C. elegans* genome database, Wormbase

The genome database for *C. elegans* is called Wormbase. The opening screen on Wormbase provides news and other information. A typical entry point is to type in the name of a gene of interest, which is the method we have used in Figure 2.23. The gene is *daf-7*, a gene affecting dauer larva formation that is used again in later chapters.

One feature of genome databases that is often confusing to newcomers is the names of the genes. Although nomenclature is discussed in Chapter 3, our introduction to Wormbase can be used to illustrate a key point. In worms, as in most model organisms, a gene such as *daf-7* was first identified, named, and mapped by the phenotype of mutant worms, often without any knowledge of the molecular identification of the gene. The *C. elegans* genome was assembled and sequenced from a variety of clones, including cosmids, but this was done

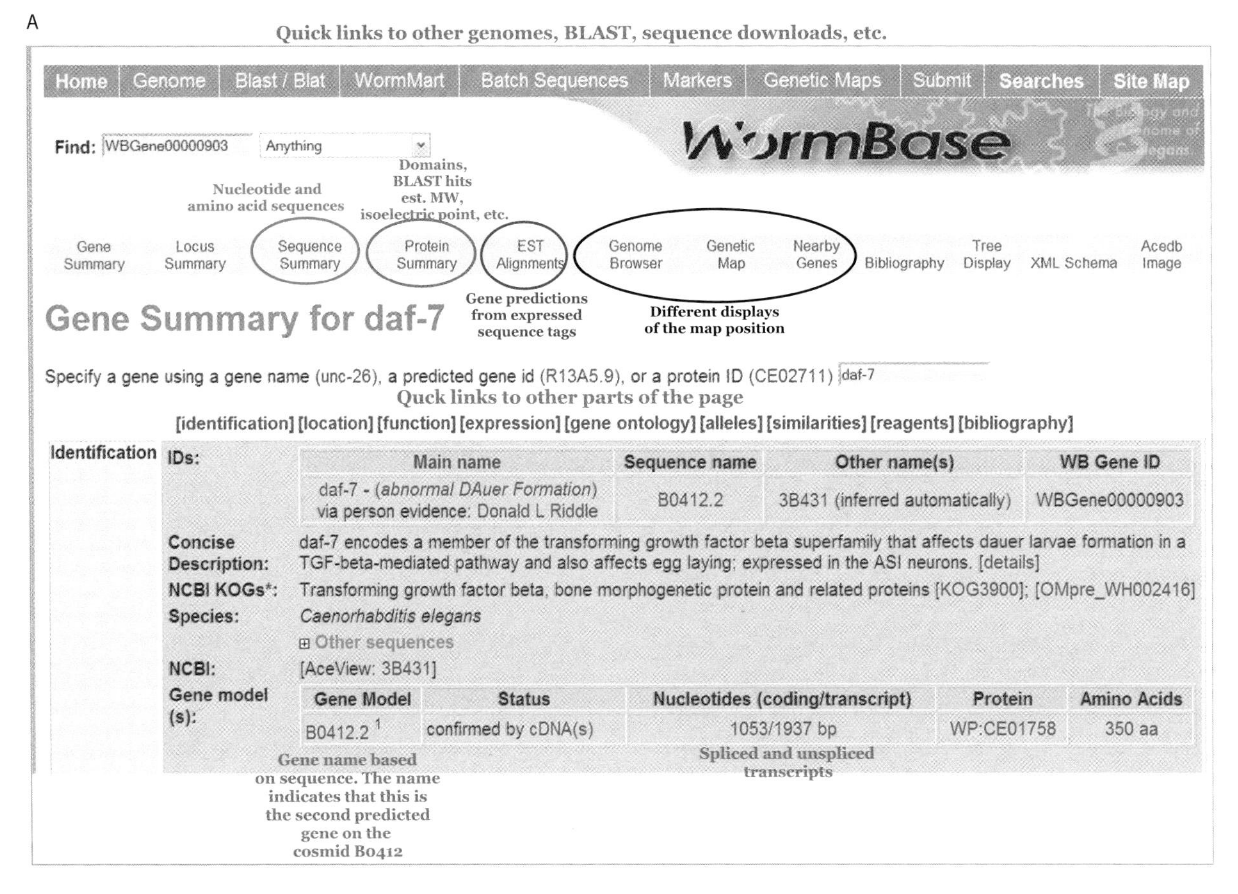

Figure 2.23 The Wormbase page for the gene *daf-7*. (A) Our annotations that do not appear on the original Wormbase page are shown in orange or with colored circles and accompanying text. The top of the *daf-7* page is shown here. Quick Links to other useful resources are shown across the top. Note the gene names *daf-7* and the gene model name B0412.2. These describe the same gene, as detailed in the text.

B

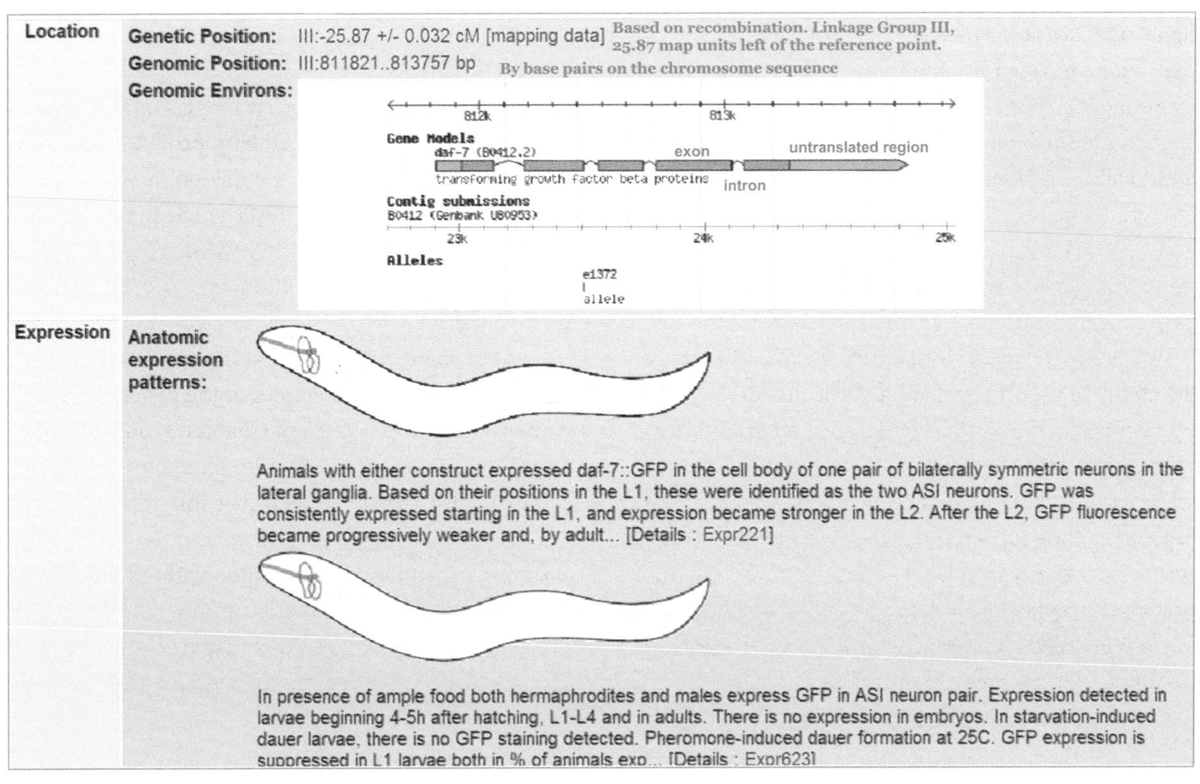

C

Function

Legacy information: e1372ts : constitutive dauer formation at 25x reversible at 15x; Type C Egl at all temperatures. ES3 (L3). NA6 (m70 m62amber n696 etc.; all alleles Egl ts Daf).
See also e1372, n696, n697, n708
[C.elegansII] e1372ts : constitutive dauer formation at 25C reversible at 15C; Type C Egl at all temperatures; dark intestine; clumpy;Srf-6 at 16C. ES3 (L3). OA11: m70, m62amb, n696, n708, m434::Tc1, etc..All alleles Egl and ts Daf. Cloned : encodes novel member of TGFbeta superfamily, possible ligand for DAF-1 receptor. [Golden and Riddle 1984; ; Thomas et al. 1993; DR; JT]
Definitions of abbreviations used in the text.

Alleles and mutant phenotype

Phenotypes: ⊞ phenotypes via RNAi

Interactions: ⊞ Interactions

Microarray Expression Data: SMD_B0412.2
193876_at
Aff_B0412.2
cea2.c.16344

Expression Cluster: cgc4489_group_10
WBPaper00025040:N2_Expressed_Genes
WBPaper00031003:24hr_muscle_depleted
WBPaper00031003:hlh_1_enriched
WBPaper00031003:total_muscle_depleted
[cgc6390]:Cluster_A

Microarray expression

Microarray "topology map" position: Mountain 1 [see Kim et al. Science (2001)]

Regulation on Expression Level: Changes localization of daf-16 [details]
Changes localization of daf-16 [details]
Positively regulates daf-28 [details]
Negatively regulates sta-1 [details]
Negatively regulates tph-1 [details]
Positively regulates daf-16 [details]
sod-3 [details]
Positively regulated by cam-1 [details]

Gene interaction

Figure 2.23 (Continued). (B) The gene structure and the expression pattern are found further down the page. The gene structure shows that *daf-7* has five exons (in pink) and both a 5′ and a 3′ untranslated region, shown in gray. The gene is transcribed from left to right, as indicated by the arrow. (C) Scrolling further down the page shows the mutant phenotype, some of the available alleles, and information about gene expression and genetic interactions.

D

Similarities	Best BLASTP matches to longest protein product (full list): (Show alignments):	Species	Hit	Description	BLAST E-value	% Length
		C. remanei	RP12288 *Note: **C. remanei** predictions are based on an early assembly of the genome. Predictions subject to possibly dramatic revision pending final assembly. Sequences available on the WormBase FTP site.*		4.2e-85	98.3%
		C. briggsae	BP:CBP13064	gene CBG24910	6.6e-82	93.1%
		A. caninum	TR:Q58J20	Transforming growth factor beta.	2.6e-55	93.4%
		D. rerio	SW:O42222	Growth/differentiation factor 8 precursor (GDF-8) (Myostatin).	2.7e-24	90.0%
		H. sapiens	ENSEMBL:ENSP00000257868	Growth/differentiation factor 11 precursor	5.5e-20	91.1%
		D. melanogaster	FLYBASE:CG16987-PB	Flybase gene name is Alp23B-PB	3.3e-15	41.1%

Reagents		
	Transgene Strains:	edIs24 mIs6 saIs8 ksIs2 saIs7
	ORFeome Project primers & sequences:	mv_AAC47389 OSTF155F1_1
	Primer pairs:	cenix:327-h8 TH:327H8 sjj_B0412.2
	SAGE Tags:	SAGE:cgcgcagctcacaaagt SAGE:atttcaaccc SAGE:atttcaaccccgtagtc
	Matching cDNAs:	EB996372 EC018447 U72883 yk1070f11.3 yk1199g10.5 EB997181 EC021930 yk885f02.3 yk1070f11.5 yk1572d07.3 EC004319 OSTF155F1_1 yk885f02.5 yk1199g10.3 yk1572d07.5

PCR primer sequences to amplify the gene

Figure 2.23 (Continued). (D) The bottom of the page shows the orthologs of *daf-7* in other organisms as identified by BLAST, ranked by the similarity to the *C.elegans* gene. Other available reagents are also shown, with links.

without relying on knowledge of the genetic names and locations of genes. The locations and structures of the genes on each cosmid were predicted by examination of the DNA sequence. Thus, when the cosmid B0412 was sequenced, several genes were predicted on it and named B0412.1, B0412.2, etc. Eventually the genetic name *daf-7* and the sequence name B0412.2 were connected, i.e. the *daf-7* gene has been cloned and the corresponding DNA sequence is B0412.2.

Wormbase and SGD have similar information for each gene but organize it in different locations on the page. In general, Wormbase places more information on the screen whereas SGD emphasizes summaries with links to other pages. With a little patience and practice, navigating these genome databases becomes much simpler. Patience and practice are needed simply because so much information is organized and displayed. Wormbase also has a link to the Worm Book, an extensive online set of review articles on all aspects of *C. elegans* genetics and molecular biology, and to the Worm Atlas, which contains remarkably detailed and instructive images of the worm.

2.4 *Drosophila melanogaster*: if you have to ask . . .

Unlike those who work with yeast or worms, *Drosophila* researchers have usually felt no need to tout the use of genetics in their organism. Working on flies was like playing the blues—if you have to ask for an explanation, you probably wouldn't get it anyway. No other experimental organism has such a long and continuous history of research involving so many geneticists for so many years. The first mutation in flies was *white* eyes, found in 1908—the mutation that persuaded scientists that genes were part of chromosomes. The first paper published in the journal *Genetics* was devoted to research on *Drosophila* and, in fact, more than half of the pages in the first two issues of *Genetics* were devoted to Bridges' classic article 'Non-disjunction as proof of the chromosome theory of heredity'. A comparison of some of the properties of *Drosophila* with other model organisms is shown in Table 2.1.

This history of research has generated a huge catalog of mutants, chromosome rearrangements, and other tools of genetic analysis in *Drosophila*. More recently, *Drosophila* has again moved to the forefront with genome sequences of 12 different species from the genus. Not many other genera have 12 well-described species, and none has genome sequences from all of them. Reducing genetic research in *D. melanogaster* to a few paragraphs is a daunting task; after all, a series of books entitled *The Genetics and Biology of Drosophila* was compiled before most molecular analysis or any genomic analysis and still ran to 12 volumes.

D. melanogaster is raised in glass or plastic vials, originally in pint milk bottles with mashed bananas, but now with commercially available fly food, which includes corn meal, agar, and yeast. At 24°C, the life cycle takes about 10 days, with complete metamorphosis from crawling larvae to flying adults (Figure 2.24). There are three distinct larval stages, called **instars**, followed by a period of pupariation; as pupae, most of the larval cells and organs degenerate to be replaced in the adult

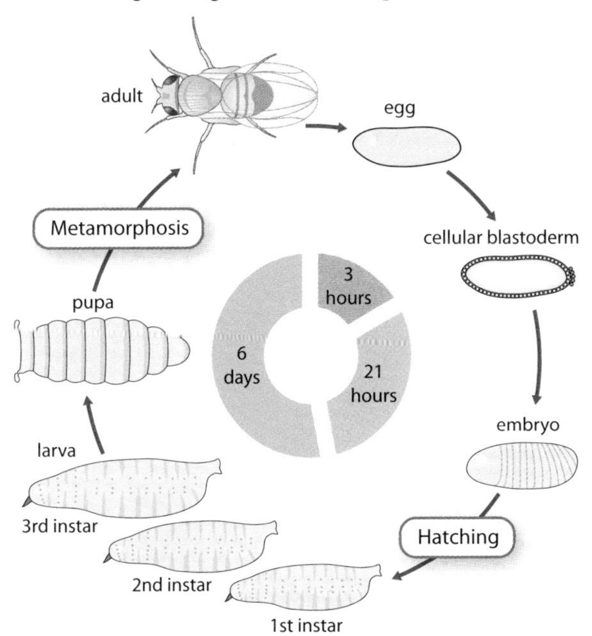

Figure 2.24 The life cycle of *D. melanogaster*. The entire life cycle of *D. melanogaster* can be completed in slightly more than 10 days, depending on the growth temperature. The three larval instars crawl and feed on their own. During pupariation, most of the larval cells die. Many adult structures arise from imaginal disks, as shown in Figure 2.25.

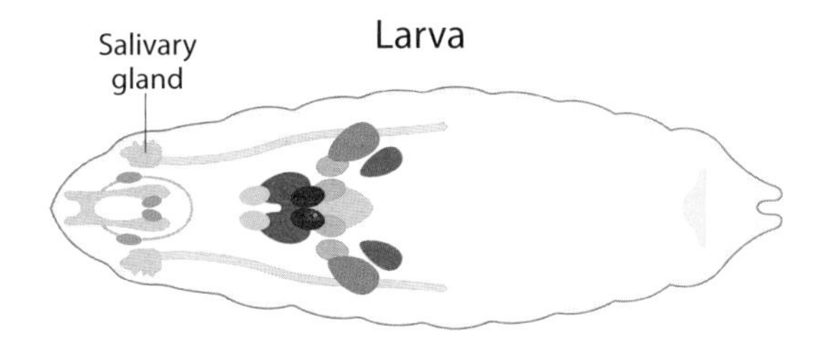

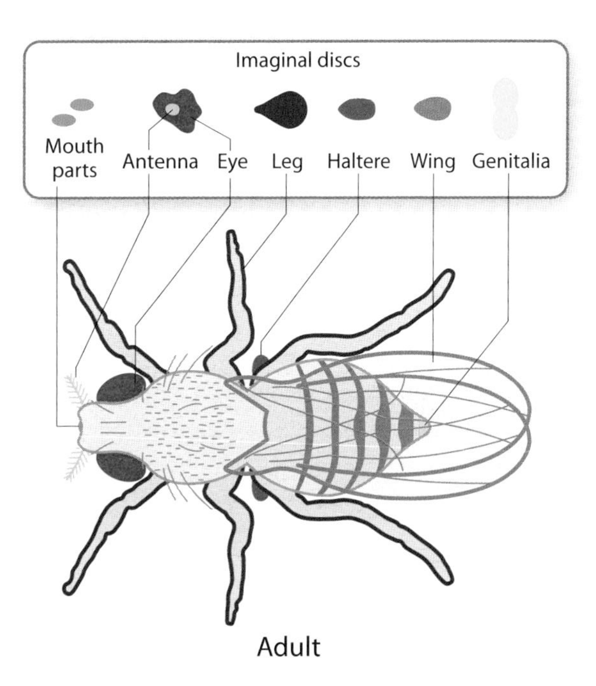

Figure 2.25 The imaginal disks in the *Drosophila* larva. These cells form no structures in the larva but differentiate during the pupa stage to form the adult structures shown.

fly. The third-instar larvae have groups of undifferentiated cells termed the **imaginal disks**; these separate pockets of cells do not degenerate but will differentiate to give rise to the mouth parts, antennae, eyes, legs, wings, halteres (the modified hind wing), and genitalia in adults, as shown in Figure 2.25. When an adult female emerges from the pupa case, or ecloses, she is unable to mate for about 12 hours. These virgin females are used for crosses, and it is common to hear fly geneticists talk about 'collecting virgins'. Mated females can store sperm for several days and continue to mate with other males, so virgin females ensure that the proper cross has been done.

Flies have an intricate external anatomy, with many bristles and hairs, pigmentation differences in the body and the eye, structures such as wing veins, sex combs, antennae, and more. Seemingly, all the anatomical features have corresponding mutant phenotypes—forked bristles, multiple wing hairs, black body, white eyes, crossveinless wings, and so on. Mutant strains are maintained by matings, often involving strains that have many different defined mutations throughout the genome.

D. melanogaster has some peculiar features

Because it is such a well-studied model organism, it is tempting to think of *D. melanogaster* as typical and other organisms as strange. Nonetheless, some aspects of *Drosophila* biology are unusual; more precisely, the biology of dipteran insects (of which *Drosophila* is a representative) affects the way that *D. melanogaster* is used as a model organism. First, unlike other model organisms, *Drosophila* has no recombination in males. In nearly all animals, the sexes have a different number and distribution of crossovers, and genetic maps are typically based on recombination in females. Dipteran insects are extreme in this regard. In *Drosophila*, males segregate their chromosomes at meiosis without chiasmata and recombination, and all crossing over occurs in females. The absence of recombination in males has been exploited by geneticists to maintain particular configurations of alleles, as shown in Figure 2.26.

Secondly, in certain tissues such as the larval salivary glands, *Drosophila* chromosomes replicate 10 or 11 times but the DNA strands do not separate. This results in very thick chromosomes known as **polytene chromosomes** (Figure 2.27). Although the underlying selective advantage for polytene chromosomes is not known for certain, the process in *Drosophila* produces many copies of the genes that encode the proteins that glue the pupa to its resting place; thus, in *Drosophila* at least, polytenization allows the production of large amounts of the glue proteins just before they are needed in development. As seen in Figure 2.27, polytene chromosomes have a distinctive and consistent pattern of light and dark bands, which are easily visible under the light microscope. The banding pattern of polytene chromosomes allowed Bridges and others to construct detailed

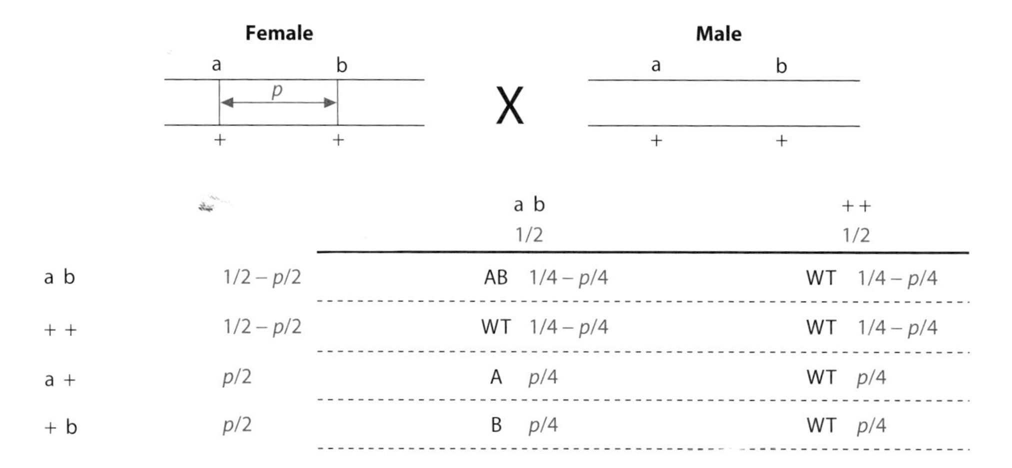

Totals

A B double mutant	$1/4 - p/4$
A not B	$p/4$
B not A	$p/4$
Wild-type	$3/4 - p/4$

Figure 2.26 The absence of crossing-over in *Drosophila* males. The diagram considers two genes, *a* and *b*, separated by the recombination distance *p* in the female. The ova made by the female are shown down the left side of the Punnett square, with the frequency of each gamete shown in red. The sperm made by the male are shown across the top of the Punnett square, and the frequency of each class of offspring is shown within the square. Of particular interest are the recombinant A not B and B not A offspring. In the absence of crossing over, each of these classes arises at a frequency of *p*/4. If crossing over occurred in the male, each of these classes would be approximately twice as frequent, assuming that *p* is small. Correspondingly, the two parental classes would be less frequent if crossing over occurred in the male.

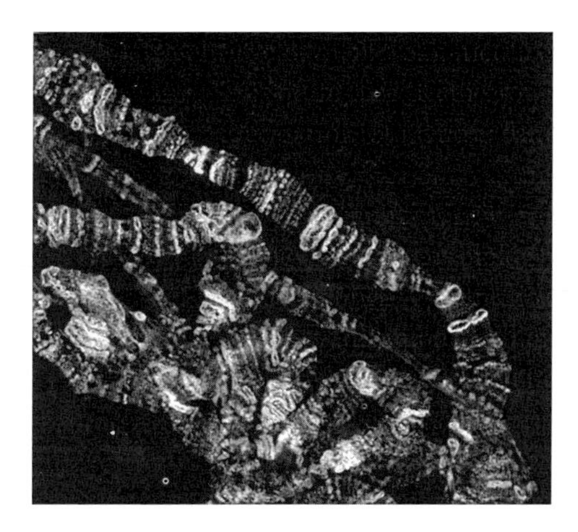

Figure 2.27 Polytene chromosomes in *Drosophila*. The chromosomes from larval salivary glands were stained with the DNA stain DAPI to illustrate the banding pattern. Each chromosome arm is labeled. Polyteny occurs when the DNA replicates but the sister chromatids do not separate.

structural maps of the chromosomes, with each of the five chromosome arms divided into 20 segments and the bands in each arm designated by a combination of letters and numbers. The breakpoints of chromosome rearrangements and the location of genes could be located with respect to these bands, and the position of the gene on the chromosome can be described by different types of coordinates. For example, the *forked* bristle locus is found at position 1-56.7 on the genetic map (i.e. on chromosome 1 and 56.7 map units from the reference point at the left tip) and at position 15F4-7 on the cytological map of polytene chromosomes.

Transformation of *Drosophila* uses a transposable element

The most significant development for *Drosophila* genetics in the molecular age has been the use of P elements as a vector for transformation. P elements are typical DNA transposable elements, and have a structure and mechanism similar to DNA elements in other eukaryotes. A complete P element, which is about 2.9 kb in length, encodes a transposase that acts on the 31 bp inverted repeats at its ends to catalyze transposition of the full element (Figure 2.28A). Like other transposable elements, most P elements are defective and cannot transpose autonomously, either because the transposase or the ends of the element have become mutated. However, an active transposase encoded at any site in the genome can catalyze the transposition of any element with the appropriate repeats at the ends, regardless of what that target element encodes.

P elements were apparently absent from the original laboratory strains of *D. melanogaster* but were present in fruit flies in the wild. *Drosophila* geneticists learned to capture wild male flies (e.g. at produce markets in exotic locations throughout the world) and mate them to laboratory females. The F_1 offspring of this cross are normal but have a number of defects in their germlines that appear in subsequent generations. This syndrome of genetic alterations, known as **hybrid dysgenesis**, includes chromosome breakage, male recombination, sterility, and, most importantly, new mutations.

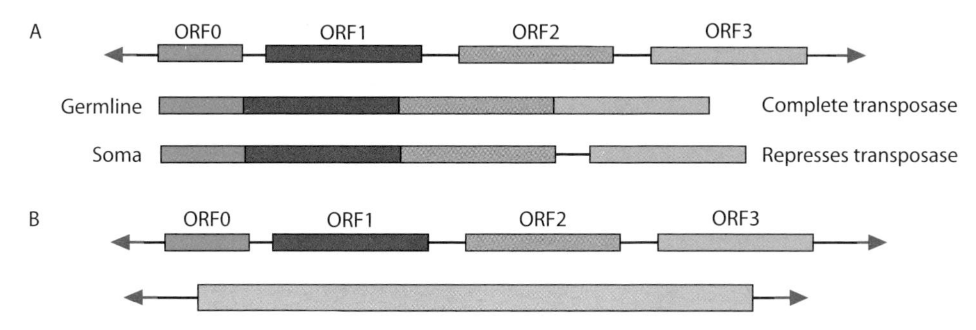

Figure 2.28 P elements in *Drosophila*. The P element is a DNA transposable element that is only mobile in the germline. A complete P element is shown in panel A. It has four open reading frames, referred to as ORF0 through ORF3, and inverted repeats at the ends, as indicated by the red arrowheads. In the germline, the introns are removed and the four exons are spliced together to make active transposase. In the somatic cells, the third intron separating ORF2 from ORF3 is not removed, generating an inactive transposase. In fact, this product represses transposition. Panel B illustrates how the P element can be used as a transformation vector. The transposase exons are removed from the P element and replaced by a gene to be transformed into the fly, shown here in light blue. The inverted repeats are retained at the ends. When this is injected into a fly, it cannot move on its own, but it can be moved by active transposase from a different P element. Integration into the genome occurs at random sites.

During the 1930s, hybrid dysgenesis became a preferred mechanism for producing new mutations at high frequency. The cross had to involve wild males and laboratory females, so paternal factors (hence the name P elements) were clearly implicated. The mechanism for hybrid dysgenesis was not resolved until the late 1970s, although the existence of a transposable element similar to what Barbara McClintock had shown in maize was one long-standing hypothesis. This hypothesis proved to be true; hybrid dysgenesis is caused by the sudden introduction and mobilization of P transposable elements.

The laboratory strains that lacked P transposable elements also lacked the ability to regulate their transposition, so P elements were able to move at an extremely high frequency in these unprotected strains. The unusual feature that the effects are restricted to the germline is explained by alternative splicing of the transposase gene encoded by the P element. The splicing pattern that yields a functional transposase occurs in the germline, whereas the splicing pattern in somatic cells results in an incomplete and non-functional transposase.

Rubin and Spradling harnessed the effects of the P element as a transformation vector by replacing the transposase gene with other DNA sequences; they retained the terminal inverted repeats as target sequences (Figure 2.28B). Since these engineered elements cannot move without a source of transposase, a complete P element encoding the transposase was also included in the transformation mixture. This DNA mixture of a complete P element and an engineered element is microinjected into the pole cells of the developing embryo; the pole cells become the germline in the adult. The injected fly has active P transposase and the engineered P element in its germline, so it is analogous to the F_1 offspring of a dysgenic cross. The P elements, both the complete element and the engineered element, insert at random throughout the genome in the offspring of the injected fly. Once inserted, the P element is inherited along with the rest of the chromosome. Thus, the insertions result in stable inheritance, unlike *C. elegans* in which the injected DNA remains extrachromosomal. On the other hand, unlike yeast, the insertions occur at random locations with respect to the sequences on the engineered elements, and targeted insertions to particular sites are very rare. If the transposase gene is subsequently outcrossed and removed from the chromosome, the engineered element becomes a stable integrant in the genome. (The concept of outcrossing is discussed more completely in Chapter 3 in our discussion of mapping newly arisen mutations.)

Many modifications to this basic procedure of P element transformation have been introduced, but the fundamental strategy has remained the same. The most significant modification has been the production of *Drosophila* strains with the P transposase integrated at known specific sites into chromosomes. The transposase does not have the terminal repeats and cannot move once inserted; these are referred to as 'wings clipped' elements. With integrated transposase, there is no need to inject a complete P element along with the engineered transformation vector. In addition, since the location of the 'wings clipped' element in the genome is known, outcrossing the newly arisen transformed line to eliminate transposase and subsequent transposition is simplified.

Sex determination in *Drosophila* involves a pathway of alternative splicing

Drosophila provides many examples of advanced genetic analysis, of which only a few will arise in this book. We use the genetic analysis of *Drosophila* segmentation as one of the principal examples throughout this book; in particular, the gene *patched* is described thoroughly as one of our main examples of genetic analysis. Sex determination in *D. melanogaster* was introduced in Chapter 1 and will be described here in a little more depth to illustrate how genetic analysis in *Drosophila* has provided insights into an important biological process in nearly all multicellular organisms.

Nearly all animals have some form of sexual dimorphism, so sex determination is a truly fundamental developmental process. Naively, as was the case with many of us in the early 1980s when this was being worked out, sex determination could be considered so fundamental that it would be expected to be highly conserved in evolution. This is not true. Sex-determining mechanisms diverge widely among animals, and many different processes are involved. Very few genes that are involved in sex determination in one species are involved in another species, with one important exception as we will note. The molecular processes involved in sex determination

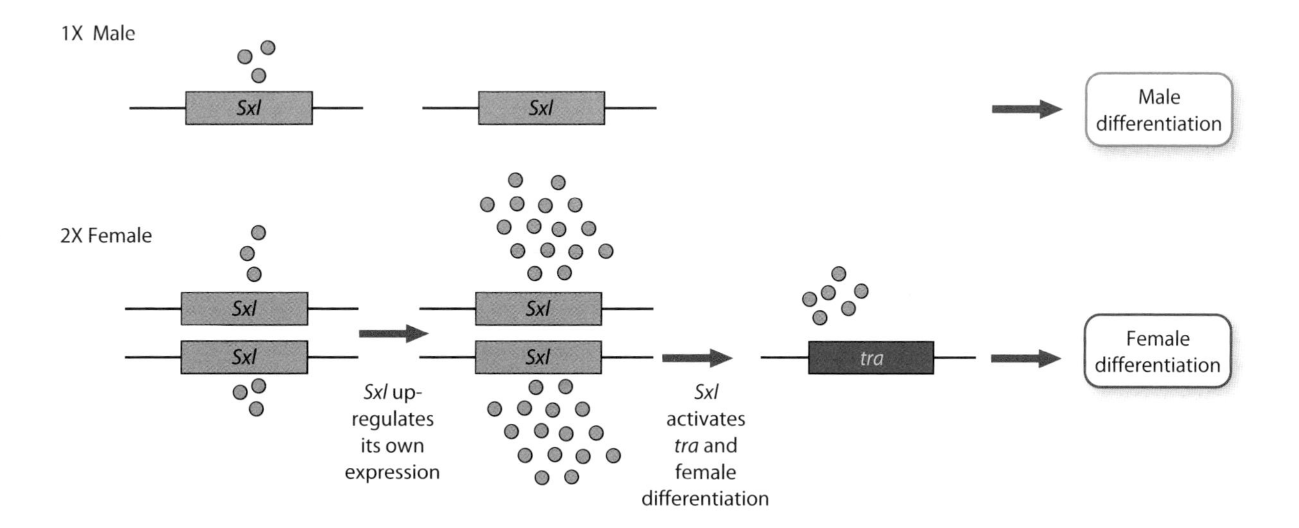

Figure 2.29 Sex determination and the number of copies of the *Sxl* gene in *Drosophila*. Sex determination in *Drosophila* depends on the number of X chromosomes; flies with one X chromosome are males, whereas flies with two X chromosomes are females. The key gene that initiates the sex determination pathway is the X-linked gene *Sex-lethal* (*Sxl*). *Sxl* is expressed very early in the embryo, and because the gene is X-linked, the expression is higher in 2X embryos than in 1X embryos, as shown on the left. In 2X embryos, the expression of *Sxl* is high enough for the gene to auto-activate and regulate its own expression. This auto-activation amplifies a twofold difference in the level of *Sxl* expression into a much larger difference between 1X and 2X embryos. The *tra* gene is a direct target of *Sxl* activation; as shown, this activation only occurs in 2X flies where *Sxl* activity is high. As a result of *tra* expression, female differentiation occurs. In the absence of *tra* expression in 1X flies, male differentiation occurs.

in *Drosophila* are used in all other eukaryotes; they are just not involved in sex determination.

In *Drosophila*, sex is determined by the number of X chromosomes, so that embryos with one X chromosome become males and those with two X chromosomes develop as females. The question of how a *Drosophila* embryo 'counts' the number of X chromosomes is itself an outstanding example of the application of genetic analysis that we can only briefly summarize here. The X-linked gene known as *Sex-lethal* (*Sxl*) is the key regulatory switch.

Sxl is expressed extremely early in *Drosophila* embryos and, because it is X-linked, its level of expression is initially different in those embryos with one X chromosome and those with two X chromosomes. The level of *Sxl* expression from two X chromosomes is sufficient to trigger female differentiation, whereas the level of expression from one X chromosome is not enough to trigger female differentiation and males result (Figure 2.29). Above a threshold, *Sxl* activates its own expression so that this relatively slight difference in expression between 1X and 2X flies becomes amplified by auto-activation, as illustrated for 2X flies in Figure 2.29. *Sxl* activation sets the switch, but what are its targets for female sex differentiation? Potential target genes were already known from among existing mutations.

Mutations that affect sex determination were found in *Drosophila* many years ago and, like sex-determining mutations in worms (which we consider in Chapter 11) and other animals, these mutations often result in **sex reversal** in one sex. For example, recessive mutations in the two genes in *Drosophila* known as *tra* and *tra2* transform 2X flies into the male differentiation pathway. (The 2X males, or more precisely the 2X pseudo-males, are not fertile in *Drosophila* since the germline is not also transformed.) Conversely, 1X *tra* or *tra2* mutant animals are phenotypically normal males.

Since the absence of Tra and Tra2 functions result in 2X animals becoming males with no obvious effect on 1X differentiation, it is inferred that these two genes are needed for normal female development but not for normal male development, at least in the somatic tissues. The genes exert their effects on all somatic tissues. The *tra* gene is a target of *Sxl* regulation, as summarized in Figure 2.29. The *tra2* gene is not a direct target of *Sxl* regulation but works with the Tra protein to regulate genes further downstream.

Sxl encodes a protein that is involved in RNA splicing, and its effects can be explained by its regulation of alternative splicing. The *Sxl* protein has numerous RNA targets, including its own transcript and that of

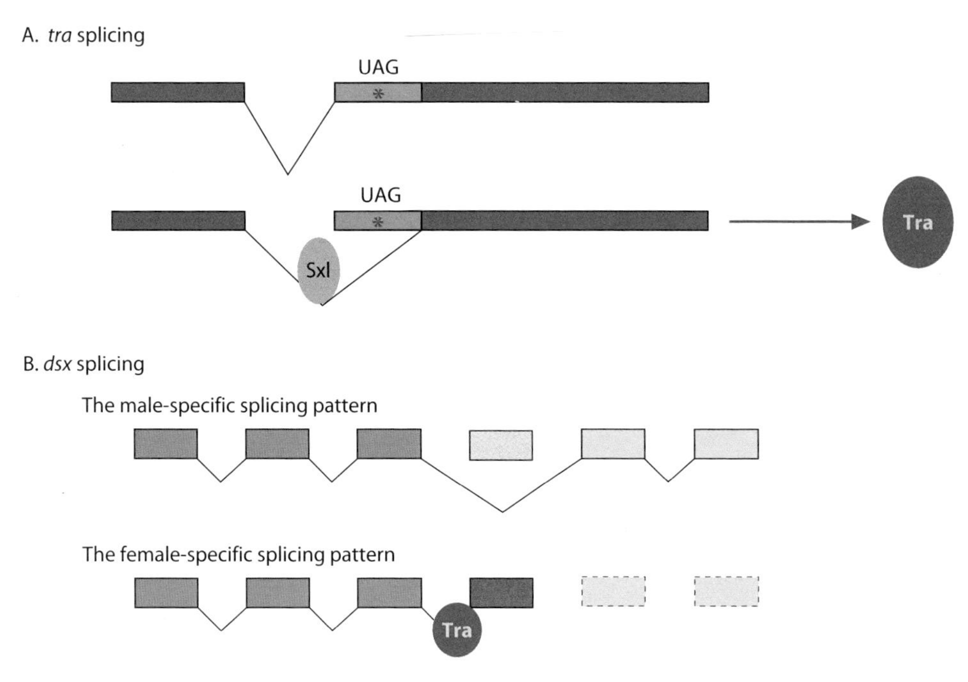

Figure 2.30 Alternative splicing and sex determination in *Drosophila*. The *tra* gene is a direct target of *Sxl* regulation, as summarized in Figure 2.29, but the mechanism of regulation only became apparent when *Sxl* was cloned and found to be an RNA splicing factor. In this figure, as in Figure 2.29, the pattern in 1X flies is shown in the upper drawing in each panel and the pattern in 2X flies is shown in the lower drawing. Regulation of *tra* gene expression in 1X and 2X flies occurs by *Sxl* regulation of its splicing, as shown in panel A. Two different splice acceptor sites are used in the second exon of the *tra* gene. In the absence of *Sxl*, i.e. in 1X flies, the more upstream splice acceptor is used, as shown in the upper drawing in panel A. This sequence of exon 2 includes a UAG stop codon, so no functional Tra protein is made. In the presence of Sxl protein, which occurs in 2X flies, this upstream splice acceptor site is blocked and a more downstream splice acceptor site is used, as shown in the lower drawing in panel A. The downstream splice acceptor results in exon 2 that does not have the stop codon so a full-length functional Tra protein is made. Only the first two exons of the *tra* gene are shown for clarity, and even in the presence of Sxl protein, some of the transcript with the stop codon is made. The *tra* gene also encodes a splicing factor, and the Tra protein directly regulates the splicing of the *doublesex* (*dsx*) gene, as shown in panel B. In this case, the alternative splice of the *dsx* gene involves the fourth exon. In the absence of the Tra protein in 1X flies, as depicted in the top drawing in panel B, the fourth exon is skipped and the first three exons of the *dsx* gene are spliced to the fifth and downstream sixth exon. In the presence of the Tra protein in 2X flies, as shown at the bottom drawing in panel B, the splice is directed from the third exon to the fourth exon and neither of the downstream exons is used. The first three exons of the *dsx* gene encode the DNA-binding portion of this transcription factor, whereas the more downstream exons encode parts of the protein that differ in their interactions with other proteins.

the *tra* gene. The structure of the *tra* gene is shown in Figure 2.30A. The *tra* gene is transcribed in both 1X and 2X flies, but the RNA splicing pattern of the second exon is different in the two sexes. In the absence of Sxl protein in 1X flies, as shown in the upper half of Figure 2.30A, an upstream splice site in exon 2 is used; this portion of exon 2 includes a stop codon so the protein produced in males is truncated at this position and non-functional. In the presence of Sxl protein in 2X flies, as shown in the lower half of Figure 2.30A, a more downstream splice occurs in exon 2 of the *tra* gene. This splice site is located after the stop codon in exon 2, so that the transcript produces a Tra protein that is full-length and functional in females. However, the regulation of splicing does not stop with Sxl protein. The Tra protein is itself a splicing factor, which also regulates alternative splicing of its downstream target genes.

Mutations in another gene, known as *doublesex* (*dsx*), have a different effect than the mutations in the *tra* and *tra2* genes. Rather than sex reversal, both 1X and 2X flies in *dsx* mutants become intersexual, with some aspects of somatic differentiation of both sexes. The 1X and 2X intersexes are nearly identical in phenotype, indicating that *dsx* is involved in sexual differentiation in both males and females. The difference between the male and female versions of the Dsx protein again lies in the pattern of its splicing, as summarized in Figure 2.30B. The presence of the Tra protein, which is only found in females because of its own splicing, results in the female-specific version, whereas the absence of the Tra protein results in the male-specific version. The role of the *tra2* gene is to function with Tra in regulating this splice but it does not encode a splicing factor itself.

Both universal and taxon-specific properties are revealed by sex determination in *Drosophila*

Although *Sxl* and *tra* have orthologs in other species that are involved in alternative splicing, these orthologs are not involved in sex determination. Alternative splicing is not the key mechanism for controlling sex determination, other than in *Drosophila* species and some other dipteran insects. However, orthologs of the *doublesex* gene are transcription factors that *are* involved in sex determination in other species such as worms and mammals. One inference is that the evolution of a sex-determining mechanism has preserved the role of Dsx and Dsx-related proteins as transcriptional regulators of downstream sex differentiation target genes. The mechanism by which the expression of *doublesex* itself is regulated has not been conserved in evolution.

In sum, *Drosophila* has modeled both universal and taxon-specific properties in the genetic analysis of sex determination. The universal properties involve alternative splicing as a means to regulate gene expression and function. The role of *doublesex* as a transcriptional regulator of sexual dimorphism in animals is also widespread. However, the involvement of alternative splicing as a means of regulating sex determination is apparently a taxon-specific property of *D. melanogaster* and some other insects.

The *Drosophila* genome database, Flybase

The *Drosophila* genome database is called Flybase. Panel A of Figure 2.31 shows the opening screen of Flybase with news and some quick links. One interesting tidbit on the opening screen for this particular day is the news about the response to a survey—65 percent of users access Flybase at least once a day, demonstrating how central these databases have beome to contemporary geneticists. The gene page for the *patched* gene is shown in Figure 2.31B. Flybase has more drop-down menus than the other genome databases, with more open space on the page. The ImageBrowse feature (Figure 2.31C) provides an extensive database of electron micrographs, photographs, and other images useful for understanding normal and mutant phenotypes in *Drosophila*. GenomeBrowse in Flybase (Figure 2.31D) shows the general structure of the genome browsers for many other organisms.

Given the extensive history of research on *Drosophila* and the wealth of information that is available, Flybase is remarkably easy to use. Information is readily accessible from the drop-down menus. Because *D. melanogaster* is used in many teaching laboratories, there are also many educational links in Flybase.

2.5 *Arabidopsis thaliana*: a weed with a purpose

Plant scientists studying the model flowering plant *Aradopsis thaliana* are no strangers to the questions 'Why do you work on a *weed*? Why not corn or tomatoes or soybeans?' This, of course, is often hardest to explain to well-meaning family members and farmers worried that you are throwing away your life and their tax money by studying a *weed*. The assumption is that studying agronomic crops is more important because of the easily comprehended relationship to crop improvement (e.g. better yields, pest resistance, abiotic stress tolerance, and nutritional value), environmental stewardship (e.g. alternative energy sources, reduced use of pesticides, and environmental remediation), and beautiful garden flowers. Along these lines, prior to the dominance of *Arabidopsis*, the central plant genetic models included both food and horticultural crops—peas, maize, tomatoes, barley, petunias, and snapdragons.

Despite the conventional wisdom, however, not all weeds are bad, at least not in the case of *Arabidopsis*. Some of its properties are summarized in Table 2.1. In

A

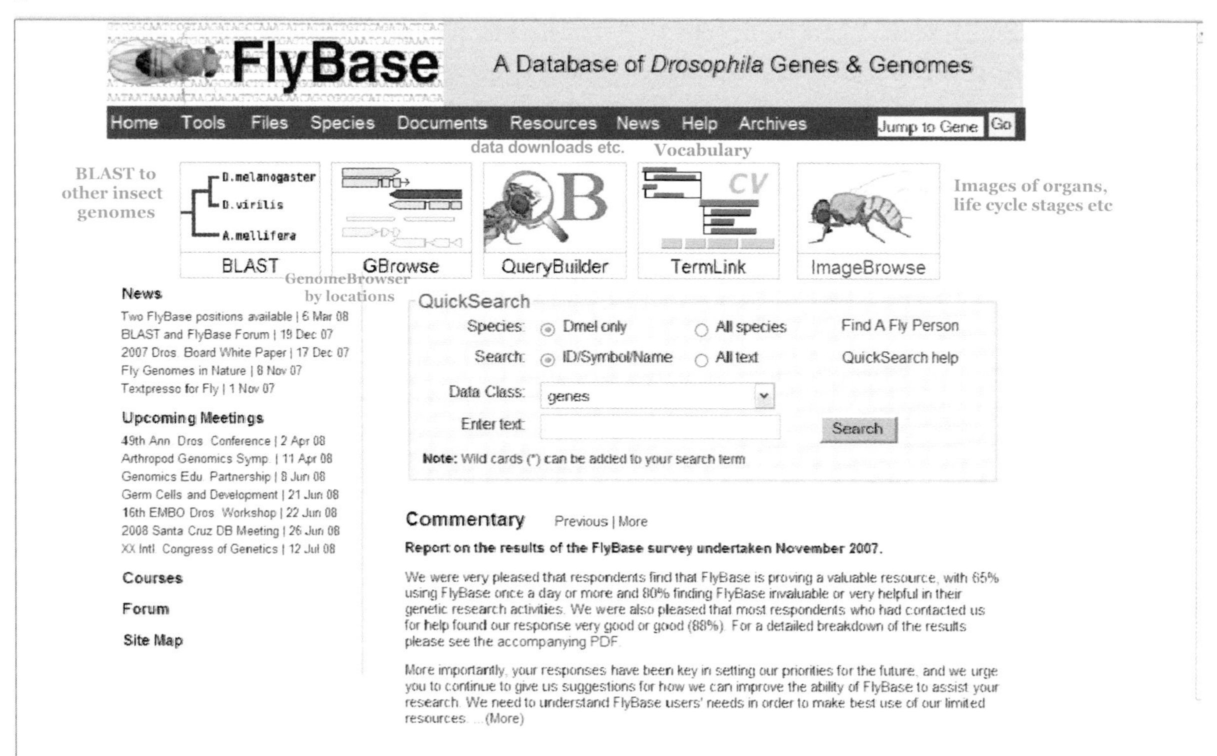

B

FlyBase | Gene Dmel\ptc

Home Tools Files Species Documents Resources News Help Archives | Jump to Gene | Go

Profile Manager | Help | Open All | Close All

General Information			
Symbol	Dmel\ptc	Species	D. melanogaster
Name	patched	Annotation symbol	CG2411
Feature type	protein_coding_gene	FlyBase ID	FBgn0003892
Created / Updated	2003-12-02/2003-12-02		
Genomic Location			
Chromosome (arm)	2R	Recombination map	
Cytogenetic map	44D5-44F1	Sequence location	2R:4,537,140..4,552,728 [+]
Map (GBrowse)	2R 4530k 4540k 4550k 4560k ptc l(2)44DEa	Decorated FastA / Get genome region / Gene region / Get FastA	

Drop down menus to get more details

- Summary Information
- Detailed Mapping Data
- Gene Model & Products
- Expression Data
- Alleles & Phenotypes
- Gene Ontology: Function, Process & Cellular Component (32)
- Sequence Ontology: Class of Gene
- Interactions & Pathways

Figure 2.31 The *Drosophila* genome database Flybase. (A) The opening screen for Flybase. Note that the database includes links to the genomes of the other sequenced *Drosophila* species. (B) The page for the *patched* gene is shown; this gene will be used for examples repeatedly throughout this book. The location of the gene shows the chromosome arm, the polytene bands, and the sequence location. Flybase has drop-down menus to provide specific detail information.

C

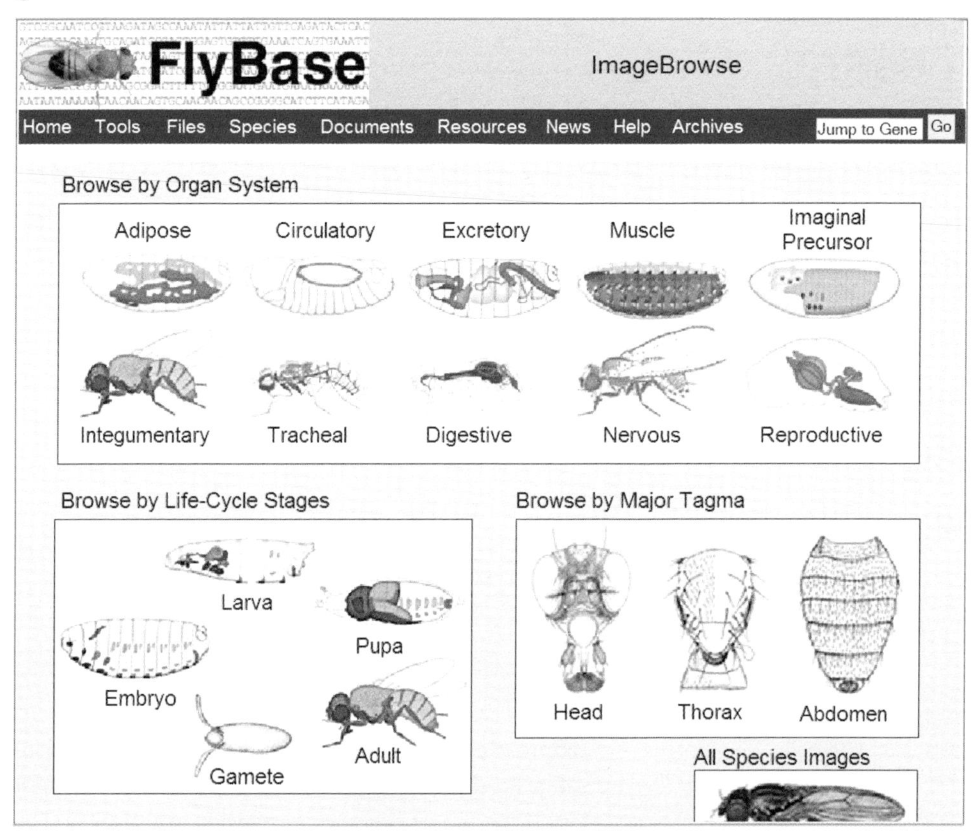

D

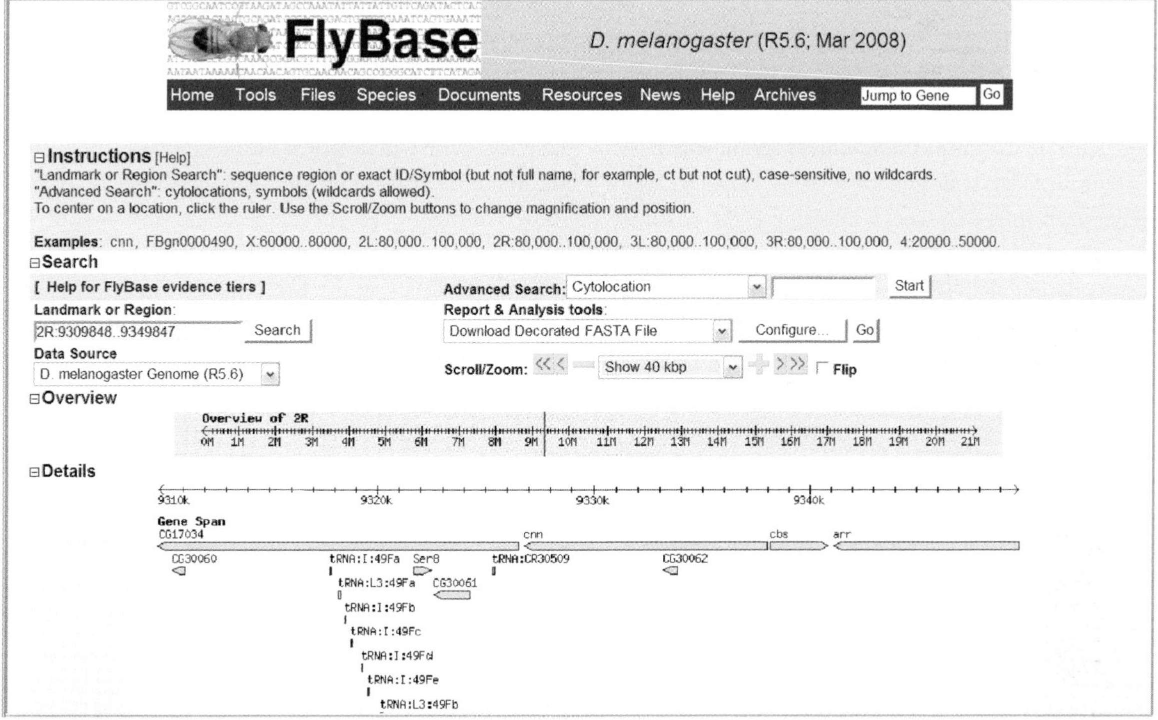

Figure 2.31 (Continued). (C) This is the opening page of the ImageBrowse link from the Flybase home page. Flybase has an extensive database of images that aid greatly in the descriptions of mutant phenotypes. (D) The GenomeBrowse feature from Flybase is similar to genome browsers for other genome databases.

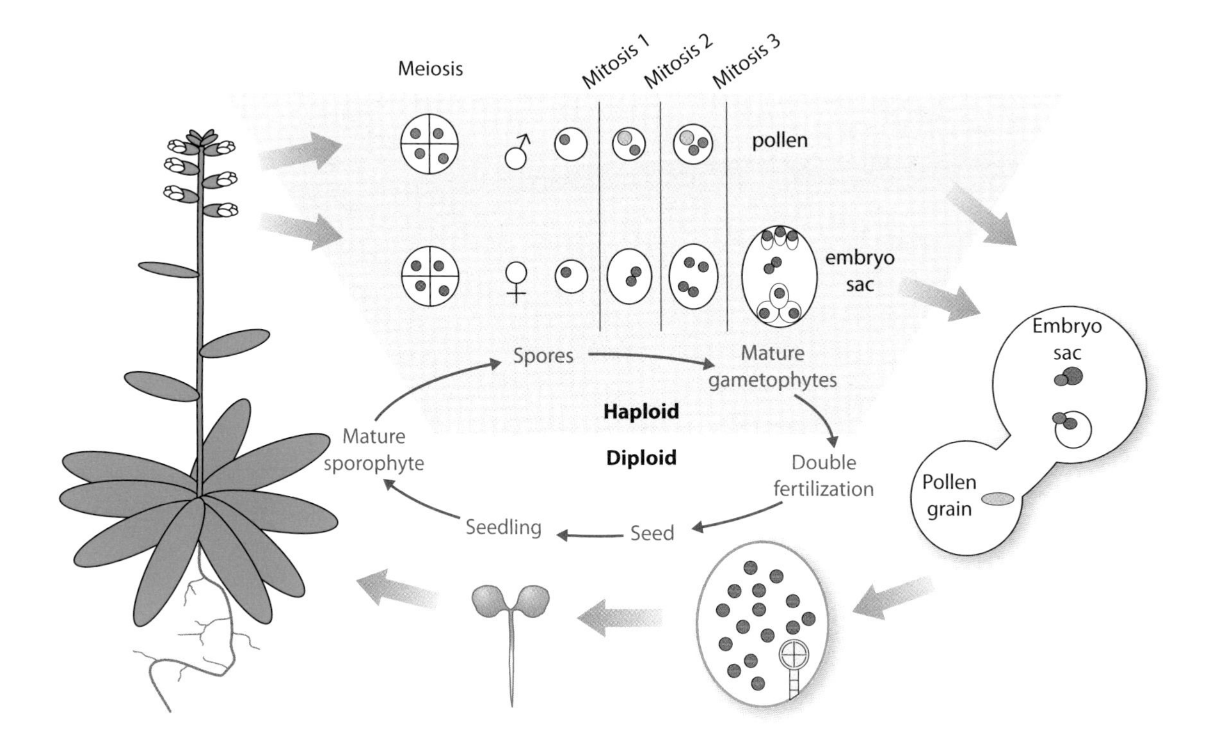

Figure 2.32 The life-cycle of *Aradopsis thaliana*. *Arabidopsis* has a life-cycle similar to other flowering plants, but it is much shorter. The complete life-cycle requires about 6 weeks, and thousands of seeds are produced per plant. The haploid phase of the life-cycle is shaded in light blue, and the diploid phase in white.

keeping with the properties of ideal model organisms described earlier and in contrast to some of the other plants mentioned above, *Arabidopsis* is small, which is important for large-scale genetic screens. It has a typical generation time of 6 weeks, allowing eight or nine generations a year, and produces up to thousands of offspring per parent plant (Figure 2.32). *Arabidopsis* is also naturally self-pollinating, which allows easy production of homozygotes, but can be manually crossed so that new genetic variants can be bred, as shown diagrammatically in Figure 2.33. Other plant genetic models suffer from one or more of the following problems that slow research progress: large size, a longer generation time, a naturally cross-pollinating reproduction, a large genome and number of genes, and difficult or labor-intensive transformation.

In addition, the *Arabidopsis* genome is small for a plant (about 125 Mb) and it was the third multicellular eukaryote, after *C. elegans* and *Drosophila*, to be completely sequenced. Recent estimates are that there are approximately 27 000 protein-coding genes, also probably small for a plant. Furthermore, since *Arabidopsis* has a natural worldwide distribution, it is well studied for many investigations of natural diversity.

As with the other model systems, *Arabidopsis* became the established model organism for the plant world in part by chance and timing, and in part by intentional choice from the possible model plants. Friedrich Laibach of the University of Bonn, Germany, published the first PhD dissertation on *Arabidopsis* in 1907, but by the 1970s there were still very few laboratories studying it. In the late 1970s and early 1980s, a handful of talented young geneticists, many from outside the plant world, were attracted to *Arabidopsis*. Chris Somerville and his wife Shauna, two of these young scientists, were in Paris in 1978 on an extended vacation after he completed his PhD, trying to decide what to do with their lives, when they settled on *Arabidopsis*. Over coffee in Parisian cafes, the same environment that produced Monod's conclusion about *E. coli* and elephants, Shauna tried to persuade Chris to begin studying plants. Both were concerned about how to address the effects of the world's growing population on food supplies and the environment. While on vacation, they read two key papers that helped convince him to study plants, and *Arabidopsis* specifically. The first was a 1977 report

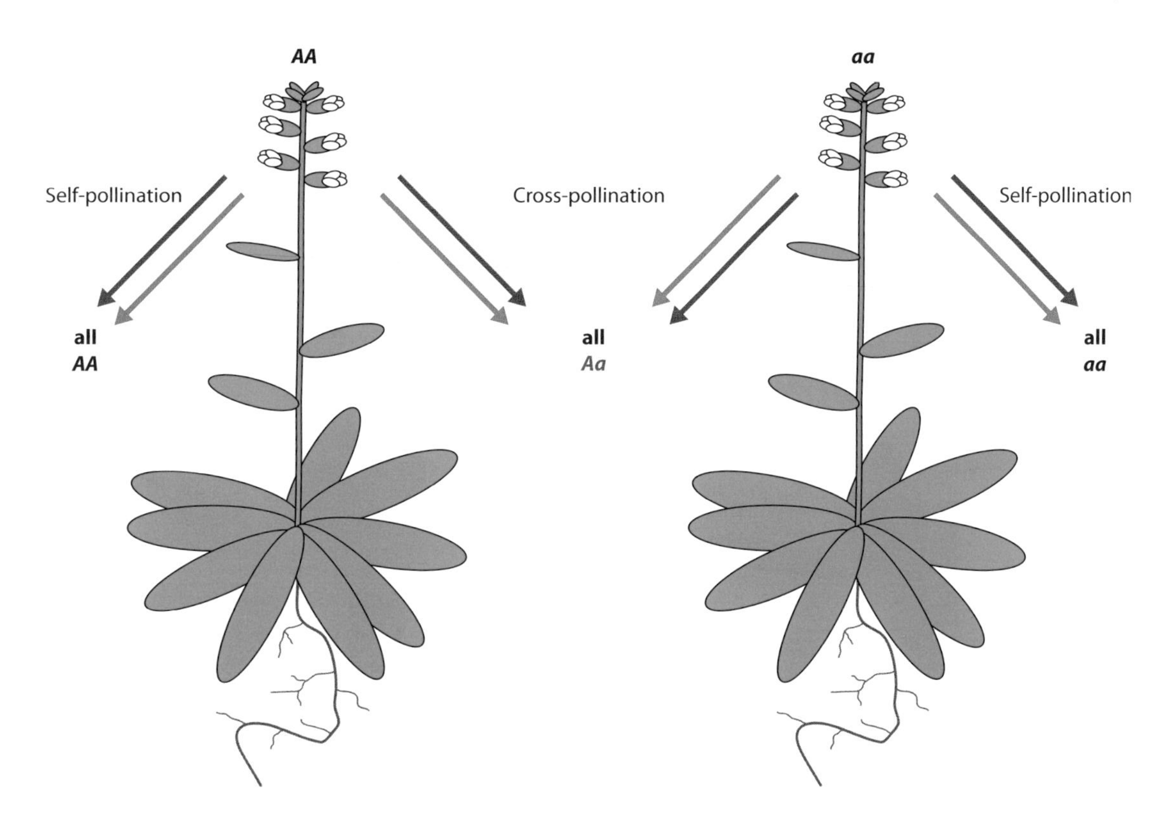

Figure 2.33 Cross-pollination and self-pollination in *Arabidopsis*. Each flower produces both pollen and eggs, and the plant is capable of both self-pollination and cross-pollination, leading to fertilization. Multiple generations of self-fertilization, as in *C. elegans*, produce lines that are homozygous at all loci. Cross-pollination is needed to perform matings.

of the first successful transformation of plants using *Agrobacterium tumfaciens* (a method described below). The second was a review by George Rédei extolling the virtues of *Arabidopsis* as a genetic model. With the help of other key converts, by the late 1980s the number of *Arabidopsis* researchers and laboratories had grown exponentially.

The annual International Conference on *Arabidopsis* Research now attracts up to 2000 scientists using *Arabidopsis* as a model for studying photosynthesis, plant hormones, plant development, abiotic and biotic stress responses, signal transduction pathways, gene families, small RNA pathways, chromatin modification, cell biology, and so on. The goal of *Arabidopsis* researchers is one day to understand completely the function of all of the genes of this one 'simple' plant—one with less DNA and fewer genes than most other flowering plants. They hope that a computer model of a 'virtual plant' will eventually be possible, so that an investigator could test the effect of a mutation, transgene, or change in gene expression *in silico*. This would allow a better understanding of other flowering plants, especially those all-important crop plants. Although it is a small weed to agronomists, *Arabidopsis* towers above other plants to geneticists.

The *Arabidopsis* life cycle is shorter than that of many flowering plants

Arabidopsis has a 6-week life cycle from germination to mature seeds. The seed itself contains a dormant embryo consisting of two cotyledons, or seed leaves, which act as an energy store for the young seedling, two immature leaf primordia, a hypocotyl, and a root. In contrast to animal embryos, the plant embryo does not have all of the organs of the final adult. Instead, plants display indeterminate tip growth. This is primarily due to the presence of a store of stem cells in meristems. The embryo contains a shoot apical meristem (SAM) at the shoot apex and a root apical meristem (RAM) at the tip of the root. As the plant grows, the SAM will give rise to the leaves, branches, inflorescences, and flowers, and the RAM will allow for the growth and branching of the root system.

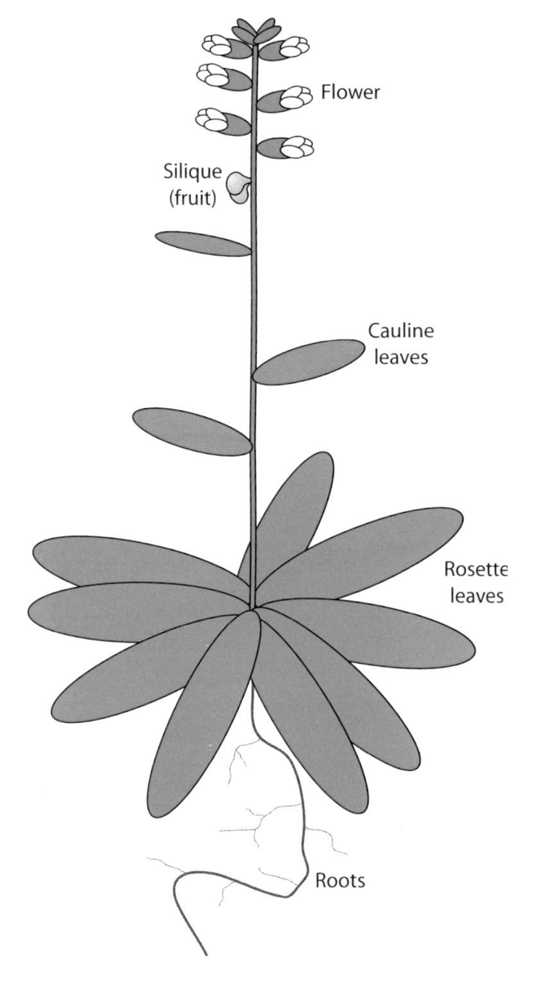

Figure 2.34 ***Arabidopsis*** **anatomy.** *Arabidopsis* plants are about 45 cm tall at maturity, and consist of leaves (rosette leaves at the base and cauline leaves on the inflorescence stem), a stem, flowers, fruits called siliques that contain seeds, and roots.

Like other higher plants, *Arabidopsis* has three major phases of post-embryonic development: a juvenile vegetative phase, an adult vegetative phase, and a reproductive phase (Figure 2.34). The juvenile vegetative phase consists of the first few leaves, and is characterized by a lack of reproductive competence and certain morphological characteristics. The juvenile stage can last for years in woody plants, and likely plays an important role in delaying the onset of flowering and hence reproduction since in the adult vegetative phase the plant is sensitized to respond to flowering signals. Complex signaling pathways and regulation by small non-coding RNAs result in major shifts in gene expression that control the transitions between these three phases.

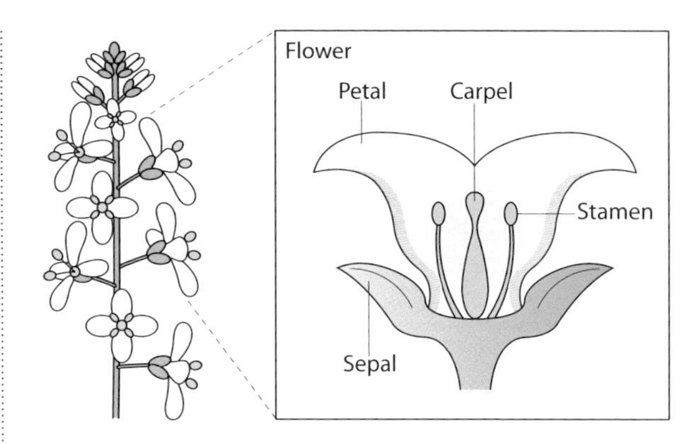

Figure 2.35 ***Arabidopsis*** **flower.** The *Arabidopsis* flower consists of four concentric radial whorls of organs: four sepals, four petals, six stamens, and one pistil made up of two carpels.

The first flowers of wild-type plants open within 3 or 4 weeks of germination. A diagram of a flower is shown in Figure 2.35. Each flower has three symmetrical whorls of organs: four leaf-like structures called sepals, four petals, six stamens, and one pistil. Some of the earliest work on *Arabidopsis* development studied genes regulating floral development, and many mutants are known that regulate the identity and structure of the floral organs. We will come back to this in a later chapter. The structure of the flower is such that the plants are naturally self-pollinating, which is advantageous to genetic studies, in the same way that self-fertilization is important to those studying *C. elegans*. Cross-pollinations are made by carefully emasculating flowers before the stamens release their pollen and then manually pollinating with pollen from another flower. Following fertilization, the pistil develops into a specialized seedpod called a **silique**, which may hold 50–100 seeds.

Transformation of *Arabidopsis* is done using *Agrobacterium tumefaciens*

One of the main reasons that *Arabidopsis* has become the dominant plant genetic model is that it is so easily transformed with DNA from other sources. Like many other dicotyledonous plants (dicots), *Arabidopsis* can be transformed by manipulating the biology of a natural pathogen—the crown-gall-causing bacterium *Agrobacterium tumefaciens*. *Agrobacterium* infects a wide range of dicots by inducing the formation of a tumor or gall at a wound site with the aid of a special plasmid called a tumor-inducing (Ti) plasmid. This plasmid contains

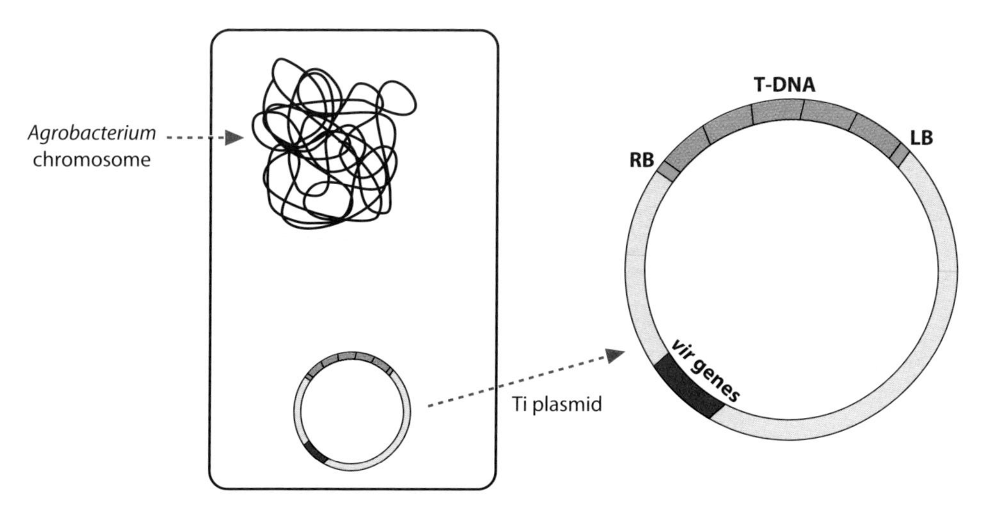

Figure 2.36 The Ti plasmid of *Agrobacterium tumefaciens*. Transformation in *Arabidopsis* takes advantage of the soil bacterium *A. tumefaciens*, which produces a tumor known as a crown gall. A complete Ti plasmid has T-DNA, which encodes the genes required for tumor formation, flanked by 25 bp repeats known as the right and left borders (RB and LB), shown here in green. Upon infection, the Ti plasmid is transferred to the plant cell, and the T-DNA randomly integrates into the plant genome. The plasmid also encodes the virulence genes (*vir*) that encode functions needed for infection and transfer of the T-DNA into the plant DNA.

a 20 000 bp region of DNA called T-DNA that is stably transferred into the genome of the host (Figure 2.36). Within this T-DNA are genes encoding biosynthetic enzymes for making the plant growth hormones auxin and cytokinin, as well as a gene for synthesizing opines, which are amino acid derivatives that the bacterium uses as a carbon and nitrogen source. The high misexpression of these hormones directs the formation of a tumor from the infected cell, which produces food for the bacteria, allowing them to multiply and flourish. Generally, these kinds of tumor are fairly innocuous to the plant, and plants can live for years in such a state. Trees that are a hundred years old may have huge crown galls.

The insertion of T-DNA into the plant genome is directed by two 25 bp repeats, called the right and left borders (RB and LB), which flank the sequence to be inserted. The Ti plasmid includes a set of *vir* genes that encode virulence factors required for the transfer; the *vir* genes are not transferred with the T-DNA. The process of transformation is shown in Figure 2.37. Geneticists replace the hormone and opine synthesis genes with any gene of interest, reintroduce the plasmid into *Agrobacterium*, and use it to infect the plant. *Agrobacterium* will then insert whatever DNA is between the RB and LB sequences into the plant DNA. However, the bacteria and hence the crown gall will not grow within the plant because of the absence of a food source. Typically, two genes are included within the modified T-DNA: the gene of interest and a selectable marker encoding antibiotic or herbicide resistance. One caveat of this process is that *Agrobacterium* inserts the T-DNA into the plant genome relatively randomly, and so multiple transgenic lines must be examined initially because of the effects of the insertion site on the expression of the T-DNA genes. The random insertion of T-DNA is analogous to the random insertion that occurs in *Drosophila* using P transposable elements, but is different from the targeted insertions and gene disruptions that can be done in yeast and mice. To date, there is still no efficient way to make directed transgenics or perform homologous recombination in *Arabidopsis* and other higher plants, although it is possible in the moss *Physcomitrella patens*.

Many plants can be transformed using *Agrobacterium*, but the process is particularly easy in *Arabidopsis*. Simply briefly dipping the immature inflorescences in a dilute *Agrobacterium* suspension is enough. In this method, the *Agrobacterium* infects the egg cells with the immature flower. As a result, some of the seeds collected from a plant transformed by a 'floral dip' will be transgenic. The transgenics can be selected from the non-transgenics at the seedling stage by testing the resistance to the selectable marker.

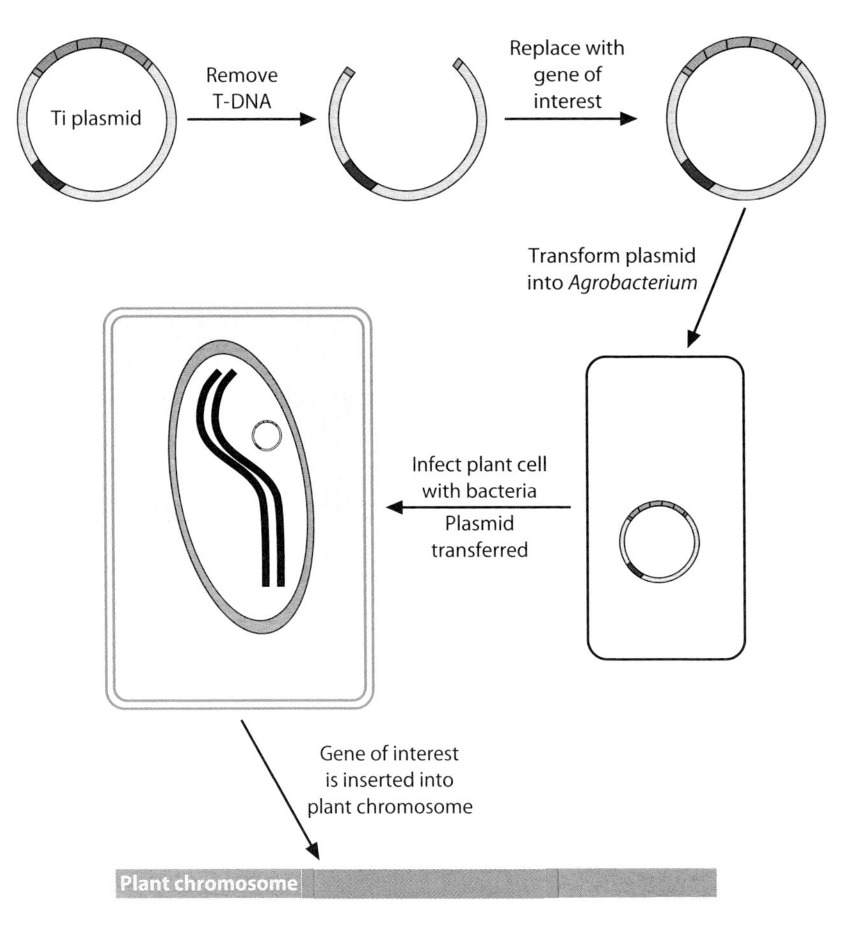

Figure 2.37 Transformation of *Arabidopsis* using modified Ti plasmids. The T-DNA is removed from the Ti plasmid *in vitro* and replaced with the gene of interest to be integrated into the plant DNA. The modified plasmid is transformed into the bacteria, which are then used to infect the plant. In *Arabidopsis*, the infection is done by dipping the flowers into a solution containing the bacteria. The bacteria transfer the Ti plasmid to the plant cell, where the region flanked by the RB and LB is integrated into the plant chromosome. Integration occurs at random. In *Arabidopsis*, the infected flowers produce transgenic seeds containing the T-DNA.

The brassinosteroid hormone biosynthetic and signaling pathways have been delineated by genetic analysis in *Arabidopsis*

Plants, like animals, have a number of hormones—key molecules produced at one location and acting on another that are required for proper growth and development. In addition to auxin and cytokinin, other plant hormones include gibberellins, ethylene, abscisic acid (ABA), and brassinosteroids. These plant hormones have incredibly wide-ranging and overlapping roles.

The last of the plant hormones to be identified were the brassinosteroids in 1979. Their name comes from the fact that they were originally isolated from the pollen of the plant *Brassica napus*. As their name suggests, brassinosteroids are steroid hormones structurally similar to the insect molting hormone ecdysone and the mammalian hormones testosterone, estrogen, progesterone, and corticosteroids. Despite the similarity, plants do not have receptors for these animal steroid hormones. (Molecular biologists have taken advantage of this to make estrogen-inducible promoter constructs for use in transgenic plants.)

Brassinosteroids have since been found in all plants tested, as well as in at least one alga. They come in many forms, but the most active endogenous form is brassinolide (BL). They have defined roles in cell elongation, vascular differentiation, senescence (the active process of plant death), repression of photomorphogenesis (light-regulated growth) in the dark, and the modulation of stress responses. Brassinosteroids regulate these

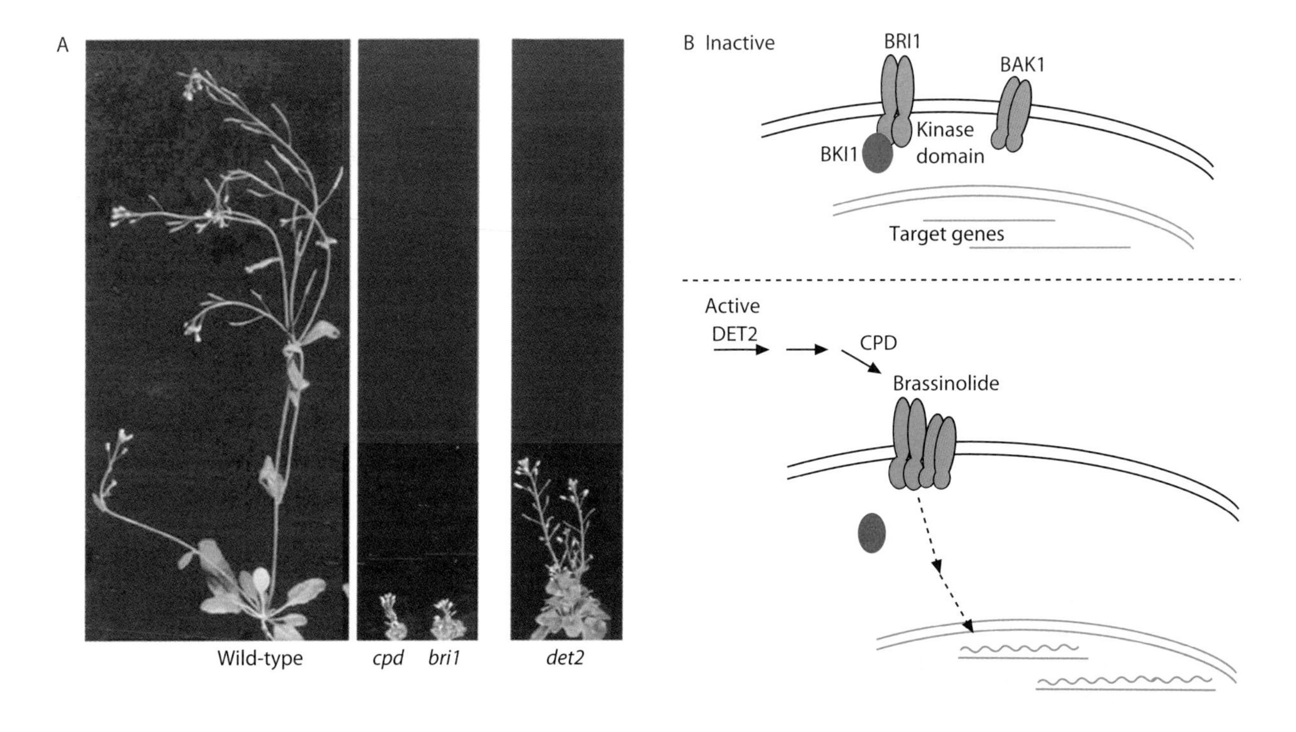

Figure 2.38 The brassinosteroid signaling pathway in *Arabidopsis*. The biological significance of brassinosteroids was proven through the use of mutants in *Arabidopsis*. (A) The phenotype of some of the dwarf mutants involved in hormone synthesis and signaling. The phenotypes of the three mutants described in the text (*cpd*, *det2*, and *bri1*) are shown. (B) In the absence of the active hormone brassinolide, the receptor BRI1 is in an inactive complex with the inhibitor BKI1, and the target genes are not transcribed. The genes *DET2* and *CPD* encode two of the enzymes involved in the synthesis of the active hormone brassinolide. The gene *BRI1* encodes the receptor itself. Brassinolide binds to the BRI1 receptor, releasing it from a complex with BKI1 and allowing BRI1 to form a complex with BAK1. In a series of signaling steps, not all of which are known, this triggers the transcription of the target genes. In *det2* and *cpd* mutants, little or no hormone is made, resulting in the dwarf phenotype shown in panel A. In a *bri1* mutant, no receptor is made and so the pathway is not active.

processes via crosstalk with other hormones, particularly auxin. Recent work has shown that the global gene expression patterns of BL- and auxin-inducible genes are significantly overlapping, which helps to explain their synergistic interaction in certain plant processes.

Mutations demonstrated the presence of brassinosteroid pathways in *Arabidopsis*

Despite various studies demonstrating the effect of exogenous BL on plant growth, many plant scientists remained unconvinced of the status of brassinosteroids as plant hormones until the identification of the first brassinosteroid mutants in *Arabidopsis* in 1996. The first brassinosteroid biosynthetic mutants were identified in screens for seedlings defective in skotomorphogenesis or etiolation (growth habit of plants grown in the dark). Normally, dark-grown seedlings display a very different phenotype from those grown in the light, characterized by an elongated hypocotyl, closed hooked cotyledons, and an absence of chlorophyll. Researchers interested in the regulation of this process screened for mutants that displayed normal light-grown growth (short hypocotyl, open cotyledons, and chlorophyll) even in the dark. The phenotypes of some of these mutants are shown in Figure 2.38A. When two of these mutants, *cpd* (*constitutive photomorphogenesis*) and *det2* (*de-etiolated 2*), were cloned, they were found to be mutated in genes predicted to encode brassinosteroid biosynthetic enzymes, as shown in the diagram of the signaling pathway in Figure 2.38B. Treating the plants with exogenous BL restored a wild-type dark-grown phenotype. Such screens are analogous to Beadle and Tatum's screens for *Neurospora* auxotrophs. These results suggested that brassinosteroids repressed photomorphogenesis in the dark. When grown in the light, the mutants displayed many other phenotypes, including severe dwarfism, dark-green leaves, delayed senescence, and poor fertility.

At the same time, other scientists were carrying out screens specifically looking for the BL receptor. While BL had been shown to enhance shoot elongation, it had also been shown to inhibit root elongation in *Arabidopsis* seedlings grown on BL-containing artificial media. The mutant *bri1* (*brassinosteroid insensitive 1*) was identified as a mutant insensitive to the root-growth-inhibiting effects of BL. Soil-grown *bri1* knockout alleles look indistinguishable from *cpd*. When *bri1* was cloned, it was shown to encode a leucine-rich repeat (LRR) receptor-like kinase. BRI1 is localized to the plasma membrane, and BL interacts directly with BRI1 via the extracellular LRR domain, activating the kinase, and setting off a downstream signal transduction cascade (Figure 2.38B). This finding was very exciting because, in contrast, animal steroid receptors bind to steroids in the cytoplasm, translocate to the nucleus, and directly bind to promoter regions, thereby acting as both receptor and transcription factor.

It was the identification of *cpd*, *det2*, and *bri1* and their pleiotropic phenotypes that convinced many plant scientists that brassinosteroids really were required for growth and development. Subsequent screens for plants displaying the severe dwarfism of these mutants have identified most of the enzymes in the brassinosteroiod biosynthetic pathway and almost an entire signal transduction pathway from BRI1 to the downstream transcription factors.

The Arabidopsis Information Resource

Research resources, including genome information, are available at the The Arabdopsis Information Resource (TAIR). Screen captures of some of the information at TAIR are shown in Figure 2.39. TAIR has a list of drop-down menus across the top of the screen. Figures 2.39A and 2.39B show a few of the links available from these menus. The Search drop-down menu (Figure 2.39A) has the links for genes, clones, mutations, microarrays, and so on that are familiar from other genome databases. The Tools and Portals drop-down menus have more detailed information and links to outside resources. The Portals menu includes a link to Education and Outreach resources, which provide the information teachers need to work with *Arabidopsis* in the classroom or the teaching laboratory. Figure 2.39B shows the results obtained on entering the gene name *bri1* in the Search box in the upper right of the home page. Note that we used the Exact Name Search because we specifically wanted information on this gene. The format of the page is familiar to those who have used National Center for Biotechnology Information (NCBI) resources such as PubMed and includes the locus name, a brief description of the gene, a link to the structure of the gene, and keywords associated with the gene. By clicking on the locus name, the images in Figures 2.39C and 2.39D appear. This provides more detailed information about the individual gene including microarray results, the size of the transcript and the predicted protein, and so on.

2.6 *Mus musculus*: mighty mouse

Scientists or not, all of us probably have yeast in our house, nematodes in the dirt in our gardens, and fruit flies near our fruit stands. We may also have *Arabidopsis* growing uninvited in our gardens or along our sidewalks. But we do not raise these organisms as pets. Alone among the model organisms, mice owe their standing to their role as house pets, particularly among those 'mouse fanciers', hobbyists who bred the animals for coat color and behavioral differences from the early days of the twentieth century. Other mammals such as dogs and cows also have important and long-standing relationships with humans, and many of them provide useful examples in Mendelian genetics. Nearly all of these other mammals are the object of substantial genome projects, but none of them has the history of genetic experimentation that comes with the mouse. Mice are among the smallest of the domesticated mammals and have the fastest lifecycle. Although many other mammals have provided important insights into the fundamentals of human genetics, none has been such a good model for human genetics as the mouse. Some of the properties of mice as model organisms for genetic analysis are shown in Table 2.1.

A

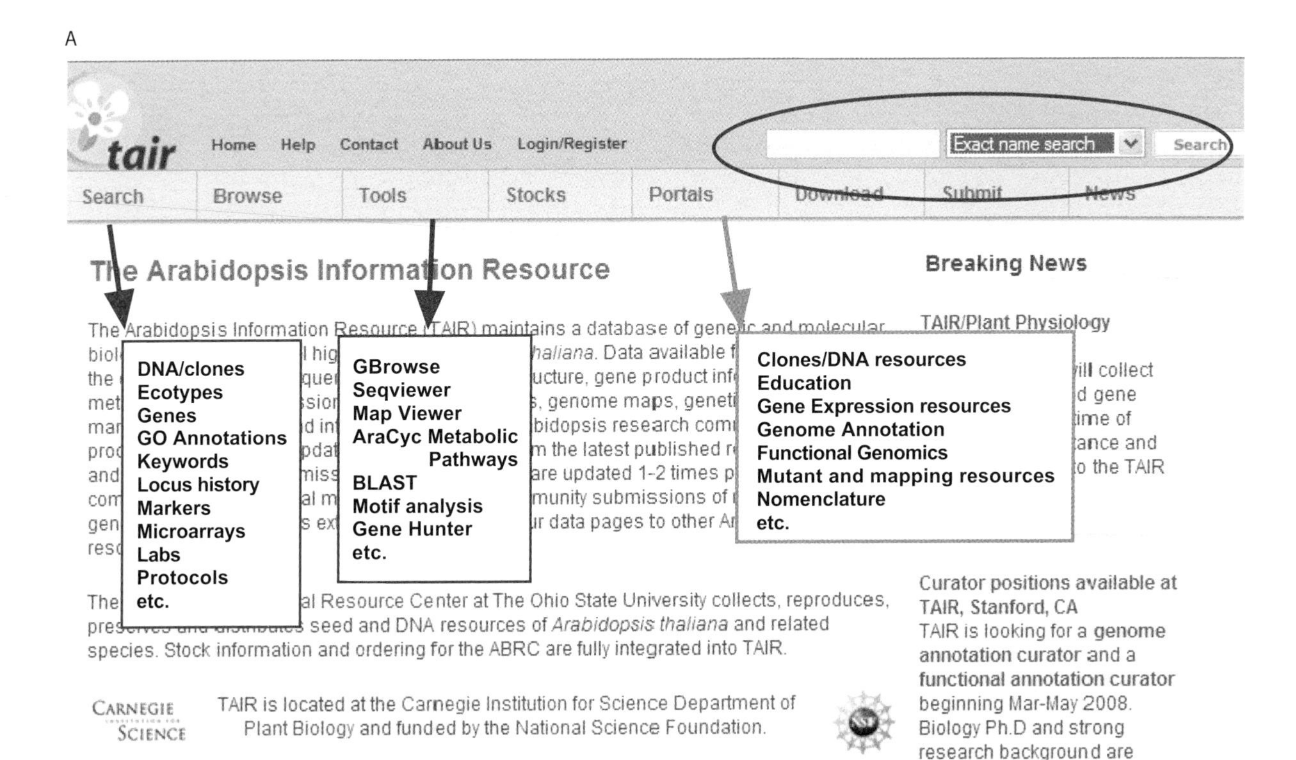

Figure 2.39 The Arabidopsis Information Resource (TAIR). (A) The opening screen has news and general information with a series of drop-down menus across the top. Some of the links from these drop-down menus are shown. TAIR has a set of links to educational resources from the Portals menus and to other genomics tools from the Tools menu. The search box in the top right corner has 'Gene' as the default setting rather than 'Exact name search' as shown here.

B

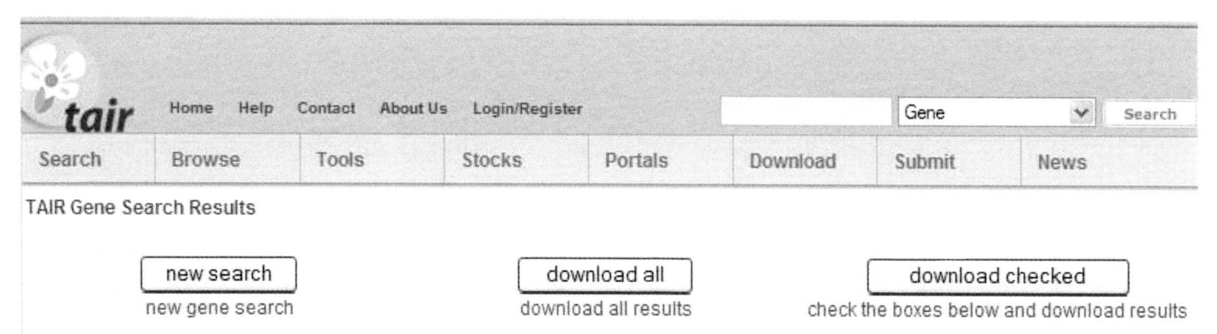

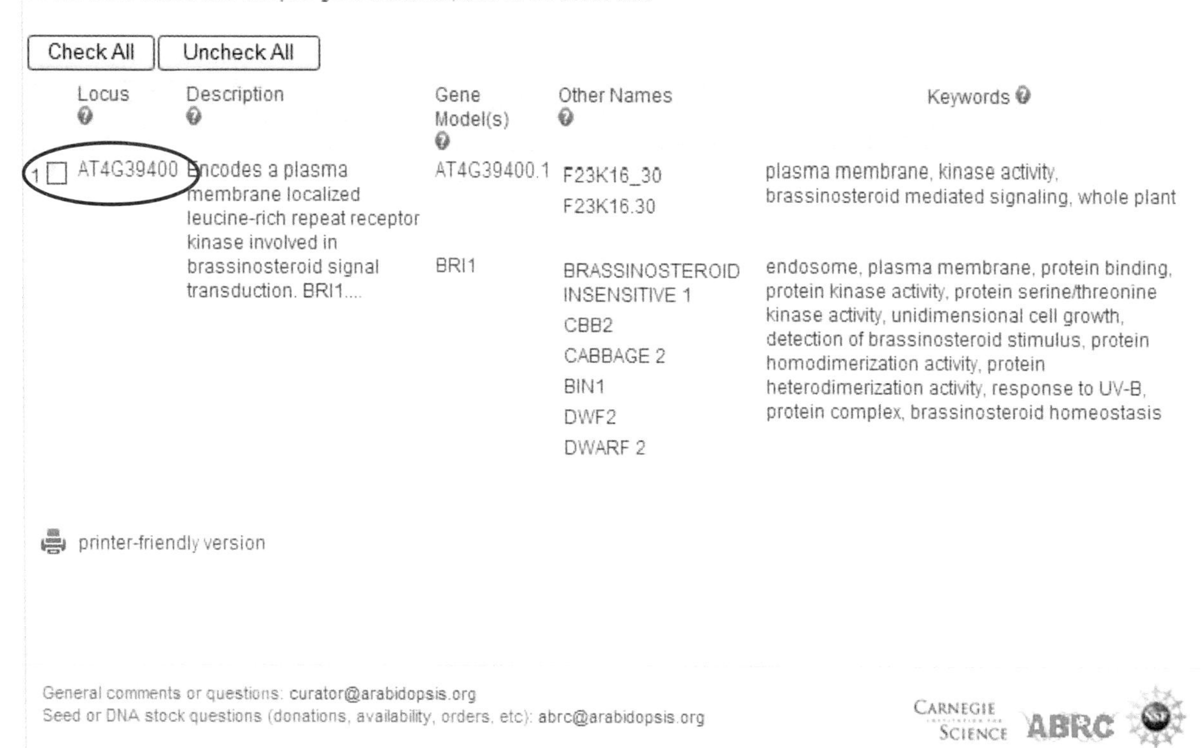

Figure 2.39 (Continued). (B) An exact name search for *bri1* gives this page of information on the gene. Most of the other information is organized by locus name.

C

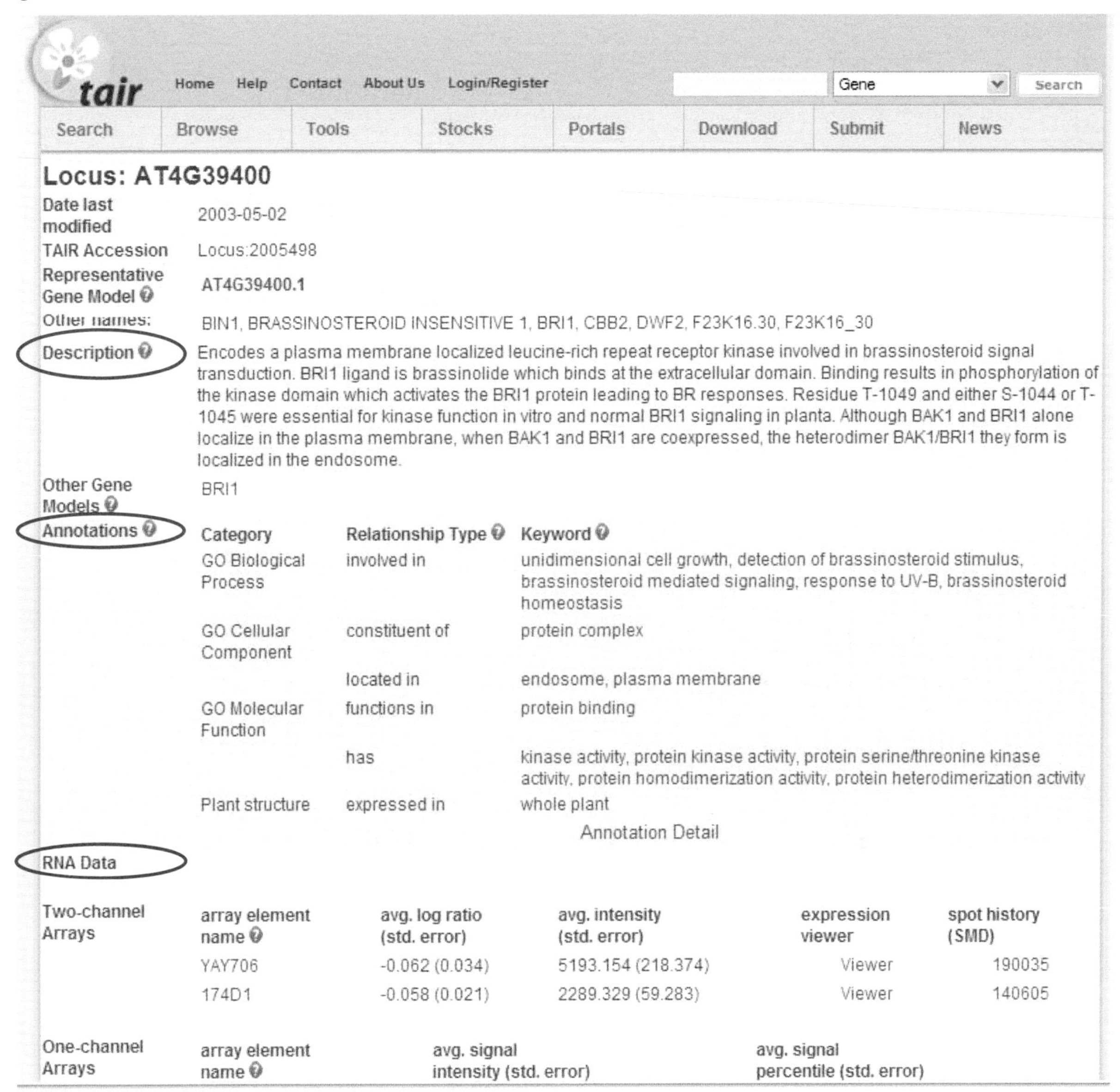

Figure 2.39 (Continued). (C) The top part of the page associated with AT4G39400 or *bri1*. Note the heading for a summary description of the role of the gene, the annotation terms in the Gene Ontology (GO) database, and the RNA data on microarray results.

Mice have the typical life cycle of placental mammals that we are familiar with as humans. Gestation takes about 21 days in mice, compared with about 38 weeks in humans, and puberty is reached in about 6 weeks as opposed to 12 or more years. Thus, a mouse life cycle is slightly more than 2 months from fertilization of an ovum until that fertilized egg can produce its own offspring.

Mouse development is described in more detail in Chapter 6. The details of mouse embryology play an important role in the production of transgenic mice and gene targeting, a technique that was recognized with the Nobel Prize in 2007. The transformation techniques in mice will be discussed in Chapter 6. For the purposes of this chapter, the important aspects are that mice can be genetically manipulated either by controlled matings

D

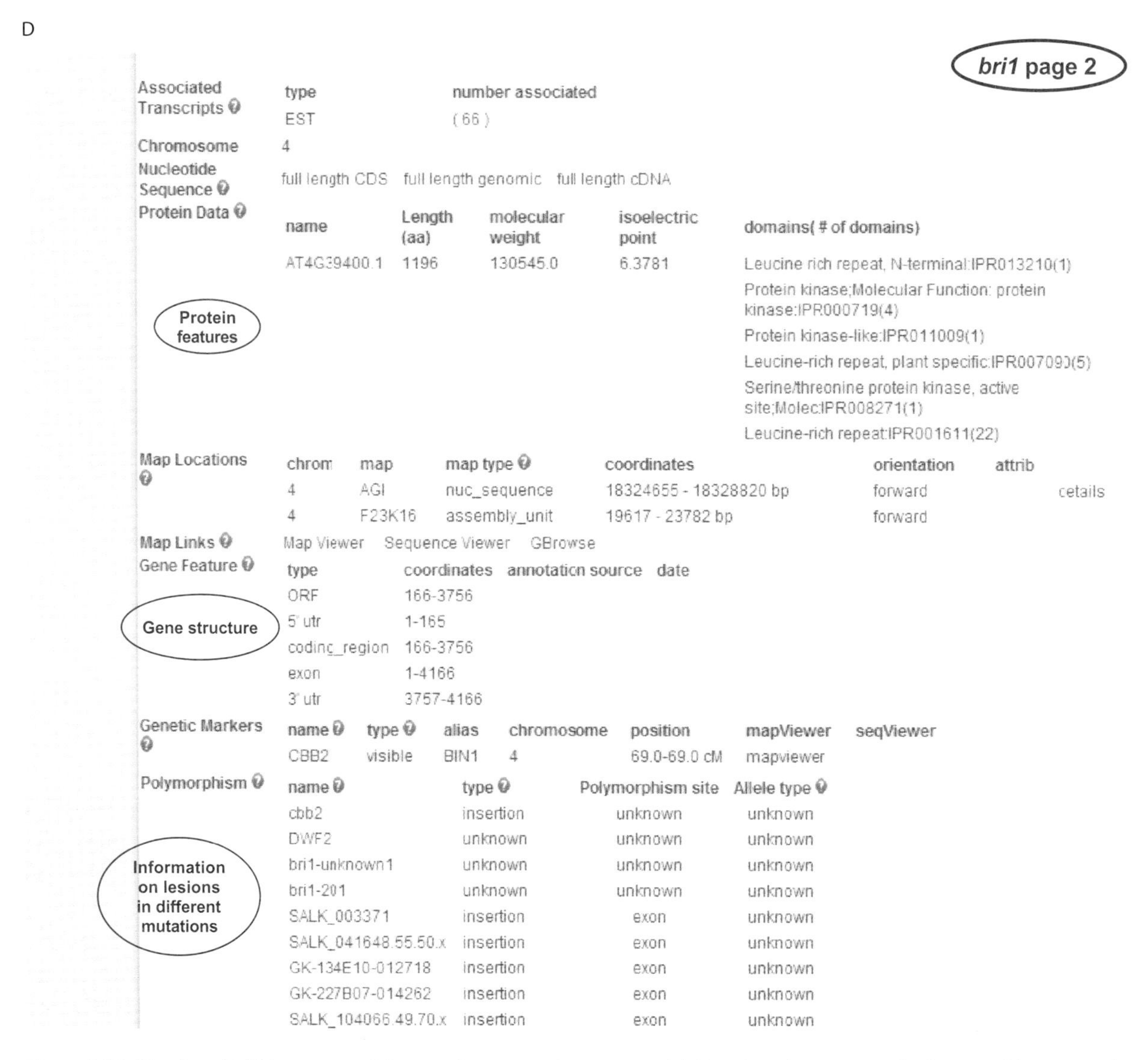

Figure 2.39 (Continued). (D) Lower part of the same locus page shown in panel C. Note the information on the features of the protein, the structure of the gene and its mRNA, and different mutant alleles of the gene. The lowest part of this page (not shown) includes information on the phenotypes of certain mutations and references.

between males and females or by the alteration of single cells. Embryonic stem cells from mice can be grown in culture, and genes can be introduced into these cells. Transformed cells that have integrated the gene stably into the chromosome by a double crossover, similar to yeast integration, are selected. These transformed embryonic cells are then used to produce an adult mouse that has the introduced DNA as a targeted integration. The introduced gene can replace or 'knock out' the function of the chromosomal gene; other manipulations of mouse genes are described in detail in Chapter 6.

It is estimated that mice and humans diverged from a common ancestor approximately 75 million years ago; mammalian phylogenic relationships are shown in Figure 2.40. The mouse genome is approximately the same size as the human genome, roughly 3×10^9 bp. The number of protein-coding genes is also about the same as humans, perhaps 30 000 (although the number is regularly being revised in both mice and humans). One estimate is that more than 99 percent of genes in the mouse have at least one ortholog in humans, although different gene families have been expanded in

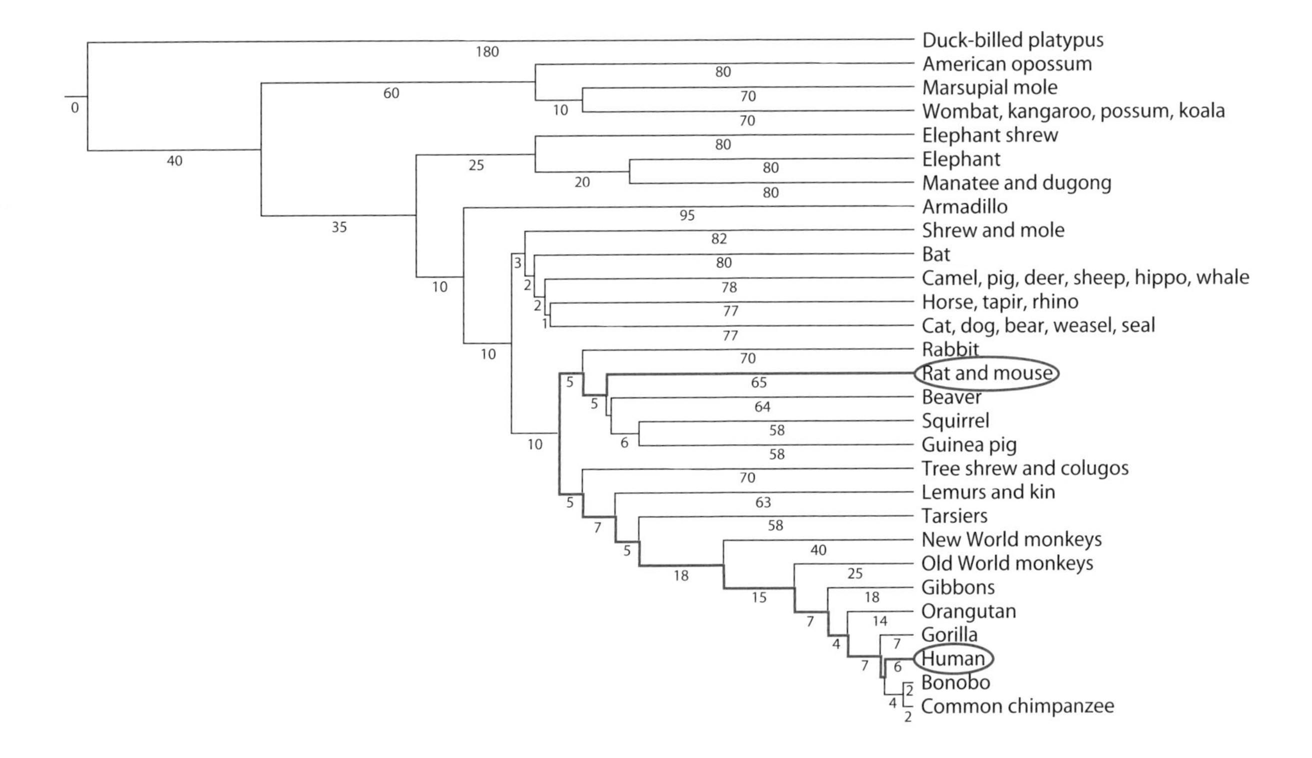

Figure 2.40 A mammalian phylogenetic tree. The taxonomic relationship between humans and mice is shown by the red line. The estimated last common ancestor of mice and humans lived an estimated 75 million years ago.

each lineage. According to a recent count, there are 118 genes in the mouse for which humans have no ortholog; many of these genes are found in other mammals as well as the mouse, so they could have been lost during primate evolution. There is extensive preservation of the linkage arrangement of genes as well, so that 80 percent or more of the mice genes are found at the same relative chromosomal location as their human orthologs. These conserved linkage arrangements are known as **syntenic blocks**, and a comparison of synteny allows a partial reconstruction of the genome rearrangements that occurred during mammalian divergence (Figure 2.41). The value to understanding diseases of having a well-studied genetic model that is so closely related to humans cannot be overstated.

Because mice were raised as pets, variations in coat color were noticed and bred. The most widely used coat colors are white, black, and brown, although genetic modifications in pattern and intensity of color are known (Figure 2.42). More than 75 genes that affect coat color are known. A classic reference book with many color pictures, *The Coat Colors of Mice* by W.K. Silvers, has been made available electronically by the Jackson Laboratory and can be accessed from the mouse genome database (http://www.informatics.jax.org/wksilvers/). Other mouse genetic markers that are used include morphological features such as eye color and size, tail structure, coat texture, and others. *Mouse Genetics* by Lee Silver, which presents a fascinating historical record of the use of mice in genetic research, can also be accessed from the Jackson Laboratory website (Figure 2.43). Each of these books is integrated with the current mouse genome database so that an investigator or a curious reader can easily compare the classic mouse-fancier mutations with the contemporary genomic analysis of these same genes.

The Mouse Genome Informatics database

Hundreds of genetic diseases in humans have a corresponding mouse model and many more correlations will undoubtedly be found in the years ahead. Mice are

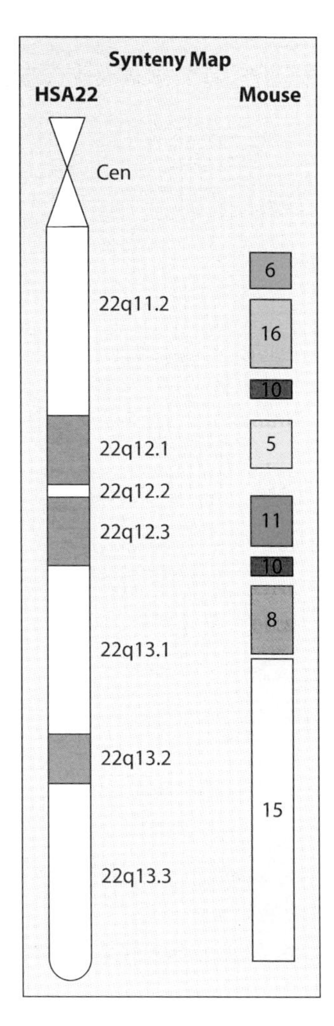

Figure 2.41 The syntenic relationship between human chromosome 22 and the mouse chromosomes. Many of the human genes found in the region 22q13.1, 13.2, and 13.3 are found in a similar linkage relationship on mouse chromosome 15, as shown in yellow. Other blocks of human chromosome 22 are syntenic with parts of other mouse chromosomes. This image and many other syntenic blocks among mammals are available from the Sanger Institute website.

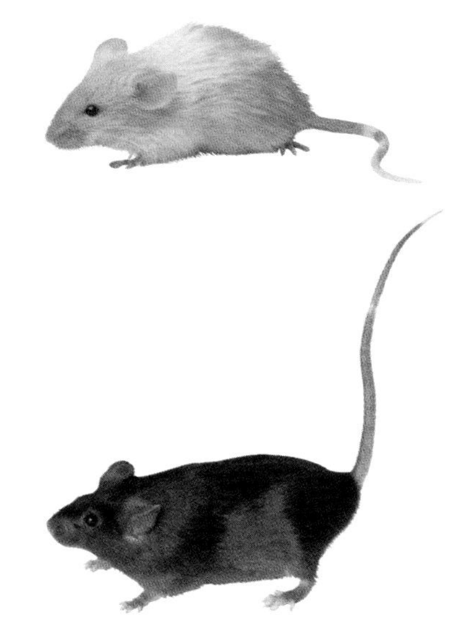

Figure 2.42 Different coat colors in mice. Mutations in more than 75 different genes that affect coat color have been identified in the mouse. The most commonly used colors are black, brown, and white.

not a perfect model for human disease—their life span is much shorter for example—but they provide far and away the best model that we have for genetic analysis of human biology. The mouse genome database, known as Mouse Genome Informatics (MGI) is shown in Figure 2.43. MGI is designated to be accessible to those with medical interests. Most pages have an FAQ section and tutorials are also common, as seen in Figures 2.43A and 2.43B. A search for the mouse *patched* gene, called *Ptch1*, is shown in Figures 2.43C and 2.43D. The detailed descriptions of individual mutant alleles, which also show their origin and the genetic background, are particularly helpful (Figure 2.43D).

2.7 Five model organisms: a summary

We have barely begun to describe genetic analysis in these five model organisms. As a summary of these model organisms, we refer again to Table 2.1, particularly the sections on the particular advantages and peculiarities of each model organism. Some biological questions are more readily explored using one model organism rather than another; often, but not always, the advantages and limitations of the organism as a model arise directly from some of its biological characteristics. For example, *S. cerevisiae* is ideal for studying nearly any property that occurs intracellularly. Although many of the fundamental properties of eukaryotic cells such as DNA replication and protein trafficking are also studied in other organisms, most of the original work was done in *Saccharomyces*, and yeast still provides a point of comparison in these other studies. On the other hand,

A

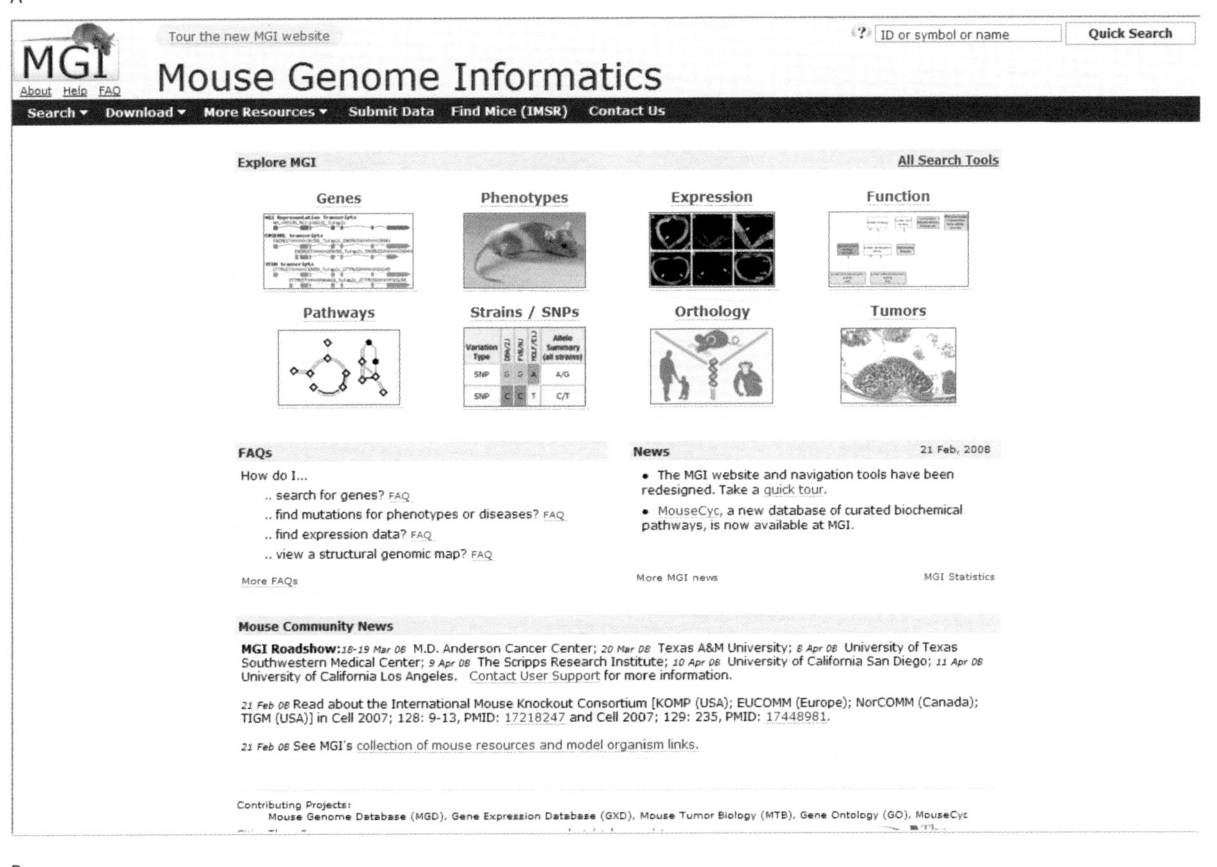

B

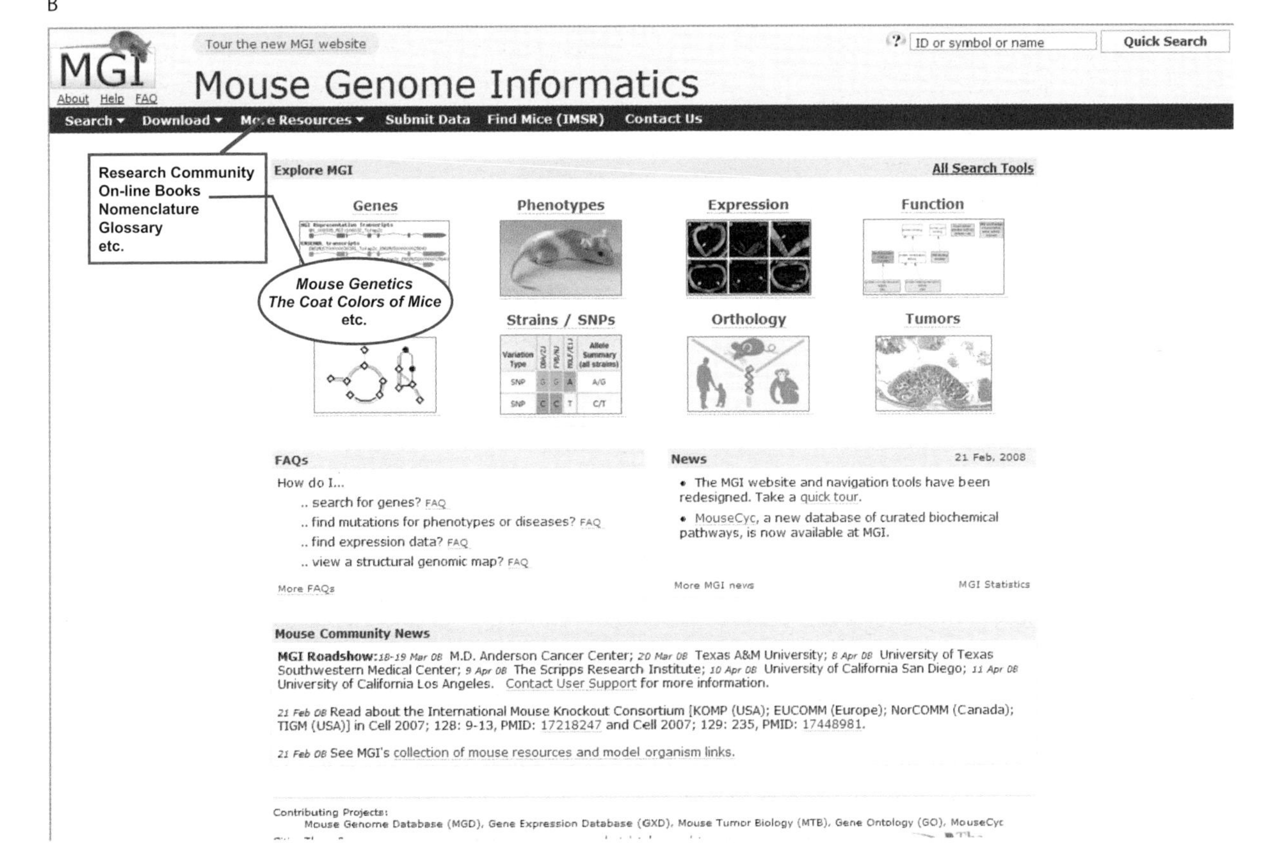

Figure 2.43 The Mouse Genome Informatics (MGI) database. (A) The home page of the MGI includes an FAQ section that is particularly helpful for new or inexperienced users. The organization of the information reflects that most investigators use mice as a model for understanding human biology and diseases. (B) The drop-down menu from More Resources has links to research laboratories, nomenclature, and, as shown here, some classic books available online.

C

MGI

Tour the new MGI website

ID or symbol or name | Quick Search

About Help FAQ

Home | Genes | Phenotypes | Expression | Function | Pathways | Strains / SNPs | Orthology | Tumors

Search ▾ Download ▾ More Resources ▾ Submit Data Find Mice (IMSR) Contact Us

Gene Detail

Your Input Welcome

Symbol	**Ptch1**
Name	**patched homolog 1**
ID	MGI:105373 — Nomenclature History
Synonyms	A230106A15Rik, Patched 1, Ptc, Ptc1
Genetic Map	Chromosome 13 36.0 cM Detailed Genetic Map ± 1 cM Mapping data(17)
Sequence Map	Chr13:63520755-63574742 bp, - strand (From NCBI annotation of NCBI Build 36) Ensembl ContigView | UCSC Browser | NCBI Map Viewer Mouse Genome Browser
Mammalian homology	human; rat (Mammalian Orthology) Comparative Map (Mouse/Human Ptch1 ± 2 cM) Protein SuperFamily: Drosophila membrane protein patched
Sequences	Representative Sequences — Length, Strain/Species, Flank genomic 19206 NCBI Gene Model | MGI Sequence Detail — 53988 C57BL/6J ± 0 Kb transcript NM_008957 RefSeq | MGI Sequence Detail — 4305 C57BL/6 polypeptide Q61115 UniProt | EBI | MGI Sequence Detail — 1434 Not Applicable For the selected sequences download in FASTA format Go All sequences(40)
Phenotypes	All phenotypic alleles(10) : Targeted, knock-out(5) Targeted, other(4) Spontaneous(1) Homozygous null embryos die by day 10 with neural tube defects. Heterozygotes are large with excess cerebellar granule cell proliferation and sometimes, hindlimb defects and medulloblastomas. Hypomorphic and spontaneous mutants have reproductive defects. Associated Human Diseases (1) Alleles Annotated to Human Diseases (2)
Polymorphisms	SNPs within 2kb(53 from dbSNP Build 126)
Gene Ontology (GO)	Process dorsal/ventral pattern formation, embryonic limb morphogenesis... Component extracellular region, integral to membrane

D

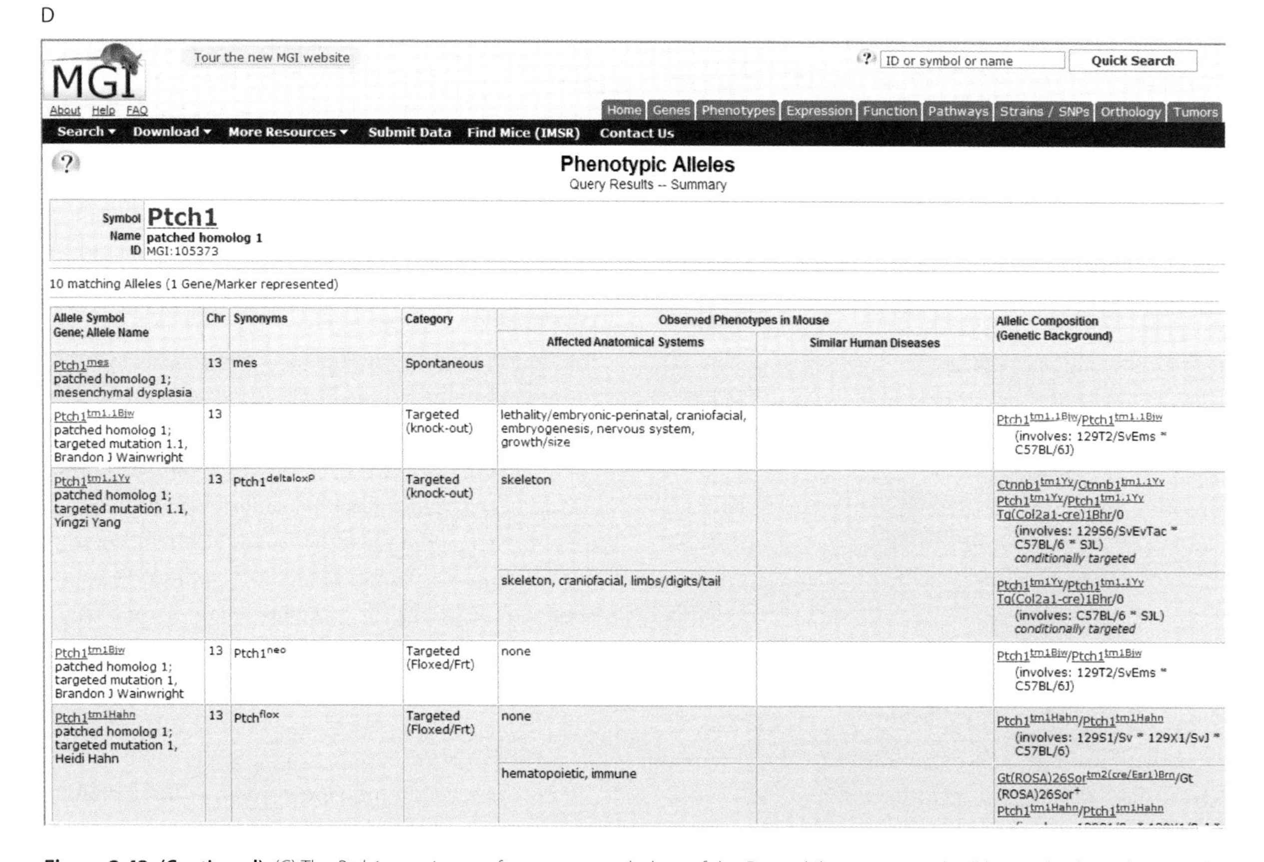

MGI

Tour the new MGI website

ID or symbol or name | Quick Search

About Help FAQ

Home | Genes | Phenotypes | Expression | Function | Pathways | Strains / SNPs | Orthology | Tumors

Search ▾ Download ▾ More Resources ▾ Submit Data Find Mice (IMSR) Contact Us

Phenotypic Alleles

Query Results -- Summary

Symbol **Ptch1**
Name **patched homolog 1**
ID MGI:105373

10 matching Alleles (1 Gene/Marker represented)

Allele Symbol Gene; Allele Name	Chr	Synonyms	Category	Observed Phenotypes in Mouse: Affected Anatomical Systems	Observed Phenotypes in Mouse: Similar Human Diseases	Allelic Composition (Genetic Background)
Ptch1^mes patched homolog 1; mesenchymal dysplasia	13	mes	Spontaneous			
Ptch1^tm1.1Bjw patched homolog 1; targeted mutation 1.1, Brandon J Wainwright	13		Targeted (knock-out)	lethality/embryonic-perinatal, craniofacial, embryogenesis, nervous system, growth/size		Ptch1^tm1.1Bjw/Ptch1^tm1.1Bjw (involves: 129T2/SvEms * C57BL/6J)
Ptch1^tm1.1Yy patched homolog 1; targeted mutation 1.1, Yingzi Yang	13	Ptch1^deltaloxP	Targeted (knock-out)	skeleton		Ctnnb1^tm1Yy/Ctnnb1^tm1.1Yy Ptch1^tm1Yy/Ptch1^tm1.1Yy Tg(Col2a1-cre)1Bhr/0 (involves: 129S6/SvEvTac * C57BL/6 * SJL) *conditionally targeted*
				skeleton, craniofacial, limbs/digits/tail		Ptch1^tm1Yy/Ptch1^tm1.1Yy Tg(Col2a1-cre)1Bhr/0 (involves: C57BL/6 * SJL) *conditionally targeted*
Ptch1^tm1Bjw patched homolog 1; targeted mutation 1, Brandon J Wainwright	13	Ptch1^neo	Targeted (Floxed/Frt)	none		Ptch1^tm1Bjw/Ptch1^tm1Bjw (involves: 129T2/SvEms * C57BL/6J)
Ptch1^tm1Hahn patched homolog 1; targeted mutation 1, Heidi Hahn	13	Ptch^flox	Targeted (Floxed/Frt)	none		Ptch1^tm1Hahn/Ptch1^tm1Hahn (involves: 129S1/Sv * 129X1/SvJ * C57BL/6)
				hematopoietic, immune		Gt(ROSA)26Sor^tm2(cre/Esr1)Brn/Gt(ROSA)26Sor^+ Ptch1^tm1Hahn/Ptch1^tm1Hahn

Figure 2.43 (Continued). (C) The *Ptch1* gene is one of two mouse orthologs of the *Drosophila ptc* gene and will be used in later chapters. The gene detail is shown here with many links that are similar to the ones used at NCBI for other genomes. Links to other browsers such as Ensembl and UCSC are also shown to provide different views and functions related to the gene. (D) By clicking on the link for 'All phenotypic alleles' on the Gene Detail page, a table of all of the mutant alleles of the *Ptch1* gene is shown. The table includes the phenotype, the origin of the mutation, the phenotype with the affected tissue or organ, and the background strain on which the allele is found.

alternative splicing and microRNA regulation do not occur in yeast, or at least are not common, so those processes are studied in other organisms.

The great advantage of *C. elegans* over other model organisms is its cellular simplicity and defined cell lineages. This is what drew geneticists to develop *C. elegans* as a model organism originally. Thus, many of the biological questions that arise from interactions between cells and that shape development have been studied productively in worms. We noted earlier the importance of genetic analysis in worms for untangling most signal transduction pathways and for recognizing the central role of apoptosis. It should also be noted that the complete connectivity of the nervous system in *C. elegans* has been described, and *C. elegans* has had a major role in integrating cellular neurobiology with simple behaviors.

One of the greatest advantages of *D. melanogaster* is that it has been used as a model organism for genetics longer than any other organism. The wealth of tools and depth of knowledge available to *Drosophila* geneticists are remarkable. It is hard to think of a significant biological question in animals, or in eukaryotes generally, that has not been studied using genetic analysis in *Drosophila*. These biological questions range from embryogenesis to ecology and evolution. However, targeted gene disruptions and gene replacement, which have contributed so much to genetic analysis in yeast and mice, are not readily available in worms and flies.

For biological questions involving either higher plants or mammals, there is one most obvious choice of model organism. The development of advanced tools of genetic analysis in *Arabidopsis* has been a great boon to plant biology, and many geneticists have discovered the wonders of plant biology from *Arabidopsis*. Mice have long been viewed as an experimental stand-in for humans, and they have usually succeeded in this role.

One feature shared by all of these model organisms is the active and interactive research communities that rely on them. The importance of a community of scientists who share reagents, strains, ideas, and criticisms cannot be overstated. Many excellent genetic models have not quite risen to the prominence warranted by their biological properties because other researchers did not adopt them readily; in a few cases, rivalries among the researchers have hampered progress on the organism. The extensive genome databases referenced in Table 2.1 and in the online supporting material are testimonies to the collaborations and energy of the researchers studying these five model organisms.

2.8 Other useful model organisms

It would be very misleading to suggest that these five widely used model organisms are the only eukaryotes used by geneticists. While it is true that these five model organisms dominate the literature, other organisms are also frequently used and deserve more than the passing mention that can be made here. A few of them are listed in Table 2.2, with examples of their utility and the names of their genome databases. Both the organisms and the examples of the biological questions they can be used to answer are selective rather than comprehensive. The list in Table 2.2 does not include some organisms that are researched principally because they are closely related to one of the five most widely used organisms. Such a list would include *Drosophila pseudoobscura* and *Drosophila virilis*, *Caenorhabditis briggsae*, and others. Table 2.2 also does not include any of the domesticated mammals whose genomes are being sequenced, such as dogs, cats, cows, pigs, rats, and so on. Many mutations have been identified in these organisms and serve as models for human biology and disease. Mutations are known and genome sequences are available for many plants of agricultural significance as well. It is also worth noting that the genomes of most of the higher apes are likewise being sequenced, although these animals are not genetic models in the same sense that other mammals are.

Are more model organisms needed?

The five eukaryotic model organisms (plus humans) probably will reveal all the *universal* properties of eukaryotes. If our goal is to determine the common underlying molecular biology of eukaryotes, then it seems at least

Table 2.2 Some other model organisms widely used for genetic analysis

Species	Type of organism	Example of biological questions	Genome database
Schizosaccharomyces pombe	Fission yeast	Cell cycle, chromosome structure	*Schizosaccharomyces pombe* GeneDB
Aspergillus nidulans	Fungus	Meiosis	*Aspergillus* Comparative Database
Neurospora crassa	Fungus	Meiosis	*Neurospora crassa* Database
Tetrahymena thermophila	Ciliated protozoa	Chromosome structure, nuclear differentiation	*Tetrahymena* Genome Database
Paramecium tetraurelia	Ciliated protozoa	Excitable cells and ion channels	ParameciumDB
Dictyostelium discoideum	Slime mold	Signaling and morphogenesis	dictyBase
Chlamydomonas reinhardtii	Green alga	Flagella and motility, chloroplast development	ChlamyDB
Physcomitrella patens	Moss	Evolution of plant development	JGI *P. patens* subsp. *patens*
Zea mays	Maize	Transposons, epigenetics, plant development	MaizeGDB
Oryza sativa	Rice	Plant development, abiotic stress, model monocot	OryzaBase Database, TIGR Rice Genome Annotation
Populus trichocarpa	Poplar	Woody plant growth and development, secondary growth	JGI *P. trichocarpa*
Danio rerio	Zebrafish	Vertebrate development, behavior	ZFIN
Canis familiaris	Dog	Behavior, disease, large mammal development	NHGRI Dog Genome Project

possible that no additional model organisms will be needed. In addition, if our goal is to have model organisms that tell us about *human biology*, an argument could be made that no additional model organisms are needed. We have the human genome sequenced, we have powerful tools for genetic analysis in mice, and we have comparative sequence from higher apes, primates, and other mammals. Perhaps that is enough.

However, understanding the universal properties of eukaryotes and applying these principles to humans were only two of the reasons that investigators rely on model organisms. If our goal for research with model organisms is to understand biology, i.e. to understand the evolutionary processes that shape the living world, these model organisms are far from sufficient. Let us consider these organisms from an evolutionary perspective. The positions of these organisms on a standard taxonomic tree of eukaryotes are shown in Figure 2.44A. (More detailed evolutionary trees and taxonomic comparisons can be observed by visiting the websites for the Tree of Life and taxonomy from NCBI.) The gaps in our tree of genomes are readily noticed. Certain phylogenetic groups, such as ascomycetes fungi, insects, and mammals, are very well represented. Other taxa are almost absent.

For example, *Arabidopsis* cannot really be expected to serve as a model for all higher plants. Woody angiosperms such as most trees, gymnosperms such as evergreens, monocotyledons and grasses, and perennial plants are dominant species about whose molecular genetics relatively little is known. Our knowledge is even more sparse when we look at the non-flowering plants, including ferns, horsetails, mosses, and the like.

Notice from Figure 2.44A some of the other phyla for which very little is known, and imagine what they could tell us. Protozoa could give us insights into some of the distinguishing features of eukaryotes, such as nuclear organization, meiosis, and more. Porifera (sponges) might tell us about the evolutionary origins of multicellular organization, Cnidaria about the origins of sensory cells, and so on. These features are highlighted in the view of animal phylogenetic relationships presented in Figure 2.44B. The list of what we may be able to learn is limited only by our imagination. Table 2.3 presents a phylum- and class-oriented view of the invertebrates, noting interesting features of some of the taxa. Many of these organisms could have become useful genetic models, but have been little studied. The reasons may be that the habitat or the growth requirements are too finicky (or simply not known), or that the life-cycle is long and complex. However, one can contemplate and dream about some of the evolutionary grandeur that we have not yet viewed.

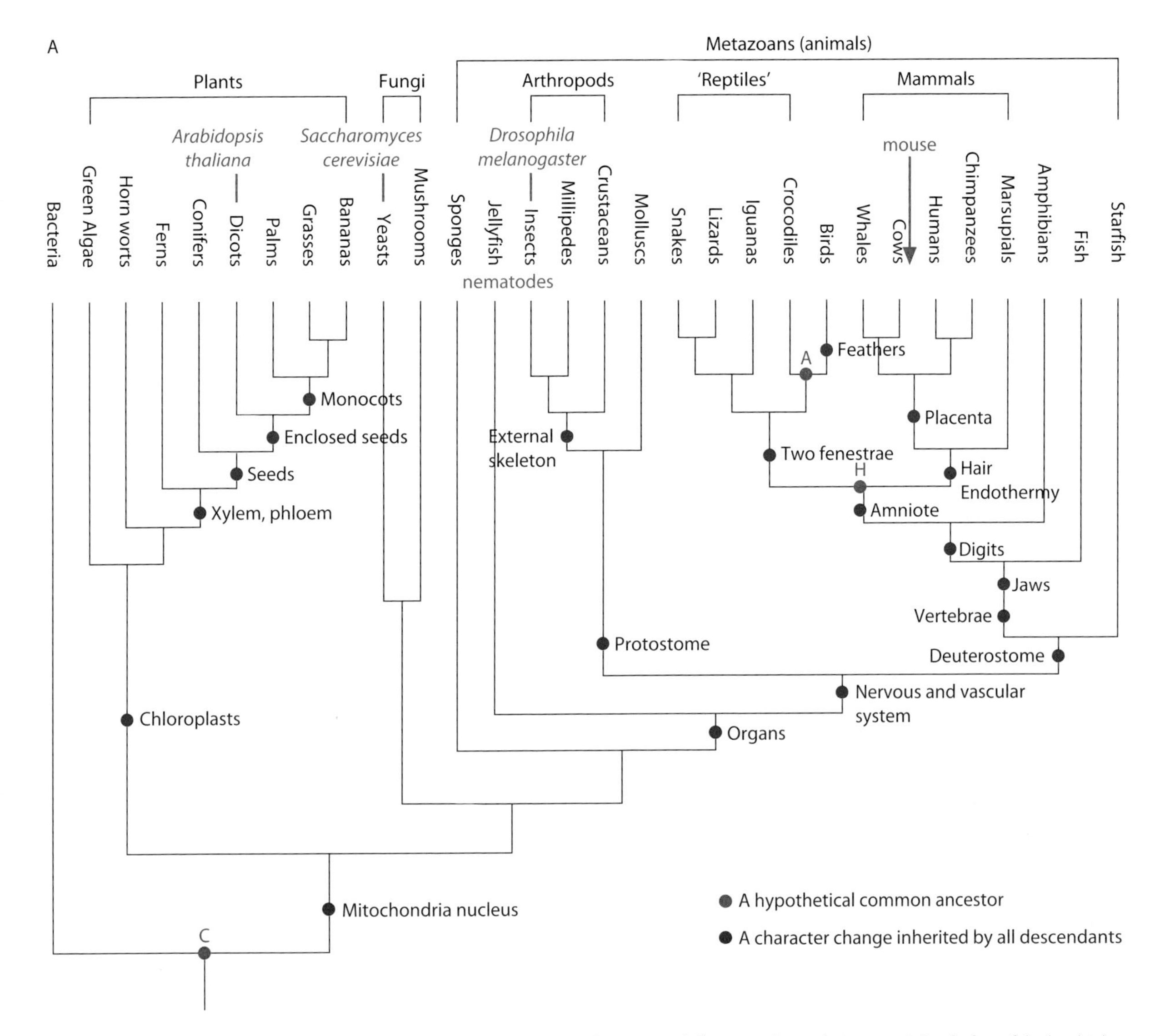

Figure 2.44 Panel A shows a conventional phylogenetic tree of some eukaryotes and illustrates the evolutionary origin of a few of the key biological characteristics. The positions of the five model organisms are noted.

Genetic analysis in a non-model organism

The limitations presented by some of these organisms may be changing so that we might not need to continue to speculate about interesting biological questions for some species. Genome sequencing has progressed far more rapidly than could have been imagined a decade ago. New genome sequencing projects are being initiated daily, and others are being completed. While most of these genomes are drawn from vertebrates on the one hand, and from prokaryotes and archeae on the other, many others are filling some of the holes in our view of life. With well-annotated genome sequences, much more comparative genomic analysis and inference of evolutionary relationships will be possible.

However, genome sequences and cloned genes provide only some of the important tools of genetic analysis, and they still do not meet all the criteria for a model organism presented earlier in the chapter. Another criterion is the availability of mutations. Because mutations require the ability to grow the organism in the laboratory and to observe phenotypes easily, this criterion is often the most difficult one for potential research organisms to satisfy. While it is unrealistic to hope that any of these other possible models will have as many genetic mutations available as our five main

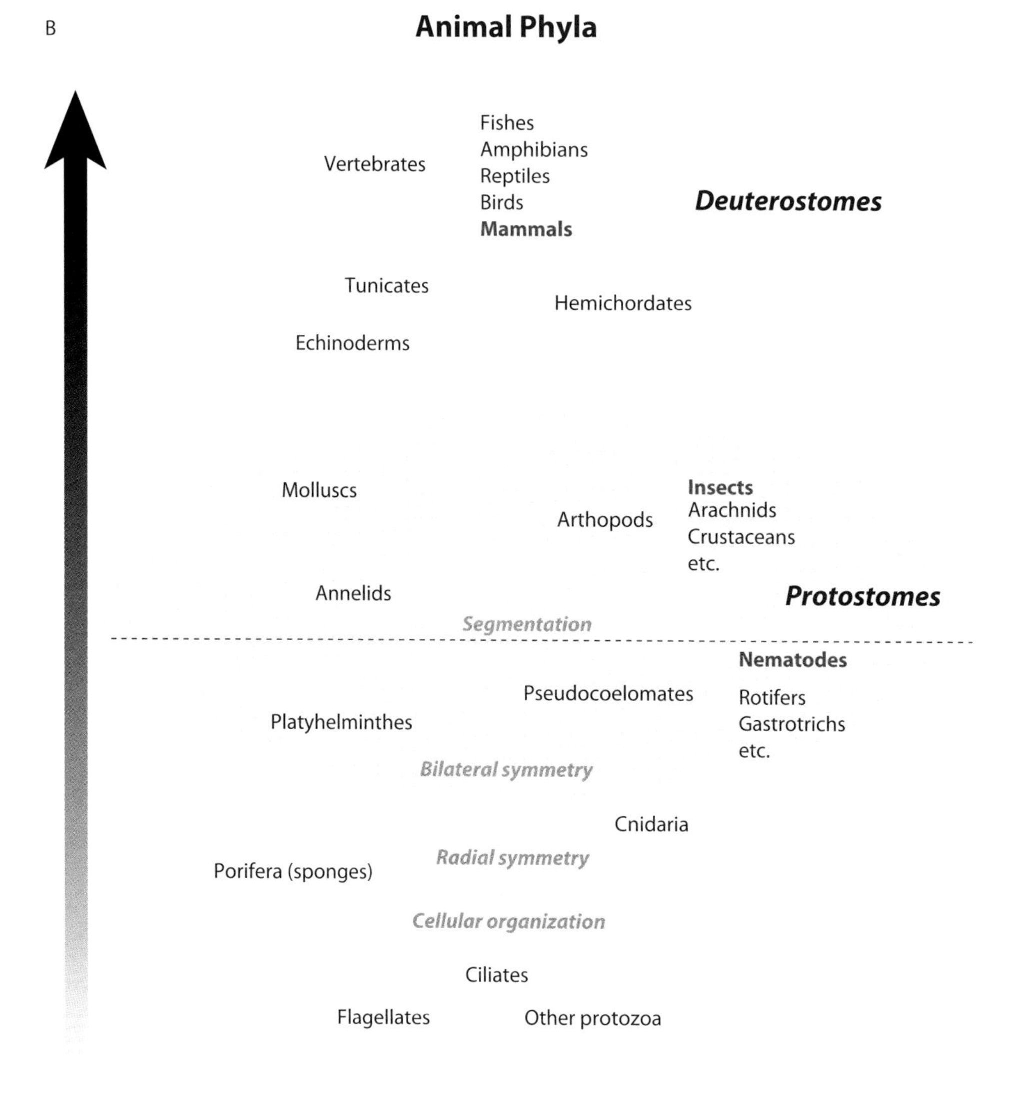

Figure 2.44 (Continued). Panel B focuses on the properties of some animal phyla, with the arrow indicating increasing biological complexity. Organisms are collected by the properties that they share, including multicellular organization, radial and bilateral symmetry, and segmentation. The two large ovals represent the conventional distinction between two animal body plans, protostomes and deuterostomes.

models, another approach to finding mutant phenotypes offers considerable promise here as well. In Chapter 7 we describe genome-wide screens involving **RNA interference (RNAi)**. While it is not quite as good as having a mutation, RNAi provides a powerful alternative for organisms in which mutations are not readily available but whose genome sequences are known.

We close this chapter on model organisms by describing one example of what genetic analysis may look like over the next few years. In Case Study 2.1 we discuss how RNAi has been used to study genes involved in regeneration in planaria. Planaria belong to the phylum Platyhelminthes, one of the under-studied taxa near the base of our phylogenetic relationships in Figure 2.44B. Planaria are famous for their ability to regenerate entire organisms from small dissected pieces, a fascinating biological phenomenon whose mechanism has not been thoroughly studied. Genetic analysis seems like a perfect method to approach this question, except that mutations are not available for planaria. As shown in Case Study 2.1, RNAi offers the possibility that any organism whose genome is sequenced can be studied by genetic analysis. In this view of life, any organism becomes a model organism for experimental genetics.

Table 2.3 Properties of some invertebrates

Group	Properties	Potential interest
Protists	Protozoa: widely diverse in body shape, movement, habitats	Evolution of eukaryotic properties such as meiosis, chromosome structure, spindle
Porifera	Sponges: multicellular but with little cell specialization	Intercellular signaling, cell differentiation
Cnidaria	Jellyfish, hydra: radial symmetry, stinging cells	Origins of excitable cells, life cycle stages are morphologically different
Platyhelminthes	Flat worms, e.g. planaria: bilateral symmetry, primitive nerve nets and behaviors, three germ layers, protostomes	Regeneration, nervous system, origins of muscles and gut, complex reproductive systems
Nematoda	Unsegmented round worms: very widespread habitats and many different species, protostomes	Many parasites of economic and medical importance, complex nervous system and behaviors, cellular simplicity, *C. elegans* is a model
Rotifera	Pseudocoelom: morphologically diverse, body is regionally specialized, protostomes	Complex sexual or asexual cycle, evolutionary position uncertain
Annelids	Segmented worms: circulatory system, protostomes	Origins of segmentation, circulatory systems
Arthopoda	Insects, spiders, crustaceans: exoskeleton, jointed appendages, very widespread, very diverse morphology and life cycles, protostomes.	Diversity of body plans and life cycles, origin of appendages, *D. melanogaster* is a model
Mollusca	Snails, clams, etc.: typically shelled, complex organ systems, complex life cycle, protostomes	Classical embryology is well understood, larval forms are morphologically distinct
Echinodermata	Starfish: exoskeleton, complex life cycles, deuterostomes	Inversion of gastrulation and body plan

Case Study 2.1

Regeneration in planaria

Genetic analysis in a non-model organism

The five model organisms discussed throughout this book have been used for experimental research for many years. Many mutations are available for each of them, as well as a complete or nearly complete genome sequence, both for these species and for some closely related species. As valuable as these organisms have been for understanding the fundamental principles of eukaryotic molecular biology, they cannot possibly represent the full range of interesting biological questions. In this Case Study, we describe a genetic approach to the study of regeneration in planaria, an organism for which no mutations are available.

Regeneration in planaria has been studied for more than a century

Planaria are flatworms, members of the phylum Platyhelminthes. The anatomy of a planarian is shown in Case Study Figure 2.1. Planaria have all three germ layers that are characteristic of higher organisms. Their body plan includes a simple nervous system, which controls behaviors such as photo-avoidance, muscles, and a gut with a single opening. Planaria live in freshwater, feed on small animals or animal debris, and average about 10 mm long depending on the species (or about 10 times the length of *C. elegans* and four times the length of *Drosophila*). No mutations have been found for any member of this phylum. Indeed, molecular and genetic approaches had been minimal until Newmark and Sánchez Alvarado began working with *Schmidtea mediterranea*, a diploid with both sexual and asexual reproduction.

Planaria are a classical model for studying regeneration using surgical procedures. In fact, T.H. Morgan studied regeneration in planaria before working on *Drosophila*. The only proliferating cells in the adult are the neoblasts, which appear to be the equivalent of a totipotent stem cell population capable of regenerating any adult structure. Upon wounding, a **blastema** of neoblast cells covered by epidermis forms, and the neoblast cells proliferate and differentiate (Case Study Figure 2.2). Dissected tiny pieces from any part of the animal are capable of forming such a blastema, and can subsequently regenerate entire animals with complete organ systems. Many physical manipulations and surgeries have been performed during the past century to demonstrate this amazing regenerative capacity, but the genes and molecules are unknown.

Regeneration genes were identified using RNAi

In order to identify and characterize genes involved in regeneration, researchers employed a process known as RNA interference (RNAi), which will be described in detail in Chapter 7. For this case study, the important aspect of RNAi is that a double-stranded RNA (dsRNA) corresponding to a gene is introduced into the organism, whereupon it reduces or eliminates the specific expression of that gene. As a result, the organism exhibits the phenotype of a mutation in the gene, although the gene itself is unaffected. Since regeneration occurs from the neoblast, RNAi was applied to a group of genes expressed in the neoblasts. The

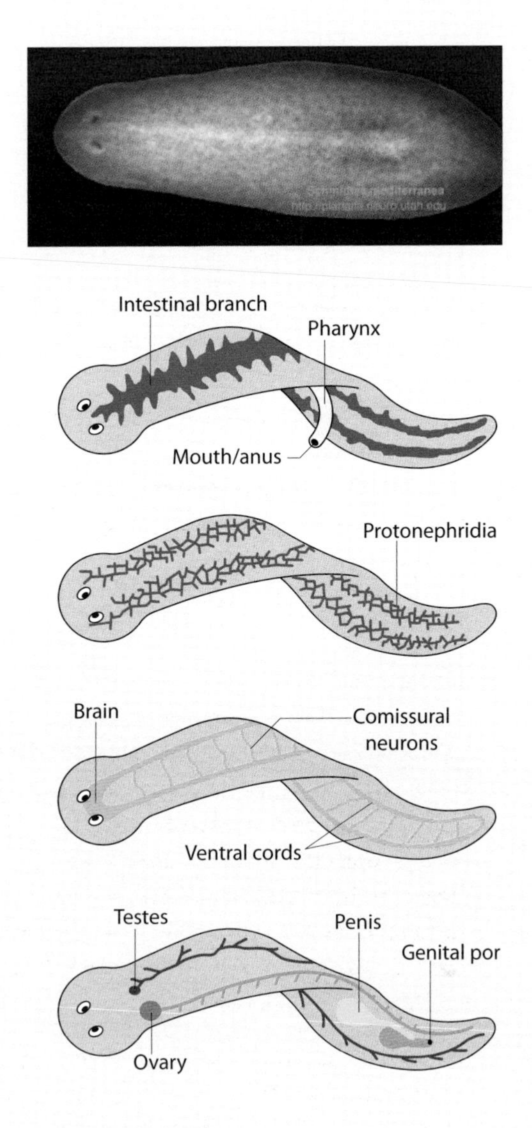

Figure C2.1 The anatomy of the planarian *Schmidtea mediterranea.*

regeneration assay is shown in Case Study Figure 2.3. A dsRNA from a gene expressed in the neoblast was mixed with a liver extract and agarose and fed to the animal. After feeding, the head and the tail were removed from the planarian, and regeneration was allowed to occur for 9 days. The process of feeding dsRNA and dissection was repeated on these 'regenerated' organisms, and the regeneration ability of the dissected cells was assessed.

RNAi assays were performed for 1065 neoblast-expressed genes, involving more than 50 000 dissected planaria. A collection of 240 different genes was found that affected regeneration, i.e. when dsRNA from these genes was fed to planaria, the regeneration process was abnormal. The regeneration defects were classified into 11 different morphological categories, including an inability to regenerate at all, alterations in the shape and growth of the blastema, and others. One of the main advantages of genetic approaches, as will be described in Chapters 3 and 4, is that little or no prior knowledge of the process is required, and unexpected results can be obtained. This was true for the defects observed in the regeneration process. For example, some genes affected only regeneration by the tail blastema and not

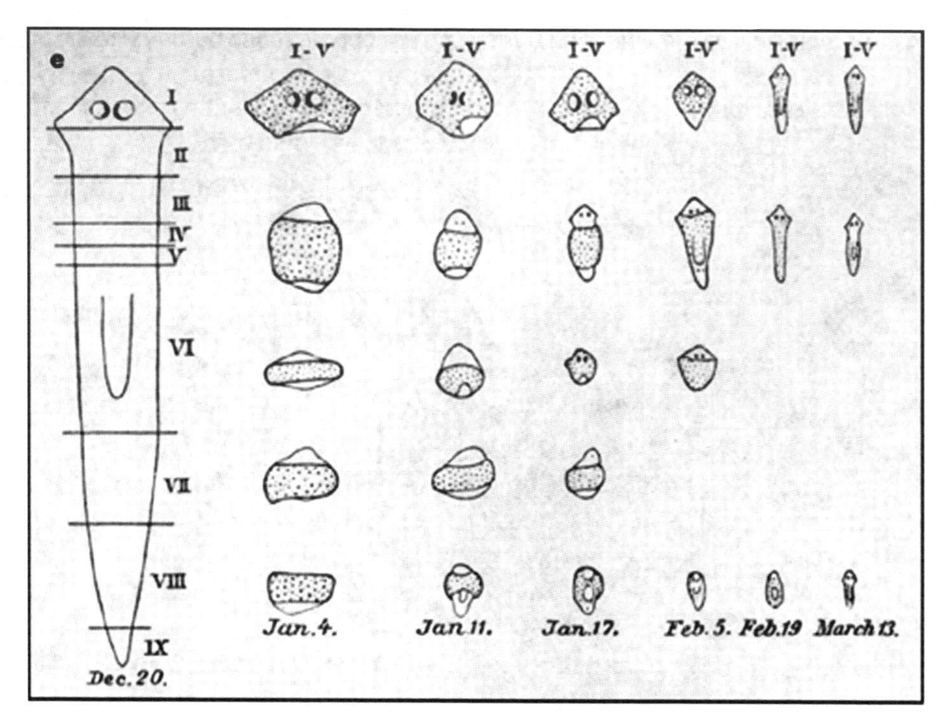

Figure C2.2 Regeneration in planaria. These drawings are from a paper by T.H. Mongan in 1898, showing the ability of aplanaria to regenerate from different fragments.

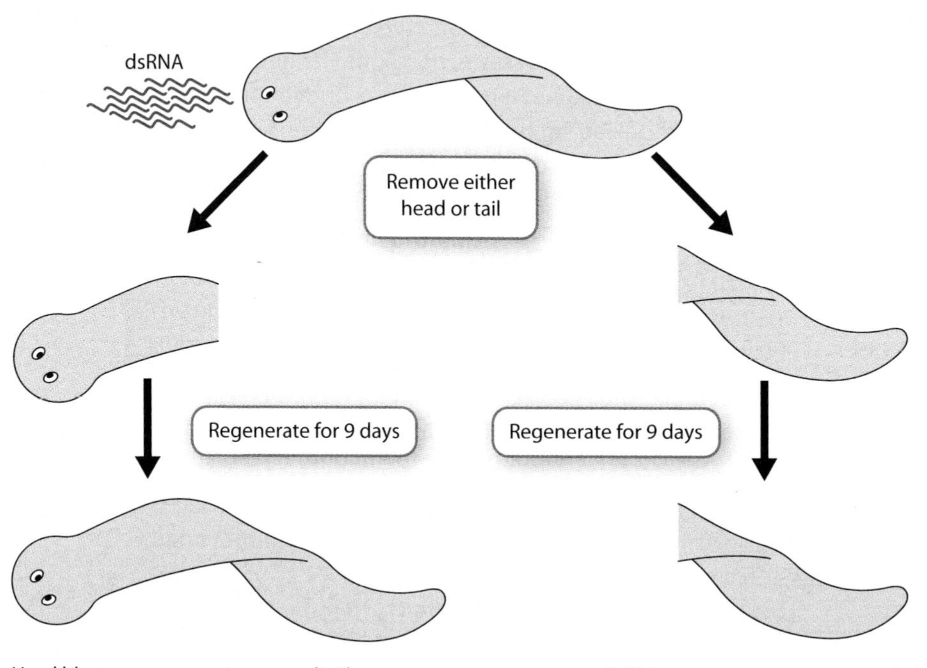

Figure C2.3 Assaying genes for their effect on regeneration using RNAi. A praparation of corresponding to an individual gene is fed to the planarian. The dsRNA knocks down repression of the gene by a process known as RNAi. Either the head or the tail is removed and the remaining cells are allowed to regenerate for 9 days. The process is then repeated. In the hypothetical example shown, the gene affects the ability of the tail blastema to regenerate a normal head but does not affect the ability of the head blastema to regenerate a normal tail.

the head blastema, suggesting that regeneration in the two ends of the animal involves some different genes. Recent further research has implicated β-catenin and genes in that pathway as defining the difference between head and tail regeneration. Other defects included asymmetric growth of the blastema and an unusual pigmentation pattern. None of these effects was suspected from the surgical manipulations.

Roughly a quarter of the genes tested (240 of 1065) have effects on regeneration, which is not very surprising since so many different cellular processes are involved in regeneration. The number of genes in the *S. mediterranea* genome is not known, but the authors estimate that the 1065 genes tested are perhaps 5–10 percent of the genome. Most of the planarian genes identified (205 of the 240) had identifiable orthologs in other organisms. This provides some clues about the wide range of molecular processes involved in regeneration. Included among the genes affecting regeneration were ones implicated in many fundamental cellular processes, including genes encoding an RNA helicase and a topoisomerase, several proteins related to signal transduction, and some related to chromosome structure. Interestingly, 38 of the regeneration genes in planaria have a human ortholog that is associated with a known genetic disease; most of these genes do not have an identified mouse genetic model for the disease, and so the distinctly related planaria could provide an applicable genetic model for humans. Thus, although the primary interest in applying genetic analysis is to understand regeneration in planaria, a taxon-specific process as described in the chapter, a second benefit could be to provide an additional model for human biology, possibly including the biology of stem cells.

Genetic analysis of regeneration in planaria required only a few tools (and a lot of patience in performing the dissections). The first of these is RNAi to provide 'mutant' phenotypes. In planaria, as in all organisms in which RNAi has been applied, the procedure for introducing the dsRNA had to be specifically developed and modified from what is done in other organisms, but the fundamental approach is the same. RNAi provides a method to knock down or eliminate the function of any gene in most multicellular animals and plants. The second necessary tool for this kind of analysis is genomic sequences or cDNA clones for the genes to be studied. For planaria, these were cDNA clones from neoblast cells. With modern molecular biology, cDNA clones from many different organisms can readily be obtained. The path for genetic analysis of remarkable biological questions has never been more inviting.

Further reading

Newmark, P.A. and Sánchez Alvarado, A. (2002). Not your father's planarian: a classic model enters the era of functional genomics. *Nature Reviews Genetics*, **3**, 210–19.

Reddien, P.W., Bermange, A.L., Murfitt, K.J., Jennings, J.R., and Sánchez Alvarado, A. (2005). Identification of genes needed for regeneration, stem cell function, and tissue homeostasis by systematic gene perturbation in planaria. *Developmental Cell*, **8**, 635–49.

Sánchez Alvarardo, A. (2004). Planarians. *Current Biology*, **14**, R737–8.

Chapter Capsule

Model organisms

Five eukaryotic species have been used as the experimental model organisms through which much genetic analysis has been done. These five organisms, *Saccharomyces cerevisiae*, *Caenorhabditis elegans*, *Drosophila melanogaster*, *Arabidopsis thaliana*, and *Mus musculus*, are easy to grow in the laboratory and have large and expanding storehouses of mutants. Complete genome sequences are available for all of them in easily accessible databases. Most of the remainder of the book will use examples from these five organisms. However, these five species represent a fraction of the organisms that have been used for genetic analysis. Furthermore, as the genomes are sequenced for even more species and as RNAi allows a form of genetic analysis without mutations, other organisms will also provide informative models for evolutionary and comparative studies.

Unit 2
Genes and mutants

The two fundamental tools of genetic analysis are a collection of mutants that affect the function of the gene and a cloned copy of the gene.

CHAPTER MAP

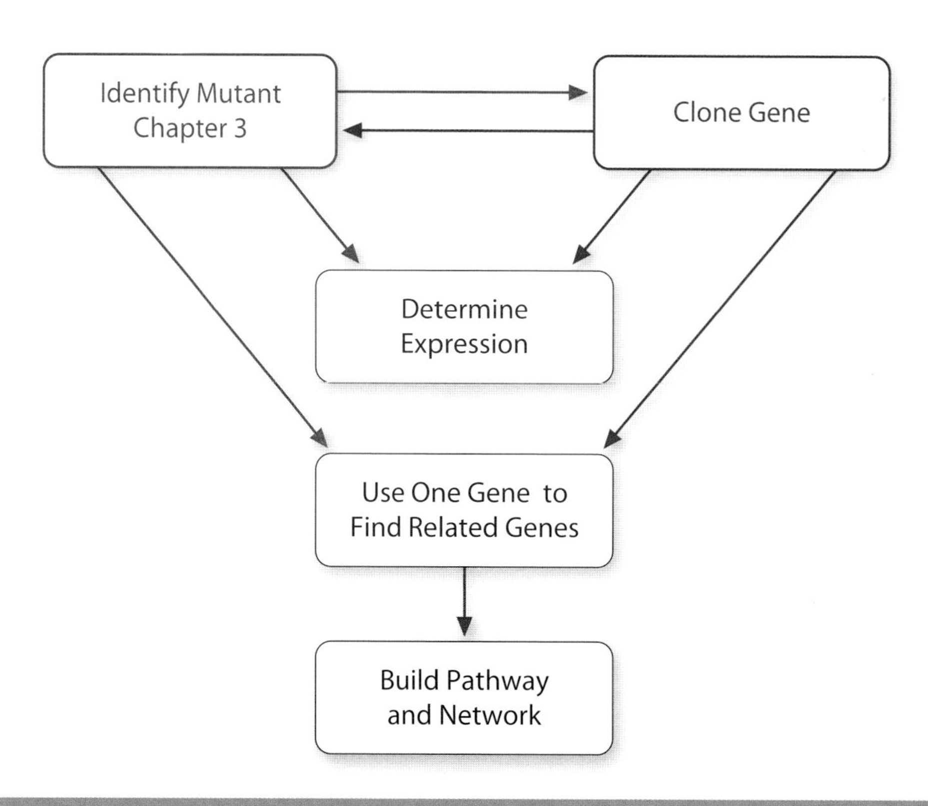
Identify Mutant
Chapter 3
Clone Gene
Determine
Expression
Use One Gene to
Find Related Genes
Build Pathway
and Network

CHAPTER 3

Identifying mutants

TOPIC SUMMARY

The process of genetic analysis often begins with a mutant phenotype. The power of mutant analysis is that the investigator does not need to know in advance what to expect when the biological process is disrupted. Mutant analysis can be summarized as follows.

- Finding mutations, often using some mutagenic agent to increase their frequency.
- Characterizing the phenotype of the mutation, including examining its phenotype under different conditions.
- Assigning the mutations to genes using complementation tests.
- Classifying the type of defect in each mutation based on its effect on the function. While most mutations are recessive and cause a loss or a reduction in the function, some interesting mutations are dominant and arise from other alterations in a gene.

This chapter describes how mutations are found and the phenotype is characterized. The next chapter discusses assigning mutations to genes and classifying the mutant defect. Chapter 5 describes some strategies by which genes are cloned, i.e. that a mutant phenotype is connected to an altered DNA sequence.

IN BRIEF

Mutations that disrupt a biological process are the crucial tools required for its dissection by genetic analysis.

INTRODUCTION

Biological processes are the outcomes of millennia of evolution. Geneticist and Nobel Laureate Max Delbruck said: 'Any living organism carries in its genome the results of a billion years of experimentation by its ancestors.' We cannot trace the long history that determines how an organism executes a certain biological process in a particular way or how it has come to employ different genes and gene products in specific ways. On the other hand, it does not take any special training in biology to see the complexity and glory of a living organism.

If we really want to understand the beauty of an organism, however, we need to analyze how it carries out its functions. Without knowing all the turns and dead ends of the evolutionary history, we cannot deduce this underlying logic. How then can we study the strategy of normal biological processes? We can infer the history by examining specimens that fail to carry out the process normally, i.e. those that have a mutant phenotype. **Mutant analysis**, the topic of this chapter, is the study of failed biological processes. Something in the normal development, physiology, or behavior of the organism goes wrong and we can see a mutant phenotype as a consequence. From that failure we infer what must occur normally—genetic analysis is about learning from mistakes.

However, not just any mistake or mutant is useful for genetic analysis. We need to know what gene has been mutated and how it has been changed. We need to find other mutations in the same gene and compare their effects. We need to use that mutation to lay the foundation for many other types of analysis.

The tool that we *explain* in this chapter is the mutated gene and its mutant phenotype. A mutant is fundamental to many of the tools that we *develop* in the rest of the book. Finding a mutant is often the first and crucial step in untangling the complex strategies of biological processes. Genetic analysis with mutants is recognizing the beauty of failures.

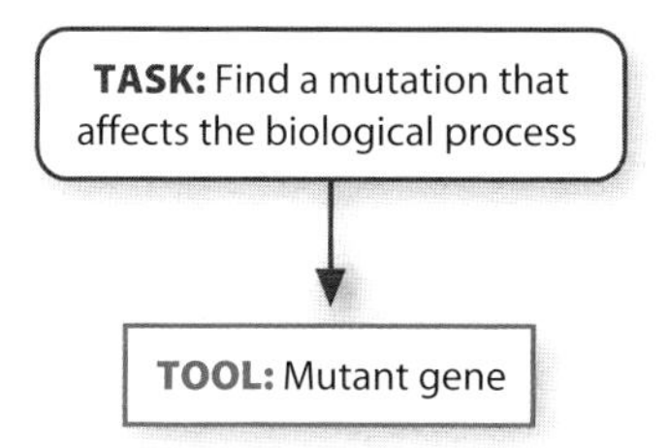

Figure 3.1 This chapter describes how to identify and characterize mutations, a crucial tool for the subsequent chapters.

3.1 Finding mutations: an overview

How does one begin to study a complicated biological process? Imagine working with one of the model organisms described in Chapter 2. You watch a fruit fly buzzing around its vial, more beautiful and subtle than you ever thought, with those big red eyes tracking your every gesture. Or you see a worm swimming gracefully on the surface of the Petri dish, attracted to some chemicals and repelled by others. What could you do to understand this or any other biological process?

Almost all geneticists who work with model organisms will answer this question with the same three-word mantra: Find a mutant. Do you want to know what determines the eye color in a fruit fly? Find a mutant with a different eye color. Do you want to understand how the worm moves so gracefully? Find a mutant that cannot move. You want to understand a biological process? Find a mutant that cannot carry it out properly.

It may seem paradoxical that the strategy for analyzing a biological process begins by observing its malfunctions. Yet that is what mutant analysis involves—disrupting the process in some way and asking what happens. In other words, breaking it in a limited and controlled way to see what happens. Or, to repeat the geneticist's mantra, 'Find a mutant.'

In this chapter we describe the procedure by which geneticists find and characterize mutants. The mutated gene and the mutant phenotype provide essential tools for every other part of genetic analysis, and topics in nearly every other chapter in the book assume that the investigator has a mutated gene to work with. Furthermore, we assume not only that the investigator has such a mutant but also that the mutant has been characterized, i.e. that the investigator has an answer to questions such as the following:

- Is the mutation recessive or dominant?
- Does the mutant phenotype change in different growth conditions?
- Does the mutation affect more than one biological process?
- Does the mutation reduce or eliminate the function of the gene? Or does it increase its function or introduce a new function?
- How many other genes with similar mutant phenotypes are in the genome?

3.2 Forward and reverse genetic analysis

It is sometimes helpful to distinguish between two types of genetic analysis, often called 'forward genetics' and 'reverse genetics'. Precise usage of these terms varies, and we are rapidly approaching a time when the experimental distinction between them is so blurred as to be meaningless. Nonetheless, the distinction is useful for our descriptions, and we will define these terms as follows. **Forward genetic analysis** or, more simply and commonly, **genetic analysis** means that the investigator has begun with an organism with a mutant phenotype and has worked towards identifying and characterizing the gene that is responsible. 'Forward' means from phenotype to gene. The mutant phenotype is what the investigator has as the starting point for the analysis and the DNA sequence of the gene has to be obtained. **Reverse genetic analysis** means that the investigator has identified a cloned gene or DNA sequence, and has worked towards identifying the mutant phenotype or understanding the biological process that the gene affects. This chapter deals with forward genetic analysis, hereafter referred to as genetic analysis. Chapters 6 and 7 more specifically address reverse genetic analysis.

Forward and reverse genetic analysis approaches are compared in Table 3.1. Figure 3.2 illustrates the differences between these two approaches using a mutant wing phenotype in *Drosophila* as one example. The greatest advantage of forward genetic analysis is that the investigator allows the organism to provide insights into the biological process. The investigator does not need to understand the biochemistry or the physiology of

Table 3.1 Forward and reverse genetic analysis: a comparison

Type of analysis	Starting material	Experimental question	Origin of mutant genes
Forward genetic analysis	Mutant phenotype	What gene causes this phenotype?	Induced by random mutagenesis
Reverse genetic analysis	Cloned gene	What phenotype does this gene cause?	Targeted to specific genes

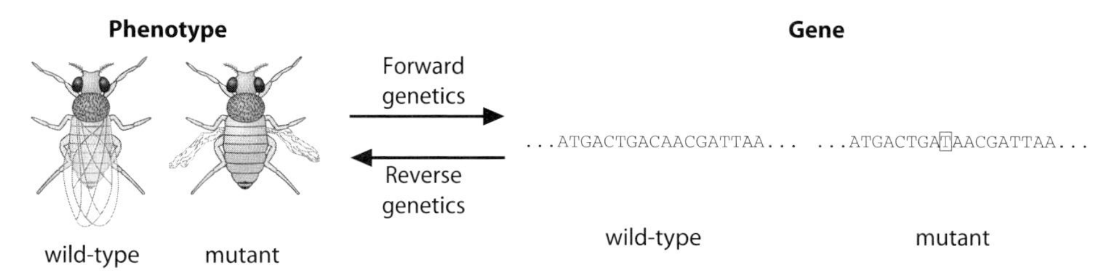

Figure 3.2 A comparison of forward genetic analysis with reverse genetic analysis. Forward genetic analysis (on the left) begins with a wild-type and mutant phenotype in the organism. In this figure, the mutant phenotype is vestigial wings in *Drosophila*. The analysis proceeds from the organism to the gene (on the right), where the sequences of the wild-type and mutant genes are compared. In this example, the mutated gene is represented by the red box around the site of the mutation. In reverse genetic analysis, the gene is known but the mutant phenotype needs to be determined. Reverse genetic analysis is described in Chapter 5. The sequence is used for illustration and is not a sequence from the *vg* gene or any of its mutant alleles.

the underlying process in advance and often will know little about the molecules that are involved. The investigator does not need to know in advance how many genes are involved or anything about their functions or relationships, although that often helps. In fact, the investigator does not even need to know what alternative phenotypes exist for genes involved in the process. The mutant phenotypes yielded by the organism will reveal those things. The mutations can occur with any gene in the genome. Genetic analysis that begins by finding mutants is an extremely powerful first step for nearly any biological problem.

Reverse genetic analysis has the advantage that the molecular function of the gene may already be known. Often reverse genetic analysis begins when a gene has been identified in one organism, in the fruit fly for instance, and the investigator wants to know what the equivalent gene does in another organism, say the mouse. Or perhaps the gene is known to be expressed in cells or a stage of the life cycle under study. Thus, the gene is presumed to be important. On the other hand, reverse genetic analysis requires more knowledge to begin than forward genetic analysis does. This will be discussed more fully in Chapter 6.

To begin our discussion of forward genetic analysis, let us consider a simple example. A wild-type fruit fly has brick-red eyes. How does this eye color come about? What other colors are possible? Does the biochemistry of eye pigmentation allow flies to have blue eyes, brown eyes, or black eyes? Does eye color have any relationship to the colors of other tissues in the fly? How can we analyze this complex biological problem?

Remember our mantra: find a mutant. To do this, a geneticist will look for flies that have eyes of any color other than the wild-type brick red. Some of the phenotypes that might be found are shown in Figure 3.3. These mutant flies will be analyzed by mating with flies of defined genotype in order to determine how many genes are involved and what has happened to those genes to give rise to this mutant eye color. The initial mutant hunt is repeated and more flies with mutant eye color will be found and analyzed by this scheme of crosses. If the investigator persists long enough, many or even most of the genes that affect eye color will be found. Then we can begin to answer questions about how the color of the eye is determined. In reality, investigators will answer many of these questions as they are collecting the mutants; the analysis of the first mutations to be

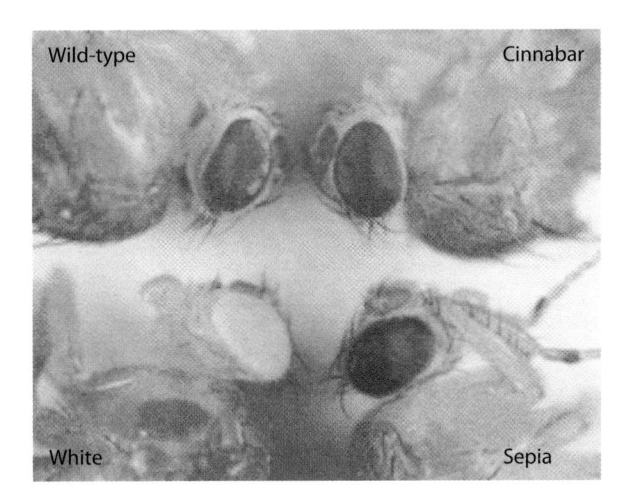

Figure 3.3 Mutant eye colors in *D. melanogaster*. More than 20 different genes have effects on eye color. Only a few of them are shown here.

found will help to determine what scheme of crosses can be used to identify and analyze additional mutations. However, it is important to recognize some of the questions that were *not* necessary for the initial analysis.

- The investigators did not ask what possible colors could be found or if they could find a particular favorite color. Instead they asked what colors were actually found in mutant flies.
- The investigators did not ask what molecular or biochemical process was being affected. They asked if the eyes were a different color. The different color could occur because the gene for one or more pigments is not being transcribed, because some enzyme involved in pigmentation synthesis is not being made properly, because the pigment granules are not being transported into the cells, or because of a host of other molecular and biochemical processes. That understanding comes later, but it is not needed in advance.
- The investigators did not have to ask what other tissues use the same pigmentation process as the eye. After finding an eye color mutant, they would see, for example, that the body color is not affected but the color of an organ called the Malpighian tubules is affected by these same mutants.

In one sense, genetic analysis is simple: start anywhere and find a mutant. But after finding one mutant, the investigator begins to ask several questions. 'What can I do to produce more mutants with similar phenotypes?' 'What do I do with my mutant once I have found it?' 'How do I know what has gone wrong in this mutant?' Genetic analysis can help to answer all of these questions.

3.3 Producing mutations

Mutations in single genes arise spontaneously as errors in DNA replication and repair, among other processes, but these events occur too infrequently for a geneticist who wants to find many mutations as quickly as possible. In Text Box 3.1, we discuss the frequency and origin of spontaneously occurring mutations in humans. Geneticists working with a model organism cannot wait for spontaneous mutations to occur. A geneticist will use some other reagent—a **mutagen**—to induce mutations at a rate much higher than occurs spontaneously. The mutagenic effect of different agents is its own active field of research with its own journals and textbooks. There is a strong and vital interaction between the fields of mutation research and environmental health, in large part because mutagenic chemicals in one organism are often mutagenic or carcinogenic in another. For our purposes, we will focus on some of the most practical aspects of laboratory mutagens rather than attempt to survey the entire field. Although many different mutagens could be used in the laboratory to induce new mutations, relatively few mutagens are routinely used by geneticists. For most model organisms, one particular mutagen is very widely used with a few others utilized for certain specific purposes. The properties of some widely used mutagens are summarized in Table 3.2. These mutagens can be divided into three primary classes: chemical mutagens, radiation, and insertional mutagens including transposable elements.

BOX 3.1 Paternal age and mutations in humans

Mutations in humans occur in DNA sequences in all cells, both germline and somatic. Mutations arise from a plethora of sources, few of which can be conclusively identified. Since the source of the mutation can rarely be identified, mutations in humans are usually thought of as arising spontaneously as part of the normal processes in the cell. For decades, geneticists have attempted to measure the rate of spontaneous germline mutations for humans, recognizing that the rate is different for different genes. Perhaps more interesting than the rate of human mutation is the recognition that for many different genes, the rate of germline mutation is very different in the two sexes—nearly all single nucleotide changes in humans arise in males.

Methods of detection

One complication in determining the origin and frequency of spontaneous mutations is that most mutations are expected to be recessive. As a result, many mutations could remain undetected in heterozygous carriers for generations. The accurate and ethical detection of these mutations is difficult. Before genome sequencing projects became established, human mutation rates were detected by two different procedures that avoided the complication of dominance.

The first of these procedures is to look for the appearance of X-linked recessive mutations arising in the sons of a woman in a previously unaffected family. One familiar example of this approach is the occurrence of hemophilia among the offspring of Queen Victoria. Being nobility, the family history is well documented and no occurrence of hemophilia is noted before the birth of her eighth child (and fourth son), Prince Leopold, Duke of Albany, in 1853. Based on a subsequent pedigree analysis, her daughters Alice and Beatrice had to be heterozygous for the disease since hemophilia appeared in the royal families of Russia and Spain who could trace their descent back to Victoria. It is likely that the mutation did not arise in Victoria herself but rather in her father, Edward, Duke of Kent, so that Victoria was heterozygous. (It is also possible that the mutation arose in her maternal grandfather, but this is less likely since none of his other descendants were affected.) Hemophilia has both advantages and disadvantages as a subject for the study of mutation: the advantage is that the disease is severe and easily detected; the disadvantage is that more than one form of hemophilia is inherited as an X-linked recessive trait. (It is not known which form Victoria's descendants had.) However, by compiling the frequency at which hemophilia arises in the sons in a previously unaffected family, an estimate of the mutation rate can be obtained.

The second and somewhat more common method of monitoring the frequency and origin of human mutations is to examine children with dominant disorders who are born to unaffected parents. Achondroplasia, a dominant form of short-limbed dwarfism, arises from missense mutations in the gene *FGFR3*. Long before the molecular lesion was known, achondroplasia was used as a means of estimating the rate of human mutation. Achondroplasia has a distinctive and easily detected phenotype, and unaffected parents with achondroplastic children would have sought the attention of medical professionals. From studies in a range of different populations over many decades of recording, the estimated spontaneous frequency of achondroplasia is approximately 1 in 30 000 live births, although both higher and somewhat lower estimates have also been made. No matter what the estimate is from a particular study, the rate of spontaneous mutation for this disorder is very high, particularly since most cases of achondroplasia arise from a limited subset of the possible mutations in the gene.

A paternal age effect on mutation rate

One important observation during the studies of the rate of human mutation is that the spontaneous occurrence of many different genetic disorders increases dramatically with paternal age, as much as fourfold between the ages of 22 and 52. Haldane noted this for X-linked diseases in 1947, and the paternal age effect on mutation rate has now been demonstrated for more than 15 different human disorders, including achondroplasia, Marfan's syndrome, and Waardenburg's syndrome. The parent of origin for a particular mutation is determined using flanking genetic markers, and the results are compelling. From studies of six different dominant genetic disorders involving 154 spontaneous mutations in three different genes, all the mutations arose as single base substitutions in the father. No maternally derived mutations were found for these genes. The inevitable conclusion is that nearly all base substitution mutations arise during spermatogenesis, and that this rate increases with the age of the father.

The paternal origin and the age effect of these mutations suggest a shared explanation. All these mutations were base substitutions, which arise during DNA replication; spermatogenesis involves many more rounds of DNA replication than does oogenesis. In fact, spontaneous occurrences of diseases that are due to small deletions in the gene rather than replication errors, such as Duchenne's muscular dystrophy and neurofibromatosis type 1, do not show a pronounced effort of parental age or paternal origin. This supports the idea that the paternal age effect contributes to base substitutions, and that this is tied to the rounds of DNA replication.

There is a pronounced difference in the number of rounds of DNA replication for sperm and ova in mammals. In women, the pre-meiotic and meiotic DNA replications in the ovary occur before birth so that all potential ova have completed the meiotic S phase years before the woman reaches sexual maturity. (This long delay between the initiation of meiosis and its completion in oogenesis is commonly hypothesized to explain the increased rate of chromosome loss and non-disjunction with increasing maternal age.) A mature ovum is the product of 22 pre-meiotic replications and one meiotic DNA replication regardless of the age of the woman. In contrast, after the onset of puberty, spermatogenesis continues throughout the life of a male. Sperm produced by a man in his mid-twenties have gone through ten times more DNA replications than any ovum does; the sperm produced by a man is in his mid-thirties have gone through approximately 500 rounds of DNA replications. Since each round of DNA replication produces a small number of base substitutions, the increased number of cell cycles can explain an increased rate of spontaneous mutation with paternal age.

Rounds of DNA replication cannot completely explain the paternal origin

However, the data also indicate that the number of rounds of DNA replication cannot be the only explanation for the paternal origin of these base substitution mutations. The difference in the rounds of division between ova and sperm in couples of comparable age is 20- or 30-fold, whereas the difference in the origin of the mutations for the three genes studied most thoroughly is 154 paternal mutations to zero maternal mutations. Some other genes show a similar but less extreme difference in the occurrence of paternal and maternal mutations, but all of them show a greater bias in paternal origin than could be explained solely by differences in the number of rounds of replication.

Several different hypotheses have been suggested to account for such an extreme difference for these three genes, including a detection bias for these genes or some other special property of these mutations and a generally decreased ability to repair mutations with paternal age. Current work on this question typically involves direct sequence analysis of individual sperm rather than pedigree analysis in affected families. Many couples are waiting until they are older to have their children and are aware of the effect of maternal age on chromosome non-disjunction. However, the effect of paternal age on the rate of spontaneous mutations is also substantial and should not be overlooked.

Further reading

Crow, J.F. (2000). The origins, patterns and implications of human spontaneous mutation. *Nature Reviews Genetics*, **1**, 40–7.

Table 3.2 Mutagens that are widely used in model organisms

Class	Mutagen	Primary effect	Type of mutation	Administration	Organisms
Chemical	Ethylmethane sulfonate (EMS)	Alkylates purines	G to A transitions	Soaking, feeding	*Drosophila*, *C. elegans*, *Arabidopsis*
	Nitrosourea (ENU)	Modifies thymidine	Transitions	Direct application to testes	Mice
	Acridine orange	Intercalation between bases	Frameshifts		Prokaryotes
Radiation	UV light	Pyrimidine dimers	Transitions, tranversions	Irradiation	Many. Sometimes combined with psolaren to potentiate the effect
	X-rays, gamma-rays	Chromosome breaks	Chromosome rearrangements	Whole-organism irradiation	Many
	Fast neutrons	Chromosome breaks	Deletions	Irradiation of seeds	Plants
Insertions	Transposable elements	Random insertion or excision	Single-gene disruptions and small deletions	Mutator strains	Many
	T-DNA	Random insertion	Single-gene disruptions	Floral dip	*Arabidopsis*

Chemical mutagens modify or replace nucleotide bases

Chemical mutagens are used in nearly all organisms and often produce a limited spectrum of single-gene mutations. They can act by chemically modifying the nucleotide base, being incorporated into the DNA during replication, or intercalating between the bases on a DNA strand. In the first two cases, the altered nucleotide base is more likely to mispair during replication than the original base.

The widely used mutagen ethyl methane sulfonate (EMS) is an example of a chemical mutagen that modifies the nucleotide base. EMS attaches ethyl groups to guanine nucleotides, resulting in a higher probability that the modified guanine will mispair with thymidine rather than its normal cytosine as shown in Figure 3.4A. EMS also affects adenine, resulting in mispairing with cytosine. Thus, if the investigator uses EMS as a mutagen, most of the mutations recovered will affect a single base, although other types of events such as chromosome breaks and small deletions are known to occur

A. Modified bases and transitions

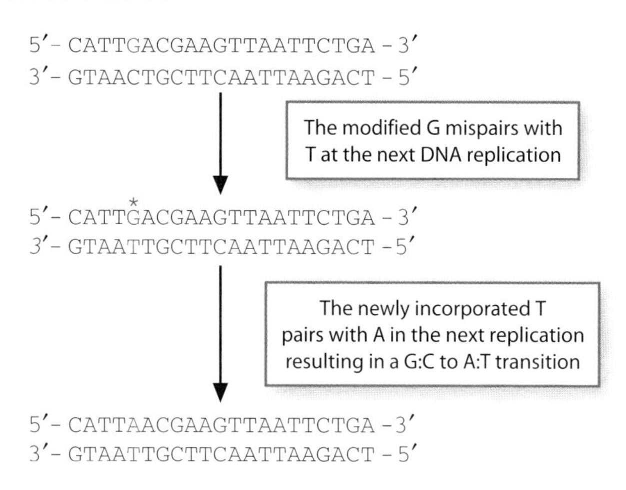

B. Base analogs and transitions

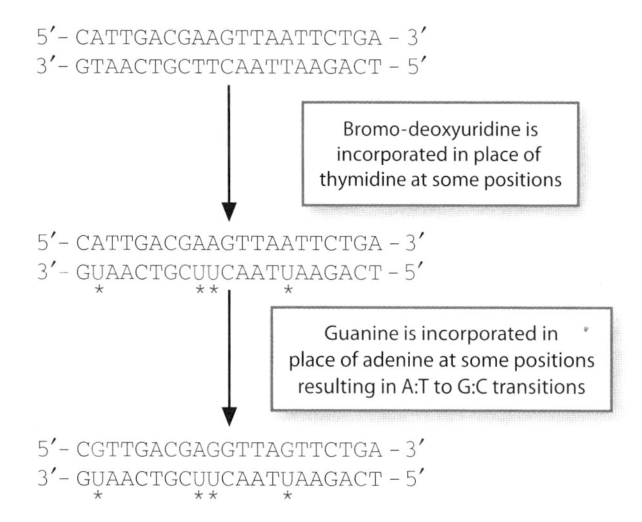

Figure 3.4 The effects of chemical mutagens. Two different effects of chemical mutagens are shown. (A) The mutagen modifies one of the nucleotide bases, in this case a guanine. The modified base can mispair in the next rounds of DNA replication, resulting in a transition. Alkylating agents such as EMS and ENU are examples of mutagens of this type. (B) The chemical is structurally similar to one of the nucleotide bases and is incorporated in place of the normal base, in this case bromo-deoxyuridine instead of thymine. The base analog is more likely to mispair than the normal base in subsequent rounds of DNA replication, resulting in a transition.

occasionally or with higher dosages. The usual mutation induced by EMS is a transition (one purine replaced by the other purine or one pyrimidine replaced by the other pyrimidine.) *N*-ethyl-*N*-nitrosourea (ENU) is an alkylating agent that is widely used in mice. ENU acts primarily on thymines, resulting in transitions of A:T to G:C or transversions of A:T to T:A. These chemicals are usually administered to the organism by ingestion, either by mixing the mutagen with the food or by soaking the organism in a solution contained the mutagen. In mice, the mutagen is often directly applied to the testes of the male and is absorbed through the skin, where the rate of mutation is roughly one mutation per gene among 700 sperm.

In contrast with chemical mutagens that modify bases, the base analogs such as bromo-deoxyuridine (BrdU) and 2-amino purine (2-AP) are incorporated into DNA during replication in place of a normal nucleotide base. The effect is shown in Figure 3.4B. BrdU is chemically similar enough to thymidine that DNA replication in the presence of BrdU results in the occasional incorporation of BrdU rather than thymidine. BrdU pairs with guanine more readily than does thymidine, resulting in a T:A to C:G transition. Similarly, 2-AP can substitute for adenine and pairs with cytosine, resulting in an A:T to G:C transition. Neither of these mutagens is widely used in eukaryotes, in part because the proofreading functions of DNA repair enzymes render them relatively ineffective.

Another class of mutagenic chemicals includes dyes such as acridine orange and ethidium bromide. These molecules are flat and stack between the nucleotides, distorting the DNA backbone. Such a distortion can result in a frameshift mutation when a base is added (or removed) during replication. Thus, although the mutagen is not altering a base in the same way as EMS or BrdU, the phenotypic effects are typically limited to a single gene.

Radiation induces chromosome breaks and structural rearrangements

The first mutagen to be demonstrated in a model organism was X-radiation, for which Muller was awarded a Nobel Prize in 1946. Most of the mutations induced by ionizing radiation like X- and gamma-rays and by fast neutrons in plants are double-stranded chromosome breaks. These breaks result in chromosomal rearrangements such as deletions, duplications, and translocations. The effects of high-energy radiation are summarized in Figure 3.5. In most uses of X- and gamma-rays, the entire organism is irradiated with doses of 1000 rad or more, or more than 20 times the dose of a typical X-ray used in medical or dental diagnosis. The mutagenic process is complex and the induction of mutations does not follow a simple dose–response curve, in part because the repair mechanisms vary in different organisms and perhaps even in different cell types. Often the rearrangements involve multiple breakpoints on the same or different chromosomes making them difficult to characterize.

Such rearrangements provide useful genetic tools for varying gene dosage or maintaining linkage arrangements by suppressing recombination. We will discuss a few of these uses later in this chapter and in Chapter 4. Occasionally an investigator will use ionizing radiation to identify single-gene mutations, but this is less common than using either chemical mutagens or transposable elements. Usually an investigator using ionizing radiation is attempting to induce chromosome rearrangements.

Fast-neutron mutagenesis is a common use of ionizing radiation in plants, and has also been used although less commonly in animals. Mutagenesis in plants is performed by exposing seeds to a nuclear fission reaction that releases high-energy fast neutrons. These neutrons induce double-stranded chromosome breaks, like X-rays, but the most frequent class of mutation induced is deletion of a few bases to several kilobases. As a result, the same mutagenic event can delete one or multiple genes.

An entirely different type of mutation arises from UV radiation. The preponderance of effects from UV radiation is the induction of pyrimidine dimers, i.e. a covalent crosslink between adjacent pyrimidines on the same DNA strand. The intra-strand crosslink affects base pairing, resulting in the mutation. This is illustrated for thymine dimers in Figure 3.6. Thus, unlike ionizing radiation, UV radiation typically induces mutations in a single gene rather than a chromosome rearrangement. Specialized repair mechanisms for UV damage have been found in nearly all cells from bacteria to mammals. These repair mechanisms are quite effective (fortunately for our health) and have themselves been

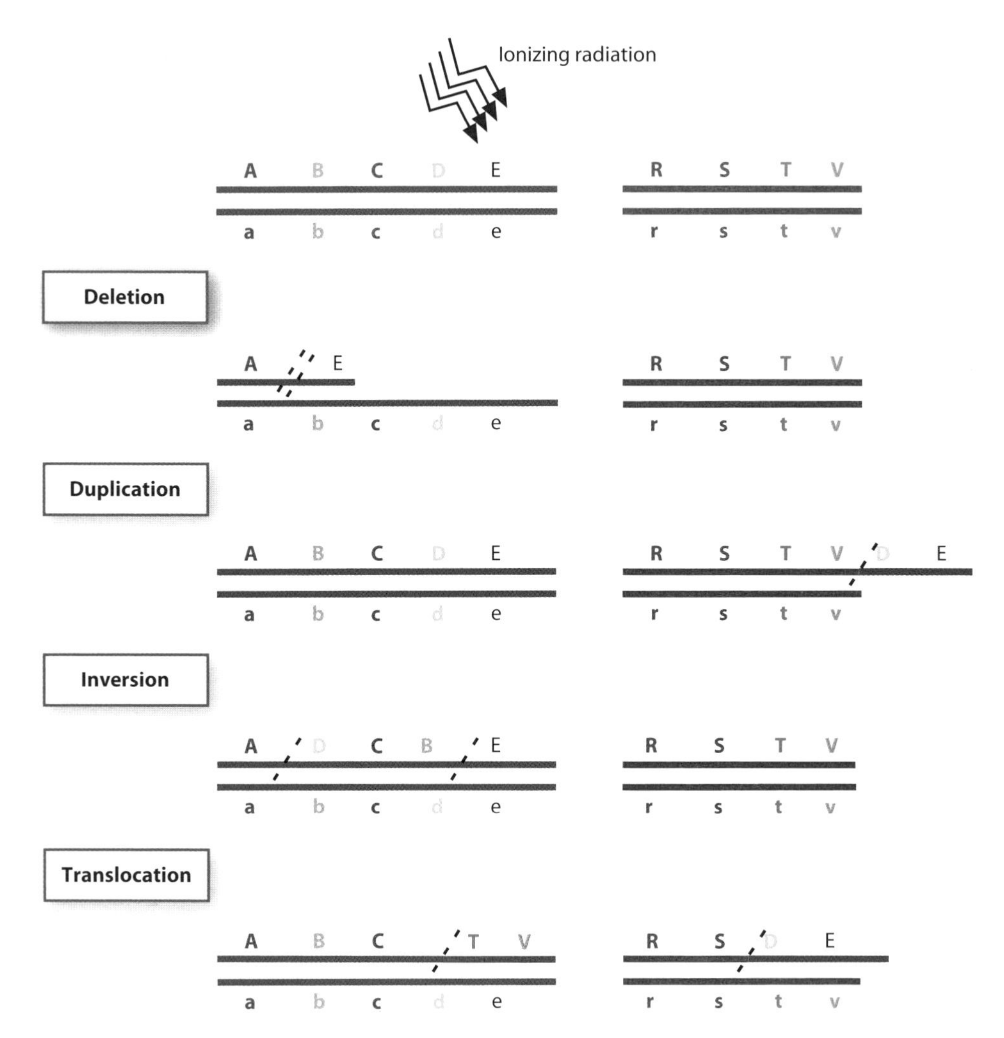

Figure 3.5 The effects of ionizing irradiation. X- and gamma-rays induce double-stranded breaks in DNA. The mutagenic effects are complex, and can occur for a single gene or part of a chromosome. Four different types of chromosome rearrangement are illustrated. The first is a deletion of part of the chromosome and the subsequent re-fusion of the two broken ends. The deletion shown here results in a chromosome that lacks or is deficient for the genes B, C, and D. Such rearrangements are also called deficiencies. The second rearrangement shown is a duplication in which a segment of the chromosome is broken and reattaches either elsewhere on the same chromosome or, as shown here, on a different non-homologous chromosome. The third rearrangement is an inversion, in which a segment of the chromosome is reversed and the ends reattached. The inversion shown here affects the genes B, C, and D. The fourth rearrangement is a translocation, shown here as a reciprocal translocation involving two non-homologous chromosomes. In the example shown, the D and E genes from the blue chromosome have been moved to the red chromosome, and the T and V genes from the red chromosome have been moved to the blue chromosome. Close analysis of the endpoints of chromosome rearrangements has found that the break events are not simple and often involve multiple rearrangements, such as small deletions or duplications at junctions of inversions.

extensively studied. The rare human disease xeroderma pigmentosum involves mutations that disrupt the UV repair mechanism, making the afflicted individuals extremely sensitive to the carcinogenic effects of sunlight. The effectiveness of the repair mechanism has limited the utility of UV radiation for laboratory induction of mutations. Although widely studied as a common environmental mutagen, UV radiation is not often used as a laboratory mutagen in the search for new mutations.

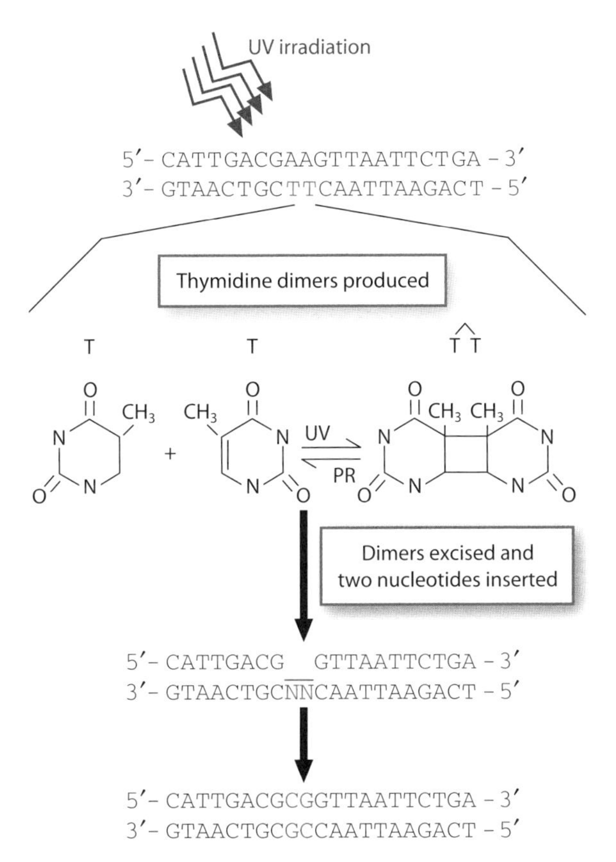

Figure 3.6 The effects of UV radiation. The mutagenic effects of UV light are distinctive and characteristic for this mutagen. UV irradiation results in covalent bonds between adjacent pyrimidines on the same DNA strand, shown here for adjacent thymines. The dimer is recognized during DNA synthesis and excised by the enzyme photolyase. This excision results in a gap in one strand, and repairing the gap can result in base substitutions. In this example, the adjacent thymines in the wild-type are replaced by a guanine and a cytosine. Because UV repair is efficient, UV radiation is not often used as an experimental mutagen.

Insertional mutagens are among the most useful laboratory mutagens

Among the most widely used mutagenic agents in current genetics laboratories are the naturally occurring insertional mutagens such as transposable elements and T-DNA insertions. Transposable elements, also known as transposons, are discrete genetic entities that move as a unit within a genome. Some transposable elements move via a DNA intermediate, whereas others move via an RNA intermediate like a retrovirus. Transposable elements can be classified into families based on nucleotide sequence similarity and their method of transposition from one site to another. They range in length from less than a kilobase to tens of kilobases, although the sizes are widely variable even among members of the same family.

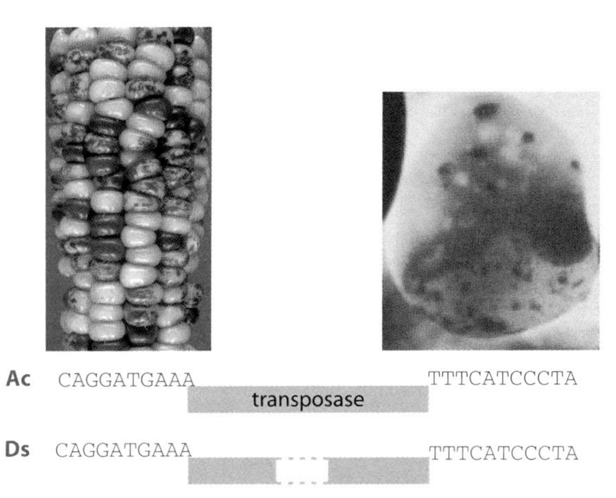

Figure 3.7 The Ac/Ds transposable element in maize. The photograph on the left shows the element inserted into the P pigmentation gene; the photograph on the right shows a kernel with the element inserted into the bronze gene. The lack of pigmentation in each figure is due to inactivation of the gene by insertion of an Ac/Ds element. Although this is classically described as a 'two-element system', molecular analysis has revealed that Ac and Ds are the same element. Ac is a complete element, with 11 bp repeats at each end and a complete transposase. Ds has the same the 11 bp repeats but has deletions that eliminate part of the transposase gene and its function. Ac elements can move Ds elements since the terminal repeats are the same.

Many different classes of transposable element have been studied, and no species of prokaryote or eukaryote is free of their effects. The first element to be recognized, the Ac/Ds transposable element in maize, is demonstrated in Figure 3.7. This element was found by Barbara McClintock who was awarded a Nobel Prize in 1983 for her studies, which began in the 1930s. There is also an extensive literature on the types of transposable elements, the mechanism by which each type moves, and their evolutionary effects, but our focus will be on their use as mutagens.

As noted in Chapter 1, transposable elements, both active and inactive, comprise a sizeable percentage of the genome sequence of multicellular organisms. The use of transposable elements as laboratory mutagens began in *Drosophila* at about the same time as McClintock's experimental studies, although the mutagenic effect was not attributed to transposable elements until decades

HISTORICAL COMMENT

Mendel's wrinkled peas were due to an insertion of a transposable element similar to Ac/Ds into a gene involved in starch biosynthesis.

later. Regulating the movement of transposable elements, particularly of the P element in *D. melanogaster*, has given rise to some of the most valuable tools in modern genetic analysis. This topic was introduced in Chapter 2; in later chapters, we describe the use of P elements as vectors for cloning genes and making transgenic flies. For now, we will limit our description to the use of transposable elements as mutagens, with P elements as our primary example.

Transposable elements are effective mutagens for inducing single-gene mutations

The effectiveness of transposable elements as mutagens rests on two key properties: first, they move as a discrete unit and, secondly, they insert more or less randomly throughout the genome. Strictly speaking, transposon insertion is not at random with respect to the DNA sequence, and each element has its own preferred target site for insertion. However, the insertion target site is usually a simple dinucleotide that is found at hundreds of places within any typical gene. Thus, with respect to genes, insertion can be considered as a random event—every gene is a target for transposable element insertion. The frequency with which a transposable element moves and produces a mutation depends on the type of element and the genetic make-up of the organism. Many laboratory strains are full of quiescent transposable elements that rarely, if ever, move. Other strains will have a transposable-element-induced mutation in a particular gene once every few thousand gametes. These strains are generally known as **mutators** because of their high rate of apparently spontaneous mutations; in reality, the mutations are occurring because of the movement of transposable elements.

Some of these mutator strains in which a transposable element is particularly active are not very fertile because transposition of a mutagenic element requires the same evolutionary balance as other mutagenic processes; it occurs sufficiently frequently that the transposable element can be used as a laboratory mutagen, but not so often that the overall fitness of the organism is affected. Most transposable elements move in all cells of the organism, and induce both somatic and germline mutations. However, one of the advantages of P elements is that they move only in the germline and not in somatic cells. Thus, even *Drosophila* strains in which P elements are very active in the germline are relatively healthy and show few somatic effects. The effects are seen in the offspring.

The means by which an investigator induces transposable elements to move depends entirely on the type of element and on the organism. With P elements, regulating the movement can be done by particular genetic crosses. In the 1930s, *Drosophila* geneticists isolated many of their 'spontaneous' mutations by using crosses that mobilized P elements, not realizing why the effect was occurring. Many of the original mutant alleles of classic *Drosophila* genes are the result of P-element insertions.

The early *Drosophila* geneticists learned that, broadly speaking, there are different classes of wild *D. melanogaster*. Decades of laboratory matings and almost two decades of engineering P elements have created a wider spectrum of laboratory types, but the historical framework is still useful. In the 1930s, when male flies caught in the wild were mated with laboratory females, the syndrome of genetic effects known as **hybrid dysgenesis** was seen in subsequent generations (see Chapter 2). This syndrome was only seen when the male came from the wild and the female had not been previously exposed to these wild males. The female was said to have an M cytotype, while the males had a P cytotype.

Modern *Drosophila* geneticists do not rely on naturally occurring versions of P elements or M cytotypes. Instead they have modified P elements *in vitro* and transformed these back into flies to control the movement of P elements. Examples of these modified P elements are shown in Figure 3.8. The mobility of a DNA-based transposable element such as a P element depends on two components: an enzyme known as a transposase, encoded by the P element itself; and its substrate, the terminal repeat sequences of the P element. If either component is defective, the element cannot move. However, a functional transposase encoded on one P element can mobilize many other P elements so long as their terminal repeats are intact. *Drosophila* geneticists have taken advantage of this by engineering strains that allow them to control P-element movement. A gene encoding transposase under the control of an inducible promoter can be inserted into the genome; the inserted gene has no terminal repeats and so the modified transposon

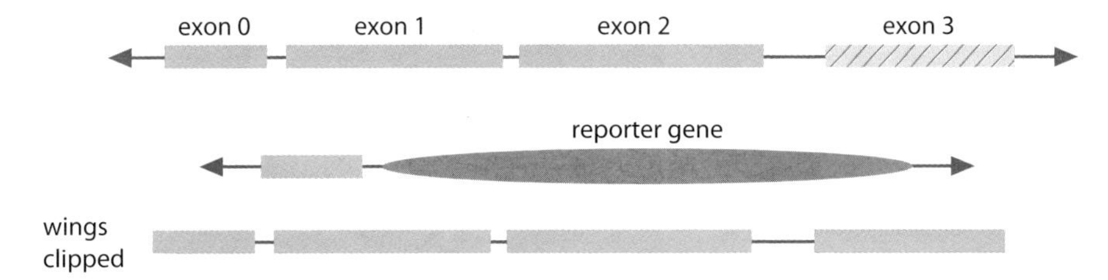

Figure 3.8 The P element in *Drosophila* with the wings clipped element. An intact P element, shown at the top, has a transposase encoded by four exons. The last exon (shown in green) is alternatively spliced, and a complete transposase is only made in the germline. P elements have been genetically engineered, with examples drawn below. The element in the middle has the transposase replaced by a reporter gene and could not move on its own. Other genes or sequences could be inserted in addition to or in place of the reporter gene. The wings clipped element at the bottom encodes functional transposase but lacks the terminal repeats needed for transposition. A wings clipped element cannot move itself, but it can move the reporter gene element if both are expressed in the same genome.

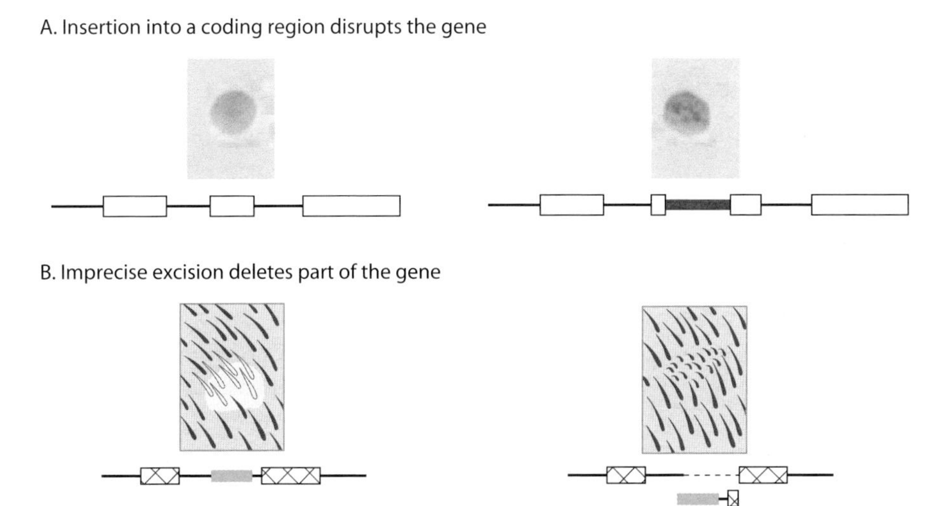

Figure 3.9 Two different mutagenic effects of transposable elements. (A) The round and wrinkled alleles that Mendel used in garden peas were the result of a transposable element (shown in red) inserted into the coding region of a gene for starch biosynthesis. The insertion of the transposable element disrupts the gene, resulting in the wrinkled allele. (B) The singed mutant allele in *Drosophila* has arisen from an imprecise excision of a transposable element. In wild-type flies, a transposable element (shown in green) is inserted in an intron, with the two flanking exons shown as the cross-hatched boxes. The gene is functional because the transposable element is inserted in an intron. Excision of the transposable element also deletes part of the neighboring coding region. The size of the deletion is exaggerated for clarity; imprecise excision usually removes fewer than 10 bp.

cannot move. Other strains have transposons with the terminal repeats (which are short and relatively easy to place on a cloned construct) but do not have the active transposase. When the two strains are crossed, the F_1 has both an active transposase and terminal repeats on which it can act, and so the cloned construct with the terminal repeats can be moved about the genome.

The mutagenic effect of a transposable element arises from either of two activities, as shown in Figure 3.9. First, the element can be inserted into a new location within a particular gene, thereby disrupting the coding region or interfering with splicing, transcription initiation, or any other aspect of gene expression. In effect, a few kilobases of foreign DNA have been inserted into a location that did not have it, and the activity of the site is disrupted. Secondly, an existing transposable element can excise imprecisely from its location within a gene. Imprecise excision induces a mutation by deleting a few nucleotides surrounding the insertion/excision site. Both effects are known to occur frequently, and both have been used to produce mutations within a gene.

Insertional mutagenesis in *Arabidopsis* offers many of the same advantages as transposon mutatgenesis. As mentioned in Chapter 2, the plant pathogen *Agrobacterium tumefaciens* inserts a piece of DNA (called T-DNA, flanked by the right and left border sequences) at random into the plant genome; this forms the basis for insertional mutagenesis in *Arabidopsis*. Investigators replace the sequences between the right and left borders with any sequence, which the bacterium will insert into the plant genome upon infection. Because the sequence inserts into the genome at random locations, insertions are often within a gene and disrupt its normal function. Each T-DNA insertion disrupts at most one gene. Like transposable elements, T-DNA insertions in exons often result in a complete loss of gene expression and function. Because *Agrobacterium* infection is an easy procedure (as discussed in Chapter 2), T-DNA insertion is a common mutagen in *Arabidopsis*, second only to EMS in its use as a mutagen.

One reason that transposable elements and insertional mutagenesis have become methods of choice for many applications is that they provide a means of cloning a gene based on its mutant phenotype. This process, known as **transposon tagging** for transposable element insertions, is described in Chapter 5. For our purposes in this chapter, the important aspect is that transposable elements and T-DNA insertions are now among the most widely used mutagens in the laboratory for model organisms. Like chemical mutagens, they typically induce single-gene events. However, they have the advantage that they produce events that are readily detected in molecular analysis and they present no possible health hazard to the investigator.

A summary of mutagenic effects

Let us return to one of our opening questions to summarize this section. What can an investigator do to produce more mutations? By choosing one type of mutagen in preference to another, the investigator has the means to induce mutations of different types. Some of the information is given in Table 3.2. If the primary goal is to obtain single-gene events, either chemical mutagens or insertional elements are effective mutagens. Chemical mutagens usually produce a small molecular alteration, typically a single base change; this can be an advantage for genetic analysis since the disruption is discrete and limited in scope. Insertional mutagenesis is frequently the most useful method if the investigator wants to use the mutation itself to clone the gene, as described in Chapter 5. If the primary goal is to induce mutations that affect multiple genes, ionizing radiation is usually the preferred mutagen.

A key question for any mutagenesis is the potency of the mutagen and the appropriate dosage or treatment. The investigator wants to find many mutations, so it is helpful to use the highest mutation rate possible. This usually means that mutations are being induced in many different genes in the genome. On the other hand, if too many mutations are induced in the genome, the organism is often rendered sterile or unable to grow. Because many different mutations are induced in any mutagenesis scheme, the investigator will need to cross the newly recovered mutant with a wild-type strain to replace most of the genome with unmutagenized chromosomes. This process helps determine that the mutant phenotype is due a single specific mutation. This process, called outcrossing, can take several generations of careful mating to wild-type. Thus, inducing many mutations in the genome increases the chances of finding a mutant phenotype but it also usually increases the amount of effort needed to isolate the new mutation of interest from any other mutations induced by the mutagenesis process.

Chemical mutagens present the largest potential safety hazard to the investigator but are also probably the most available and the simplest to use. No chemical mutagen is entirely safe for the investigator to use, so very potent mutagens are often avoided for safety reasons. One reason for the popularity of EMS as a chemical mutagen is that it breaks down in soapy water, making it safer to use than some of its more potent relatives. On the other hand, the related alkylating agent methyl methanesulfonate is not an effective mutagen because it lacks the potency of EMS. Safety and potency are also related to the DNA repair mechanisms, since all biological organisms have evolved sophisticated and diverse DNA repair mechanisms.

All of these methods result in mutations in random targets because any gene in the genome presents an

DEFINITION

outcrossing: Outcrossing is a protocol of matings that replaces mutagenized chromosomes with chromosomes that have not been exposed to a mutagen.

appropriate target. In a typical mutagenesis experiment, mutations in many different genes are induced without regard to the effect of the gene. Other methods that target specific genes based on their DNA sequence are also widely used and very effective. These are described in the reverse genetics approaches discussed in Chapters 6 and 7.

Finding mutations

Having discussed the ways that an investigator can induce mutations at a high frequency, we turn to the genetic methods that are used to find and characterize mutations that affect a biological process. Often the process of finding new mutations is referred to as a mutant hunt. If populations of mutagenized organisms are examined for a mutant phenotype, the hunt is usually called a mutant search. If some method is used to favor the recovery of mutants of one type, the hunt is called a mutant selection. The searches and selections that geneticists use to find mutations have personal styles in the same way that any artistry reflects its artist. Screens and selections may involve simple or complicated mating strategies, subtle or obvious phenotypes, easy or arduous searches. There is beauty in a well-defined mutagenesis hunt that allows the investigator to find exactly what he/she is searching for with minimum effort. For our example, we will outline a simple mutagenesis scheme and then add the embellishments later. Case Study 3.1 describes a more realistic scheme that was used to find mutations affecting segmentation in *Drosophila* embryos.

For our hypothetical mutagenesis scheme, we return to our questions about *Drosophila* eye color. A simple scheme is shown in Figure 3.10. To reveal the logic of eye pigmentation, a population of wild-type flies is

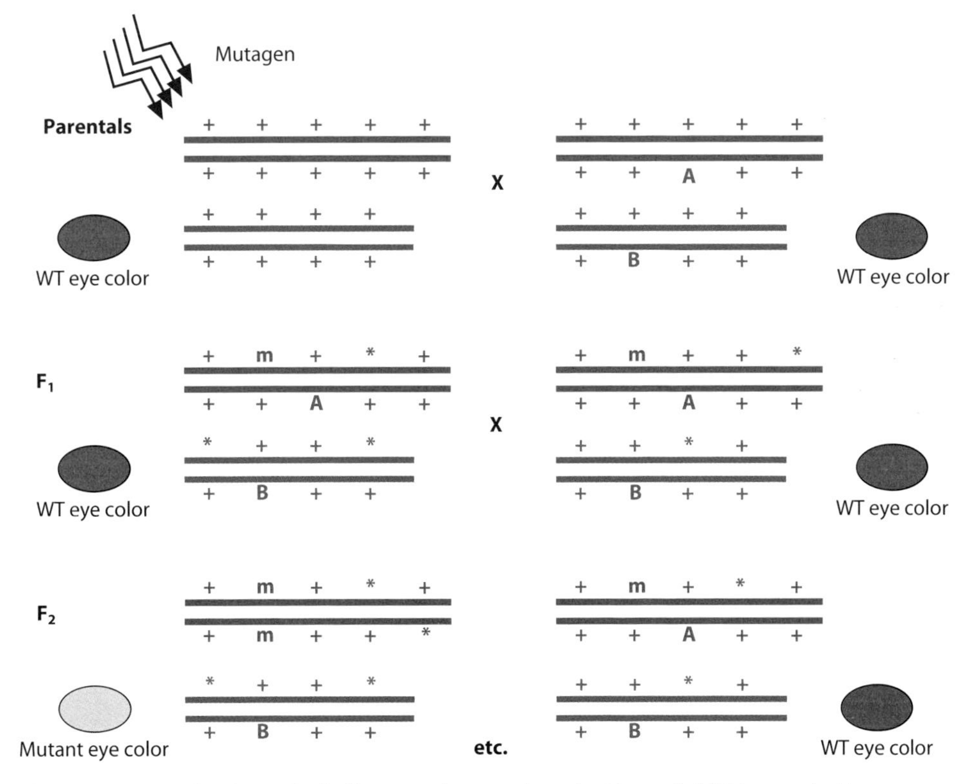

Figure 3.10 A simple scheme for finding eye color mutations. A wild-type fly (WT) is treated with a mutagen and mated to a fly with wild-type eye color but with other genetic markers. The other genetic markers are represented by A and B on two different chromosomes. Because most mutations are recessive, the F_1 flies are expected to have wild-type eyes but will be heterozygous for different mutations affecting eye color. In this diagram, the eye color mutation, represented by *m*, is on the same chromosome as the A genetic marker but on the other homolog. The F_1 flies are mated among themselves to produce an F_2 generation, some of which have a mutant eye color, shown here in pink. Many mutation screens will need to go to an F_3 generation in order to produce homozygotes. Other mutations that do not affect the eye color are represented by asterisks (*) and will be outcrossed, as in Figure 3.11. This scheme would be laborious, and much cleverer schemes involving fewer F_1 matings are more likely to be done.

Case Study 3.1

Part 1: Find a mutant: segmentation in *Drosophila* embryos

Many possible examples of the application of mutant analysis to a complex biological problem could be used as a case study for this chapter. However, it is likely that no example has been more successful than the study of segmentation in *Drosophila* embryos. When Nüsslein-Volhard and Wieschaus began the analysis in the late 1970s, no one would have been so bold or prescient as to predict publicly that it would not only unlock the strategy for embryogenesis in flies but would also provide insights for embryology in nearly all other animals. Even more remarkable, no one could have dreamed that it would provide key insights into the underlying biology of some of the most common forms of cancer. Nüsslein-Volhard, Wieschaus, and their collaborators combined remarkable knowledge, observation, dedication, and hard work with the power of genetics in a study that few have matched. The story has been told often, and it deserves its place in genetics textbooks. Because the results of their work are well known and intellectually compelling, it is tempting to skip over the details of how they reached their conclusions. Our goal in this case study is to show the use of the application of genetic analysis in the context of this broader project. That is, while they may have begun by saying 'Find a mutant!', much more went into the project.

A summary of *Drosophila* development

The most visible aspect of an insect embryo is that it has segments—14 of them in *D. melanogaster*, although segments in the head are fused and harder to distinguish than those throughout the main body. A picture of a *Drosophila* embryo is shown in Case Study Figure 3.1. Each segment has a characteristic appearance and location along the body axis, and several give rise to the distinctive structures in the adult. For example, the three thoracic segments (T1, T2, and T3) give rise to the three pairs of legs; in dipteran insects such as *Drosophila*, T2 also produces the wings, whereas T3 produces an evolutionary derivative of the hindwing known as a haltere. E.B. Lewis spent a lifetime studying mutants that affected segment identity and produced homeotic transformations of one segment or structure into another. When Nüsslein-Volhard and Wieschaus began their investigation, Lewis had published a highly influential paper in *Nature* (Lewis 1978), which laid out a strategy by which the homeotic genes specified the identities of individual segments.

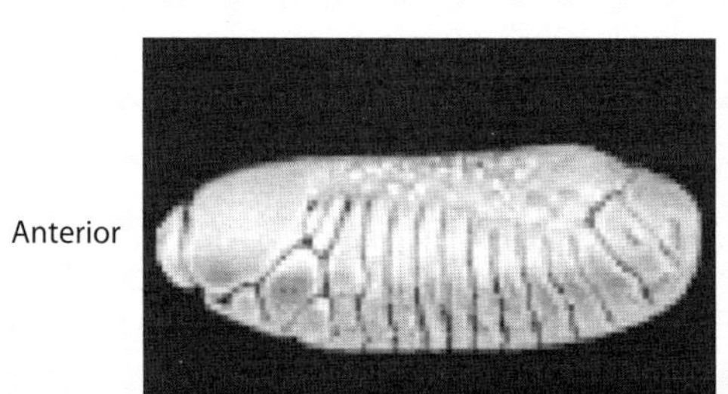

Figure C3.1 **Scanning electron micrograph of a *Drosophila melanogaster* embryo.** The anterior is to the left and the posterior is to the right. The positions and the structure of the segments are evident throughout the mid-region (thorax) and posterior (abdomen).

However, the origin of the segment pattern itself was mysterious. Segmentation patterning was widely believed to be connected to the determination of the body axis in the embryo, in particular the anterior–posterior axis. Physical manipulations of cells and nuclei including transplantation, ligation, and ablation had been done in *Drosophila* and other insects but these could provide only the broadest framework of how segments arose in their proper positions. For a more detailed understanding, mutations were required.

It was also clear from both simple observation and physical manipulations that the single-cell embryo already had a polarity derived from an underlying polarity in the oocyte. The nucleus of the single-cell *Drosophila* embryo divides rapidly without cytoplasmic division—eight divisions in about an hour to generate a syncytium with roughly 1000 nuclei in a common cytoplasm. The nuclei migrate to the periphery of the egg and divide four more times before boundaries form between them to make individual cells around the edge of the embryo. Thus, during the early stages before cellularization occurs at the periphery, any gene product (either mRNA or protein) that is free to diffuse in the cytoplasm might affect all of the nuclei in the embryo. The progression from the cellular blastoderm stage with the nuclei arrayed around the periphery of the embryo to a segmented embryo had been thoroughly described by scanning electron microscopy and other methods, but the underlying process of segmentation itself was not understood at all.

Find a mutant: mutagenesis

In order to understand the genetic logic of segmentation, Nüsslein-Volhard and Wieschaus set out to find mutants. In contrast with what many other geneticists had done previously, they were not content to find a few new mutations in a few new genes. Instead, they set out from the very beginning to find as many genes affecting segmentation as possible. Nüsslein-Volhard and Wieschaus used EMS as their principal mutagen and concentrated almost exclusively on recessive mutations. Every gene is a potential target for EMS-induced mutations, and most of the mutations are single base changes that result in a loss or reduction of gene function. Their initial mutagenesis strategies made three important assumptions about the genes they might find, all of which proved to be true.

1. Mutations that affect segmentation in embryos are very likely to be lethal, such that the embryo cannot progress to adulthood or become fertile. Therefore, they needed to establish methods to maintain their mutant strains as heterozygotes.
2. The genes would map throughout the genome, so procedures would be needed that allowed them to balance heterozygotes at many different loci.
3. Some of the mutations would exhibit a maternal effect.

This last assumption needs a little more explanation.

Maternal-effect mutations involve gene activity in the mother

Maternal-effect mutations had been observed in other organisms as well as *Drosophila*, and the explanation for the phenotype arises from an understanding of embryology. In insects, as in most invertebrates, the cytoplasm and cellular components of the early embryo are derived from the oocyte. Proteins and mRNA molecules that the mother has deposited in the oocyte will be used after fertilization to direct or sustain early embryonic development.

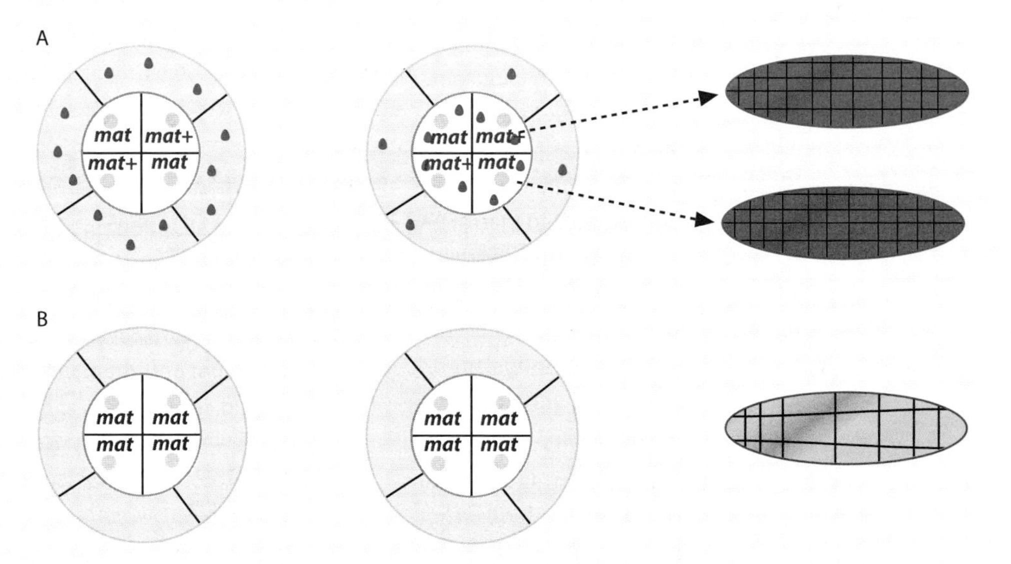

Figure C3.2 Maternal-effect mutations. The diagram shows two developing oocytes surrounded by somatic cells of the ovary in light blue. These somatic support cells are called nurse cells in insects, but cells with similar functions are found in most invertebrates. The somatic cells secrete a gene product *mat*, indicated in red, which is localized to the developing oocytes without regard to the genotype of the oocyte itself. The gene product is encoded by the *mat+* gene. The presence of this red gene product is necessary for normal embryonic development after fertilization. The developing oocytes are fertilized by the sperm from a heterozygous male in each example. In panel A, a heterozygous *mat*/+ mother makes oocytes that are *mat+* or *mat*, each of which receives the normal *mat+* gene product from the somatic nurse cells. As a result, these embryos will develop normally regardless of their own genotypes, and even those embryos that are *mat/mat* in genotype have a normal phenotype. In panel B, a homozygous *mat/mat* mutant mother does not make the red gene product and so all her offspring will be mutant regardless of their own genotype; even those that receive the *mat+* allele from the sperm and are *mat*/+ have a mutant phenotype.

Thus, a mutation in one of these genes will have a delayed phenotype. This is shown diagrammatically in Case Study Figure 3.2.

Consider a heterozygous mother of genotype *mat*/+, where *mat* is a recessive mutation in a gene that affects the early embryo. Because the female has a wild-type phenotype, all her oocytes will have the full complement of gene products that she deposits, regardless of the genotype of the individual oocyte. Both oocytes that inherit the *mat* allele and those with the + allele can support normal embryogenesis, regardless of the paternal contribution from the sperm, because they have received the *mat* gene product from their mother. Therefore, the father could also contribute a gamete with a *mat* mutation in it, but this will not matter. Even a *mat/mat* embryo has a wild-type phenotype, again because the wild-type *mat+* gene product has been deposited by the mother. However, when this *mat/mat* embryo itself makes oocytes, it cannot make the product of the *mat+* gene, and its oocytes will lack this gene product. The resulting embryos will exhibit a mutant phenotype that arises because of the lack of *mat* gene product from their mother. In effect, we are seeing the mutant phenotype skip a generation.

Some genes with maternal effects had already been identified in *Drosophila* before Nüsslein-Volhard and Wieschaus began their study. The phenotypes of such maternal-effect mutations suggested that maternally contributed gene products are responsible for the major body axes of the embryo—anterior vs. posterior and dorsal vs. ventral. (For some genes, this effect is independent of the paternal allele so that any embryo from a *mat/mat* mother has a mutant phenotype regardless of its own genotype.) Although several different types of mutations were found in their screens, Nüsslein-Volhard and Wieschaus focused on those that had effects in the larvae, postulating that these mutations would identify genes that interpret maternally deposited signals.

Further reading

Jürgens, G., Wieschaus, E., Nüsslein-Volhard, C., and Kluding, H. (1984). Mutations affecting the pattern of the larval cuticle in *Drosophila melanogaster* II. *Roux's Archives of Developmental Biology*, **193**, 283–95.

Lewis, E.B. (1978). A gene complex controlling segmentation in *Drosophila*. *Nature*, **276**, 565–70.

Nüsslein-Volhard, C. and Wieschaus, E. (1980). Mutations affecting segment number and polarity in *Drosophila*. *Nature*, **287**, 795–801.

Nüsslein-Volhard, C., Wieschaus, E., and Kluding, H. (1984) Mutations affecting the pattern of the larval cuticle in *Drosophila melanogaster* I. *Roux's Archives of Developmental Biology*, **193**, 267–82.

Wieschaus, E., Nüsslein-Volhard, C., and Jürgens, G. (1984). Mutations affecting the pattern of the larval cuticle in *Drosophila melanogaster* III. *Roux's Archives of Developmental Biology*, **193**, 296–307.

mutagenized with any convenient mutagen such as EMS or a P element. The mutagenized flies are mated to a genetically identifiable strain to produce an F_1 generation, which is then mated among itself to produce an F_2 generation. The F_2 generation might then need to be mated to produce an F_3 generation that is homozygous for the chromosomes that have been mutagenized. (This is somewhat easier to do in self-fertilizing organisms such as *Arabidopsis* and *C. elegans* since homozygotes are easily found in the F_2 generation.) The genetic markers used in the unmutagenized parent, represented by A and B in Figure 3.10, play the important role of distinguishing the mutagenized and unmutagenized chromosomes in subsequent generations. Because most mutations are recessive, all (or nearly all) of the flies in the F_1 generation are expected to have the wild-type brick-red eye color. Recessive mutations will begin to be seen when they become homozygous in the F_2 and subsequent generations, and the mating scheme is done such that flies are homozygous for a new mutation. (Case Study 3.1, Part 1 provides some more specific details about how one particular mutant hunt was done.) The investigator can screen through these flies looking for any whose eye color is different from wild-type. The difference in phenotype is due to a mutation in a gene that affects eye color. Dominant mutations, which are far less common, can be identified in the F_1 generation. Because the vast majority of mutations are recessive, we will concentrate on working with them.

The newly induced eye color mutant is mated to an appropriate strain for further characterization. 'Appropriate' in this context depends on what exactly the investigator wants to do next. For example, as suggested by the asterisks in Figures 3.10 and 3.11, the mutant fly is likely to have more than one mutation in its genome, only one of which affects eye color. Thus, one appropriate mating is to outcross the mutant fly to a strain with unmutagenized chromosomes to replace particular chromosomal regions from the mutant strain with the same region from the unmutagenized strain. One outcrossing scheme for our hypothetical eye color mutant is shown in Figure 3.11. The outcrossing mating schemes can be devised to replace every region of the mutagenized genome with its corresponding unmutagenized region—every region, that is, except the one with the newly arisen eye color defect. Because outcrossing replaces mutagenized genes and chromosomes with unmutagenized ones, the outcrossed individuals may grow better, be more fertile, or have

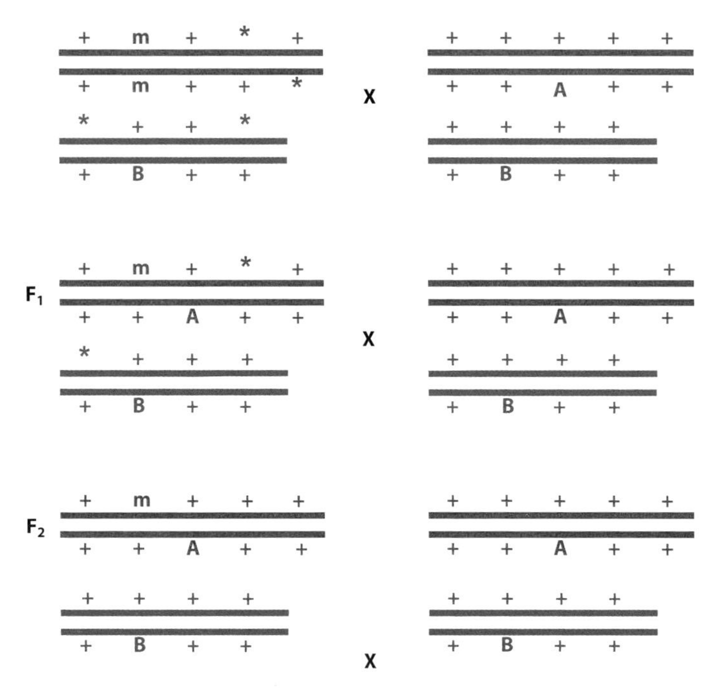

Figure 3.11 Outcrossing to eliminate unlinked mutations. A simple outcrossing scheme is shown (more elaborate ones are likely to be used). The goal of the outcrossing scheme is to replace every region of the chromosomes that has been mutagenized, except for the newly arisen mutation, with the corresponding region from an unmutagenized chromosome. Outcrossing is often done as part of the procedure for mapping mutations. In the diagram, the unwanted mutations to be outcrossed are represented by asterisks (*) and the mutation of interest is designated *m*. Genetic markers, represented here by *A* and *B*, are used to monitor the mutagenized and unmutagenized chromosomes. In each generation, the mutagenized chromosomes, which have no other genetic markers in this example, are replaced by chromosomes that were not mutagenized, which in this example have the genetic markers. Each generation or round of outcrossing replaces one or more regions of the chromosomes and removes some of the background mutations.

a slightly different phenotype than the original mutant strain.

Another appropriate set of matings, often done during the outcrossing procedure, will map the mutation to a region of the genome. Once the mutation is mapped, more precise outcrossing schemes can be devised to eliminate extraneous mutations that are closely linked to the eye color defect. Other appropriate matings, such as tests for dominance, mutation strength, and complementation, are done as part of the analysis of the mutation, as described in Chapter 4.

Mutation hunts can encounter some known challenges

Imagine that an eye color defect has been recovered from a mutagenesis procedure and outcrossed so that it is likely to be the only mutation remaining in the strain. An investigator could stop the mutagenesis scheme there, characterize this one mutation and its corresponding gene, and have an idea about a single step in the eye pigmentation process in *Drosophila*. The method for finding mutations is easy (although the outcrossing can be a bit tedious) and the results are clear. However, if the procedure were really this uncomplicated, a university would probably not give graduate degrees in genetics. Complications do arise and geneticists have developed methods to deal with them. These methods are sometimes referred to as 'genetic tricks', but that name suggests that the geneticist is doing something by sleight of hand. The strategies are in fact rooted in our understanding of basic genetics, and most have been used in one form or another for many decades. We will pose these complications as a series of questions, something like FAQs for mutagenesis schemes.

How can I reduce the number of F_1 and F_2 individuals that have to be examined?

The importance of this question can be illustrated by some arithmetic. Mutation rates vary widely depending on the mutagen used, its dose, the gene under investigation, and many other factors. A *typical* mutation rate for a typical gene with a typical chemical mutagen may be about one in 2000; that is, from the mutagenized organism, one gamete in 2000 will have a recognizable mutation in any specific gene. (The mutation rate can be 100-fold, or even 1000-fold, less than this, so the rate of one mutation in 2000 gametes should be considered as illustrative.) Therefore, in a simple scheme such as the one we described, the investigator needs to have about 2000 F_1 individuals to find one mutation.

In fact, since mutations occur at random with respect to the gene, many more F_1 individuals will have to be examined to have a high probability of having a mutation —possibly 15 000 or 20 000. Because this is an imaginary scheme, we can imagine that we are lucky and only have to worry about 5000 individuals. Now each of these 5000 F_1 indivdiuals needs to be put through the mating scheme to make an F_2 generation that is homozygous for a mutation, and so a certain amount of labor will be involved. The scheme may require an F_3 generation as well. This labor is offset somewhat by the fact that many different genes can probably be mutated to give an eye color defect, such that mutations can be recovered once every few hundred F_1 flies.

Almost no one will be satisfied with only one new mutation, and the investigator will want to look at tens of thousands of F_1 flies to find mutations. A generation time in *Drosophila* is about 12 days and multiple generations are required, and so this can easily consume months of an investigator's time. A person does not have to do such a scheme very often before wishing for a means to reduce the number of F_1 and F_2 individuals that have to be examined. Many very clever methods have been developed to enrich the population for ones with a mutation of interest at the expense of ones with no mutation. The best of these methods is a genetic **selection**.

In a selection, only the mutation of interest is capable of growth or reproduction. It has similarities to a weed-killer, which selects *against* a weed and lets the flowers grow. Many different biological properties can form the basis for a selection scheme. For example, antibiotic or nutritional selections have been widely used in prokaryotic genetics, and similar procedures using nutrition, drug sensitivity, or drug resistance can be applied to many eukaryotes. We will describe one type of selection using an example from yeast (Figure 3.12). For example, normal yeast cells cannot grow in the presence of the chemical 5-fluoro-orotic acid (5FOA). To find mutations that confer resistance to 5FOA, the investigator can mutagenize the yeast cells, allow them to grow for a generation or two, and then plate them on media containing 5FOA. A standard agar plate can easily hold a few thousand yeast colonies but, since 5FOA kills normal cells, hundreds of thousands of yeast cells can be plated onto a single agar plate containing 5FOA without overgrowth. In fact, nothing will grow unless a mutation conferring resistance to 5FOA has been induced, so the selective agent is killing all non-mutant cells. Even if the mutations to resistance are quite rare, the strength of the selection allows the investigator to find them easily with one or two Petri plates.

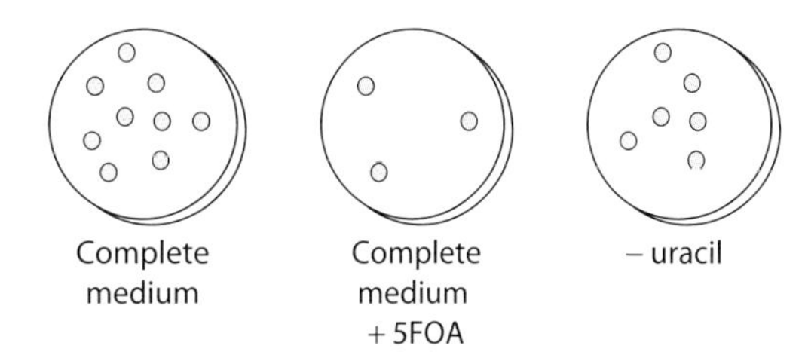

Figure 3.12 Selection and counter-selection in yeast. Yeast cells grow on complete medium. When cells are replica-plated onto plates containing the toxin 5FOA, only a few yeast cells grow. 5FOA is metabolized into a toxic product by normal cells in a process that requires the uracil biosynthetic enzymes. The only cells that can grow on 5FOA are auxotrophic mutants for uracil biosynthesis. When normal cells are replica-plated onto plates lacking uracil, the cells that grew on 5FOA do not grow because they cannot synthesize uracil, whereas those that could not grow on 5FOA are able to synthesize uracil and grow.

This same example can also be used to illustrate another type of selection. Upon analysis of the resistant mutants, the mutations that confer resistance to 5FOA are found to be alleles of the gene *ura3*. The wild-type product of URA3 is necessary for uracil biosynthesis. Uracil auxotrophy provides another means of selection, since, if uracil is omitted from the growth medium, the *ura3* mutant cells cannot grow and only $URA3^+$ cells will be found. These two conditions make it possible to set up a **counter-selection** for either $URA3^+$ or *ura3* mutant: on medium with 5FOA and all other nutrients,

URA3+ cells will be killed and only *ura3* mutant cells will grow, whereas on medium with no uracil or 5FOA, only the URA3+ cells can grow. The counter-selection is illustrated by comparing the growth in Figure 3.12. Nutritional selections of this type are commonly used in yeast genetic analysis.

A few selection schemes in multicellular organisms can be based on nutrition or drugs, as the yeast *ura3* scheme is. Other selections can be based on drugs that affect the nervous system or on some other chemical reagent, and many other types of selection can be done using hormones, nutrition, and the ability to respond to a chemical stimulus. However, we can make a more general example of selection by thinking of the *URA3* gene as a **selectable marker** as shown in Figure 3.13A. For example, suppose that we construct a plasmid that includes the *URA3* wild-type gene and a copy of our gene of interest. (Different types of yeast plasmid vectors are described in Chapter 2.) The yeast cell itself has a mutated *ura3* gene, and is a uracil auxotroph. The cells are transformed with the plasmid, and the transformants are grown in the absence of uracil. The only cells that will grow are those with the plasmid. Thus, by using uracil auxotrophy as the basis for our selection, we have also been able to select for the presence of our gene of interest without the need to monitor its effect or phenotype directly.

Selection with linked lethal mutations

The use of selectable markers in this way can be extended to include many other types of genes beside nutritional markers. The general principle is that the phenotype

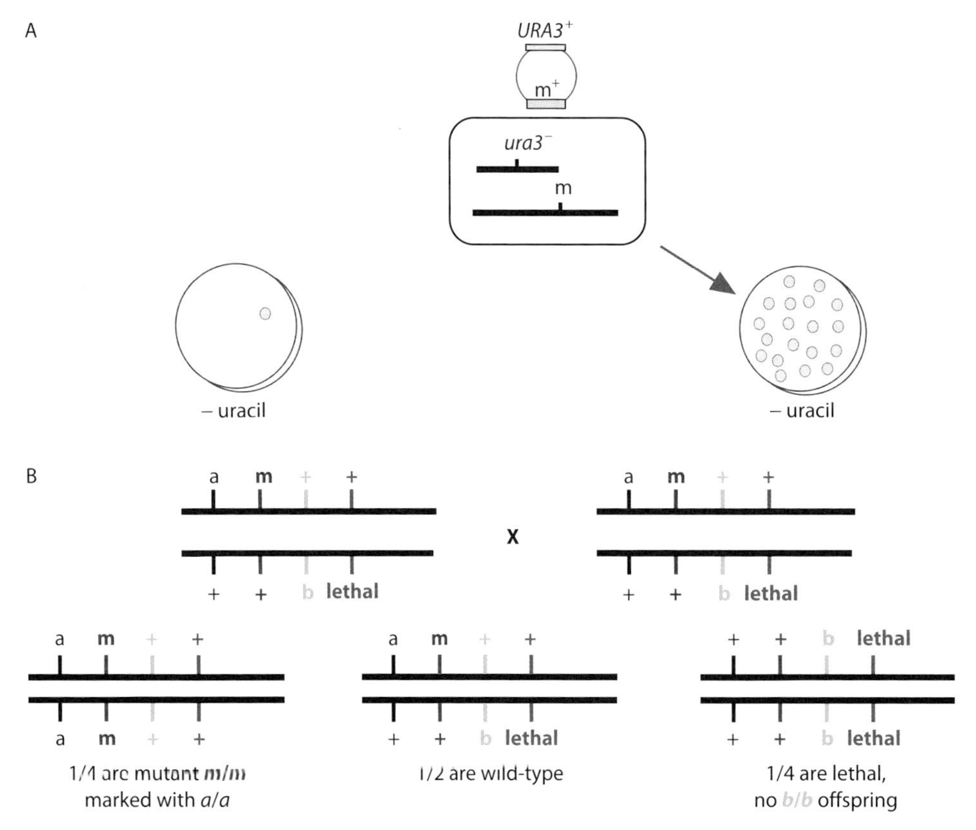

Figure 3.13 Selectable linked markers on plasmids and chromosomes. In panel A, *ura3*− cells cannot grow in the absence of added uracil. These cells are also mutant for the gene of interest *m*. A plasmid that has the wild-type *URA3*+ and the wild-type *m*+ gene is transformed into *ura3*− *m*− cells. Selection for the ability to grow in the absence of added uracil will also have the *m*+ allele. A similar principle is used for selecting or screening for linked mutations on chromosomes in diploids. In panel B, a heterozygote is shown that has the *a* marker closely linked to the *m* mutation of interest. The other homolog has the *b* marker and lethal mutation. When the heterozygotes are mated, as in strain maintenance, the *b/b* offspring die because of the linked lethal mutation, the *a/a* offspring are also homozygous for the mutation of interest *m*, and the heterozygotes are wild-type.

of one gene serves to monitor the presence or absence of many other linked genes whose phenotypes are not actively being examined. This principle is widely used in multicellular organisms such as *Drosophila* and *C. elegans* by employing lethal mutations as selectable markers on the unmutagenized chromosome. This is illustrated in Figure 3.13B. In our eye color mutant hunt, a quarter of the flies are homozygous for the parental non-mutagenized chromosome and are of no interest in the screen; in fact, their presence makes it more difficult to find the mutations with eye color defects. Suppose that, before the mutant hunt has begun, a recessive lethal mutation is crossed onto the marked chromosome in the non-mutagenized parent, as in Figure 3.13B. Now, all of those offspring who inherited only the non-mutagenized chromosome will die, making it easier to spot the ones with eye color defects. One of the most common means of reducing the number of offspring that have to be examined is to include a lethal mutation or some other selectable marker in the mating scheme.

What if the mutation I am interested in has a lethal phenotype?

In the previous example, a lethal mutation was used as a selectable marker to reduce the number of F_1 or F_2 individuals that need to be examined. Suppose that the mutation of interest is itself lethal when homozygous. This is not a rare occurrence. By some estimates, as many as three-quarters of genes in *C. elegans* and *Drosophila* detected by a simple genetic screen have a recessive lethal phenotype. As a result, when the newly arisen mutation becomes homozygous in the F_2 or F_3 generation, the mutant dies and is lost from further analysis. In other words, the investigator needs to be prepared to deal with lethal mutations.

Two general approaches have been used to avoid the complication of lethal mutations. The first approach is to employ conditions, especially temperature regimens, under which recessive homozygotes can survive. These are known as **conditional mutations**, with the specific example of **temperature-sensitive mutations**. The second approach is to maintain the mutations as heterozygotes with other genetic markers. The procedure of maintaining a mutation in heterozygous condition is known as **balancing** the mutation. Both approaches are widely used; let us explore each in turn.

Conditional mutations

Temperature-sensitive conditional mutations were first employed with the bacteriophage T4. In this genetic system, as in haploid eukaryotes, maintaining the mutation as a heterozygote is not feasible. The solution in many haploid organisms is to conduct the mutant hunt at a growth temperature slightly higher than the normal temperature used for the organism. Mutants found at the high temperature are then grown at a temperature slightly below the normal growth temperature. In many cases, mutations that do not grow at the high temperature will grow and exhibit a wild-type or nearly wild-type phenotype at the lower temperature.

This approach has been so successful that, even in diploid organisms in which it is possible to maintain mutations as heterozygotes, temperature-sensitive conditional mutations are widely used. The temperature at which the mutant phenotype is seen is termed the **restrictive temperature**. The temperature at which the wild-type phenotype is seen is called the **permissive temperature**. Often these temperatures differ by only a few degrees: the permissive temperature for *C. elegans* is 15°C, whereas the most widely used restrictive temperature is 25°C. Not all temperature-sensitive mutations are lethal, although this is the use we are considering. The key is that the mutant phenotype is seen under a specific set of growth conditions, in this case temperature. The two concepts are illustrated in Figure 3.14 using the vestigial wing phenotype in *Drosophila*.

In most cases, the restrictive temperature is the high temperature, so that the mutations actually have a heat-sensitive phenotype. For this reason, when the term 'temperature-sensitive' mutation is used with no further explanation as above, it can be assumed that the restrictive temperature is the high temperature and the permissive temperature is the low temperature. Such

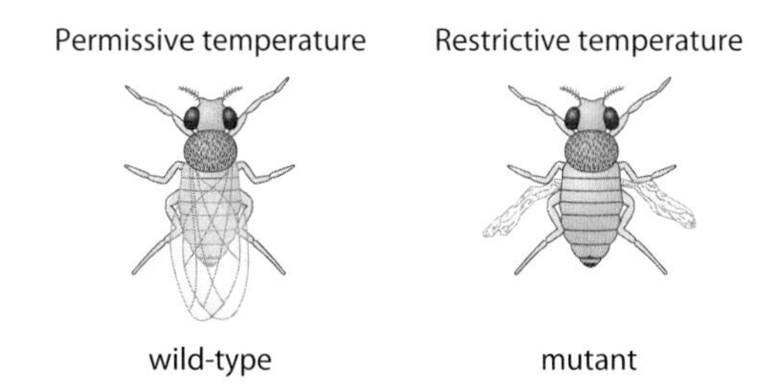

Figure 3.14 Temperature-sensitive mutant phenotypes in *D. melanogaster*. At the permissive temperature, the fly has wild-type wings. At the restrictive temperature, the fly has vestigial wings.

temperature-sensitive mutations are often designated by the notation 'ts'. Exceptional cases have been described in which the restrictive temperature is the lower temperature and the permissive temperature is the normal growth temperature for the organism. These mutations are called 'cold-sensitive', designated by 'cs'.

It is easy to imagine how a temperature-sensitive effect might occur by recalling that the three-dimensional shape of a protein depends on the temperature, and that the function of a product depends on its three-dimensional shape. Thus, that a mutation slightly destabilizes the structure of the protein may have no effect on the function until the protein is placed under higher-temperature conditions—not so high that the native protein is destabilized but high enough that an already slightly destabilized protein becomes non-functional. With a non-functional protein, a mutant phenotype can occur. Under lower permissive temperature conditions, the protein is stable and functional, and so a wild-type phenotype is observed. Although this is an easy way to think about temperature-sensitive mutations, this explanation has not been verified for most temperature-sensitive mutations and is clearly an oversimplification for some Nonetheless, it provides an appropriate intellectual framework for thinking about temperature-sensitive mutations.

Temperature-sensitive mutations have also been widely used as a means of estimating the time at which expression of a gene is required. This type of analysis will be described in Chapter 9 when we discuss temperature shift experiments to analyze the time of gene activity. For our purposes in this chapter, the important point is that temperature-sensitive mutations allow the investigator to maintain a lethal mutation as a homozygote simply by growing the mutant strain at the permissive temperature. As a practical matter, many mutant hunts are routinely conducted at the restrictive temperature characteristic for that organism so that temperature-sensitive mutations can be recognized if they occur. Although not all genes are found to have temperature-sensitive alleles, many genes do; it is worth taking advantage of an easy change in growth conditions to find such mutations.

Balanced heterozygotes

Since most mutations are recessive, lethal mutations can be maintained as heterozygotes in diploid organisms; the homozygous lethal individuals are produced when two heterozygotes are mated. A generalized genotype for such a balanced heterozygous state is shown diagrammatically in Figure 3.15, but the ingenuity of generations of geneticists has greatly modified this generalized example. In the example, the lethal mutation is on the same chromosome as the recessive markers *a* and *c*, while the recessive markers *b* and *d* mark the other homolog. The lethal mutation could have become linked to *a* and *c* because the original mutagenesis scheme was carried out with an *a*/*a*; *c*/*c* homozygous individual. This multiply heterozygous individual has a wild-type phenotype because all the mutations are recessive. When two heterozygotes are mated, a quarter of the offspring will be homozygous for both *b* and *d*, which will mark these as not carrying the lethal mutation of interest. Like the parents, half of the offspring will be heterozygous for all four markers and the lethal mutation, which allows the investigator to recognize and keep the balanced heterozygote for future experiments. A quarter of the offspring will be homozygous for the *a* and *c* markers, which the investigator recognizes as being homozygous for the lethal mutation. In other words, because the mutation results in a lethal phenotype, the absence of *a c*/*a c* homozygous offspring indicates the presence of the mutation.

This generalized and simplified example illustrates several important features of using balanced heterozygotes, which other modifications have addressed. Notice that the effectiveness of this method is affected by recombination and linkage. If the lethal mutation is unlinked from the marker mutations used to balance it, in our case the genes *a* and *c*, it will quickly be lost. For this reason, most screens that involve looking for lethal mutations are directed at genes at a particular location in the genome, and different genetic markers are used for different regions of the genome. Even when many different marker mutations are available, recombination defines a limitation of this approach, and it is limited to intervals of a few map units in practice.

The early *Drosophila* geneticists developed a set of **balancer chromosomes** to circumvent the problem with recombination. Certain chromosome rearrangements, particularly inversions, act as **crossover suppressors**: the gametes with a crossover are aneuploid and give rise to inviable offspring. These crossover suppressors are the basis for balancer chromosomes. One example of how this happens is illustrated by the inversion in

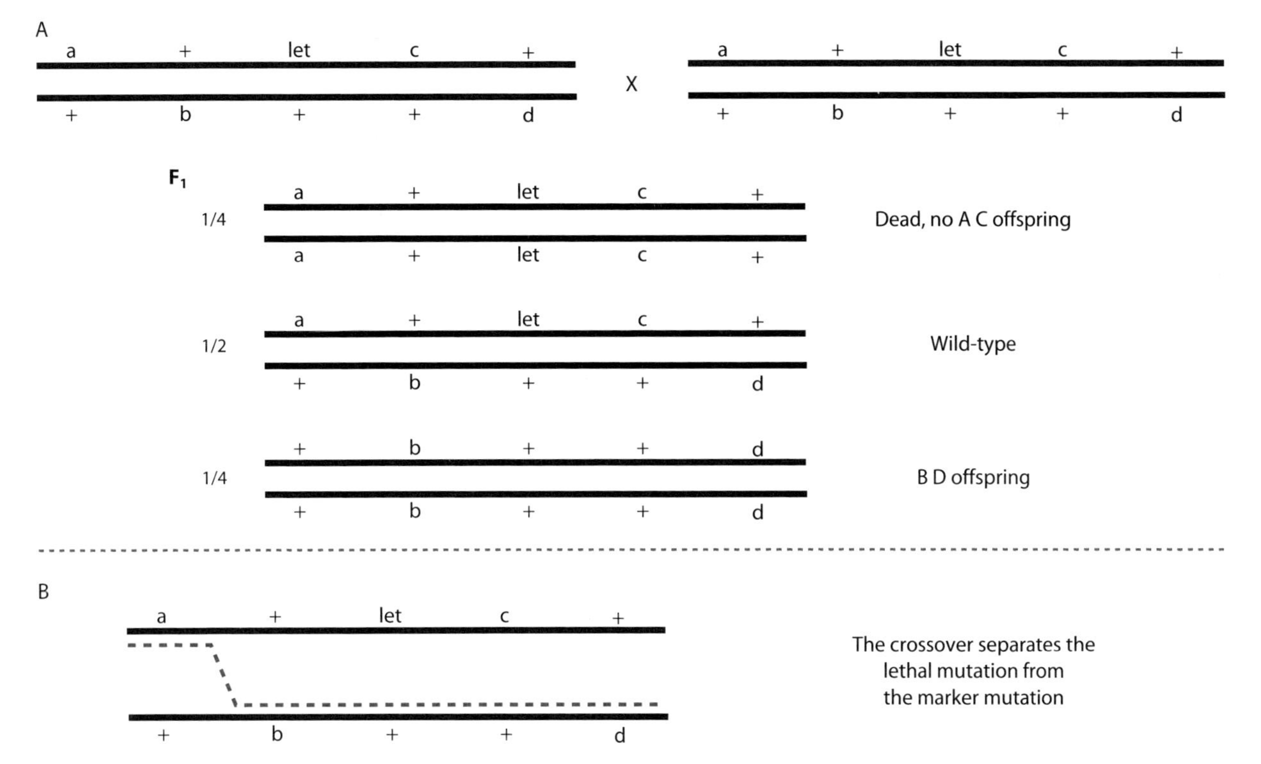

Figure 3.15 Maintaining a lethal mutation as a balanced heterozygote. Because many mutations have a lethal phenotype, they are maintained in heterozygotes, with one general scheme shown here. The lethal mutation is symbolized 'let', and the genes *a*, *b*, *c*, and *d* are recessive marker mutations used to monitor the chromosomes. The lethal mutation was induced on a chromosome that had the *a* and *c* markers on it. The *b* and *d* markers are on the other homologous chromosome. Because all the mutations are recessive, this individual has a wild-type phenotype. In panel A, the heterozygotes for the lethal mutation are mated to produce the F_1 offspring. The phenotypes of the flanking marker mutations are used to identify the various genotypes. The homozygous lethal individuals, which have the phenotypes of interest in this case, are recognized either because of their lethality or by the absence of the A and C phenotypes. The individuals that do not carry the lethal mutation are recognized by the B and D mutant phenotypes. The heterozygotes used to maintain the lethal mutation have a wild-type phenotype. Panel B shows the limitation of balancing lethal mutations in this way. A crossover separates the lethal mutation of interest from one or more of the markers used to monitor its presence on the chromosome. Because of this effect, balancing heterozygous mutations are limited to very small intervals or require a method to reduce or eliminate recombination. This is shown in Figures 3.16 and 3.17.

Figure 3.16—even if an inversion heterozygote has normal meiotic pairing, a crossover in the inverted region results in meiotic products that are dicentric, acentric, or aneuploid. Dicentric or acentric chromatids cannot complete meiosis and are lost, whereas aneuploid gametes result in inviable offspring if they are not also lost. Thus, if a lethal mutation is placed as a heterozygote opposite an inversion chromosome, it can be balanced over the length of the inversion and the effect of recombination is greatly reduced. Most of the balancer chromosomes that are used in *Drosophila* consist of multiple complex inversions and other rearrangements that extend over the entire chromosome arm or the entire chromosome. Since there are three chromosomes in *D. melanogaster*, an investigator can use one of these rearranged chromosomes to balance heterozygotes for a nearly a third or more of the genome at once. In Case Study 3.1, Part 2 (finding mutations affecting embryonic development), only three different balancer chromosomes were needed in order to screen most of the genome since the balancers for each autosome included the entire chromosome. Mutations in genes located at opposite ends of the same chromosome were recovered in the scheme using these balancers.

Drosophila geneticists have refined the use of such balancer chromosomes even further. Widely used versions of these chromosomes have one or more recessive lethal mutations on the same chromosome as the inversion so that the rearrangement homozygote also does not survive. This use of lethal mutations to select again the balancer chromosome homozygotes parallels

DEFINITION

'Balancing' a mutation refers to maintaining it as a heterozygote.

A. Normal and inversion chromosomes

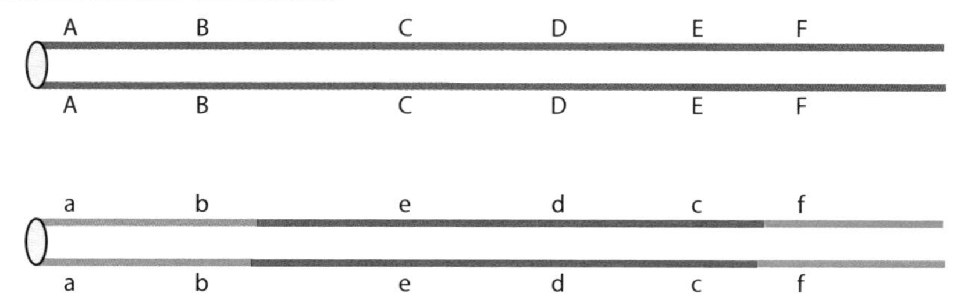

B. Pairing requires an inversion loop

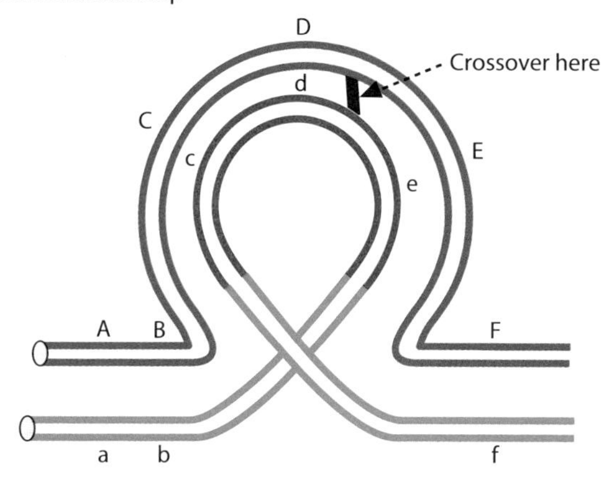

C. The recombinant chromatids are acentric or dicentric

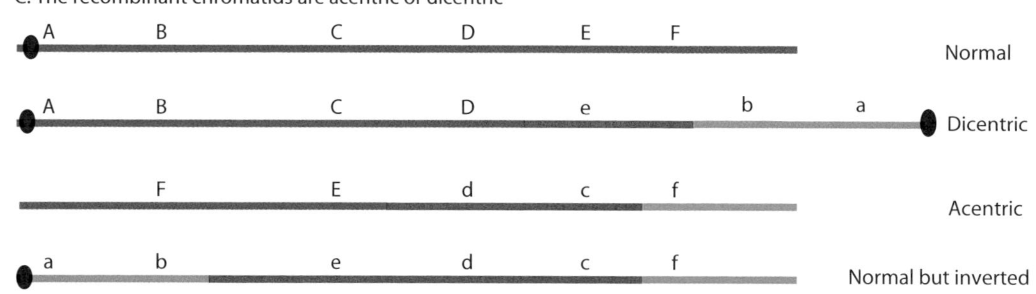

Figure 3.16 Inversions and crossover suppression. A normal and an inverted chromosome, each with two chromatids, are shown in panel A. The inverted region is indicated in red. In order for these chromosomes to pair in meiosis, one chromosome has to bulge outward and the other needs to loop on the inside, as shown in panel B. These inversion loop structures have been seen in some plants and in *Drosophila* polytene chromosomes, and have been inferred to occur in other organisms. A crossover occurs in the inverted region, as shown by the arrow. When these chromosomes segregate at meiosis, only the non-crossover chromatids are normal, as shown in panel C. The chromatids that crossed over will be either dicentric or acentric, neither of which will segregate normally in meiosis. Thus, there is an absence of crossover gametes and the genes *A* and *F* will always segregate together, as if crossing over between them is suppressed.

what was described above for lethal mutations and selections. Some balancer chromosomes also have one or more dominant markers on the rearranged chromosome that are themselves homozygous lethal. An example with Curly wings is shown in Figure 3.17. Thus, in these cases, the only surviving animals are the heterozygotes between the balancer rearrangement and the mutation being studied, providing a selection method to make stock maintenance much simpler. The balancer chromosomes used in Case Study 3.1 had both dominant visible markers and recessive lethal mutations.

Although the use of balancer chromosomes is most advanced in *Drosophila*, similar methods for maintaining lethal mutations as heterozygotes have been used in other organisms including worms and mice. Most of these methods bear some resemblance to the more sophisticated methods used in *Drosophila*. Balancer chromosomes are described in more detail in Text Box 3.2.

Case Study 3.1

Part 2: Find a mutant: segmentation in *Drosophila* embryos

The use of balancer chromosomes

Three different papers detail the mating schemes that were used to find these mutations, each paper addressing genes from a different chromosome (Nüsslein-Volhard *et al.* 1984; Jürgens *et al.* 1984; Wieschaus *et al.* 1984). Nüsslein-Volhard, Wieschaus, and their collaborators used different balancer chromosomes for the X chromosome, for chromosome 2, and for chromosome 3. The procedure was approximately the same for each chromosome, but the genetic strains differed. For example, the screen that identified mutations on chromosome 2 used the balancer chromosome CyO, an inversion that includes most of the second chromosome and is marked with the dominant mutation Curly wings (*Cy*), as shown in Case Study Figure 3.3. In the chromosome 2 scheme, males homozygous for the mutations cinnabar (*cn*) eyes, brown (*bw*) eyes, and spotted (*sp*) were fed on EMS to induce mutations during spermatogenesis. These males were mated to the females that were heterozygous for the CyO inversion and a multiply marked chromosome.

The F_1 flies with newly induced mutations balanced over the inversion were recognizable because they have curly wings and a distinctive eye color. A single F_1 male fly was then back-crossed with a female from the CyO heterozygous strain to produce a balanced F_2 strain. At this point, the mutagenized copy of chromosome 2 is heterozygous but is not yet known to have any mutations of interest. By using an unmutagenized F_1 female, the F_2 males will have an unmutagenized X chromosome. Chromosome 3 may still contain newly arisen mutations (and, in fact, some were found), but these are eliminated in subsequent outcrosses and mapping experiments. The subsequent crosses also determined that the mutations were recessive, and that each mutant phenotype was due to a single mutation rather than to two mutations acting synthetically or synergistically. Each F_2 line was mated with its siblings and the F_3 larvae were examined for mutations.

The balancer chromosome in this scheme plays several important roles. First, a mutation that arises anywhere on chromosome 2 is kept as a heterozygote. If a mutation is found, siblings heterozygous for the balancer are used to maintain the mutation. Mutations that were subsequently mapped to opposite ends of chromosome 2 could be maintained without the need for any preliminary limitation on the map location. With a balancer chromosome this effective, approximately 40 percent of the *Drosophila* genome is screened at the same time. Secondly, the heterozygous siblings needed to maintain the mutation are easily identified by the dominant marker Curly wings on the balancer chromosome. Thirdly, the balancer chromosome is itself lethal when homozygous, although at a different developmental stage than the mutations of interest. As a result, in the F_2 and subsequent generations when flies heterozygous for the CyO balancer are mated, the CyO/CyO embryos (expected to be a quarter of the total) do not hatch, so that the fraction of mutant heterozygotes among the offspring is increased. The CyO balancer allowed them to test and establish nearly 4600 lines for embryonic lethal mutations. From these, 321 lines were determined to have an embryonic lethal mutation on chromosome 2.

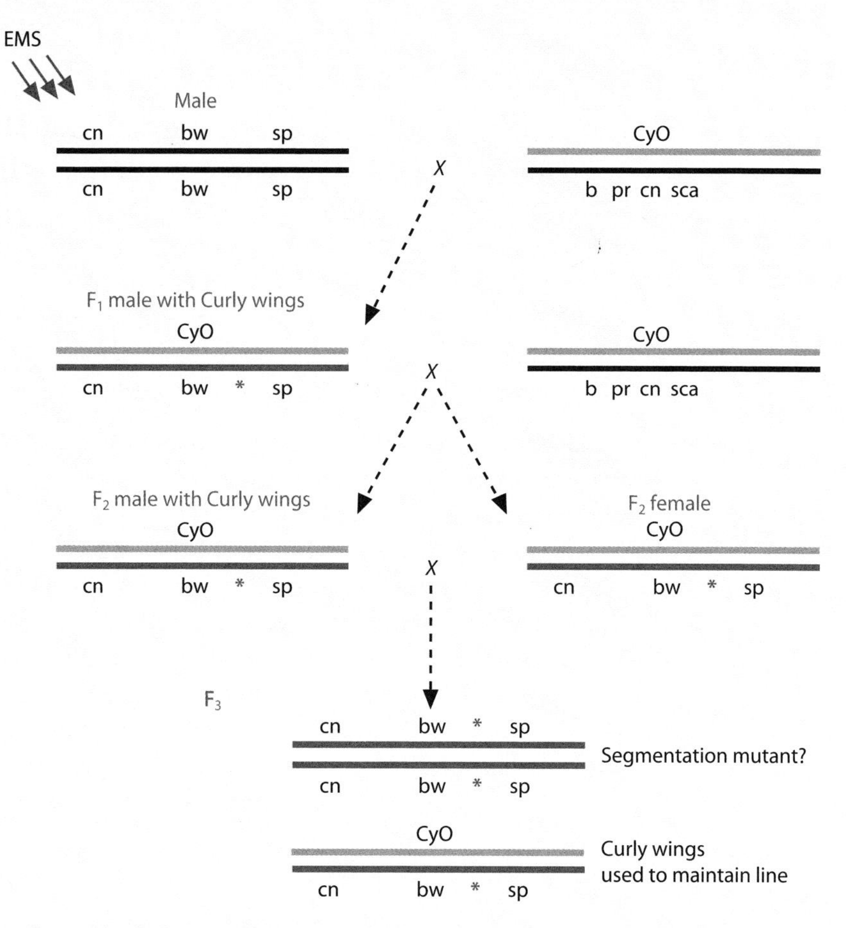

Figure C3.3 The balancer chromosome scheme used for chromosome 2. A male homozygous for the recessive marker mutations *cn*, *bw* and *sp* is mutagenized and mated to a female that is heterozygous for the CyO balancer chromosome. The mutagenized chromosome is shown in red and the balancer chromosome in orange. The newly induced mutation is indicated by a red star. The CyO balancer is marked with the dominant mutation Curly wings (*Cy*) so that any individual heterozygous for the balancer chromosome has a Curly wings phenotype. An individual F_1 male that is heterozygous for the mutagenized chromosome and for the CyO balancer is mated to a female with the balancer chromosome. This produces F_2 males and females that are heterozygous for the same mutagenized chromosome and the CyO balancer. These are mated to produce an F_3 generation, which includes flies that are homozygous for the mutagenized chromosome to be tested for new segmentation mutants. The CyO heterozygotes are used to maintain the newly arisen mutation without recombination. The F_3 generation will also include CyO/CyO homozygotes but these die for unrelated reasons and can be distinguished from the newly arisen segmentation mutant.

Different balancer chromosomes were used for the schemes involving the X chromosome and chromosome 3, but the strategy was the same. The balancer chromosomes allowed them to test nearly the entire length of the chromosome as one unit, without mutations being lost through recombination. For the X chromosome, 122 lines with lethal mutations were established; for chromosome 3, 198 mutant lines were established. One of the crucial features of this mutagenesis scheme was the ability to isolate and maintain as many lethal mutations as possible, which is only possible with the balancer chromosomes available in *D. melanogaster*.

Many flies were grown: a temperature-sensitive selection against the unmutagenized flies

In order to find as many mutations as they did, Wieschaus, Nüsslein-Volhard, and their colleagues cultured and examined hundreds of thousands of flies. The amount of effort was formidable, but they were able to use a genetic selection to keep the numbers rather more manageable. Prior to their work, several dominant temperature-sensitive lethal mutations (referred to as DTS) had been identified by other geneticists. One of these temperature-sensitive lethal mutations was crossed onto each chromosome being tested, on the normal and non-inverted homolog in the unmutagenized mother. These temperature-sensitive lethal mutations provided the means for a very useful selective strategy.

It is important to realize that, in finding mutations in *Drosophila*, individual flies often need to be mated. Mating of individual flies requires that the females be virgins; the DTS mutations provide an efficient means of recovering virgin females. The F_1 flies were mated and then shifted to the restrictive temperature. At the restrictive temperature, the parents with the DTS mutation die, so that no special effort is required to collect virgin females at each generation. The temperature-sensitive lethal mutation also ensures that only the balanced heterozygotes are maintained in the F_2 generation, since the siblings with the non-inverted chromosome die. Thus, this particular genetic selection was used to reduce the number of flies that had to be examined for mutations rather than to select specifically for the mutations of interest. Nonetheless, the selection also allowed the investigators to maintain and examine many more mutant lines than would otherwise have been possible.

Segmentation mutations: summary so far

The account of finding and analyzing segmentation mutations in *Drosophila* will be continued in the next chapter. For descriptive purposes, we have divided the genetic analysis of segmentation into sections, which might suggest that the procedures were done sequentially. This is not entirely accurate. While Nüsslein-Volhard and Wieschaus were finding the mutations, they were continuing with all of the other steps in the analysis, and the results of each mutant hunt informed the next one. Nonetheless, if we think of this as a stepwise process for clarity, at this step the investigators have a very large number of mutations maintained as heterozygotes. They have some preliminary mapping data arising from the use of the balancer chromosomes, i.e. they have been able to assign the mutations to a chromosome. They have a broad idea of what types of mutant phenotypes are being found. The next steps in the genetic analysis of segmentation are described in Case Study 4.1 in Chapter 4.

Further reading

Jürgens, G., Wieschaus, E., Nüsslein-Volhard, C., and Kluding, H. (1984). Mutations affecting the pattern of the larval cuticle in *Drosophila melanogaster* II. *Roux's Archives of Developmental Biology*, **193**, 283–95.

Nüsslein-Volhard, C. and Wieschaus, E. (1980). Mutations affecting segment number and polarity in *Drosophila*. *Nature*, **287**, 795–801.

Nüsslein-Volhard, C., Wieschaus, E., and Kluding, H. (1984). Mutations affecting the pattern of the larval cuticle in *Drosophila melanogaster* I. *Roux's Archives of Developmental Biology*, **193**, 267–82.

Wieschaus, E., Nüsslein-Volhard, C., and Jürgens, G. (1984). Mutations affecting the pattern of the larval cuticle in *Drosophila melanogaster* III. *Roux's Archives of Developmental Biology*, **193**, 296–307.

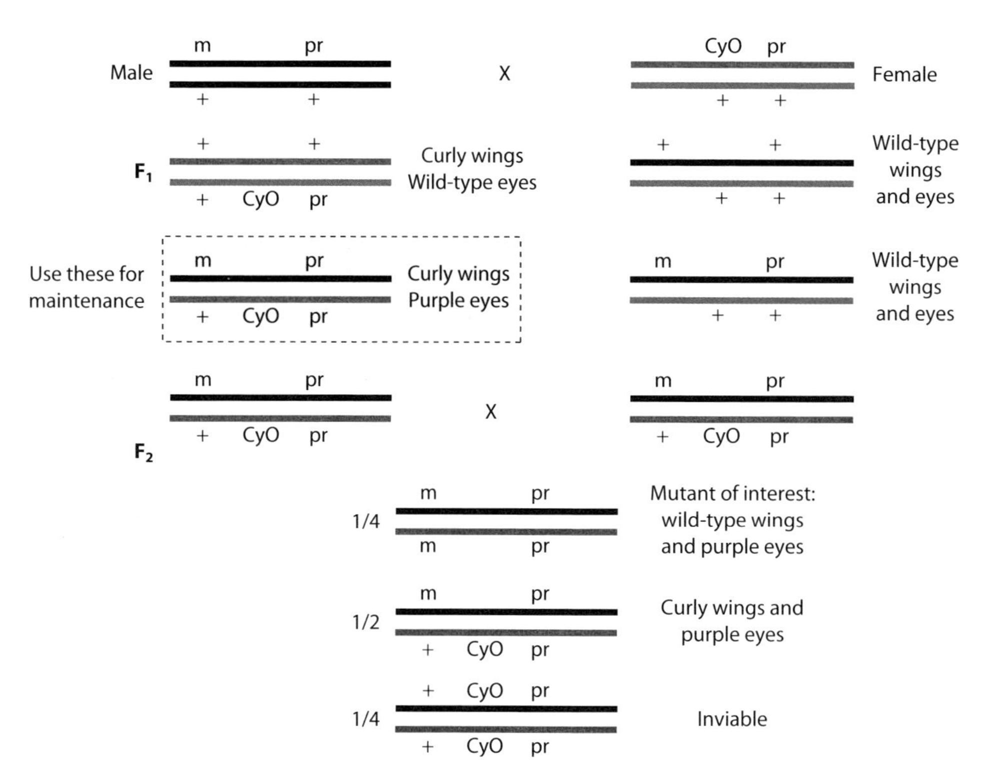

Figure 3.17 Using a balancer chromosome. One widely used balancer chromosome in *Drosophila* is called CyO. This chromosome has multiple inversions that suppress recombination, the dominant marker Curly wings (*Cy*) and the recessive marker purple eyes (*pr*), among other genetic markers. A heterozygous male that has the mutation of interest (*m*) linked to the *pr* gene is mated to the balancer heterozygous female. Since males do not recombine in *Drosophila*, *m* and *pr* will remain linked in this cross. Four classes of F_1 progeny will result, as shown. The ones of interest for further analysis and maintenance are recognized because they have curly wings and purple eyes. These flies are mated to each other to produce the F_2 offspring, as shown. The ones with purple eyes and normal wings are also homozygous for the mutation of interest. Even if *m* is only loosely linked to *pr*, the two genes always segregate together because of the CyO balancer. The flies with curly wings and purple eyes are like the parents and can be used to maintain the stock. The ones that are homozygous for the CyO chromosome are inviable because *Cy/Cy* homozygotes die.

BOX 3.2 Balancer chromosomes and genetic screens

The main advantage of identifying mutations is that the mutant phenotypes do not have to be anticipated by the investigator. However, mutant screens present some challenges as well, including the realization that many mutants are non-viable. Such mutations have to be maintained in heterozygous strains with easily visible markers nearby on the chromosome, an arrangement referred to as a **balanced heterozyote**. As described in the chapter, this is an effective means of maintaining the mutation, but a crossover between the visible marker and the mutation being balanced means that the mutation will be lost. Thus, the region that can be balanced in a typical normal chromosome is quite small and limited by the amount of recombination.

The *Drosophila* geneticists H.J. Muller and Alfred Sturtevant found a solution to this problem more than 80 years ago with the first of many balancer chromosomes. Balancer chromosomes are widely used in *Drosophila* genetics and have also been useful tools in mutant screens in both *C. elegans* and mice. A balancer chromosome allows the geneticist to maintain any mutation, including lethal mutations, in a heterozygous state. In addition, the balancer chromosome allows the geneticist to identify genes within a particular region of the genome, establishing a preliminary map position that can be refined by more precise methods.

Properties of balancer chromosomes

A good balancer chromosome has two essential characteristics. First, it must greatly reduce or eliminate crossing over in a region of the genome being investigated. With crossover suppression, the balanced mutation will not be lost even if the visible marker is not close by on the chromosome. Secondly, individuals homozygous for the balancer need to be distinguishable from heterozygotes, which in turn need to be distinguishable from the homozygous non-balancer mutants.

Crossover suppression

The first of these characteristics, crossover suppression, has often been solved using inversion chromosomes, as it was in *Drosophila*. From cytological observations in maize, it is generally believed (but rarely demonstrated) that an inversion heterozygote can form a loop structure upon meiotic pairing (Figure 3.16). A crossover that occurs within this loop will produce gametes with duplication and deletion chromosomes, including chromosomes that lack a centromere (known as acentric fragments) and chromosomes that have two centromeres (dicentric fragments). Both acentric and dicentric chromosomes will fail to complete meiosis normally, and thus will not contribute to viable gametes. Since only the non-recombinant chromatids will appear in viable gametes and the subsequent offspring, an inversion effectively reduces or eliminates crossing over.

Inversions are the most widely used chromosome rearrangement for balancer chromosomes, but reciprocal translocations have also been used for this purpose, particularly in *C. elegans*. Translocation heterozygotes are usually, if not always, much less fertile than inversion heterozygotes, and so inversions are advantageous if they can be found. The most widely used balancer chromosomes in *Drosophila* have multiple inversions induced by repeated irradiation over many years. These multiply inverted chromosomes eliminate crossing over in a large region of the chromosome. In fact, most of the euchromatic portion of the *D. melanogaster* genome can be balanced by one of three balancer chromosomes, one for the X chromosome and one for each of the two autosomes. (The tiny heterochromatic fourth chromosome has very few genes and does not recombine even in wild-type flies, so no balancer is required and it is typically ignored in mutagenesis screens.) No other organism has balancer chromosomes for all parts of its genome as *Drosophila* does. The ability to use only three different balancer chromosomes to screen the genome reflects both the years of genetics research using *Drosophila* and the small number of chromosomes.

Distinguishing marker mutations

The second necessary characteristic for a balancer chromosome is the ability to distinguish the balanced heterozygote, which is the genotype of interest for strain maintenance, from either of the two homozygotes. In *Drosophila* balancer chromosomes, this is usually accomplished with a dominant marker on the inversion chromosome. Some of the dominant markers that have been used are themselves homozygous lethal or the inversion also has other recessive lethal mutations present on it. For example, the balancer chromosome described in Case Study 3.1 is marked with the dominant marker Curly wings as well as with several recessive lethal mutations. The balanced heterozyote has curly wings and is easily identified. An example is shown in Figure 3.17. The homozygotes for the balancer inversion die because of the recessive lethal mutations, so they need not be considered further. The offspring that do not have curly wings will be ones with the mutation being studied. Such a system is particularly effective because *Drosophila* is maintained in culture tubes by random mass matings rather than matings of an individual pair of flies. Therefore, if the mutation of interest is lethal, which is a standard use of balancer chromosomes, the two homozygous classes will die and only the balanced heterozygote will survive. The balanced mutant line can be maintained in culture by simply dumping the contents of one vial into a fresh culture vial without the need to separate males and females or perform individual matings. The only surviving flies are those that are heterozygous for the balancer and the mutation of interest.

A more recent innovation has been to insert the green fluorescent protein (*gfp*) gene or a modified derivative of GFP onto the balancer inversion. (The use of *gfp* as a reporter gene is described in more detail in Chapter 8.) The fluorescence arising from GFP can be identified in early embryos long before any adult visible phenotype such as curly wings can be spotted. Even more important, the embryos can be sorted automatically by fluorescence in a flow cytometer and the three genotypic classes can easily be identified and separated from one another at a very early stage. Since balancer chromosomes are used to maintain lethal mutation stocks, it is very helpful to be able to distinguish the inviable embryos arising from the effects of the mutation from those arising from the effects of the balancer chromosome.

Balancer chromosomes in *C. elegans*

The most widely used inversion balancer chromosome in *C. elegans* does not have a dominant marker and does not have an associated lethal phenotype. Instead, the inverted chromosome is marked with a visible recessive mutation that results in a dumpy (short and fat body shape) phenotype and a second recessive mutation that results in paralysis. The configuration is shown in Text Box Figure 3.1. The homozygous balancer worms are then both dumpy and paralyzed, and are very easy to identify. Often the other non-inverted homolog has been marked with a recessive visible mutation that results in uncoordinated movement. Thus, the homozygotes for the non-inverted chromosome, which will also be homozygous for the mutation under study, will also have a distinctive phenotype, and the heterozygotes between the balancer inversion and the visible

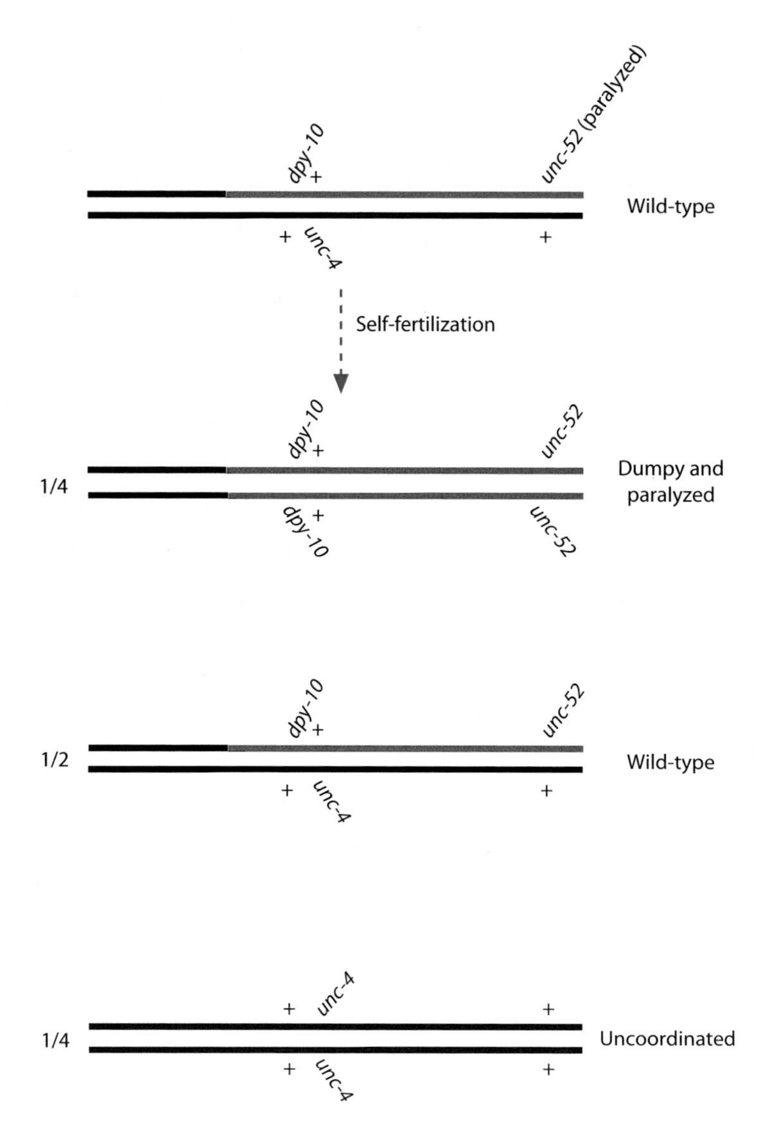

Figure B3.1 A balancer chromosome in *C. elegans*. One widely used balancer chromosome in *C. elegans* is an apparent inversion on linkage group II, shown here in red. The balancer chromosome has two recessive markers called *dpy-10* and *unc-52*; a *dpy-10/dpy-10* worm is short and fat (called dumpy) while an *unc-52/unc-52* worm is paralyzed. The genes are loosely linked in the wild-type, but because of the chromosome rearrangement (depicted in red) no recombination occurs between them in the balancer strain. A heterozygote for the balancer chromosome produces progeny that are dumpy and paralyzed (homozygous for the balancer chromosome), wild-type (heterozygous for the balancer), and homozygous for the other homolog (marked here with the recessive mutation *unc-4*), resulting in a worm that moves but is uncoordinated. Worms homozygous for *unc-4* are easily distinguishable from those homozygous for *unc-52*. Because *C. elegans* is maintained in the laboratory by picking individual worms rather than by matings between males and females *en masse*, the balancer chromosome homozygotes do not need to be inviable in order to be useful.

marker but non-inverted chromosome will be wild-type in both morphology and movement. One reason that a dominant mutation is not needed on the balancer chromosome is that worm cultures are usually propagated by allowing a single hermaphrodite to self-fertilize rather than by random mating as in a population of flies. A marker gene such as GFP could be helpful to mark either the inversion chromosome or the non-inverted chromosome, but such a construct has not been made.

Balancer chromosomes in mice

Inversion balancers have also become widely used in mice for large mutagenesis screens. The first balancers to be used in mice had very large inversions that did not completely suppress recombination over their entire length and did not have an easily scored visible marker. These chromosomes represented only a small portion of the mouse genome and thus were not appropriate for large-scale mutagenesis screens of the genome.

More recently, a set of inversions appropriate for use as balancer chromosomes has been engineered using the Cre–*lox* system. The Cre–*lox* system of targeted recombination is described in more detail in Chapter 6, so our review here will be very brief. The Cre recombinase enzyme is derived from bacteriophage P1 and will induce targeted recombination at the chromosomal sequences known as the *loxP* sites. The *loxP* site consists of two 13 bp inverted repeats flanking an 8 bp core sequence. Because the core sequence is unique, its orientation confers a polarity on the *loxP* sites. As shown in Text Box Figure 3.2A, a recombination event between two *loxP* sites in the same orientation will cause a deletion of the chromosomal region between the two *loxP* sites. On the other hand, if the *loxP* sites are in opposite orientation, as in Text Box Figure 3.2B, the recombination event will produce an inversion of the chromosomal region. Cre recombinase can catalyze recombination between two *loxP* sites separated by several million base pairs, so this system can induce quite large inversions. The strategy is to introduce *loxP* sites into particular locations in the genome in embryonic stem cells, as described in Chapter 6.

The first use of this method introduced the *loxP* sites into the *Wnt3* region and into the gene for a p53-related protein, two genes that are separated by 24 map units on chromosome 11. When Cre recombinase was expressed, a recombination event occurred between the two *loxP* sites and the entire 24 map unit interval was inverted. Although the *loxP* endpoints could have been inserted at many different places in the genome, the *Wnt3* gene is needed for embryonic survival, so its disruption by *loxP* confers the recessive lethal phenotype on inversion homozygotes. In addition, the vector used to target *loxP* to the *Wnt3* gene also contained a dominant pigmentation marker, the *agouti* gene under the control of the keratin promoter. Thus, homozygous inversion mice die because the essential *Wnt3* gene is disrupted, and heterozygous balanced mice can be distinguished from their non-inverted litter-mates by the presence of the dominant pigmentation transgene.

Subsequent uses of the Cre–*lox* system to make balancer chromosomes have not always inserted the *loxP* site into an essential gene so that balancer homozygotes are viable. Mice are typically maintained by mating a single pair rather than random mating among a population so, as with *C. elegans*, an inversion balancer in which the homozygotes can be distinguished from the heterozygotes is adequate for most purposes. In fact, if the mutation being balanced opposite the inversion is itself lethal, it is an advantage to have a homozygous balancer that is viable so that the only non-viable embryos are those that arise from the mutation being studied. Some of the recent balancer chromosomes on chromosome 4 have used insertion into the tyrosinase gene as a suitable visible marker because the pigmentation differences arising from heterozygous and homozygous tyrosinase mutations can be distinguished. It should also be possible to use GFP genes as a dominant visible marker, as has been done in flies.

Further reading

Hentges, K.E. and Justice, M.J. (2004). Checks and balancers: balancer chromosomes to facilitate genome annotation. *Trends in Genetics*, **20**, 252–9.

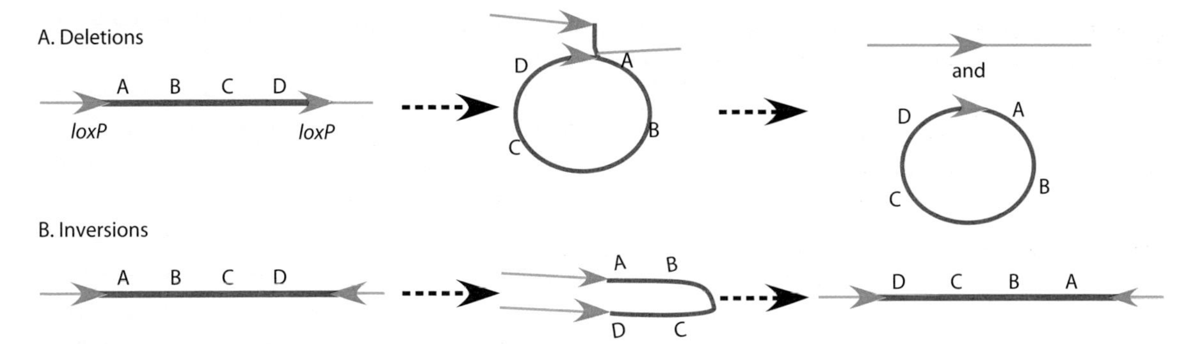

Figure B3.2 Targeted recombination and inversions in mice. Inversions that can be used as balancer chromosomes in mice are produced using the site-specific recombination of the Cre–*lox* system. The Cre enzyme catalyzes recombination at the specific sites known as *loxP*, shown here as blue arrowheads. The direction of the arrowhead indicates the orientation of the *loxP* site, which determines the type of chromosome rearrangement that is produced. In panel A, the *loxP* sites have the same polarity so that a crossover between them deletes the genes between the sites. In panel B, the *loxP* sites have opposite polarity. When these pair and recombine, the sequences between them are inverted. More information about the use of Cre–*lox* in mice is given in Chapter 6.

What if the mutation affects multiple different phenotypes?

Many mutations affect biological processes other than the one being studied. For example, many of the eye color mutations in *Drosophila* also affect the pigmentation of the Malpighian tubules, a separate organ involved in excretion. This effect, in which a mutation affects several different tissues, is, known as **pleiotropy**. Pleiotropy often arises because the same gene products are used in two or more different tissues, for example in both the eye and the Malpighian tubules. The phenotype reveals an aspect of underlying physiology that may not have been anticipated in the original conception of the genetic screen, one of the advantages of this type of genetic analysis.

Although it often reveals interesting phenotypic relationships, pleiotropy can also present some less interesting alternatives. Pleiotropy sometimes reveals trivial relationships between phenotypes. For example, mutations that cause paralysis may interfere with eating simply because the animal cannot move to find food. The gene itself does not affect eating or mouth parts or digestion; starvation is an indirect pleiotropic effect of the inability to move.

Often the pleiotropic relationship is more subtle than expected—both white eyes and yellow body in *Drosophila* have pleiotropic effects on male fertility, in part because female flies preferentially mate with wild-type males. A pleiotropic effect can sometimes obscure another interesting and unexpected effect of the gene. For example, a mutation affecting sex determination in humans was nearly overlooked because of its effects on the developing kidney, a serious medical issue in newborns. Subject to these few cautions, pleiotropic mutations at their most interesting can provide subtle and unsuspected insights into development or physiology. Because few genes work in only one tissue or one developmental process, pleiotropic phenotypes are the rule rather than the exception.

What if the mutant phenotype is only seen in some of the homozygous mutant individuals?

The best-studied mutations are those in which every homozygote exhibits a mutant phenotype. However, there are many other examples in which some mutant homozygotes have a wild-type phenotype; in these cases, the mutant phenotype is observed by examining a population of mutants rather than a single individual. For example, certain recessive mutations in *C. elegans* result in the formation of blisters on the cuticle. This blistered phenotype is illustrated in Figure 3.18, in which all the worms are homozygous for the mutation. A homozygous blistered hermaphrodite will have offspring that do not show the blistered phenotype, although they are homozygous for the mutation. When one of these non-blistered hermaphrodites is allowed to self-fertilize, a fraction of its offspring will have the blistered phenotype, although all of them are homozygous for the mutation. The effect is often expressed quantitatively: if 30 percent of the homozygotes are found to be blistered, the mutation is said to be 30 percent **penetrant**. It can also be expressed qualitatively as mutations that exhibit incomplete penetrance.

The molecular basis for mutations that are less than 100 percent penetrant is not well understood. In some cases, it may be that the mutant allele does not completely eliminate the function of the gene and the residual function is sufficient to allow some individuals to execute the function of the gene normally. Mutations that reduce but do not eliminate the function of the gene, known as hypomorphic mutations, are described in Chapter 4. In these cases, the reduced penetrance of certain alleles may be indicative of slight stochastic variations in gene expression. A threshold level of gene activity may be necessary for wild-type function. Those individuals that have, by chance, slightly more activity than the functional threshold will have a wild-type phenotype, whereas those that have slightly less than the functional threshold have a mutant phenotype. For some other genes, reduced penetrance may reflect a complex interaction with other genes or some aspect of the environment. Such an explanation could be advanced when even null alleles of the gene (as defined in Chapter 4) are less than 100 percent penetrant. Mutant phenotypes with reduced penetrance are discussed in more detail in Chapters 10 and 12, although it should be emphasized that molecular explanations are elusive.

Incomplete penetrance can be seen with either recessive or dominant mutations although, for unknown reasons, it appears to be somewhat more common for dominant mutations. Tourette's syndrome in humans

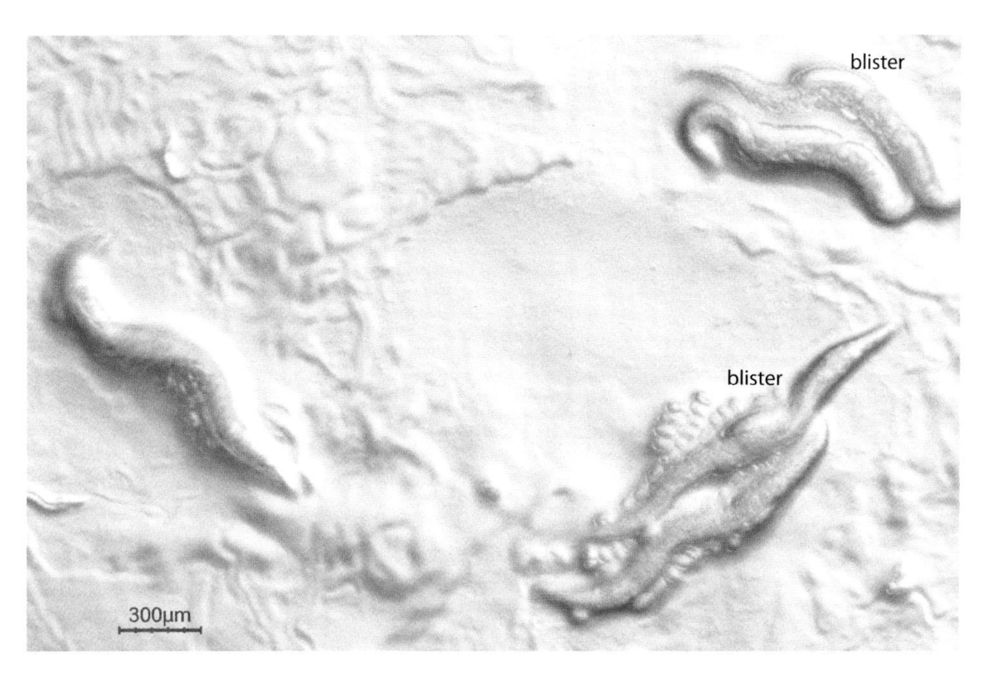

Figure 3.18 Penetrance and expressivity. All of these worms are homozygous for a recessive mutation that results in blisters forming on the cuticle. Several worms with blisters are indicated. Although all worms are homozygous mutants, not all of them have blisters and several worms without blisters can be seen. In other words, the mutation is not 100 percent penetrant. In addition, not all of the worms have the same number of blisters or have blisters in the same location on the cuticle. This means that the mutation has variable expressivity. Nearly all mutations have some degree of variable expressivity.

is often cited as an example of incomplete penetrance. Tourette's syndrome is a neurological disorder characterized by repetitive and involuntary tics, such as rapid eye blinks, grimances, and vocalizations. The symptoms typically appear in early adolescence and often disappear by early adulthood. Although environmental influences are clearly important in the exact symptoms (e.g. the disorder is much more common in males than in females), the syndrome is inherited. Pedigrees from many families are consistent with an autosomal dominant trait, i.e. since only one Tourette's 'allele' is needed to show the disorder, one parent with Tourette's will often have one or more children who also have the syndrome, regardless of the genotype and phenotype of the other parent. The disorder also exhibits incomplete penetrance, so that an affected boy may have an unaffected father but have paternal uncles who were affected. Based on this inheritance pattern, one explanation is that the father carried the allele for the disorder but did not exhibit symptoms; the disorder has incomplete penetrance.

It is also well known that many, possibly all, mutations have a somewhat variable mutant phenotype; that is, even though the organism is mutant, not all the mutants with the same genotype look exactly the same. A population of mutants is often more variable in phenotype than a population of wild-types. This is the phenomenon of **variable expressivity**. If we return to our example of mutant worms with blistered cuticles in Figure 3.18, not all of the blistered mutants have the same number of blisters or have the blisters in the same location. Similarly, those individuals with Tourette's syndrome have a range of symptoms. Variable expressivity is rather simply explained by realizing that mutations cause the reduction, loss, or change of some biological function. Just as each unhappy family is unhappy in its own way, each mutant organism will reflect the loss or change of function in its own way. Variable expressivity is discussed in more depth in Chapter 9 in the consideration of gene activity, but a detailed explanation for any particular effect is often lacking.

What if no mutant phenotype is observed?

One of the most striking and humbling aspects to arise from genome-wide mutant screens (described in

Chapter 7) is that many genes appear to have no mutant phenotype. Thus, conventional genetic screens would not have detected the presence of these genes. These mutations with no mutant phenotype are one of the main topics discussed in Chapters 10, 11, and 12, so we will defer the discussion of this effect until those chapters. As discussed in those chapters, such mutations are currently a very active field of research. Part of the recent interest in these mutations is precisely the recognition that some key biological processes could have been missed by standard genetic analysis with single-gene mutations, as described in this chapter. If there is no mutant phenotype, the investigator does not realize what he/she has missed.

SUMMARY

Mutant hunts are a standard tool in genetic analysis

These few examples present only some of the challenges encountered with nearly any scheme to find new mutations. Every mutant hunt has its own peculiar features, and it is often those very features that characterize the cleverest and most successful procedures. These examples present some of the common elements of mutant hunts, but in the end the geneticist will have to let the organism show the direction and even the solution to the challenges. Finding a mutant may be easier said than done, but it remains one of the basic tools of genetic analysis. However, this is only the beginning. Once the mutants have been found, it is necessary to assign them to genes and to determine the nature of the defect. These are the topics for the next chapter.

Chapter Capsule

Finding a mutant

Genetic analysis often begins by finding a mutant phenotype. Some of the key questions in finding mutants for genetic analysis can be summarized as follows.

Which mutagen will I use?

- Chemical mutagens often affect only a single gene or even a single base.
- Radiation usually produces chromosome rearrangements.
- Transposable elements and other insertional mutagens affect single genes and make molecular analysis easier.

Which procedure will I use to search for mutants?

- Screens involve looking through populations of mutagenized individuals to find the mutants.
- Selections provide the means to search for mutants among large populations.
- Lethal mutations can sometimes be maintained as homozygotes under particular environmental conditions, such as low temperature. Lethal mutations can also be kept as stable heterozygotes with balancer chromosomes.

CHAPTER MAP

A flowchart of genetic mutant analysis. The process begins by finding mutants, as described in Chapter 3. The genes responsible for the mutant phenotype are mapped (on the left) and tested against each other for complementation (on the right). As genes are identified and mapped, the types of alleles are classified for the phenotypic effect. Ultimately, the gene is cloned, as described in Chapter 5. Mapping, complementation, and classification are interconnected processes, with the results of each experiment affecting the order of the next steps.

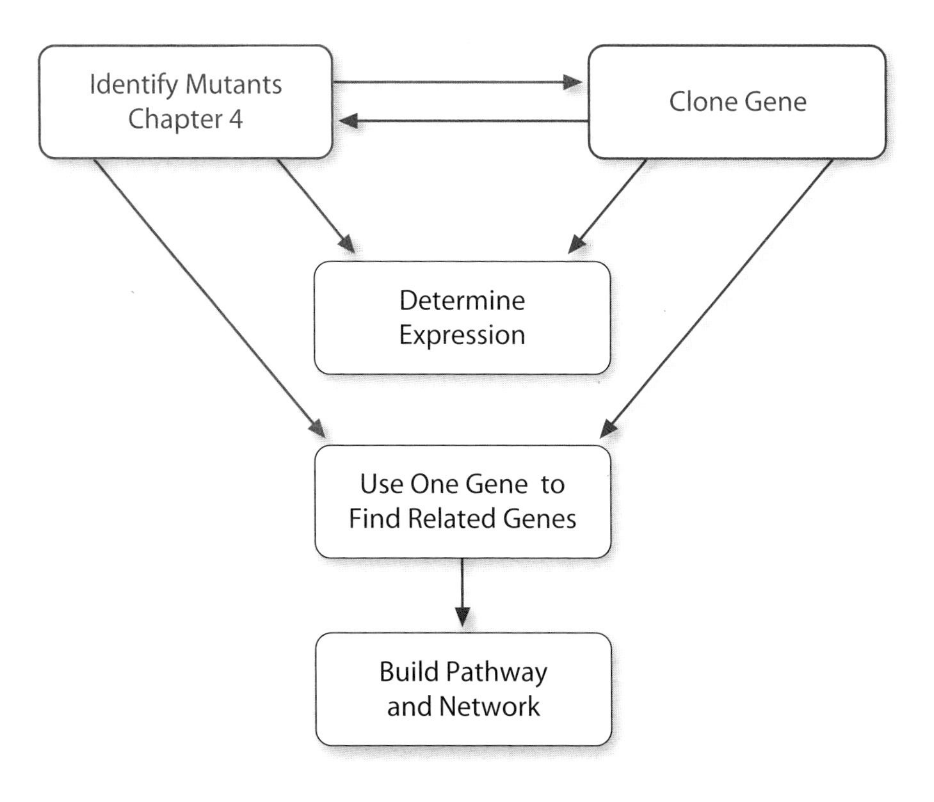

CHAPTER 4
Classifying mutants

TOPIC SUMMARY

Once mutants are found, the next steps in genetic analysis include:

- Assigning the mutations into genes using complementation tests.
- Mapping the locations of the genes.
- Classifying the type of defect in each mutation based on its effect on the function. While most mutations are recessive and cause a loss or a reduction in the function, some interesting mutations are dominant and arise from other alterations in a gene. These include over-production of the normal gene product and expression in the wrong tissue. Other dominant mutations might prevent the normal products from functioning.

These processes of classification do not usually occur sequentially because results obtained from one set of experiments inform the next steps in the characterization.

IN BRIEF

Mutations are mapped and assigned to genes, and the nature of the functional defect is determined.

INTRODUCTION

A large collection of mutants is an appropriate starting point for the genetic analysis of many biological processes. However, this is only the beginning of genetic analysis because the mutants must also be classified. For example, which of the mutations are different alleles of the same gene, and thus affect the same function? Which, if any, of the mutations eliminate the function of the gene, and which reduce the function but do not eliminate it? Do any of the mutations alter the function more subtly or even over-produce the normal function?

Sorting through the many mutations that arise from a thorough mutagenesis scheme takes time and judgment. Nearly all geneticists have collections of mutants that have never been studied thoroughly, possibly because they are similar in phenotype to other mutations in the same gene. In this chapter we describe how a geneticist assigns mutations into genes and maps the genes to chromosomes. We also describe how geneticists analyze the effect of a particular mutation on the function of a gene.

4.1 Classifying mutants: an overview

While it is intriguing and even exciting to find a mutant disrupted in a particular biological process, the next critical step is to identify which gene has been mutated. Only when the mutated gene has been identified is the investigator able to begin analysis of the function of the normal gene. That is, the investigator first has to find the mutant, as described in the preceding chapter. The next steps are to identify the gene that is affected by that mutation and determine how the gene has been affected by the mutation. This strategy for classifying mutations is presented in the flowchart at the beginning of this chapter. All of the steps in characterizing a mutant depend on the results from other steps.

Identifying which gene has been mutated typically involves a procedure of **mapping** the mutation to a particular location on a chromosome. The mutation is compared with other mutations that are similar in phenotype and map position to determine whether these mutations are alleles of the same gene, a process known as **complementation testing**. Often, as suggested in the flowchart, mapping and complementation are done as a coordinated effort so that only one allele of a gene has to be mapped and only mutations that map to the same location are tested for complementation.

Having a collection of different mutations that define the gene, the investigator needs to determine what each mutation has done to the normal function. That is, has the function been completely eliminated or only reduced? Most mutations are recessive to wild-type and have a loss of gene function, either total or partial. Alternatively, does the mutation result in over-production of the gene product or in a gene product that interferes with the normal function? This is typically, although not always, what happens in the unusual mutant alleles that are dominant to wild-type; these are known as gain-of-function mutants. Part of the process of classifying the type of mutant will almost certainly occur during the processes of mapping and complementation testing. For example, determining whether a mutation is recessive or dominant is fundamental to every other step in the classification processes. Usually the investigator finds out whether the mutation is recessive or dominant during the mutant isolation process described in Chapter 3. If this is not determined during isolation, it will almost certainly be done in the initial matings performed for mapping.

Having a well-mapped gene with many mutant alleles whose functional changes are known sets the stage for most of the analysis that follows in the remainder of this book. For example, having these mutant alleles and the map positions is one strategy that the investigator can use to clone the gene and determine its normal molecular function (described in Chapter 5). Well-characterized mutant alleles can be used for examining the activity of the gene and inferring the time and tissues in which it is active (described in Chapter 9). Mutant alleles for one gene also provide one resource needed to find additional genes that affect the same biological process (Chapter 10) and for determining how the activities of these genes affect and interact with one another (Chapters 11 and 12). A collection of well-characterized mutant alleles is the most fundamental tool in genetic analysis.

4.2 Mapping mutants to genetic locations

A mutant can be mapped using recombination

After the schemes in Chapter 3, the investigator may have many different mutants with similar or related phenotypes. These mutants have to be assigned to genes to determine which ones affect the same function and which ones affect different but potentially related functions.

Each gene lies at a particular position on a chromosome, known as a **genetic locus**. The most widely used procedure for assigning a mutation to a gene is to assign it to a genetic locus, i.e. to **map** the mutation to a location on a chromosome using segregation and recombination. The distinction between a gene and a genetic locus is often semantic rather than biological because genes are so closely identified with their location in the genome. Mapping mutations involves basic principles of Mendelian genetics, and a general procedure is summarized in Figure 4.1.

Suppose that the newly identified mutation *m* is being mapped with respect to a known genetic marker *a*; in this example, both *m* and *a* are recessive to wild-type, alive, and fertile. The map position of *a* is known, and in fact many different genetic markers will be used in the course of this experiment. Thus, our notation *a* refers to both an individual genetic variant at a particular known location and genetic variants in general whose known locations can be used in mapping. The marker *a* need not be a gene, and in fact, in most contemporary mapping, the genetic markers are molecular variants such as single-nucleotide polymorphisms or sequence-length polymorphisms rather than phenotypic variants such as white eyes or uncoordinated movement. The logic of the procedure is essentially the same no matter what type of variant is used for a mapping marker; the only differences are the methods used to assess the genotype.

The *m/m* mutant is mated to the *a/a* marked strain to make F_1 heterozygotes of genotype *a*/+;*m*/+. Since both *a* and *m* are recessive in our example, the F_1 heterozygotes are wild-type in phenotype. The F_1 heterozygotes are mated to produce an F_2 generation. If *a* and *m* assort independently (as did Mendel's markers), then 1/16th of the offspring will be *a/a;m/m* double mutants. If *a* and *m* are linked, by definition they will not segregate independently and the fraction of *a/a;m/m* double-mutant offspring will be less than 1/16—more precisely, the fraction will be $p^2/4$, where *p* is the probability of a crossover between *a* and *m*. (To anticipate a potential

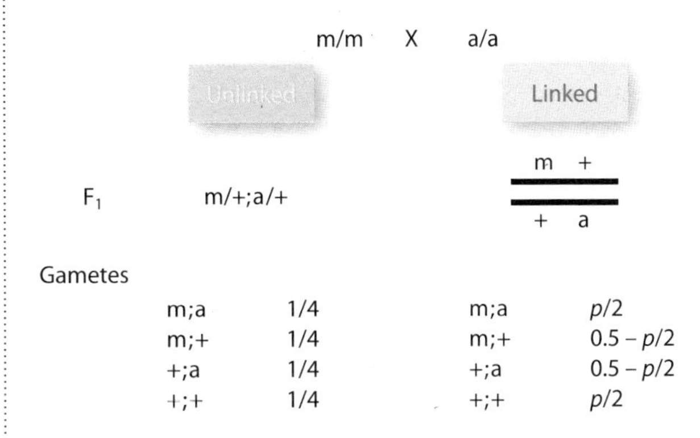

Figure 4.1 Distinguishing unlinked from linked mutations. In this diagram, *m* represents a recessive mutant allele to be characterized and *a* is a recessive marker variant with a known map location. The mutation and the marker are crossed to make an F_1 heterozygote. If *m* is not linked to *a*, as shown on the left, the four types of gametes will occur with equal frequency. If *m* is linked to *a* by a map distance *p*, the two recombinant gametes *m;a* and +;+ will be less frequent than the two parental gametes. These gamete frequencies can be determined by a test cross or an F_2 ratio.

question about the number 4 in the denominator, if the probability of a crossover between *a* and *m* is denoted by p, then only $p/2$ gametes will be *a;m* double mutant; a similar fraction of $p/2$ will be +;+. In a diploid organism, the probability that *both* gametes are recombinant is $p/2 \times p/2$.)

This method of determining linkage requires that the investigator is able to determine the difference between 1/16th and less than 1/16th of the offspring, which requires examining enough F_2 offspring to have statistical significance. Mapping using the ratio of progeny in the F_2 generation as we just described is easy to do in self-fertilizing organisms such as *C. elegans* and *Arabidopsis*. To make linkage easier to detect, a test cross of *a*/+;*m*/+ to the *a*/*a*;*m*/*m* double mutant can also be used. In this case, a quarter of the offspring are expected to be double mutants if the two genes are unlinked, and less than a quarter if the genes are linked. Test crosses are frequently used because fewer offspring are needed to determine the difference between linked and unlinked genes if a quarter are expected to be double mutants. Alternatively, the investigator could examine only the *a*/*a* offspring and determine whether a quarter of them are also *a*/*a*;*m*/*m*. Although the fraction of double-mutant offspring will differ depending on the exact cross used, the principle of random segregation of unlinked genes is the same as inferred by Mendel.

If enough markers with known genetic locations are tested, linkage to a chromosome or a chromosomal region can be accurately determined. A mutant is usually mapped to a genetic interval between two markers with information about the relative distance from the mutant to each flanking marker. The challenge in these experiments is to find enough markers with known map locations and to have the patience to test them all. Typically this is simplified by using a mapping strain that has markers on many different chromosomes so that numerous genetic regions can be tested simultaneously. Because of their frequency and distribution in the genome, molecular polymorphisms are the linkage markers of choice for nearly all model organisms. We return to the use of molecular polymorphisms in Case Study 5.1 in Chapter 5 with the cloning of the cystic fibrosis gene in humans. We also discuss the process of association mapping in Chapter 5 (Text Box 5.1). Although association mapping is a recombination-based procedure to locate a gene within a genetic interval, it is closely associated with attempting to clone the gene; therefore, we have covered the procedure in that context.

Mapping genes in humans presents significant challenges for the general scheme outlined above. Not least of these are that crosses are not controlled (fortunately) and the number of offspring is very small. Other methods for mapping genes in humans are described in Text Box 4.1.

Our example with the hypothetical mutation *m* assumed that no prior linkage information was known about the newly arisen mutation. In the preceding chapter, we described balancer chromosomes as a means of maintaining mutations as heterozygotes. Balancer chromosomes are also useful for determining a map position. If some system of heterozygous balancers is used in the original mutagenesis scheme, the newly arisen mutation is already assigned to a chromosome or a chromosomal region. It will then need to be mapped more precisely within that region, which can be done using recombination with genetic markers. Again, with enough markers and patience, the map location can often be determined to less than a map unit.

Recombination also locates genes with respect to one another

Mapping a gene to a region of less than a map unit in size is an estimable achievement, but we will point out in Chapter 5 that this is unlikely to identify one specific mutated gene at that site in the genome. A map unit on the genetic map corresponds to hundreds of thousands or even several million base pairs of DNA sequence; an average gene in an invertebrate or *Arabidopsis* has a size in the order of thousands or possibly tens of thousands of base pairs. Therefore, many, many genes might be found in the same one map unit genetic interval on the chromosome.

It is often helpful or even necessary to place the new mutant with respect to other genes in the same region, even if the exact recombination distance between two genes is not known. For positional cloning, as described in Chapter 5, the investigator wants to work with the smallest feasible genetic interval, with markers close to the mutant on each side. One method of determining the location of one gene with respect to two known

BOX 4.1 Mapping genes in humans

Almost as soon as genes began to be mapped to chromosomes in *Drosophila*, similar attempts were made with humans. Genetic mapping in humans faces a number of particular challenges, including the long generation time and the small number of offspring. Furthermore, humans have 24 linkage groups (including the Y chromosome), many more than is found in most model organisms. Even if two genes were shown to be linked by pedigree analysis, it was difficult to assign them to a chromosome. In addition, much of the most familiar genetic variability that is observed in humans arises from multifactorial traits such as height and blood pressure rather than from single-gene traits; other than blood groups, most of the single-gene traits are relatively rare genetic diseases. Geneticists attempting to map a human trait could (and sometimes do) assign the trait to different map positions because the trait has an unexpectedly complex inheritance.

Despite these difficulties, many genes have been placed on the human genetic map using recombination and segregation data. Among the earliest ones to be mapped were X-linked traits such as red–green color-blindness and hemophilia, but many autosomal linkage arrangements between blood antigens and disease traits were also uncovered. The impetus for these studies is the hope that one can understand the biochemical and molecular basis for a human disease, leading to improved detection and treatments. Recombination and segregation analysis in humans has been revolutionized by the use of molecular polymorphisms, including restriction fragment length polymorphisms (RFLPs) and single-nucleotide polymorphisms (SNPs). The use of these markers is described in Chapter 5 in the context of using its map position to clone a gene, a process referred to as **positional cloning**. Molecular polymorphisms as mapping markers also allowed the development of sophisticated methods for **association mapping**, as described in more detail in Chapter 5. Association mapping is a procedure based on recombination and segregation in populations rather than individual families, and is now the most widely used method to map genetic variants in humans. With greatly improved genetic maps, mapping genes by recombination in humans has been simplified, although it will never be as easy as mapping a gene in *Drosophila* or *C. elegans*.

Recombination and segregation analyses to map a gene rely on detectable phenotypic differences between alleles to recognize particular genotypes. Other methods have also been used to map genes in humans, taking advantage of other properties of genes. In this text box, we briefly describe a few other methods that are used for mapping genes in humans. These methods are also used in other organisms, but it makes sense to describe their application to human genetics.

Somatic cell hybrids map genes based on expression

In the late 1960s and early 1970s, the ability to grow human cells, particularly fibroblasts, in culture was well established and stable cell lines had been produced for humans and many other mammals. It was also recognized that cells of two different mammals (humans and a rodent such as the Chinese hamster) can be fused in culture by a variety of different methods. The resultant cell, called a **heterokaryon**, initially has two nuclei with the complete chromosomal complement of each species. However, human chromosomes are lost with each cell division. The reason for the preferential loss of human chromosomes is not entirely clear, although the slower cell cycle in humans could be a contributing factor. The loss is not completely random since some human chromosomes are lost or retained more often than others, but different cell lines retain different human chromosomes.

Using these properties, new somatic cell lines were established that had the complete set of rodent chromosomes and a subset of human chromosomes. For each cell line, the retained human chromosomes were identified by cytological banding. With a bit of patience and collaboration, human genetics laboratories could have a bank of cultured cell lines that would allow them to assign a human gene to its appropriate chromosome. Many such collections of cell lines are in existence.

The mapping procedure relies on gene expression in these cell lines. An example is shown in Text Box Figure 4.1. Suppose that we wanted to map the gene encoding the enzyme acid β-glucosidase (GBA) because mutations in this gene result in Gaucher disease. Based on pedigree analysis, the gene was known to be autosomal. Various cell lines could be assayed for the presence of GBA enzyme using enzyme assays, gel electrophoresis, antibodies that react with the enzyme, or any of several other methods. All cell lines will have the rodent GBA since they have the complete set of rodent chromosomes. However, the human version of the enzyme may have a slightly different K_D, molecular weight, or other property that allows the geneticist to distinguish the human version of the enzyme from the rodent version. As different cell lines are tested for the presence of human GBA, a correlation could be established between the presence of one human chromosome and the presence of human GBA. By correlations of this type, the human gene was placed on chromosome 1. This use of gene expression

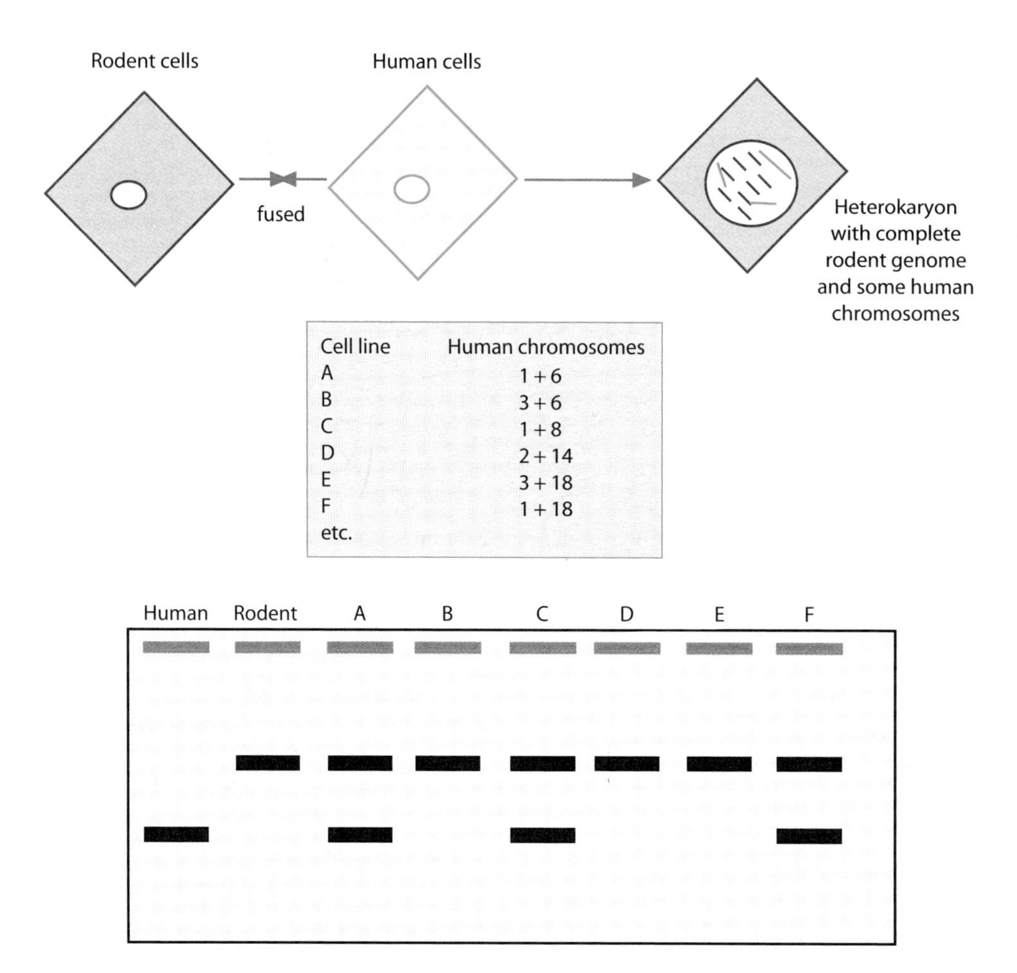

Figure B4.1 Mapping human genes by somatic cell hybridization. Human cells in culture are fused with cells from a rodent such as a mouse or Chinese hamster. The nuclei fuse to form a heterokaryon and stable cell line with a complete set of rodent chromosomes and some human chromosomes. Each cell line is assayed for the production of the human enzyme. In the hypothetical example shown here, the human and mouse enzymes differ in their electrophoretic mobility, and six cell lines are assayed to show the correlation between human chromosome 1 and the human enzyme. For GBA, 52 cell lines were tested by an antibody precipitation assay.

differences between humans and rodents has been an exceptionally good method of assigning human genes to their chromosome. The combination of this method with the linkage arrangements observed by recombination and segregation (many of which could be observed but not assigned to a chromosome) led to a much more complete human genetic map in the early 1980s.

As described here, somatic cell hybrids would assign a gene to a chromosome but may not be able to assign it to a particular locus. It was a logically straightforward step to induce or employ segmental aneuploids including deletions, duplications, and translocations, as are used for mapping in *Drosophila* and other organisms. In these methods, the human chromosomes are fragmented, often by irradiation but other methods have also been used, so that the rodent–human hybrid cell line now has only particular segments of the human karyotype rather than entire chromosomes. Otherwise, the procedure is similar to what is described in the preceding paragraphs, although now the location is limited to a smaller region of the chromosome rather than the entire chromosome. By working back and forth between the locations obtained by hybrids and segregation data, particularly from molecular polymorphisms, the human genetic map was greatly refined even as the human genome sequencing project was beginning.

Fluorescence *in situ* hybridization maps genes based on DNA sequence

Although these procedures were developed somewhat later than the somatic cell mapping methods, hybridization methods are also widely used in concert with somatic cell lines. Fluorescence *in situ* hybridization (FISH) avoids the potential complication of finding an expression difference between human and rodent proteins, and maps a gene using a cloned DNA sequence itself. Thus, FISH can be used if a gene has been cloned but no assay has been developed for its gene product. The DNA sequence to be mapped, which need not encode a protein, is copied *in vitro* using a fluorescently labeled nucleotide. Many different fluorescent labels with different emission spectra have been developed for this purpose. The fluorescently labeled DNA sequence makes the probe for hybridization, and probes of about 1000 bp can be routinely used. Metaphase chromosomes are fixed, much as is done for chromosome banding, and the fluorescent probe is hybridized to the spread chromosomes. The location of the hybridization signal establishes the physical position of the DNA probe in the genome.

FISH is a very widely used method for many procedures in addition to mapping. With two- or three-color FISH, several probes (each with a different label) can be placed with respect to each other. As more sequences are used for FISH probes, the probes themselves are useful for recognizing complex chromosome rearrangements, and many other applications. One visually striking technique is termed multicolor FISH or **chromosome painting** (Text Box Figure 4.2). In this method, all of the probes that come from one chromosome are labeled with the same dye, and probes from different chromosomes are labeled with different dyes or combinations of dyes. When hybridized to a chromosome spread, each chromosome is 'painted' with probes of the same color. Such paints are particularly helpful for examining tumor cells. Many tumor cells are aneuploid, and most have complex chromosome rearrangements, such as deletions, duplications, and especially translocations. These can often involve more than two chromosomes, and the chromosomes may be barely recognizable by normal banding patterns. Chromosome painting can be used to characterize these rearrangements in more detail. This can be important in determining whether a certain breakpoint is often associated with the same type of cancer. It can also be used to determine whether chemotherapy or radiation therapy has eliminated all of the tumor cells.

Both of these methods demonstrated and relied on the conservation of linkage arrangements among mammals, known as **synteny**. As a result of evolution, genes that are genetically linked in one organism are very likely to be located close together in a related organism. Thus, if two genes were shown to be closely linked in one mammal, it was reasonable to postulate that they were also linked in other mammals, which could then be demonstrated (or not) by FISH or other methods. Synteny also made the development of the human genetic map more rapid.

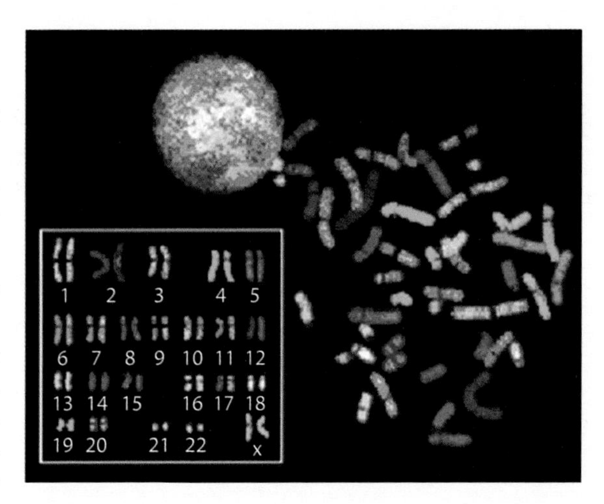

Figure B4.2 Fluorescence *in situ* hybridization (FISH) is used for whole chromosome painting. As described in the text box, a series of probes that are specific to a chromosome are fluorescently labeled. Each chromosome is given a different combination of fluorescent labels so that each chromosome is a different color. Although one- or two-color FISH is used for mapping, chromosome painting is applicable to characterizing complex chromosome rearrangements, syntenic relationships between species, and other events involving multiple chromosomes.

As more genomes are sequenced, the conserved linkage of genes allows syntenic blocks to be constructed, and the complex rearrangements that occurred during evolution to be inferred. Chromosome painting has been used to find syntenic blocks between related species. Thousands of syntenic regions have been found between the human and mouse genomes, some of them as short as a few million base pairs. The largest syntenic block is the mammalian X chromosome, recognized by Ohno in the 1960s as being largely conserved among placental mammals. Even among these syntenic regions, small rearrangements such as inversions, duplications, and deletions are observed. A chromosome map, like the organism itself, is the product of complex evolutionary forces, and it bears the marks of its history.

Physical maps and recombination maps

In this text box we have focused on two methods for mapping genes in humans. Both of these methods result in a physical map, i.e. they demonstrate the location of the gene on the chromosome but they do not show the recombination distance between the genes. Most model organisms have both a physical map arising from methods such as these and from genome projects, and a genetic map arising from recombination data. The relationship between physical and genetic maps is interesting and complex. A quick summary is that, whereas the order of genes is necessarily the same, the observed 'distance' between them may be quite different. Two genes that are 5 million bp apart (a physical distance) may be less than one map unit or more than 20 map units apart (a genetic distance), depending on the organism and the location in the genome.

We are familiar with different ways of measuring distance in our everyday world. If asked, 'How far is it from town A to town B?', we may answer 'Thirty miles'—a physical distance. Or we may answer in terms of driving distance, recognizing that driving 30 miles may take as little as 30 minutes or as much as several hours, depending on the location and the road conditions. Each of these is a useful measure of the distance but they are useful in different ways. Similarly, both physical and genetic maps provide useful information about gene locations, but they provide different information. A physical map provides the molecular position of the gene on a chromosome or DNA molecule. A genetic map measures the probability that the linkage arrangement between alleles at two genes will be inherited together.

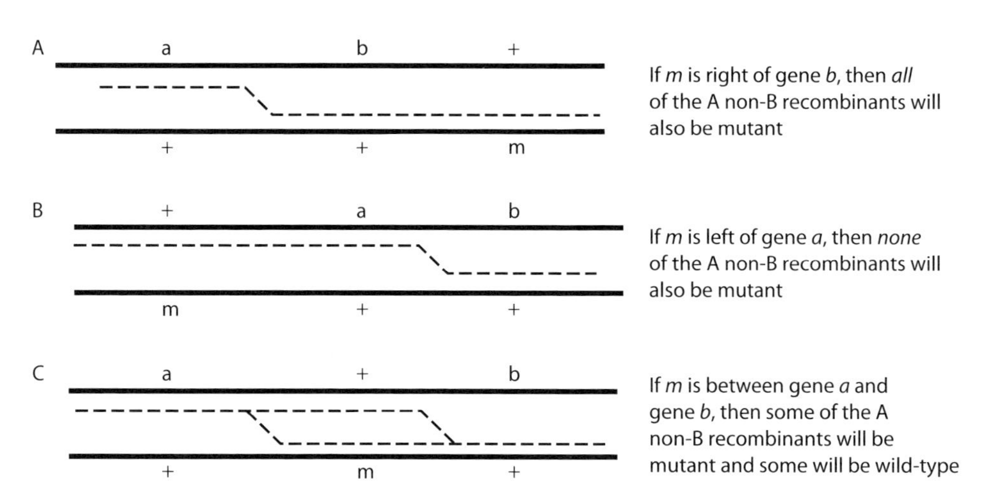

Figure 4.2 Determining the position of a mutation with respect to other linked genes. In this diagram, *m* represents a recessive mutant allele to be mapped, and *a* and *b* are the recessive markers at two locations with known map positions. The *m* allele has been shown to be linked to the *a* marker by a procedure similar to that shown in Figure 4.1, but its position with respect to the *a* and the *b* markers is not known. The three possible locations for *m* are shown; it could be to the right of the *b* marker (panel A), to the left of the *a* marker (panel B), or between the two markers (panel C). Crossovers between the *a* and *b* markers are identified as A non-B recombinants, and analyzed for the presence of the *m* allele, ignoring the possibility of rare double crossovers. If *m* is to the right of marker *b*, as in panel A, all the recombinants between *a* and *b* will also have the *m* allele, whereas if *m* is to the left of the *a* marker, as in panel B, none of the recombinants between *a* and *b* will have the *m* allele. If *m* is between the two markers, some recombinants will have the *m* allele and some will not, as shown in panel C.

genetic markers is shown in Figure 4.2. In this example, the two markers are the familiar variants *a* and *b*, and *m* is the mutant to be located. The locations of the markers *a* and *b* are known, as is the distance between them. A recombination event is detected between *a* and *b* (which is easy if some form of selection can be used so that only recombinants are examined), and the recombinants are then tested for the presence or absence of the new mutant *m*. Even if only a few recombinants between the markers *a* and *b* are detected, the position of *m* with respect to *a* and *b* can usually be determined. The procedure that locates the mutant *m* with respect to the markers in this way can be repeated with other markers.

Although mapping of this type may not determine the exact location of a gene, the relative position or order with respect to other markers can be useful. We illustrate this with an analogy. Suppose that you ask friends for directions to a restaurant. If your friends had the information, they could give you the precise GPS coordinates; likewise, with genome projects (as described in subsequent chapters), a gene can be positioned to the exact base pair coordinates once you know where it is. However, your friends are more likely to give you the relative position of the restaurant with respect to other landmarks: 'The restaurant is past the gas station but before the shoe store.' With mapping genes, as with directions to restaurants, a relative location may be enough to help you find your way. If a gene can be positioned between two other landmarks whose location is known, it can then be located more precisely by other means. We return to this topic in Chapter 5 when we discuss ways that an investigator can use a map position to clone a gene.

A mutation can be mapped using deletions and duplications

When a mutant has been mapped to a particular chromosomal region, it is sometimes easier to determine its more precise position using chromosome rearrangements such as deletions and duplications, at least in organisms such as *Drosophila* and *C. elegans* where many deletions and duplications are known. (Many more duplications and deletions have been found and analyzed in *Drosophila* than in any other model genetic organism, an advantage arising from a century of genetic analysis.)

Figure 4.3 summarizes the structures of deletion and duplication chromosomes schematically. A **deletion chromosome**, also known as a **deficiency chromosome**, is missing all of the genes in a contiguous piece of the chromosome. It is often drawn as a chromosome with a gap or a hole in it, although in reality the two endpoints are fused together to make a new and shorter chromosome. A **duplication chromosome** has an additional fragment of the chromosome elsewhere in the genome. Deletion chromosomes almost always have to be maintained as a heterozygote (often with a balancer chromosome as described in Chapter 3) because homozygous deletions that affect more than a few genes are nearly always lethal. As a heterozygote, the deletion strain will have one copy of every gene that is missing from the deleted region, the copy on the other homologous chromosome.

A **duplication strain** will have three (or possibly four) copies of every gene present within the duplicated region; two of these copies are the ones present on the normal chromosomes, and the additional copy (or copies) is present on the duplication. When compared with deletions, duplications are nearly always alive and fertile as heterozygotes (i.e. with three total copies of the region) and often alive as homozygotes (with four total copies of the region) since, to a cell, having too much of a gene and its product is typically less deleterious than having too little. Duplications may have a mutant phenotype but, if so, it is often not very severe, such as slower growth or reduced fertility. Both deletions and duplications are called **segmental aneuploids**,

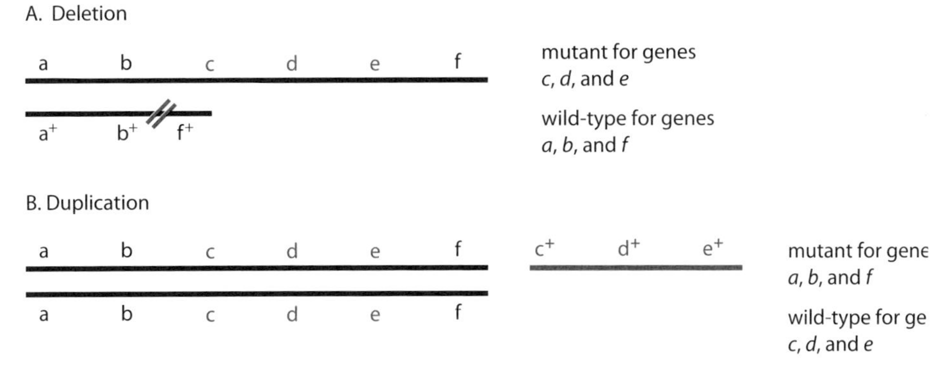

Figure 4.3 The structures of deletions and duplications. The deletion chromosome shown in panel A lacks the genes (or regions) *c*, *d*, and *e*, as shown in red, and can be considered as having a null mutant allele for these three genes. The diploid shown will exhibit a mutant phenotype for genes *c*, *d*, and *e*, but will have a wild-type phenotype for genes *a*, *b*, and *f*, which are heterozygous. The duplication shown in panel B has wild-type alleles for genes *c*, *d*, and *e*. The diploid shown will have a mutant phenotype for the genes *a*, *b*, and *f*, which are homozygous for the recessive mutant alleles, but will have the wild-type phenotype for genes *c*, *d*, and *e*. The duplication is shown as free and unattached, but it may be attached to one of these chromosomes or to another chromosome. Deletions and duplications have one or three copies of these genes, a genotype known as segmental aneuploidy.

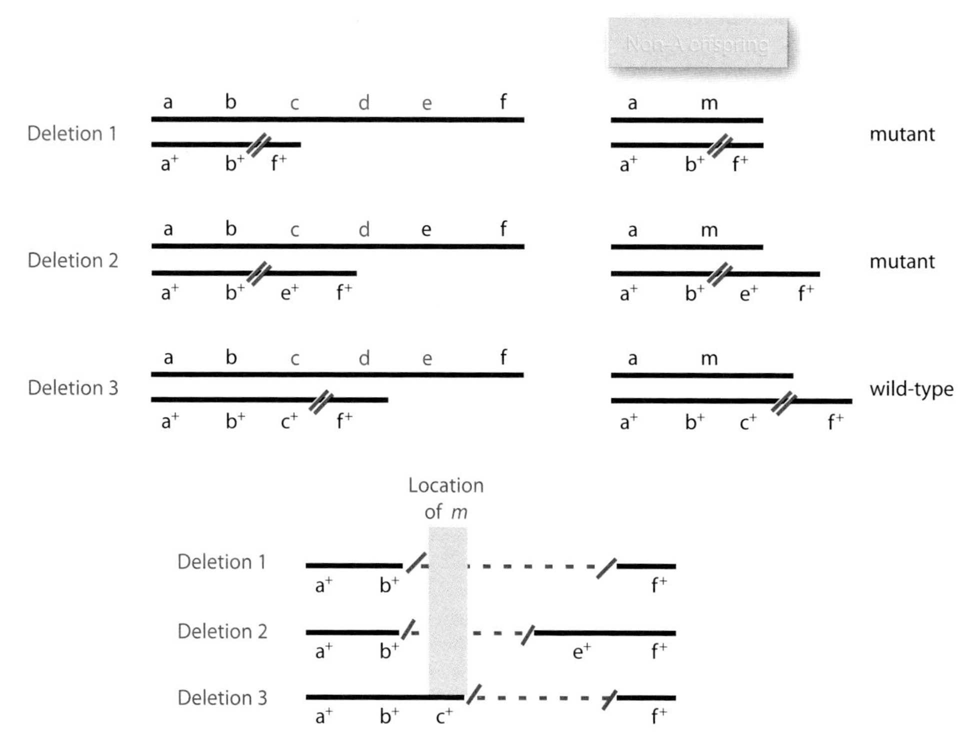

Figure 4.4 Mapping with deletions. The mutation *m* is mapped using three different deletions, referred to as deletions 1, 2, and 3. The deletions remove different but overlapping segments of the chromosome. The mutation is linked to a marker known not to be deleted by the deletion, in this case the marker allele *a*. A deletion strain, typically maintained as a heterozygote as shown here, is mated to the marker mutant strain and the F_1 progeny are examined. As shown here, only the non-A progeny are examined since these are the ones with the deletion. In the results illustrated, the mutant *m* is deleted by deletions 1 and 2 but not by deletion 3. This places the *m* mutation in the region shown by the pink box, approximately the same region as the *c* gene. A complementation test would be used to determine whether the mutation *m* is an allele of the *c* gene.

being monosomic (having one copy) or trisomic (having three copies) for a particular defined segment of a chromosome.

Segmental aneuploids provide a convenient means of mapping a gene within the duplicated or deleted chromosomal segment. Segmental aneuploids for many regions of the genome are available from the genetic stock center for each model organism. The procedure for mapping using a series of deletions is shown in Figure 4.4. In a simple form of deletion mapping, a strain that is doubly mutant for a known marker *a* and the mutant to be mapped *m* is crossed to the deletion strain. The deletion being tested is known to delete the *a* marker and so the presence of F_1 offspring with an *a* mutant phenotype is expected from the cross. If these progeny with the *a* mutant phenotype also exhibit an *m* mutant phenotype—whatever the *m* mutant phenotype might be—the deletion chromosome must not carry the wild-type allele of the *m* locus, and the *m* locus lies within the deleted region. On the other hand, if these *a* mutant progeny do not exhibit an *m* mutant phenotype, the deletion must have the wild-type allele of the *m* gene and the *m* locus does not lie in the deleted region. By using a set of partially overlapping deletions, the *m* locus can often be positioned using relatively few matings.

The procedure for mapping with duplications is conceptually similar, with the following difference. Deletions are the genetic equivalent of a knockout mutation of a gene and will fail to provide the wild-type function for (or fail to complement, as we discuss below) any mutated gene; thus *m/deletion* strains are mutant. In contrast, duplications are usually designed to carry the wild-type allele of any gene that they include, sometimes abbreviated as *Dp(m+)*. Thus, a duplication strain will provide the wild-type function for the mutated gene and complement the mutation. This can be described in terms of the genotype and phenotype: *m/m* has a

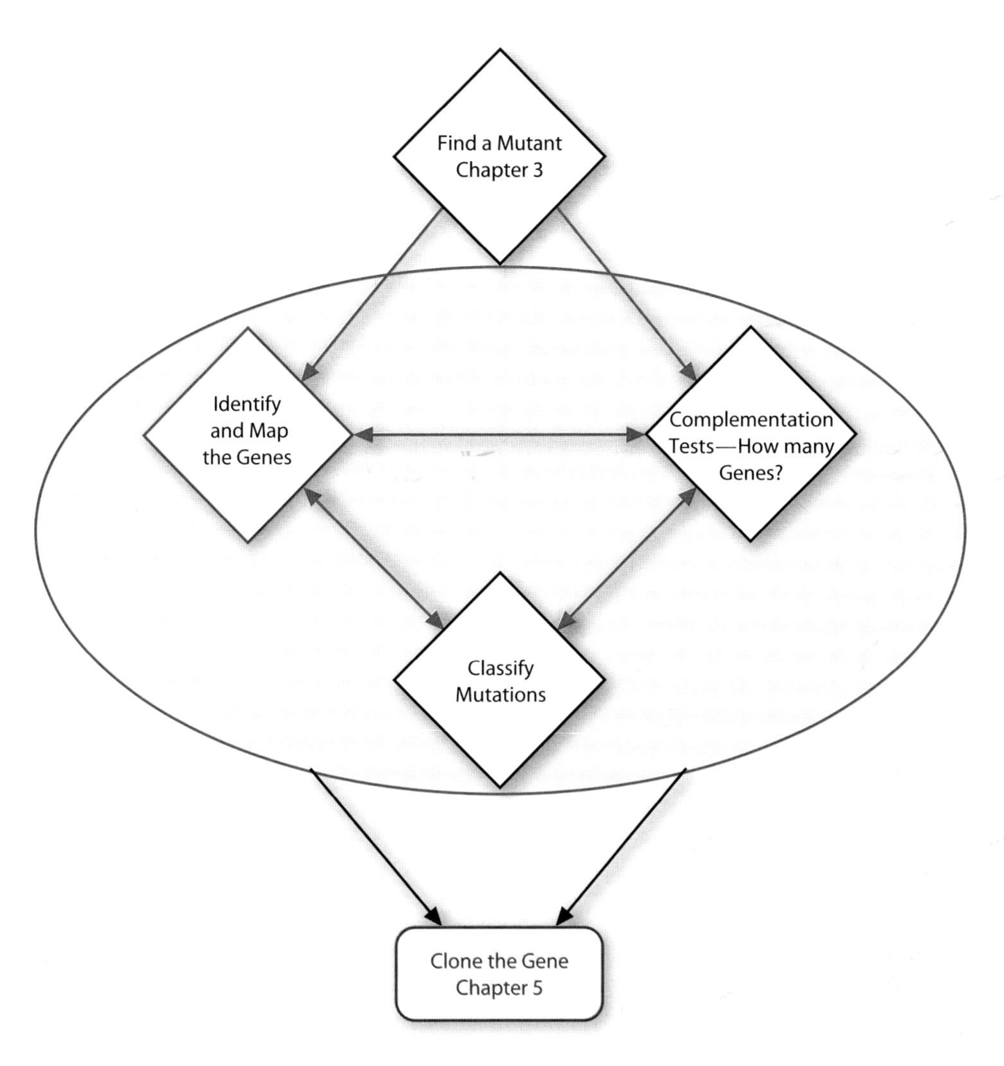

Figure 4.5 Flowchart of genetic mutant analysis. Mutations have been identified and mapped, as indicated by the red box and lettering. Complementation tests are used to determine how many genes are involved.

mutant phenotype, whereas *m/m;Dp(m+)* includes the wild-type allele on the duplication and so has a wild-type phenotype. By testing a series of overlapping duplications, some will provide the wild-type function and some will not; the *m* locus can be positioned with the chromosomal region as result. Often duplications and deletions can be used together to define the position most carefully.

As will be described below, duplications and deletions are used to classify the type of mutant that is being characterized as well as for mapping the mutant. Duplications and deletions allow the investigator to vary the number of wild-type (and mutant) copies of the gene, which, as we shall see below, is a key aspect of determining what type of mutant allele is being studied.

Let us summarize where we are in our process of identifying genes that affect a biological process. Our progress in the analysis is shown diagrammatically in Figure 4.5. We have now collected a number of mutations with a range of phenotypes, but all of them affecting the biological process that we are analyzing. The mutant strains have been outcrossed sufficiently that we have confidence that the mutant phenotypes are due to a single mutation rather than to multiple different mutations in the same strain. We have mapped the mutations with respect to the location of other known genes. The map location provides one of the most important strategies for cloning the gene, as described in Chapter 5. However, we do not yet know how many different genes are represented by the mutations we have collected, we have not yet named these genes, and we have not yet worked out how the mutation affects the function of the gene and the biological process. Each of these steps will be considered in turn.

4.3 Complementation tests to assign mutants to genes

Many mutants with similar phenotypes are usually found in a mutant hunt. These mutants could be different mutant alleles in the same gene, and thus affect the same function, or could be mutant alleles in different genes that have related functions. The goal of complementation testing is to determine how many different genes have been identified by mutations.

The **complementation test** is one of the fundamental principles by which genes are defined. Like the map position, complementation testing also provides a strategy for cloning genes that we describe in Chapter 5. First explicitly described by Seymour Benzer using rII mutations in the bacteriophage T4, complementation tests are now used in all diploid organisms to answer the following question:

> **Are these two mutations alleles of the same gene or are they in two different genes?**

In order for a complementation test to be performed, both mutations must be recessive to wild-type. This is the only requirement for complementation testing. On the other hand, it only makes sense to do a complementation test if the two mutations map to the same location since mutations on different chromosomes cannot define the same gene. In fact, because it is usually easier to do a complementation test than it is to map a gene, it sometimes makes sense to perform complementation testing before mapping so that fewer mutants need to be mapped. The procedure for a complementation test is shown in Figure 4.6.

Let us define *m1* and *m2* as two different and independent mutations that have similar mutant phenotypes and that map to the same location on the chromosome, at least so far as our mapping procedure can resolve their positions. Thus, there may be one gene at this locus, and *m1* and *m2* are two different mutant alleles of this gene. Alternatively, there may be two genes with related functions in this same region of the chromosome, with *m1* being a mutation in one gene and *m2* being a mutation in the other gene. To distinguish between these possibilities, an *m1* strain is crossed to an *m2* strain to make a hybrid offspring that is *m1/m2*. (The shorthand notation for writing genotypes can be a little confusing at this stage. If *m1* and *m2* are mutant alleles of the same gene, then *m1/m2* is the proper way to write the genotype. If *m1* and *m2* are mutant alleles of two different genes, the genotype would be written *m1/+;m2/+*.) The phenotype of this F_1 hybrid is examined.

According to the complementation test, if the *m1/m2* hybrid has a *mutant* phenotype, *m1* and *m2* are defined as *failing to complement* and thus are *alleles of*

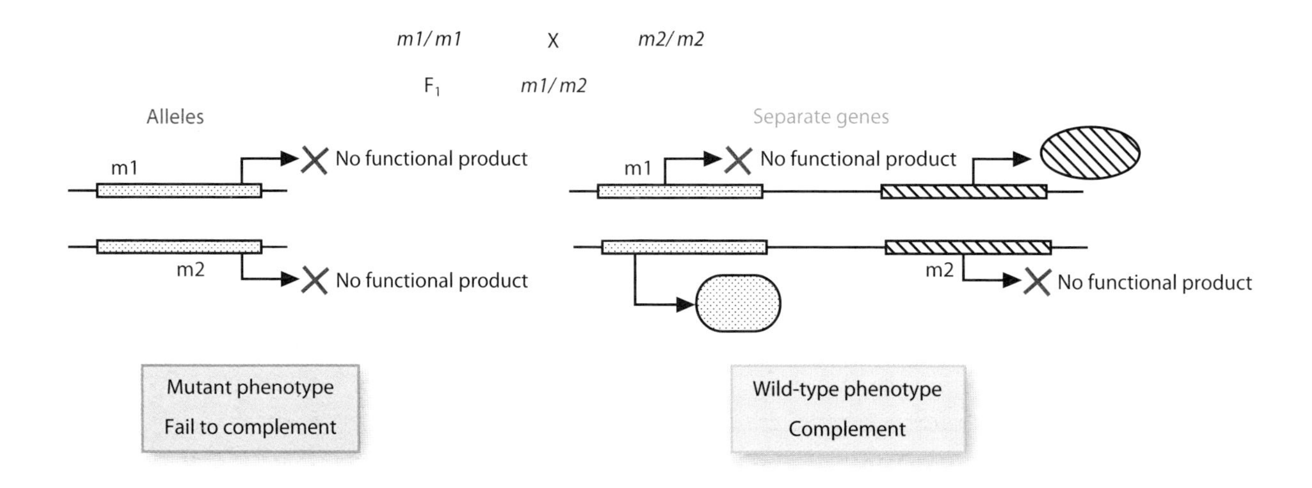

Figure 4.6 Complementation tests to determine if two mutations are alleles of the same gene. Both *m1* and *m2* are recessive mutations. If they are two different alleles of the same gene, as shown on the left, the heterozygote between them will not make a functional product and will have a mutant phenotype. The mutations fail to complement. Conversely, if *m1* and *m2* are alleles of two different genes, as shown on the right, the *m1* mutant strain will provide the wild-type *m2* function and the *m2* mutant strain will provide the *m1* wild-type function. The heterozygote will have a wild-type phenotype, and the mutations complement.

the same gene. On the other hand, if the *m1*/*m2* hybrid has a *wild-type* phenotype, *m1* and *m2* are defined as *complementing* each other and thus are *alleles of different genes*. In effect, if the two mutations are alleles of two different genes, the *m1* strain is capable of providing the wild-type function that the *m2* strain lacks and the *m2* strain is capable of providing the wild-type function that *m1* lacks—the *m1* mutant completes or complements the defect in *m2* and vice versa. Mutations that fail to complement each other are assigned to the same complementation group.

Complementation tests are not a perfect method of determining if two mutations are alleles, but they are easy to do and are reliable indicators in most cases. In a few cases, two mutations can complement each other but still prove to be alleles of the same gene because of some unusual property of the protein. This phenomenon is known as **intragenic complementation**. In particular, a protein with two independent functional domains can sometimes be mutated in one domain without affecting the function of the other domain. If *m1* and *m2* affect different domains, they might complement since each is providing one mutant and one functional domain for the protein. Intragenic complementation has also been seen when the protein functions as a multimer. In cases involving intragenic complementation, the proof of allelism often comes from a third mutation that fails to complement each of the other two. If both *m1* and *m2* fail to complement *m3*, then *m1* and *m2* are inferred to be alleles of the same gene even though they complement each other. Proof of allelism can also be obtained when the gene is identified and the mutations are sequenced.

In contrast to the situation when two alleles of the same gene complement each other, mutations might fail to complement even when they are not alleles of the same gene. For example, mutations that map to different chromosomes clearly cannot be alleles of the same gene, and yet may fail to complement each other. This describes a phenomenon known as **non-allelic non-complementation** in which mutations in two different genes fail to complement. This is discussed in Chapter 10 when we describe gene interactions. Whereas intragenic complementation often indicates something about the protein product of the gene, non-allelic non-complementation often provides information about the functional relationship between two genes. Nonetheless, with these cautions in mind, complementation tests are usually the standard for defining allelism and determining whether two mutations affect the same gene and the same function.

Complementation tests can be used to screen for new mutant alleles of a gene

Complementation tests are so fundamental to our definition of genes that they can be used themselves as an experimental tool. In Chapter 5, we describe how the principle of a complementation test is used as one strategy to clone a gene and confirm that the correct gene has been cloned. Another way that complementation tests are used as an experimental tool is in a screen or selection to find new alleles of a gene that the investigator wants to study in more detail. After all, one mutation in a gene may be interesting, but many more mutations in the same gene are much more informative in understanding the biological process.

For example, suppose that we have a mutation that results in vestigial wings in *Drosophila* and we want more alleles of this same gene. Of course, we could continue to mutagenize the flies, collecting mutations that affect wing shape, mapping them, and performing complementation tests to determine if they are alleles of the *vestigial* gene that we most want to study. However, once the investigator has decided to study one particular gene in more detail, a mutagenesis procedure known as an **F_1 screen** or a **non-complementation screen** can be performed to specifically identify alleles of this one gene. The strategy is shown diagrammatically in Figure 4.7. In this case, the mutagenized flies are mated immediately with a strain that carries a known mutation in the gene, the *vestigial* wing mutation in this example. The F_1 offspring are examined for wing defects, in particular for vestigial wings. Most of the F_1 offspring will have wild-type wings because they have no mutation in the *vestigial* gene. However, a few of the F_1 offspring will have vestigial wings because a new mutation in the *vestigial* gene has been induced in the mutagenized male parent. That is, the new mutation induced in the male will fail to complement an existing mutation inherited from the female and will then be a new allele of the same gene.

One significant advantage of such a screen is that the mutations are screened in the F_1 generation rather than in the F_2 or F_3 generation, saving the time and effort

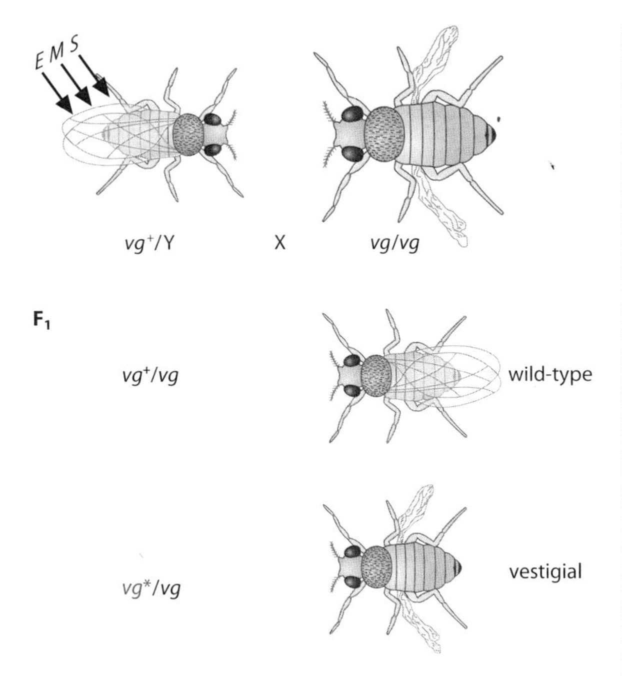

Figure 4.7 An F_1 or non-complementation screen illustrated for the *vestigial* wings gene in *Drosophila*. A wild-type male is treated with a mutagen (EMS) and mated to a female that is homozygous for the *vestigial* (*vg*) mutant allele. The parental female could be homozygous as shown here or could be heterozygous with a balancer chromosome. Most of the F_1 flies will have wild-type wings since they have inherited a wild-type vg^+ allele from the father. A few of the F_1 progeny will have a newly induced mutation in the *vg* gene, shown in red as vg^*, which will fail to complement the *vg* allele from the mother, resulting in a mutant fly.

of breeding and examining a second generation that includes few mutations of interest. In addition, mutations with only a subtle effect can be detected because they fail to complement the original mutation; this effect is described in more detail below in the discussion of an allelic series. By placing a mutation with a weak phenotype *in trans* with a mutation with a strong phenotype, the mutant phenotype of the weak mutation becomes more obvious. Furthermore, the new mutations found by a non-complementation screen do not need to be mapped. In short, an F_1 screen is one of the best methods of finding new alleles of any gene of interest.

In this chapter, we continue our case study on *Drosophila* segmentation mutants that we began in Chapter 3. The combination of complementation tests and mapping procedures used for these genes is described in Case Study 4.1, Part 1. Because 577 mutations were placed into 137 genes by these methods, efficient and informed procedures for mapping and complementation were crucial.

The total number of genes that could be found by a mutagenesis procedure can be estimated by statistical methods

As the mutants from a large collection are assigned to genes using complementation tests, an investigator will want to estimate how many additional genes can be found from a particular mutant hunt. That is, he/she will want to know if it is worthwhile continuing the same screen or selection to look for more mutants, or if the procedure will simply provide yet more mutant alleles of genes that have already been identified. The question of whether one has **saturated** for a screen is a complicated statistical problem. Some aspects of this problem are discussed in Text Box 4.2. Briefly stated, a key consideration in the estimate is the fraction of genes that have only one identified mutant allele and the fraction that have more than one mutant allele. From this, it is possible to make a rough estimate of the number of genes that remain to be identified. The general concept is illustrated in Figure 4.8 for a collection of balls in a series of bins.

Consider a simple version of this problem as it relates to mutations. Suppose that an investigator collects 10 mutants and learns from complementation tests that they fall into two genes, each with five mutant alleles. Intuitively it seems likely that the next mutant to be identified by the same method will also be an allele of one of these two genes. The number of genes affecting the biological process that can be identified by this method is then probably close to these two genes. It is possible that there are many more genes that could be identified by a modification of the screen or selection but, if the goal is to find more genes, continuing with this same scheme is probably not worth the effort.

In contrast, suppose that the investigator collects 10 mutants, all of which complement each other and define 10 different genes. Now, it is very hard to know whether the next mutant will be a new allele of one of the existing genes or define yet another new gene since no genes have been mutated more than once. No matter what the result with the next mutant, the number of genes that can be identified by this screen is clearly not saturated and it seems very likely that many more genes remain to be found. In Case Study, Part 2, we describe how the *Drosophila* geneticists estimated that they were nearing saturation for their mutagenesis scheme.

Case Study 4.1

Part 1: Find a mutant: segmentation in *Drosophila* embryos

Classifying mutants: complementation tests

A total of more than 640 different mutations affecting segmentation in *Drosophila* embryos, mapping to three different chromosomes, were found in the extensive mutant hunts described in Case Study 3.1. The next step for Nüsslein-Volhard, Wieschaus, and their colleagues was to determine the total number of genes that these mutations represent using complementation tests. To do this, they separated the mutants by chromosome and by phenotype.

For example, among the 321 mutations on chromosome 2, five phenotypic categories were used: anteroposterior defects, dorsoventral defects, holes in the cuticle, general differentiation defects, and head defects. Subgroups were also used within each category. These were subjective phenotypic categories that arose from the investigators' experience with and knowledge of mutant *Drosophila* embryos. Since it was not feasible to perform pairwise complementation tests among all 321 lethal mutations expeditiously, separation into categories reduced the labor and time. Nothing other than the mutant phenotype suggested that these were meaningful distinctions since different mutations from the same gene might have fallen into different categories. The researchers are candid about the success and limitations of this approach:

> Several mutations were tested as members of more than one subgroup. In general, this strategy proved useful and in many instances our predictions of allelism were correct. (Needless to say, our abilities in this matter improved greatly during the course of the experiment.)...In other cases, the initial classification was less successful. Several mutant phenotypes had more than one prominent feature and what later turned out to be allelic mutants had initially been distributed among two or three different categories. (Nüsslein-Volhard *et al.* 1984, p. 272)

Problematic classifications are described, and it is noted that complementation tests were performed among genes that proved to map close together to be sure that the mutations belonged to separate genes. The results are tabulated in Case Study Table 4.1. In the end, 259 mutations on chromosome 2 were assigned to 48 different genes with more than one allele, four mutant lines were found to have mutations in more than gene, and 13 mutations were in the sole allele of a gene. In sum, the complementation tests defined 61 genes on chromosome 2, with 48 genes having more than one allele. The number of alleles per gene ranged from 1 to 18, with an average of 4.5 mutant alleles per gene. These numbers will be useful when we consider the question of how many genes could have been identified by this approach. Similar strategies identified 198 mutations in 45 genes on chromosome 3 and 101 mutations in 20 genes on the X chromosome. Overall, for the three chromosomes, the number of mutations per gene ranged from 1 to 20, with an average of 4.5 mutant alleles per gene.

Table C4.1 Distribution of mutations affecting segmentation

Number of alleles	Number of genes			
	On X	On chr. 2	On chr. 3	Total genes
1	13	13	13	39
2	1	13	5	19
3	7	7	4	18
4	2	8	6	16
5	1	3	7	11
6	3	5	1	9
7	1	1	1	3
8	1	1	2	4
9	2	5		7
10			1	1
11			1	1
12	1	1		2
13		1		1
14			1	1
15		1	1	2
16				0
17		1		1
18		1		1
19				0
20			1	1

Each row is the number of genes identified with that number of alleles, i.e. 13 genes on the X chromosome had one allele, one gene on the X chromosome had two alleles, seven genes had three alleles, and so on. In total, 137 genes were identified.

Classifying mutants: map positions

More than 120 genes affecting the segmentation pattern in *Drosophila* embryos were identified in these screens, located in many different regions of the genome. Each of these genes had to be mapped and tested with previously identified genes that mapped in the same location. Both of the mapping methods described in the chapter (recombination distances and segmental aneuploidy) were used to assign these new genes to a chromosomal location. The method used for recombination mapping on chromosome 2 is summarized in Case Study Figure 4.1; similar methods were used for the genes on chromosome 3 and a slightly modified version was used for the genes on the X chromosome. Most genes were also mapped using the many deficiencies and duplications that are available to *Drosophila* geneticists.

Even upon re-reading these papers after more than two decades, the immediate impression is the remarkable amount of work that went into finding these mutations, classifying them into genes, and mapping the genes. Nearly every genetic method available in *Drosophila* was used in these experiments. Most readers will probably skip quickly to the pictures of the mutant phenotypes and the tables of the genes since those tell us what the flies reveal about the strategy for segmentation. But the lessons for us in these papers lie in the details of the methods that were used, an example of how thorough a genetic analysis of a biological process can be.

Further reading

Hooper, J.E. and Scott, M.P. (1989). The Drosophila *patched* gene encodes a putative membrane protein required for segmental patterning. *Cell*, **59**, 751–65.

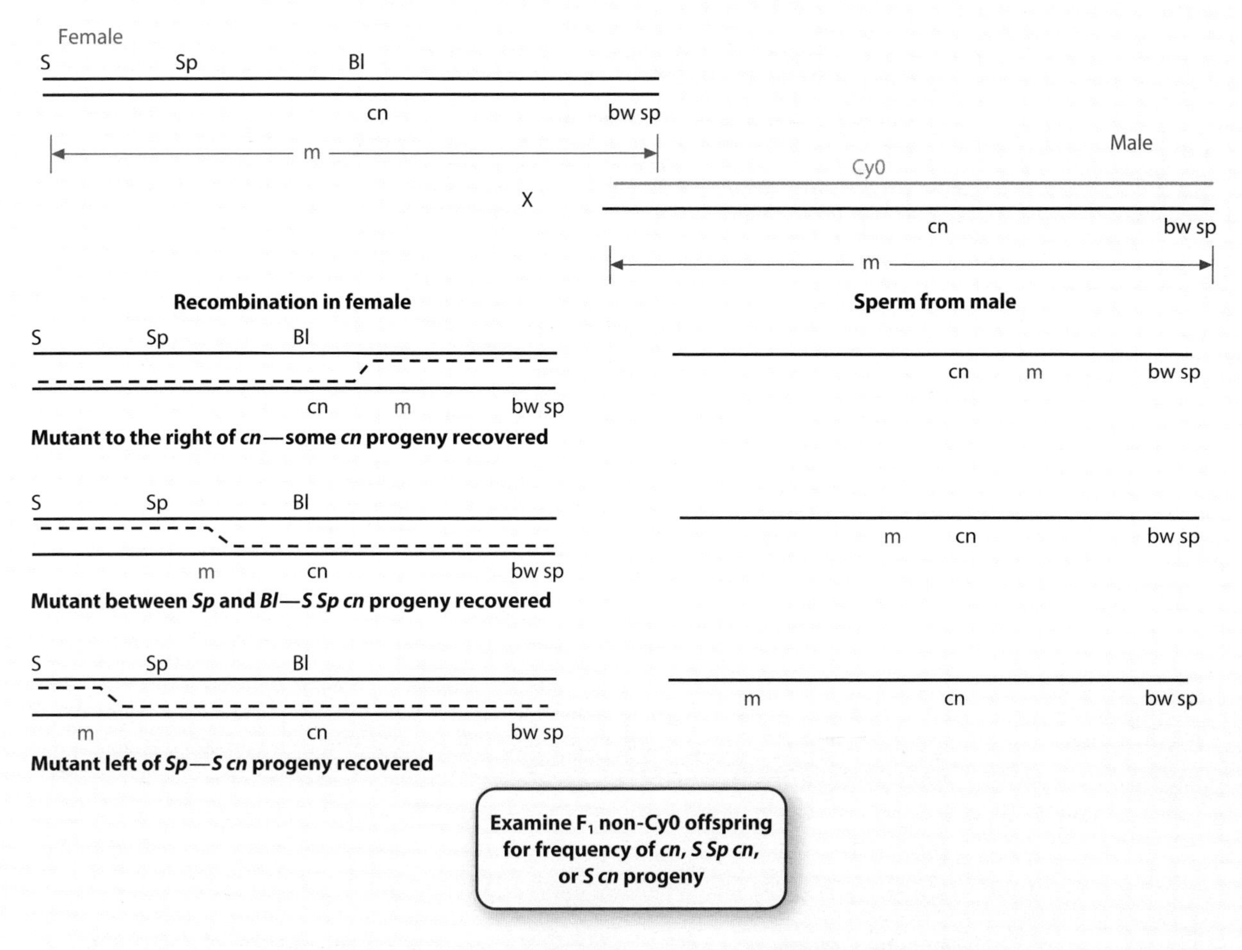

Figure C4.1 Mapping by recombination on chromosome 2. A mutation *m* was mapped by the following cross. The *m* mutation was induced on a chromosome marked with the recessive markers *cn*, *bw*, and *sp*. Two different mutant alleles of the same gene were used, one in each parent, indicated here by the different colors for *m*. The location of *m* is not known, as indicated by the line in the parents. The genotypes of the parents are at the top of the figure, with the relative location of the mapping markers indicated. The mapping mutations on the other homologous chromosome in the female are the dominant mutations *S*, *Sp*, and *Bl*. Recombination occurs in the female parent to separate *cn* from *m*. The non-Cy0 F_1 offspring are examined because these have inherited the *m* allele from the male parent. The *m* mutation is found in one of three intervals on the chromosome, as shown. If *m* is located to the right of *cn*, some of the F_1 progeny will have a *cn* phenotype but will not have any of the dominant markers. If *m* is located to the left of *cn* in the interval between *Sp* and *Bl*, the F_1 progeny will have an *S Sp cn* phenotype. If *m* is located to the left of *cn* in the interval between *S* and *Sp*, some of the F_1 progeny will have an *S cn* phenotype. The relative frequency of each of the phenotypes of F_1 progeny places the *m* mutant into one of these three intervals.

Jürgens, G., Wieschaus, E., Nüsslein-Volhard, C., and Kluding, H. (1984). Mutations affecting the pattern of the larval cuticle in *Drosophila melanogaster*. *Roux's Archives of Developmental Biology*, **193**, 283–95.

Nüsslein-Volhard, C. and Wieschaus, E. (1980). Mutations affecting segment number and polarity in *Drosophila*. *Nature*, **287**, 795–801.

Nüsslein-Volhard, C., Wieschaus, E., and Kluding, H. (1984). Mutations affecting the pattern of the larval cuticle in *Drosophila melanogaster*. *Roux's Archives of Developmental Biology*, **193**, 267–82.

Wieschaus, E., Nüsslein-Volhard, C., and Jürgens, G. (1984). Mutations affecting the pattern of the larval cuticle in *Drosophila melanogaster*. *Roux's Archives of Developmental Biology*, **193**, 296–307.

BOX 4.2 Estimating the number of genes

A mutant screen is considered to be saturated when all the genes that *can* be identified by that particular screen *have* been found. We cannot know the total number of genes, and the genes that have been identified will inevitably represent a sample of those that could be found. Mutations are isolated and assigned to genes, and the frequency with which a gene is mutated is used to estimate how many additional genes could be found. Therefore **saturation**, i.e. estimating the number of genes that can be found by a particular mutagenesis procedure, is inherently a statistical concept.

We can illustrate this approach with an analogy. Suppose that I own an MP3 player, but I do not know how many songs are stored on it. I want to know how many songs are in the library by measuring the frequency at which a given song is played. Like many MP3 players, mine has a function that will select and play a song at random from my library and will count how many times that song has been played. However, the random selection on my hypothetical MP3 player is different from most commercial players in that songs are returned to the library once they are played. (Most actual MP3 players generate a playlist when the random shuffle feature is selected but a given song will appear on the list only once. Thus, once a song is played, it is effectively removed from the library until a new playlist is generated. In statistical terms, this is referred to as sampling without replacement. On my hypothetical MP3 player, we are using sampling with replacement.) Therefore, with my player, the same song could be played several times while I listen, and might even be played two or three times in a row.

Now let us think about what will happen. The first song will necessarily be one that has not been played before. The second song could be the same as the first one or it could be a new song that has not been played before. Likewise the third song could be the same as one of the first two songs or it could be a different song. The process continues indefinitely, until the person listening to the song list is satisfied that all of the songs have been heard, or at least that he/she can estimate how many songs are stored on the MP3 player. Some songs will have been heard many times, other songs only a few times, and perhaps some songs will not have been played at all. Nonetheless, by knowing how often a song has been played and making some assumptions about how the song selection process occurs, it is possible to estimate how many songs are on the MP3 player. The procedure is saturated when additional song selections will not provide additional information about the total number of songs in the library.

We can express this situation mathematically. Assume that there are N songs in the library. The probability that the first song is one that has not been played previously is 1. The probability that the second song is the same as the first one is $1/N$, and the probability that it will be a different song that has not been played is $1 - 1/N$. We can express this more generally if we define h as the number of songs that have been played before each selection is made. With each selection, the probability of playing a song that has not been heard previously is $(N - h)/N$.

Let us illustrate this with some numbers. On the first selection, no songs have been heard previously so $h = 0$, and the probability of playing a song that has not been played before is $N/N = 1$ regardless of the size of N. Suppose that $N = 3$, i.e. there are three songs in my music library. After the first selection, $h = 1$ because one song will have been played. The probability that the second selection will play a different song is $(3 - 1)/3 = 2/3$. If the second song is different from the first song, $h = 2$. The probability that the third selection will play a different song is $(3 - 2)/3 = 1/3$. In other words, as the number of selections increases, h approaches N, and so $N - h$ approaches zero: most selections will pick a song that has already been heard. Although we do not know N, we can estimate it by using h and the change in h as the number of songs played increases. This is the statistics behind the concept of saturation.

Comparing the MP3 player's shuffle with a mutagenesis screen

Before considering the statistics of this situation more fully, we need to compare it with a mutagenesis scheme. Each selection of a song corresponds to finding a mutation, and the songs in the library represent the genes that could be mutated. In genetic terms, this means that every mutation can be identified as an allele of some gene. In addition, the songs are selected without regard to what songs have been played previously; this indicates that mutations can be found in any gene without an intentional bias introduced by the investigator. Mutations are being induced at random with respect to the genes. Thirdly, with each song selection, the probability increases that that the song will be one that has been heard previously. In genetic screens, a complementation test is used to determine whether the new mutation defines a new gene or a new allele of a previously existing gene. At the point of saturation, all songs will have been heard previously. In genetic terms, this is equivalent to demonstrating that all new mutations are alleles of a gene that has previously been identified.

Each of these assumptions seems like a realistic approximation of a genetic screen. However, selecting a song by a random shuffle is far less effort than finding a new mutation and identifying which gene it defines. With the MP3 player, one could simply keep listening to songs until all of them have been heard, some of them many times. With genetics, it makes sense to estimate the number of genes well before all of them are identified, if only to estimate the size of the remaining task.

The assumption of equal mutation rate per gene

Our MP3 example makes the important assumption that each song has the same probability of being played next. This assumption is reasonable for an MP3 player that selects songs based on the title. In genetic terms, this corresponds to the very important assumption that all genes have the same probability of being mutated by the procedure. All genetic data indicate that this assumption is an oversimplification that can result in a significant distortion in the estimate of gene number. Consider the data from mutations on chromosome 2 affecting *Drosophila* segmentation as displayed in Case Study Table 4.1. Even without knowing much about frequency distributions, it seems very unlikely that the genes have the same probability of being mutated when one gene is hit 18 times, another one is hit 17 times, and a third is hit 15 times, and yet 13 genes are hit only once. Intuitively, it seems clear that the genes hit so much more frequently must somehow present a better target for mutagenesis.

There are many biological ways that this apparent high rate of mutability could happen. Most mutations alter base pairs, and not all genes are comprised of the same number of base pairs. Perhaps one gene is physically larger and thus presents more possible targets for mutation. This effect has been observed. Spontaneous mutations to muscular dystrophy in humans are much more frequent than other genes because the affected gene is one of the largest genes in mammals. This effect of gene size may not be recognized until a gene has been cloned.

Another very likely explanation is that some mutant phenotypes are easier to see than others. In genes with an easily detected mutant phenotype, both hypomorphic and amorphic alleles are spotted, whereas genes with a subtle mutant phenotype might be detected only by amorphic alleles or sometimes be overlooked all together. This effect was seen for mutations affecting segmentation. The investigators note that the five genes with the greatest number of alleles also have strong and easily detected mutant phenotypes, making it easier for the investigator to recognize them.

Yet another possibility is that mutations in that gene are not easily found by the particular mutagen used, in this case EMS, which primarily induces single base changes. The investigators note that the only allele of one gene is a small deletion rather than a single base change. Different mutations definitely do produce a different spectrum of mutations, but generally this effect is not very important for saturation estimates. Although nearly every mutagenesis scheme performed has found this bias in the distribution of mutations, it is rarely explained and all these possibilities likely play some role. However, the statistical models typically require an assumption that all genes are equally mutable, and nearly all calculations for saturation are made with this assumption in place.

Saturation and the *Drosophila* mutagenesis screen

Having examined the underlying assumptions that went into our model, what did the investigators observe and how did they determine if they were near saturation? They plotted the number of new mutations found and the number of new genes found as a function of the number of mutagenized lines. This plot is reproduced in Case Study Figure 4.2. In other words, they are showing the change in $N - h$ (the total number of genes, which is not known, minus the number of hit genes) as a function of the number of mutations or tosses. The data show that roughly 5000 chromosomes were scored and the number of mutations increased linearly with the number of chromosome tested. The number of genes increases until about 1500 chromosomes have been tested and about 75 mutations have been found. Then the number of genes increases much more slowly than the number of mutations, such that among the final 1500 chromosomes almost no new genes were found. This indicates that nearly all genes that could be found by this method probably have been found.

In making these estimates of the number of genes, the investigator is forced to make some assumptions that are clearly incorrect. These assumptions and the use of some statistical methods to estimate the number of genes are described in Text Box 4.2. Estimates of the number of genes based on mutant frequency are probably best regarded as a rough approximation. One of the significant advantages of genome-wide screens, as described in Chapter 7, is that they avoid estimates of gene number based on the statistical principles of saturation.

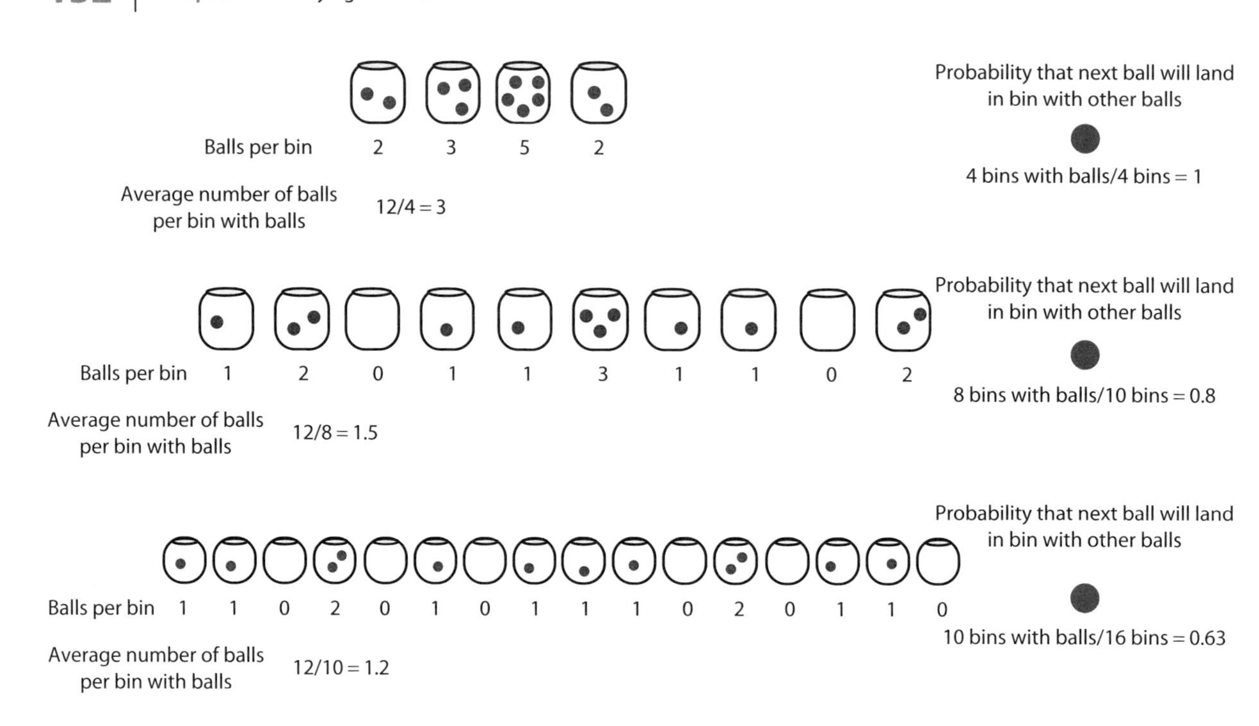

Figure 4.8 Saturation as a means of estimating the number of genes. Each line of this diagram shows a number of bins with 12 red balls distributed throughout the bins. In the line with four bins, every bin has at least one ball in it, and so the probability that the next ball (shown in blue on the right) will land in a bin with a ball already in it is 1. As the number of bins increases, the average number of balls per bin will decrease. Furthermore, the number of bins with no balls in them increases, and so the probability that the next blue ball will land in a bin with other balls also decreases. Thus, the fraction of bins with no balls in them can be estimated from the number of balls per bin and the probability that the next ball lands in a bin with a ball already in it. In a genetic screen, the balls represent the mutations and the bins the genes that can be identified. The number of genes is not known because the number of genes that have not been mutated (the bins with no balls) is not known. However, as the number of mutations increases, the number of alleles per gene increases and the frequency with which a new mutation is an allele of a gene that has already been identified decreases. A genetic screen is said to be **saturated** when new mutations no longer identify new genes.

Gene names are usually, but not always, based on mutant phenotypes

As the investigator is identifying mutations, the mutations will probably have laboratory names such as A28C or 42-3-D or some system that only the investigator can understand. Once the mutations have been assigned to a newly identified gene, the gene has to be named and the mutation is given an allele designation. It is probably safe to say that no other topic in genetics has more potential for confusion than gene names and allele designations.

The potential for confusion arises from many sources. First, different organisms have different systems of nomenclature. Gene names in *Drosophila* consist of one to four letters and rarely involve a number (e.g. *w*, *ptc*, *Antp*), whereas gene names in *C. elegans* must be three letters long and always include a number (e.g. *tra-1*, *dpy-10*). In *Aradidopsis*, gene names are usually three letters but can vary from two to six letters. They may include a number or even two numbers (such as *bri1-1*) in which the second number refers to the allele. Secondly, many of the nomenclature systems are products of a long history of research in the organism. A gene name that made sense to an investigator working on a particular mutant allele years ago may be quite obscure to a modern investigator. For example, the *Notch* gene in *Drosophila* encodes a receptor involved in cell–cell signaling; orthologous genes are found in nearly all animals. The gene name arises because the first mutation resulted in notches in the wing margin, a relatively insignificant defect in light of all of the other roles of this gene. However, the gene name indicates its history. Thirdly, a gene name reflects a particular viewpoint on the nature of genes. This last point will provide our brief guide to how genes are named.

DEFINITIONS

The questions of when to use italics, capital letters, or superscripts are also completely different for different model organisms. For guidance in an important but slightly arcane topic, the interested and patient reader is directed to the nomenclature section of the online database for each specific organism referenced in Chapter 2.

Case Study 4.1

Part 2: Find a mutant: segmentation in *Drosophila* embryos

Classifying mutants: saturation and the number of genes

With such an impressive collection of mutations, how do investigators know when they are done? It is always possible to run the mutagenesis screen one more time, find more mutants, and characterize them. There is always the suspicion that the next mutant could be the most interesting one in the entire collection or that it will identify a crucial gene that has never been seen before. On the other hand, each new mutation requires additional effort that could be spent more productively on some of the other aspects of genetic analysis that we cover in this book. At some point, most of the mutations are new alleles of previously identified genes and add little new information. This is known as **saturation**—when all the newly identified mutations are alleles of previously identified genes and no new genes can be found by this method. How can an investigator recognize when this point of saturation has been reached?

Among the other attributes of their mutagenesis screen, Nüsslein-Volhard and Wieschaus came closer to saturation than most other geneticists had done or would have done. One of the greatest features of this work is the extent of the collection of mutants and the number of genes. They have provided stimulating and informative research projects for many geneticists for many years. The investigators themselves discuss the degree of saturation of their efforts. Some of the assumptions that are made in estimating how many genes could be found are discussed in Text Box 4.2.

In order to determine whether they were near saturation, the investigators plotted the number of new mutations found and the number of new genes found as a function of the number of mutagenized lines. This plot is reproduced in Case Study Figure 4.2. The data show that roughly 6000 chromosomes were scored and the number of mutants increased linearly with the number of chromosomes tested. In contrast, the number of genes increased until about 1800 chromosomes had been tested and about 50 different genes had been found. Then the number of genes increased much more slowly than the number of mutations, such that almost no new genes were found among the final 1500 chromosomes. This indicates that nearly all genes that could be found by this method probably have been found. We can examine this conclusion with the benefit of hindsight. More than two decades have passed since these studies were published, and few, if any, additional segmentation genes have been found. The goal of using genetic analysis to find all of the genes affecting segmentation in *Drosophila* was met.

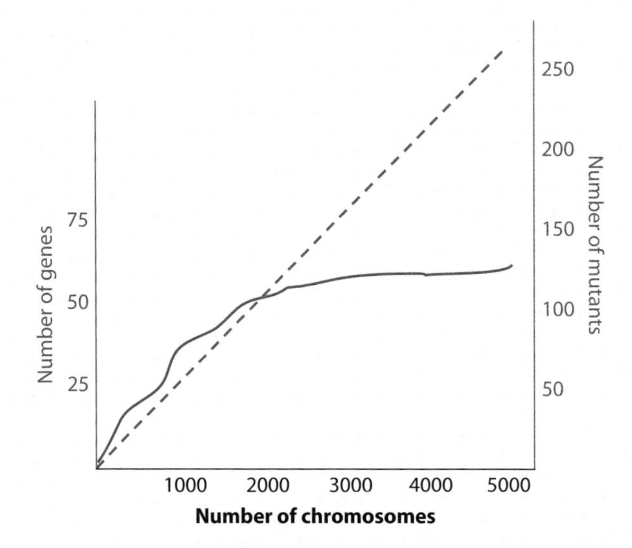

Figure C4.2 Estimating the number of segmentation genes on chromosome 2. One way of estimating if a mutant screen has been saturated is to compare over time the number of mutants recovered with the number of genes. The data for the analysis on chromosome 2 has been redrawn from Nüsslein-Volhard *et al.* (1984). The number of mutants recovered is shown in blue and increases almost linearly as more chromosomes are tested. That is, when 1000 chromosomes had been tested, approximately 60 mutants had been found. When 3000 chromosomes had been tested, approximately 170 mutants had been found. The number of genes (shown in red) does not show a similar increase. A linear increase with the number of chromosomes is observed until about 1800 chromosomes had been tested and about 50 genes had been found. Testing approximately 4000 more chromosomes identified only about 12 more genes because new alleles were found for genes that had been previously identified.

Further reading

Jürgens, G., Wieschaus, E., Nüsslein-Volhard, C., and Kluding, H. (1984). Mutations affecting the pattern of the larval cuticle in *Drosophila melanogaster. Roux's Archives of Developmental Biology*, **193**, 283–95.

Nüsslein-Volhard, C. and Wieschaus, E. (1980). Mutations affecting segment number and polarity in *Drosophila*. *Nature*, **287**, 795–801.

Nüsslein-Volhard, C., Wieschaus, E., and Kluding, H. (1984). Mutations affecting the pattern of the larval cuticle in *Drosophila melanogaster. Roux's Archives of Developmental Biology*, **193**, 267–82.

Phillips, R.C., Roberts, I.J., Ingham, P.W., and Whittle, R.S. (1990). The *Drosophila* segment polarity gene *patched* is involved in a position-signalling mechanism in imaginal discs. *Development*, **110**, 105–14.

Wieschaus, E., Nüsslein-Volhard, C., and Jürgens, G. (1984). Mutations affecting the pattern of the larval cuticle in *Drosophila melanogaster. Roux's Archives of Developmental Biology*, **193**, 296–307.

In an era of genetics that predated genome projects or even molecular biology, genes were recognized solely by mutant alleles. This began with one of the first genes to be named, the *white* gene in *Drosophila*, and continues until today. The gene was called *white* because flies with a mutation in the gene had white eyes. Thus, the gene was named for its *mutant* phenotype—in effect, for what happens when the normal function is missing or altered.

This method of naming genes is contrary to the way everyday objects are named. For example, when I am driving and want my car to stop, I press a particular pedal on the floor and the car slows down. The normal function of this pedal and its associated parts is to brake the car's momentum, and we call the pedal the brake after its normal function. However, with genes, especially in the earlier days of genetics, the normal function was usually unknown; the best clue to the normal function was to see what phenotype resulted when the function was disrupted. Thus, when this particular gene was disrupted, it was observed that the fly had white eyes.

The normal function of the *white* eye gene is not to make white eyes—it is to make wild-type red eyes. However, we name a gene for what happens when the gene malfunctions, *not* what happens when the gene functions normally: the gene is called the *white* gene and the protein that it makes is called the White protein. If we were to use a similar system to describe the brake pedal, we would name it for what happens when it is non-functional; thus, it would be called *car rolls forward*, or *no stopping*, or some other name indicating that the car fails to slow down and stop.

This system of nomenclature has a particular type of confusion when human genes are named. Sometimes we speak of 'disease genes' when we actually mean that a known disease arises when the gene is *mutated*. The cystic fibrosis gene (called *CFTR* and discussed in more detail in Chapter 5) does not cause cystic fibrosis. The normal function of the gene is to transport chloride ions across the membrane. It is a *mutation* in the *CFTR* gene that results in cystic fibrosis—the gene is named for its mutant phenotype.

This type of short-hand confusion is rampant, particularly when the news media report that a gene for some behavioral trait such as schizophrenia or Alzheimer's disease has been identified. It must be recalled that the gene is being named (or nicknamed) for the phenotype that arises when the gene is altered. The normal function of the gene is not to cause schizophrenia or Alzheimer's disease any more than the normal function of the *white* gene is make white eyes. However, we may not know the normal function of the gene; we only know what happens when the gene is altered. Thus, the gene is named for its mutant phenotype.

The potential for confusion becomes exacerbated by geneticists' own inconsistency in conferring gene names. The inconsistency arises in part from the way a particular gene was first found and studied. Not all genes are now recognized first by their mutant phenotype, especially since biochemical and molecular genetics and genomics are used in addition to traditional mutant analysis to find genes. In humans, the β-globin gene is very well studied. Notice how the gene is named—for its polypeptide product and not for the diseases that arise from mutations. Mutations in the β-globin locus are very well studied, and literally hundreds of naturally occurring mutations are known in humans. These mutations result in a group of diseases called the β-thalassemias, which include sickle-cell anemia. But the gene is not called the thalassemia gene or *thal-1* or something similar. It is named instead for its normal function, and is called the β-globin gene. The same is true for many genes that encode metabolic enzymes or familiar proteins such as β-tubulin or collagen. Occasionally one such gene will be named for its mutant phenotype. One particular collagen gene in *C. elegans* is called *rol-6* because mutations in the gene result in a worm with a rolling corkscrew movement. However, this is not the rule; many other collagen genes in *C. elegans* are named *col-* for their normal protein products. In addition, in an era in which genomes are sequenced and genes are found by computational analysis, many members of gene families have been named on the basis of their predicted functions and their similarities to genes in other organisms.

Thus, genes are named for their mutant phenotypes most of the time but not always, and there is no simple universal rule. Unfortunately, it is something that must be learned for each gene individually. One small satisfaction is that each gene name records something of the history of discovery for the gene, and so learning the name often allows the investigator to learn something of the background by which the gene was identified. This is usually a small and private satisfaction, more useful than being able to name all the members of a little-known rock band or all the popes of the Roman Catholic Church perhaps, but not a conversation topic that arises in social discourse with non-biologists.

Does this exhaust the ways that gene names can cause confusion? Unfortunately, it does not. Most genes are evolutionarily conserved and have orthologs in other species. For example, the cyclin B protein is found universally among eukaryotes. A gene performing the same function in different organisms will often have a different name. In *Schizosaccharomyces pombe*, the cyclin B protein is encoded by a gene called *cdc2* (*cdc* stands for cell division cycle, the mutant phenotype that is defective when the gene is altered). In *Saccharomyces cerevisiae*, the cyclin B protein is encoded by a gene called *cdc28*. The proteins have the same function—in fact, the wild-type version of the *S. pombe cdc2* gene can complement a *cdc28* mutation in *S. cerevisiae* and vice versa—but the gene names are different for historical reasons. The relationship between the two genes was not known until well after the mutations had been identified and characterized. To make matters worse, *S. cerevisiae* does have a gene named *cdc2*, but it encodes a different protein and not cyclin B. As cyclin B genes have been found in other organisms, they are often called *cdc2* genes although not usually because of the mutant phenotype in that organism.

Fortunately, an initiative called the Gene Ontology (GO) Project is attempting to bring some coherence to gene names and gene functions. We return to discussing the GO Project in Chapter 8. In its own words, the GO Project is 'developing...structured or controlled vocabularies' for describing the functions of genes. The GO Project organizes genes and gene products based on three criteria: the molecular function, the cellular component, and the biological process. By separating the names of the genes from their function or their mutant phenotype, the GO Project allows searches and correlations based on biological and molecular criteria. For example, a search on 'cdc2' finds the biological process of cell-cycle checkpoint (with more examples and links), the molecular function of a cyclin-dependent protein kinase, and the cellular component cyclin-dependent protein kinase holoenzyme complex. It also finds the gene in other organisms and provides the related gene names.

A different project, called HomoloGene, administered by NCBI, identifies homologous genes in different organisms and organizes them together in one place. It also allows the investigator to learn what an orthologous gene is called in other organisms, and provides links to the appropriate databases. Although the confusion over gene names probably cannot be completely eliminated, both the GO Project and HomoloGene are useful tools for reducing the potential confusion from different genetic nomenclatures used in different organisms.

Our progress in characterizing mutants is summarized in Figure 4.9. We have identified mutants, mapped them, and sorted them into genes based on complementation tests. By carefully analyzing the number of mutations per gene as each new mutant is identified, it is possible to estimate how many genes are yet to be found. The next step is to classify the mutants according to how they affect the function of the gene. Some of this classification has already been done (e.g. determining whether a mutant is recessive or dominant), but we now describe it in a more structured way.

4.4 Classifying mutants

A mutation disrupts the normal function of a gene. But exactly how does it disrupt the function? Does the mutation completely eliminate the function or merely reduce it? Does it over-produce the normal function of the gene, resulting in a mutant phenotype? Perhaps it produces the normal product but at the wrong time or in the wrong tissue, thereby causing a mutant phenotype. For most contemporary molecular biologists, these questions may best be answered by a molecular analysis of the DNA sequence (as described in Chapter 5) and the expression pattern of the gene product, either RNA or protein (as described in Chapter 8). From such analyses, one could determine if the mutation produced a stop codon, resulted in the substitution of an amino acid, altered a splice site, deleted or altered an upstream regulatory sequence, or caused some other change in

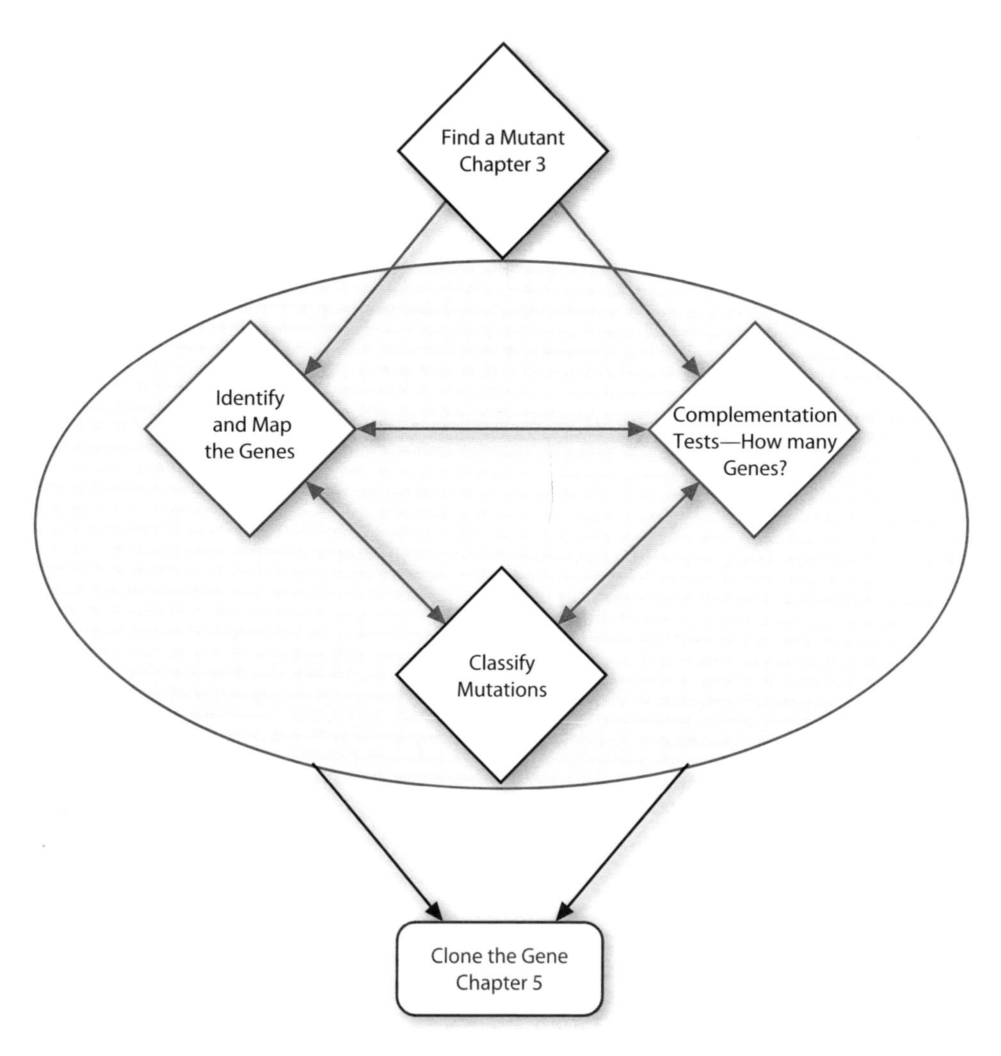

Figure 4.9 Flowchart of genetic mutant analysis. Once the mutations have been identified and mapped, they are placed into genes by complementation tests. Often these processes take place at the same time. The next step is to classify the mutations by how they affect the normal function of the gene.

the expression of the gene. The investigator will want to know this information for any mutation in any gene being studied.

Geneticists working in the days before molecular analysis faced these same questions and developed methods to answer them, at least in part. These methods typically involve duplications and deletions to vary the copy number of either the wild-type or the mutant allele and then observing the mutant phenotype. Thus, this classification may occur during the process of mapping. The classical work involving this concept was published by H.J. Muller in 1932, and many of his terms and ideas are still in use, although a few have been clarified or renamed. Our description will draw heavily on that of Muller while attempting to combine his ideas with what we now know about gene function.

Two key points need to be made at the outset. First, Muller was working on *D. melanogaster*, a diploid organism, and our description will also draw on diploids. Secondly, recall that most genes are named and described in terms of the *mutant* phenotype rather than the wild-type function. Most mutations are recessive and reduce or eliminate the function of a gene rather than confering a new function. Recessive mutations are often termed **loss-of-function** mutations. Thus, a strong mutant allele has a more severely mutant phenotype than a weak mutant allele of the same gene. The strong allele probably has less of the wild-type function remaining than does a weak allele, but the comparison is made involving the mutant phenotype, and the strong mutant allele eliminates more of the function than does the weak mutant allele. This important point will be illustrated with our examples.

Recessive mutations generally have a loss of or a reduction in the normal function of the gene

In classifying mutations, the first test is one of the simplest—to determine whether the mutation is recessive or dominant to wild-type. By this stage of classifying mutations, mapping genes, and performing complementation tests, the investigator will have already learned this. Most mutations are recessive to wild-type, at least when the phenotype is compared at the level of the organism. That is, if ***m*** is a mutant allele, the heterozygote ***m***/+ has the same wild-type phenotype as the +/+ wild-type homozygote. In the context of gene function, this result implies that the organism is producing enough of the gene product that even a single wild-type dose is enough to carry out the normal function. From an evolutionary perspective, dominance of the wild-type allele makes superficial sense: a single mutation in a single gene occurs regularly, so there has been selection for a level of gene expression that is insensitive to the loss of a single dose. The concept of dominance, which has interesting evolutionary and physiological implications, is discussed in more detail in Text Box 4.3. Exceptions are certainly known in which one mutant allele is enough to disable the gene so that the heterozygote has a mutant phenotype; such genes, termed haplo-insufficient, are discussed below.

With recessive mutations, the next question is whether the mutation eliminates the normal function of the gene or reduces it but retains some residual wild-type function. Muller referred to mutations that eliminate the normal function as amorphic mutations, the name reflecting the fact that the gene had no detectable wild-type function. A more common contemporary usage is to refer to these as **null** mutations, and amorphic and null are synonyms. We employ these terms interchangeably so that the reader becomes familiar with both usages.

Mutations that retain some residual normal function but have a reduction in function were referred to as hypomorphic mutations, where 'hypo' reflects the lower function compared with wild-type. Hypomorphic mutations are often referred to as leaky mutations, but this term can have somewhat different meanings in different contexts and so we prefer to use the older term hypomorphic mutations or hypomorphs.

Muller devised a simple test using deletions to distinguish amorphs from hypomorphs. His strategy is shown diagrammatically in Figure 4.10. The test is based on the following logic. A deletion completely lacks the genetic locus; therefore, it certainly has no remaining gene function and must be an amorphic mutation. If the mutation has the same phenotype as a deletion in these genetic tests, the mutation is also considered to be amorphic. That is, if the *m*/deletion strain has the same mutant phenotype as the *m*/*m* strain, *m* must be an amorphic mutation since replacing one copy of it with a deletion makes no difference to the mutant phenotype.

However, if the *m*/deletion strain has a more severely mutant phenotype than the *m*/*m* strain, then *m* is not behaving like the deletion and *m* must have some amount of residual wild-type function. *m* is then defined to be a hypomorphic mutation. Text Box 4.4 describes hypomorphic and null alleles of the *microphthalmia* gene in mice, and Case Study 4.1, Part 3, describes null and hypomorphic alleles of the *patched* gene in *Drosophila*.

Recall that a more severe mutant phenotype is defined in terms of the difference from wild-type. For example, perhaps the *m*/*m* strain arrests its growth at some particular developmental stage, whereas the *m*/deletion strain arrests its growth at an earlier developmental stage. By this test, the *m* mutation is hypomorphic. Another test for hypomorphs is to increase the dose of the mutant allele. If *m*/*m* is mutant, then the partial trisomic *m*/*m*/*m* with three mutant doses should be less severely mutant than *m*/*m*, and *m*/*m* should be less severely mutant than *m*/deletion. If the phenotype becomes less severe (or more like the wild-type) with each additional copy of the mutant allele, *m* is a hypomorphic mutation. On the other hand, increasing or further decreasing the dose of an amorphic mutation will have no effect on the mutant phenotype.

DEFINITIONS

amorphic: An amorphic or a null mutation eliminates the activity of a gene.

hypomorphic: A hypomorphic mutation reduces but does not completely eliminate the activity of a gene.

BOX 4.3 Thoughts on dominance

One of Mendel's key insights was that certain alleles are dominant to other alleles. Dominance is such a basic principle of genetic analysis that we can take it for granted. However, the existence of dominance suggests some interesting principles about evolution and biochemistry.

First, let us make sure that we are defining dominance correctly. The dominant phenotype is the one that is observed in a heterozygote. Note that this definition says nothing about being wild-type, advantageous, normal, or frequent; it refers solely to the phenotype of a heterozygote. It is also helpful to consider what phenotype is being examined. The familiar example of sickle-cell anemia can serve to illustrate this. If we are considering the general health of the individual at normal atmospheric pressure, sickle-cell anemia (Hb-S) is recessive to the presence of normal globin (Hb-A). However, by using isoelectric focusing gel electrophoresis, both the Hb-S and the Hb-A allele can easily be seen, and the alleles may be considered co-dominant. By sequencing the DNA, the alleles will also be seen as co-dominant, or more generally, the concept of dominance does not arise. Dominance depends on both the heterozygote being examined and the particular phenotype being tested. For this reason, terms such as 'semi-dominant' or 'intermediate dominant' are only helpful when we define what phenotype is being examined. Dominance is not only a property of the allele; it is also a property of the phenotype.

What does dominance tell us about the evolution of gene expression and basic biochemistry? Most fundamentally, it demonstrates that the organism makes a surplus of most gene products. Most mutant alleles are recessive to wild-type. When the amount of a gene product is halved by a null mutation, the phenotype of the organism is usually unaffected. Genomic tests in which every gene in the genome is systematically knocked out, as described in Chapter 7, have shown that the vast majority of mutations have no effect as a heterozygote with wild-type. The few exceptional cases typically involve a gene that is dose sensitive, and eliminating one copy of the gene produces a mutant phenotype. Dominant loss-of-function or haplo-insufficient alleles are notable because they are uncommon. We have become accustomed to observing dominance, and so we may overlook this insight into gene expression.

The advantages of dominance

How could such a biochemical system with dominant alleles have evolved? This question is more complicated than it may first appear. Making twice as much of a product as required could be considered quite wasteful. Having two alleles making the same (if slightly varied) product is redundant. Why does the cell devote energy and other resources to making proteins in excess of what it needs?

The key to this question lies in the phrase 'what it needs'. The amount of a gene product that is needed is not constant, peaking at certain times and in certain tissues and declining in others. Furthermore, since there are natural environmental fluxes, which we can rarely simulate in the laboratory, the amount of a gene product that is needed is an unknown function of the particular environmental changes. Perhaps natural selection has favored organisms with more than one allele, or even more than one copy of a gene, as being able to cope with environmental changes.

Even more important than environmental flux is the effect of mutation. Mutations occur for all genes in all cells at nearly all times. Most mutations reduce or eliminate the function of a gene. Dominance reflects the insurance that natural selection has provided against the effect of random mutation. The term **robustness** has been used in many fields, including genetics, to indicate the situation when a change in one component does not change the overall system. To express this thought more carefully, dominance is an indication of robustness against environment and genetic changes. Organisms cannot plan for change, of course, but genotypes that are buffered against the effects of environmental and genetic changes would have had a selective advantage. The topic of robustness will arise again in Chapter 12 when we consider redundancy in gene function as well as in allele function.

Dominance and trisomic syndromes

In discussing dominance in these terms, we are sidestepping another interesting problem, namely the existence of trisomic syndromes. After all, if a twofold surplus of a gene product is an evolutionary advantage, why do trisomic genotypes or duplications, a threefold surplus, often have a mutant phenotype?

We noted in Chapter 3 that mutant phenotypes arising from a duplication are much less common than mutant phenotypes arising from a deletion. Having too much of a gene product is clearly less of a disadvantage than having too little of it. In systematic searches of the *Drosophila* genome in the 1970s using duplications and deletions, very few genes were found to have a mutant phenotype when present in three or four copies rather than the standard two copies. This suggests that increasing gene dosage (and probably gene expression) by 50 percent, or even 100 percent, probably matters only for those few genes that are the most dose sensitive.

The data indicate that a mutant phenotype arising from the over-expression of one particular gene is not very common. Yet we know that trisomic individuals have distinctive phenotypes, so having three copies of an entire chromosome clearly does have a detrimental effect. Two explanations, which are not mutually exclusive, are posited to account for the trisomic effect. First, the phenotype might arise principally from those few genes that are dose sensitive in the range of two versus three copies. A few such genes have been identified or postulated for the phenotypes in trisomy-21 individuals. For example, one gene on chromosome 21 in humans encodes the β-amyloid protein that has been associated with Alzheimer's disease; many trisomy-21 people experience an early onset of Alzheimer's disease. Other effects of aneuploid syndromes can also be attributed to a few dose-sensitive genes.

A second explanation is that the trisomic phenotype arises from a cumulative effect of the hundreds, or even thousands, of genes whose expression is increased even slightly. Supporting this interpretation is the fact that most of the phenotypes seen in trisomy-21 individuals cannot be mapped to a particular locus on the chromosome. This result suggests that the increased expression of hundreds of genes simultaneously exceeds the ability of organism to buffer against genetic change.

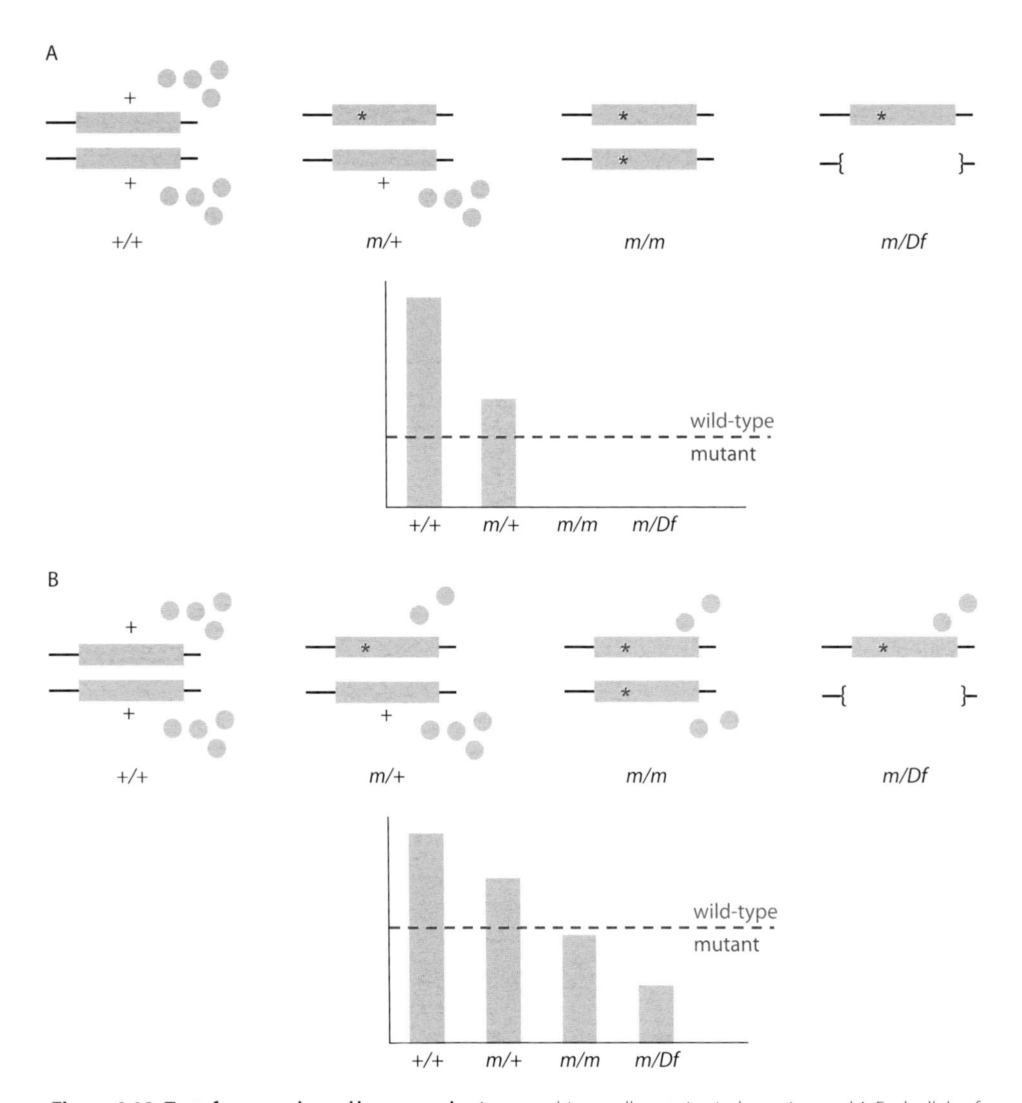

Figure 4.10 Tests for amorphs and hypomorphs. An amorphic or null mutation is shown in panel A. Each allele of the wild-type gene produces an amount of the protein, shown here as the four gray balls and summarized in the bar graph below for gene activity. An amorphic mutation eliminates the function of the gene. A heterozygote makes only half as much gene product, but since the heterozygote has a wild-type phenotype, the amorphic allele is inferred to be recessive. This is shown by the location of the dotted line in the bar graph. The dotted line in these graphs represents the functional threshold between a wild-type and mutant phenotype. A homozygote for the null allele makes no functional gene product and is mutant. When the null allele is placed opposite a deletion or a deficiency (Df) of the gene, no change is seen in the phenotype; the null or amorphic allele behaves like a deficiency in this test. A hypomorphic mutation is shown in panel B. In this case, the mutation only reduces the function of the gene but does not eliminate it. As a mutant homozygote, *m/m* has a mutant phenotype but retains some functional activity of the gene. This can be inferred when the hypomorphic mutation is placed opposite a deficiency. In this case, the deficiency has less activity than the hypomorphic mutation and the mutant phenotype becomes more severe.

BOX 4.4 Microphthalmia mutations in mice—an allelic series

Mutations that affect coat color in the mouse were among the earliest and easiest mutations to be studied in mammals; many of these genes are pleiotropic and affect other processes in addition to coat color and pigmentation. Mutations in the *microphthalmia* (*mi*) gene were first described in the 1940s and affect both coat color and, as the name suggests, eye development. Hearing loss is also frequently associated with these mutations, although it is not always reported as a key aspect of the phenotype. All of these affected tissues have a common cellular origin in melanocytes—cells responsible for pigmentation in the skin, eyes, and ears, as well as for cochlear functions in the ear. As we describe below, some alleles also have what appear to be ectopic effects in other tissues. Mutations in the human ortholog of the gene are responsible for Waardenburg syndrome type IIA, a dominant syndrome characterized by pigmentation alterations and hearing loss. Several studies have correlated the molecular lesion of mutations with the phenotype, and the gene provides examples for the topics discussed in the main chapter. Both recessive and dominant mutations have been identified. This text box describes an allelic series among the recessive mutations. Some of the dominant mutations are discussed in Text Box 4.5.

Although the gene was identified based on its mutant phenotypes, it is easier to explain the mutant defects in light of the molecular biology of the gene and its predicted protein. The *mi* locus is predicted to encode a transcription factor of the basic helix–loop–helix leucine zipper (bHLHZip) class. The protein is known as MITF, a name also now applied to the gene. bHLHZip proteins form homo- or heterodimers, usually with another bHLHZip protein, via the leucine zipper. Subsequent studies of MITF indicate that it functions as a homodimer, although it can form heterodimers with some closely related bHLHZip proteins *in vitro*. The basic helix regions of the dimeric protein, joined together by the loop region, bind to specific DNA sequences to activate transcription. The two-domain structure of these proteins, with the dimerization domain dependent on the leucine zipper and the DNA binding domain dependent on the basic helix–loop–helix region, is relevant to understanding their mutant phenotypes. Since functions of the protein family have been correlated with particular structural features that can be recognized in the amino acid sequence, it was possible to infer the functional defect from the molecular lesion. We will discuss four of the recessive mutations. Their effects are summarized in Text Box Figure 4.3.

- ***eyeless-white*** The most severe mutant phenotype among the recessive mutants is found with *microphthamia eyeless-white* (the gene symbol is $Mitf^{mi\text{-}ew}$). Since all the mutations we are discussing are alleles of the same gene, the alleles will be called by the superscript name, such as *eyeless-white*. Homozygotes for *eyeless-white* have a white coat with the eyes reduced and almost absent. Molecular analysis indicated that the mutation is an intragenic deletion that removes most of the basic helical region of the protein. Thus, the mutant protein, if it is made at all, would be unable to bind DNA to activate transcription. As expected from this lesion, the *eyeless-white* mutation behaves as a null allele and both pigmentation and eye development are knocked out.
- ***cloudy-eyed*** A slightly less severe mutant phenotype is seen in the mutant *cloudy-eyed*. This mutant also has a white coat, but it does form eyes. The eyes are typically reduced in size and cloudy white in pigmentation. Thus, although neither mutant has pigmentation in the coat, eye development in *cloudy-eyed* is not as severely affected as in the *eyeless-white* mutation. The *cloudy-eyed* mutation has a stop codon that is predicted to truncate the protein without a leucine zipper. Based on studies in other bHLHZip proteins, the leucine zipper is required for high-affinity dimerization but retains the sequence specificity in DNA binding, albeit with a lower affinity than wild-type. By inference, this protein may be able to bind some of the regulatory sequences and provide residual function, and the hypomorphic phenotype is consistent with this inference. One possible interpretation is that, in wild-type development, pigmentation requires more of the functional protein than does eye development. If this is true, a partly functional protein will have a more severe effect on pigmentation than on eye development.
- ***vitiligo* and *spotted*** Two other recessive mutations have less severe mutant defects. The *vitiligo* mutant mouse has relatively normal pigmentation in both coat and eyes when young, but gradually loses coat pigmentation as it ages; old adult mice are nearly white and experience retinal degeneration. The mutant phenotype arises from a missense mutation in the first helix of the dimerization domain that is postulated to result in unstable protein dimers. It may be that the instability results in the age-related change in the phenotype. In contrast with the other recessive mutations we have discussed, the *spotted* mutation has no visible phenotype in coat or eye color, although a close inspection shows that it has reduced numbers of the pigment-producing melanocytes. The spotted mutation affects a splice acceptor site and appears to result in alternative splicing. When the *spotted*

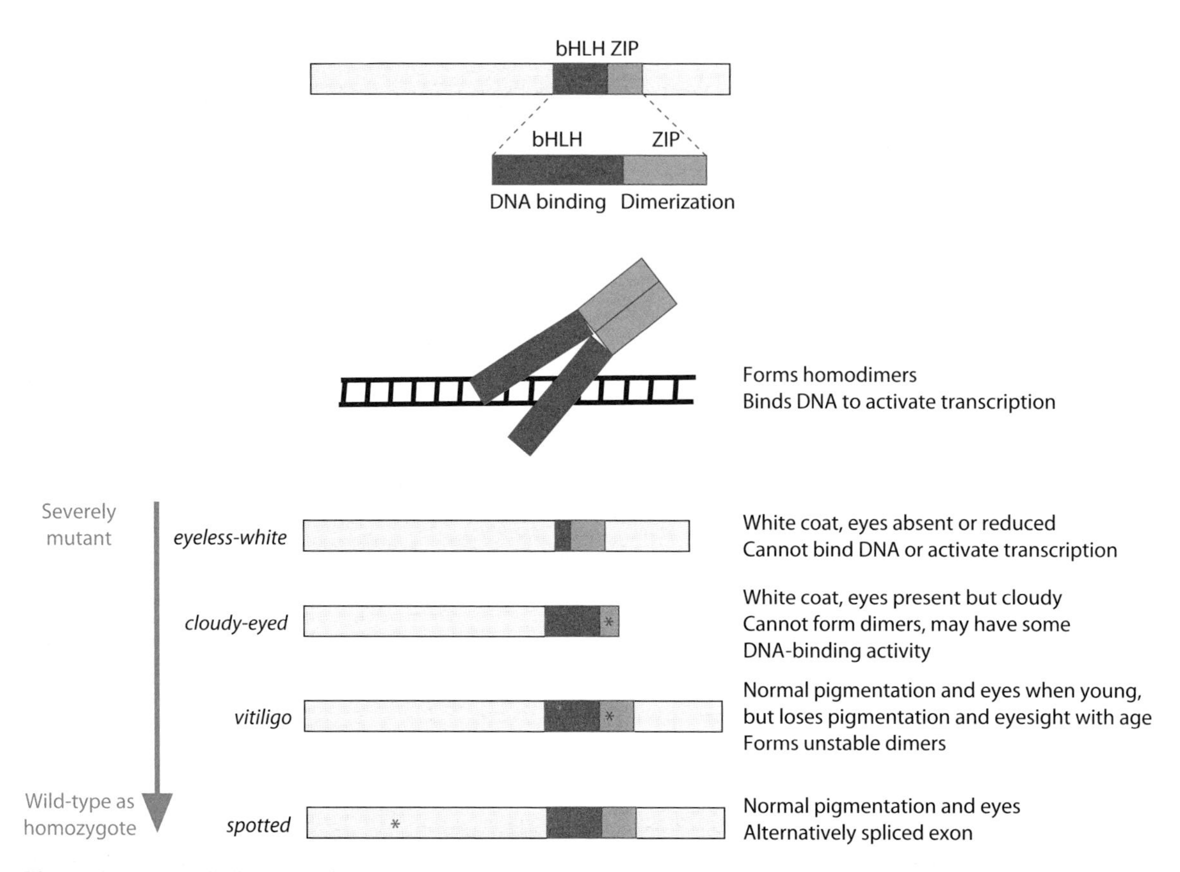

Figure B4.3 An allelic series for recessive alleles of the microphthalmia-associated transcription factor (Mitf) in mice. The *Mitf* gene encodes a basic helix–loop–helix (bHLH) leucine zipper (ZIP) protein which normally forms homodimers. Dimer formation depends on the ZIP domain, whereas DNA binding requires the bHLH domain. Many mutants are known, of which four recessive alleles are shown. The *eyeless-white* allele has a deletion of the bHLH domain and is apparently a null mutation. The other alleles are hypomorphic, with varying effects on both coat pigmentation and eye development. The *spotted* mutation has wild-type pigmentation and eye development but shows a spotted coat phenotype when heterozygous with the *eyeless-white* mutation.

mutation is placed as a *trans* heterozygote with a null mutation in the *Mitf* gene, such as *eyeless-white*, the animal has white spots on its coat. Thus, *spotted* is also a hypomorphic mutation, retaining enough activity for normal function when homozygous but not when the other allele is providing no activity. The effect on coat color with normal eye development also reinforces the view that pigmentation and coat color require more functional activity from the gene than does eye development.

The pleiotropic effects of these mutations help us to place them in a tentative allelic series, with *eyeless-white* being the most severe and *spotted* the least severe, with *cloudy-eyed* and *vitiligo* being intermediates. We have presented only some of the phenotypes and alleles for this gene. Many other alleles are known, and the mutant phenotype can be correlated with the development of cells derived from the neural crest, which includes both the melanocytes involved in pigmentation and cells involved in eye development. The interested reader is referred to the mouse genome database at http://www.informatics.jax.org/

Further reading

Steingrimsson, E., Moore, K.J., Lamoreux, M.L., *et al.* (1994). Molecular basis of mouse *microphthalmia* (*mi*) mutations helps explain their developmental and phenotypic consequences. *Nature Genetics*, **8**, 256–63.

Case Study 4.1

Part 3: Find a mutant: segmentation in *Drosophila* embryos

Classifying mutants: types of allele

Because of the mutagen and the methods used to find them, nearly all of the mutations recovered in these screens were recessive: among the more than 20 000 mutations recovered of all types, fewer than 40 proved to be dominant. None of the dominant mutants was tested further. Since most of the genes had many recessive mutant alleles, the investigators were able to compare the phenotype of different mutations in the same gene. That is, they could attempt to construct an allelic series for some genes. In addition, by mapping the mutations using deletions, the investigators could also compare the phenotype of a deletion with the phenotype of a mutant allele—Muller's test for hypomorphs and amorphs as described in the chapter.

The information for the genes on chromosome 3 is both informative and typical of the results. Among the 32 genes with more than one mutant allele, the mutants have the same phenotype in 17 cases. Thus, no allelic series could be constructed for those 17 genes. Six of these 17 genes were tested with deletions, and in all cases the phenotype of m/deletion was the same as phenotype of m/m. Thus, for these genes, the phenotype arising from these mutations was considered amorphic or null. For 15 other genes, mutant alleles had different phenotypes and an allelic series could be constructed for several of these genes. Thirty-five mutations in five of these genes were tested with deletions, with the result that 19 were hypomorphic, 15 were amorphic, and one could not be readily classified. In sum, about two-thirds of the mutations were amorphic and one-third were hypomorphic.

In addition, all of the mutants were tested for temperature sensitivity by comparing the phenotype at the permissive temperature of 18°C and the restrictive temperature of 29°C. Eleven temperature-sensitive mutations were identified for nine different genes. Thus, about 5 percent of the mutations (11 of 198) were temperature-sensitive, and about 20 percent of the genes (9 of 45) had temperature-sensitive mutant alleles.

Classifying mutants: gaps, pair rules, and segment polarity

The range of mutant phenotypes found among these genes was remarkably diverse. Some mutants appeared to be missing entire blocks of the embryo, whereas others had missing or disorganized regions. When a person who does not routinely study *Drosophila* embryos looks at the array of mutant phenotypes, no apparent logic emerges. However, Nüsslein-Volhard, Wieschaus, and their collaborators did recognize a pattern in the defects, and were able to divide the mutant phenotypes into three main categories: gap genes, pair-rule genes, and segment polarity genes. The categories are summarized in Case Study Figure 4.3. One category of mutants is missing large regions of the embryo, resulting in embryos that are often much smaller. That is, these mutants have gaps in their segmentation pattern, and the genes are collectively called the **gap genes**. The inference was that these genes are needed for the organization of broad regions of the embryo.

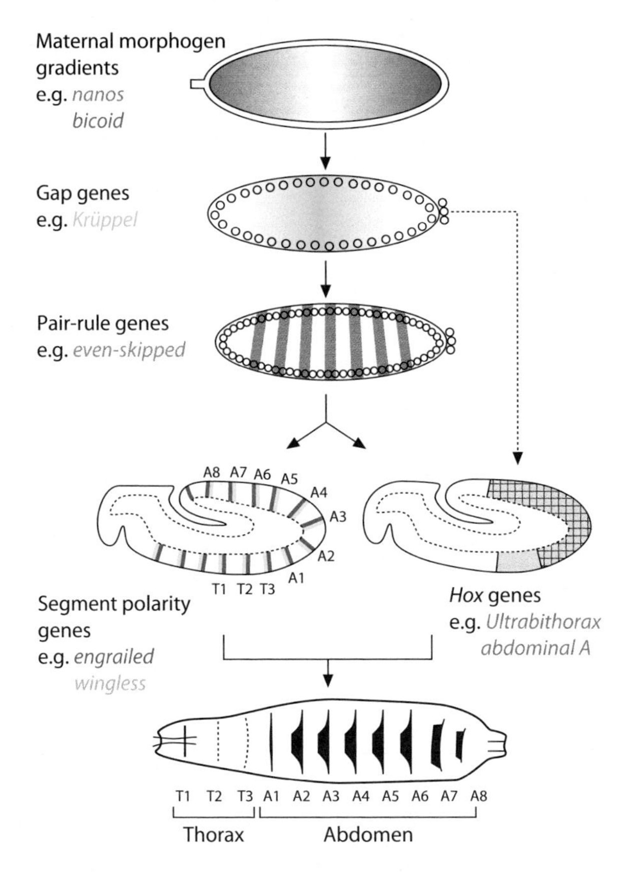

Figure C4.3 Gap genes, pair-rule genes, and segment polarity genes. The initial polarity of the embryo is established by the mother in the ovum with maternal-effect genes. Regions of the embryo are specified by the actions of the gap genes, of which *Krüppel* is an example. These regions are further subdivided by the actions of the pair-rule genes, including *even-skipped.* These subdivisions result in the establishment of 14 segments. The polarity within these segments is determined by the segment polarity genes, including *engrailed*, *wingless*, and *patched.*

A second group of genes has probably the most unexpected mutant phenotype, in that they affect particular combinations of segments, often a pair of segments or alternate segments but typically only seven of the 14 segments. These genes are collectively called the **pair-rule genes**. The inference from this mutant phenotype is that the broad regions of the embryo defined by the gap genes are further subdivided by the pair-rule genes, so that each segment is specified by a combinatorial code of gap and pair-rule gene expression.

A third group of genes affects every segment, often by altering the polarity or orientation of the segment. This group, the largest number of genes found, is known as the **segment polarity genes**. The inference from this group is that each segment is specified by the expression of some combination of gap and pair-rule genes, but the orientation of structures within each segment is specified by the segment polarity genes.

The *patched* gene

One of the segment polarity genes was named *patched* (abbreviated *ptc*) because the mutant phenotype has patches of bristles; to the untrained eye, the mutant phenotype as shown

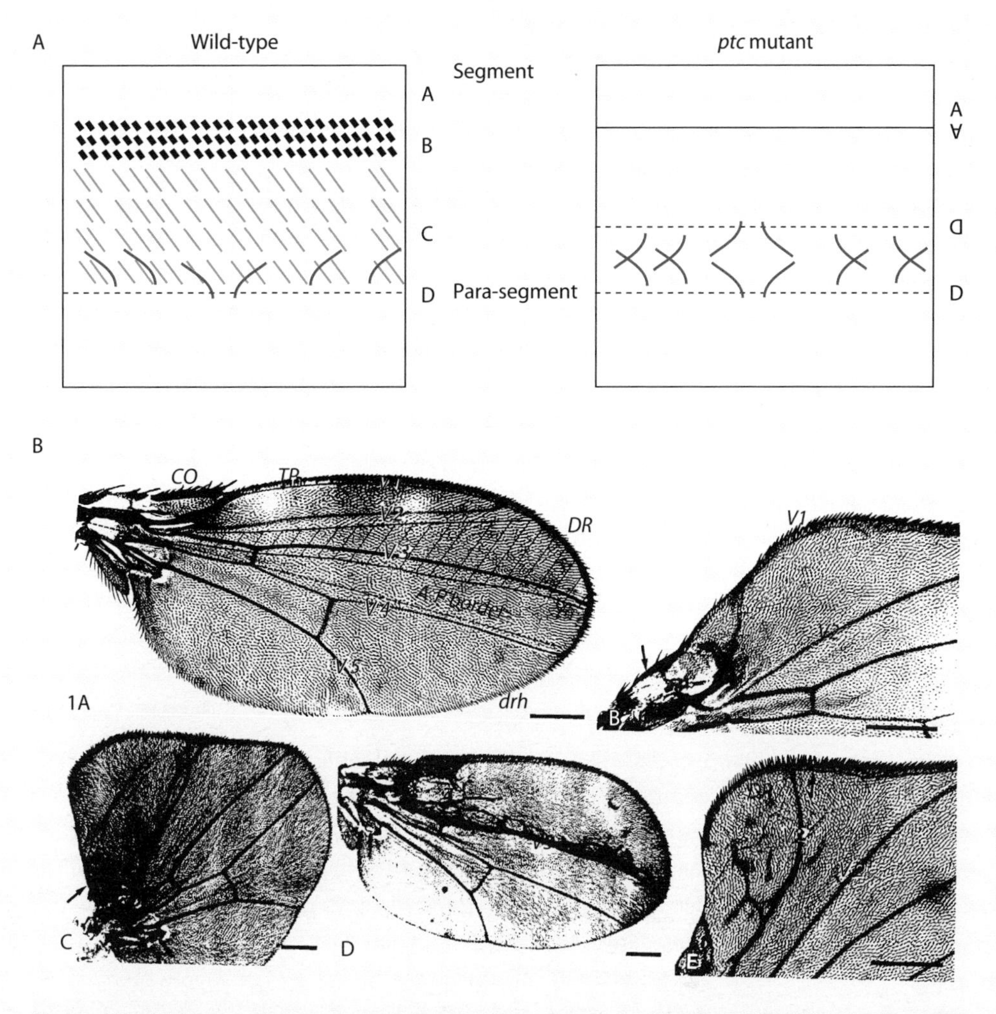

Figure C4.4 The *patched* mutant phenotype in embryos and in the wing. Panel A shows the effect of *ptc* mutants on embryonic segmentation. The drawing depicts the dorsal side of the third thoracic segment in *Drosophila* embryos. Anterior is to the top, posterior to the bottom. The wild-type segment can be divided conceptually into four zones, each with a characteristic pattern of bristles and hairs. These are shown in different colors for clarity. The most anterior portion of the segment has no hairs or bristles, and the region posterior to the para-segment boundary has no hairs or bristles. The *ptc* mutant phenotype can be interpreted as a deletion of zones B and C, with a mirror image duplication of zones A and D. There are two segment and two para-segment boundaries. This has been redrawn and modified from a diagram in Hooper and Scott (1989). Panel B shows the effect of *ptc* mutants on wing morphology. A wild-type wing (top left) and various *ptc* mutant alleles or combinations of *ptc* mutant alleles are shown. Redrawn from Phillips *et al.* (1990).

in Case Study Figure 4.4 looks a little like a man who shaved with a dull razor using a poor mirror, with patches of bristles remaining at unusual places. However, the pattern and orientation of the remaining bristles in *patched* mutants reveal that the mutant shows deletions of structures in adjacent segments with mirror-image duplications of remaining structures. The mutants have twice the normal number of segment boundaries, with reversed polarity around them. Since *patched* will be used as an example in other parts of the book, we will describe some of the characterization here in a little more detail.

In the original screens, 12 mutant alleles of *patched* were found, including one temperature-sensitive allele and one other hypomorphic allele that affected fewer structures than the null alleles. The *patched* gene mapped to a location that corresponded to that of a gene called *tufted*, found more than 30 years earlier but not extensively studied. *tufted* mutations are viable, with variable effects on the wing and lesser effects on bristles and structures in the head, eye, and antenna. In the wing, *tufted* alleles cause deletions of some structures including wing veins, sensory organs, and bristles, with duplications of other structures in the anterior compartment of the wing, including other veins and bristles; in severe examples, the wings are much shorter than wild-type and are shaped liked paddles.

The *tufted* mutations and the *patched* mutations map to the same location and have some similarities in their effects on bristles, although at different times in development and in different tissues. Given the similarities, it was logical to test whether these were different alleles of the same gene. When complementation tests were done between the *tufted* mutations and the *patched* amorphic mutations, the mutations failed to complement for both the embryonic lethality and the wing defects, demonstrating that these are alleles of the same gene. The *tufted* alleles are hypomorphic, whereas the *patched* alleles are more likely to be null. These classes of mutant phenotypes show an advantage of having an allelic series for a gene. If the phenotype of the *tufted* mutations was considered without knowledge of the *patched* mutation phenotype, the important function of the gene in embryo segmentation could be overlooked. Conversely, in the absence of the *tufted* hypomorphic alleles, the effects of the gene on adult tissues would have been missed.

In subsequent studies using an antibody to monitor the levels of Ptc protein, a method described in Chapter 8, the null alleles had no detectable amounts of the wild-type protein in the wing, whereas a *tufted* hypomorphic allele had greatly reduced amounts of the protein; this was reduced even further in a hypomorphic *tufted*/null *ptc* heterozygote. Thus, our description of hypomorphic and amorphic alleles in Figure 4.10 holds true in this case. The gene name *tufted* was dropped in favor of the name *patched*, and the original *tufted* alleles were designated alleles of *ptc*.

Although the original screens for all segmentation genes found 12 alleles of *patched*, other investigators subsequently studying the *patched* gene itself conducted more mutant screens to find even more alleles. Additional *patched* alleles, including some that were induced by P elements, were also found by an F_1 screen. These alleles were used to clone the gene by transposon tagging, described in Case Study 5.2. The F_1 screens found about a dozen more alleles of *patched*, including viable alleles (similar to the *tufted* alleles) and embryonic lethal alleles. The viable alleles were used to examine the effect of *patched* on the wing and on other adult structures. The analysis of these mutations on wing development will be described in Chapter 9, and will also be important in some of the experiments discussed in Case Study 11.1. More than 140 mutant alleles of *patched* have now been identified.

Find and classify a mutant: a triumph for genetic analysis

It may be hard for a contemporary reader to understand the excitement that surrounded this mutant collection when it was published in 1984. Here at last were representative alleles of most, if not all, of the genes that affect segmentation in *Drosophila*, a distinctive but mysterious biological process. Now at last we had the genetic tools to analyze this process. The papers on individual genes began appearing quickly afterwards and scarcely slowed over the next two decades as orthologs of these genes were found in other organisms. Each of these 120 genes has its own interesting story to tell, and similar accounts could be made for the mutant analysis of many other biological processes. In subsequent chapters, we focus on the *patched* gene because most of the tools of genetic analysis were used in studying it. The bristle pattern may have been reminiscent of a poorly shaven man, but the gene has much more to tell us.

Further reading

Hooper, J.E. and Scott, M.P. (1989). The *Drosophila patched* gene encodes a putative membrane protein required for segmental patterning. *Coll* **59**, 751–65.

Jürgens, G., Wieschaus, E., Nüsslein-Volhard, C., Kluding, H. (1984). Mutations affecting the pattern of the larval cuticle in *Drosophila melanogaster. Roux's Archives of Developmental Biology*, **193**, 283–95.

Nüsslein-Volhard, C. and Wieschaus, E. (1980). Mutations affecting segment number and polarity in *Drosophila. Nature*, **287**, 795–801.

Nüsslein-Volhard, C., Wieschaus, E., and Kluding, H. (1984). Mutations affecting the pattern of the larval cuticle in *Drosophila melanogaster. Roux's Archives of Developmental Biology*, **193**, 267–82.

Phillips, R.C., Roberts, I.J, Ingham, P.W., and Whittle, R.S. (1990). The *Drosophila* segment polarity gene *patched* is involved in a position-signalling mechanism in imaginal discs. *Development*, **110**, 105–14.

Wieschaus, E., Nüsslein-Volhard, C., and Jürgens, G. (1984). Mutations affecting the pattern of the larval cuticle in *Drosophila melanogaster. Roux's Archives of Developmental Biology*, **193**, 296–307.

Null and hypomorphic alleles of a gene can be placed in an allelic series

Both amorphic and hypomorphic mutations provide useful information in understanding the function of the gene. The amorphic mutation phenotype reveals the stage or tissue that *first* requires the gene product. Often the true function of a gene can best be inferred from its amorphic mutation when the function is absent. The interpretation of mutant phenotypes is discussed in more detail in Chapter 9. Certainly when we discuss gene interaction tests in Chapters 10 and 11, it is important to know the type of allele and to use null mutations whenever possible.

On the other hand, the amorphic phenotype for an essential gene cannot reveal functions of the genes that occur later in development. For essential genes, hypomorphic mutations are more useful. Hypomorphic mutations may have enough residual wild-type function to accomplish an early stage but not enough to carry out a later step in development. This same logic works in terms of tissues or organs. This is illustrated in Figure 4.11, for a hypothetical insect. Imagine that the wild-type function of a particular gene is needed for both wing and leg development, but that the wing requires much more of this function than does the leg. A hypomorphic mutation may develop legs (perhaps with defects) but have defective wings because of the difference in the requirement of the two appendages whereas an amorphic mutation will probably fail to develop either wings or legs normally. This concept is developed more thoroughly in Chapter 9 in the discussion of inferring gene activity from mutant phenotypes.

In the course of acquiring many mutant alleles of a gene, the investigator will sometimes be able to place them in a phenotypic order, from the most severely affected (the strongest allele) to the least severely affected. This is called an **allelic series** of mutant strengths; realizing the strength of a mutant refers to how much of the normal function has been lost rather than how much has been retained. An allelic series for the *white* gene in mice is shown in Figure 4.12 and described for the *microphthalmia* gene in Text Box 4.4 for the recessive alleles and Text Box 4.5 for the dominant alleles. Many of the mutations involved in segmentation in *Drosophila* have also been placed in an allelic series. We will return to allelic series in different contexts throughout this book because having a series of slightly differing mutant phenotypes can be used to understand the functions and interactions of a gene. Some of the best examples of allelic series can be found at Online Mendelian Inheritance in Man (OMIM), which can be accessed

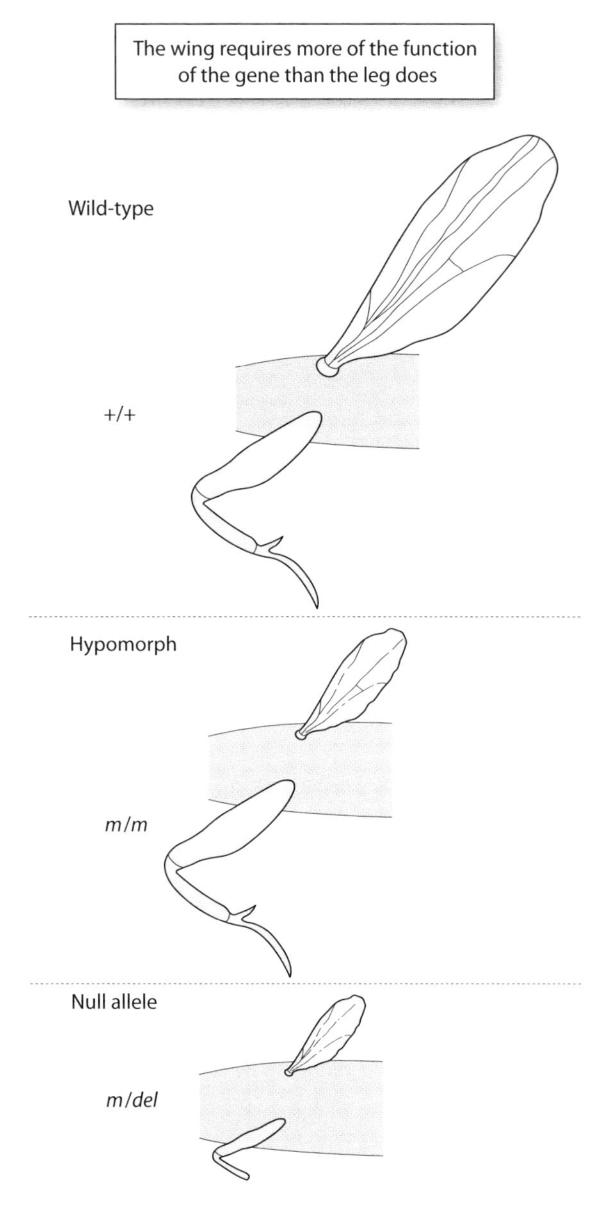

Figure 4.11 Hypomorphic mutations affecting different tissues. The wing and the leg of a hypothetical insect are shown. Mutations of the gene in question affect the wing more strongly than the leg; perhaps more of the gene product is needed for wing development than for leg development. A hypomorphic mutation, shown in the center, has reduced wings but normal legs. When the hypomorphic mutation is placed opposite a deletion, as shown at the bottom, both limbs are reduced in size.

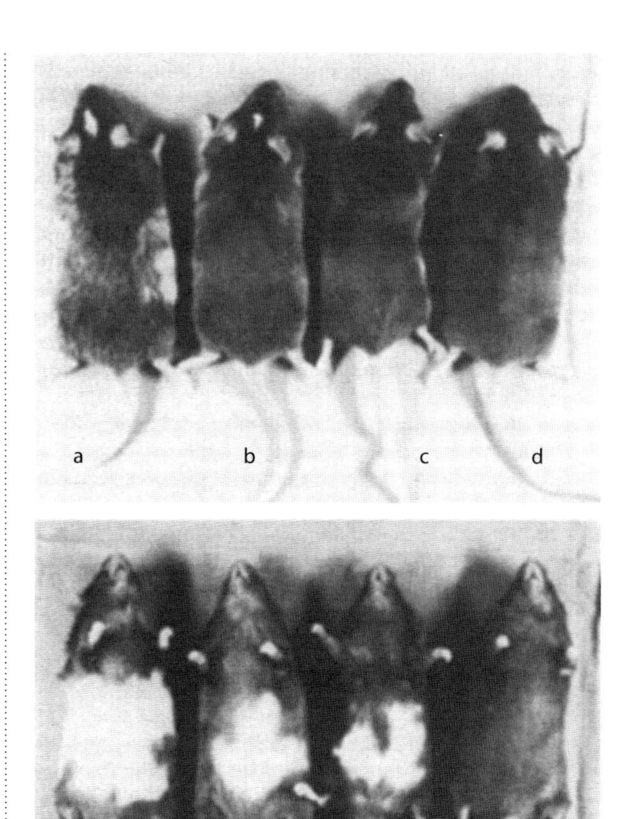

Figure 4.12 An allelic series of the *white* gene in mice. The amount of white pigmentation increases in more severe mutant alleles, so that the mutant labeled 'd' on the right has an almost normal coat color. Many alleles of the *white* gene are known, of which four are shown here.

via the NCBI web site at www.ncbi.nlm.nih.gov/sites/entrez. This remarkable database describes the phenotypes and molecular lesions of the tens of thousands of different allelic variants that have been described for human genes.

These historical definitions of mutants do not require any particular knowledge of the molecular basis for the mutation and, in fact, many different molecular lesions could be seen as amorphic or hypomorphic mutations. As a general rule, hypomorphic mutations are likely to be **missense mutations** in which one amino acid is substituted for another that is slightly less functional. This generality does not work in reverse because many missense mutations are amorphic if, for example, the substituted amino acid is at the active site of the protein. Null mutations can arise from missense mutations, nonsense mutations, insertions, small deletions, splicing defects, and many other types of molecular lesions.

BOX 4.5 Microphthalmia mutations in mice—dominant mutations

In addition to the recessive mutations described in Text Box 4.4 that eliminate or greatly reduce gene function, the *microphthalmia* gene (*Mitf*) is also defined by a number of dominant or semi-dominant mutations. That is, the dominant mutations at the *Mitf* locus (m^{Dom}) have a mutant phenotype in m^{Dom}/+ heterozygotes but a more severe and slightly different phenotype when homozygous. As in Text Box 4.4, we will refer to the mutations by their superscript names. For example, mice heterozygous for the *Oak Ridge* mutation have a diluted coat color, paler pigmentation on the ears and tail, and a white streak on the belly or head (Text Box Figure 4.4). In homozygous mice, the coat is completely white and the eyes are either very small or absent. The *white* mutation has an overall similar phenotype: a diluted coat color and paler pigmentation in heterozygotes, with homozygotes being white with small and unpigmented eyes. Each of these mutations affects amino acids involved in DNA binding but not in dimerization, i.e. the mutant protein is predicted to be able to form a dimer with the normal protein. However, the dimeric proteins with one normal subunit and one mutant subunit cannot bind DNA or activate transcription. Therefore, the mutation has a dominant-negative or antimorphic effect and impairs the function of the normal

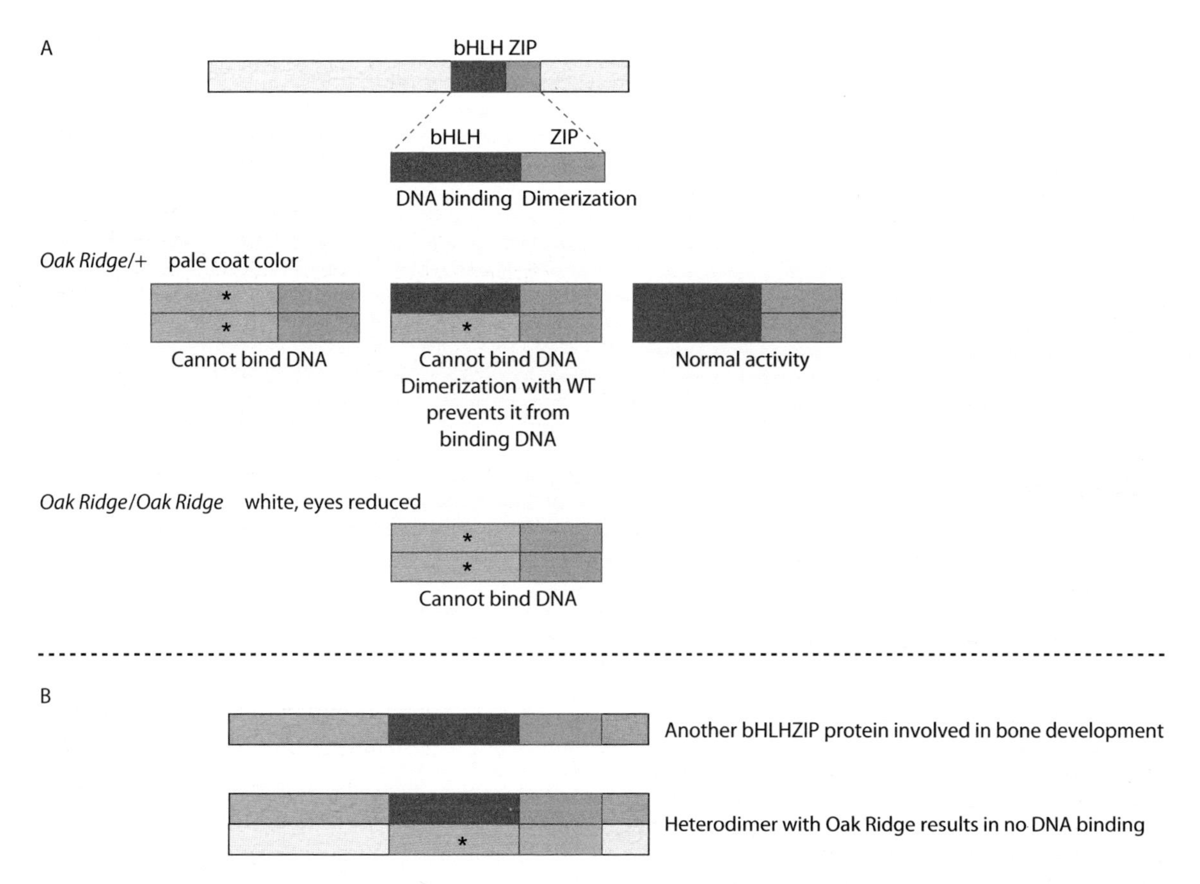

Figure B4.4 Dominant alleles of microphthalmia-associated transcription factor (Mitf). Numerous dominant alleles of the *Mitf* gene are known, one of which is the *Oak Ridge* mutant. The *Oak Ridge* mutation is a transition in the basic region that affects DNA binding. As shown in panel A, a heterozygous mouse makes both normal and mutant Mitf protein, and has some residual gene activity from the homodimers of the normal protein. The normal protein and the Oak Ridge mutant protein can form a dimer but cannot bind DNA, so the mutation has a dominant-negative effect. Panel B suggests one way that the *Oak Ridge* mutation could have a neomorphic effect on bone formation. Another bHLHZip protein is needed for normal bone formation. Expression of the Oak Ridge protein in the same cells could result in the formation of heterodimers between the normal bone growth transcription factor and the mutant Oak Ridge protein. The heterodimer cannot bind to DNA because of the *Oak Ridge* mutation, and so the activity of the bone growth transcription factor is blocked.

protein. Some residual function remains in the heterozygotes because the mutant and normal subunits are both made, so some dimers will have two normal subunits and function normally.

At least some of the dominant mutations also exhibit a gain of function or neomorphic effect that may arise from ectopic expression of the gene. Both the *Oak Ridge* and *white* mutations described in the preceding paragraph result in bone overgrowth or osteopetrosis. Bone growth phenotypes are not observed in mutants that completely eliminate gene expression or in some other dominant mutants. One explanation lies in the postulated effect of these mutations on dimerization and DNA binding. Many bHLHZip proteins can form heterodimers as well as homodimers, and the balance between these two forms is important for expression for the proper target genes. It is possible that a bHLHZip protein is needed for osteoclast function and bone development. One postulate is that these dominant *mi* mutations can form heterodimers with a bHLHZip protein needed for bone growth. Formation of these (putative) heterodimers would block the normal function of the osteoclast bHLHZip protein since the heterodimer cannot bind DNA or it binds to DNA sequences other than its normal binding site. The interpretation that this is a neomorphic activity arises from the failure to see effects on bone growth in other mutants, so there must be some unusual property of these two alleles. In fact, a related bHLHZip protein known as Tfe3 is expressed in osteoclasts and is needed for their differentiation. Although the two proteins can form heterodimers *in vitro*, they appear to form only homodimers *in vivo*, so the neomorphic effect might arise from these unusual heterodimers. However, the effects are rather complicated, and other explanations are also possible.

Microphthalmia mutations and Waardenburg syndrome

Microphthalmia mutations have been used as a model for understanding Waardenburg syndrome in humans. The syndrome has complex and diverse effects on hearing, skin and hair pigmentation, and eye development, similar to what is seen in microphthalmia mutants. At least three distinct types exist, known as types I, II, and III, and at least four different genes are responsible for subtypes IIA, IIB, IIC, and IID. Mutations in the human *MITF* gene have been associated with cases of Waardenburg syndrome type IIA. Waardenburg syndrome is inherited as a dominant disorder, and so it seems logical to suggest that the dominant alleles of the mouse *Mitf* gene could provide insights into the human disorder. However, this hope has proved to be overly simplified. Most of the human mutations are not dominant-negative alleles, as is the case for the mouse dominant alleles. Rather, three of the eight studied human mutations are splicing defects that would reduce or eliminate the function of the gene. One possible explanation is that the MITF protein is more dose sensitive in humans than in mice, so that mutations that reduce or eliminate the function of the gene are haplo-insufficient. The complexity of these phenotypes demonstrates both the simplicity and the limitations of compiling an allelic series in the absence of molecular data about the gene and its product.

Further reading

Steingrimsson, E., Moore, K.J., Lamoreux, M.L., *et al.* (1994). Molecular basis of mouse *microphthalmia* (*mi*) mutations helps explain their developmental and phenotypic consequences. *Nature Genetics*, **8**, 256–63.

Conditional mutations are often hypomorphs

In Chapter 3, we described temperature-sensitive mutations and suggested an explanation for how they could result in slightly unstable proteins. By this explanation, temperature-sensitive mutations are predicted to behave as hypomorphic mutations. This is generally true, and it is frequently observed that m^{ts}/m^{ts} has a temperature-sensitive mutant phenotype, whereas m^{ts}/deletion or m^{ts}/null has a non-conditional mutant phenotype. This is also the reason why most F_1 non-complementation screens are done by finding mutations that fail to complement a null allele, since this will result in a more severe mutant phenotype than using a hypomorph for the complementation test.

Case Study 4.1, Part 3, recounts some of the examples of different types of mutant alleles encountered for genes affecting segmentation in *Drosophila*. Because many of the mutations were mapped using deletions, determining whether an allele is null or hypomorphic was often done as the mutations were mapped and complementation tests were performed. This helps to illustrate the interrelatedness of the processes of mapping, complementation testing, and classifying mutations. The last section of the case study also introduces the gene *patched*, one of the genes that will serve as a recurring example throughout the book. As described in the case study, investigators studying *patched* specifically isolated a number of new alleles using an F_1 non-complementation screen. These alleles proved to be important when the gene was subsequently cloned.

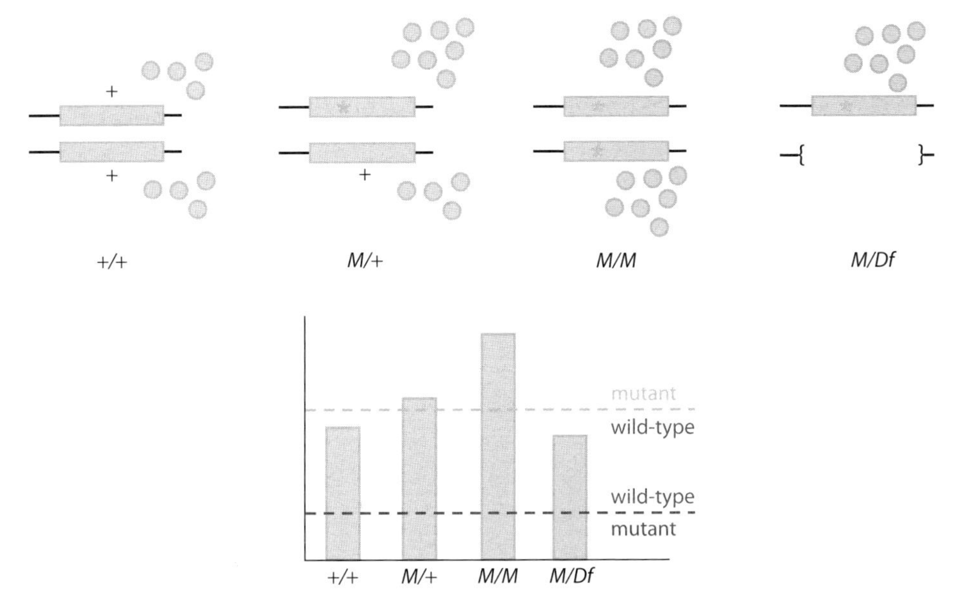

Figure 4.13 Test for hypermorphic mutations. A hypermorphic mutation, represented by *M* and the green star, over-produces the gene product or the gene function. The over-production results in a mutant phenotype in both heterozygotes and homozygotes. A deletion that removes the wild-type copy of the gene may restore the wild-type phenotype. Compare this figure with the recessive amorphic and hypomorphic mutations in Figure 4.10.

Dominant mutants generally arise from over-producing a normal function

Although most mutations are recessive, mutations that are dominant to wild-type are found in some genes. Some human genetic syndromes are primarily recognized from dominant mutants. For example, the form of dwarfism known as achondroplasia is dominant to wild-type, as are syndromes such as polydactyly (extra digits), Huntington's disease, and several cancer syndromes. Dominant mutations, or the gene that they affect, often have some unusual feature rather than simply reducing the function of the gene as is the case with recessive mutations. In fact, the molecular explanation for dominance is different in each of the examples listed.

Most dominant mutations are thought of as **gain-of-function** mutants, but a precise definition of 'gain of function' is rather elusive. Muller divided dominant mutants into three categories, although not all of his terms are widely used nowadays. The most frequently encountered type of dominant mutant is the class Muller referred to as hypermorphic mutations; many geneticists nowadays call these **over-producers**. These mutants produce the normal gene product, usually in the same cells as the wild-type gene, but the mutant phenotype results in the production of too much gene product. For example, the hypermorphic mutation may eliminate a site or sequence needed for repression, and the gene activity is over-produced as a result. The site or sequence needed for repression could occur at any level of gene expression. Hypermorphic mutants could include a mutation that results in high levels of transcription or the failure to degrade the transcript; thus, the mutation affects the level of stable RNA being produced. In either case, too much gene product is being made, and the activity of the gene is over-produced. Hypermorphic mutants could also include a mutation that results in a protein that is not degraded or that fails to interact with a repressor protein such as a kinase or phosphatase; thus, the mutation affects the level or the activity of the protein product from the gene. At the level of mutational analysis, all of these mutants have roughly the same effect and the gene product is over-produced.

Muller's test for identifying a hypermorphic mutation was to vary the mutant and the wild-type doses of the gene. His strategy is shown in Figure 4.13. The test is essentially the same as that used for amorphs and

DEFINITION

hypermorphic: A hypermorphic mutation over-produces the activity or function of a gene.

hypomorphs but the expected results are quite different. Suppose that *M* is a dominant mutation, perhaps one that causes notches in the margin of a wing in Drosophila. Thus, *M*/+ has a more severe mutant phenotype than does +/+, which has no mutant phenotype at all. If *M* is a hypermorphic mutation, then *M*/*M* should have a more severe mutant phenotype than does *M*/+, with more or deeper notches in the wing margin. The partial trisomic *M*/*M*/*M* (if it can be examined) should be even more severely mutant than *M*/*M*. That is, each additional dose of the mutant allele makes the phenotype more severely mutant because the mutation is causing an over-production of the function of the gene.

A similar and important supplemental test can be done by decreasing the copies of the wild-type allele by using a deletion. If *M*/+ is mutant and *M* is a hypermorphic mutation, then *M*/deletion should be less severely mutant than *M*/+; the amount of gene product has been diminished by replacing the wild-type allele with a deletion. Imagine this in terms of simple arithmetic. If the + allele contributes one unit of functional product, then the normal diploid has two units. (We are not attempting to define a functional unit in this example.) The *M* allele that over-produces the gene product could be thought of as making, say, two functional units by itself. Thus, the *M*/+ will have three units and be mutant, and the *M*/*M* will have four units and be even more severely mutant. However, the *M*/deletion strain will have two functional units, with both coming from the *M* allele and none from the deletion chromosome. This is the same level as the wild-type, and so it should be less severely mutant than *M*/+ and, in our contrived example, may even resemble the wild-type.

Many dominant oncogenes in vertebrates are the result of hypermorphic mutations, often ones in which the function of the protein cannot be repressed. For example, the normal cellular and the oncogenic versions of the Src proteins differ at their C termini. In the normal cellular protein, phosphorylation at a tyrosine residue near the C terminus inactivates the protein. The activity of the oncogenic version of the protein is over-produced because it lacks this tyrosine.

Another example is found in the *ras* oncogene in humans and its orthologs in other organisms. The Ras proteins are part of a well-known signaling cascade that is used in a variety of processes in different organisms. The Ras protein binds guanine nucleotide, such that Ras is active when bound with GTP but inactive when bound with GDP. Hydrolysis from GTP to GDP requires interaction with a GTPase activating protein (GAP). Mutations that replace the glycine at position 12 in Ras prevent the interaction with GAP and thus prevent hydrolysis to GDP. As a result, Ras is constitutively bound to GTP and active, so Ras activity is increased. This residue is conserved across many species, and mutations at this position in other species are also hypermorphic.

In the examples of dominant mutations discussed so far, the hypermorphic phenotype arises from a mutation that makes the protein insensitive to inactivation by another protein. Hypermorphic mutations can also arise from other forms of over-production of an active protein. Examples of hypermorphic mutations are known in which the protein is not degraded and thus persists, in which the gene is over-transcribed because of a mutation in a regulatory region, and in which the gene has become duplicated at its site in the genome. In all cases, the over-production of a normal form of the protein results in a dominant mutant phenotype. By definition, all of these over-producer mutations are hypermorphic.

Hypermorphic mutations are also encountered when the expression level of a gene is manipulated experimentally. For example, the coding region of a gene can be placed under a transcriptional regulatory region that expresses at a high level. When this is reintroduced into the organism, as described in Chapter 2, the gene is over-expressed and a mutant phenotype is sometimes observed.

Dominant hypermorphic mutations may not exist for all genes because the biological process is relatively insensitive to the dose of that particular gene product. Hypermorphs are analogous to having too much of a good thing. For some good things, such as health and happiness, having too much is not a problem, and does not produce an aberrant phenotype. For some other good things, such as money, having too much may be a problem in some circumstances, and an unusual phenotype can arise. For yet other good things, such as back copies of *Genetics*, having too much is definitely a problem once the accumulation reaches a critical level. We return to this question about the level of gene expression and gene activity in Chapters 8 and 9.

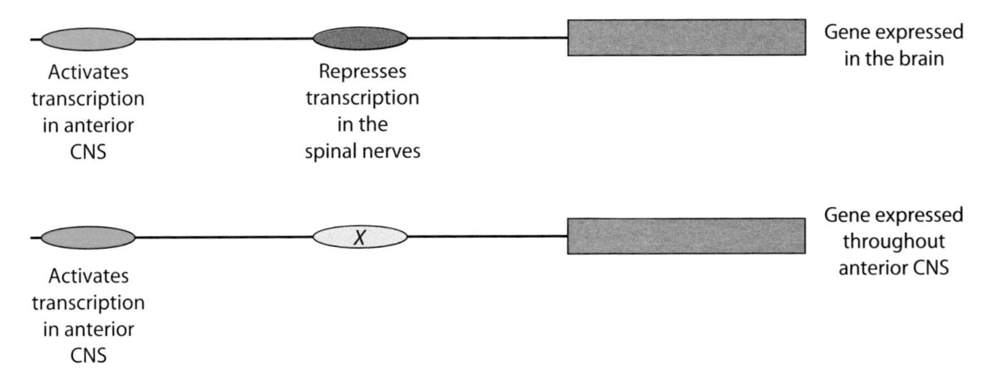

Figure 4.14 Neomorphs and ectopic expression. This hypothetical gene, whose expression is controlled by two upstream regulatory regions, is expressed in the brain in the wild-type. The regulatory region in green is responsible for activating transcription in the activator central nervous system (CNS). The regulatory region in red is responsible for repressing transcription in the spinal nerves. The combined effect of these regulatory regions in the wild-type is that the gene is expressed in the brain. A mutation that affects the red regulatory region might eliminate its repressive effect, resulting in ectopic expression of the gene throughout the anterior CNS. Neomorphic mutations result in a novel phenotype and ectopic expression is a common explanation.

Mutations producing an unexpected function can also be dominant

Muller recognized two other categories of dominant mutants, which he called neomorphs and antimorphs. Although the concepts are still useful, the terms have largely been replaced by other descriptors. Neomorphs are mutations that result in a **novel** function or novel phenotype. As a strict definition of novel phenotypes, this would seem to be uncommon because it suggests that the gene has somehow altered its nature or acquired some new properties. However, many neomorphic mutations cause the **ectopic** expression of the gene in a cell, tissue, or developmental stage when it is not usually expressed. Figure 4.14 illustrates one way that a neomorphic mutation could arise.

With this recognition of ectopic expression, neomorphic mutations are not so rare. The *Drosophila* homeotic mutation *Antennapedia* is a well-known example. The Antp protein is normally expressed in the foreleg, and if the locus is deleted the forelegs fail to develop. In the *Antennapedia* mutant, the Antp protein is expressed in the antenna and a foreleg arises at an ectopic location. Thus, a phenotype that to Muller may have looked like a novel function of the gene is due to expression of the gene in a novel tissue, giving rise to a neomorphic phenotype. One of the common Antp neomorphic mutations arises from mutations in the regulatory region, such that the gene is expressed in the wrong tissue.

An example in humans is illustrated by Burkitt's lymphoma, a solid tumor of B lymphocytes. The vast majority of cases of Burkitt's lymphoma are characterized by a specific reciprocal translocation between chromosomes 8 and 14 that fuses the regulatory region of the antibody heavy-chain genes on chromosome 14 with the protein-coding region of the *myc* gene, normally found on chromosome 8 (Figure 4.15). The Myc protein functions in regulation of the cell cycle but is not normally expressed at high levels in B lymphocytes, in contrast to antibody heavy-chain genes. The translocation places the *myc* gene under the control of the transcriptional enhancers of the antibody heavy-chain gene, resulting in unusually high and unregulated expression in a cell type that does not normally express the Myc protein. Other cases of Burkitt's lymphoma are due to translocations that place the *myc* gene under the control of other antibody genes located on chromosome 2 or 22. In other words, Burkitt's lymphoma is due to a dominant neomorphic mutation in the *myc* gene.

Antimorphic mutations **antagonize** the wild-type function of the gene; these mutations are most easily thought of as **poisons** of the normal function. Although they are dominant mutations, antimorphs behave like a loss of function in the gene rather than a gain of function. For this reason, the term antimorph has largely

DEFINITIONS

Neomorphs: Neomorphic mutants result in novel activities of a gene, often as a result of ectopic gene expression.

Antimorphic: Antimorphic or dominant-negative mutations poison or antagonize the normal function of the gene.

A

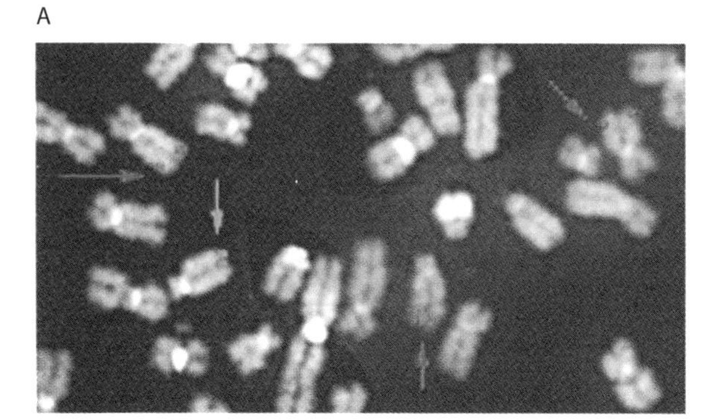

B

Figure 4.15 Burkitt's lymphoma arises from a neomorphic allele of the c-*myc* gene. Panel A is a two-color fluorescence *in situ* hybridization of metaphase chromosomes, with the *myc* gene identified by the red probe and the IgH gene shown in green. The *myc* gene is present in three copies, two on its normal locations of the chromosome 8 homologs and a third that has been translocated to chromosome 14 near the IgH gene. The translocation is shown diagrammatically on the right. The dominant mutant effect arises because the translocated *myc* gene is ectopically expressed in the immune system, resulting in a novel activity.

been replaced by the somewhat more descriptive synonym of a **dominant negative** mutation.

A dominant negative mutation is often indicative of a protein product that forms dimers or multimers. One type of dominant negative mutation is illustrated in Figure 4.16. For example, many receptors form homodimers to trigger a downstream signaling cascade upon ligand binding. A dominant negative mutation can arise that allows the binding of the ligand but not the signal transduction to the downstream molecules. The mutant receptor acts as a dominant mutation in that it will form a dimer with the normal receptor and bind the ligand, but it will prevent the function of the normal copy of the receptor. In other words, a heterozygote has a mutant phenotype because the mutant subunit in the receptor dimer is preventing the normal receptor from functioning normally.

The *Shaker* gene in *Drosophila* is characterized by antimorphic mutations. The wild-type *Shaker* locus encodes a polypeptide that assembles as a multimeric potassium ion channel protein in the cell membrane. The presence of one defective subunit interferes with normal channel assembly and function, and thus results in the *Shaker* mutant phenotype. Another example is seen with mutations in some of the collagen genes of *C. elegans*. Collagen proteins form a highly organized bundle; the absence of one copy of the gene (i.e. an amorphic mutant) is rarely detrimental because the normal protein is still made from the wild-type allele and assembles normally. Thus, amorphic mutations are recessive. However, a mutated protein can assemble with the wild-type protein and interfere with the normal function of the entire bundle. The antimorphic or dominant negative mutation acts as a loss-of-function mutation because it is interfering with the normal gene product, often in the assembly of a multimeric protein. Curiously, since they antagonize the normal function of the gene, many antimorphic mutations depend on the presence of a wild-type allele for their mutant phenotype; homozygotes for antimorphic mutations may have no mutant phenotype at all or may have a different mutant phenotype.

Dominant negative mutations are generally uncommon among the mutants recovered in a mutant hunt. However, investigators frequently produce a dominant negative mutation *in vitro* by mutating one domain but not the protein interaction or dimerization sites. Such a mutant has the advantage of behaving like a loss-of-function or even a null allele, while being dominant at the same. Because the mutated version is dominant, it is not necessary to eliminate the wild-type copy of the gene as well to see the effect; in fact, the mutant effect often depends on having the wild-type copy present.

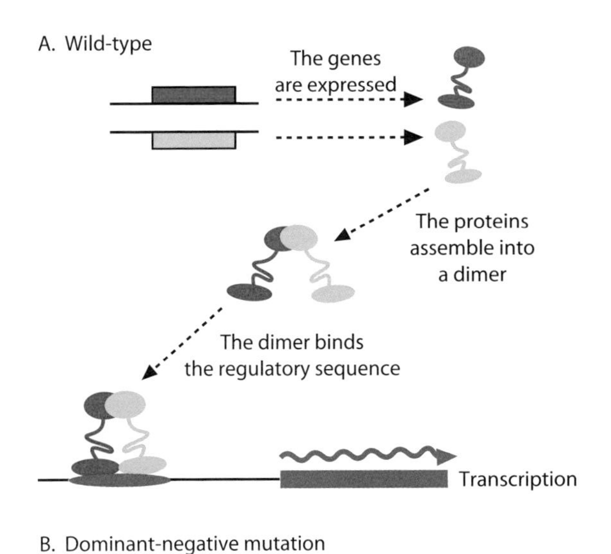

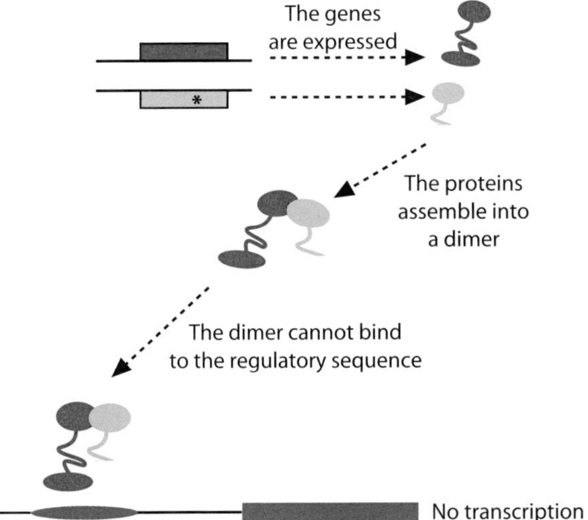

Figure 4.16 Dominant negative mutations or antimorphs. These are mutations that interfere with the normal wild-type function. One type of example is shown here, using a transcription factor and its target gene (in blue). In the wild-type (panel A), each allele is expressed and makes a protein; for clarity the two alleles and their protein products are shown in red and green. The two proteins form a dimer by interactions in one domain of the protein, and the dimeric protein binds the regulatory region to trigger transcription of the target gene. In a dominant negative mutation (panel B), each allele is expressed and the protein products assemble into a dimer. However, the mutant allele (shown in green) lacks the DNA-binding domain so that although it can form a normal dimer with the wild-type allele, it cannot bind DNA and cannot trigger transcription.

Haplo-insufficient mutants are dominant and define dose-dependent genes

Although dominant gain-of-function mutations such as those described above are the exceptions, most mutations reduce or eliminate the normal function of a gene rather than creating a novel function or over-expressing a normal function. This is true simply because there are many ways for something to go wrong, whether with a gene or any other function. Most loss-of-function mutations are recessive because the remaining wild-type allele provides enough gene activity to be able to carry out the normal function. However, for a few genes, the organism requires both wild-type copies for the normal function. This situation is shown diagrammatically in Figure 4.17. A mutation that eliminates gene function in such a dosage-sensitive gene is dominant to wild-type, and the gene is considered to be **haplo-insufficient**, i.e. a single functional copy of the gene is inadequate to carry out the normal function.

Not many genes have such a dose-sensitive mutant phenotype but some important examples are seen with genes implicated in cancers. For example, many cancers have mutations in the gene for p53, a transcriptional regulator; mice that are heterozygous for an inherited mutation in p53 are predisposed to tumors, particularly lymphomas. Why does a mutation that knocks out p53 function produce a dominant phenotype? Two possible explanations are illustrated in Figure 4.18.

The first idea postulates that the phenotype arises from a combination of germline and somatic mutations in the same gene, a two-hit hypothesis for the gene. With this hypothesis, the inherited *p53* mutation is expected to be recessive, so all of the cells in the affected individual are *p53/+*. Under the two-hit hypothesis of cancer, the high rate of tumors occurs because the one remaining wild-type allele in these *p53/+* heterozygous cells has been knocked out by a somatic mutation, inactivating the *p53* gene and resulting in a *p53/p53* phenotype. If this hypothesis provided the complete explanation, the tumor cells would be predicted to be homozygous mutant, with one inherited mutation and one somatic mutation. Because the inherited mutation and the somatic mutation are independent events, they are expected to be different molecular lesions. Although this hypothesis may explain the origin of some of the tumors, it clearly does not explain all of them. In at least some cases, the tumor cells have retained the wild-type *p53* allele and do not have a somatic mutation in the

DEFINITION

haplo-insufficient: Haplo-insufficient mutants are dominant loss-of-function mutations that reveal dosage-sensitive genes.

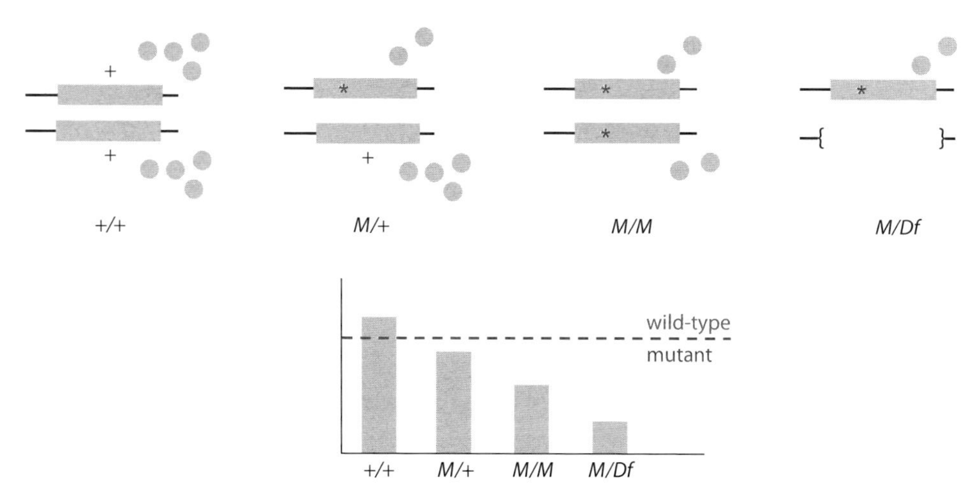

Figure 4.17 Haplo-insufficent mutations arise in dose-dependent genes. A haplo-insufficient mutation is a dominant loss-of-function mutation, reducing or eliminating the amount of the gene product. The mutation is dominant because the threshold of gene activity needed for wild-type function is very close to the amount made in normal diploids. A reduction results in a mutant phenotype, with a further reduction resulting in a more severe mutant phenotype. Compare this figure with Figure 4.10. Although the effect of the mutation is the same, haplo-insufficient mutations are dominant because the gene is dose dependent.

gene. Thus, an alternative explanation is that the *p53* gene is dose dependent or haplo-insufficient. With this hypothesis, one mutation in the gene reduces its activity to a level that is below what is needed for normal function. The mutant 'predisposes' the mice to tumors with no second genetic event needed. In effect, the *p53* mutant is dominant, but it has reduced penetrance as discussed earlier. The amount of gene activity has been lowered to close to the threshold needed for normal function, and in some cells the activity of the gene falls below this threshold.

A similar explanation has been posited for cancers arising from mutations in the *PTEN* gene and may be true in other oncogenes as well. The dose dependence of certain genes will be discussed further in Chapters 8 and 9 in the context of gene expression and gene activity Haplo-insufficient mutations differ from hypermorphic or neomorphic mutations in that they are loss-of-function mutations rather than over-producing mutations. They are different from antimorphic mutations in that they do not require the presence of a wild-type allele and, like hypomorphic alleles, haplo-insufficient mutations produce a more severe phenotype when they are homozygous. Thus, haplo-insufficient mutations are similar to null or hypomorphic mutations except that they are dominant because of the specific dosage requirements for that gene.

Systematic analysis of *Drosophila* chromosomes using deletions and duplications suggests that relatively few genes are haplo-insufficient for overall morphological phenotypes, but they do not rule out the possibility that a gene may be dose sensitive in particular cells. In fact, analysis of hypomorphic mutations in an allelic series indicates that this is almost certainly true. The sensitivity of the phenotype being monitored in the mouse experiments, i.e. cancer occurring in particular tissues, may allow the effects of subtle changes in gene dosage to be observed. This may suggest that genes that are dose sensitive are more likely to be oncogenic when mutated.

Summarizing the effects of mutations with an analogy

The effects of these mutations can be illustrated by an analogy. Consider two workers in the same office performing the same task; these workers are analogous to the two alleles of the gene. The absence of one (an amorphic mutation) does not interfere with the function of the office if the remaining worker (the wild-type allele) is able to carry out all of the functions. Even if one worker is lazy or ill and does less work (a hypomorphic mutation), the other worker is able to carry out the normal functions of the office. These are analogous to recessive mutations.

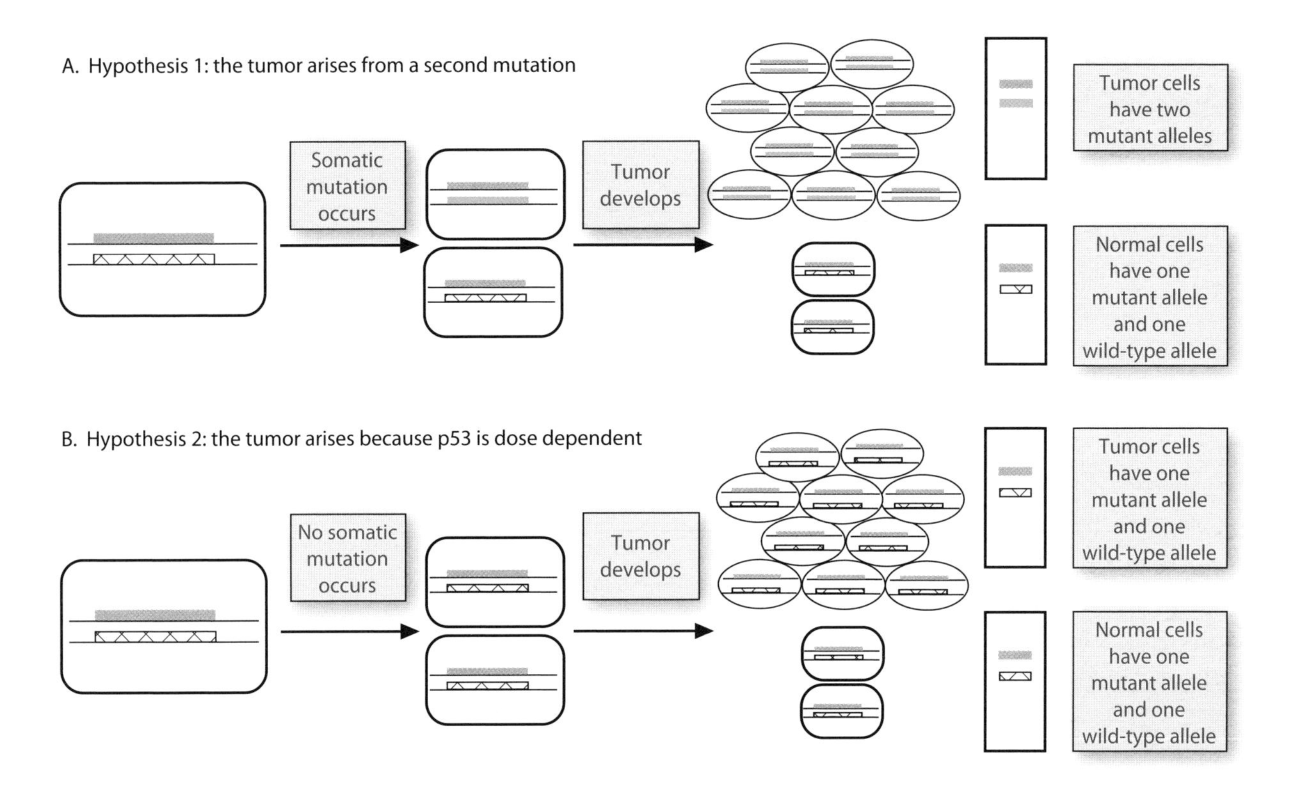

Figure 4.18 p53 has haplo-insufficient mutations. A cell with two copies of the *p53* gene is shown on the far left. One copy, shown in blue, has an inherited mutation, and so the mouse that is heterozygous for the *p53* mutation and wild-type has a high rate of tumors. Two explanations are advanced for the increased susceptibility to tumors in *p53* heterozygotes. Hypothesis 1, shown in panel A, is that another mutation occurs in the *p53* gene during somatic growth. This second mutation is shown in green. This cell is now homozygous mutant for *p53*, which results in a tumor, depicted by the proliferation of the rounded light brown cells. Under this hypothesis, tumor cells will have two different *p53* mutations, as shown on the far right, whereas normal cells will have the inherited *p53* mutation and a wild-type allele. Hypothesis 2, shown in panel B, is that *p53* is dose dependent and the inherited allele is haplo-insufficient. No second mutation occurs, and the tumor develops because the one mutant allele of *p53* has reduced the amount of *p53* activity below the required threshold. Under this hypothesis, tumor cells and normal cells will have the same genotype, with one mutant allele and one wild-type allele. Some of the tumors in the *p53* heterozygous mice are found to have a normal *p53* allele, supporting the second hypothesis.

But what if one office worker is absent and the other one cannot keep up with the work requirements? Problems arise because the function of the office is not occurring normally. This is analogous to a haplo-insufficient gene, in which both alleles must be active for the normal function of the gene to occur.

Suppose instead that one worker becomes hyper-active or over-caffeinated and produces vastly more work than usual; the extra output can create a problem in the flow of work through the office. This over-active worker could be considered a hypermorphic mutation. The mutant phenotype may actually result because other biological processes (other offices in our analogy) cannot keep up with the increased output, or the increased output itself may be the problem. In either event, in a hypermorphic mutation, the mutant phenotype arises from an increased activity from one allele, and the effect is dominant to the normal function.

The same office worker analogy can be used to explain other dominant mutants. In a second situation, the worker begins carrying out activities that he/she does not normally do, perhaps by being transferred to a different office; this is what happens with a neomorphic mutant or a gene that is being ectopically expressed. Having a worker who does not normally work there creates a problem in that new office. In a neomorphic mutation, the problem arises because of the ectopic function of the gene, i.e. a gene functioning in a tissue or at a time when it does not normally function.

The worst situation from the perspective of the office functions may arise when a worker becomes disruptive and prevents his/her coworker from functioning normally. This one disruptive worker is not only failing to perform his own work but is also reducing the function of the normal workers as well. This worker is analogous to an antimorphic or a dominant negative mutant—not only does he/she perform no useful work but he/she also prevents his/her colleague from working.

Often the only way to restore normality to an office with one of these dominant workers is to remove him/her from the office. Likewise, as we will see in Chapter 11 in the discussion of suppressor mutants, one method of overcoming the effect of a dominant mutant is to find second mutations that eliminate its effect and restores a normal phenotype.

SUMMARY

A mutant provides a crucial starting point for genetic analysis of a biological process

Mutations are the foundation tool of classical genetic analysis. Geneticists spend much of their intellectual and physical energy in finding and classifying mutants because they provide one of the best ways of understanding the normal function of a gene. Since many geneticists and molecular biologists define nearly all of their current activities in terms of cloned genes and genomes, it may be surprising to realize that genetics was a sophisticated field even before the advent of molecular biology. Mutational analysis as described in Chapters 3 and 4 has the advantage that very little prior understanding of the biological process is required. We describe molecular analysis of genes in the subsequent chapters, but this is often understood in terms of knowing a mutant phenotype. Few geneticists would do a molecular analysis without also working with a mutant, and every mutant needs to be understood in molecular terms. But ask a geneticist how to analyze a complex biological process and the answer is likely to consist of three words: 'Find a mutant.'

Chapter Capsule

Classifying mutants

Genetic analysis often begins by finding a mutant phenotype, but the next steps are to characterize the types of mutants that have been found. Some of the key questions in classifying mutants for genetic analysis can be summarized as follows.

How many genes are represented by my mutants?

- Mapping to chromosomal locations can be done by recombination or by chromosome rearrangements.
- Complementation tests are used to determine whether mutations are alleles of the same gene.
- Statistical methods can be used to estimate how many more genes have not yet been detected.

What kinds of mutations have I found?

- Recessive mutations usually represent a loss of function and could be:
  - amorphic or null mutations in which the normal function is eliminated
  - hypomorphic mutations in which the normal function is reduced.
- Dominant mutations usually represent an unusual function or gain of function and could be:
  - hypermorphic mutations that over-produce the normal function
  - neomorphic mutations that introduce a novel function, often by ectopic expression
  - antimorphic or dominant negative mutations that antagonize or interfere with the normal function
- Haplo-insufficent mutations are dominant mutations that arise from the loss of function in a dose-sensitive gene.

- **Unit 1 Genes and genomes**
- **Unit 2 Genes and mutants**
  - 3 Identifying mutants
  - 4 Classifying mutants
  - **5 Connecting a phenotype to a DNA sequence**

    This chapter describes how a gene is cloned, often by beginning with a mutant phenotype. The cloned gene and the mutant phenotype are crucial tools for many other types of analysis described in subsequent chapters.
  - 6 Finding mutant phenotypes for cloned genes
  - 7 Genome-wide mutant screens
- **Unit 3 Gene activity**
- **Unit 4 Gene interactions**

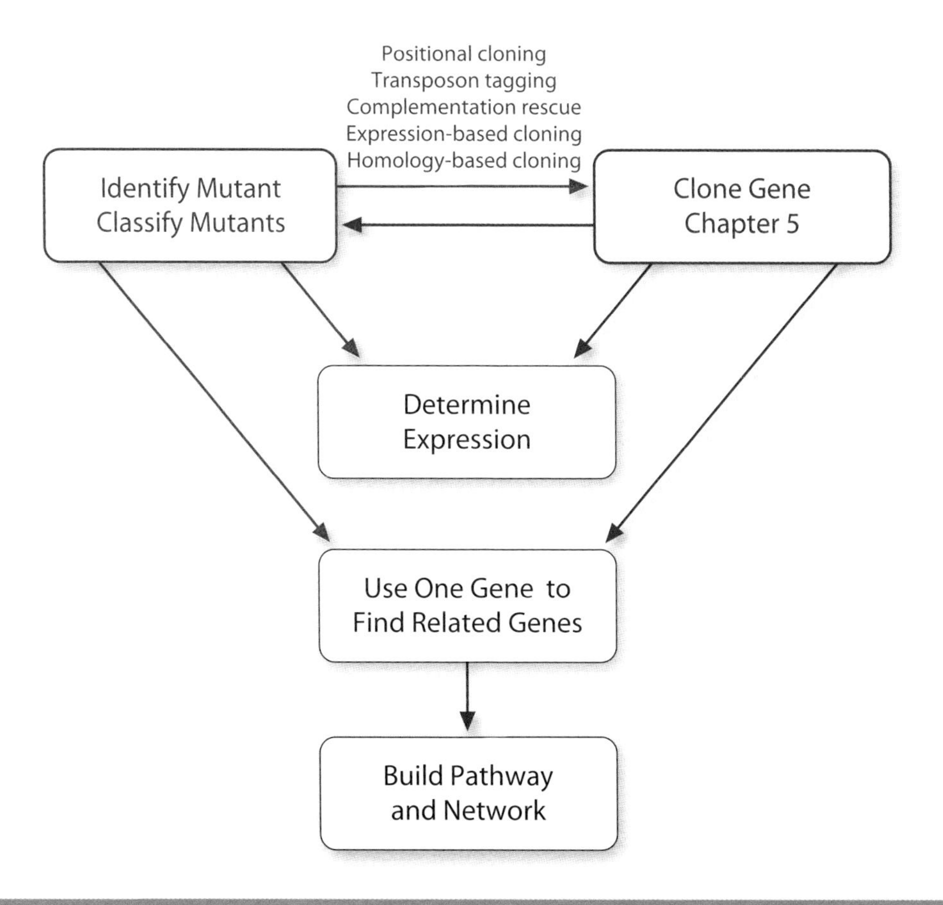

5

CHAPTER 5

Connecting a phenotype to a DNA sequence

TOPIC SUMMARY

Once the investigator has identified mutations in a gene that affect the biological process being studied, one of the next steps is to clone the gene. The initial approach to cloning a gene is often based on one of five properties that all genes share:

- map position
- mutant alleles
- complementation
- pattern of expression
- evolutionary descent or homology.

IN BRIEF

The properties of a gene inferred from Mendelian and population genetics provide strategies that can be used to clone it.

INTRODUCTION

A mutant phenotype cries out for an explanation. Which gene is mutated in a fly with no wings or a worm that rolls like a corkscrew, and what is the nature of that defect? What is the normal function of the protein encoded by that mutant gene? In order to answer these questions and many others, the gene must be **cloned**, i.e. the mutant defect seen in the cell or organism must be connected to the change in a specific gene and gene product.

One of the beauties of genetics is that it is possible to make the connection between the mutant phenotype of a cell, animal, or plant and a change in the DNA sequence. In this chapter, we describe some of the ways to make that connection between a phenotype and a DNA sequence. We will see how the very properties of genes discovered by geneticists working before we even knew that genes consisted of DNA (i.e. the geneticists mentioned in Chapter 1, such as Mendel, Morgan, Muller, and Beadle) can be used to identify the corresponding DNA sequence. Many different approaches to the problem of cloning a gene have been tried and shown to work. Our aim is to explain them in the context of the known properties of genes.

The tool that we *explain* in this chapter is the cloned gene (Figure 5.1). A cloned gene is fundamental to many of the tools we *develop* in the rest of the textbook, including the analysis of gene expression and finding new genes. A mutant phenotype can demonstrate what process the underlying gene affects, but it is essential to clone the gene if we want to know what it actually does. In other words, it is essential to clone a gene to know its molecular and biochemical functions in the process altered in the mutant.

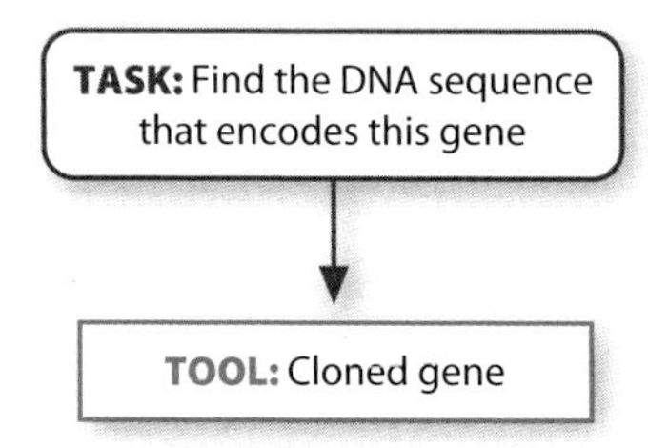

Figure 5.1 Developing the tool. The cloned gene.

Table 5.1 Definitions of a gene

Sub-discipline of genetics	Typical symbols	Defining quality	Related concepts
Mendelian	*A/a*, w^+/w	Inheritance in generations	Linkage, allelism, dominance
Population	*p* and *q*	Inheritance in historical time	Allele frequencies, selection, drift
Molecular	...ATCGGCTAAT...	DNA sequence	Transcription, replication, expression

5.1 Cloning genes: an overview

The field of genetics combines several very different concepts under one scientific heading. In each of the sub-disciplines of genetics, the 'gene' has a slightly different meaning. A few of these ideas are collected in Table 5.1. For example, we can talk about Mendelian genetics (also known as classical genetics or transmission genetics); in this sub-discipline, the gene is often a symbol such as ***A*** or w^+ and is defined by its pattern of inheritance in a family over a few generations with known parents. Included under this heading are the principles of linkage, allelism, dominance, and so on. If the term 'genetics' makes you think of Punnett squares or linkage maps, you are thinking about Mendelian genetics.

We can also talk about evolutionary or population genetics. The focus of this sub-discipline is the frequencies of the different alleles within a population of organisms and how these frequencies change over time. In population genetics, the frequencies of two alleles are generally symbolized by ***p*** and ***q***. In addition to allele frequencies, this field also considers the concepts of mating patterns, selective advantages, migrations, and so on. If the term 'genetics' makes you think of Hardy–Weinberg equilibrium or natural selection, you are thinking about population or evolutionary genetics. Both classical and population genetics were intellectually rich fields of science before Watson and Crick worked out the structure of DNA and even before DNA had been conclusively shown to be the genetic material (the biochemical constituent of the abstract concept of the gene). There may be a tendency among some scientists to disregard these pre-molecular fields, but such an attitude would miss much of the beauty and power of genetics.

We can also talk about molecular genetics. This is the field in which the gene is defined by its DNA sequence and its RNA and protein products. Included in this sub-discipline are the concepts of the cloned gene, gene expression and activity, transcription, translation, replication, and so on. No reader of this book will need to be persuaded that molecular genetics is also a fantastically rich field of science. The great charm of genetics is that all three of these seemingly different views of genetics can be integrated. In fact, not only *can* they be integrated, but they *must* be integrated in order to understand how genetics is used to analyze a biological problem fully. An allele can be described and analyzed in terms of an eye color variation (a Mendelian phenotype), its frequency in the population (an evolutionary concept), or a DNA sequence (a molecular concept). All of these concepts are relevant to our understanding of genetics.

One way in which the many concepts of genetics come together is represented in the molecular identification of genes—gene cloning, which is the topic of this chapter. This is a large topic, and every gene that has been cloned comes with its own history of false leads and unexpected results. No molecular geneticist will forget the thrill (and possibly the frustration) of the first gene he/she cloned. Those of a certain age can tell you where they were when they read about the cloning of the human globin gene, the cystic fibrosis gene, or the bithorax complex genes. Most could tell you about a time when they scratched their head and wondered how some particular DNA sequence (the molecular definition of the gene) could explain some particular mutant phenotype (the classical definition of the gene). The intellectual excitement of genetics often comes from connecting classical genetics to molecular genetics and even to population genetics. Population genetics becomes particularly important when studying human genetics because many of the

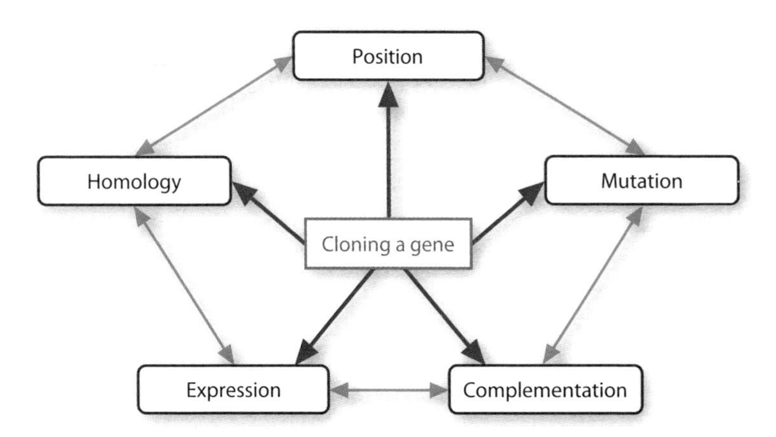

Figure 5.2 A summary of the strategies described in this chapter for cloning a gene. Each of the five properties is characteristic of a gene, and each forms the basis for at least one significant initial strategy in isolating the DNA sequence of a gene. Just as any of these can serve as the initial cloning strategy, all of the others can serve as confirmation that the correct gene has been cloned. For example, a gene may be cloned by its position on the chromosome but confirmation that one has cloned the right gene could come from the sequence of mutations or from expression patterns.

tools of classical genetics, such as large numbers of progeny from controlled crosses between largely homozygous individuals varying at only one or a few loci, are not applicable.

If each cloned gene has its own story, how can the genetic anthology of all genes be compressed into one chapter? We will take the following approach. Each section will begin with a classical property of a gene that applies to all genes. Then we will describe how that particular property of genes has been used as a primary strategy for cloning them. This is not a comprehensive list of the ways in which genes have been cloned, nor is it an instruction manual that will allow a person to clone a particular gene. Rather, it is an effort to show how all these properties of genes have provided the insights needed for connecting genes to DNA sequences.

The properties of genes that are used to organize this chapter are as follows and are summarized in Figure 5.2.

1. **Position** A gene maps to a particular locus on a chromosome. For example, the *white* gene of *Drosophila* maps to the X chromosome at position 1.5, or 1.5 map units from the reference point at the left tip of the chromosome. The fact that a gene can be mapped to a particular locus is the principle that defines the strategy known as **positional cloning**, and it is one of the underlying reasons why genome sequencing projects have been undertaken.
2. **Mutation** A gene exists in many different forms known as alleles or mutants. For example, the *white* gene has alleles that give red eyes (the wild-type phenotype), white eyes, ivory eyes, pale yellow eyes, and many other phenotypes. The positional cloning strategies mentioned above require alleles with different phenotypes, but the molecular basis of these alleles is not important. As discussed in Chapter 3, the kind of mutagen chosen for a large-scale mutagenesis determines the classes of mutation (point mutations, deletions, insertions, or rearrangements) that are seen. Some cloning strategies use the molecular basis of the mutation to clone the gene. The most common examples of this involve mutants with large insertions of known sequence, such as transposons. Using a transposon insertion to clone a gene is known as **transposon tagging.** In addition to their roles in positional cloning and transposon tagging, mutations are essential for confirming that the investigator has cloned the correct gene.
3. **Complementation** A gene is defined by complementation. As discussed in Chapter 4, when characterizing an organism carrying two heterozygous recessive mutations, a wild-type phenotype (complementation) indicates that the two mutations are located in two different genes, while a mutant phenotype (non-complementation) reveals that the two mutations are found in the same gene. In classical terms,

non-complementation is the critical test for functional allelism. For example, a fly that is heterozygous for two different *white* recessive alleles will have white eyes if both mutations are alleles of the *white* gene. The key to these complementation tests of recessive mutations is that at least one functional copy of a gene is required for a wild-type phenotype. In the molecular analysis of genes, this property has been used in the strategy known as **transformation rescue**, where each clone in a gene library is tested for its ability to complement the mutant phenotype in a transgenic organism. The ability of a clone to complement the phenotype suggests that it may encode the gene that is non-functional in the mutant.

4. **Expression** A gene has a specific pattern of expression. A mutant phenotype can suggest where the gene is likely to be expressed. For example, while phenotypic descriptions of mutations in the *white* gene usually focus on eye pigmentation, these mutations may also affect other pigmented structures in the fly, such as the Malpighian tubules. Therefore, whatever the function of the *white* gene, it must at least be expressed in the eye and the Malpighian tubules. This overlap between the phenotype of a mutant and the expression pattern of the mutated gene has led to a number of different gene-cloning strategies using gene expression patterns. These approaches are collectively termed **expression-based cloning**.
5. **Homology** A gene is the unit of evolutionary descent or homology. For example, the *white* gene in *Drosophila* encodes a type of protein known as an ABC transporter, a type of protein found in all organisms. This molecular similarity of genes gives us the underlying principle of **homology-based cloning**.

The two case studies in this chapter should also be used to understand how these five properties are used together to clone a gene. Cloning a gene consists of two steps: finding candidate genes and sorting through the candidates to find the correct one. Some of the strategies are useful for *finding* candidate genes but are unlikely to pinpoint exactly one gene; the most useful methods for finding candidate genes are positional cloning (property 1) and expression-based cloning (property 4). Other strategies are more useful for identifying and *confirming* the correct gene from the list of candidates; the most useful methods for confirmation are allelic differences (property 2) and complementation (property 3). Homology (property 5) occupies the middle ground and is helpful but not definitive as both an initial cloning strategy and a way of confirming that the cloned gene is the right candidate. Although these properties of genes are generally more useful for one of the two steps of gene cloning, any of them can be used as the first step (finding candidate genes), and all of them are important for confirming that the right gene has been cloned.

Investigators who are trying to clone a gene (particularly if they are trying to clone it before a competitor does) are almost certain to use more than one of these approaches and are probably going to use almost all of them. One strategy may provide several candidates, which can then be narrowed down using one of the other strategies. For example, if one has obtained the DNA sequence of a region of the chromosome containing the gene of interest, using the position (property 1) as the first approach, it is very likely that there will be more than one gene at that location. Therefore, it is almost certainly necessary to show that the gene at that region has DNA sequence changes in mutant individuals (property 2), can complement the mutant defect (property 3), is expressed in a time and place that makes sense for the phenotype (property 4), and/or encodes an evolutionarily conserved protein whose deduced function makes sense for the phenotype (property 5). And when all five properties converge on one gene? It is time to start writing the paper telling its story.

5.2 Property 1: a gene maps to a particular locus on the chromosome (positional cloning)

Genes were first mapped to a specific location on a chromosome in the early days of *Drosophila* genetics, and genetic maps have been the atlas for geneticists ever since. Therefore, it is not very surprising that one of the properties most widely used for cloning a gene has been its map position. The underlying logic is simple: gene

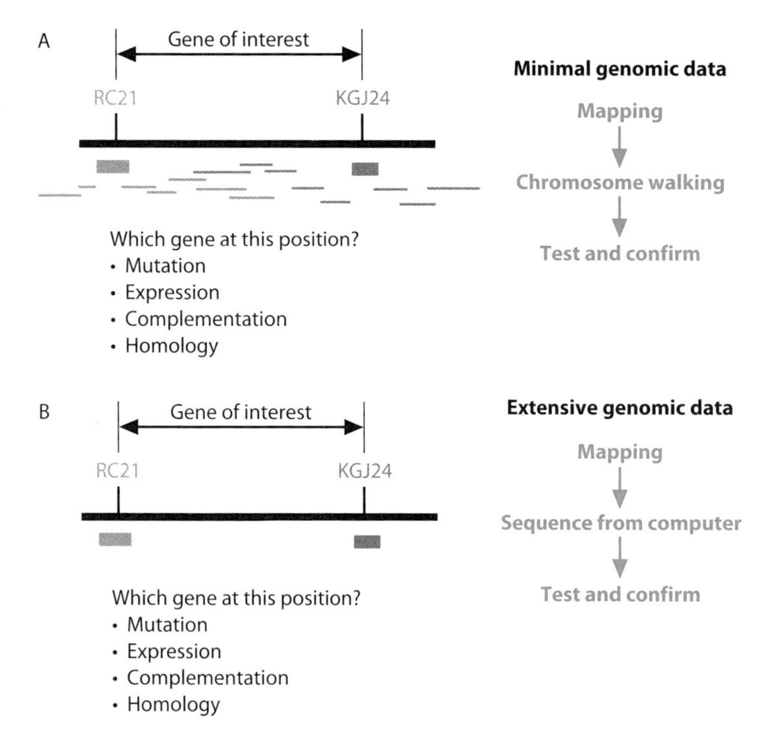

Figure 5.3 Using position as the initial strategy. The gene of interest is mapped between two previously cloned molecular markers, the polymorphisms RC21 and KGJ24. If minimal genomic information is available, as in panel A, the region between the two markers is isolated by chromosome walking or a similar strategy. These clones are tested for the other properties of genes such as mutation. If extensive genomic information is available, as in panel B, the chromosome walking step is replaced by retrieving the DNA sequence from the computer.

B is known to map between loci *A* and *C*; both locus *A* and locus *C* have been cloned; therefore, if the region between loci *A* and *C* can also be cloned, it must include gene *B*. This strategy of positional cloning is shown in Figure 5.3. The positional cloning approach is a major impetus for genome sequencing projects. Indeed, for the early genome projects, whether bacteria or worms, the progression from genetic map to genome project was more gradual than abrupt. Therefore, it can be difficult to decide exactly when the genetic map became the genomic project.

Locating a gene with respect to cloned markers and genes

Let us examine each of the two primary requirements for positional cloning: locating a gene between two molecularly defined markers, and being able to clone the region between the two markers. We will start by discussing the types of molecularly defined flanking markers. Remember that for a marker to be used in recombination mapping, it must exist in two different alleles or polymorphisms in the mapping population. In the most traditional sense, the flanking markers might be other cloned genes, or genes that were cloned after the mapping had been done. These are often the most convenient flanking markers to use for many inbred laboratory organisms, such as yeast, flies, worms, or *Arabidopsis*, especially as more genes are mapped and cloned. As discussed below, genes are not the only flanking markers that can be used and, in fact, are often not the best flanking markers to use in contemporary genetic mapping. However, since genes were the phenotypic markers used for creating the earliest genetic and molecular maps, we consider them first. Mapping a gene to the region between two cloned markers is by conventional recombination mapping, which even allows the relative position between the markers to be estimated. Suppose that the distance from gene *A* to gene *C* is 0.5 map units from recombination mapping and 100 kb from molecular mapping, and that recombination experiments have placed gene *B* about two-thirds of the distance between *A* and *C*. Gene *B* can then be estimated to lie closer to *A* than to *C* in molecular distance.

The primary goal of this stage of positional cloning is for the cloned markers at the margins of the flanking region to be as close to each other as possible, simply because there will be far fewer candidate genes in a region of 100 kb than a region of 1000 kb. On the other hand, it is more challenging to find closely linked markers and to map genes with respect to them because recombination events will be less frequent within short genetic distances. Much of the effort in genetic mapping has been directed at increasing the resolution of maps, i.e. finding cloned markers separated by the smallest realistic intervals.

An insight that was crucial for efforts in positional cloning was that the cloned markers do not need to be genes. Any type of molecular polymorphism, such as a restriction fragment length polymorphism, a single-nucleotide polymorphism, or a short sequence-length polymorphism, works as well as a cloned gene for this purpose. The procedure for mapping using these molecular markers is exactly the same as that for conventional phenotypic markers. The only difference is that molecular markers have a molecular phenotype, such as the size of a band on a gel, the presence or absence of a restriction site, or the DNA sequence itself.

Furthermore, in most organisms, non-coding DNA (which is less likely to be subject to selection) has a far greater frequency of sequence polymorphisms than coding DNA and thus provides the best source of phenotypes to use for mapping. These sequence polymorphisms are found when comparing two different wild-type strains, such as two plants or worms found in different geographical locations and brought into the laboratory. With current methods for DNA sequencing, single-nucleotide polymorphisms (known as SNPs and pronounced 'snips') are among the most widely used markers for mapping.

Cloning the region between two molecular markers

The second of the two requirements for positional cloning is the ability to clone the region between the cloned markers. Cloning the region of interest has often been a laborious process involving chromosome walking in a library. Chromosome walking refers to the process of using a portion of one cloned fragment as a probe to find overlapping clones. Then those overlapping clones are themselves used as probes to find the next set of overlapping clones. The process is repeated to provide a step-by-step molecular path along the chromosome. One general procedure for chromosome walking is shown in Figure 5.4, but many alternative procedures have been devised to reduce the distance that has to be walked and to determine the orientation of the walk. Fortunately, genome projects depend on the assembly of overlapping clones for the entire genome of a species, a procedure that is frequently automated. Chromosome walking is now largely an unlamented memory for most researchers working in model organisms.

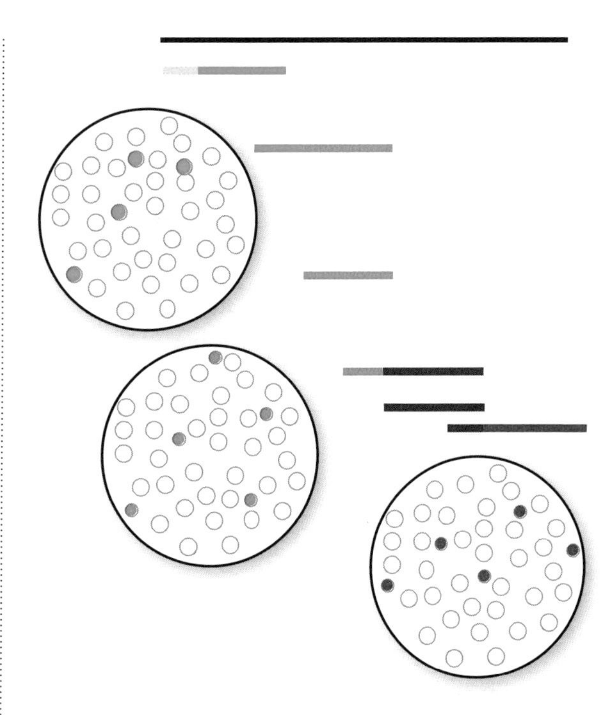

Figure 5.4 A generalized procedure for chromosome walking. A clone at one site, shown here in green, is used to probe a library, depicted as colonies on a Petri dish. The hybridizing sequences are picked and purified. One of these is then used as a probe for a second step, shown here in light blue. The process is repeated (with the probe in purple and then the one in red) to find overlapping clones that span a region.

Other strategies for cloning genes based on position have also been used occasionally, particularly before genome sequencing projects were developed. Among these other position-based strategies were the microdissection of *Drosophila* polytene chromosomes and flow sorting of mammalian chromosomes. In all cases, the underlying goal has been to *place* the gene at a particular location in the genome and to isolate the

DNA from that location. Genome projects have made positional cloning the most widely used strategy for cloning a gene in nearly all well-studied organisms. The investigator maps the gene of interest with respect to as many other cloned genes and/or molecular polymorphisms as possible, thus defining a relatively small interval; nowadays, mapping is usually the slowest step in the cloning process for most genes. Genome databases (introduced in Chapter 2) are then used to identify candidate genes in the region, and strategies using the other four properties of genes help to determine which candidate gene is of greatest interest.

Positional cloning is especially useful when working with organisms of known pedigree. Most genetics studies in natural populations, including the human population, use an alternative to positional cloning called **association mapping**, which does not require known lineages or crosses. In Text Box 5.1 we consider the use of association mapping for identifying human disease genes and loci responsible for natural variation in *Arabidopsis* found in the wild. Although the steps for determining the map position depend somewhat on the gene of interest and the organism, the underlying strategy based on the gene's position on the chromosome remains the same. There is a certain satisfaction in realizing that map positions were the evidence from a century ago which convinced biologists that genes are on chromosomes; this same evidence is now the most widely used strategy to connect a phenotype to a DNA sequence in cloning a gene.

5.3 Property 2: a gene has different alleles (transposon tagging)

Whether it causes round or wrinkled peas, white or red eyes, or some other phenotype, a gene is recognized because it has different alleles. A phenotypic difference between alleles implies an underlying molecular difference, although the exact nature of the molecular difference may not be known for some time after the gene is described. Nowadays, the underlying molecular difference between alleles is most commonly used to confirm that the investigator has cloned the correct gene rather than to obtain candidate genes; in fact, the molecular basis for allelic differences, usually determined by sequencing the gene, is often the critical step in confirming that one has cloned the correct gene. In some cases, however, the underlying molecular difference can be used as the initial strategy for cloning a gene. Here, we will describe a common cloning method that relies on the molecular difference between alleles—a technique called **transposon tagging**. This method has been used in many different organisms, but probably most frequently in *D. melanogaster*.

Transposon tagging in *Drosophila* relies on the properties of the P transposable element. P elements were described in Chapter 2 as a means of producing transgenic flies and in Chapter 3 as a method of producing mutations, but they also provide a means of cloning a gene if the mutant phenotype has arisen from the insertion of a P element. In a transposon-tagging procedure, the P element is mobilized and mutant flies are identified in a subsequent generation. The screen for mutations is exactly the same as for any other mutant screen, as outlined in Chapter 3, but the mutation in these flies results from the insertion of a P element, and the P element can be cloned. If one extracts DNA from the mutant flies and isolates the P element together with some flanking DNA, a piece of the gene of interest has been cloned and can be used as a probe to find the rest of the gene. Thus, the gene has been *transposon tagged* by its mutant phenotype. The procedure is illustrated in Figure 5.5.

The success of using P elements for transposon tagging has led to modifications that make it simpler to mobilize and identify P elements in the genome. For example, in Chapter 3 we described how a P element encoding a transposase but lacking the terminal repeats (a wings clipped element) has been integrated into the *Drosophila* genome to provide a stable source of transposase. The introduced element can then be marked with a reporter gene rather than the transposase, which makes it easy to distinguish from other transposable elements already present in the genome. (Reporter

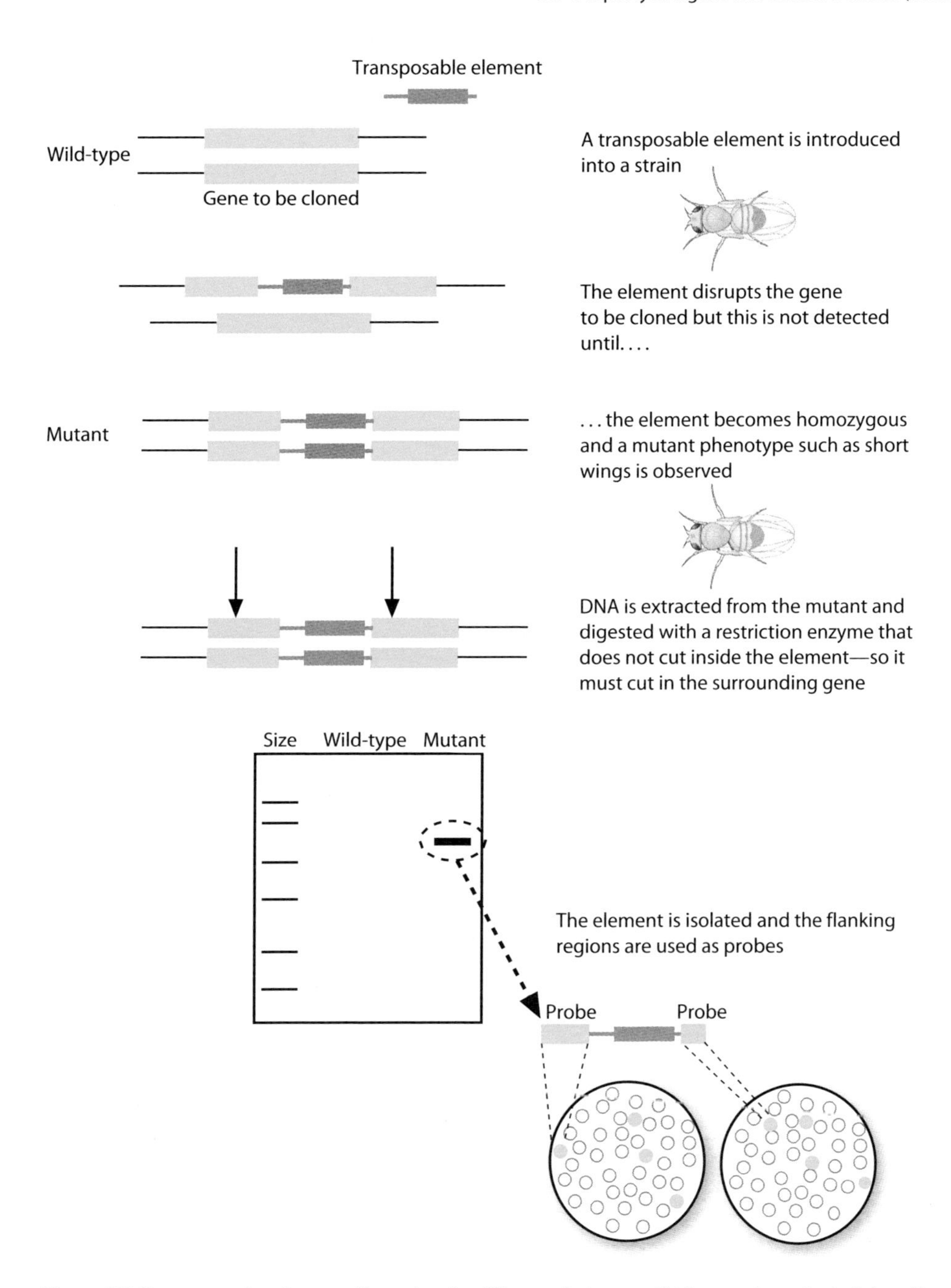

Figure 5.5 Transposon tagging uses the molecular difference between wild-type and a mutant allele as the initial strategy for cloning. A transposable element, shown in pink, is introduced into a normal strain and integrates throughout the genome. One insert is in the gene of interest, shown in yellow. A mutant phenotype is observed when the transposable element becomes homozygous; for example, the fly has short wings. DNA is extracted from the mutant animal and digested with a restriction enzyme that does not cut within the element. The digested DNA is run on a gel and probed with the element; only the mutant has an inserted element. This fragment is purified and the regions flanking the element are isolated. These flanking regions correspond to part of the disrupted gene and can be used to probe a library to obtain the entire gene.

BOX 5.1 Association mapping: human disease genes and natural variation in *Arabidopsis*

As we emphasize throughout this book, there are many advantages to using model laboratory organisms. Gene cloning based on a genetic map is much easier in model species because their genomes have been completely sequenced, inbred strains homozygous at all loci can be compared, and the genotypes of the parents in a mating can be controlled. Most of these advantages do not occur in natural populations. Is it possible, then, to use map-based cloning in instances where individuals are heterozygous at many loci, controlled matings are not possible, and there are few progeny from each mating? This is the situation when doing genetics in humans and when using natural variation to identify genes related to a particular process.

A technique called **association mapping** is a recombination-based technique that can be used to clone genes in natural populations. Association mapping relies on the concept of positional cloning, discussed in the chapter, but the approach is somewhat different. The easiest way to understand association mapping may be to consider it in parallel with positional cloning. In positional cloning, a genetic strain or line containing a mutation of interest is crossed with a different strain that lacks the mutation. In addition to the difference at the locus to be cloned, many other molecular polymorphisms also differ between the two strains. The map locations of these other polymorphisms are known or can readily be inferred. As a result, an F_1 hybrid between these two strains will be heterozygous for both the mutation of interest and the other polymorphisms. Because the state of each of these polymorphisms in the parent strains is known, linkage relationships in the F_1 generation are established. In the subsequent F_2 generation, linkage between these polymorphisms and the mutant phenotype is used to determine the position of the mutation of interest. The closer this mutation is to one of these other loci, the more strongly linked it will be. Thus, the mutation and the polymorphism will segregate together.

Haplotypes and linkage disequilibrium

In natural populations, including the human population, the linkage relationships are not known *a priori* in the absence of a pedigree, and many loci in each individual are heterozygous. (In population genetics terms, there is a high degree of **heterozygosity**.) Association mapping identifies genes based on the assumption that the particular gene mutation affecting the phenotype of interest originally arose on one chromosome in one individual in the population. As shown in Text Box Figure 5.1, this mutation arises in a specific neighborhood of

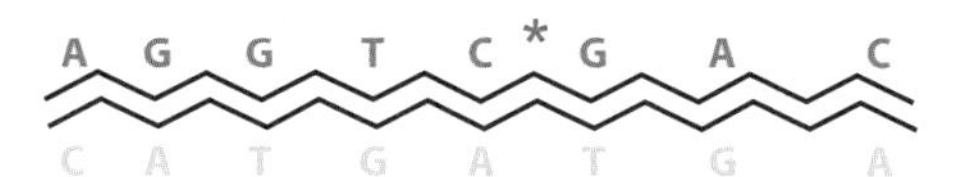

Figure B5.1 The origins of a haplotype. A region of the genome is shown from a diploid individual, such as a human. Single-nucleotide polymorphisms (SNPs) are found throughout this region, which may be tens or even hundreds of kilobases in length. Because of genetic linkage, the SNPs on the top chromosome, shown in blue, will tend to be inherited as a unit, as will the SNPs shown in green on the bottom. The block of polymorphisms that is inherited as a unit is known as a **haplotype**. A new disease-causing mutation, indicated by the red star, arises on one of the two chromosomes. Individuals who inherit this disease-causing mutation will also be likely to inherit the other polymorphisms of this haplotype. Thus, by monitoring the inheritance of the blue SNPs, the investigator can monitor and map the disease-causing mutation.

other polymorphisms. In other words, the appearance of this mutation at a specific site on one particular chromosome establishes a specific linkage relationship with the surrounding polymorphisms in this founding individual. The mutation and the specific alleles at polymorphic loci linked to it form a **haplotype**. As the mutation is passed on to its progeny, its progeny's progeny, and so on, the mutation and the polymorphism will tend to co-segregate, and the frequency at which the mutation and a polymorphism co-segregate indicates how closely linked they are on the chromosome.

This tendency of linked alleles to segregate together in a population is called **linkage disequilibrium**. The term 'disequilibrium' reflects the fact that these polymorphisms will not be found at Hardy–Weinberg equilibrium with respect to each other. In a population in Hardy–Weinberg equilibrium, the segregation of alleles at two loci is expected to be independent of linkage and depends on their allele frequencies. Thus, all combinations of alleles (i.e. genotypes) will be seen at the frequencies predicted from the frequencies of the individual alleles. In linkage disequilibrium, however, the presence of an allele at one locus is an indicator of the presence of a specific allele at a linked locus. As a result, a haplotype can also be defined as a set of polymorphisms that exhibit linkage disequilibrium.

Mapping in a population

In association mapping, the investigator maps a gene by screening a population for polymorphisms that are in linkage disequilibrium with the mutant phenotype. Essentially, the

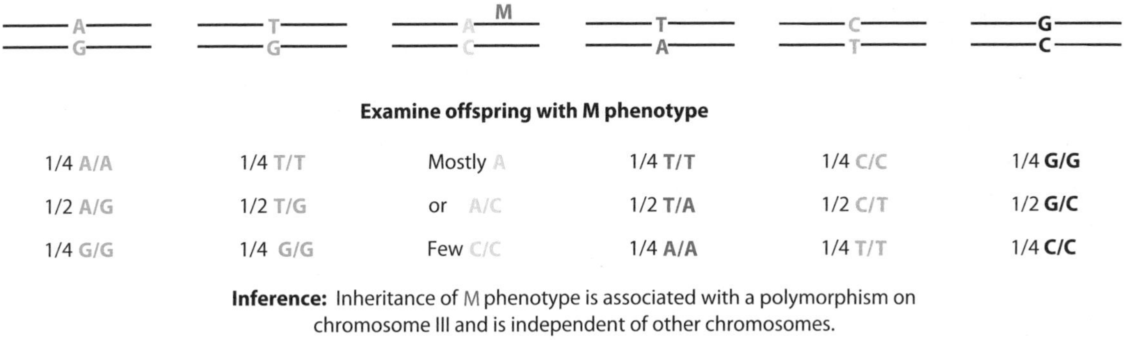

A/G T/G M A/C T/A C/T G/C

Examine offspring with M phenotype

1/4 A/A	1/4 T/T	Mostly A	1/4 **T/T**	1/4 C/C	1/4 **G/G**
1/2 A/G	1/2 T/G	or A/C	1/2 **T/A**	1/2 C/T	1/2 **G/C**
1/4 G/G	1/4 G/G	Few C/C	1/4 **A/A**	1/4 T/T	1/4 **C/C**

Inference: Inheritance of M phenotype is associated with a polymorphism on chromosome III and is independent of other chromosomes.

Figure B5.2 Association mapping in *C. elegans*. A SNP is located on each of the six chromosomes in worms, shown in different colors, and can be used for rapid mapping. A new mutation *M* arises on chromosome III and is mapped by observing the segregation ratio of the six polymorphisms in the subsequent generation. The M offspring are assayed for the presence of the six SNPs. The 1:2:1 ratio is characteristic of unlinked genes, whereas most of the M mutants will have the A/A or A/C polymorphism on chromosome III.

geneticist is identifying the haplotype in which the mutation originally arose by looking at these linked polymorphisms.

Before considering an example from a natural and complex population, let us see how the concepts of association mapping apply to a positional cloning problem in *C. elegans*, as shown diagrammatically in Text Box Figure 5.2. Assume that you have identified a new mutant phenotype in *C. elegans* that you want to map and the mutation is homozygous. You mate this homozygous strain to a different homozygous strain that is polymorphic across the genome compared with the first strain. In F_2, you assign the trait to a chromosome by examining the association in a large population of the mutant phenotype with one polymorphism on each of the six chromosomes. As shown in Text Box Figure 5.2, linkage equilibrium between the mutant phenotype and a marker is seen for markers on five of the six chromosomes: a quarter of the mutants are homozygous for one allele of the marker, half of the mutants are heterozygous, and a quarter are homozygous for the other allele. However, one allele of the marker on another chromosome is seen to be over-represented in the mutant individuals. Thus, this locus and the mutation of interest are in linkage disequilibrium. The mutation has now been mapped to a chromosome. The region of the chromosome carrying the mutation can then be narrowed down using the level of linkage disequilibrium to additional markers.

Our analysis of linkage disequilibrium in this example was made easier by the controlled mating of parents of known genotypes, and the same principles can be applied to natural populations. The key to this kind of approach is having enough polymorphisms to consider. Any type of polymorphism could be used for an association study, but the most commonly used ones are **single-nucleotide polymorphisms (SNPs)**. These are often monitored by their ability to create or eliminate a restriction site, although they can also be scored by sequencing or by hybridization to microarrays (as described in Chapter 8); hybridization to microarrays offers the most rapid identification of SNPs. Because haplotypes are comprised of many linked polymorphisms rather than individual SNPs, assaying one or a few key SNPs in a region of high linkage disequilibrium can take the place of analyzing hundreds or even thousands of individual sites. These key polymorphisms are referred to as **tag SNPs**. The concept of a tag SNP is shown diagrammatically in Text Box Figure 5.3. These tag SNPs have made it possible to overcome the problem of testing the large number of regions of linkage disequilibrium in humans.

Identifying common haplotypes in the human genome is one of the primary goals of the Human HapMap project. Although the effort is still in progress, we are beginning to get a perspective on how many different polymorphisms will be needed for thorough association studies. One current estimate

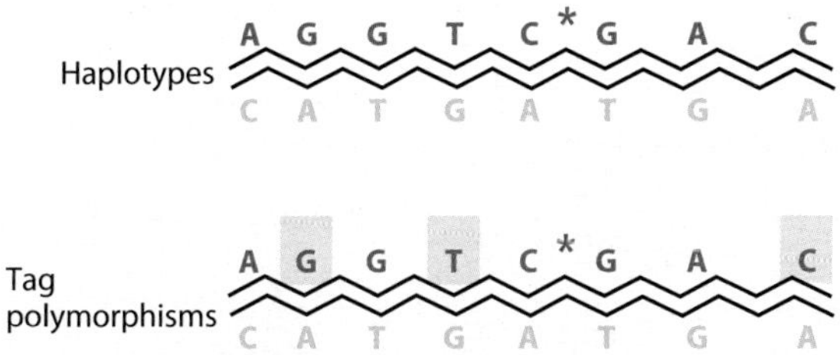

Haplotypes
A G G T C * G A C
C A T G A T G A

Tag polymorphisms
A G G T C * G A C
C A T G A T G A

Figure B5.3 Monitoring tag polymorphisms. The same haplotypes as in Text Box Figure 5.1, with eight SNPs, are shown here. Rather than monitor all eight SNPs, an investigator could monitor the three tag polymorphisms shown by the blue shading to infer the inheritance of the entire haplotype and the disease-causing mutation.

is that association studies for most of the genome can be provided by about 500 000 tag SNPs. Although still a large number, monitoring 500 000 tag SNPs is much more feasible than testing all of the estimated 10 million SNPs with two or more alleles present in the human population at a frequency greater than 1 percent. The number may be reduced further as we understand more about the genome and the history of human populations. Linkage disequilibrium for some regions of the genome provides useful information over distances as great as 50–100 kb. (Human chromosome 16 is approximately 100 Mb, so even a linkage disequilibrium for 100 kb is a very small percentage of a chromosome.) However, linkage disequilibrium is affected by the local rate of recombination, which varies by at least a factor of 10, by the evolutionary history of the population, and by chance events. Thus, in some regions, linkage disequilibrium is not informative over more than a few kilobases. For such regions, association mapping for disease traits will be especially difficult unless other tag SNPs can be identified in at least some populations. Nonetheless, for much of the genome, the HapMap project provides great promise for identifying genes that affect some of the most important aspects of human health and for developing improved therapies for some of our most common diseases.

Association mapping of complex traits

For diseases inherited as a single trait with high penetrance (i.e. diseases in which the mutant phenotype shows up in nearly all individuals with the mutation), positional cloning using a few large pedigrees may be sufficient to provide at least preliminary mapping data. Positional cloning of this sort was used to clone a gene for a recessive disease such as cystic fibrosis (see Case Study 5.1) and for a dominant disease such as Huntington's disease. Association mapping could also have been used to clone these genes, but it is most valuable for working with complex traits. Many human traits, such as dyslexia, depression, diabetes, and hypertension, have a clear genetic component but are not inherited as single-gene traits. Traits that are affected by many genes and the environment are known as **complex traits**. Since many complex traits have a quantitative phenotype (such as blood sugar levels), the genes that contribute to them are called quantitative trait loci (QTLs). Since complex genetic traits influence our health and life span, identifying one or more associated QTLs may provide therapeutic strategies, diagnostics and prognostics, and basic biological information.

Association mapping has also been used in *Arabidopsis* and other model species to identify both single-gene loci and the QTLs responsible for phenotypes in natural populations. *Arabidopsis* has a wide geographical distribution, and it is possible to examine plants found in different locations (**ecotypes**) to identify genetic loci that have evolved to fit a particular niche. For example, a recent study used association mapping with 95 different ecotypes to identify genes involved in flowering time and pathogen resistance. As in the HapMap project in humans, researchers are continuing to characterize SNPs as well as short sequence-length polymorphisms in additional ecotypes, increasing the genetic resources for association mapping studies. Because *Arabidopsis* is self-fertilizing (as described in Chapter 2), each *Arabidopsis* ecotype is highly inbred and homozygous across the genome. As the ecotype is homozygous and self-fertilizing, its genotype only needs to be determined once, but it can be assayed repeatedly, each time for a different phenotypic trait (Aranzana *et al.* 2005).

Association mapping has two important advantages over traditional positional cloning. First, the mapping scheme is greatly simplified because there is no need for crosses or pedigrees. Secondly, the linkage relationships are being studied many generations after the original mutation appeared. As a result, generations of meioses have reduced the linkage disequilibrium everywhere except at those loci closest to the gene of interest. In the original *C. elegans* example, only one recombination event occurs on each chromosome per generation, and so the region of the genome showing some level of linkage disequilibrium with respect to the mutation is one whole chromosome. Thus, it is easy to see how the resolution after a single meiosis would be much lower than that after multiple meioses, when the more distant regions of the chromosome should reach linkage equilibrium with respect to the mutation of interest.

The greatest drawback of association mapping is that the basis for the phenotype being studied is not initially known, and often the many separate loci that contribute to the phenotype are segregating. The analysis of phenotypes in natural populations is also made more difficult by the inability to control fully the environment of the experiment. Thus, it can be challenging to separate the effects of genetics on the phenotype being studied from those of the environment. The effect of the environment on the expression of complex traits has been particularly high. Association mapping data can also be skewed by the presence of an underlying population structure and the absence of random mating. Nonetheless, most genetic mapping in human populations is done using association mapping, and the techniques continue to be refined.

Further reading

Aranzana, M.J., Kim, S., Zhao, K., *et al.* (2005). *PLoS Genetics*, **1**, e60.

genes are described in more detail in Chapter 8.) By separating the transposon and transposase to different loci, the modified P element can be distinguished from other P elements in the genome by the reporter gene, which facilitates outcrossing and isolation of the P element.

Beyond *Drosophila*, transposon tagging has been used to clone genes in *C. elegans*, mice, *Arabidopsis*, and many other organisms. The essential feature of the process is to have one strain in which the transposable element is present but inactive because the transposase is absent, and a second strain in which the element is absent but there is a transposase. The transposable element is mobilized by crossing the two strains and the resulting mutations are cloned using the transposable element as the molecular probe. The major limitation of transposon tagging is that the resultant mutant needs to be extensively outcrossed to remove extraneous elements, which can be a time-intensive step. The exact number of generations of outcrossing depends on the mating scheme, the activity, the number of transposable elements in the background, and the number of linkage groups in the organism. This situation is no different from the scheme used for outcrossing any new mutation (as described in Chapter 3), except that the investigator must be particularly careful to eliminate all of the extraneous elements that can be found by the same molecular probe, as these give rise to confusion about which gene has been tagged.

Transposon tagging has certainly been the most widely used mutation-based strategy for gene cloning, but it is not the only one. In *Arabidopsis*, T-DNA insertions can also be used to tag and clone genes. Furthermore, some cancer-related genes in mammals were cloned on the basis of the chromosome rearrangements that have breakpoints within the gene. The overall strategy in these cases is similar to that of transposon tagging: a molecular lesion in the gene provides the starting point for cloning the gene. In Chapter 8, we will also describe how microarray technologies can be used as a genomic approach to identify specific deletions and SNPs causing a mutant phenotype. Despite these advances, mutation-based strategies are still not widely used as an initial approach for cloning genes; the advent of genome sequencing projects has made positional cloning the strategy of choice for most cloning projects. Nonetheless, the molecular basis for allelic differences is still very widely used, and it is probably the single most important property for confirming that one has cloned the right gene.

To use this property to confirm that a candidate gene is the right one, the candidate gene is first sequenced from the wild-type strain and from several different mutant strains. If the correct gene has been cloned, the sequences will show different changes in each of the mutant alleles—missense mutations, nonsense mutations, splicing mutations, deletions, insertions, and so on. For model organisms that are derived from highly inbred strains such as flies, worms, and *Arabidopsis*, even a single base change in the coding portion of the gene can confirm that the investigator has cloned the correct gene. For organisms that do not have highly inbred laboratory strains, including humans, the investigator needs to be careful to distinguish a mutation that alters the function of a gene from a naturally occurring polymorphism that has little or no functional impact on the gene. For this reason, the case for having cloned the right gene becomes more persuasive with each mutant allele that is sequenced and shown to have a molecular lesion in the gene. This is also true when working with model organisms.

5.4 Property 3: a gene is defined by functional complementation (transformation rescue)

As described in Chapter 3, the functional definition of a gene has been the complementation test: two mutations are in different genes if they can complement each other, restoring the wild-type phenotype, and they are in the same gene if they cannot complement each other and restore wild-type function to the organism. A complementation test and its interpretation are shown in Figure 5.6. In a similar way, if the wild-type allele of a gene is introduced into a mutant organism, the wild-type copy will provide the normal function lacking in the mutant and will restore it to wild-type growth, movement, color, or whatever. In laboratory jargon, the

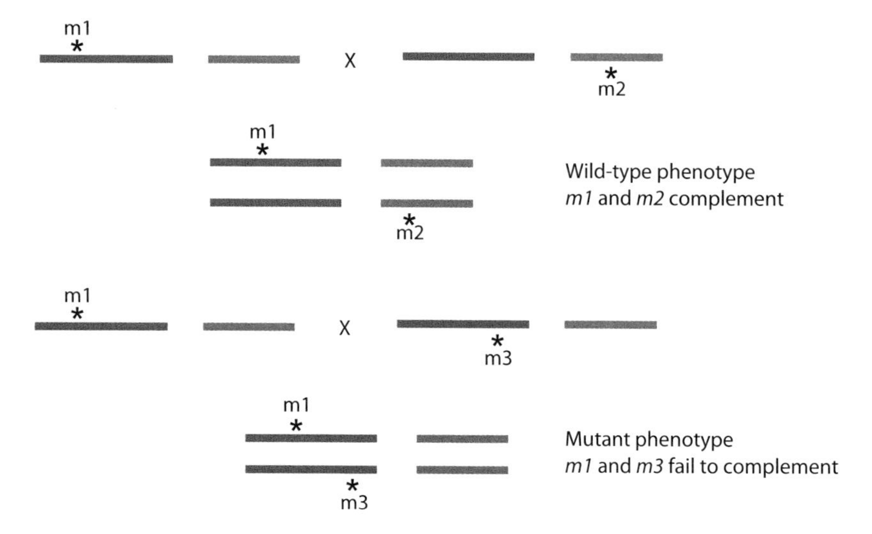

Figure 5.6 Complementation tests are the functional definition for a gene. Three recessive mutations, designated *m1*, *m2*, and *m3*, have similar mutant phenotypes and are tested for allelism. When the F_1 heterozygote between *m1* and *m2* is examined, a wild-type phenotype is observed. Thus, *m1* and *m2* complement each other and define different genes—each mutation is supplying the wild-type complement of the function that the other mutant lacks. When the F_1 heterozygote between *m1* and *m3* is examined, a mutant phenotype is observed. Thus, *m1* and *m3* fail to complement and are alleles of the same gene because they cannot provide the function that the other lacks.

wild-type allele has *rescued* the mutant phenotype. This functional complementation, also called **transformation rescue**, is used as both a screen and a selection at the initial step of cloning a gene and as confirmation that one has cloned the correct gene.

An example of how functional complementation can be used to clone a gene comes from auxotrophic mutants of the yeast *Saccharomyces cerevisiae*. Recall that auxotrophs are mutants that have lost the ability to synthesize some necessary nutrient and thus require the addition of that particular nutrient to survive (Figure 5.7). A yeast cell with a mutation in the *ura3* gene cannot grow in the absence of exogenous uracil. A library of plasmids containing fragments of the yeast genome is transformed into the *ura3* mutant cells, which are then plated onto medium lacking uracil. The only cells that can grow to produce colonies are those in which the wild-type *URA3*$^+$ gene has been introduced and where it is expressed to make a functional product. Although only a small number of cells acquire the appropriate plasmid, the powerful selection for growth in the absence of uracil ensures that many plasmids can quickly be assayed to identify the correct gene. The complementing plasmid is isolated from the cells that grow, and the complementing wild-type *URA3*$^+$ gene is cloned as a result.

Many genes have been cloned by functional complementation, particularly in unicellular organisms, such as the yeasts *S. cerevisiae* and *Schizosaccharomyces pombe*, where library transformation is straightforward and powerful selections are available with nutritional markers. The strategy has been less widely used in other organisms, at least as the initial approach for cloning a gene, because selections are not as easily available and transformation techniques with plasmid libraries are more difficult.

However, transformation rescue has been used frequently in multicellular organisms to narrow down a list of candidate genes in order to confirm that one has cloned the correct gene. For example, imagine that a positional cloning strategy (property 1) has yielded a collection of three or four cosmids that must contain the gene (a cosmid contains approximately 40 kb of DNA on average). It is not feasible for an individual investigator to sequence each cosmid and look for a mutant lesion. In worms, flies, and *Arabidopsis*, 10 (or more) different genes might be encoded in a region of 40 kb, and it is simply too laborious to sequence each gene to look for a mutation. Instead, transformation rescue (i.e. complementation) can be used. Each cosmid is introduced into the mutant organism and the transformants are tested for rescue of the mutant phenotype.

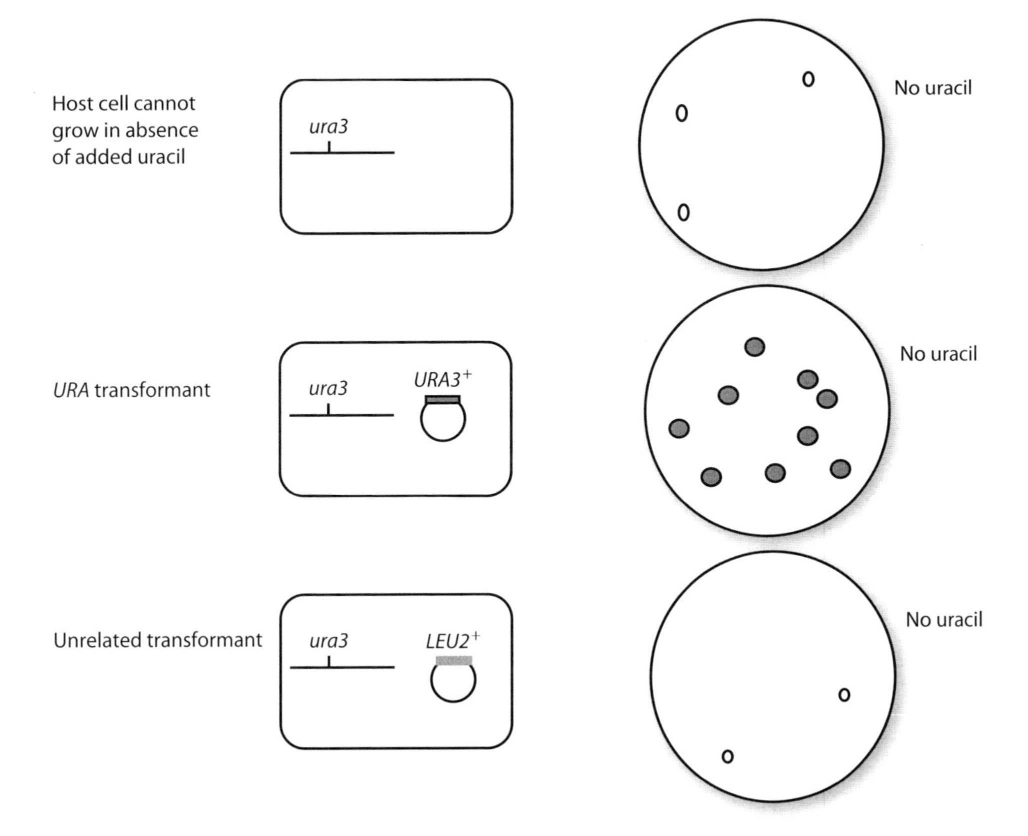

Figure 5.7 Using complementation to clone the *URA3*⁺ gene in yeast. A yeast strain with a *ura3* mutation cannot grow in the absence of added uracil. Plasmids are transformed into the *ura3* mutant cells; only those plasmids with the functional *URA3*⁺ gene can provide the wild-type function and rescue the mutant phenotype. An unrelated gene such as *LEU2*⁺ does not complement the defect. The complementing plasmid is isolated in order to clone the *URA3*⁺ gene. This strategy is referred to as transformation rescue.

The complementing cosmid can be subcloned into smaller plasmids until a smaller rescuing fragment is defined (property 3). In other words, the investigator is finding the region that provides functional complementation. The rescuing fragment is now sequenced from both wild-type and mutant organisms to find the gene with a molecular lesion in the mutant organism (property 2). In this way, complementation is being used to reduce the number of candidate genes to a more manageable number than position alone could manage, and the molecular defect in the mutant allele is used after complementation to confirm that the correct gene has been found. This common strategy is illustrated in Figure 5.8.

Complementation is also used to confirm that a gene cloned by another method is the correct one. In fact, most studies cloning a gene responsible for a mutant phenotype confirm that the gene is correct by complementation of the mutant phenotype with a wild-type copy in a transgenic organism or the presence of multiple alleles with defined molecular mutations in the same gene, as described above. A paper reporting the cloning of a new gene is rarely published without at least one of these two independent confirmations (in addition to the molecular basis of the mutation itself) that the real gene has been cloned.

In addition to its use for reducing the number of candidate genes, complementation is also widely used to provide transformation markers for other transgenic experiments. In *Drosophila*, for example, the P element used for transformation often includes the wild-type version of the *rosy* (*ry*) gene, and the recipient flies are mutant for *ry*. This allows the investigator to determine which flies have the transposable element by screening for wild-type eye color against a background of *rosy* flies or by a selection involving the xanthine dehydrogenase enzyme encoded by the *ry* gene. In these and all other

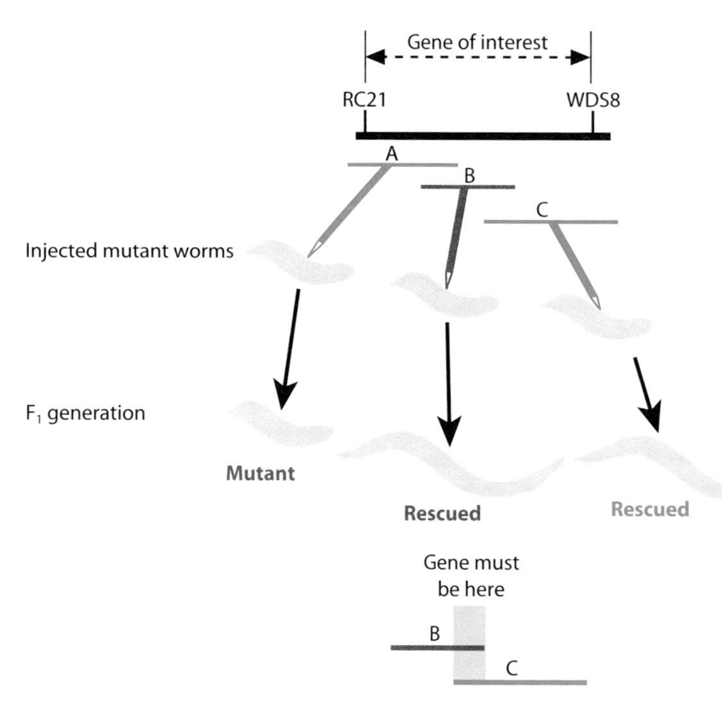

Figure 5.8 Using transformation rescue to test candidate genes. In *C. elegans*, a mutant that results in a short, fat phenotype has been mapped between the polymorphisms RC21 and WDS8, a region that is spanned by three cosmids, A, B, and C, shown here in different colors. Each cosmid is microinjected into a mutant worm. Both cosmid B and cosmid C rescue or complement the mutant defect and give a phenotypically normal worm in the next generation, but cosmid A does not. Therefore the wild-type copy of the gene must lie in the region where cosmids B and C overlap, shown in yellow.

cases that we have described in this section, it is essential that the mutation is recessive to the wild-type allele that provides the rescue.

It is worth noting that all of these complementation strategies are also widely employed in bacteria and bacteriophages and none was used first on eukaryotes. Where convenient methods have been available to screen or select for the transformants, complementation by transformation rescue has provided a very powerful strategy to clone genes and confirm the gene is correct. Investigators rely on it when using bacterial or yeast plasmids for many other molecular biology experiments, usually without thinking about complementation tests. That may be the highest praise for the strategy of complementation: it works so well that most scientists use it routinely and rarely think about it.

5.5 Property 4: a gene has a specific pattern of expression (expression-based cloning)

A gene or its mutant form has a phenotype that affects certain tissues, cells, or processes at designated times. With what we know about mutant phenotypes, it seems almost trivial to be reminded that a gene has a particular expression pattern. The pattern or level of gene expression has provided many different cloning strategies, some of them quite elaborate. The strategies for expression-based cloning can be divided, somewhat arbitrarily, into three categories:

- approaches using the expressed RNA
- approaches using the expressed protein
- approaches using the phenotype of the organism itself.

We will first describe some common expression-based cloning strategies that use RNA or protein, and then describe an approach using mutant organisms known as enhancer traps. The brevity of our discussion of RNA-expression- or protein-expression-based strategies should be understood as merely an overview of the many sophisticated and clever methods that have been used. A more extensive discussion of some of the methods involved in examining gene expression and in cloning genes is given in Chapter 8.

RNA-expression-based cloning

The most commonly used methods for expression-based cloning rely on RNA. In fact, RNA-based cloning strategies are probably second only to positional strategies in how often they are used to clone genes. The fundamental and original procedure for expression cloning using RNA is to extract total mRNA from a tissue type or developmental stage that is known to transcribe the gene at high levels. The RNA is then reverse transcribed into cDNA, which can be cloned to make a cDNA library. cDNA libraries from different tissues or developmental stages are a valuable resource and have often been the subject of informal exchanges at meetings: 'We had no luck finding it in the [first] library but got it from the [second] library.' cDNA is used for these experiments rather than the mRNA itself because it is more stable and more easily manipulated. The cDNA library will contain many genes that are expressed in the tissue, including genes that are expressed in many other tissues as well. Therefore, it provides many candidate genes, and so additional screening methods are required to find the particular gene of interest. Because the abundance of a particular clone in a well-made cDNA library approximates the abundance of the transcript in a particular tissue or stage, a strategy such as this is particularly useful for finding the genes that are transcribed at the highest level.

A further advance with such libraries is to use a subtractive procedure to enrich for stage-specific or tissue-specific transcripts. The technique is summarized in Figure 5.9. The mRNA or cDNA from the tissue of interest is hybridized in solution to cDNA or RNA from another tissue. The double-stranded hybrids represent genes that are expressed in both tissues, whereas the single-stranded transcripts represent tissue-specific expression. By isolating the cDNAs corresponding to transcripts that are not expressed in both tissues, the investigator can specifically enrich the cDNA library for tissue-specific expression. Such subtractive procedures are desirable in situations where other screening methods are difficult. They are rather more laborious, but subtractive procedures are useful to find less abundantly transcribed genes and to reduce the number of potential candidate genes.

A more contemporary genomic approach to expression-based cloning comes from the use of microarrays (Figure 5.10). We describe microarrays in more detail in Chapter 8, so this will be a brief introduction. In a two-channel microarray such as the one shown in Figure 5.10, gene-specific oligonucleotides corresponding to thousands of genes are spotted robotically onto a support (e.g. a silicon or glass slide) in a specific pattern, such that each location in the array corresponds to a defined gene. RNA is isolated from one tissue, stage of the life cycle, or treatment condition. The RNA is then reverse transcribed and fluorescently labeled to make labeled cDNA (or cRNA—complementary RNA), which is hybridized to the array. The fluorescence pattern is scanned, and the intensity at each spot is converted to a numerical expression value. In the case of two-channel arrays, two different samples are co-hybridized to the same array. The expression from the two samples is distinguished by labeling them with different fluorophors—Cy3 or Cy5. Single-channel microarrays are more reliable than two-channel arrays because the probes are directly synthesized on the surface of the array. Here, only one sample is hybridized to each array, and so for comparing two samples the process is repeated on an identical microarray chip with RNA from a different tissue or stage.

In either case, the results from the different samples are compared to identify differentially expressed genes, with particular interest in transcripts found in only one of the samples—the same kinds of transcripts we were considering in subtractive hybridization. The position of each oligonucleotide spot on the microarray slide is observed for expression differences, and the gene from which the oligonucleotide sequence is derived is known. This information allows an easy correlation between an expression pattern and a gene sequence, so that the gene can be cloned.

Cloning based on RNA expression will identify a large number of candidate genes, for example, genes

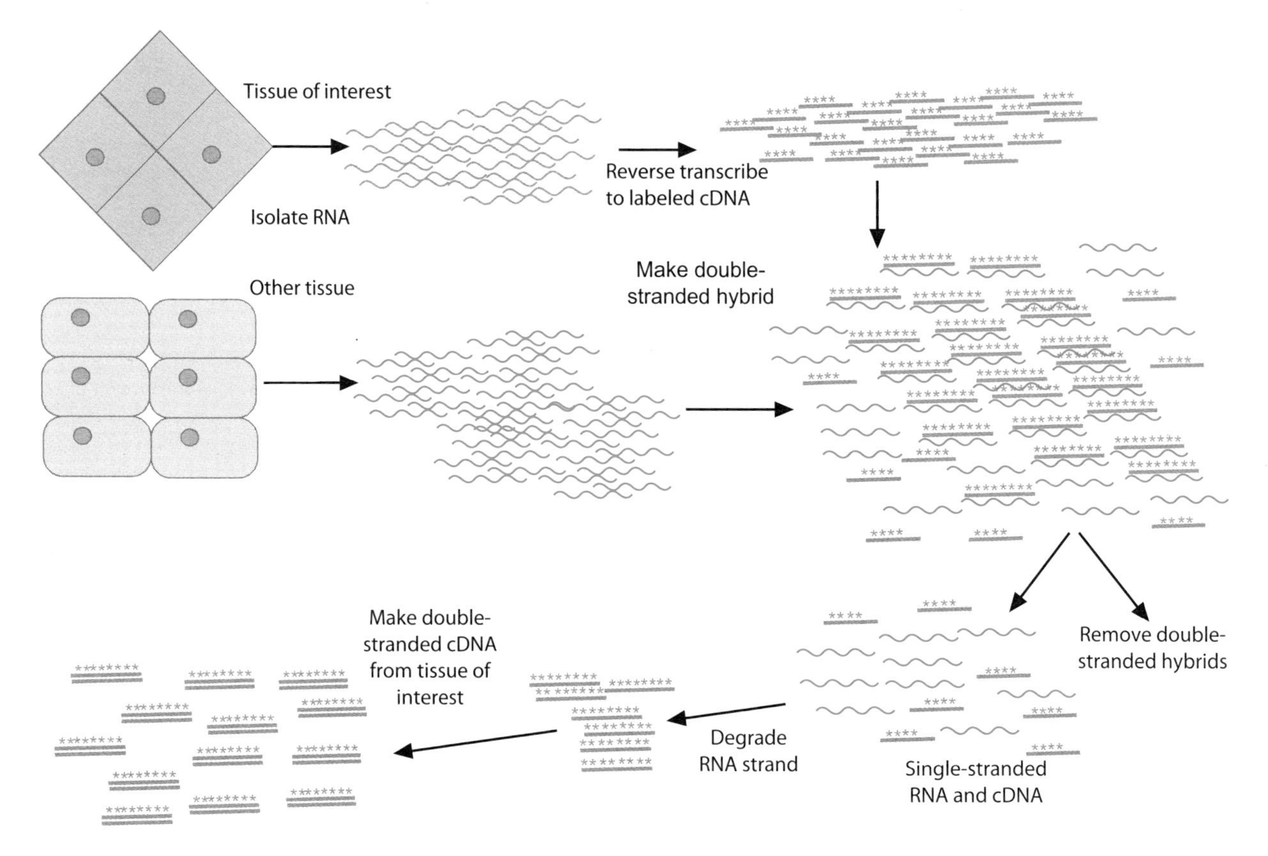

Figure 5.9 A subtractive hybridization procedure used in expression-based cloning. The goal of this experiment is to clone genes specifically expressed in the tissue of interest, shown here in red. RNA is isolated from this tissue as well as from another tissue, shown here in green. The RNA from the tissue of interest is reverse transcribed into cDNA with a label. The cDNA from the tissue of interest and the RNA from the other tissue are hybridized to each other, making a double-stranded (ds) cDNA–RNA hybrid. Note that an excess of RNA is used from the other tissue so that all the cDNA from the tissue of interest is found in the ds hybrid. The ds nucleic acid is separated and removed, leaving the RNA and the cDNA. The RNA is degraded and only the cDNA from the tissue of interest is retained. This is cloned and analyzed.

expressed differently in tumor cells and normal cells of the same type. Often, particularly for an expression profile from a microarray, there are too many expression differences to be able to clone one individual gene of interest. On the other hand, the greatest advantage of RNA-expression-based cloning is that it can be used in situations when other methods are not feasible. For example, if one is working with an organism in which the genetic map is incomplete or the genome is not being sequenced, expression-based cloning can be the best strategy for identifying developmentally important genes. This has been the case for many organisms including sea urchins, frogs, and snakes. It is the best method of compiling a large collection of candidate genes and provides the material for follow-up experiments using other properties of genes.

Protein-expression-based cloning

Protein-based cloning methods are not so common in contemporary molecular genetics, but we include them briefly for completeness. Proteins have been purified from crude cell or tissue extracts for many decades, and dozens of *in vivo* purification schemes have been used. One example relevant to cloning involves proteins that bind DNA, including many transcription factors. A technique for identifying DNA-binding proteins uses an electromobility shift assay (EMSA), also called a gel shift assay. In this case, a crude protein extract is incubated with a DNA sequence, and the DNA-binding proteins are isolated on the basis of their ability to shift the location of the DNA sequence in an electrophoresis gel. An example of an EMSA gel is shown in Figure 5.11. The binding protein is extracted from the gel, and a portion of its amino acid sequence is determined. This sequence is used to synthesize an oligonucleotide or a pair of polymerase chain reaction (PCR) primers to provide a probe for

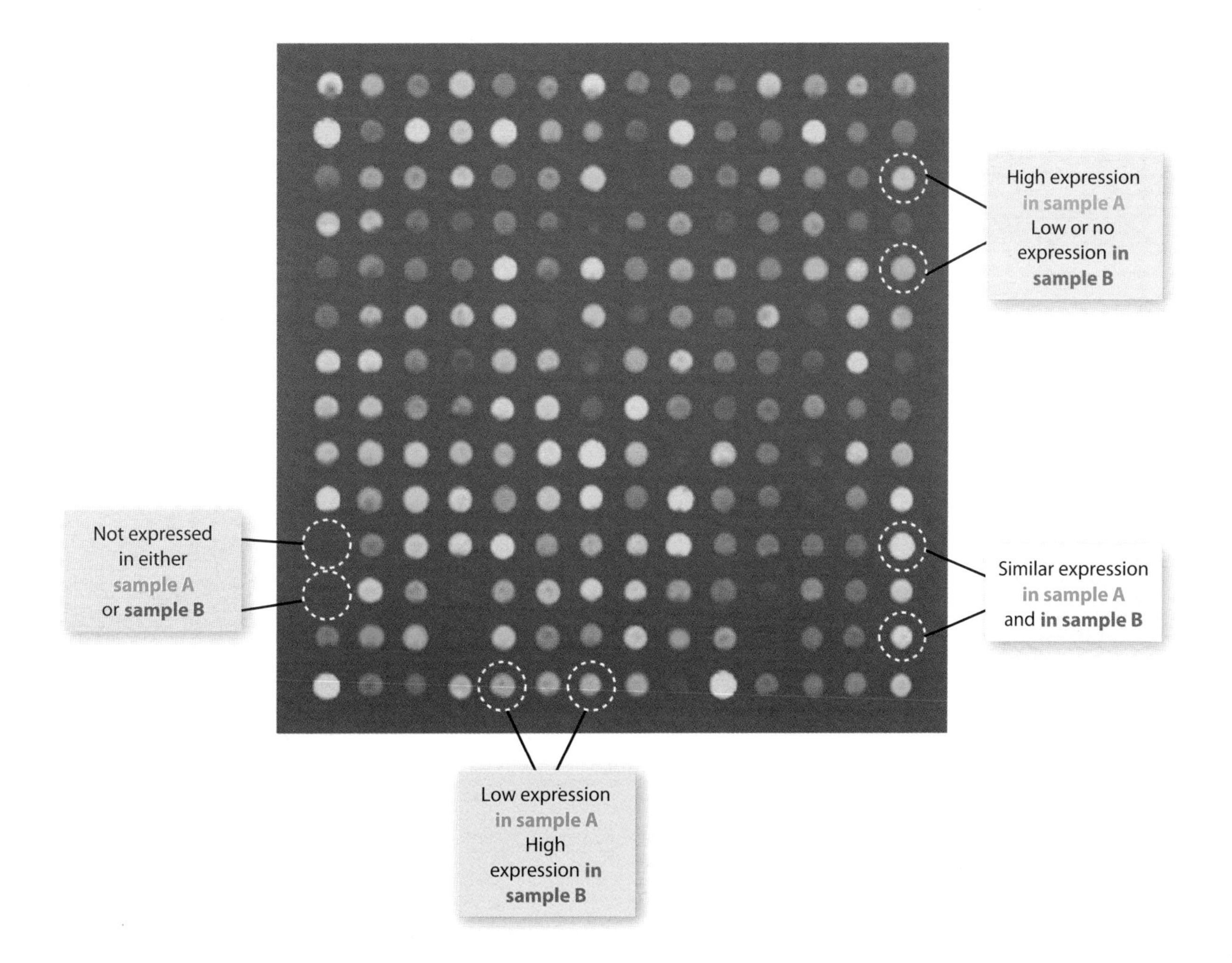

Figure 5.10 A microarray can be used as an RNA expression cloning strategy. Sequence probes specific to different genes are fixed to a support in a defined array. This array shows 14 × 14 or 196 different genes. This is a two-channel array, as described in Chapter 8. RNA from sample A has been made into cDNA and labeled in green. RNA from sample B has been made into cDNA and labeled in red. The cDNAs are hybridized to the array. Genes expressed at high levels in one sample but at low levels in the other sample will appear as spots of one color, as indicated here. Genes expressed in both samples at similar levels will appear as yellow spots, and genes not expressed in these samples are black. Since genes occupy known positions on the array, genes expressed in one or the other sample can readily be identified and cloned.

identifying the rest of the gene from a cDNA library, in a manner similar to homology-based cloning (described below), or to derive a nucleotide sequence conceptually and search the genome sequence computationally. Although our example describes DNA-binding proteins, almost any protein purification scheme can be adapted into a cloning strategy as long as some amino acid sequence information can be obtained from the purified protein.

Although examples of genes that have been cloned using protein expression can be found, this strategy is used only infrequently in some very specific cases. The most fundamental limitation is that the techniques for protein purification and sequencing are not as easy as those for nucleic acids. Because this limitation is significant, most investigators will prefer to use a nucleic-acid-based cloning strategy if available.

Phenotypic-expression-based cloning: enhancer traps

Using isolated RNA or protein—the macromolecules expressed by a gene—as a means of cloning the gene is conceptually simple, even though the actual procedures may be quite sophisticated. Cloning a gene using its morphological expression pattern in a whole organism is subtler, but can be equally effective. One widely

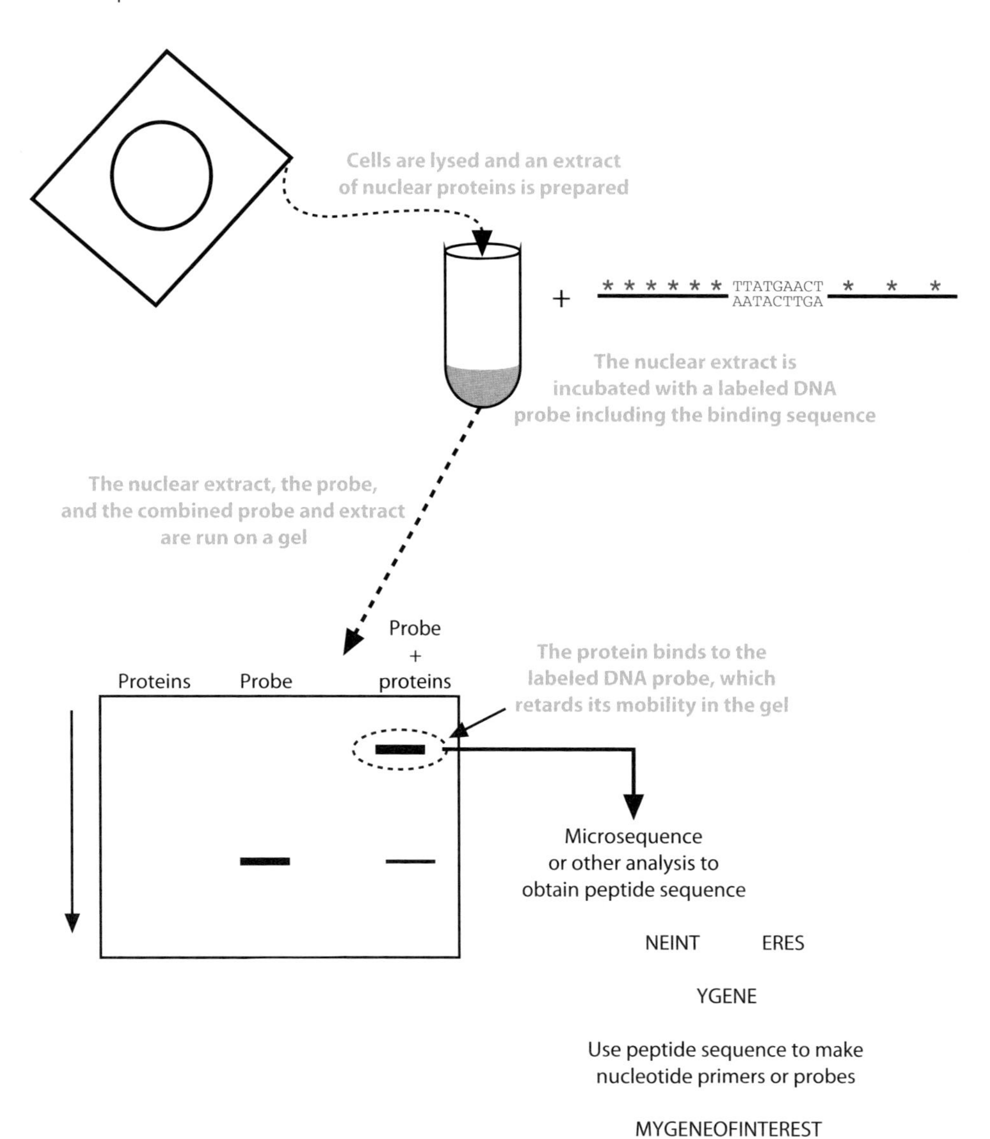

Figure 5.11 A protein-based expression cloning strategy based on electrophoretic mobility shift assays. Gel shift assays have been widely used to clone DNA-binding proteins. A crude protein extract is prepared from cells that are thought to be making the protein of interest. This crude extract is incubated *in vitro* with a labeled DNA probe that includes the binding sequence for the protein. The protein–DNA mix is then run on a gel, as shown. The binding of the protein to the DNA sequence retards the mobility of the labeled DNA sequence probe; notice that the band in the 'probe + proteins' lane is higher in the gel than the 'probe' alone. The protein is partially purified from this binding complex, and the amino acid sequences of the peptides are obtained.

used cloning strategy using morphological phenotype is known as an **enhancer trap**. Enhancer traps are also used to analyze the transcription pattern of a gene, as described in Chapter 8. The technique is the same in each case, but the goal is slightly different; we will discuss the use of enhancer traps in cloning a gene.

In order to understand the strategy, it is important to understand the name 'enhancer trap'. It refers to the fact that a gene's enhancer, i.e. the flanking chromosomal region that regulates the expression pattern of a gene, is being used to trap or clone the gene's coding region. This name can be a source of confusion because the enhancer itself is not usually being trapped. Rather, the presence and function of an enhancer is used to clone the gene under its control. The procedure will be described as it is done in *Drosophila*, where it has been most widely used; related procedures have also been used in many plants and animals.

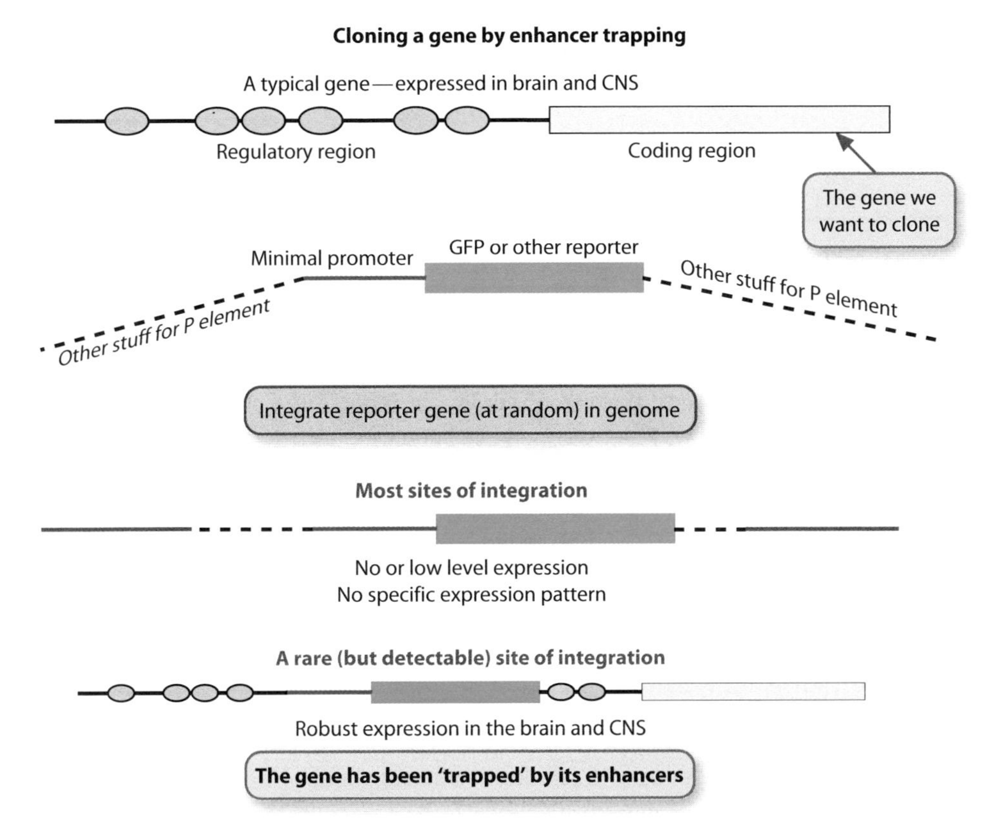

Figure 5.12 An enhancer trap is a strategy for cloning genes based on the phenotypic expression pattern. The gene to be cloned is expressed in the brain and central nervous system. A reporter gene construct is integrated at random into the genome. At most sites, little specific expression is observed. However, if the reporter gene integrates into the enhancer region of a gene expressed in the brain and central nervous system, there will be strong expression in those tissues. The surrounding region is cloned to find the gene at that locus expressed in the brain and nervous system.

An enhancer trap strategy is outlined in Figure 5.12. It has two essential components: a reporter gene and a method to produce random integration of transgenic reporter genes into the genome. As will be discussed further in Chapter 8, a reporter gene, such as the *lacZ* gene from *Escherichia coli*, which makes β-galactosidase, encodes a protein whose expression can easily be assayed. In *Drosophila*, P elements provide the vector for such random integration events. A reporter gene with a minimal promoter is introduced into flies and is allowed to integrate throughout the genome. At most integration sites, the reporter gene will either not be expressed or will only be expressed at low levels and in non-specific patterns. However, at some sites the reporter gene will be integrated by chance near an enhancer element, whose presence is inferred from the specific pattern of expression of the reporter gene.

For example, suppose that the reporter gene in one particular transgenic line is expressed only in the eye primordium during the second larval stage. The pattern of expression indicates that the site of integration is close to an enhancer that directs expression in that specific pattern. Therefore, there must be a gene nearby that is normally expressed in the eye primordium during the second larval stage. The function and identity of this inferred gene is completely unknown; its presence is detected or trapped by its enhancer. The transgenic line with the appropriate expression pattern is painstakingly outcrossed to eliminate all other integrated P elements that may be influencing expression. Then the region of the chromosome is cloned using procedures similar to positional cloning, as described above, and the corresponding gene is cloned and analyzed. Alternatively, in the genomics age, PCR can be used to amplify the

Drosophila DNA at the insertion site, which can then be sequenced. The sequence can be entered into BLAST to identify the region of the genome having the insertion, and the neighboring gene(s) can be identified and then cloned.

Enhancer trapping has several attractive features as an initial strategy for cloning genes. As with all of the cloning approaches in this category, genes are cloned using simply their pattern of expression. It is not necessary to know the function, map location, or mutant phenotype of the gene in order to clone it. For an enhancer trap, it is also not necessary to have mRNA, protein, or a functional assay corresponding to the gene product in order to clone the gene. On the other hand, enhancer traps are labor-intensive, particularly when the outcrossing procedure is considered. Most enhancer trap experiments have been done by several groups working together, perhaps interested in genes with different patterns of expression. Large collections of transgenic fly strains with different integration sites and different patterns of expression have been maintained as a collective enterprise, which reduces the need for an investigator to begin the protocol from the beginning.

Summary

These summaries only begin to describe the diversity of the cloning strategies that are grouped together as expression-based cloning, using the fourth property of genes. As a rule, expression-based cloning is much better at identifying candidate genes than at pinpointing the exact gene of interest. Expression-based approaches are also very helpful when used in combination with other approaches. In Case Study 5.1 on the human disease cystic fibrosis, we describe how an expression-based strategy, namely a cDNA library isolated from sweat glands, was a crucial step in sorting through candidate genes identified by positional cloning.

5.6 Property 5: a gene is the unit of evolutionary descent (homology)

The fifth and final property of genes that has been used in cloning is that they are units of evolutionary descent, i.e. most of the genes found in one mammal, such as the human, are also found in all other mammals. Furthermore, many of the genes found in humans have a recognizable nucleotide or amino acid sequence similarity to genes found in *Drosophila*, *C. elegans*, yeast, and even *Arabidopsis*. In other words, each gene will have orthologous genes in other organisms, and that evolutionary relationship will often result in sequence similarity that can be used to clone the corresponding genes in those other organisms. For example, in Case Study 5.2 on the *Drosophila* gene *patched*, we describe how homology with the *Drosophila* gene was used to clone orthologs of the *patched* gene from other insects, chickens, and ultimately mice and humans.

This is an appropriate point at which to define and discuss some terms used to describe the relatedness of genes. By definition, genes derived from a common ancestral gene are homologous. **Homology** is a description of evolutionary history, and as such is a **qualitative** term: genes are either homologous or they are not. Some molecular biologists might use an expression such as '40 percent homology' to describe two genes, but this is a faulty use of a qualitative concept. Genes can no more be partially homologous than a woman can be partially pregnant. The false concept, which probably reflects careless language rather than a misunderstanding of the principles, arises because homology itself cannot be measured. Homology of two genes or two proteins is inferred from **sequence similarity**. The degree of sequence similarity can readily be assessed quantitatively (e.g. by using BLAST) and rigorously tested for statistical significance. BLAST and its use for inferring homology from amino acid sequence similarity are described in Text Box 5.2.

It is often useful to distinguish two types of homologous genes. An organism often has multiple copies of a particular gene with identical or related functions. For example, *C. elegans* has four actin genes and *S. cerevisiae* has two α-tubulin genes (as described in more detail in Chapter 10). Closely related genes within one genome

Case Study 5.1

Positional cloning of the cystic fibrosis gene in humans

Cystic fibrosis (CF) is one of the most commonly occurring genetic diseases, particularly among people of European descent. Although the frequency of the disease varies among populations, approximately one in 2500 Caucasian babies has CF. The principal symptom is an accumulation of mucus in the lungs, which leads to difficulty in breathing and an increased susceptibility to bacterial infections. Most affected individuals also have defects in exocrine secretion in the pancreas; ducts in other organs (such as the vas deferens in the testes) can also be affected. Some symptoms of the disease have been familiar for decades, in particular the unusual secretion of chloride ions in sweat. As recently as 50 years ago, most individuals with CF lived only into their teens, although with current diagnosis and treatments, the average life expectancy is now about 30 years.

CF is inherited as an autosomal recessive trait. From the frequency of affected children, it is estimated that about 4 percent of Europeans are heterozygous for the disease but have no symptoms. Because of its high frequency and simple pattern of inheritance, CF has been the focus of a substantial amount of research. Cloning the gene and characterizing the molecular lesion that causes the disease were a *tour de force* of human genetic analysis.

Identifying candidate genes

The gene was first mapped to the long arm of chromosome 7 (7q) by pedigree analysis and subsequently placed in a region of about 1500 kb, with a large number of polymorphic molecular markers mapped to either side. In a Herculean effort involving some novel cloning strategies that need not concern us here, this entire 1500 kb region was cloned into a set of 49 phage and cosmid clones. Thus, the first step in the cloning process relied on the map position of the *CF* gene. This step is summarized in Case Study Figure 5.1.

This cloning process occurred more than a decade before the human genome project was completed, at a time when few details were known about the structure of the human genome. Once the region of the chromosome that included the *CF* gene was delimited by mapped markers and cloned, it was necessary to define the genes in the region. This was itself an arduous process. Two properties of all genes, as well as one characteristic of mammalian genes, played a significant role in the annotation of genes in this region. The first key property that guided investigators was *homology* to other mammalian DNA sequences. Homology was established by hybridization of each part of the 1500 kb region with DNA from other mammals. Because coding sequences are more highly conserved than non-coding sequences, the regions of hybridization with other mammals served as a first approximation to the locations of exons.

The second key property used to find candidate genes was the *expression* of the *CF* gene in affected tissues. Investigators isolated mRNA and probed cDNA libraries from tissues such as tracheal epithelium, lung, sweat glands, and pancreas, i.e. the tissues and organs known to be affected in CF children. The third useful property for finding genes was the recognition

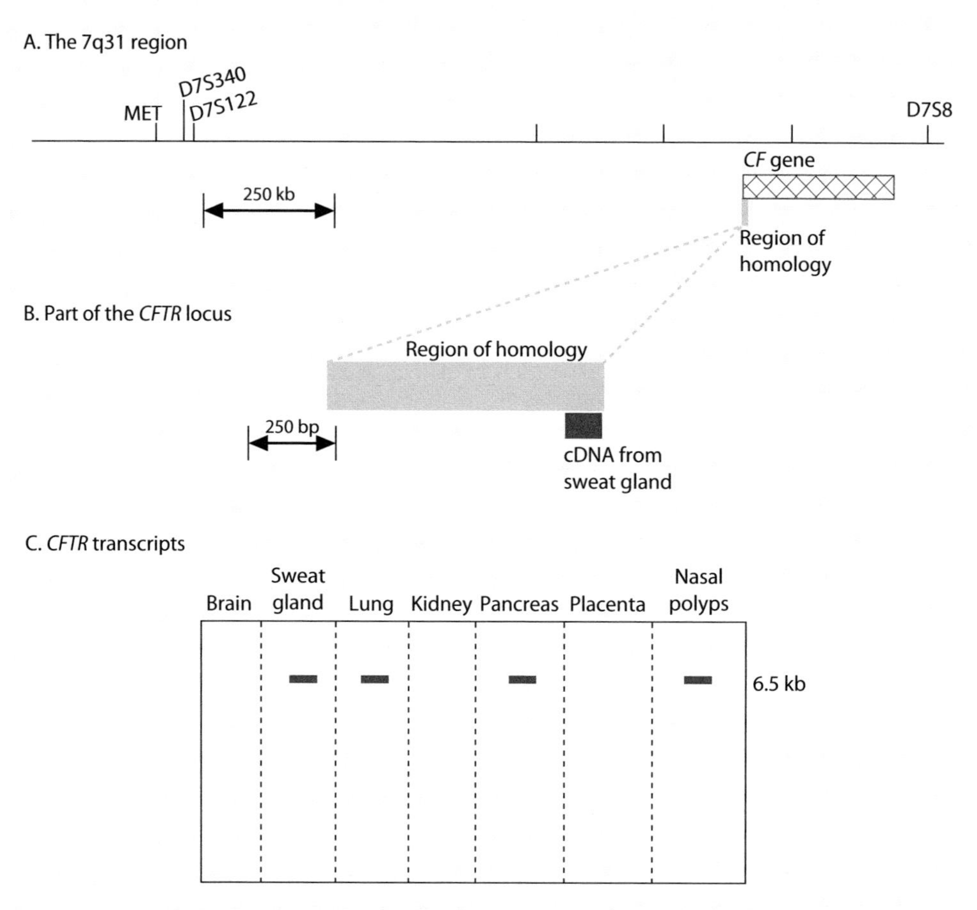

Figure C5.1 Positional cloning of the human *CF* gene (CFTR). (A) The polymorphisms closest to the *CF* gene on chromosome 7q31 are shown, and the region between them was cloned. This region was probed by hybridization for nucleotide sequence similarity to other vertebrates. The region of hybridization and inferred homology found with other vertebrates is shown in green, with the location of the *CFTR* gene shown for reference. The scale bar is 250 kb, so the entire region is approximately 1.5 Mb and the region of homology is approximately 1 kb. (B) An expansion of the region of homology, with the small fragment of exon 1 of *CFTR* that was used as a probe indicated in blue. This exon was found by probing a cDNA library from sweat glands of a healthy individual. The scale bar represents 250 bp. (C) The fragment of exon 1 was also used to probe RNA blots from a variety of human tissues for transcripts of the gene. Only some of the tissues are shown, for illustrative purposes. A transcript is detected in the tissues known to be affected by CF such as sweat glands, lungs, and pancreas, but not in tissues that are not affected by CF, such as brain and kidney. Simplified from Rommens *et al.* (1989).

that most mammalian genes have a CpG island—a cluster of CG dinucleotides—at their 5′ end. This gave the investigators more insight into the structure of all of the candidate genes in this region of the chromosome, not just the *CF* gene.

From these approaches, the investigators expected to find a candidate gene at the right location on the chromosome and expressed in the appropriate tissues. No one of these methods was guaranteed to identify all of the genes in the region; accurate gene identification remains one of the more formidable tasks in genome analysis. In fact, a careful reading of the papers on CF illustrates the limitations of the approaches that they used since analyses using expression patterns implicated several false candidates for the *CF* gene. The correct gene was found only after repeated and exhaustive screening of a cDNA library prepared

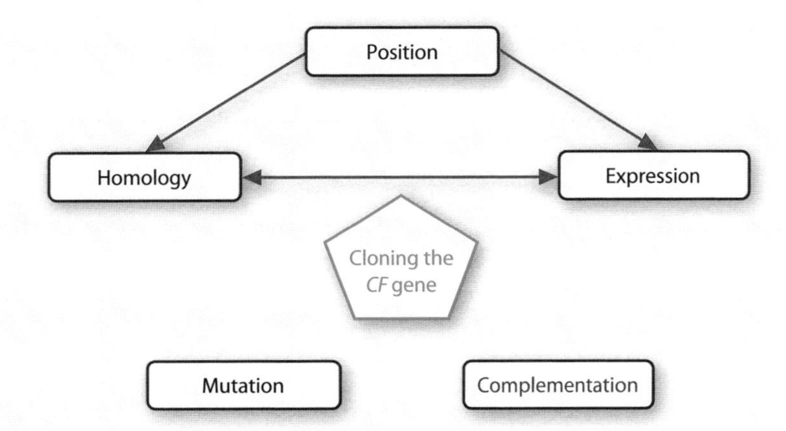

Figure C5.2 Summary of the strategy to clone the human gene responsible for CF. The gene was first mapped to a locus on chromosome 7q. The DNA from this region of the chromosome was isolated and analyzed for candidate genes. Candidate genes were recognized by a combination of homology and expression pattern.

from the sweat glands of a normal individual. The strategy used to arrive at this candidate *CF* gene is summarized in Case Study Figure 5.2.

The CFTR protein and gene

What could be said about the protein encoded by the candidate *CF* gene, called *CFTR*? CF is associated with defects in secretion, and suggestions that the gene might encode a membrane-channel protein predated its molecular or biochemical analysis. In particular, the secretion of chloride ions mentioned earlier was considered a strong hint that the *CF* gene may be involved in secretion, either as the ion-channel protein itself or as a regulatory protein. The inferred amino acid sequence of the candidate CFTR protein supported the conclusion that it could be encoded by the *CF* gene. When compared with other protein sequences, the predicted CFTR protein has extensive sequence similarity to membrane-channel proteins (Case Study Figure 5.3). This was certainly reassuring evidence that the candidate gene *CFTR* was the actual *CF* gene.

But how could the investigators be certain that they had found the gene that really causes CF? Map position alone, even for the carefully mapped *CF* gene, identified candidates but did not limit the number to only one gene. Homology searches and expression pattern analysis refined the search for candidate genes but still did not pinpoint the exact gene. The inferred amino acid sequence of the protein is consistent with the predicted role of the CFTR protein, but, by itself, this would not have persuaded critics that the investigators had found the *CF* gene. So what property would provide the most unambiguous evidence that the investigators had found the *CF* gene?

Mutations in the *CFTR* gene

For proof that they had cloned the causative gene for CF, the investigators turned to mutations. The gene is defined by the disease that occurs when it is mutated. If investigators had indeed found the gene responsible for a disease, individuals with the disease would have mutations in the candidate gene, and unaffected individuals from families with no history of the disease would usually be homozygous for the wild-type allele. In principle

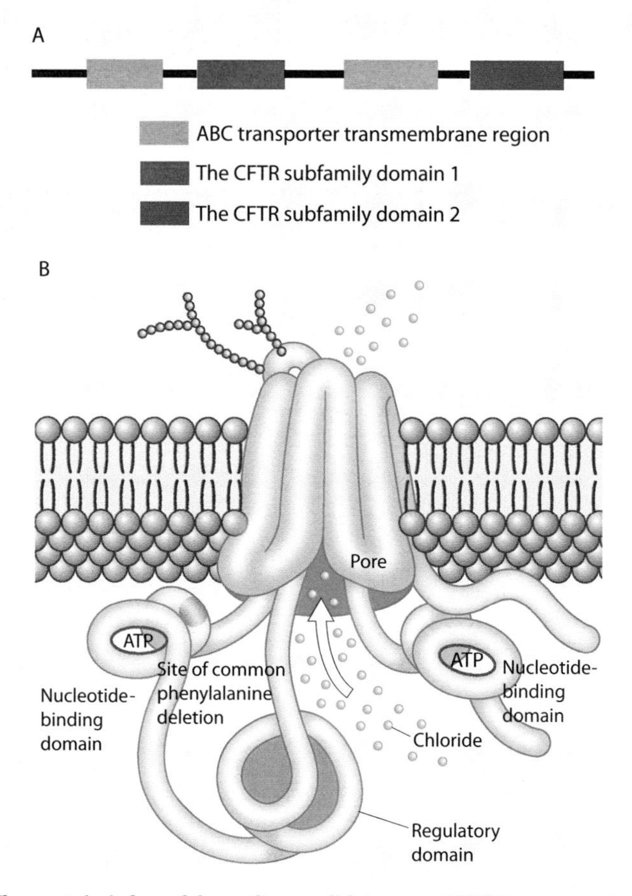

Figure C5.3 The protein inferred from the candidate gene *CFTR* has a structure predicted for ion-channel-related proteins. Panel A shows the domain structure of the CFTR protein from the Conserved Domain Database. The presence of the two ABC transporter transmembrane domains was strong evidence for the biochemical role of CFTR. The CFTR subfamily domains are blocks of amino acids that are found in many CFTR orthologs but not in all ABC transporter proteins. Panel B shows a predicted three-dimensional structure of the CFTR protein in the membrane. The location of the common ΔF508 mutation is shown on the intracellular domain.

then, one could simply sequence the candidate gene from an affected individual and from an unaffected individual and look for sequence differences. Some sequence differences between these people will be naturally occurring polymorphisms but at least should identify the gene.

For a rare genetic disease (such as phenylketonuria, which affects 1 in 17 000 newborns), an unaffected individual from a family with no history of the disease can reasonably be expected to have a wild-type allele of the gene. Thus, sequence differences in this person can be attributed to natural polymorphisms rather than to causative mutations.

But even this conceptually straightforward approach was more challenging for the *CF* gene because unaffected heterozygotes are relatively frequent. Thus, any person in the population might well have one mutated copy of the gene. Furthermore, although frequent, heterozygotes may not be detected because the disease is recessive. A person is known to be a carrier only if he/she has an affected child. Suppose that two individuals, both of whom are heterozygous carriers for CF, marry and have children. The probability of a healthy child is 3/4. Even with two children, the probability is 9/16 that neither child is affected, and so the carrier status of the parents remains undetected in many families. In short, undetected carriers for CF are not rare. Thus, it may not be informative simply to compare the sequence from the candidate gene with the sequence from another person; the other person could be a carrier.

Since the investigators could not make an assumption about *unaffected* families, they turned to a more comprehensive analysis of *affected* families. Unaffected parents who have an affected child must be heterozygotes and must have one wild-type allele of the *CF* gene. The investigators looked not only at the putative *CF* gene (*CFTR*) itself but also at nearby molecular polymorphisms that flank the gene. These polymorphisms allowed them to establish a correlation between the inheritance of other linked molecular markers and the inheritance of CF, which could then be used to determine both normal and mutant alleles of the gene. Previous studies had indicated that many cases of CF occur on the same chromosomal background, which suggested that many children with CF have inherited the exact same molecular lesion. This suggestion was directly confirmed. Upon sequencing the candidate *CF* gene *CFTR* from heterozygous carriers, investigators discovered that 68 percent of the CF chromosomes had a 3 bp deletion that removes a phenylalanine at residue 508 in the predicted CFTR protein. The frequency of this one mutation, designated ΔF508, is discussed in more detail in the section below entitled 'The population genetics of cystic fibrosis'.

None of the normal chromosomes has this mutation, and no recombination has been observed between it and the inheritance of CF, so inheritance of this mutation completely co-segregates with the inheritance of the disease. This analysis confirms that a mutation in the *CFTR* gene is responsible for CF. In subsequent studies, the *CFTR* gene has been sequenced from other children with the disease, particularly those whose symptoms differ from the classical CF disease phenotype and those from other ethnic backgrounds. These children also have mutations in the *CFTR* gene, although usually not the ΔF508 mutation. To date, more than 100 different mutations in the *CFTR* gene have been observed in patients with CF, but ΔF508 still comprises about 70 percent of CF mutations, and more than that in some populations.

CF and positional cloning

In this case study, we have seen how investigators used genetic map position, homology to other genes, and expression pattern to clone a candidate gene for CF. The analysis of mutant alleles confirmed that the candidate gene *CFTR* is the gene that is actually responsible for the disease. This strategy is summarized in Case Study Figure 5.4.

The cloning of the *CF* gene is rightly recognized as one of the landmark events in human genetics. However, a current reader of the original papers from 1989 would probably be struck by how much laboratory procedures have changed. A comparison of the procedures used in the classic papers and a current similar analysis is presented in Case Study Figure 5.5. Unlike current papers, much of the description in the classic papers is devoted to the recombination analysis used to position the gene and to procedures for chromosome walking to clone the entire region. Many more polymorphisms have now been determined and close genetic map locations are now easier to obtain.

Even more significantly, positional cloning is now done by searching sequences on the computer, so the entire process of finding clones across the region is made largely

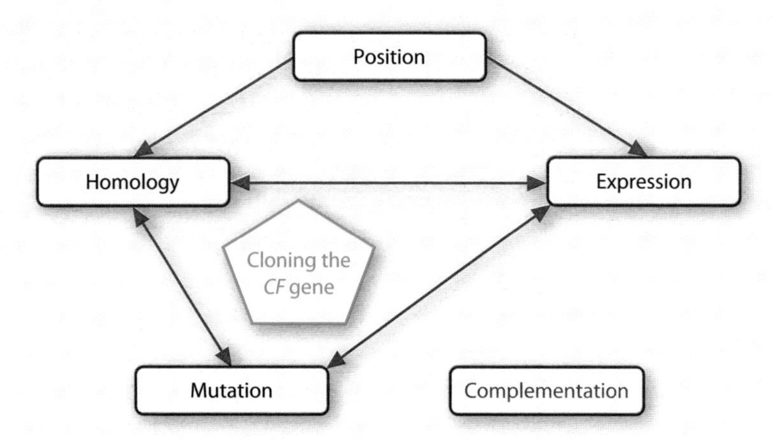

Figure C5.4 A summary of the strategy used to confirm that the candidate gene *CFTR* is responsible for CF. Compare this figure with Case Study Figure 5.2. The candidate gene identified by the procedures shown in Case Study Figures 5.1 and 5.2 was sequenced from known carriers of CF and from affected individuals to determine the molecular lesion. These individuals had a detectable molecular change in the candidate gene, confirming that this gene is responsible for CF.

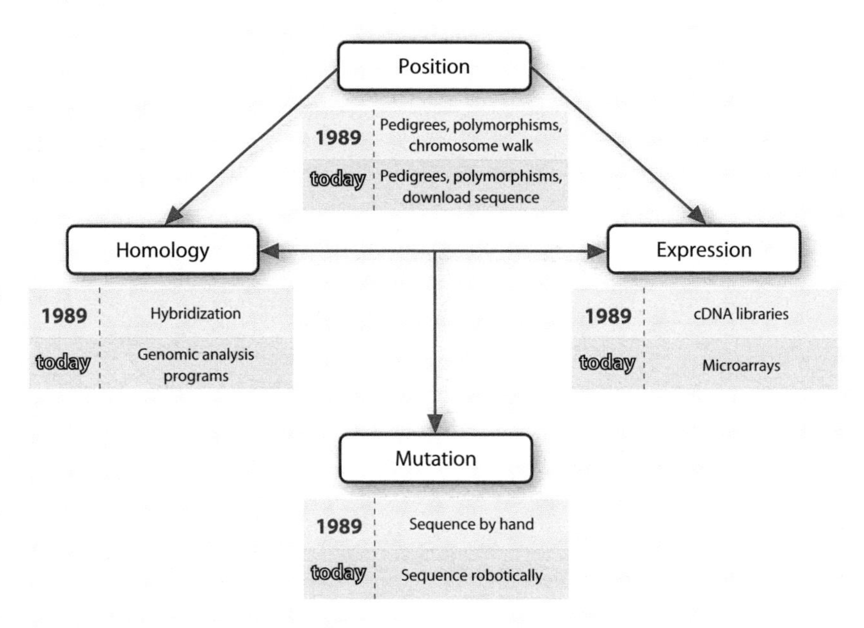

Figure C5.5 A comparison of the procedures used to clone the *CFTR* gene with current procedures. The essential strategy for cloning the gene is very likely to be similar to what was done in 1989, but the Human Genome Project and related endeavors have changed the experimental techniques. Positional cloning remains the primary approach for many disease-related genes. More polymorphisms are available for mapping, and the need to perform a chromosome walk to obtain the DNA spanning the region has been replaced by the ability to download the sequence from a database. The use of hybridization to infer homology has been replaced by genome analysis programs. Likewise, expression analysis is done by microarrays rather than by probing cDNA libraries. The key confirmation of the cloning strategy remains the molecular lesion associated with a mutation. In this case, manual sequencing has been replaced by sequencing machines and tiling microarrays. Nonetheless, these are changes in methodology and not in the underlying strategy that was originally used to clone the *CFTR* gene.

unnecessary. For a current investigator, the entire region of the genome has been sequenced and information on sequence similarity to other mammals is available at the press of a few computer keys. Thus, the need to do hybridization with the DNA of other mammals to find the highly conserved regions is also eliminated. Likewise, the structure of most human genes has been predicted, and the gene predictions, although not perfect, include the location of CpG islands, splice sites, and termination sites, so all these steps are also already completed. In addition, for many human genes, expression patterns have been determined by microarrays (see Chapter 8) or other methods, so the protocol for finding mRNAs or cDNAs from the appropriate tissues is also shorter. Predicted amino acid sequences have been obtained for nearly every gene, and orthologs for many human genes have been identified in other species. This is the level of information available for many genes in the human genome, regardless of whether the gene is known to be connected to a human genetic disease. The procedure now is to map the disease gene as closely as possible using well-characterized molecular polymorphisms, and then to find transcripts with the predicted expression pattern to identify candidate genes.

If the genetic times have changed so much, why does the cloning of the *CF* gene still deserve a place of honor in a genetics textbook? Several reasons come to mind. First, just as for CF, it is still essential to sequence candidate genes from affected individuals (or their parents) to confirm that one has cloned the right gene. That crucial part of the analysis has not changed, and CF shows why it is necessary. Although it is true that sequencing methods and availability of data have dramatically changed this part of the analysis, the only true confirmation that one has cloned the right gene is to identify mutations that cause the disease.

In addition, a current reader will realize how much more difficult it is to clone a human gene when the gene does not cause a well-described syndrome or the disease does not have a simple Mendelian inheritance pattern. CF is one of the genetic diseases that we know best, both in the clinic and in the research laboratory; the analysis would have been much more difficult for a more obscure disease gene or for a disease with a multifactorial pattern of inheritance. The final reason for describing the cloning of the *CF* gene is that the high quality of the research itself—not only the quality of the data but, more importantly, the quality of the logic—has been an underlying guide in the Human Genome Project. The cloning and analysis of the *CF* gene set the standard by which the cloning of other human genes and the Human Genome Project itself has been measured.

The population genetics of CF: Why is CF so common among Europeans?

The prevalence of the ΔF508 mutation is an interesting story in genetic analysis by itself—in this case, in population genetics. From Hardy–Weinberg equilibrium considerations, it seems implausible that a mutation that results in a severe disease could accumulate to a high frequency, yet as many as 4 percent of Europeans are carriers for the disease. Five different forces are known to affect allele and genotype frequencies in a population and any one of them (or some combination of forces) could be responsible for this high frequency. One of these forces, non-random mating, affects the frequency of genotypes but not the frequency of an allele itself. The other four—mutation, migration, genetic drift, and selection—affect the allele frequencies. In typical natural populations, certainly including natural populations of humans, these five forces occur together and their individual effects can be difficult to separate. However, it is instructive to examine the prevalence of CF among Europeans, as an example of the effects of the five forces, by examining what each one predicts and comparing that with what is seen for CF.

The effects of non-random mating

We will begin by discussing the possible effects of non-random mating, also known as inbreeding or consanguinity. Non-random mating does not affect the frequency of the allele itself and could not explain the high frequency of the ΔF508 mutation. However, non-random mating, which in humans takes the form of first cousins having children together, will increase the frequency at which recessive alleles become homozygous, thereby increasing the frequency of a recessive disease.

It seems unlikely that consanguinity plays a significant role in the frequency of the disease because CF often arises in offspring of unrelated parents. On the other hand, one of the hallmarks of inbreeding is that the allele is 'identical by descent' as shown in Case Study Figure 5.6; that is, as it is derived from a common ancestor to both parents, the allele will have the same molecular lesion. The highly prevalent allele ΔF508 often occurs on the same haplotype or chromosomal background (haplotypes are described in Text Box 5.1).

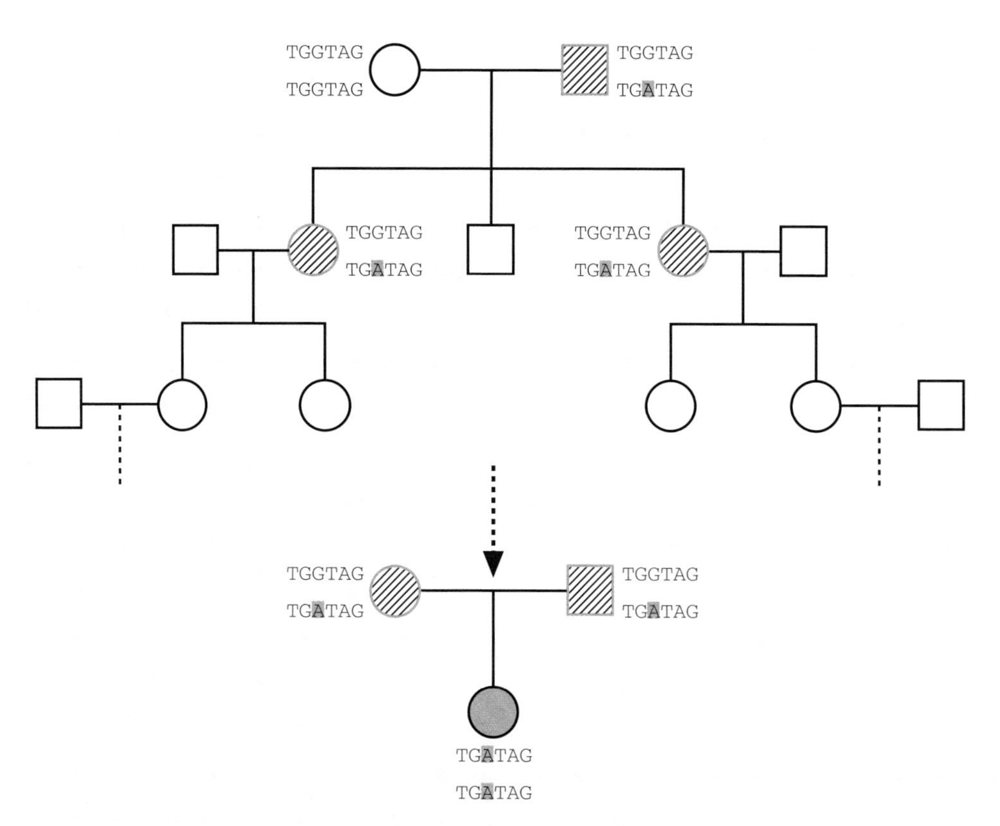

Figure C5.6 Inbreeding and alleles identical by descent. Heterozygotes for a recessive condition are shown as cross-hatched and the affected person is in blue. The affected girl is homozygous for the same mutation, the G to A transition in the man in the first generation. The man passed this mutation on to two of his daughters and several generations later, two heterozygotes have an affected child. Her alleles are identical by descent from the common ancestor of her parents, although her parents may not realize that they have a common ancestor. Because human populations have been small throughout our history, many apparently unrelated people have common ancestors.

Thus, although the present high frequency of CF is almost certainly not due to non-random mating in modern populations, the structure of human populations in our historical past might have resulted in consanguinity, so that the allele that arose on this chromosome became identical by descent.

Migration and genetic drift

Several other theoretical possibilities can be used to account for the high frequency of CF. It could be that, for some reason, the site in the gene is subject to recurrent mutation. This is not a likely explanation for the high frequency of CF. If high mutation rate were a likely explanation, each occurrence of the mutation would be expected to be a new molecular lesion and to occur against a new chromosomal background. If recurrent mutation were the correct explanation, affected individuals would not be expected to have the same molecular lesion and would certainly not share flanking polymorphisms. This is discussed in Text Box 3.1 on rates of human mutation. However, this explanation does not work for CF, because the majority of CF cases among European populations are due to one mutation, and the majority of these cases also share polymorphisms near the *CFTR* gene. This shows that the ΔF508 mutation is not a recurrent mutation, and so we must turn elsewhere to explain the allele frequency.

A more likely explanation is that the high frequency of the ΔF508 allele is due, at least in part, to random effects known as **genetic drift**. That is, the allele might have arisen in one or a few people when the population in Europe was small. As a result of the small populations, some groups could have a high frequency of an otherwise rare allele, following a process analogous to sampling error, as suggested by Case Study Figure 5.7. From these individuals, the mutation evidently spread throughout Europe. Also, since the effective population size was small at that time, consanguinity may have occurred, making the effect of genetic drift more pronounced. First cousins have a pair of common grandparents. Even if mating of first cousins was rare, individuals who are more distantly related (such as sharing a common great-grandparent) could both be carriers of the same allele.

Genetic drift is often accompanied by **migration**, another force that affects allele frequencies. If two small populations are genetically separated by geographic or cultural barriers, the genetic differences between the populations will increase, whereas the diversity within a population will decrease. Both theoretical and geographical evidence support the hypothesis that a combination of genetic drift and migration has been an important force in the frequency of the disease. First, throughout much of our history, interbreeding populations *have* been small, with no more than a few hundred people; thus, the population structure lent itself to genetic drift. Secondly, the frequency gradient of the ΔF508 allele is lowest in Turkey and highest in Denmark, consistent with the known migration pattern of groups of people across Europe. Some combination of these two factors may have led to the high frequency of the ΔF508 allele.

Selection and the CF mutation

However, an even more intriguing effect was probably also at work in the historical past that determines the high frequency of the ΔF508 allele today Although in the past homozygotes for the mutation died without reproducting, the ΔF508 allele appears to have conferred a selective advantage on heterozygotes, as summarized in Case Study

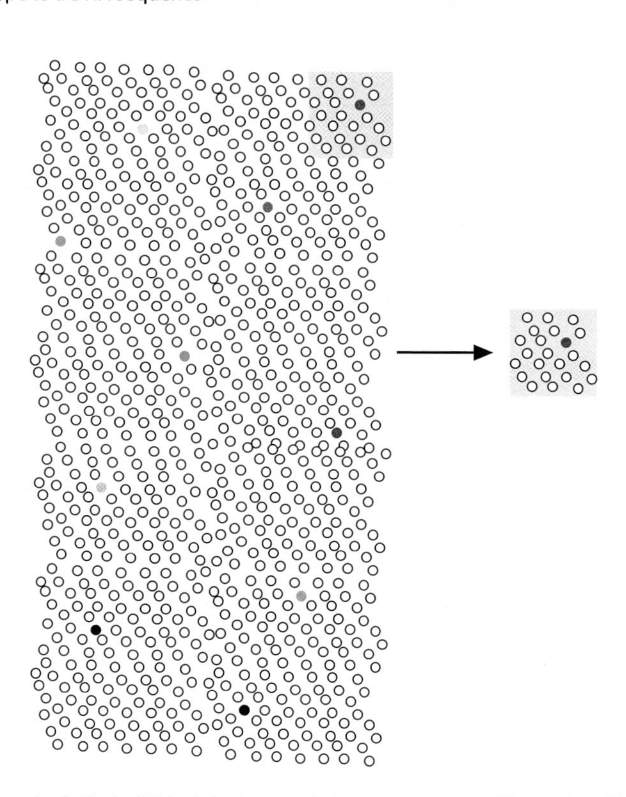

Figure C5.7 Genetic drift. Individuals in the population are represented by circles; filled circles indicate individuals who carry a particular allele, with different colors representing different alleles of the same gene. In the original large population, the 'red' allele is relatively rare. In a subsequent generation, only a small fraction of this initial population reproduces, as indicated by the shaded box. This may be because of a population bottleneck that eliminates most of the initial population, a founders' effect by which this small group forms its own population, or a sampling effect when the population size is small. In the new population, the red allele is comparatively common but others, such as the green, yellow, and orange alleles, are completely absent.

Figure 5.8. Numerous studies in animal models have indicated that heterozygotes for ΔF508 have increased resistance to several water-borne bacterial diseases, including typhoid fever and cholera. The history of European people includes a history of infectious diseases, and so it is possible to imagine how the heterozygote advantage of ΔF508 could have led to the high frequency of the disease by selection.

In this model, the mutation arose at random in one individual and increased in frequency as a result of genetic drift in small populations. These populations spread across Europe,

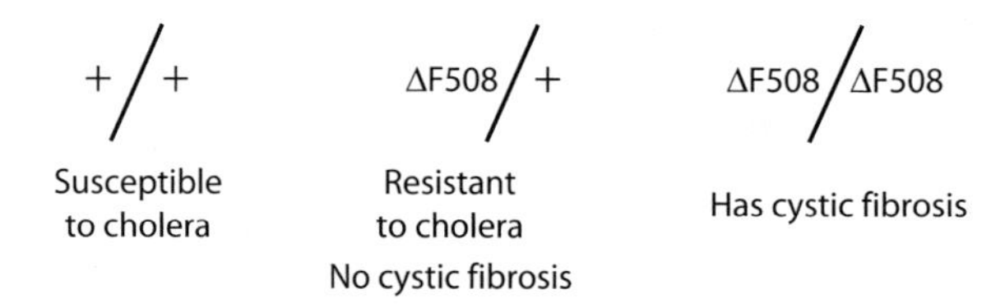

Figure C5.8 Selective advantage in heterozygotes and CF. Individuals who are heterozygous for the ΔF508 allele appear to have a selective advantage over both homozygotes. When water-borne diseases such as cholera and typhus were common, those with the ΔF508 allele may have been resistant. However, a homozygote for this allele has CF. Because the heterozygote has a selective advantage over either homozygote in an environment with water-borne pathogens, both alleles are maintained in the population.

carrying their mutation into regions where water-borne infectious diseases were common. When people with the same mutation migrated to arid regions, in which water-borne diseases are rare, the mutation conferred no selective advantage and either was eliminated or remained at a low frequency. However, in the presence of infectious diseases, the mutation was advantageous in the heterozygous state. It is now present at high frequency, even though it is selected against in homozygotes.

Allele frequencies in our genome can be compared to a scrap book containing fragments of our long-forgotten past. We may look at an allele now in the way we look at a ticket stub and wonder where it came from and why it was kept. However, as geneticists studying human populations encounter these scraps, they can often reconstruct a little more of our shared history, reminding us again of where we came from.

Further reading

Kerem, B., Rommens, J.M., Buchanan, J.A. *et al.* (1989). Identification of the cystic fibrosis gene: genetic analysis. *Science*, **245**, 1073–80.

Riordan, J.R., Rommens, J.M., Kerem, B., *et al.* (1989). Identification of the cystic fibrosis gene: cloning and characterization of complementary DNA. *Science*, **245**, 1066–73.

Rommens, J.M., Iannuzzi, M.C., Kerem, B., *et al.* (1989). Identification of the cystic fibrosis gene: chromosome walking and jumping. *Science*, **245**, 1059–65.

are referred to as **paralogs**. Paralogs commonly arise by gene duplication (among other possible mechanisms), so that paralogs belong to the same gene family, and it is frequently helpful to compare paralogs to understand the functions of a gene within an organism. In contrast, similar genes in different species that are present in both because the gene existed in a common ancestral species are referred to as **orthologs**. The actin genes from *C. elegans* are orthologs of the actin genes in yeast, fruit flies, *Arabidopsis*, and humans. The functions of the actin genes in these different organisms are similar but not identical; all five organisms use actins in cell division, but yeast and *Arabidopsis* do not use them for muscle contraction. The functional and evolutionary relationships among the genes are inferred from the similarities and differences in sequence similarity.

Sequence similarity or inferred homology is also used for cloning a gene, either an ortholog or a paralog. Suppose that an α-tubulin gene in the form of a cDNA has been cloned from chickens, possibly by expression-based cloning or by using the known amino acid sequence to infer and identify a corresponding nucleotide sequence. The cloned chicken gene can be used to find orthologs in other organisms.

The chicken cDNA clone is used as a probe for Southern blots or in library screens for α-tubulin genes from other organisms, such as mice. The cross-hybridizing sequence from the mouse, identified because it has sufficient nucleotide sequence similarity to the chicken gene, is extracted from a gel or a library and itself used as a probe to find the remainder of the mouse α-tubulin gene. The procedure works because of the sequence similarity of the genes, which is a result of their inferred homology or descent from a common ancestral tubulin gene.

Homology-based cloning was very widely used when gene cloning was still a relative novelty, and it is still useful for organisms whose genomes have not been extensively studied. In contemporary examples that use homology and hybridization as an initial cloning strategy, it is far more common to use a PCR-based strategy initially to find the related sequences. An example is shown in Figure 5.13. To do this, the investigator examines orthologs from as many organisms as possible to find the most similar regions of the protein's amino acid sequence. PCR probes are prepared corresponding to that region and used to amplify and clone the portion of the gene. It is often useful to design the PCR primers with nucleotide degeneracy at certain positions

Case Study 5.2

Cloning the *patched* gene from *Drosophila*

Mutations in *patched* affect the segmentation pattern in *Drosophila*

One of the great triumphs of genetic analysis has been the use of mutants to understand the embryonic development of *Drosophila melanogaster*. The mutant screens used were introduced in Chapters 3 and 4. In this case study, we will see how one gene found in these screens, known as *patched* (*ptc*), was cloned. Two different groups, working separately but in communication with each other, cloned the *ptc* gene from *Drosophila* at about the same time using different approaches. Most of the general properties of genes discussed in this chapter were used in the cloning and molecular analysis of *ptc*. We will also describe how the cloned *ptc* gene from *Drosophila* was used to clone *ptc* orthologs from other species, including humans. A study of *ptc* provides an in-depth analysis of a gene that proved to be interesting in development and important in disease, but it is not an unusual example. Many other genes from flies or worms or yeast could have been used in this case study. However, our story is about *ptc*.

As the mutants that affected the pattern of segmentation in *Drosophila* embryos were analyzed, their phenotypes were seen to fall into three broad categories, known as 'gap' genes, 'pair-rule' genes, and 'segment-polarity' genes. The gap genes have mutant phenotypes that affect broad regions of the embryo, the pair-rule genes affect every alternate segment in the embryo, and the segment-polarity genes affect the orientation and development of every segment in the embryo.

The *ptc* gene falls into the segment-polarity category and thus affects every segment in the fly embryo. In *ptc* mutants, the middle portion of each segment is transformed into structures characteristic of the borders of the segment, and ectopic segment borders form. This phenotype is shown in Case Study Figure 5.9. Although most *ptc* mutations are lethal in early development, a few mutant alleles and allelic combinations survive to later stages of development, when other defects become evident. Like many other segment-polarity genes, *ptc* affects structures in both the embryo and the adult fly, including the development of the adult wing. The wing defects will be described in more detail in Case Study 9.2 in Chapter 9. Although the mutant phenotypes are complex, the effects on boundaries of the segments and of the wings suggest that *ptc* may be involved in cell signaling. However, in order to discover its actual role, the *ptc* gene had to be cloned.

Approach 1: *ptc* was cloned on the basis of its map position

As described in the main chapter, positional cloning is one of the most widely used methods for cloning a gene, particularly in organisms whose genomes have been extensively studied or sequenced. The *ptc* gene was cloned about a decade before the *D. melanogaster* genome was sequenced, but an extensive genetic map was available. *ptc* was mapped by conventional recombination analysis and deletion mapping (as described in Chapter 4) to a region on the right arm of chromosome 2. The overall procedure is shown in Case Study Figure 5.10. As the gene was mapped between two deficiency (Df) breakpoints, probes from this

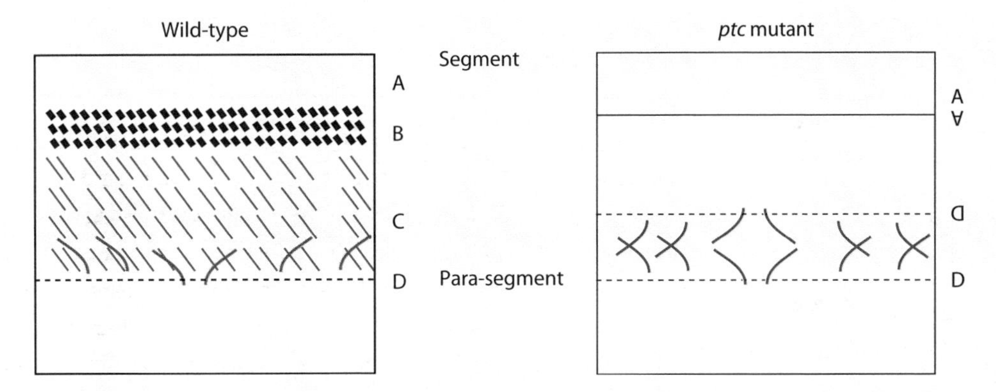

Figure C5.9 The *patched* mutant phenotype. The drawing shows the dorsal side of the third thoracic segment in the *Drosophila* embryo: anterior is at the top and posterior at the bottom; the segment boundary is a solid line and the para-segment boundary is a dashed line. The wild-type segment can be divided into four zones conceptually, each with a characteristic pattern of bristles and hairs. These are shown in different colors for clarity. The most anterior portion of the segment has no hairs or bristles, and the region posterior to the para-segment boundary has no hairs or bristles. The *ptc* mutant phenotype can be interpreted as a deletion of zones B and C, with a mirror-image duplication of zones A and D. There are two segment and two parasegment boundaries. Redrawn and modified from a diagram in Hooper and Scott (1989).

region were tested against the deficiency strains and the position of the gene relative to the breakpoints was identified. A chromosomal walk identified a 95 kb region that had to include *ptc*, as shown at the top of Case Study Figure 5.10. Mapping with molecular polymorphisms narrowed the *ptc* region slightly, but there are many genes in a chromosomal region of this size in *Drosophila*. Positional cloning had identified candidate genes but had not, by itself, found the *ptc* gene.

The next step in the cloning process took advantage of the molecular lesion in *ptc* mutant alleles. DNA was isolated from 15 different *ptc* mutants and probed with different portions of the *ptc* chromosomal region. The technique had the sensitivity to detect mutant lesions greater than about 100 bp but not smaller lesions such as single base changes. One allele induced by X-rays had a molecular alteration that was detected with a particular probe (the other 14 alleles have molecular lesions that were too small to be detected in this way). This probe found no lesions in other parts of the *ptc* chromosomal region, and no other probe found alterations in any other part of the *ptc* region. In addition, a reversion of a *ptc* mutation also involved sequences in this same region. The location of both the X-ray induced allele and the reversion allele is shown in the pink boxes in the middle panel of Case Study Figure 5.10.

Thus, from among the candidate genes determined by map position, the investigators used allelic differences between the wild-type chromosome and an X-ray-induced mutant allele to identify the candidate gene most likely to be *ptc*. However, this did not prove that the investigators had found the actual *ptc* gene, because an X-ray-induced rearrangement allele could still be affecting more than one gene. To confirm that the candidate gene was *ptc*, the investigators used a third property of the gene—its expression pattern.

Some alleles of *ptc* have a temperature-sensitive phenotype, and this can be used to give an approximate time of gene expression. (Temperature-sensitive alleles are described in Chapter 3, and estimating the time of gene activity by temperature-shift experiments is described in Chapter 9.) This analysis suggested that *ptc* gene activity is needed during

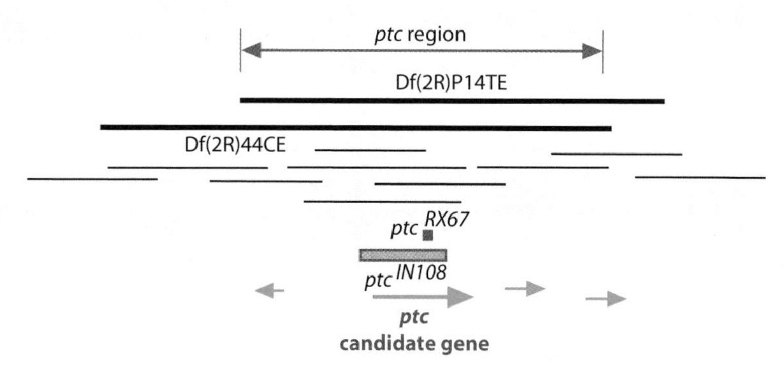

Figure C5.10 Positional cloning of the *Drosophila ptc* gene. *ptc* was mapped to a 95 kb region on chromosome 2 (blue) using deficiency (deletion) strains, and placed between the right endpoint of deficiency Df44CE and the left endpoint of deficiency DfP14TE (thick black lines). Cosmid and phage clones that span the region were isolated by probing libraries from wild-type flies (thin black lines); the clones shown are illustrative and do not represent the actual clones used. DNA from different *ptc* mutant strains were tested for mutations in the sequences corresponding to the overlapping clones from this region. The two *ptc* mutant alleles *RX67* and *IN108* had detectable differences when tested with the clones that included the region in red. These mutant lesions indicated a probable locus for the *ptc* gene. Transcripts from the appropriate developmental time for *ptc* expression were located by Northern blotting (green). Four transcripts are shown; others have been omitted for simplicity. One transcript includes the locus identified by the *ptc* mutant lesions, identifying this gene as the best candidate to be *ptc*. In addition, this transcript is altered in the *ptc* mutants. The drawing is not to scale. The original information can be found in Hooper and Scott (1989).

a window of 4–8 hours after egg-laying. Therefore, the investigators isolated mRNA from embryos that were 4–8 hours old and used Northern blots to place transcription units onto the chromosomal region. Five transcripts expressed during the time window were found in the *ptc* chromosomal region (blue arrows in Case Study Figure 5.10). By itself, expression analysis helped to find candidate genes but did not clearly point to one gene. However, the X-ray-induced *ptc* allele mentioned above showed a clear alteration in only one of the five transcripts. Thus, the combination of the expression analysis and the molecular lesion in this allele showed that the correct gene had been cloned and assigned to the *ptc* locus.

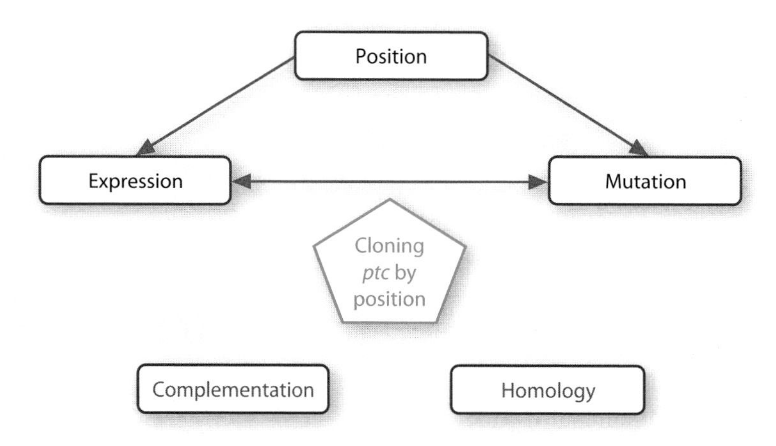

Figure C5.11 In this approach, the *ptc* gene was first mapped to a particular locus on the chromosome and that region of the chromosome was cloned. The expression pattern and mutational lesions were used to identify, from among the candidate genes at that position, the correct gene as *ptc*.

In this approach, three different known properties of the *ptc* gene—its map location, its alleles, and its expression—were used to clone the gene and to confirm that the investigators had cloned the correct gene. Case Study Figure 5.11 summarizes the relationship among the properties of the *ptc* gene that were used in its cloning. The cloning and confirmation did not use transformational rescue by the wild-type sequence, a fourth property of the *ptc* gene, although that was widely available at the time of the analysis and could have been done if needed. This approach to cloning *ptc* demonstrates both the labor and the insight that was needed to clone genes before the genome had been completely sequenced.

Positional cloning in *Drosophila* now

Suppose that one were to clone a gene like *ptc* now, once the genome has been sequenced. What would change? All of the *Drosophila* genome is already available in mapped clones, so the chromosomal walk, a labor-intensive step, is eliminated. Probing mutant strains for their molecular alterations is simplified because only regions predicted to contain genes need to be tested. In addition, probing different genes is likely to be done by PCR amplification and sequencing or by tiling microarrays (described in Chapter 8), which allow more sensitivity in detecting small changes than does hybridization. If microarray or other expression data are available for each predicted gene, as is the case for *Drosophila*, expression analysis by RNA isolation is either greatly simplified or has already been completed. However, these are changes in technique and not in strategy. They reduce the time and labor involved in the cloning process but not the fundamental ideas used to clone the gene. The same known properties of genes are likely to be used now for cloning a gene from *Drosophila* as were used to clone *ptc*.

Approach 2: *ptc* was cloned by transposon tagging

Meanwhile, in another laboratory....

One of the many advantages of *D. melanogaster* is the existence of a method of mutagenesis known as hybrid dysgenesis. A more complete description of hybrid dysgenesis is given in Chapters 2 and 3, but fundamentally it involves the mobilization of a particular class of transposable element known as P elements. A dysgenesis cross often results in mutations because P elements have moved into a gene and disrupted it. One can think of P-element mutagenesis in the same way as any other mutagenic element. A key advantage of using a transposable element as the mutagen is that the P-element transposon is the molecular allelic difference that allows the disrupted gene to be cloned. This relies on the second property of genes described in the chapter: different alleles have different molecular lesions.

This method, known as transposon tagging, was used to clone the *ptc* gene in an F_1 screen. (F_1 screens are described in Chapter 4). Case Study Figure 5.12 shows the procedure used in transposon tagging. Because *ptc* is homozygous lethal, a *ptc* heterozygote with suitable flanking markers was used as one of the parent strains. The other parent produced gametes with random mutations induced by P elements. The F_1 offspring were necessarily heterozygous for these newly arisen mutations and for the parental chromosome with *ptc* on it.

P elements insert throughout the genome, so an F_1 fly is heterozygous for P-element-induced mutations in many different genes. If one of the newly arisen mutations inserts into the *ptc* gene, it fails to complement the *ptc* allele: one allele is the original *ptc* mutation inherited

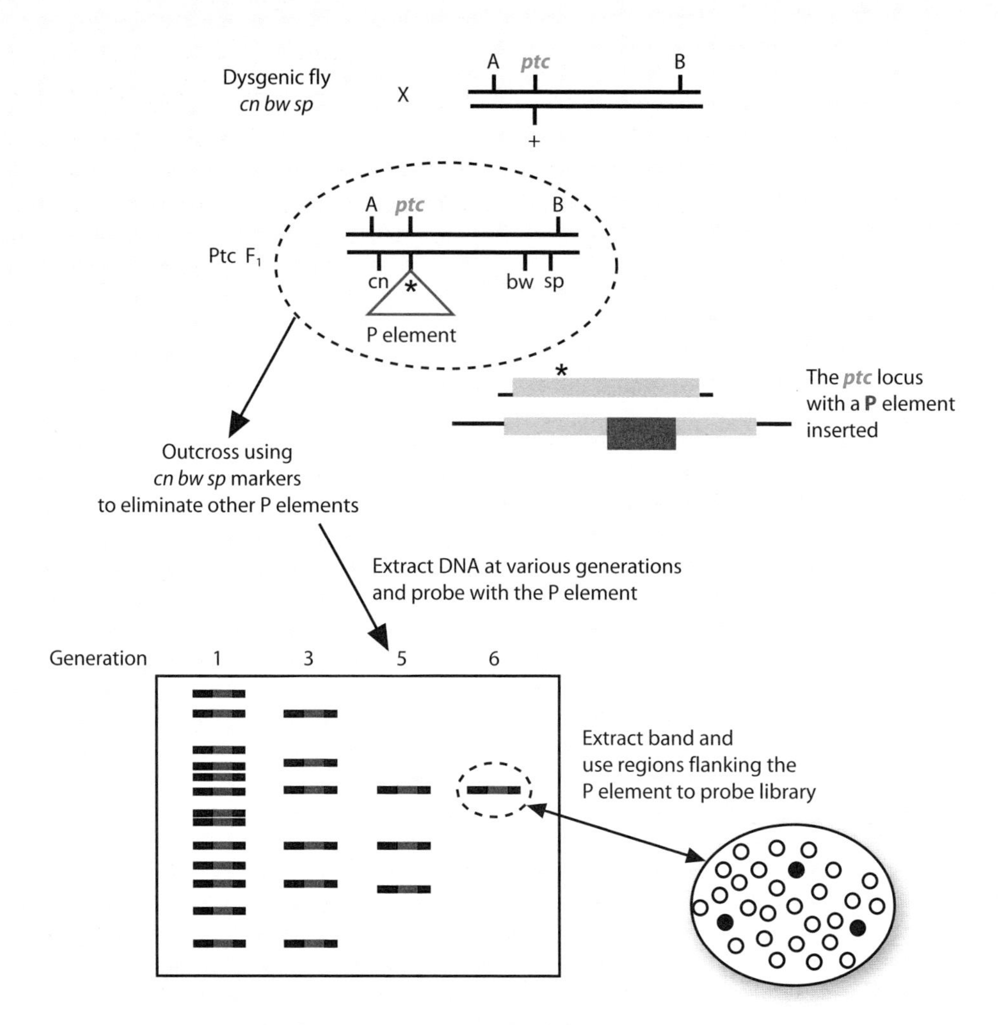

Figure C5.12 Cloning the *Drosophila ptc* gene by transposon tagging. The *cn, bw,* and *sp* mutations are recessive markers linked to *ptc* and are used to distinguish the mutagenized dysgenic chromosome from the other chromosome. The parent on the right has a *ptc* mutation and dominant markers nearby. If the dysgenic fly on the left makes a gamete with a P element inserted into the *ptc* locus, the F_1 fly will show a *ptc* mutant phenotype (the *ptc* locus is in blue and the P element in red). The F_1 *ptc* fly is outcrossed and each generation is probed for the presence of P elements. In Drosophila, this can be done by using *in situ* hybridization to polytene chromosomes, but in this example the more general case is shown in which the presence of the transposable element is monitored by Southern blots. DNA is extracted from the flies in each generation and digested with a restriction enzyme that does not cut inside the transposable element. The number of generations needed to reduce the presence of extra transposable elements depends on the linkage of the elements and the outcrossing scheme that is used. In the example, flies in generation 6 had only one P element remaining. This element and its flanking DNA are extracted from the gel, and the flanking regions are used as a probe for genomic libraries. Given that the fly had a *ptc* mutant phenotype, the flanking DNA must contain part of the *ptc* locus.

from the parent and the other allele is the mutation newly induced by the P element. Thus, the F_1 fly has a *ptc* mutant phenotype. Because *ptc* is a recessive mutation, all of the other F_1 flies that do not have a P element inserted at the *ptc* locus have a wild-type phenotype, so the *ptc* mutants can be spotted easily against this background.

Although the *ptc* homozygote died, the newly induced P-element allele was recovered and maintained using the marked heterozygote (the exact details of this process need not concern us here). The newly arisen *ptc* mutation was outcrossed to wild-type and, in each generation, the *ptc* mutant flies were picked and maintained. This procedure helped to stabilize the mobile P element. In addition, given that the dysgenic fly had P-element insertions at multiple sites in its genome, the outcrossing procedure removed the P elements that were not tightly linked to *ptc*. The number and distribution of the remaining P elements could be determined by *in situ* hybridization of a labeled P-element probe to polytene chromosomes. In an idealized example, the only P element remaining in the genome would be the one inserted at the *ptc* locus. In organisms in which *in situ* hybridization to polytene chromosomes is not possible, a similar procedure can be done by Southern blotting. This is the procedure shown diagrammatically in Case Study Figure 5.12.

Because the mutation in *ptc* is due to the insertion of defined DNA, the investigators could isolate the complete *ptc* gene. They isolated DNA from the *ptc* mutant fly and constructed a genomic library, which they probed with a P element; this allowed isolation of not only the transposon but also the flanking region of the *ptc* gene. The region had one transcript, and it is expressed during embryogenesis when the earliest defect in *ptc* mutants is observed.

The transposon tagging approach relied primarily on the molecular lesion induced by P-element mutagenesis, i.e. on the ability to detect the molecular difference between wild-type and mutant *ptc* alleles. The properties used to clone the *ptc* gene by this approach are shown diagrammatically in Case Study Figure 5.13. Additional confirmation came from the map position and the expression pattern of the gene.

Thus, both groups of investigators used the same three properties of genes: the molecular lesion associated with alleles, the map position, and the time of expression. Their approaches began from different starting points but fortunately they ended with the same candidate gene, providing even more confidence that both groups had cloned the right gene.

How has transposon tagging as a method of cloning a gene been affected by the availability of sequenced genomes? Using transposon tagging to clone a gene was one of the most popular and important methods before genome sequences were available. Although transposon tagging is still used, positional cloning is more widely used for organisms with

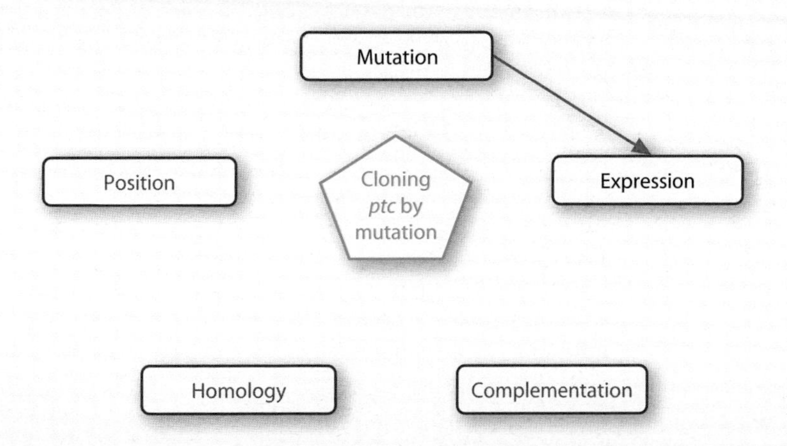

Figure C5.13 The second approach to cloning the *ptc* gene involved using the molecular lesion of a *ptc* mutation by transposon tagging. The mutation specifically identified the *ptc* gene, which was further confirmed by the expression pattern. Compare this scheme with Case Study Figure 5.2, in which the map position provided the initial information for cloning the *CF* gene.

sequenced genomes. The molecular difference between alleles is now widely used as the proof that the right gene has been cloned, but not so often as the starting point for the cloning process.

The *Drosophila ptc* gene was used to clone orthologs from other organisms

Since the initial work in *Drosophila*, *ptc* has become one of most thoroughly studied of the segment-polarity genes. We will return to experiments involving *ptc* in later chapters, but for now we will limit our discussion to the type of protein it encodes and the methods used to clone *ptc* orthologs from other organisms. At the time the gene was cloned from *Drosophila*, no protein similar to Ptc had been found in other organisms. The structure of the Ptc protein was inferred from a hydropathy analysis, which predicted a number of hydrophobic stretches. The length of these hydrophobic stretches suggested that they were membrane-spanning regions, several of them long enough to cross the membrane more than once. It seemed likely that the Ptc polypeptide crossed and recrossed the plasma membrane and could be acting as a receptor protein. The predicted structure is consistent with a postulated role in cell signaling, as was inferred from the mutant phenotype.

Cloning *patched* from other insects using homology

One of the methods used to clone *ptc* from other animals was homology—using the property that genes are the units of evolutionary descent. Direct searches of mammalian genomes for sequences with nucleotide similarity to *Drosophila ptc*, using Southern blots for example, ran the risk of finding many functionally unrelated sequences that happened to share some sequence similarity with the *ptc* DNA sequence. Even worse, because the genetic code is degenerate, a mammalian Ptc protein might share amino acid similarity with *Drosophila* but have very little nucleotide similarity. Thus, not only might the investigators clone the wrong gene, but they might also fail to clone the correct gene if it exists.

Instead, the investigators worked their way to the mammalian *ptc* gene via homology in organisms more closely related to *Drosophila*. They used the cloned fruit fly gene to clone orthologs from a mosquito, a butterfly, and a flour beetle. In these cases, the nucleotide similarity among the genes in different insects made it possible for the investigators to use hybridization to find the related gene. Aligning these sequences allowed them to identify the portions of the *ptc* gene that are the most highly conserved, at least among insects. PCR primers were designed that would amplify the conserved portion of the gene and these were used to clone homologs from a mouse.

Using the expression pattern of *patched* to clone the gene from the mouse

Another property of genes was also important in cloning homolog(s) of the *ptc* gene from the mouse. Even with primers from the most conserved sequences of the gene, the evolutionary distance between insects and mice could have been too great to find the correct gene. Thus, to help ensure that they would clone mouse ortholog(s) of *ptc*, the investigators used information about *ptc* expression. They knew that *Drosophila ptc* is co-expressed with the *hedgehog* (*hh*) gene, and so it was reasonable to expect that *ptc* and *hh* orthologs are co-expressed in mice as well. (The relationship between *ptc* and *hh* is described in more detail in Chapter 11.)

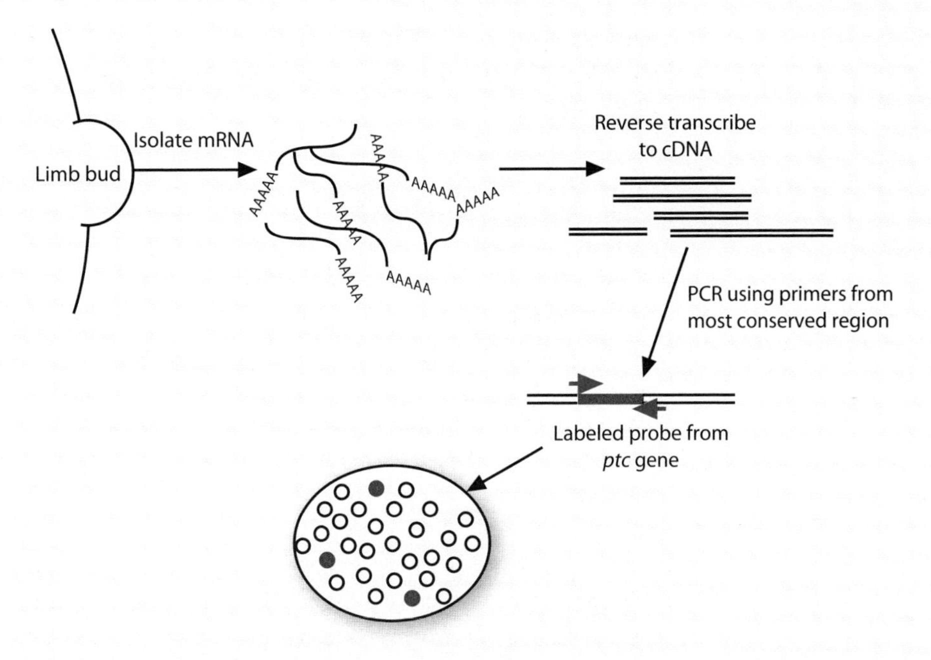

Figure C5.14 Cloning of *Ptc* from mice using the expression pattern. Homologs of *Drosophila ptc* were inferred to be expressed in the developing limb bud because of the relationship between *patched* and *hedgehog*. The mouse ortholog of *hedgehog* is expressed in the limb bud, so it seemed likely that *patched* was expressed there as well. Limb buds were dissected from embryos and mRNA was extracted and used to make cDNA. The cDNA was used as the template for PCR, with the primer sequences derived from the regions of high sequence identity among insects. The PCR product provided the probe for a cDNA library that yielded the complete mouse *Ptc* gene.

In mice, an *hh* ortholog known as *Shh* is expressed at a high level in the developing limb bud. With this in mind, the investigators isolated mRNA from the limb bud and performed RT-PCR on this mRNA using the primers from the most conserved region of the *ptc* gene derived from the insect sequences; the procedure is shown diagrammatically in Case Study Figure 5.14. This process gave them a unique product, which was similar in sequence to part of the *ptc* gene from insects. This portion was used as a probe to isolate an entire gene from mice; this gene was called *Ptc* (it is now known as *Ptch1*). The mouse protein Ptc had the same predicted hydrophobicity profile as the insect protein, giving the investigators confidence that they had cloned the correct gene. This strategy of expression cloning with homology is shown in Case Study Figure 5.15. A large body of work since its isolation has confirmed that *Ptch1* is one of the the mouse orthologs of insect *ptc*.

The *patched* gene in humans

What about humans? Do we make a version of the Ptc protein? Certainly if the gene is present in mice and *Drosophila*, it seems very likely to be found in humans as well. Just as two groups working independently cloned the gene from Drosophila, two groups also cloned an ortholog of the *ptc* gene from humans (called *PTCH1*), using somewhat different approaches.

Beginning with the mouse gene

The first group relied on the cloned mouse gene to clone the human ortholog of *Ptch1*, thus depending on the strategy of **homology**. Using the cloned human gene as a probe for *in situ*

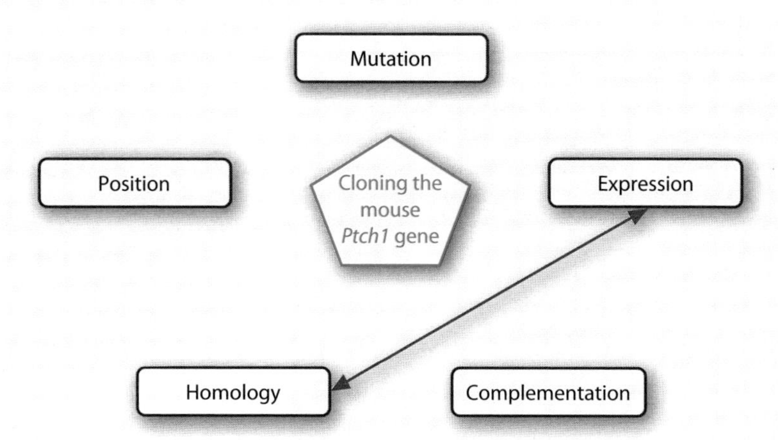

Figure C5.15 The strategy for cloning the mouse *Ptch1* gene. The investigators began with the expression pattern of *Shh* to infer the expression of the mouse *patched* gene in the developing limb bud. They used the region of highest similarity between the *patched* gene in insects and chickens to develop PCR primers for RT-PCR on the RNA from the limb bud. Thus, they combined the strategies of expression pattern and homology to clone the mouse *patched* gene *Ptch1*.

hybridization to chromosomes, they then identified the map location of the human gene, placing it at 9q22.3.

However, knowing the *position* of the gene in humans does not describe its function. What does the human gene *Ptch1* actually do? Now the investigators turned to the property of **mutational lesions**, and in particular to human genetic disorders. Several human genetic diseases have been mapped to the 9q22 region, including a relatively rare disease known as nevoid basal cell carcinoma syndrome (NBCCS). This is a syndrome marked by developmental abnormalities of the face and recurrent tumors, including basal cell carcinomas, a form of skin cancer.

Based on the expression pattern of the *Ptch1* gene in mice, mutations in the human *Ptch1* gene appeared to be a good candidate for the basis of NBCCS. When the gene was sequenced from NBCCS patients, many of the individuals had mutations in the human *Ptch1* gene. Thus, this approach to cloning the human gene relied on the homology of the gene with the mouse, followed by the map position, expression, and finally mutational lesions. This approach is summarized in Case Study Figure 5.16.

Beginning with the disease gene

A second group used a different approach to show that mutations in the human *PTC* gene are a cause of NBCCS. In the course of studying the disease itself, these investigators had mapped the causative mutation to a region on 9q22 and had isolated DNA from the region. The DNA sequence was examined for candidate genes, i.e. genes whose expression pattern or homology suggested that they might be the underlying cause of the disease. Among the candidates was the *PTCH1* gene. Knowing the effect of *ptc* mutations on *Drosophila* development, the investigators decided that it was a good candidate and therefore tested NBCCS patients for mutational lesions in the human *PTCH1* gene. Several individuals affected with NBCCS had mutations in *PTCH1*. Thus, this approach began with the map position of the disease phenotype, then used expression and homology, and finally examined mutational lesions to identify the correct gene. This is summarized in Case Study Figure 5.17.

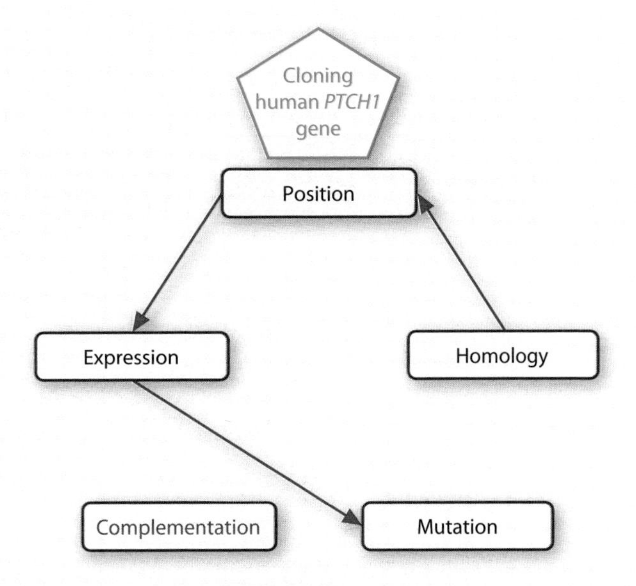

Figure C5.16 One research team used homology with the mouse *Ptch1* to clone the human ortholog of *Ptch1*. The human gene was placed on the map by fluorescence *in situ* hybridization to chromosomes. Genes at this map position with a disease phenotype affecting the same tissues as *Ptch1* affects in mice were considered as candidates for the human *PTCH1* gene. The gene was sequenced from individuals with the disease syndrome NBCCS and mutational lesions were identified.

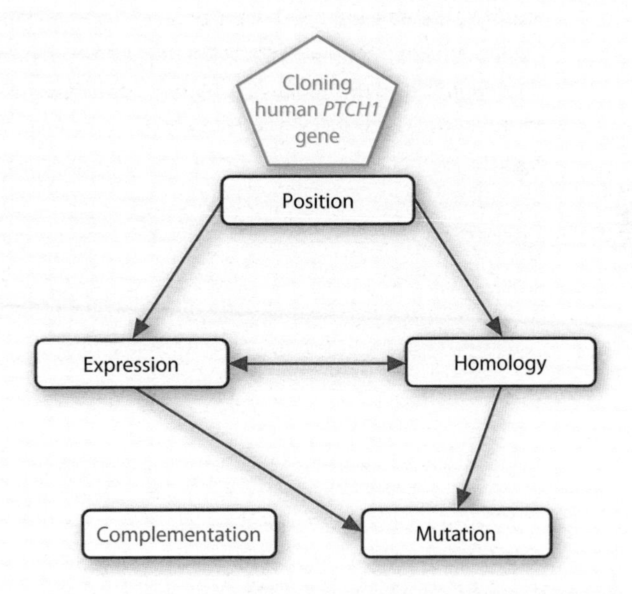

Figure C5.17 Another research group was studying the NBCCS syndrome and had mapped the disease to the location that included the human *PTCH1* gene. Through a combination of expression pattern and homology to the mouse gene, the *PTCH1* gene was considered a likely candidate for causing NBCCS. The *PTCH1* gene was sequenced from individuals with the disease syndrome and mutational lesions were identified.

The role of the *patched* gene in humans

It is always satisfying to investigators to identify the molecular basis for a genetic disease, or to learn that the gene they have studied from a model organism could have an important human ortholog. NBCCS is a relatively rare cancer syndrome, but it has an intriguing connection to the most common form of cancer in humans—skin cancer. Most cases of skin cancer, including basal cell carcinomas, are the result of sporadic events rather than an inherited disposition to cancer. Identifying the genetic mutation for a sporadic cancer is an exceptionally difficult process. However, knowing that *PTCH1* mutations could account for the inherited disposition to basal cell carcinomas in NBCCS, the two groups who had cloned human *PTCH1* also wondered whether any of the sporadic cases of basal cell carcinoma might be due to somatic mutations in *PTCH1*. To test this idea, the investigators sequenced the *PTCH1* gene from basal cell carcinoma cells in different patients. The postulate proved to be true—at least some sporadic cases of basal cell carcinoma are due to the mutations in the *PTCH1* gene.

One specific case was particularly illustrative. The patient had had a basal cell carcinoma removed from his cheek; the *PTCH1* gene was sequenced and a mutation was found in the cancer cells. In fact, the mutation in *PTCH1* was apparently due to the formation of a pyrimidine dimer. As discussed in Chapter 3, geneticists have known for decades that pyrimidine dimers are caused by UV exposure, such as would occur with sunlight on exposed skin cells. Here, then, was a direct connection from the effect of a known mutagen (UV light) on a known gene (*PTCH1*) to a well-studied phenotype, basal cell carcinoma.

When Nüsslein-Volhard and Wieschaus isolated and characterized *ptc* mutations, they were probably not thinking about skin cancer. Even the two groups who cloned the *ptc* gene from *Drosophila* were probably not thinking about skin cancer. In fact, you can search their papers in vain for any reference to cancer. Nonetheless, the connection exists. We have focused on the *ptc* gene (both in this case study and elsewhere in the book) not because of its role in human cancer but because it is an example of the many different genes that have been cloned and analyzed in *Drosophila* and other organisms. We described *ptc* because many different approaches have been used in its analysis. Many other genes might have been described in this way, including others with connections to human disease, simply because the tools used for *ptc* are generally applicable.

Further reading

Goodrich, L.V., Johnson, R.L., Milenkovic, L., McMahon, J.A., and Scott, M.P. (1996). Conservation of the *hedgehog/patched* signaling pathway from flies to mice: induction of a mouse *patched* gene by Hedgehog. *Genes & Development*, **10**, 301–12.

Hahn, H., Wicking, C., Zaphiropoulous, P.G., *et al.* 1996. Mutations of the human homolog of Drosophila *patched* in the nevoid basal cell carcinoma syndrome. *Cell*, **85**, 841–51.

Hooper, J.E. and Scott, M.P. (1989). The Drosophila *patched* gene encodes a putative membrane protein required for segmental patterning. *Cell*, **59**, 751–65.

Johnson, R.L., Rothman, A.L., Xie, J., *et al.* (1996). Human homolog of *patched*, a candidate gene for the basal cell nevus syndrome. *Science*, **272**, 1668–71.

Nakano, Y., Guerrero, I., Hidalgo, A., Taylor, A., Whittle, J.R., and Ingham, P.W. (1989). A protein with several possible membrane-spanning domains encoded by the *Drosophila* segment polarity gene *patched*. *Nature*, **341**, 508–13.

BOX 5.2 Inferring homology from sequence similarity

The chapter describes how the biological principle of homology is used as one strategy to clone genes. Homology is defined as sequences or structures that are derived from a common ancestral sequence or structure. In this text box, we discuss how homology is inferred from the similarity of macromolecular sequences; in particular, we look at the similarity of amino acid sequences as evidence for homology. Homology itself is not observed, since descent from a common ancestor occurs over evolutionary time. Two structures or sequences are inferred to be homologous by examining and evaluating the similarities and differences in the currently existing structures or sequences. To reiterate a point made in the chapter text, homology is a statement of a qualitative relationship: two sequences are either homologous or they are not. On the other hand, similarity between sequences can be defined quantitatively. These concepts are illustrated in Text Box Figure 5.4.

The process of examining and evaluating the relationship between two sequences is often done by a family of programs collectively called BLAST. Versions of BLAST, including versions for amino acid sequences (BLASTP) and nucleic acid sequences (BLASTN), are running on servers worldwide. Many users log on to BLAST through the servers at NCBI, Bethesda, MD, which also has outstanding tutorials on the different versions of BLAST. Our overview of BLAST is most specifically applicable to amino acid comparisons (i.e. BLASTP), but the principles are similar for the other versions. BLAST is not the only set of programs used to compare amino acid or DNA sequences. The FASTA programs have similar functionality but the underlying computational approach is slightly different; the FASTA family of programs is also widely used.

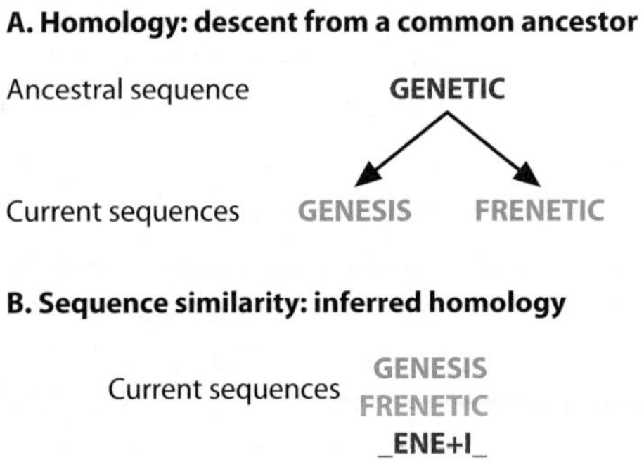

A. Homology: descent from a common ancestor

Ancestral sequence GENETIC

Current sequences GENESIS FRENETIC

B. Sequence similarity: inferred homology

Current sequences GENESIS
FRENETIC
ENE+I

Figure B5.4 Homology and similarity. The amino acid sequence GENETIC in panel A becomes mutated over time and amino acid substitutions arise. Thus, two species that share this common ancestor will have orthologs of the original protein but the current sequences will not be identical to each other or to the ancestral sequence. Since descent from a common ancestor can rarely be directly observed, homology is usually inferred from sequence similarity in the current sequences, as shown in panel B. Although the sequence in the extinct ancestor cannot be known for sure, the similarity between the two extant sequences can be used to infer their common origin. In the bottom sequence, identical amino acids are shown and the conservative change between S and T is represented by the symbol +. From the current sequences, it is not possible to know whether S or T was the amino acid in the ancestral sequence. The residues that do not align and are not similar are represented by lines.

Defining an alignment

BLAST compares two sequences by first aligning them and then evaluating the similarities in the alignment. Buried beneath that simple sentence is a remarkably powerful and complex program. An investigator uses BLAST or one its relatives and probably never thinks about the computer algorithm and the sophisticated mathematical rigor that serve as its foundation. Even more fundamentally, one rarely thinks about the evolution of macromolecules, particularly of amino acid sequences, that BLAST implicitly tests. However, it is worth recalling that every BLAST search is a test of specific hypotheses about evolution. In a different context, we could discuss the computer algorithm and the statistical principles, but we will only touch on those here. Our primary focus will be on the biological assumptions and hypotheses.

First, then, what is meant by an **alignment**? An alignment can be considered as a point-by-point comparison of two sequences. This concept is not unique to biological sequences. When the detectives in a police story use voice recognition to compare a recorded message with a speaker or when bird watchers compare the song of an unseen bird with a recorded song, they are doing an alignment problem. A word search of a document for the string of letters 'ene' is an alignment problem. The program uses some method to match the initial 'e' in the query string with each letter in the document (i.e. the database). If an 'e' is matched, it then attempts to match the 'n', then the 'e', and so on until the entire string is matched. Such a point-by-point alignment search becomes a laborious and computationally intensive process for a long query string or a large document. Thus, a quicker method is to treat 'ene' as a unit and search not just for the 'e' but for the 'en' combination, or even to use the 'ene' combination as a query string. This is one of the key principles that allow both the BLAST and the FASTA families

A. Global alignment	B. Local alignment	C. Local gapped alignment
GENETIC_	**GENETIC**	**GENE_TIC**
GENETICS	**HETEROGENEITY**	**HETEROGENE ITY**

Figure B5.5 Global and local pairwise alignments. Panel A shows a global alignment in which each position in the two sequences has to be matched. The final position in the top sequence requires a gap, and could result in a penalty when the alignment is scored. Panel B shows a local alignment with no gaps. Only the positions of similarity or identity are aligned, and other positions are ignored. Mismatches occur for the last three positions but these would be ignored in the scoring. An even better local alignment is obtained when a gap is allowed, as in panel C. The inclusion of the gap allows the T to be aligned in the two sequences, and the last two positions are ignored. Since we cannot know if this position had an insertion in the lower sequence or a deletion in the top sequence, such a position is called an indel (insertion/deletion).

of alignment programs to search so quickly. BLAST performs a **pairwise alignment**, comparing one sequence (the query) with another (the target). The final result of a BLAST search is a table with each of the sequences and a corresponding alignment score. Each sequence in that table has been tested against the query by a pairwise alignment.

Two different types of pairwise alignment are important. In a **global alignment**, the sequences are matched from end to end. In the conceptually simplest version, the aligned sequences are then scored at each position in the string, with each match or mismatch awarded a certain number of points. The number of points awarded for each match or mismatch is determined by the scoring matrix, as described in detail below. The total number of points for the alignment is tallied to determine the quality of the alignment. Thus, in a global alignment, mismatches are penalized (relative to matches) wherever they occur in the string.

Local alignments stand in contrast to global alignments. In a **local alignment**, the regions of highest identity or similarity are aligned and scored. Mismatches outside the region of local similarity are ignored. Then the different local similarities within the aligned sequences are tallied to determine the quality of the alignment. These two concepts are illustrated in Text Box Figure 5.5.

These types of alignment rely on different assumptions about the evolution of proteins. In a global alignment, proteins are assumed to evolve as units, whereas a local alignment recognizes that proteins have evolved with localized regions of high similarity, such as domains or, more generally, a block. Within the block of amino acids, such as a homeodomain or a kinase domain, the amino acid sequences of the two proteins are quite similar; outside the block of amino acids, the proteins may be similar or quite different. BLAST performs local alignments; in fact, the name stands for **B**asic **L**ocal **A**lignment **S**earch **T**ool. Some examples of protein alignment blocks are shown in Text Box Figure 5.6.

Just as our earlier example discussed searching for the string 'ene' rather than each letter separately, BLASTP subdivides the query sequence of amino acids into short 'words'; the default setting is that each word consists of three amino acids. It searches for each of the three amino acid words in the target sequence. When it finds a match, the program extends the alignment in each direction, scoring each position in the alignment string and tallying the total number for that local alignment. Once the total score for a local alignment falls below a certain threshold value, BLAST abandons the attempt to extend the alignment. It then tallies each of the local alignment scores and gives a total score to the sequence that represents the sum of the local alignment scores. Text Box Figure 5.7 shows the process of compiling an alignment score. Having done that sequence, the program then moves on to the next sequence and repeats the process, recording the alignment score for each sequence that it tests. Currently, a BLASTP search of the entire database of non-redundant sequences tests the alignment of the input query sequence against more than 3×10^7 other sequences. (Perhaps this will make you a little more patient in waiting for BLAST to return your results.)

Scoring the alignment

How does BLASTP evaluate the similarities in the alignment? Each match or mismatch along the string is given a value derived from the **scoring matrix**. The scoring matrix is perhaps the most biologically interesting aspect of an alignment search. It is derived from observed amino acid substitutions when comparing known proteins. This important point can easily be overlooked—the scoring matrix uses the properties of extant functional proteins. That is, the changes and substitutions in these amino acid sequences are ones that have been evolutionarily 'accepted' and have not been eliminated by natural selection. Many other amino acid changes undoubtedly arose in these sequences during evolution but the protein (or the

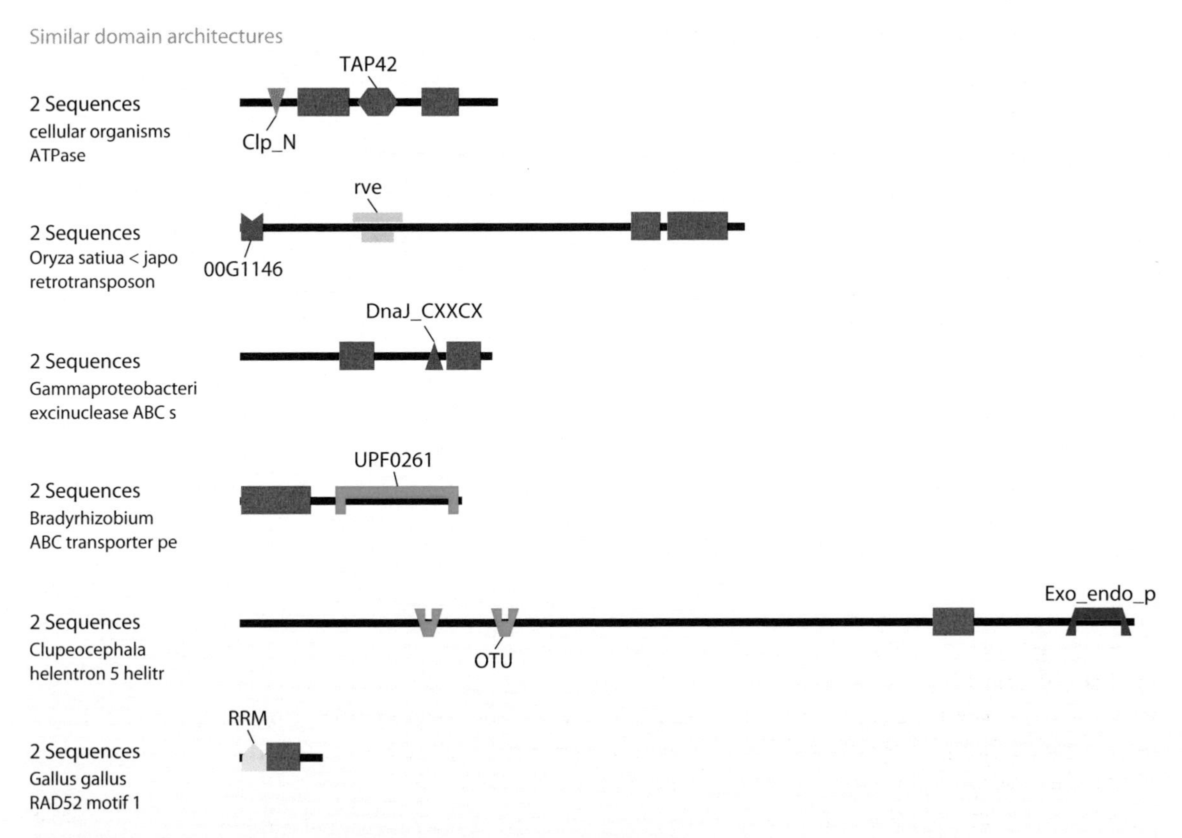

Figure B5.6 Examples of protein sequence blocks. The figure is a screen capture from the CDART database, with different domains represented by colors and shapes. All of the proteins share the ATPase domain represented by the red rectangle. This domain has become combined with different domains during protein evolution. Local alignment programs such as BLAST assume that proteins evolved in blocks of amino acids without attempting to identify the function of the block.

GENESIS
FRENETIC
ENE+I
5 5 4
-2 6 1 -1

Score: -2+5+6+5+1+4-1 = 18

Figure B5.7 Scoring an alignment. The same sequences are used as in Text Box Figure 5.5, with a score beneath each position. The scores are written in staggered lines for ease of reading. At each position in the local alignment, the match or mismatch is evaluated. Each of the positions is tallied to give a total alignment score. For example, each position in which the E has remained an E in both sequences is given a score of 5, while the replacement of the S with a T results in a score of 1 at that position. The scores are derived from the BLOSUM62 matrix, as described in Text Box Figure 5.8. The first and last positions are included here for clarity. Since the scores are negative for these positions, BLAST might ignore the first and last positions to give a total score of 21 instead of 18. The statistical significance of such scores is considered in Text Box Figure 5.9.

organism with the altered protein) was not functional and did not survive. The most widely used matrices are called the BLOSUM matrices (e.g. BLOSUM62 or BLOSUM45), so we will describe these. (The explanation is only slightly different for an earlier set of matrices known as the PAM series.)

To derive the BLOSUM matrix, local regions of high identity were found for known proteins. Each local region of high identity is called a **block**, which gives rise to the 'blo' in the BLOSUM name. For example, to derive the BLOSUM62 matrix, amino acid blocks with 62 percent or more identity were compiled. Then, within that set of identities, the other non-identical amino acid substitutions were recorded and tabulated. Since these amino acid blocks are found in naturally occurring functional proteins, the other 38 percent of the substituted amino acids represent evolutionarily accepted mutations. In other words, the underlying assumption of the scoring matrix is natural selection: Which amino acid substitutions have occurred in evolution and have not interfered with the function of the

	C	S	T	P	A	G	N	D	E	Q	H	R	K	M	I	L	V	F	Y	W
C	**9**	−1	−1	−3	0	−3	−3	−3	−4	−3	−3	−3	−3	−1	−1	−1	−1	−2	−2	−2
S	−1	**4**	1	−1	1	0	1	0	0	0	−1	−1	0	−1	−2	−2	−2	−2	−2	−3
T	−1	1	**4**	1	−1	1	0	1	0	0	0	−1	0	−1	−2	−2	−2	−2	−2	−3
P	−3	−1	1	**7**	−1	−2	−1	−1	−1	−1	−2	−2	−1	−2	−3	−3	−2	−4	−3	−4
A	0	1	−1	−1	**4**	0	−1	−2	−1	−1	−2	−1	−1	−1	−1	−1	−2	−2	−2	−3
G	−3	0	1	−2	0	**6**	−2	−1	−2	−2	−2	−2	−2	−3	−4	−4	0	−3	−3	−2
N	−3	1	0	−2	−2	0	**6**	1	0	0	−1	0	0	−2	−3	−3	−3	−3	−2	−4
D	−3	0	1	−1	−2	−1	1	**6**	2	0	−1	−2	−1	−3	−3	−4	−3	−3	−3	−4
E	−4	0	0	−1	−1	−2	0	2	**5**	2	0	0	1	−2	−3	−3	−3	−3	−2	−3
Q	−3	0	0	−1	−1	−2	0	0	2	**5**	0	1	1	0	−3	−2	−2	−3	−1	−2
H	−3	−1	0	−2	−2	−2	1	1	0	0	**8**	0	−1	−2	−3	−3	−2	−1	2	−2
R	−3	−1	−1	−2	−1	−2	0	−2	0	1	0	**5**	2	−1	−3	−2	−3	−3	−2	−3
K	−3	0	0	−1	−1	−2	0	−1	1	1	−1	2	**5**	−1	−3	−2	−3	−3	−2	−3
M	−1	−1	−1	−2	−1	−3	−2	−3	−2	0	−2	−1	−1	**5**	1	2	−2	0	−1	−1
I	−1	−2	−2	−3	−1	−4	−3	−3	−3	−3	−3	−3	−3	1	**4**	2	1	0	−1	−3
L	−1	−2	−2	−3	−1	−4	−3	−4	−3	−2	−3	−2	−2	2	2	**4**	3	0	−1	−2
V	−1	−2	−2	−2	0	−3	−3	−3	−2	−2	−3	−3	−2	1	3	1	**4**	−1	−1	−3
F	−2	−2	−2	−4	−2	−3	−3	−3	−3	−3	−1	−3	−3	0	0	0	−1	**6**	3	1
Y	−2	−2	−2	−3	−2	−3	−2	−3	−2	−1	2	−2	−2	−1	−1	−1	−1	3	**7**	2
W	−2	−3	−3	−4	−3	−2	−4	−4	−3	−2	−2	−3	−3	−1	−3	−2	−3	1	2	**11**

Figure B5.8 The BLOSUM62 scoring matrix. These are the log odds for the replacement of each amino acid by another among polypeptide blocks with an average identity of 62 percent. For example, a cysteine (C) in one polypeptide in the block usually remains as a cysteine among other polypeptides in the block, as seen by the score of 9. Substitution of alanine (A) for cysteine occurs at approximately the frequency of alanine in general, as seen by the 0, whereas all other substitutions for cysteine are found less often than predicted by chance, as seen by the negative numbers. The numbers are based on observed polypeptide blocks rather than on hypotheses based on the properties of the amino acids.

protein? These comparisons and substitutions are complied into a 20 × 20 (for each of the 20 amino acids) table, with a percentage in each cell of the table. Among blocks of amino acid sequences that are 62 percent identical in sequence, how often is an alanine in one block still an alanine in another block? How often is that alanine observed to be replaced by a valine or a glutamine or a histidine or a cysteine? Which amino acids are the least likely to change and why? Which amino acids are often substituted for one another and which are only rarely substituted for one another?

The percentages in this table are then converted to a log-odds score using the frequency of each amino acid and the substitution of each one for another. (The advantage of using log odds is that the probabilities can be tallied by adding the natural logarithms rather than multiplying the numbers.) The resulting scoring matrix, the actual BLOSUM62 matrix, is a table of these log-odds. The BLOSUM62 matrix is shown in Text Box Figure 5.8.

The numbers in the matrix are interpreted as follows. The overall frequency of an amino acid in the database of protein sequences is the probability that it will be found by chance at any position in a block. If the substitution of one amino acid for another occurs as often as predicted by chance, the score is 0. For example, from examining the BLOSUM62 scoring matrix in Text Box Figure 5.8, asparagine (E) replaces arginine (R) about as often as predicted from the overall frequency of asparagine. Although it may be a somewhat simplistic interpretation, this could be considered a neutral value for the substitution—the replacement of asparagine for arginine has had very little positive or negative functional effect on proteins overall. This is an average value for all of the proteins being compared, so it includes residues in proteins when the replacement of the asparagine for arginine has occurred more often than expected by chance as well as ones when the replacement has occurred less often than expected by chance.

If an amino acid replaces another more often than expected by random substitution, the score in the BLOSUM is greater than 0. The higher the score, the more likely those amino acids are to be found at corresponding positions in the string. For example, an arginine (R) in one protein usually corresponds to an arginine in the other proteins in the block (the score is 5), but it is replaced by lysine (K) or glutamic acid (Q) more often than predicted by the chance occurrence of these amino acids. On the other hand, if the substitution occurs less often than expected by chance, i.e. it is not a tolerated substitution, the score will be less than 0. Thus, the replacement of arginine with an uncharged amino acid such as isoleucine (I) or valine (V) occurs very rarely in functional proteins. These values based on observed substitutions generally agree with our intuition about proteins—replacing a charged amino acid with an uncharged one is likely to impair the function of the protein, whereas replacing it with another charged amino acid may not have such dire functional consequences.

The interested reader is invited to imagine his own scoring matrix before looking further at the BLOSUM62 matrix in Text Box Figure 5.8. How closely did your intuitive scoring matrix resemble the actual scoring matrix? For example, it is easy to imagine that cysteines are highly conserved since they are often involved in crosslinking. Thus, most cysteines will remain the same and a mismatch of some other amino acid with a cysteine will probably incur a substantial mismatch penalty. As expected from your intuition, a match of cysteine (C) with another cysteine in the alignment is given a score of 9, i.e. cysteine remains as cysteine at a frequency of e^9 or approximately 8000 times more likely than that expected if substitutions occurred at random. Similarly, valine (V) and isoleucine (I) can often replace each other in functional proteins, but replacement of either by a glutamic acid (Q) is very unlikely. Although we use terms like 'very unlikely', we could be highly quantitative: in naturally occurring proteins, among blocks of amino acids that are 62 percent identical, a substitution that replaces valine with glutamic acid occurs at e^{-2} or 13.5 percent as often as expected by random substitution. In other words, this is relatively rare. If the alignment being tested matches a valine with a glutamic acid, the other parts of the alignment need to be very similar to give an appropriately high score.

Interpreting the score

So far, we have discussed how the alignment is done and how a score for the alignment is computed. But what does that score mean? What was meant by 'an appropriately high score' in the preceding paragraph? We now turn our attention to the statistical basis that allows us to interpret the score and draw meaningful conclusions about the alignment. The score from the alignment of two sequences is compiled by adding all the local alignment matches and mismatches. (The scores are added because they are based on logarithms; adding the logarithms corresponds to multiplying the probabilities; assuming that the matches at each position are independent of one another.) The higher the score, the more similar are the sequences. The maximum score depends on the similarity of the two sequences being compared, their amino acid composition, and their length. But what alignment score is high enough to conclude that the proteins are genuinely related? That is, how can we convert the similarity score to an inference about their possible homology or descent from a common ancestor? Here BLAST tests a specific but hidden statistical hypothesis.

Imagine a protein of consisting of *n* amino acids, whose amino acid composition is the same as the average amino acid composition of known proteins. The amino acid sequence of the protein is then shuffled to generate a huge database (or a population in statistical terms) of random sequences, all with the same average amino acid composition. Our query protein is then aligned against each sequence in this large population and the similarity score is computed for each alignment. The similarity scores are plotted as a histogram, i.e. as a probability distribution of scores. One such histogram is shown in Text Box Figure 5.9. Although, for clarity in presentation, this particular histogram uses a different alignment program and a smaller protein database, the interpretation is the same for BLASTP and larger databases.

Suppose that our alignment gives a similarity score of 125. We can then ask: 'How many alignments in this random population of sequences give a similarity score of 125 or higher?' Any alignment in this population of sequences is necessarily a random alignment since the proteins used to create it are hypothetical strings of amino acids. If we see no similarity scores that are greater than 125 (the score our two proteins had) in this random population, we can say that our two proteins have **excess similarity** or more similarity than is expected by chance. Thus, we can infer that it is very likely that their similarity is not random and that they are in fact derived from a common ancestral sequence.

Fortunately, we need not generate a population of random amino acid sequences each time we want to do a BLASTP search. It has been shown mathematically that the similarity scores arising from this population of random sequences follow a known frequency distribution known as the extreme value distribution, which we will not describe. However, we can use the known statistical properties of the extreme value distribution to calculate exactly the expected probability of a score greater than or equal to any given score in a database of the size that was searched. In the BLAST results page, this is shown in the column marked 'E', the probability of an alignment score equal to or greater than the score in this database. The expected value is the probability that a match with this score will arise between two proteins by chance; the lower the expect value (E), the less likely that the match is due to chance. The database

FASTA searches a protein or DNA sequence data bank
version 3.4t25 Nov 12, 2004
Reference : W.R. Pearson & D.J. Lipman PNAS (1988) 85:2444–2448
Query library /usr/people/www/fasta_tmp/1036649 vs Z library
Database : PIR

```
opt E()
< 20 778 0: ==
22 0 0: one = represents 479 library sequences
24 1 0: =
26 13 6: *
28 23 64: *
30 191 390: *
32 904 1506: == *
34 3081 4084: ======= *
36 7590 8388: ================ *
38 13754 13863: ============================*
40 20577 19337: ========================================*==
42 26390 23637: =================================================*======
44 28690 26074: ======================================================*=====
46 28660 26557: =======================================================*====
48 26661 25425: =====================================================*==
50 23713 23201:================================================*=
52 20334 20397: ==========================================*
54 16711 17423: =================================== *
56 13665 14553: ============================= *
58 10989 11948: ======================= *
60 8635 9679: =================== *
62 6886 7759: =============== *
64 5470 6171: ============*
66 4153 4877: ========= *
68 3311 3836: ======= *
70 2515 3006: ======*
72 1978 2349: ====*
74 1537 1832: ===*
76 1150 1426: ==*
78 908 1108: ==*
80 713 860: =*
82 512 658: =*
84 449 521: =*
86 344 403: *
88 273 312: * inset = represents 4 library sequences
90 233 242: *
92 200 187: * :======================================*
94 161 145: * :===================================*===
96 101 112: * :========================= *
98 83 87: * :====================*
100 82 67: * :===============*====
102 61 52: * :============*===
104 57 40: * :=========*=====
106 34 31: * :=======*=
108 37 24: * :=====*====
110 32 19: * :====*===
112 28 14: * :===*===
114 25 11: * :==*====
116 11 9: * :==*
118 8 7: * :=*
>120 109 5: * :=*======================
statistics sampled from 60000 to 282721 sequences
Expectation_n fit: rho(ln(x))= 4.5918+/-0.000184; mu= 11.7296+/-0.010
```

is centered on an expect value of 1, i.e. a score of 1 indicates that at least one randomly generated sequence in a database of this size will have this score. Thus, two sequences that give an expect value of e^{-20} are almost certainly derived from a common ancestor since this degree of similarity is very unlikely to be due to a chance relationship. In a typical probability distribution, two values cannot be considered to be significantly different unless the expect value is less than 0.05, but most users of BLAST are much more conservative than this, and would regard an expect value of e^{-5} or less to be the cut-off for a significant relationship; some investigators are uncomfortable with expect values greater than e^{-10}, although this is quite conservative. A BLASTP results page with some annotations and explanations is shown in Text Box Figure 5.10.

BLAST and related alignment programs have changed molecular genetics as profoundly as tools such as thermal cyclers and automated sequencing robots. As with any tool, it is important that we understand its uses and potential misuses. This is far from a complete and rigorous description of how to use BLAST effectively, but perhaps it will serve as a reminder that when we hit that BLAST! button, we are testing specific hypotheses about homology and evolutionary relationships.

Figure B5.9 (opposite) A histogram of alignment scores. The amino acid sequence from the myelin protein in rats was aligned with the PIR protein database using the program SSEARCH. The statistics of using this program and this database are the same as for using BLASTP against the non-redundant database at NCBI, but the histogram of scores is easier to display and understand. There are three columns on the left. The first column is the score for the alignment, grouped in intervals of two. The second column (labeled opt at the top) is the number of sequences in the database that returned that score from the alignment search. The third column (labeled E() at the top) is the expected number of scores from a database of random sequences, as described in the text box. For example, look at the line with the score of 52. A total of 20 334 proteins returned an alignment score of 51 or 52 in the database. The expected number of sequences with this score is 20 397, which is very close to the observed number. Thus, proteins in the database that returned an alignment score of 51 or 52 occur as frequently as random sequences, so these are not statistically significant similarities. These data are plotted as a histogram. The equals sign (=) in this database represents 479 protein sequences in the database, and the asterisk (*) represents the number of sequences in the random database. Note that the scale changes for scores greater than 88, and = represents four sequences in the database. The number of observed sequences closely agrees with the number of expected sequences for scores below 120. However, only five sequences in the random database are predicted to score above 120 and the observed number of sequences is 109. These sequences have statistically significant 'excess similarity' and are inferred to be homologous to myelin.

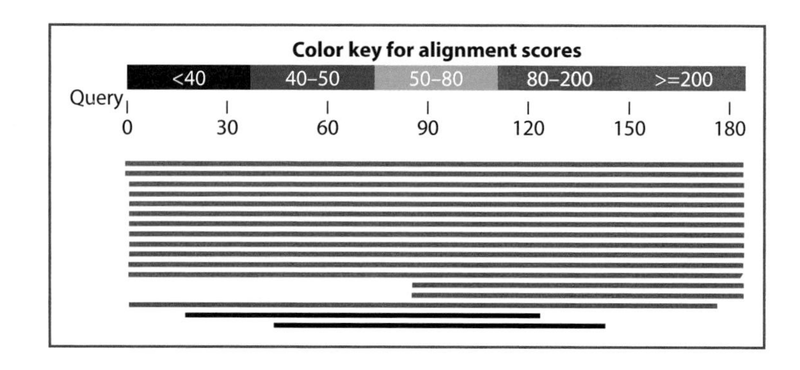

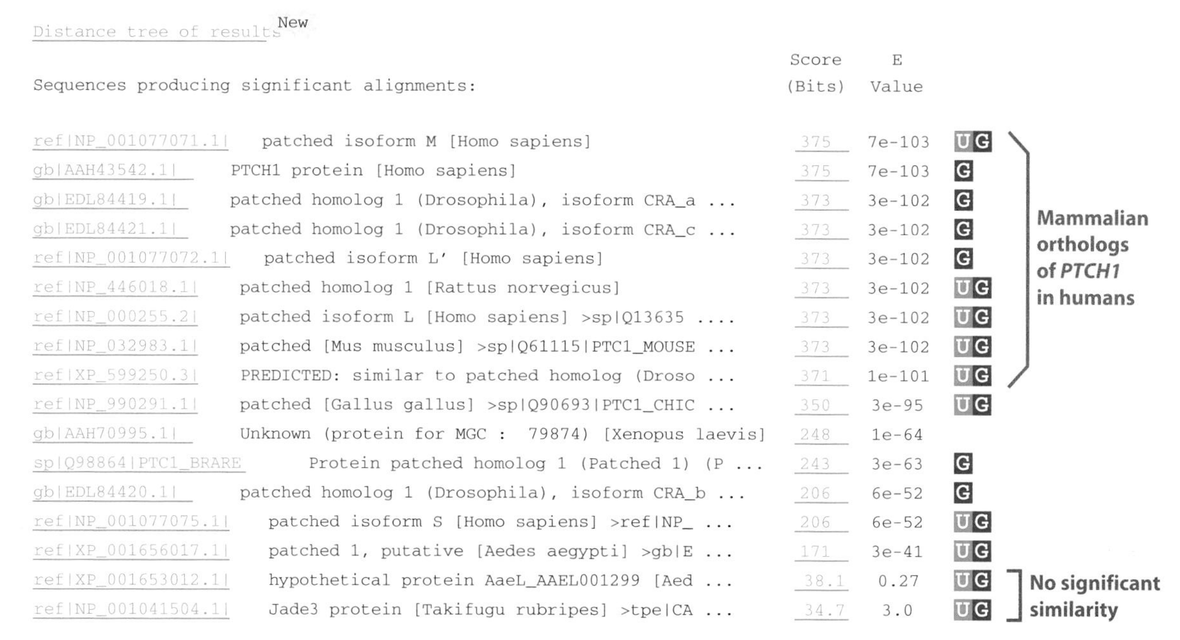

Distance tree of results New

Sequences producing significant alignments:		Score (Bits)	E Value	
ref|NP_001077071.1|	patched isoform M [Homo sapiens]	375	7e-103	UG
gb|AAH43542.1|	PTCH1 protein [Homo sapiens]	375	7e-103	G
gb|EDL84419.1|	patched homolog 1 (Drosophila), isoform CRA_a ...	373	3e-102	G
gb|EDL84421.1|	patched homolog 1 (Drosophila), isoform CRA_c ...	373	3e-102	G
ref|NP_001077072.1|	patched isoform L' [Homo sapiens]	373	3e-102	G
ref|NP_446018.1|	patched homolog 1 [Rattus norvegicus]	373	3e-102	UG
ref|NP_000255.2|	patched isoform L [Homo sapiens] >sp|Q13635	373	3e-102	UG
ref|NP_032983.1|	patched [Mus musculus] >sp|Q61115|PTC1_MOUSE ...	373	3e-102	UG
ref|XP_599250.3|	PREDICTED: similar to patched homolog (Droso ...	371	1e-101	UG
ref|NP_990291.1|	patched [Gallus gallus] >sp|Q90693|PTC1_CHIC ...	350	3e-95	UG
gb|AAH70995.1|	Unknown (protein for MGC : 79874) [Xenopus laevis]	248	1e-64	
sp|Q98864|PTC1_BRARE	Protein patched homolog 1 (Patched 1) (P ...	243	3e-63	G
gb|EDL84420.1|	patched homolog 1 (Drosophila), isoform CRA_b ...	206	6e-52	G
ref|NP_001077075.1|	patched isoform S [Homo sapiens] >ref|NP_ ...	206	6e-52	UG
ref|XP_001656017.1|	patched 1, putative [Aedes aegypti] >gb|E ...	171	3e-41	UG
ref|XP_001653012.1|	hypothetical protein AaeL_AAEL001299 [Aed ...	38.1	0.27	UG
ref|NP_001041504.1|	Jade3 protein [Takifugu rubripes] >tpe|CA ...	34.7	3.0	UG

Figure B5.10 BLAST results. This is a screen capture from a BLAST search using the human *PTCH1* gene, an ortholog of the *Drosophila patched* gene. The results have been edited for illustrative purposes. The figure at the top shows the distribution of sequences that produced alignments. The color code is based on the similarity score, and the length of the line is based on the extent of the alignment within the sequence. Mammalian *PTCH1* orthologs have high similarity over the entire length of the sequence, as expected. The pink line is the gene from the mosquito *Aedes aegypti*, with significant but lower similarity over the length of the sequence. The black sequences appear in the list of similar sequences, but the similarity scores and E values indicate that this similarity is not statistically significant so there is no evidence that these sequences are related to *PTCH1*.

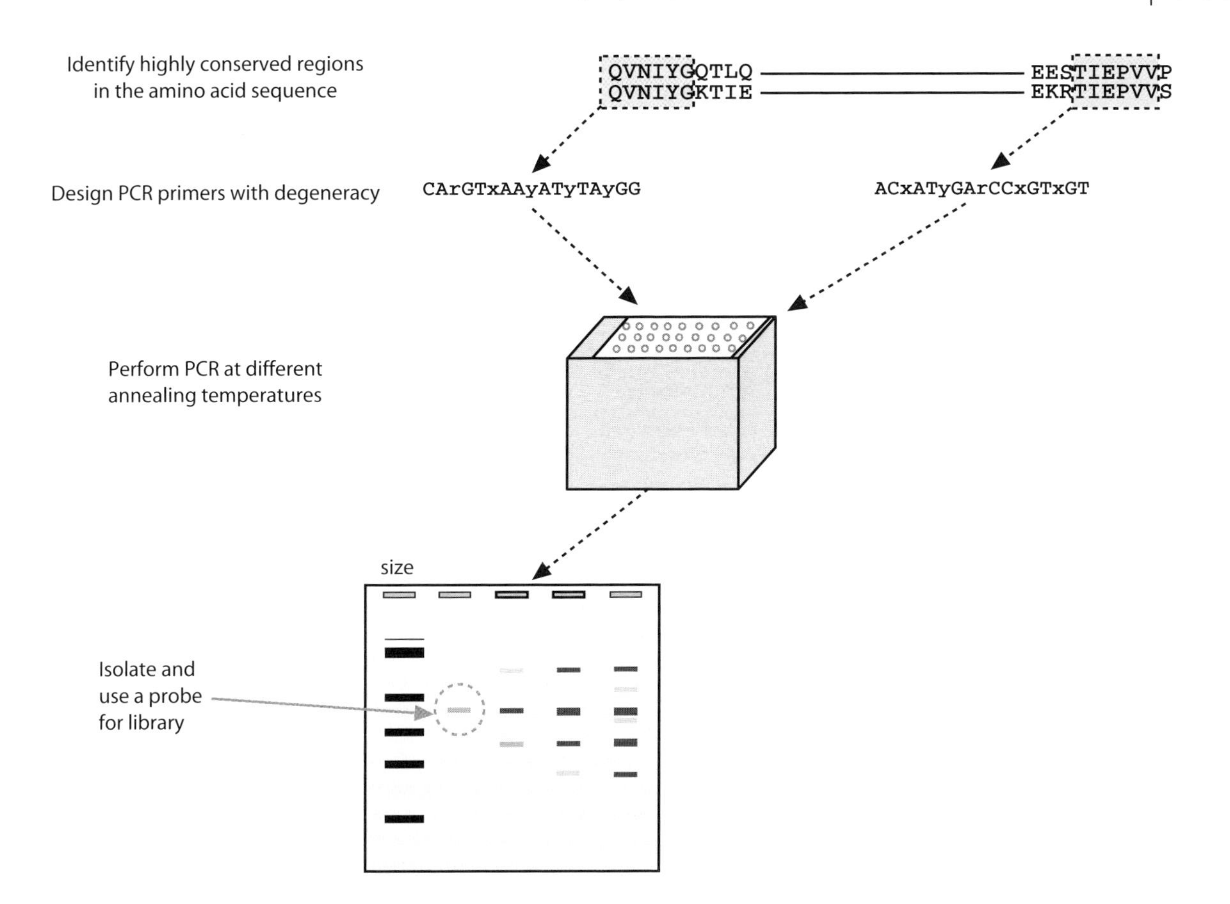

Figure 5.13 Using nucleotide homology to clone a gene. The first step is to align amino acid sequences of related genes and identify the regions of highest sequence similarity. A conceptual translation is used to produce PCR primers with degeneracy at positions that are uncertain. The degenerate primers are actually a pool of primers with slightly different sequences. In the primer sequences indicated, r indicates purine, y a pyrimidine, and x any of the four nucleotides. PCR is done with reactions at lower annealing temperatures, which allow some mismatches between the primers and the DNA sequence. The reactions are separated on a gel, and the best candidate amplification product is isolated and used as a probe in the library.

to account for the degeneracy of the genetic code. The conserved portion of the protein or gene sequence from the species being investigated is then used as a probe in a Southern or cDNA library to find the remainder of the gene. Alternatively, the initial PCR product can be used as the primer in a PCR-based technique called RACE (rapid amplification of cDNA ends) to clone the entire transcript.

Because they are based on nucleotide sequence similarity, hybridization procedures for homology-based cloning of orthologs work best for proteins or domains that are highly conserved during evolution and for closely related organisms. They are much less effective for distantly related genes because more nucleotide changes have occurred. In addition, even among related species, the redundancy of the genetic code can be an obstacle given that hybridization cannot distinguish between changes in the third codon position and changes in other bases. Thus, there may be amino acid identity when the underlying nucleotide sequence is different. Conversely, nucleotide similarity may arise even when the amino acids are completely different. Both AAG and GAA share two out of three nucleotides with AAA, but both AAA and AAG encode lysine whereas GAA encodes glutamic acid. Hybridization of nucleic acids cannot distinguish between the nucleotide sequences encoding these different amino acids.

With appropriate cautions and controls, homology-based cloning is a useful first approach and is still widely used. However, the method for finding homology has completely changed. For most applications when working with an organism whose genome has been fully or even partially sequenced, the technique for finding sequence similarity is first to perform computer searches of the sequence databases. The underlying cloning principle has not changed, but the first step in finding sequence similarity is a BLAST search that takes a few minutes, rather than a sequence hybridization procedure that requires a few days or even weeks.

Furthermore, whereas hybridization relies on nucleotide similarity, BLAST searches can use nucleotide or, better, amino acid similarity. This allows the investigator to circumvent problems presented by the redundancy of the genetic code. In addition, as described in Text Box 5.2, the similarity detected by a BLAST search is mathematically rigorous and its significance can be analyzed by well-grounded statistical methods, unlike the approximations based on hybridization results. Most importantly, it is possible to use BLAST to search many more sequences much more quickly than could ever by done in hybridization or PCR experiments. Thus, BLAST and its relatives provide a vastly superior method of identifying and cloning genes than sequence hybridization could ever offer. Nonetheless, whether it is detected by BLAST, PCR, or hybridization, the same property of genes—that they are units of evolutionary descent—is being used as an initial cloning strategy.

SUMMARY

Connecting a phenotype to a DNA sequence

The goal of this chapter has been to collate the many approaches that have been used to clone genes by using five common properties of genes as an organizing principle. These five properties were described during the earliest days of genetics, in some cases even before we knew that genes consisted of DNA sequences. The five properties do not represent all of the strategies that have been used to clone genes; for example, in Chapter 10, gene interactions are described as another strategy for cloning genes. Approaches to cloning genes continue to develop, and new techniques will probably be used in the near future. Nonetheless, these five properties will remain, and strategies for cloning genes will undoubtedly continue to be built around them.

For situations in which extensive genomic information is available, most cloning strategies will begin with either the map position or homology to other genes (properties 1 and 5). In one typical situation, the investigator will map the gene to a position on the chromosome and use the other properties of genes to determine which gene at that position is the gene of interest. A second typical situation is that a gene has been cloned for one organism, and the investigator will want to find the ortholog in another species (property 5). Expression-based cloning (property 4) is also widely used, particularly in strategies relying on microarray analysis. As described above, expression-based cloning is often the best initial strategy when very little is known about the genetic map or the genome. Properties 1, 4, and 5 are the best methods of finding candidate genes simply because they require so little preliminary information. Using mutational lesions (property 2) and complementation (property 3) provides far more useful strategies for identifying and confirming the specific gene from a set of candidates.

Having a cloned gene provides one of the essential tools for the analyses described in the rest of the book. In this book, we have taken the organizational approach that cloning the gene often follows finding the mutant phenotype (the other essential tool for genetic analysis of biological questions), as described in Chapters 3 and 4. Placing the topics in this order reflects the historical sequence in which these two essential tools were connected: mutant phenotypes were usually found before genes were cloned. However, in some experimental situations, the cloned gene was available before a mutant phenotype was known. In these situations, it has been necessary to reverse the historical sequence and to work from the cloned gene towards the mutant phenotype. These approaches, sometimes called reverse genetics, are the topics of the next two chapters. But in whichever order they are acquired, a mutant phenotype and a cloned gene are the two essential tools for any further genetic analysis.

Chapter Capsule

Cloning a gene

Five properties that define Mendelian genes have been used to clone genes and to confirm the connection between the Mendelian gene and the molecular gene. Some properties are most appropriate for identifying many candidate genes; other properties are best suited for confirming that the specific gene of interest has been cloned.

Most candidates

A gene is expressed in a particular pattern	expression cloning
A gene has a map position	positional cloning
A gene is the unit of evolutionary descent	homology-based cloning
A gene is defined by complementation	transformation rescue
A gene has different alleles	mutational lesions

Most specific

A typical cloning strategy is to identify candidate genes by mapping a mutant phenotype or by looking at genes transcribed in a particular pattern. The candidate genes can be further delimited by complementation tests or another property until one or a few candidate genes are sequenced from different mutants and a molecular lesion is found.

CHAPTER MAP

- ▶ **Unit 1 Genes and genomes**
- ▼ **Unit 2 Genes and mutants**
    - 3 Identifying mutants
    - 4 Classifying mutants
    - 5 Connecting a phenotype to a DNA sequence
    - **6 Finding mutant phenotypes for cloned genes**

      This chapter describes how a geneticist identifies a mutant phenotype for a cloned gene that affects a biological process. This is the reverse of the processes described in Chapters 3, 4, and 5. The mutant and the cloned gene are crucial tools for the types of analysis described in subsequent chapters.
    - 7 Genome-wide mutant screens
- ▶ **Unit 3 Gene activity**
- ▶ **Unit 4 Gene interactions**

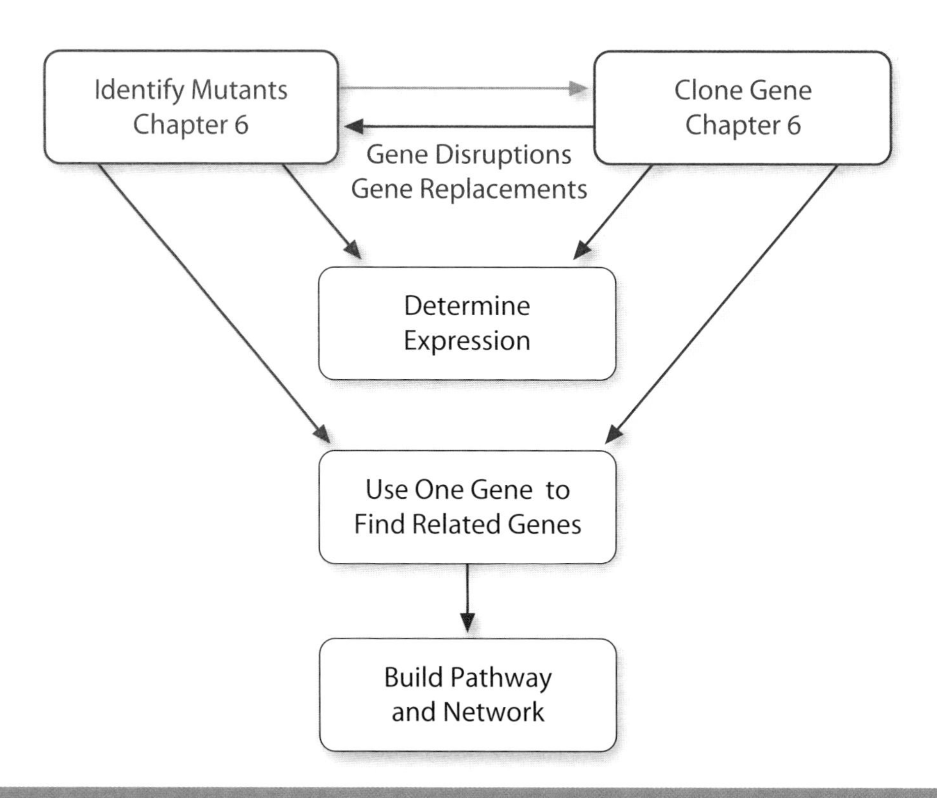

6

CHAPTER 6

Finding mutant phenotypes for cloned genes

TOPIC SUMMARY

The process of genetic analysis often begins with a mutant phenotype, as discussed in Chapter 3. Molecular biological approaches also allow us to initiate genetic analyses by using a cloned gene before a mutant phenotype is known. Even when the gene has been cloned first, it is still important to find a mutant phenotype, in effect to 'work backwards' from the DNA sequence to the mutant phenotype in order to determine its biological function. These genetic approaches are sometimes called **reverse genetics**. This chapter focuses on two of the most powerful techniques of reverse genetics—the processes used to create gene disruptions and gene replacements. There are two requirements for gene disruptions and gene replacements:

- the ability to integrate the cloned gene at its chromosomal site into individual cells by homologous recombination
- the ability to produce an organism from the genetically altered cell.

In a gene disruption, the target gene is knocked out by replacing it with a cloned donor sequence, while in a gene replacement the target gene is replaced with a copy of the gene having a defined molecular alteration. Gene targeting for reverse genetic analysis is an especially powerful analytical tool for studies in *Saccharomyces cerevisiae* and mice.

IN BRIEF

Targeted insertions can be used to find mutant phenotypes when the gene has already been cloned.

INTRODUCTION

With advances in molecular biology, genes can be molecularly cloned before a mutant phenotype has ever been observed. For example, it is possible to clone a gene based on its expression pattern or its sequence similarity to a gene in another organism. Yet, as we noted in Chapters 3 and 4, it is important to know a gene's mutant phenotype in order to understand its role in a biological process. In this chapter, we describe how to find a mutant phenotype for a gene that has been previously cloned by a molecular method.

The topic of this chapter is sometimes referred to as **reverse genetics**—an experimental tool that complements those discussed in Chapters 3, 4, and 5. In Chapters 3 and 4, the analysis began with a mutagenized population of wild-type organisms, from which a mutant phenotype was identified and characterized. The mutant could be found without knowing anything about the molecular or biochemical function of the gene. In Chapter 5, we described strategies used to clone genes, several of which depend on having a mutant phenotype. Genetic analysis that begins with a mutant is referred to as 'forward genetic analysis' or simply 'genetic analysis'. Genetic studies of this type have the advantage that very little needs to be known in advance about the biological process.

In many cases, however, the investigator *will* know something about the biological process before attempting to find a mutant. In particular, one or more of the genes that are involved in the process may have already been cloned. The current chapter—and the topic of reverse genetics—completes the connection between the previous chapters by asking: What happens when an investigator has cloned a gene but has no idea what the mutant phenotype is? How can an investigator begin with a cloned gene and identify a mutant phenotype?

The task of this chapter is to use the cloned gene to identify a mutant phenotype. The tool in this chapter is the same as the tool in Chapter 3—the mutant phenotype. The task and the tool are summarized in Figure 6.1. However, unlike Chapter 3 in which the mutations are introduced as random changes throughout the genome, the mutant in this chapter is engineered as a directed change in a particular gene. Nonetheless, the goal is the same: **Find a mutant.**

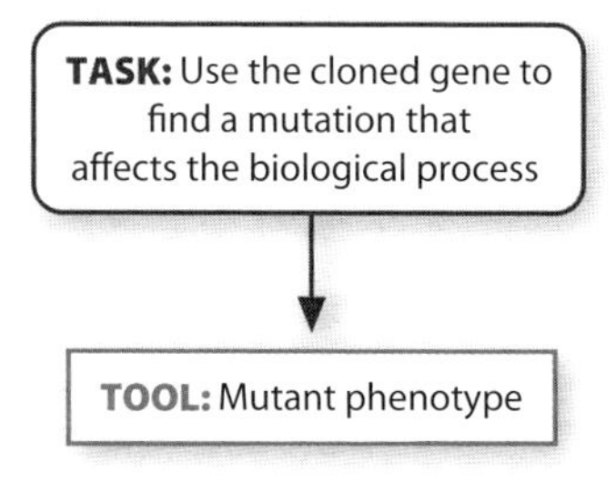

Figure 6.1 The task and the tool. The task is to use the cloned gene to find a mutant phenotype that can be used as a tool for further genetic analysis.

6.1 Overview of reverse genetics, targeted gene disruption, and targeted gene replacement

It is often possible to clone a gene before its mutant phenotype is known. Let us consider two experimental situations in which a gene has been cloned first such that a reverse genetics approach is needed to find a mutant phenotype. For example, a gene can be cloned based on its expression pattern. Imagine that the investigator is interested in genes involved in the function of the brain in a mouse and has used a microarray or other expression method to find genes expressed in the brain. (Such methods are described in more detail in Chapter 8.) How can he/she learn about the functions of those genes and determine if the expression pattern is biologically relevant? By now, we know the geneticist's answer—**Find a mutant**. However, unlike the situation described in Chapter 3, the investigator is beginning with the knowledge of at least one gene that is expressed in the correct tissue. Will he/she really need to begin with a wild-type organism and find mutants at random before identifying the one that corresponds to the gene that has already been cloned? There must be a more direct method.

In another example, perhaps the investigator has used the procedures described in Chapter 3 to identify a gene in *Drosophila* that affects brain function, and he/she has gone on to clone the orthologous gene from a mouse, as described in Chapter 5. An obvious question is to ask if the gene plays a similar role in the mammalian brain as it does in the insect brain, despite the vast difference in the complexity of the two organs. A mutant phenotype in the mouse can help to answer this question.

In both examples, the forward genetics question 'What genes affect the function of the brain?' has become one of reverse genetics: 'How does this one particular gene affect the function of the brain?' It may be possible to find the mutant phenotype associated with this one gene by screening a mutagenized population of wild-type organisms using the procedures described in Chapters 3 and 4, but these methods do not take advantage of the previous knowledge of the one particular gene. In addition, the genetic methods described in Chapter 3 will prove inefficient in this case because they will find mutations in many other genes in addition to the one of interest. Identifying these other genes is very important, but this is not the direct goal of the present study and it may take a long time to sort through all the phenotypically related mutants before finding the specific one having a mutation in the gene of interest.

Reverse genetics addresses different questions than forward genetics

Because the starting point for the investigation is different for forward and reverse genetic analysis, both the question to be answered and the scope of the answer are different. We summarize the difference between forward and reverse genetic analysis in Figure 6.2, which is the same figure used in Chapter 3 (Figure 3.2). Reverse genetic approaches as described in this chapter will not identify all of the genes that are involved in a particular biological process. Instead, the analysis is limited by the information that the geneticist has in advance. On the other hand, the only genes that are investigated are those that are likely to be important to the biological process. Thus, reverse genetic methods do not identify all of the genes involved in a process, but they do ensure that the investigator is working with at least one relevant gene.

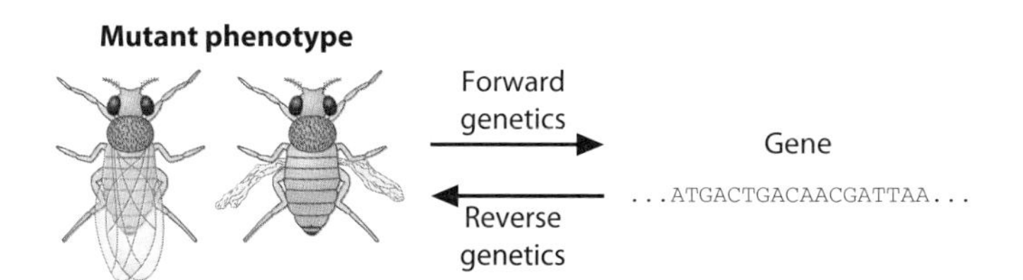

Figure 6.2 A comparison between forward and reverse genetics. In the forward genetic analysis on the left (as described in Chapters 3 and 4), the mutant phenotype is used as a starting point to clone a gene. In reverse genetics, as described in this chapter, the cloned gene is the starting point to find a mutant phenotype.

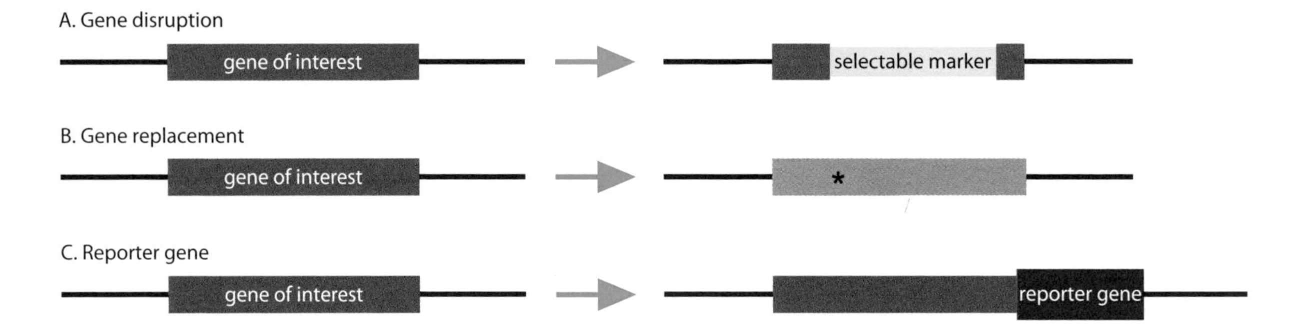

Figure 6.3 Three uses for targeted insertions. Homologous recombination can be used to target a cloned gene to a site on the chromosome. Three uses are described in this chapter. (A) The gene of interest is disrupted by another gene sequence, often something that can be used for selection. Examples of such selectable markers include nutritional genes such as *URA3* in yeast and drug resistance genes such as *neo*^{R} in mice. (B) The gene of interest is replaced by an altered version of the gene, with a particular mutation for example, as indicated by the asterisk. The gene is not knocked out but may be reduced in function, as indicated by the lighter color. In mice, such gene replacements are often referred to as 'knock-in' experiments. (C) A reporter gene has been introduced as a fusion with part of the normal gene, preserving the regulatory region of the target gene. Examples of reporter genes include *lacZ* and *gfp* whose expression is monitored by colorimetric assays.

Because the investigator has a gene sequence as a starting point and wants to identify its mutant phenotype, it is often possible to define the type of mutation that is made in a mutant. This is especially true for reverse genetic approaches that involve targeted gene replacements. As was discussed in Chapters 3 and 4 for induced mutations in random genes, there are different types of mutant alleles, and each type can tell us something different about the function of the gene. This is also true in reverse genetic analysis. For example, using sequence information, the gene can be specifically disrupted or knocked out to produce a null allele. Alternatively, the regulatory region of the gene can be altered so that the gene is expressed ectopically (a neomorph) or at unusually high levels (a hypermorph). With only a few modifications in the cloned gene, hypomorphic alleles can also be produced by targeted gene replacement. In fact, by making a range of specific changes to the cloned gene *in vitro* before reintroducing it into the organism, an allelic series of defined mutations can be created. The characterization of the mutants is much the same as that described in Chapter 4. The difference is the manner in which the mutants were obtained.

Reverse genetics is done in all of the model organisms, but it dominates genetic analysis in yeast and mice where targeted gene replacement is a standard procedure. In these organisms, targeted gene replacement has allowed geneticists an unprecedented ability to manipulate and explore the functions of genes. With these approaches, the scope of biological analysis changes from cloning a gene using a mutant to using the cloned gene to find a mutant.

In reverse genetics in yeast and mice, a cloned gene is inserted at a specific site in the genome

When beginning a reverse genetics study in yeast and mice, the investigator has a cloned gene that is thought to be involved in the biological process under investigation. However, the investigator does not know the gene's precise function and wants to use the mutant phenotype to infer the function. The cloned gene is altered *in vitro* by disruption with an unrelated piece of DNA (such as a selectable drug or nutritional marker), a specific local change in the coding region, or a reporter gene inserted under the same regulatory region. These changes are shown diagrammatically in Figure 6.3.

The first requirement for efficient reverse genetic analysis by gene replacement is a method for integrating a cloned gene at a specific targeted site in the genome. This can be done by homologous recombination, as shown diagrammatically in Figure 6.4. A cloned fragment of DNA to be integrated is introduced into the cell, typically as linear DNA. Homologous recombination between sequences on the cloned DNA fragment and identical sequences on the chromosome integrates the DNA fragment precisely into the chromosome. Often the site of integration will be the locus of the normal chromosomal gene, so that the chromosomal copy of the

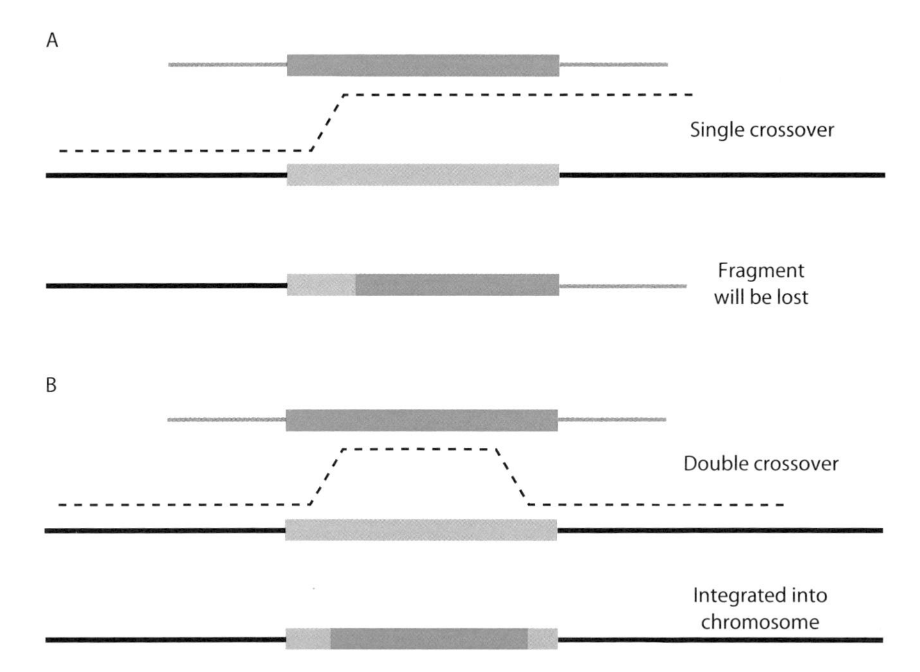

Figure 6.4 Double crossovers and targeted integration. The region of homology is shown in green, with dark green on the linear donor fragment and light green on the chromosome. Homology can rely on sequence similarity in any part of a gene or its regulatory region. (A) A single crossover will yield a chromosome fragment that will be lost. (B) A double crossover stably integrates the donor sequence into the chromosome. The amount of homology needed is only about 100 bp in yeast but is much higher in metazoans. In gene targeting in the mouse, regions of homology are usually about 8 kb.

gene is disrupted or replaced by the cloned copy. These two alternatives correspond to targeted gene disruption and targeted gene replacement, respectively, and are the most common reverse genetics approaches in species capable of efficient homologous recombination.

In one of the most common targeted gene disruption procedures, a specific piece of DNA, such as a selectable marker, is integrated within the cloned gene under study, thereby producing a null mutation. As shown in Figure 6.4, homologous sequences between the linearized plasmid and the yeast or mouse chromosome direct a crossover at a precise site on each side of the cloned fragment, creating a double crossover. (By first linearizing the plasmid DNA, it is guaranteed that a specific double crossover will be required for the cloned gene to be stably integrated into the genome. A single crossover on only one side will produce two broken DNA molecules that will be unstable or degraded by the cell and lost. If a circular plasmid had been used, only one crossover would be required for stable integration, but the entire plasmid would be integrated and the gene would not be replaced.) The site of recombination, and hence the region that is integrated, depends on the homologous sequences chosen, and so a specific location can be targeted. Crossovers and integration of the DNA fragment occur precisely, without leaving gaps in the sequence or introducing a new sequence.

It is likely that all eukaryotic organisms can carry out site-specific recombination in this manner because the mechanism is directly related to the recombination that occurs during meiosis. Therefore, in principle, any organism for which DNA fragments can be introduced into the nucleus could be used for this kind of reverse genetics approach. In reality, the rate of recombination is too low in most organisms for this approach to be useful, and the introduced DNA fragment will rarely be integrated and will eventually be lost. However, *S. cerevisiae* has a much higher rate of recombination than multicellular organisms, and so targeted integration by recombination has been a routine procedure in yeast for decades. The ability to carry out gene replacements has been one of the tools that makes yeast such a powerful model organism.

Targeted gene disruptions can be used in organisms with a lower rate of recombination, such as mammals, if an appropriate selectable marker is integrated during the recombination. That is, even if the rate of homologous recombination is very low, cells with the appropriate disruption or insertion can be selected using drug resistance markers and grown in culture. This approach has been used effectively to perform reverse genetics in mice. After a brief description of some of the techniques used in yeast, we will devote most of the chapter to reverse genetics approaches in mice.

Reverse genetic approaches of this type are not as widely used in *C. elegans*, *Drosophila*, or *Arabidopsis* as they are in yeast and mice. In *C. elegans*, homologous recombination techniques are problematic because injected DNA is rarely integrated into the genome; rather, the injected DNA is quasi-stably transmitted as an extrachromosomal array, as discussed in Chapter 2. Furthermore, the rate of recombination between the introduced DNA and the chromosome is low, and so targeted insertions rarely occur although methods to increase the rate of integration have been devised. In *Drosophila* and *Arabidopsis*, DNA fragments can readily be integrated into the genome but, as discussed in Chapters 2 and 5, the most widely used methods for carrying out this integration involve P elements and Ti plasmids, respectively. P elements and T-DNA insertions are *randomly* integrated into the genome, and are not targeted to specific genes or sites. In addition, unlike yeast and mice, neither worms nor flies can be propagated from single cells grown in culture, which again makes reverse genetics approaches more challenging. *Arabidopsis* can be regenerated from a single cell grown in culture, but the rate of homologous recombination is extremely low. The real advantage of higher-efficiency homologous recombination in *Arabidopsis* would be for use in targeted gene replacement experiments where specific gene sequences could be altered.

This is not to say that reverse genetics approaches are never adopted in these three organisms. However, rather than creating targeted changes at a specific location, these methods involve random gene disruptions by deletion or insertional mutagenesis of whole populations. The location of each disruption is determined using gene-specific PCR. For example, deletions are made by excision of transposable elements or irradiation in a population. The location of the deletions is not known, but individuals from the population are rapidly screened using PCR primers for a gene of interest to find one worm, fly, or plant with the desired deletion. Similarly, a random gene disruption by P-element transposition in *Drosophila* or T-DNA integration in *Arabidopsis* can be connected to a single gene by PCR. To improve the ability of researchers to do reverse genetics in *Arabidopsis*, the insertion sites of thousands of independent T-DNA lines have been sequenced, and it is now possible to find T-DNA-mediated gene disruptions in most genes using the *Arabidopsis* genome database TAIR and order it from a stock center. Similar projects have been done with P-element insertions and deletions in *Drosophila* and with radiation-induced deletions in worms.

Even with these alternative reverse genetics approaches, there have been continued attempts to do targeted gene disruptions in these other model organisms. The real advantage of higher-efficiency homologous recombination in these species would be for targeted gene replacement experiments, where specific gene sequences could be specifically altered, something that cannot be done using random large deletion or integration approaches. One approach that has been taken with *Drosophila* to target gene disruptions for reverse genetics is described in Text Box 6.1. Throughout the majority of the chapter, however, our attention will be on yeast and mice.

BOX 6.1 Gene targeting in *Drosophila*

Targeted gene insertions have not been as important in *Drosophila* or *C. elegans* genetics as they have in yeast and mice. Nonetheless, because targeted insertions have many advantages for genetic analysis, efforts have been made to target insertions in both organisms. To date, more success has been obtained in *Drosophila* but gene targeting is not widely used in either organism. The strategy in *Drosophila* begins by inserting sequences at random locations using P elements to create flies with particular transgenes. The flies with these different transgenes are crossed to create strains that have multiple transgenes, and interactions between the transgenes target insertion to a particular locus.

Three transgenes are required. The first has the site-specific recombinase Flp, which is expressed under an inducible heat shock promoter. Flp catalyzes recombination between two specific sites known as the FRT sequences, which do not occur naturally in the *Drosophila* genome. The interaction between Flp and the FRT sites provides the specificity needed for targeting. The second transgene has the sequence-specific endonuclease *Sce*I, also expressed under an inducible heat shock promoter. The role of this endonuclease is to cleave a circular molecule *in vivo* to provide a linear DNA donor sequence. Linear DNA recombines at a higher rate than does circular DNA, and is said to be more **recombinogenic**. These two transgenes are inserted together in one parental strain of flies. The transgenic flies are normal, but when subjected to a heat shock of 38°C for 1 hour, both enzymes will be expressed. Such a strain can be used as a parent for any targeted insertion by this method, regardless of which gene is targeted.

The cloned gene to be inserted is introduced in the third transgene. The cloned gene being targeted is flanked by FRT sites and includes a recognition sequence for *Sce*I endonuclease. The arrangement is shown in Text Box Figure 6.1A. This third transgene is inserted into the chromosome via P

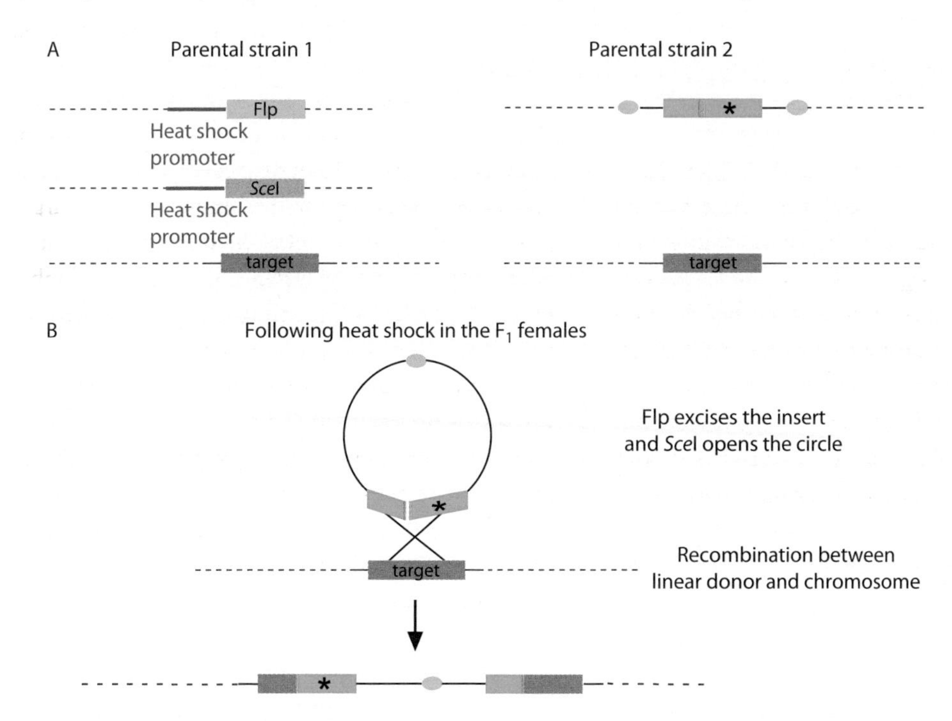

Figure B6.1 A gene targeting strategy in *Drosophila*. The target gene is shown in blue. In panel A, three different transgenes are used in two different parental strains. The first strain has inserts of Flp recombinase (green) and *Sce*I endonuclease (orange) expressed by a heat shock promoter. The other strain has the mutated donor copy of the target gene (with the asterisk) flanked by the FRT sites (green) needed for Flp recombinase and the *Sce*I recognition site (orange). The two strains are crossed to produce the F_1 flies in panel B. In panel B, the F_1 female flies are heat shocked so that both Flp and *Sce*I are expressed. The expression of Flp excises the donor gene as a circle, and the expression of *Sce*I cleaves the circle into a linear molecule capable of recombination. Recombination between the donor and the chromosome disrupts the chromosomal copy. Only one of the possible recombinant forms is shown, and other recombination products were found.

elements and is expected to insert at random sites in the genome. The two fly strains are crossed to make flies with all three transgenes.

When female progeny from this cross are subjected to heat shock, the Flp recombinase catalyzes the excision of the donor sequence, which is predicted to form a circular molecule at least some of the time. This is shown diagrammatically in Text Box Figure 6.1B. The heat shock also induces expression of the *SceI* endonuclease, which cuts the circle to produce a linear molecule in which the donor gene is disrupted. The linear molecule is recombinogenic and can undergo homologous recombination with the chromosomal copy of the gene. When this occurs, the chromosomal copy of the gene is replaced by the disrupted donor gene. This process occurs at the highest frequency when sequence homology between the donor and the chromosomal genes is about 8 kb, as is also seen in mice. The overall frequency of targeted insertions, not all of which occur by this process, is estimated to be approximately one targeted insertion per 500 gametes produced by the transgenic female.

Many modifications of this technique are possible to create specific mutations as well as to disrupt specific genes; these are the equivalent of the knockout and knock-in mutations in mice. One of the major limitations of this technique is that the transgenic strain constructions require more effort than a simple P-element insertion at a random location. In addition, many other types of recombination events have been observed besides the disruption shown in Text Box Figure 6.1. With the rich history and power of *Drosophila* genetics, targeted insertions appear to be a useful but not yet essential tool.

Further reading

Rong, Y.S. and Golic, K.G. (2000) Gene targeting by homologous recombination in *Drosophila. Science*, **288**, 2013–18.

Capecchi, M.R. (2000). Choose your target. *Nature Genetics*, **26**, 159–61.

6.2 Targeted gene insertions in yeast

Gene disruptions have been done in *S. cerevisiae* for more than 20 years, so the techniques for integration have been very well studied, and the applications for targeted insertions are limited only by the investigators' ingenuity. It is not an exaggeration to say that the linkage arrangements and gene locations in yeast have been completely reconstructed for the convenience of the investigator. Genes can be placed downstream of different regulatory regions that allow controlled expression, specific mutational changes can be engineered as needed, linkage and recombination can be altered, and so on. Cloned DNA is readily introduced by transformation, many nutritional markers are available for selection, and only about 100 bp of sequence homology are needed on either side of the DNA to be integrated for the recombination to occur between the introduced cloned gene and the targeted chromosomal gene. Yeast offers the geneticist more control over the type of mutation studied than any other eukaryotic organism. Many of the experiments that we describe in other chapters in this book depend on the ability to target yeast genes to specific locations, to disrupt specific genes, and to introduce specific genes. To cite one example that we will develop in more detail in Chapter 7, targeted insertions have made it possible for yeast geneticists to engineer deletions of every gene in the genome.

A simple example of a gene disruption for reverse genetics is provided by the tubulin family of genes. We will make use of these tubulin mutants again in Chapter 10, so it is worth describing how one of them was made by a gene disruption. Recall that tubulin is a dimeric protein, with an α-tubulin subunit and a β-tubulin subunit that assemble precisely (Figure 6.5A). In yeast, there are three genes, two α-tubulin genes named *TUB1* and *TUB3* and one β-tubulin gene named *TUB2*. Mutations in both *TUB1* and *TUB2* were initially identified in forward genetic screens by their increased sensitivity to the drug benomyl, which binds to microtubules. Based on the sequence homology with tubulin genes in other organisms, the β-tubulin gene was cloned and then altered *in vitro* by cloning a nutritional marker such as *URA3* in frame into the coding region of the *TUB2* gene (Figure 6.5B), and the altered version was targeted to replace the chromosomal gene *in vivo*. The cloned *TUB2* gene with the inserted *URA3* gene was transformed as linear DNA into *ura3* mutant yeast so that homologous recombination could occur between the chromosomal *TUB2* gene and the introduced gene.

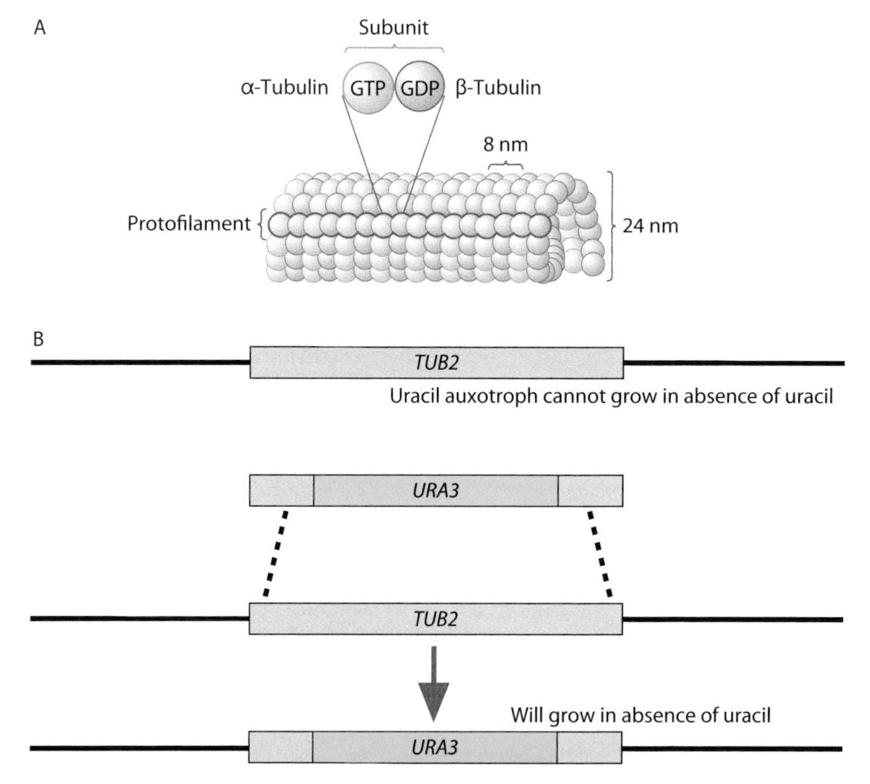

Figure 6.5 The *URA3* gene inserted into the *TUB2* gene. (A) The tubulin protofilament consists of α-tubulin and β-tubulin. In yeast, α-tubulin is encoded by the *TUB1* and *TUB3* genes, while β-tubulin is encoded by the *TUB2* gene. (B) The *URA3* gene is cloned into the middle of the *TUB2* gene. This is transformed into a *ura3* mutant that cannot grow in the absence of uracil. A double crossover between the disrupted *TUB2* gene and the chromosomal *TUB2* gene, indicated by the broken lines, replaces the chromosomal gene with the disrupted gene. This disruption can be selected by the ability to grow in the absence of uracil.

A double crossover replaced the chromosomal *TUB2* gene with the disrupted gene, identified by the ability of the yeast strain to grow in the absence of added uracil and confirmed by subsequent molecular analysis. This procedure is shown diagrammatically in Figure 6.5B. Although the frequency of the double crossover is low, perhaps one cell in a few thousand, the strong selection for uracil prototrophy allows these infrequent insertions to be found.

Modifications of this general procedure abound, but the essential strategy has not changed in more than 20 years. The success of this technique in yeast parallels attempts to carry out similar analyses in multicellular organisms. The most numerous examples of reverse genetic analysis using targeted gene disruptions are found in mice.

6.3 Targeted gene knockouts in the mouse

Embryonic stem cells can give rise to mouse embryos

In thinking about how reverse genetics might be done in a mammal, it is useful to make a comparison with yeast. Yeast can be grown as single cells in culture, and colonies arise from those single cells. Therefore, the stable genetic modification of only one cell can result in the genetic alteration of a large group of yeast cells through asexual reproduction. Since a population of cells can arise from a single altered cell, it is possible to perform selections for molecular events even if they occur rarely.

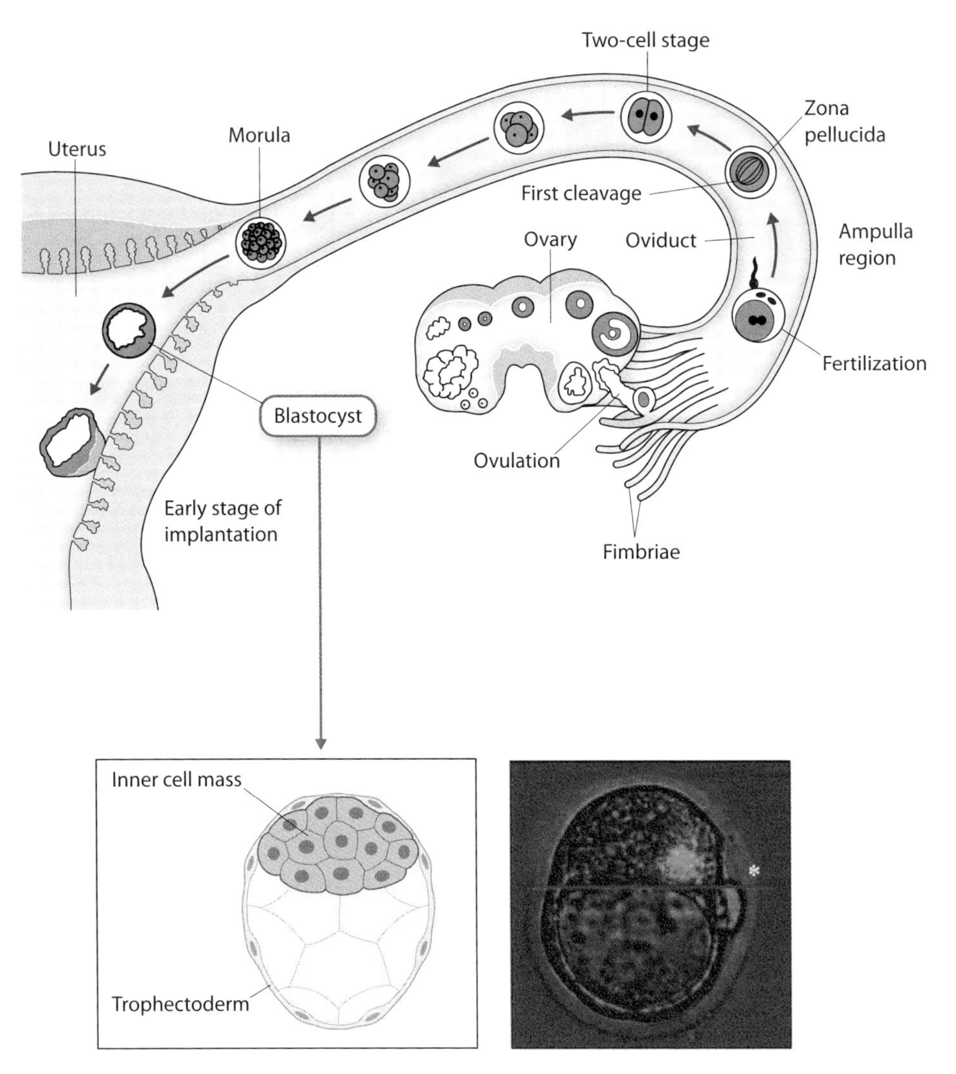

Figure 6.6 Mouse embryogenesis and the generation of the inner cell mass. Ovulation releases an oocyte that moves down the oviduct where it can be fertilized. After fertilization and the completion of meiosis, the embryo begins to divide mitotically. The dividing embryo reaches the blastocyst stage as it enters the uterus, approximately 3–4 days after fertilization, and implantation in the uterine wall follows. An isolated blastocyst is shown at the bottom diagrammatically and as a light micrograph. A large cavity called the blastocoel forms with the inner cell mass at one side. The cells of the inner cell mass will give rise to all of the cells of the embryo and adult mouse. These cells can also be isolated and grown in culture as embryonic stem (ES) cells. Gene targeting occurs in these ES cells in culture.

A similar principle is true for mice, but to understand the technique for reverse genetics in mice, we first need to briefly describe mouse embryogenesis, illustrated in Figure 6.6.

In mice, as in all placental mammals, an ovum is released from the ovary and fertilized as it moves down the oviduct, where it begins a series of cell divisions. At about the eight-cell stage, the cells rearrange themselves with a corresponding change in cell contacts in a process known as compaction. Prior to compaction, individual cells termed blastomeres can be separated from one another. Shortly after compaction, a cavity known as the blastocoel forms in the embryo, a characteristic of the blastocyst stage. Implantation into the uterine wall occurs shortly thereafter.

The embryo itself develops from the inner cell mass arrayed in a layer along the top of the blastocoel at the blastocyst stage. In a mouse, the inner cell mass forms at about 3.5–4.5 days post-coitum. The inner cell mass can be isolated from blastocysts and grown in culture

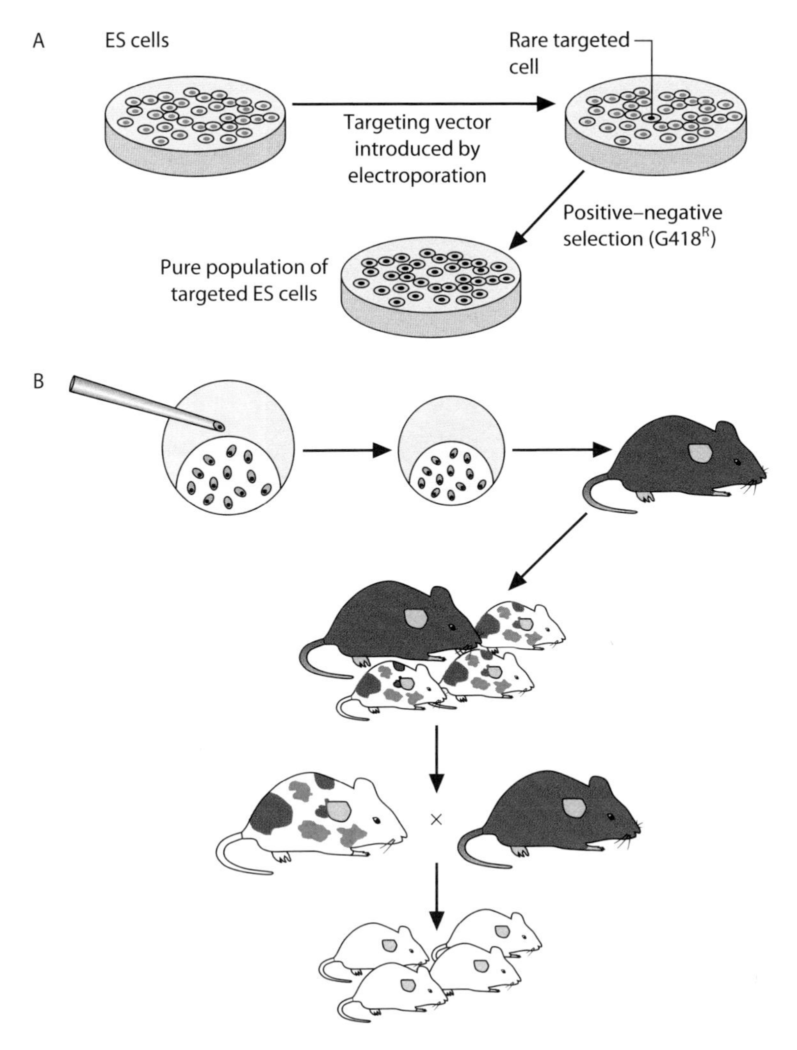

Figure 6.7 ES cells and chimeric mice. ES cells are used for targeted insertions with a selection for particular genetic events (the selection is shown in Figure 6.8). The selected ES cells can be assayed molecularly to ensure that the recombination of interest has occurred. Those cells with the proper insert are injected into the blastocele of an isolated blastocyst, and the injected blastocyst is implanted into a pregnant female host. In this example, the ES cells were derived from a white mouse and the recipient blastocyst was derived from a brown mouse, but other coat color combinations are often used. The pregnant female gives birth to a litter with some mice from her own embryos and others from the chimeric embryos. Mice derived from the chimeric embryos are recognized by the variegated coat color, and are presumed to be chimeric in all cell types. The chimeric mice are mated to produce a mouse that is derived entirely from the ES cells with the targeted insertion. Here, those mice will be white. (Redrawn from Cappechi, M.R. (2005). *Nature Reviews Genetics*, **6**, 507–12.)

indefinitely. The process is shown diagrammatically in Figure 6.7. Such cells are termed embryonic stem (ES) cells. In addition, ES cells injected into the blastocoel at this stage can also become incorporated into the inner cell mass and thus can contribute to all the tissues of the mouse that develop from this blastocyst. In order to produce transgenic mice, embryos at the blastocyst stage are isolated from a pregnant female, ES cells are introduced into an isolated blastocyst by injection, and the blastocyst is then inserted into another pregnant female (or a hormonally treated pseudo-pregnant female). The blastocyst will implant into the uterine wall, and a mouse that includes some of the engineered and injected cells will arise. Because there are very few cells in the inner cell mass and at this stage these cells can still give rise to any cell type in the adult, an introduced cell has the potential to develop into any structure in the mouse, including the germline. Thus, the mouse

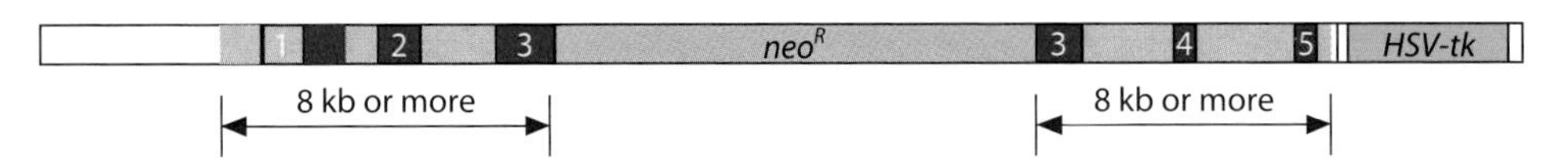

Figure 6.8 A targeting vector used for a gene disruption. The gene is shown in blue, with light blue representing introns and the dark blue exons numbered. The neo^R gene has been inserted into the third exon. The vector also includes the herpes simplex virus thymidine kinase gene (*HSV-tk*). The *HSV-tk* gene confers sensitivity to ganciclovir, whereas the neo^R gene confers resistance to neomycin or its more stable analog G418. At least 8 kb of homology is used for recombination between the targeting vector sequence and the chromosomal gene.

developing from such an embryo will be a **chimera** of cells of more than one genotype. (We return to the topic of genetic chimeras in Chapter 9 when we discuss their utility in understanding gene activity.)

In summary, the key cellular and developmental principles that make transgenic mice possible are as follows.

1. Cells of the inner cell mass can be isolated and grown in culture as ES cells.
2. ES cells can be reintroduced into a developing embryo at the blastocyst stage and will become incorporated into the inner cell mass.
3. The blastocyst can be reintroduced into a female and a chimeric mouse will develop with cells of both types.

Thus, when a gene is stably inserted into the genome of an ES cell grown in culture, a mouse may be produced with that gene insertion. This meets the first criterion for reverse genetics—that an organism can be grown from a single altered cell or a small number of altered cells.

ES cells can be targeted for gene insertion using both positive and negative selections

The second criterion for efficient reverse genetics and gene disruptions is that the introduced sequences can be targeted to specific sites in the genome. Cloned genes are typically introduced into ES cells by electroporation of linear DNA because large populations of cells can be treated simply and simultaneously. Again, just as in targeted gene replacements in yeast, integration depends on homologous recombination between the introduced donor DNA fragment and the chromosomal sequence. In contrast to yeast, however, extensive sequence homology is required for recombination to occur; most homologous recombination constructs have at least 8 kb of homology with the target sequence on either side of the DNA to be integrated. Furthermore, unlike yeast, the rate of homologous recombination in mice is very low, so a selection scheme is required to identify the ES cells with the proper insertion.

The relative inefficiency of the biological recombination process has prompted mouse geneticists to introduce both a positive and a negative selection, also called a counter-selection, as defined in Chapter 3. The targeting vector for this counter-selection process is shown in Figure 6.8. The target gene under study is disrupted *in vitro* by the insertion of a drug selection marker, typically the neomycin-resistance gene (abbreviated neo^R) that has been placed in an exon. The disrupted target gene is cloned into a targeting vector that includes the thymidine kinase (abbreviated *tk*) gene from a herpes virus. Each of these genes confers a specific drug response: neo^R confers *resistance* to neomycin and its more stable analog drug G418, and the *tk* gene confers *sensitivity* to the drug ganciclovir. Note that these drug responses are opposite to one another and make it possible to select both for and against the cloned donor gene. The targeting vector is introduced into ES cells that are diploid for a normal copy of the target gene and that are naturally resistant to ganciclovir.

The introduced targeting vector will encounter one of three fates in ES cells, as summarized in Figure 6.9. In most cells, the targeting vector will not become integrated into the chromosome at all, as shown in Figure 6.9A. The ES cells with no insertion are sensitive to G418, and so will not grow if G418 is included in the culture medium. Thus, there is a positive selection *for* integration by using the neo^R gene and G418, allowing even rare integrations of donor sequences to be detected.

On the other hand, the targeting vector can be inserted to produce G418 resistance by either non-homologous recombination or the desired homologous recombination as shown in Figures 6.9B and 6.9C. In

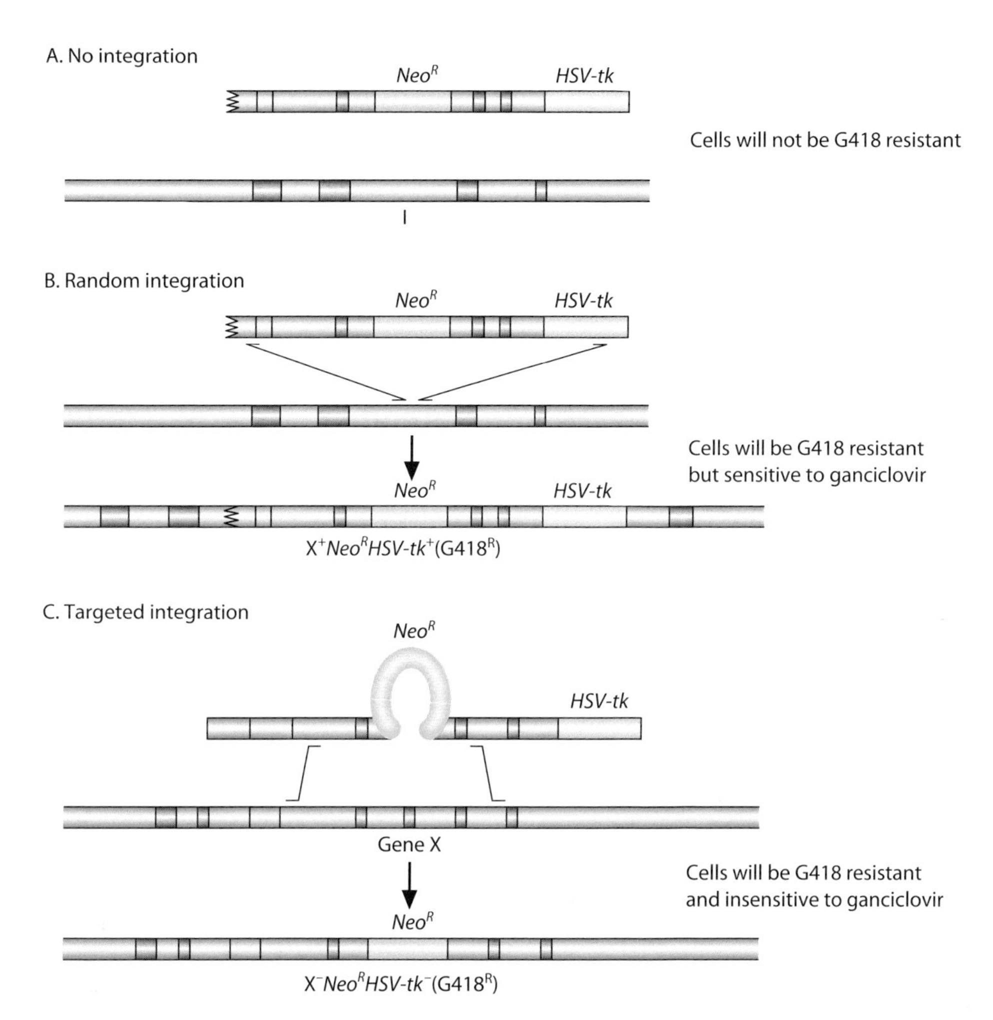

Figure 6.9 Counter-selection for targeted integrations. (A) If the vector fails to integrate into the chromosome in the ES cells, the cells will not grow in the presence of the neomycin analog G418. On other hand, G418-resistant cells can arise from either random integration with non-homologous recombination as in panel B or by targeted integration with homologous recombination as in panel C. These ES cells can be distinguished by a selection based on ganciclovir. The non-homologous events also integrate the *HSV-tk* gene, which confers sensitivity to ganciclovir, whereas homologous recombination excludes the *HSV-tk* gene and the cells are insensitive to ganciclovir. (Derived from Cappechi, M.R. (2005). *Nature Reviews Genetics*, **6**, 507–12, Figure 4.)

non-homologous recombination (Figure 6.9B), the targeting vector can be inserted into the chromosome at random sites, probably by a break repair or non-homologous end-joining mechanism. Such an insertion of the entire targeting vector by non-homologous recombination will confer not only resistance to G418 but also sensitivity to ganciclovir by the *tk* gene that is included on the vector. These untargeted events, which may be as much as 99 percent of the G418-resistant cells, can be selected *against* by including ganciclovir in the culture medium so that no insert of the *tk* gene can grow.

The counter-selection with these two drugs ensures that only those ES cells that inserted the donor gene by homologous recombination events between the disrupted target gene on the targeting vector and the chromosomal copy of the gene will be able to grow and divide. These are the very cells that are needed to construct a chimeric mouse with the engineered gene disruption. The ES cells that grow and divide in culture when subjected to this selection are isolated and introduced into a blastocyst embryo. As described above, these ES cells will become part of a new mouse.

There is yet one more screen in the gene replacement process—one that occurs among mice born in the same litter. When the chimeric blastocyst is inserted into the surrogate mother, it is not the only developing

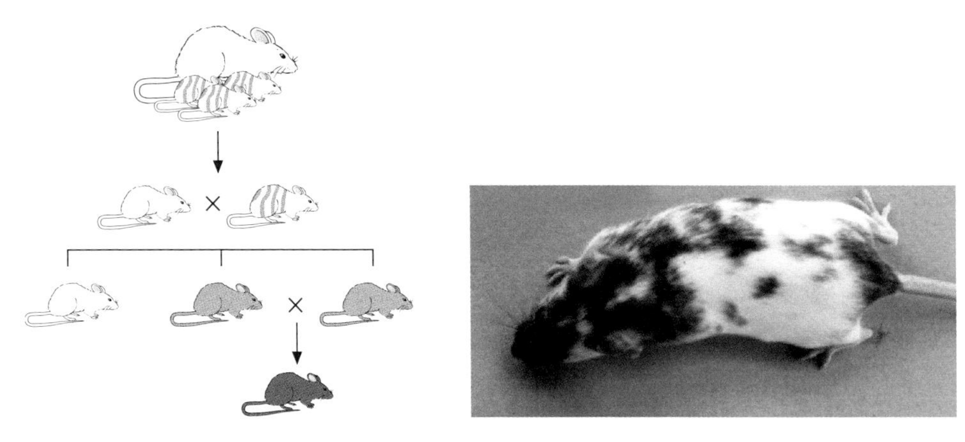

Figure 6.10 Coat colors indicate chimeric mice. Many ES cell lines are derived from the inner cell mass from brown mice; the injected blastocyst may be either black or white, as shown here. A mouse that arose from a chimeric inner cell mass will have a variegated coat color with both brown and white in different patterns. These chimeric mice can be bred to produce offspring that are derived entirely from the engineered ES cells, indicated by the brown coat. A PCR assay on cells from the tail is often used to confirm that the mouse is homozygous for the insert.

embryo in her uterus. Embryos of her own genotype are also developing. How can one tell which of the newborn mouse pups is a chimera with tissues derived from the injected transgenic ES cells and which pups are normal mice? For this screen, the geneticist uses some of the earliest mutations studied by mouse fanciers—the coat color mutations. As shown in Figure 6.10, the ES cells are often derived from a brown (or agouti) mouse and introduced into blastocysts that have been isolated from a black female. The chimeric mice will have black and brown coats, whereas those arising from a non-chimeric wild-type inner cell mass will give rise to entirely black mice. The exact pattern of black and brown in the mouse is not important, but the extent of brown fur can be helpful in recognizing the success of the process. A mouse that has only a few or very small patches of brown fur may consist primarily of black non-engineered cells, whereas one that has more or larger patches of brown fur is likely to have more cells arising from the engineered ES cells. The coat color itself does not matter to the biology of the mice but acts as a marker of chimerism. These chimeric mice are also likely to be chimeric in their germline and thus will give rise to two different types of gametes—those derived from ES cells with the disrupted gene and those derived from ES cells with the normal gene.

The chimeric mice are bred and a homozygous strain is derived in which all of the offspring are brown, which means that all of these cells arose from the engineered ES cells in a previous generation. The knockout can be confirmed by performing PCR or a similar assay on a sample of transgenic cells obtained from a snip from the tail for example.

The ability to make gene knockouts has changed mouse genetics

Until gene knockouts were available, mouse geneticists relied on the types of mutant screens described in Chapter 3. These worked very well, and many experiments still begin with random mutagenesis. However, the ability to target mutations to specific genes has added a new dimension to mouse genetics, and this general procedure has been used to create thousands of different knockout mutations in mice. Case Study 6.1 describes how the mouse ortholog of the *patched* gene, introduced in the case studies in Chapters 4 and 5, was targeted for disruption.

Once the mutation has been generated by this reverse genetics method, characterization of the mutant phenotype is done as described in Chapter 4 for other mutations. Many knockout mutations result in a lethal phenotype, which requires that the mutation be maintained as a heterozygote; some of the balancer chromosomes used for maintaining heterozygotes are also described in Text Box 3.2 in Chapter 3. Conditional mutations in which the gene is knocked out only in certain tissues or at certain times are described below and provide another method of investigating genes with widespread lethal effects.

Case Study 6.1

Patched knock-out mutations in mice

The *patched* gene in *Drosophila* was introduced in Chapter 4 as one of the segment-polarity genes that affect the orientation of each segment in the embryo. Subsequent analysis in *Drosophila*, to be described elsewhere in the book, showed that the *patched* gene was involved in a wide range of important development events through its interactions with *hedgehog* and other genes. The mouse *patched* (*Ptch*) gene was cloned using homology to *patched* in other vertebrates and its expression in the same tissues as *hedgehog*, a method described in Chapter 5. These experiments suggested that the mammalian *patched* gene encoded an important signaling protein during development and played a role in some human cancers, but they were not designed to reveal the full range of *patched* effects. For that, the investigators needed to make a gene knockout.

The mouse gene has 15 exons, and a complete genomic clone was obtained by screening a mouse library. A portion of the gene was disrupted *in vitro* with a *lacZ* gene and the *neo*^{R} gene, replacing part of the first exon and all of the second exon as shown in Case Study Figure 6.1. This part of the *Ptch* gene was cloned into a targeting plasmid vector that included the thymidine kinase (*tk*) gene. Thus, the two drug markers needed for both positive and negative selection of the disrupted cells were included by standard procedures. In addition, the presence of the *lacZ* gene served as a reporter for *Ptch* gene expression. (Reporter genes are described in more detail in Chapter 8.) This targeting plasmid was linearized and electroporated into ES cells derived from a brown mouse. The ES cells were grown in culture in the presence of G418 (the positive selection for integration) and ganciclovir (the negative selection against random integration). Three independent clones were expanded into colonies, and these were injected separately into blastocysts isolated from a black mouse. The injected blastocysts were introduced into black females and brought to term

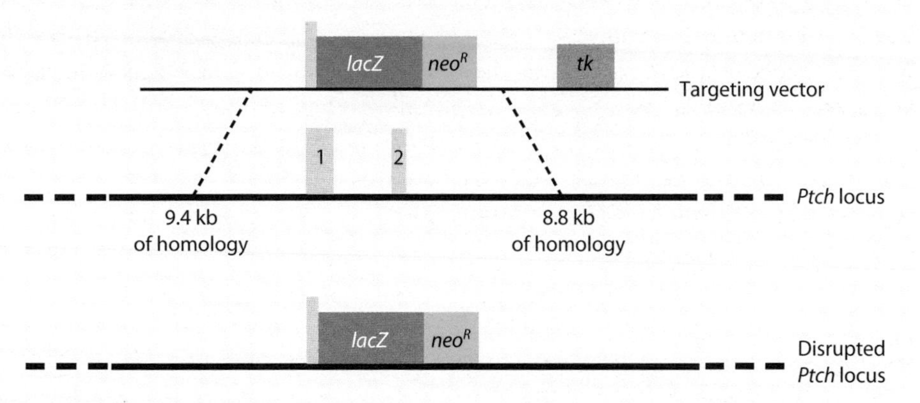

Figure C6.1 Disruption of the mouse *Ptch* gene. The targeting vector has 9.4 kb of the mouse *patched* gene on the left side and 8.8 kb on the right side to allow homologous recombination. It replaces part of the first exon and all of the second exon (shown as numbers in the light gray boxes) with the *lacZ* reporter gene and the *neo*^{R} gene. It also includes the thymidine kinase (*tk*) gene, which confers sensitivity to ganciclovir. The vector was introduced into ES cells where the *neo*^{R} and ganciclovir-insensitive cells were selected. By blotting and PCR techniques, these ES cells were shown to have the *Ptch* gene disrupted as shown. The *Ptch* gene has 15 exons. The drawing is not to scale.

as chimeric brown and black mice. The chimeric mice were then bred among themselves to make heterozygotes, and the heterozygotes were mated with the hope of finding a strain homozygous for the *Ptch* knockout. In fact, no knockout homozygotes were recovered from these matings, indicating that *Ptch* is an essential gene for mouse development. The homozygotes begin to show morphological defects at 8 days of embryogenesis and die by day 9 or 10 with neural tube and heart defects.

Nonetheless, because the mutation is recessive, the *Ptch* knockouts could be maintained as heterozygotes, and the *lacZ* reporter gene allowed the investigators to monitor the expression of the gene. Some of the results are shown in Case Study Figure 6.2. The expression pattern is complicated and dynamic, reflecting the widespread signaling role of *Ptch* during embryogenesis. However, in the heterozygote, expression is seen in the neural tube of 8-day-old embryos, as well as in other tissues. This expression pattern is consistent with the mutant defects seen in the homozygote.

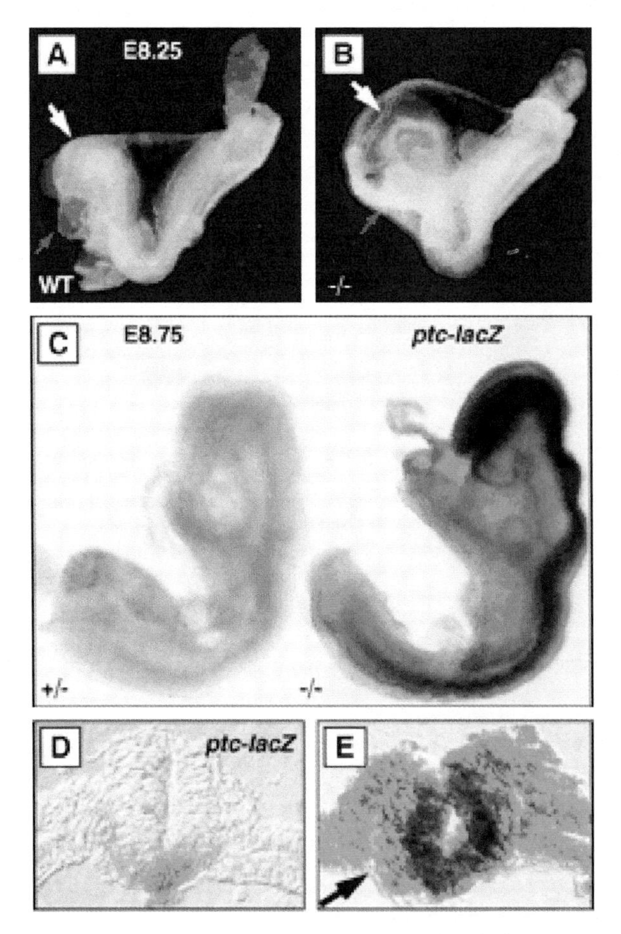

Figure C6.2 **Expression of the *ptc* gene in wild-type and *Ptch* embryos.** The gene disruption included the *lacZ* reporter gene, detected by staining for β-galactosidase, as shown here. Panels A, C, and D are heterozygous for the *Ptch* mutant and show the expression of the gene in wild-type mice. Panels B and E and the right side of panel C show expression in homozygous mutant *Ptch* embryos. Notice that *Ptch* expression in the homozygous mutant expands over a wider region of the embryo, indicating that Ptch is a negative regulator of its own expression. This is particularly noticeable surrounding the neural tube in panels D and E. (Goodrich *et al.*, 1997)

One of the interesting complexities of the *patched* gene is that the Ptc protein in *Drosophila* is a negative regulator of *patched* gene transcription, thereby decreasing its own expression. If Ptch behaves the same way in mice, the homozygous mutant animals should show *Ptch* expression in many additional tissues. That is, because the wild-type gene represses its own transcription, the mutant may exhibit ectopic expression as a consequence of lacking autoregulation. The reporter gene allowed the investigators to compare Ptc expression in *Ptch* mutant homozygotes and heterozygotes. The *Ptch* gene is expressed much more broadly in a homozygous embryo than in a heterozygote, as seen by the blue staining in Case Study Figure 6.2. Note that tissues in the embryo that do not stain in the heterozygote are stained in the homozygous mutant. Clearly, the type of autoregulation of *patched* gene expression that occurs in *Drosophila* also occurs in mice.

Although the heterozygote embryos developed normally, defects were seen in some heterozygous adults. About 7 percent of the heterozygotes died prematurely, and most of those that were autopsied had distinctive cerebellar or other tumors. Although such tumors are rare in humans, mutations in the human *patched* gene have been implicated in their formation, and the tumors in the mouse heterozygotes are histologically similar to the human tumors. In addition, a few *patched* heterozygotes had extra digits (polydactyly) or fused digits (syndactyly), indicating a broad and dosage-dependent role for *patched* in limb development.

It is worth noting that mice are now known to have two *Ptch* orthologs, and that this gene disruption targeted one of the two genes, the *Ptch1* locus. The second patched ortholog, the *Ptch2* gene, has also been disrupted, and is also lethal with severe developmental consequences. In general, the two genes are expressed in different tissues, and the *Ptch2* gene disruption shows greater defects in the hematopoetic system than does the *Ptch1* gene.

Further reading

Goodrich, L.V., Milenkovic, L., Higgins, K.M., and Scott, M.P. (1997). Altered neural cell fates and medulloblastoma in mouse *patched* mutants. *Science*, **277**, 1109–13.

In contrast to these severe effects of lethal mutations, many knockouts result in no obvious mutant phenotype, as has also been observed for mutations in other organisms. These mutations were mentioned in Chapter 3 and will be a major topic in Chapters 10 and 12. Often another gene knockout is required in these cases, and only the double mutant shows a mutant phenotype because of issues of functional redundancy. Knockout mutations are expected to be null alleles with a complete loss of gene function because the sequence of the targeted gene has been disrupted by the neo^R gene. As noted in Chapter 4, null alleles are important for recognizing the earliest time at which a gene exerts its effect, but it is often helpful to have other types of alleles as well and to generate an allelic series of mutations. Most knockout mutations are recessive, but some are dominant loss-of-function alleles, including dominant negative mutations and mutations in haplo-insufficient genes. Modifications in how the gene is cloned into the targeting vector have allowed mouse geneticists to make different types of alleles. Some of these modifications are described in the next section.

Genes can also be targeted by Cre–*lox* or other recombination systems

Targeted gene replacements in ES cells occur at the low frequency of homologous recombination but can be detected because of the strong positive and negative selections that are imposed. A higher frequency of recombinants with much less sequence homology with the target can be produced using site-specific recombination systems. These site-specific recombination systems have allowed geneticists to replace the chromosomal gene with an altered version and to make reporter gene constructs. A common application of site-specific recombination involves the Cre–*lox* system. Cre is a recombination enzyme that integrates the bacteriophage P1 genome into the bacterial host chromosome. Cre, as with many other such recombinases, uses specific sequences as its recombination substrates, in this case the sequences known as *loxP* sites. A *loxP* site, as shown in Figure 6.11, is asymmetric, with a unique 8-base core sequence flanked by two 13-base inverted repeats. Thus, the extent of required sequence homology on each side of the sequence to be integrated is reduced from 8 kb for homologous recombination to 34 bp for Cre–*lox* mediated recombination.

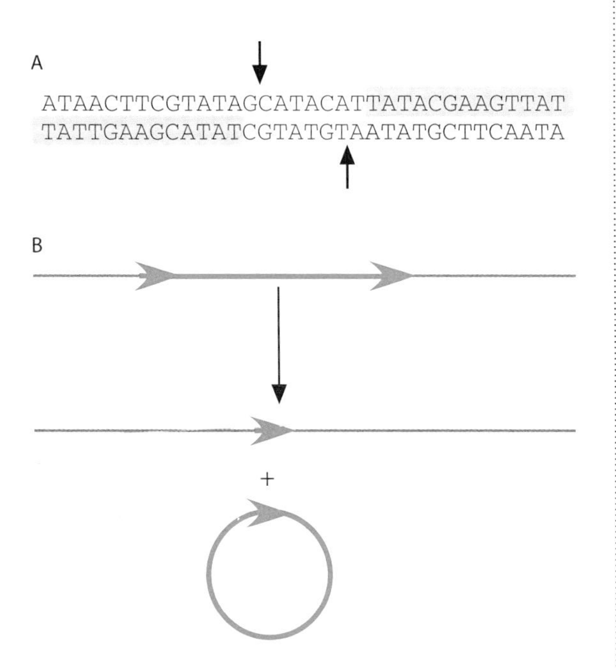

Figure 6.11 The Cre–*lox* system for site-specific recombination. (A) Cre recombinase recognizes the specific sequence of the *loxP* site shown and cleaves it at the sites indicated by the arrows. Note that the *lox* site is palindromic, as illustrated by the shaded boxes. The center core sequence is asymmetric and confers directionality to a *loxP* site, as indicated by the blue arrowheads in panel B. (B) A sequence shown in green flanked by *loxP* sites is the target for Cre-mediated recombination. The crossovers between the *loxP* sites result in the deletion of the sequence between the two sites as a circular molecule. Since both the reactants retain a *loxP* site, Cre-mediated insertion can also occur in the reverse reaction.

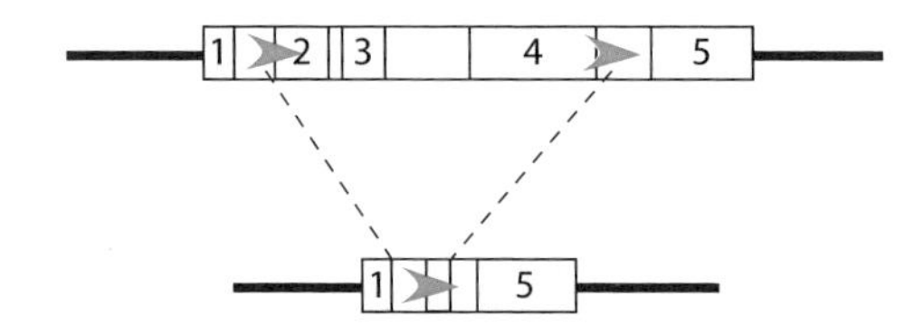

Figure 6.12 Generating a deletion using Cre–*lox*. A gene with two *loxP* sites in the same orientation is shown at the top. The *loxP* sites are shown as blue arrowheads at the beginning of exon 2 and the end of exon 4, as in Figure 6.11. The gene with *loxP* sites is said to be 'floxed'. When Cre is expressed, recombination occurs between the *loxP* sites and deletes the sequences in between. In this example, most of exon 2 and all of exons 3 and 4 are deleted.

The orientation of the *loxP* core sequence confers a direction to the recombination event, specifying the orientation in which the gene is inserted. The targeting vector used for Cre–*lox* techniques includes the *loxP* sites, but the procedure for inserting the altered gene is essentially the same as used for a standard gene knockout. Two limitations with this method are that the *loxP* sites must be included in the donor gene and Cre recombinase must be expressed in the genome of the host mouse. However, mouse strains with integrated Cre recombinase have now been developed and the *loxP* sites can be inserted anywhere in the genome using the homologous recombination procedures discussed above. More significant than these limitations is the advantage that the presence of the *lox* sites in the targeted gene allows for a number of additional manipulations.

Consider the situation when the *lox* sites are placed in direct orientation with respect to each other, as shown in Figure 6.12. When Cre catalyzes a crossover between the two sites, the sequence between the *lox* sites is deleted. (A sequence that is flanked by *lox* sites is sometimes said to be 'floxed'.) Thus, by assiduous placement of the *lox* sites, particular and limited portions of a gene can be targeted for specific deletion. Deletions of this type generate a true null allele, an advantage over gene disruptions using exogenous DNA, such as the *neo*^{R} gene, to interrupt the gene.

Tissue-specific mutations can be made by regulating Cre expression

Another significant advantage is that such a targeted deletion will only occur when Cre recombinase is present; when Cre is absent, the gene will not be deleted and will have normal or nearly normal function. Therefore, the specificity of Cre recombinase allows an additional level of manipulation when making deletions. In particular, by specifically controlling the expression of Cre, it is possible to delete the target gene in particular cells only rather than throughout the entire mouse. This is an important development with genes that have widespread lethal effects. Mouse strains have been made in which the *Cre* gene has been integrated under the control of regulated or tissue-specific promoters. The combination of the floxed gene with the specific expression of *Cre* provides a way of studying gene function in a much more defined manner.

For example, a gene such as *PTEN* that is essential in early embryonic development may have specific and additional functions in adult tissues, such as the pancreas, as well. With a gene knockout, these later functions cannot be studied because the knockout will die at the earliest stage; however, these effects can be studied using a conditional knockout in which the gene is deleted only in certain tissues. An example is shown in 6.13.

In order to study the effect of the *PTEN* gene specifically in the pancreas, the gene was replaced with a floxed version in which a recombination between the *lox* sites deleted most of the coding region. Another strain of mouse was made in which the Cre enzyme was placed under the control of a cell-type-specific promoter so that it was only made in specific cells in the pancreas; in the *PTEN* experiment, the *Cre* gene was under the control of the promoter from an insulin gene. Both strains of mice were viable and fertile even as homozygotes, since neither the presence of the *lox* sites nor the expression of the Cre protein is detrimental by itself. The two mice are mated to produce offspring with the *Cre* gene expressed in cells in the pancreas and the *lox* sites flanking the region of interest in the *PTEN* gene. The gene is then deleted specifically and exclusively in the pancreas, allowing the investigator to determine the effects of the mutation in one cell type or organ without the compounding effects of PTEN deficiency in other parts of the mouse at other times in development.

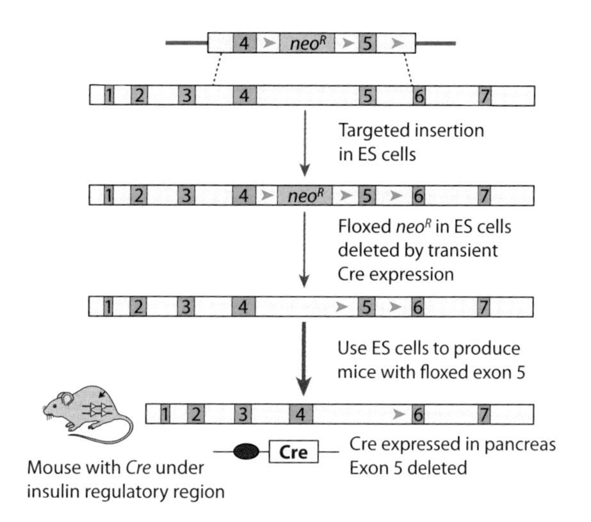

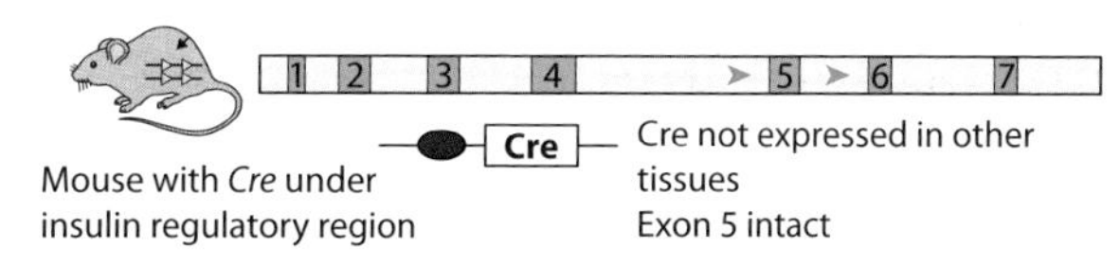

Figure 6.13 Cre–*lox* deletion of *PTEN* in the pancreas. A targeting vector is inserted in the *PTEN* gene by homologous recombination. In the targeting vector, both the *neo^R* gene in the intron and exon 5 of the *PTEN* gene are flanked by *loxP* sites. The actual *PTEN* gene has nine exons but, for clarity, only seven are shown. With transient Cre expression in the ES cells, the *neo^R* gene is deleted. These events can be detected in ES cells by appropriate molecular probes. The ES cells are used to make a transgenic mouse strain in which the *Cre* gene is expressed under the regulatory region from an insulin gene. Under the control of this regulatory region, Cre is expressed only in the pancreas, and exon 5 is deleted in those cells of the pancreas to produce a tissue-specific gene deletion. In other cells, Cre is not expressed so the mouse can survive to be analyzed for the role of PTEN in pancreatic cells.

Knock-in mutations replace the coding region with an altered function

Another important use of the Cre–*lox* system occurs when the normal coding region of a gene is replaced by an altered gene. These have been nicknamed **knock-in** experiments because the altered function is *introduced* rather than knocked out. In these cases, the gene is not deleted or disrupted but instead is replaced by a specific and defined molecular defect. Since most genetic diseases are not due to a deletion or a gene disruption, a knock-in mutation is often the best method of producing an appropriate model and testing the functional importance of particular portions of a gene. An example of one type of knock-in experiment is provided by the mouse ortholog of the human *CFTR* gene (Figure 6.14). As described in Case Study 5.1 in Chapter 5, the common mutation giving rise to cystic fibrosis in humans is a three-nucleotide deletion in the *CFTR* gene, a mutation called ΔF508 because a single phenylalanine

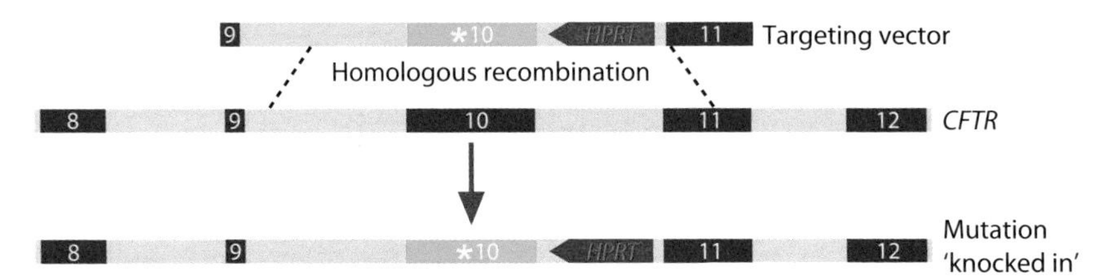

Figure 6.14 Knock-in of the *CFTR* ΔF508 allele. Most cases of cystic fibrosis are due to a specific three-base deletion in exon 10. This mutation was made *in vitro* in the mouse *CFTR* gene, shown in green with an asterisk to indicate the mutation. The mutant gene was cloned by conventional molecular methods into a targeting vector, as shown at the top. In this particular case, the selection marker was not the *neo*^{R} gene but the *HPRT* gene, which can be selected in HAT medium. The *HPRT* gene was cloned into an intron in reverse orientation. It was hypothesized that the expression of this gene would not interfere with normal *CFTR* expression in lung cells. Homologous recombination inserted this construct into the chromosomal *CFTR* gene in ES cells, as shown. These ES cells were used to make chimeric mice that carried the mutant *CFTR* gene.

residue is deleted. A gene disruption that knocked out the gene may not provide a realistic mouse model for this mutation and the disease it causes. In order to create a mouse model that mimicked cystic fibrosis, the normal coding region of the gene has been replaced by a version with this three-nucleotide deletion made *in vitro*. Several different procedures have been used to create this knock-in. One method is to make the desired alteration in the gene *in vitro* with flanking *loxP* sites. When introduced into ES cells expressing Cre, this engineered allele can replace the normal allele of the gene to produce a mouse model with the same mutation found in human patients. Similar procedures have been used for many genetic diseases for which a common disease mutation is known.

The same procedures can be used to make other types of hypomorphic alleles of a gene. This is particularly useful for genes in which the null allele created by a gene knockout is lethal, but for which a hypomorphic allele, sometimes called a 'knockdown' mutation (in keeping with the other nicknames), is not. Genes can also be engineered so that the resulting protein will have modifications in specific residues. For example, many proteins are comprised of several different domains or pro-tein motifs, such as a kinase domain, a DNA-binding domain, or a phosphorylation site. The effects of these motifs can be studied separately using mutations that affect only one of them. As discussed in Chapter 4, an allelic series of hypomorphic and null alleles is useful to see the range of effects that arise from one gene rather than simply the null effects. Dominant alleles, such as constitutively active hypermorphic mutations and dominant negative mutations, can also be created in this way.

Knock-in experiments are also used to make reporter gene constructs in which *lacZ* or another reporter coding region replaces the coding region of the normal gene. With a reporter gene of this type, the normal expression can easily be monitored. Furthermore, since the reporter gene is expressed under the control of the regulatory region of the gene it has replaced, the normal expression pattern of the gene is revealed. This can be quite complex. Since the mutation made by the reporter construct is likely to be recessive with regard to gene expression and function, a heterozygote will have a normal phenotype while the reporter gene expression will still be easily monitored. Although the procedure did not use a knock-in strategy, the *patched* gene disruption described in Case Study 6.1 also included the *lacZ* reporter gene, whose expression was important for understanding the complex roles of *patched* in mouse embryogenesis.

Another powerful use of the Cre–*lox* technology places the Cre protein under the control of a germline-specific promoter and uses it to remove the *neo*^{R} selectable marker. It is important to realize that once the G418 selection has been applied to find the ES cells, the *neo*^{R} gene is no longer needed. An example of this use is shown in Figure 6.15. In this technique, the gene is altered *in vitro* with the *neo*^{R} gene flanked with *loxP* sites and introduced into ES cells by standard procedures. The ES cells are placed into blastocysts from a mouse that has the Cre gene expressed only in the germline, for example during spermatogenesis. The chimera is

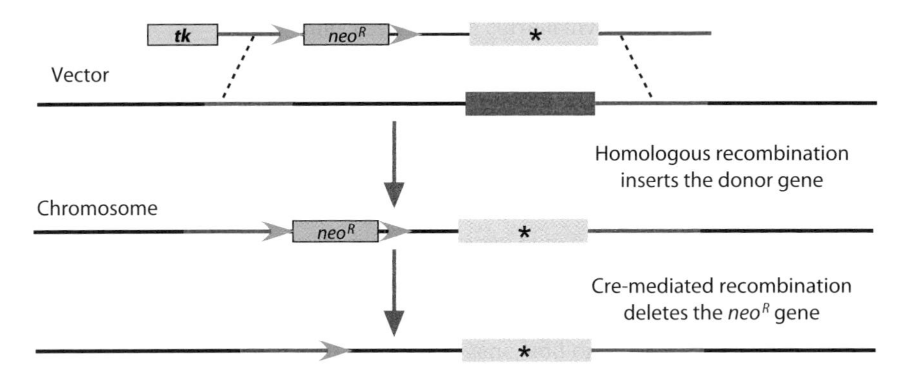

Figure 6.15 A Cre–*lox* knock-in experiment. The targeting vector with the mutated donor (shown in light red with an asterisk to illustrate the mutation) has the *neo^R* gene flanked by the *loxP* sites. Homologous recombination at the sites indicated by the broken lines knocks in the mutated gene in ES cells. The ES cells are then used to make a transgenic mouse by the standard procedure, except that the host blastocyst is derived from a mouse that expresses Cre in the germline. When the chimeric mouse becomes sexually mature, Cre is expressed and the *neo^R* gene is excised in the germline. All subsequent progeny of this mouse will have the knocked-in gene but will not have the potential complication of the *neo^R* gene. The *Cre* gene is often expressed under the control of a regulatory region from a testis-specific gene, so excision of *neo^R* occurs during spermatogenesis.

detected by normal counter-selection methods, but the *neo^R* gene has been deleted in the sperm once the male mouse reaches sexual maturity. The F_1 and succeeding generations will have the deletion or the floxed allele without complications arising from the exogenous *neo^R* gene.

I hear you knocking…

Other site-specific recombination techniques besides Cre–*lox* can be used for these techniques, including Flp recombinase from yeast and its 48 bp FRT target sites. Flp–FRT and Cre–*lox* can also be used in tandem in a procedure known as recombinase-mediated cassette exchange. Although the number of applications of this method has been limited so far, the technique could be an effective method of deleting different parts of genes or altering gene expression in different cell types at different times. A hypothetical example is shown diagrammatically in Figure 6.16. The Cre–*lox* system

Figure 6.16 Recombination-mediated cassette exchange. A combination of two different site-specific recombinases allows increased flexibility in gene manipulations. The hypothetical gene encodes a protein with two domains, one encoded by exons 2 and 3 and another encoded by exon 4. The Cre enzyme works on the *loxP* sites shown as blue arrows flanking exons 2 and 3. The Flp enzyme works on the FRT sites shown as red diamonds flanking exon 4. A transgenic mouse is made in which the gene has both recombination sites. Expression of Flp in one tissue will specifically delete exon 4 and the domain it encodes, whereas expression of Cre in another tissue will specifically delete exons 2 and 3 and the domain they encode. The experiment is similar if regulatory regions or enhancers are flanked rather than parts of the coding region. The reactions leave recombination sites, either *loxP* or FRT, in the gene for potential insertions with other gene cassettes.

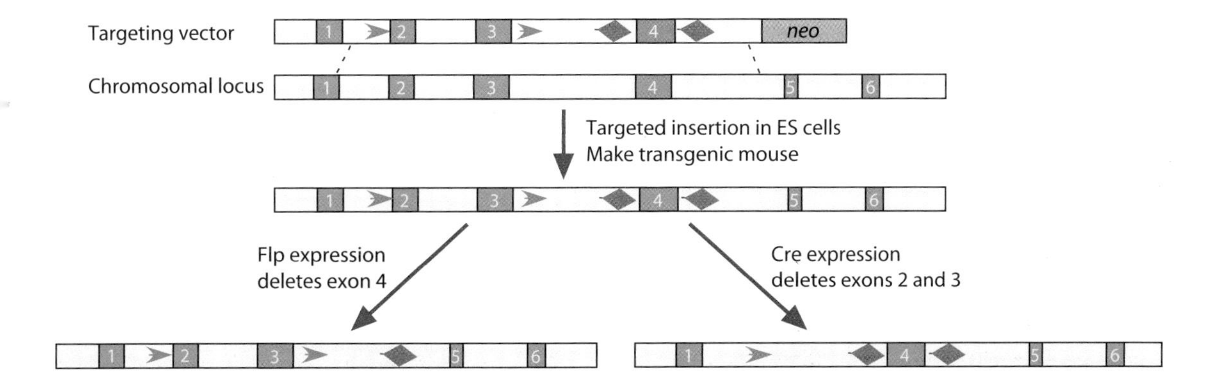

is used to delete the exons that encode one domain (e.g. a DNA-binding domain), whereas the Flp–FRT system is used to delete the exons encoding another domain (e.g. an interaction domain). Thus, by regulating expression of Cre and Flp in the mouse, the same gene replacement can yield two different mutant alleles. Another potential use involves the two recombinase systems simultaneously. By flanking both the donor gene and the target gene with one *loxP* and one FRT site and expressing either recombination enzyme, the donor gene can replace the target sequence, as in other gene knock-in experiments. A key benefit is that the *loxP* and FRT sites remain in the target gene where other mutated forms of the gene can be inserted by taking advantage of these recombination cassettes. This should make it simpler to introduce different types of knock-in mutants once the first one has been made.

SUMMARY

Reverse genetics allows specific types of mutations to be made and analyzed

The 'awesome power of yeast genetics', a phrase often associated with the late Ira Herskowitz, owes much of its truth to the ability to manipulate genes *in vitro* and target their reintroduction into the genome. It sometimes seems that it is possible to do nearly any experiment imagined using yeast genetics. Instead of relying on mutations occurring at random sites in the gene, as in Chapter 3, reverse genetics in yeast can produce the desired mutation and determine its effect. The power of targeting the mutated gene is that the investigator can replace the genomic copy rather than having the genomic copy still present in the background. Thus, the only version of the gene is the one that the geneticist has altered and introduced.

Our ability to manipulate genes and the genome in the mouse will probably never reach the level of sophistication in yeast, in part because the mouse is a more complex organism than yeast. Nonetheless, reverse genetics approaches dominate mouse genetics. Since the mouse is the most thoroughly researched model organism for understanding human genetic diseases, flexible and efficient gene targeting methods are regularly being developed. High-throughput genome-wide screens for gene disruptions in the mouse are under way, and it is certain that this technology will continue to evolve and improve. The 'knocking' is only going to get louder.

Chapter Capsule

Reverse genetics

Genetic analysis relies on having a mutant phenotype, but it may begin with a cloned gene rather than a mutant organism. In reverse genetic analysis, a cloned gene is used to identify a mutant phenotype for the gene and thereby determine the function of the gene. In the most powerful reverse genetics techniques, the cloned gene is disrupted or altered *in vitro* and used to produce a mutant phenotype by reintroducing the cloned donor gene into cells. By homologous recombination between the cloned donor gene and the chromosome target gene, specific mutations can be introduced into the genome and a mutant organism can be produced. The process of reverse genetics by targeted gene disruption or targeted gene replacement is summarized in Figure 6.17.

Procedure for gene replacement and disruption

1. A cloned gene is altered *in vitro* with a selectable marker

2. The altered gene is reintegrated into cells by homologous recombination and gene disruptions are selected

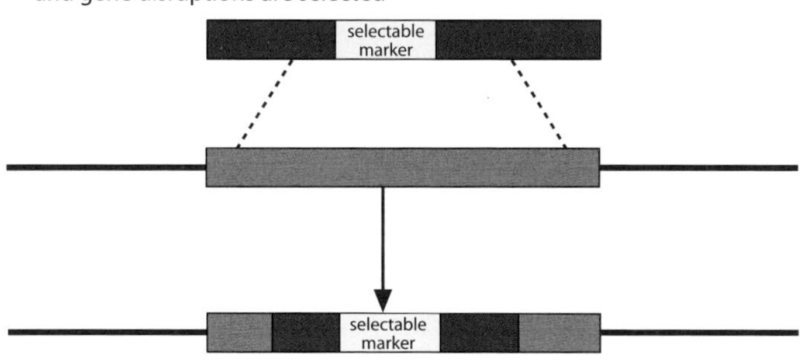

3. Cells with the disrupted gene are grown into organisms

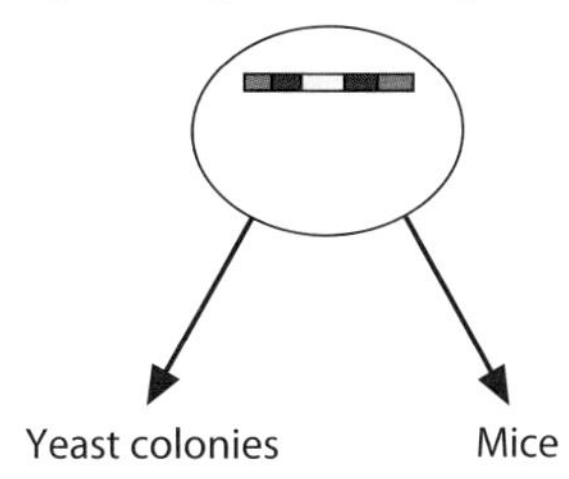

Figure 6.17 Chapter capsule. The general procedure for creating an organism with a gene replacement or gene disruption is outlined. Although the procedure could in principle be applied to many organisms, its greatest uses have been in yeast and in mice.

CHAPTER MAP

▶ **Unit 1 Genes and genomes**

▼ **Unit 2 Genes and mutants**

3 Identifying mutants

4 Classifying mutants

5 Connecting a phenotype to a DNA sequence

6 Finding mutant phenotypes for cloned genes

7 Genome-wide mutant screens

This chapter describes how a geneticist uses genomic sequence information to find mutant phenotypes for all cloned genes. The mutant and the cloned gene are crucial tools for the types of analysis described in subsequent chapters.

▶ **Unit 3 Gene activity**

▶ **Unit 4 Gene interactions**

7

CHAPTER 7

Genome-wide mutant screens

TOPIC SUMMARY

Genome-wide mutant screens attempt to determine the mutant phenotype for every gene in the genome, regardless of its possible biological function. Two methods are widely used for genome-wide screens:

- deletion of genes by targeted homologous recombination, used primarily in yeast
- disruption of a gene's expression by RNA interference, used in many multicellular organisms.

Genome-wide screens have identified unsuspected functions for many additional genes, even for biological processes that have been thoroughly studied by traditional gene-by-gene approaches. Genome-wide screens are rapidly becoming the method of choice for finding mutant phenotypes for many biological processes in many different organisms.

IN BRIEF

The availability of genome sequence information allows investigators to test every gene in the genome for its mutant phenotype.

INTRODUCTION

For decades, geneticists have searched genomes for mutants. The preceding chapters have described how mutations are induced and how the mutants are examined for phenotypic consequences. One by one, the genes are cloned. The cloned genes themselves can be used to generate gene disruptions and gene replacements. Little by little, gene by gene, genomes have yielded their functional information. The pace of this type of genetic analysis has sometimes been slow and the path is often meandering, but the scenery has usually made the route worthwhile.

However, in the past 10 years, as genomes have been sequenced, all that has begun to change. When the DNA sequence of a genome is determined, all of the genetic information that underlies every biological process has been revealed. The implications of this are enormous. Do we want to study complex developmental programs, simple or sophisticated behaviors, responses to pathogens, or other complicated biological questions? All of the genetic information is available, waiting for us to use it. With hundreds of genomes being sequenced each year, genetics has begun to move from being about *gathering* information to being about *processing* information.

Despite these changes, the established tools of genetic analysis such as mutant phenotypes, expression assays, and interaction experiments remain the tools through which we make sense of the information that genomics provides. The power of genome sequences has turbocharged genetic analysis and allowed us to examine many biological processes in ways that we could not have done previously. But the fundamental principles of the analysis do not change.

In this chapter, we introduce the concept of genome-wide genetic analysis. Genome-wide analysis tests every gene in a sequenced genome for its effect on a biological process. In later chapters, we will describe genome-wide analysis of gene expression and gene interactions and lay them side by side with the more traditional gene-by-gene approaches. In this chapter, having just described gene-by-gene mutant analysis, we describe genome-wide analysis of mutant phenotypes. Some of the technology that has been used in these genome-wide mutant screens has been innovative, even revolutionary.

The use of genome-wide analysis for mutant screens is still in its earliest stages. Yet the results, even at our current preliminary stage, have already been remarkable. In this chapter, we will describe some of the techniques being used to perform genome-wide mutant screens and some of the results that have arisen from these mutant screens. The methods can be used for any biological process, and many new experimental options are now open to us. Genome-wide mutant analysis in this chapter can identify every gene in the genome using targeted mutations. As summarized in Figure 7.1, the task is to identify the genes involved in a biological process; the tool is still the mutant phenotype, the same as we used in Chapters 3 and 6. The technology has changed but the principle has not: **Find a mutant**.

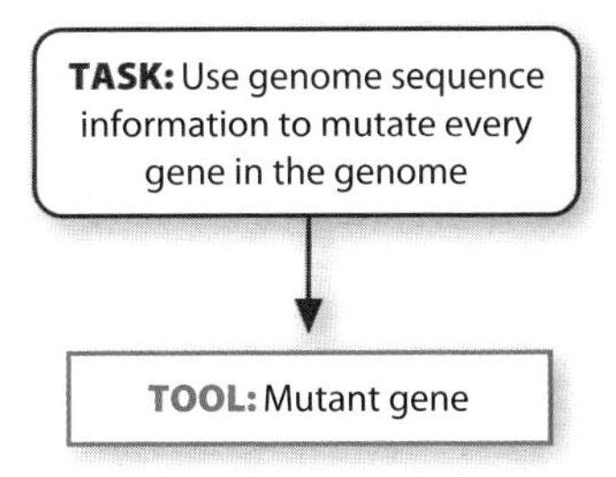

Figure 7.1 This chapter describes how genomic information is used to produce mutations in every gene in the genome.

7.1 Genome-wide mutant screens: an overview

Identifying mutations in every gene in a genome is an enormous task. The yeast *S. cerevisiae* has an estimated 6200 protein-coding genes, the fruit fly *D. melanogaster* about 15 000, the nematode *C. elegans* about 18 500, *Arabidopsis* about 27 000, and the mouse about 30 000. Generations of geneticists have been identifying mutations in these organisms, and only a fraction of genes in any of them had been mutated—at least, until genome-wide screens were developed. The development of genome-wide mutant screens has identified new genes involved in biological processes and has made it possible to consider questions that were unapproachable even 10 years ago. Only a few screens that are truly 'genome-wide' have been completed, but many more are in progress.

Before beginning to describe how genome-wide screens are carried out and what we have learned from them so far, we need to define what they are. The term genome-wide is commonly used to describe these procedures but the term is not very well defined. A genome-wide screen is one in which the investigators use genome sequence information to design specific molecular tools for the analysis of all or nearly all of the identified genes. Our definition lays out the basic requirements of a genome-wide screen: a sequenced genome for which most of the genes have been identified, and molecular probes that are specific to each gene.

Genome-wide mutant screens have two significant advantages compared with standard genetic screens using randomly induced mutations. The first is that the gene has already been cloned before the mutant phenotype has been identified. The process of connecting a mutant phenotype to a DNA sequence, as we described in Chapter 5, now happens immediately. The second is that *all* of the genes are being tested at the same frequency without regard to a suspected physiological function. This is in contrast to what happens in traditional genetic screens or reverse genetic analysis. Neither of these other methods has the capability to test *all* of the genes.

A comparison of traditional forward genetic screens, reverse genetic screens, and genome-wide screens is shown in Figure 7.2. A forward genetic screen begins with a single mutant organism, as shown at the top of the figure, a reverse genetic analysis begins with a single mutated gene, as shown in the middle, and a genome-wide screen, as shown as the bottom, begins with mutants in every cloned gene.

We previously introduced the concept of **saturation**—identifying all of the genes that can be found by a particular mutant screen. We described the process of saturation in terms of an MP3 player with a random play feature (Text Box 4.2). By playing songs at random and counting how often each song has been played, the listener can estimate how many songs are in the music library, even before all songs have been heard. However, this is only an estimate. The tally of how often a song has been played will follow a probability distribution, so there is no guarantee that all songs have been heard. By chance, some songs will be heard often, while others will be heard rarely, if at all.

Likewise, in a traditional genetic screen based on randomly induced mutations, the frequency at which the same gene is mutated allows the investigator to

DEFINITION

genome-wide: A genome-wide screen is one in which the investigators use genome sequence information to design specific molecular tools for the analysis of all or nearly all of the identified genes.

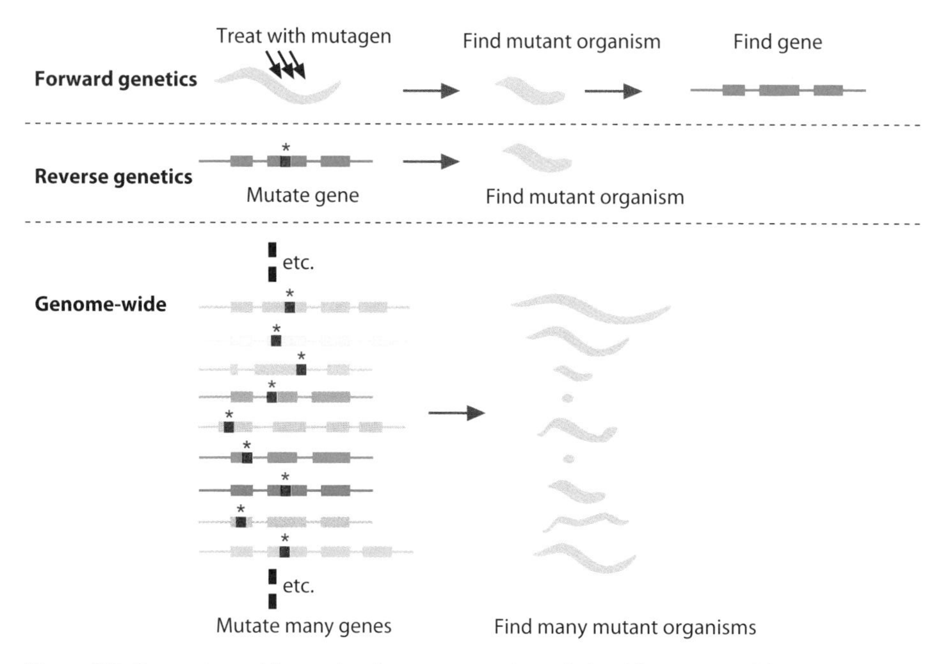

Figure 7.2 Comparison of forward and reverse genetic analysis with genome-wide mutant screens. In forward genetics, as described in Chapters 3 and 4, a wild-type organism is treated with a mutagen, mutant organisms are found, and the gene is cloned. Neither the mutant phenotype nor the gene is known in advance. In reverse genetics, as described in Chapter 6, the gene is known and is mutated *in vitro*. The mutant gene is introduced back into the organism and produces a mutant phenotype. In genome-wide screens, all genes are altered *in vitro*. The mutated genes are introduced back into organisms and the phenotype is observed.

infer how many genes could be found. We illustrate this in Figure 7.3 with a hypothetical process affected by the four genes shown. Among the first thousand mutant samples tested, one of the genes (the blue one) is mutated. In the second thousand samples, a second gene (the pink one) is mutated, and so on for 5000 samples. But again, the estimate of the number of genes is based on statistical methods and assumptions about the underlying probability distribution, so certain genes will almost certainly be missed. In Figure 7.3, one gene is found three times, one gene is found twice, and one gene is found once; but one gene (the one in yellow) is not mutated at all and would be missed.

A genome-wide screen tests every gene but does not use statistical sampling to do so. It mutates every gene based on the knowledge of its DNA sequence. In our analogy with an MP3 player, a genome-wide screen is equivalent to playing each song in the library in order. No song is heard more than once, but all songs will be heard. Similarly, in a genome-wide screen, no gene will be mutated more than once, but all genes will be mutated. This is shown at the bottom of Figure 7.3. Each gene is mutated and the mutants are divided into pools, which are then tested for their mutant phenotype.

Genome-wide screens can identify new genes even in well-described biological processes

The confidence that every gene has been mutated in a genome-wide screen presents a major advantage, which can best be illustrated with an actual example. *S. cerevisiae*, like all organisms, repairs damage to its DNA induced by environmental agents. The process of DNA repair has been studied very thoroughly and cleverly by traditional mutant analysis, and many yeast mutants that cannot repair some type of DNA damage are known. All of the genes identified by traditional genetic analysis are cloned, and orthologs of the cloned genes have been identified in many other organisms. Many of the orthologs have themselves been tested for a role in DNA repair, either by mutant analysis or by biochemical assays. Although there is always more to be learned, DNA repair can be considered a thoroughly studied biological process.

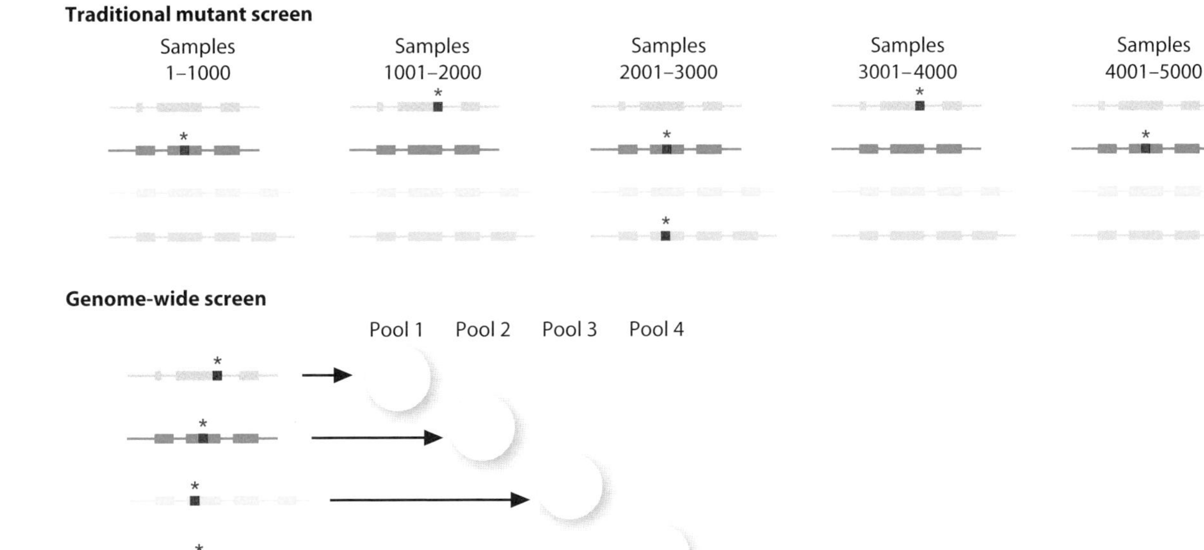

Figure 7.3 Comparison of saturation methods in traditional mutant screens and genome-wide screens. Four different genes are involved in this hypothetical process, as indicated by the four different colors on the left. In the traditional mutant screen, the first thousand samples yield a mutant in the blue gene, the second thousand samples yield one in the pink gene, and so on. After 5000 samples, the blue gene has been mutated three times, the pink gene twice, and the green gene once; the yellow gene has not been mutated at all. Thus, the role of the yellow gene in the process has not been detected and it remains unknown. In a genome-wide screen, all genes are mutated by directed methods first, without regard to what process they may affect. Each gene is tested in a pool. Because all genes are mutated, the role of the yellow gene in this process would be detected. Furthermore, no gene is mutated more than once.

However, in a recent example of a genome-wide screen, *every* gene in the yeast genome was individually mutated and *every* mutant was tested for its effect on DNA repair. Most of the mutants that were found to affect DNA repair were alleles of genes that had already been identified by traditional gene-by-gene analysis. (It is reassuring to know that the genome-wide screen also finds the genes that have been so thoroughly studied.) But new genes involved in DNA repair were also found. Specifically, mutants in three genes that had never been identified by traditional gene-by-gene analysis were found to have observable defects in DNA repair. Furthermore, each of these genes has orthologs in other species including humans, suggesting that they are fundamental to the process. This is an exciting example of the power of genome-wide screens. By being able to test mutants in all possible genes, we learned that we did not know all the parts of a process that has been thoroughly analyzed in many different species. Other examples will be given throughout the chapter.

Although the steps of a genome-wide screen differ in each organism and each screen, genome-wide analysis can be summarized as follows, and is shown in flowchart form in Figure 7.4.

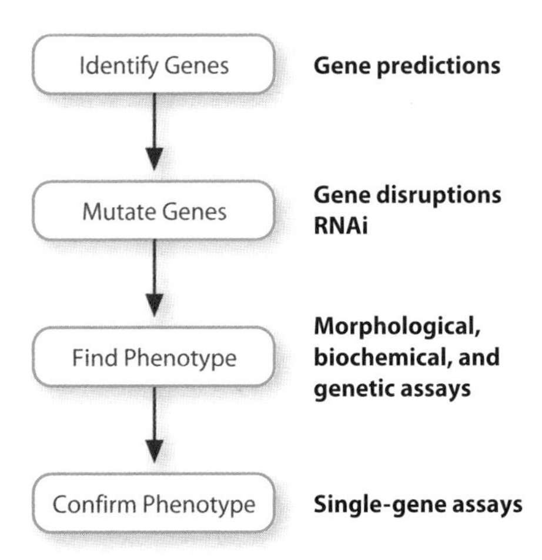

Figure 7.4 Flowchart of a typical genome-wide screen. The first step is to identify all genes in the genome, using gene prediction methods. Then genes are mutated, either by gene disruptions or by RNAi. Thirdly, phenotypes are found by a range of different assays. Finally, the role of single genes involved in the process is confirmed and refined.

1. **Identify** all of the genes to be mutated.
   For the largest scale of genome-wide screens, this set includes all of the genes in the genome. However, this large goal is often subdivided into more manageable smaller goals. For example, the set of genes to be mutated could be all genes expressed in a particular pattern, such as those expressed in lymphomas, or all of the genes of a particular functional class such as G-protein-coupled receptors.
2. **Develop** methods for disrupting or perturbing genes.
   Most of the chapter will be spent describing these methods, including targeted gene deletions in yeast and RNAi screens in many organisms.
3. **Examine** the mutants for phenotypic effects.
   The team of investigators may want to look for all mutants that have a particular phenotype or they may want to look for as many phenotypes as possible for each mutant.
4. **Confirm** and follow-up on the mutants that are found, often with gene-by-gene approaches.

We will consider each of these in turn, with most of our attention on the first three steps. Confirming the mutant phenotype, the fourth step, is similar for genome-wide screens and for traditional gene-by-gene methods, and so it is addressed in other chapters in the book.

7.2 Identifying the genes to be mutated

It may seem obvious, but the first requirement for a genome-wide mutant screen is to have a genome that has been sequenced. However, our definition requires going beyond a sequenced genome—the genes themselves must also be identified. In the jargon of genomics, the genome must be **well annotated**. This means that the coding regions of genes must be identified and the corresponding protein sequences must be inferred. These protein sequences must have been compared with other sequences (by BLAST or a related tool) to infer functions based on similarity. Furthermore, the locations of exon–intron boundaries must be identified or at least predicted by some reliable method.

Annotation of a genome eventually goes well beyond these basic parameters to include the location of repeat elements, inferences about regulatory regions, comparisons with the linkage arrangement in related species, and so on. However, the annotation needed for a genome-wide screen is more fundamental. Where in the genome are the genes? What is their likely structure and what can be inferred about their function?

Predicting the structure and location of genes from nucleotide sequences, sometimes referred to as **gene-finding**, can be relatively simple or quite complicated depending on the gene and the species. Some of the computational and experimental methods that are used for gene prediction are described in Text Box 7.1. The structures of predicted genes are continually being refined, and these gene predictions form the basis of genome-wide screens.

What set of the annotated genes will be tested in a genome-wide screen? The ultimate, if laborious, goal is to test all genes in the genome without regard to location, expression, and hypothesized functions. Such a project will certainly provide the most complete information, and has been achieved for several prokaryotes, *S. cerevisiae*, and some other organisms. However, for a variety of reasons, it is often necessary or helpful to test a **subset** of the genes initially. The subset to be tested can be defined by either arbitrary or functional means. We have summarized some of these in Figure 7.5.

For example, some genome-wide screens in *C. elegans* have tested all genes on each chromosome without regard to their postulated functions, following the same strategy for subdividing the genome as traditional genetic screens by relying on balancer chromosomes, as described in Chapter 3. Since functional criteria are not the reasons for inclusion in the screen, the mutant could have any imaginable mutant phenotype or even a wild-type phenotype. As we will describe later in the chapter, recognizing mutant phenotypes is one challenge associated with these methods. In Figure 7.5, these genes are depicted on the same chromosome.

Many genome-wide screens have tested subsets of genes based on some functional criteria. For instance, microarray data (described in Chapter 8) often

BOX 7.1 Gene predictions

For many molecular biologists who learned their craft before automated sequencing and versatile molecular biology software were in widespread use, predicting the location and structure of a gene was their first genomics problem. Raw DNA sequencing data has no landmarks, no capital and lower case letters to distinguish exons from introns, no obvious starts and stops. All these characteristics are supplied by the investigator. The same methods that were used more than decade ago to find and predict the structure of a gene are still the foundation of most contemporary gene prediction programs, although the methods are automated and are far more sophisticated and accurate than an individual working on an X-ray sequence film over a light box with pencil and paper.

As a genome is sequenced, the investigators scan the sequence with one or more gene recognition programs to identify the coding regions of likely genes. The genome sequence data are annotated to show these predicted genes, many of which will initially have no confirmation other than the computer prediction. The computer predictions for gene recognition, also called gene predictions and gene finding, are extremely good (and improving), but they remain far from perfect. This is one reason why genome annotation projects will use more than one gene prediction program, and that genome annotation is never complete. It is also one reason why investigators should understand the source of gene prediction programs and verify the predictions experimentally.

The prediction of protein-coding regions begins from one of two different perspectives, based on either experimental data or *ab initio* predictions about the nature of genes; most programs end up using a hybrid of the two different approaches. An *ab initio* approach uses the known features of genes as a set of 'first principles' of gene structure, as described below. In this text box, we describe approaches to finding protein-coding regions in the genome. Identification of regulatory regions from raw sequence data is a much more complex problem and is beyond the scope of this book.

Experimental data: ESTs and cDNA

Experimental data that are particularly used for gene predictions are expressed sequence tags (ESTs) and cDNA clones. Both of these are derived from RNA, and so they represent expressed regions of the genome. Let us consider each of these in turn.

ESTs are randomly generated and sequenced cDNA fragments, which may number in the hundreds of thousands. That is, RNA is isolated from the tissue or individual, reverse transcribed in bulk to make cDNA fragments, and sequenced. Each EST represents a portion of a mature transcript, and so a comparison of the EST sequence with the genomic sequence identifies part of an mRNA. This comparison allows the investigator to find the coding regions in the genome. Just as significantly, the ESTs accurately identify exon/intron splice junctions and untranslated regions at the beginning and ends of transcripts. For example, the genomic sequence can be aligned with a large database of ESTs that come from different parts of many different transcripts—possibly every transcript expressed in a cell type if enough ESTs are generated. Some of these ESTs will overlap in sequence with each other. From piecing together these overlaps, the structure of the transcribed gene can be inferred. Text Box Figure 7.1 uses a screen capture from Wormbase with an alignment of ESTs in a region of the genome of *C. elegans* showing how the ESTs are used to infer a gene structure. ESTs correspond to transcribed regions of the genome, which, for a mammalian genome, may be only a few percent of the total genome sequence. Thus, they provide the locations of genes and information about the structure of genes.

Further refinement and experimental confirmation of EST assembly can be done by producing a **cDNA clone**. The differences between cDNA clones and ESTs arise from the scale. An individual investigator working on one gene will look for a full-length cDNA by carefully extracting mRNA from cells or tissues that are known or believed to be transcribing the gene. A full-length cDNA provides the complete sequence of the transcript, showing all of the intron/exon splice boundaries as well as the untranslated sequences at each end of the transcript. From this, the investigator can infer the amino acid sequence and size of the predicted protein and can identify splice variants if more than one cDNA is found; the full-length cDNA is also used for many other expression-based experiments. However, in a genome with tens of thousands of genes, it is extremely laborious to isolate a full-length cDNA for each gene. In contrast, ESTs are isolated from all transcripts expressed at the same time, with no attempt to find full-length transcripts for any gene, and so the isolation is fast. The assembly of the ESTs allows many of the same inferences as a full-length cDNA but does not conclusively prove that a predicted gene structure exists since portions of the transcript may have been missed in the process.

A recent addition to the process of gene-finding with experimental data is to use a type of microarray known as a **tiling array**. Tiling arrays are described in Chapter 8, but the basic technique involves testing every part of the genome for the presence of a corresponding transcript. This involves small sequences from each DNA strand, and is only possible because microarrays can

A

```
catatcaattgacctctaaacttggtatataaaagcaacgtgtctgataaagaaaaaccg
acgagaaaacaacATGAAATCTATTATCATTTTATCAGTTTTCTTAGTTTGTTCCGCGTT
GGCTGCCAGTTCTGATAAGGATACTGAATCAAAAGTGGCCAGATATTTGAAATCCTGTGG
AGTTATCACTTCATTGAAGACTACTAAATGTGTTAAGgtgggttttataaaattgttcaa
gcaggcttcaattgtcatgattgctcatgcagttcttattttggtggaatttcaaacatg
gagatttatcacagtagtcctttttttgttgtacttgctgcatttcgaactatttccgtca
ttttaataaaaaaaacacttttagAGAAGCGATGAAGTAGATAAGGAGCTCAAAGCTCTC
GTAAAGAGCAAAGACAAGGATCTCTCAAAAGTCAAGGCTTCGTGCAAAGACACCATGgtg
ggtaacatcagtgatattaaaaacttctcaaatgtgtttcagGAATGCATTAAAGAGCTG
AAATGTAAACCACTCGAAGAAGATCACAAGGAGACTCTCGACTTCTGTGGTCGTGCTCTT
TTCCAGGACGAGTTCAAGGATTGCAGCAAGAAATTGCAGGCATTGAAGAAAAGTGATAAG
GATGCTGCAAAGTGTTTGGAAGATTTTAAATCAGAGACAAAGAAGTCGACAAAGGACACT
TGCACATTTTTGAAGGGCTCGAAAGATTGTATCAAGGCTCATATTAAGAAAGAGTGCGGG
GATGATAAGTTGCAGGGATGGCAGAAGgtaagcggatgatgtcttagaaatcggtttagg
cttcggcgcaggccttaaacttagatgcttaggcttaggcttagagagcctcaaactttc
attttcatttccaccaacttttctaaatttctatttttcagTTTGCTCAAGACTCGTTCA
AAGAGAACGAATGCGAGAAGGCAATTGGAGCAAAATGGAACTAActtcattaaatgttct
tcgatttcttgttccttgttacagtaaccctgcaaaactgcaataattaatttaaaagtc
tttaaaattccggttgagtacccattcctccctcctacattcttcatttcttcaaagggt
ttctctccaggtaccaataaaatttta
```

B

```
aaaaccgacg agaaaacaac ATGAAATCTA TTATCATTTT ATCAGTTTTC TTAGTTTGTT CCGCGTTGGC TGCCAGTTCT

GATAAGGATA CTGAATCAAA AGTGGCCAGA TATTTGAAAT CCTGTGGAGT TATCACTTCA TTGAAGACTA CTAAATGTGT
TAAGgtgggt tttataaaat tgttcaagca ggcttcaatt gtcatgattg ctcatgcagt tcttattttg gtggaatttc
aaacatggag atttatcaca gtagtccttt tttgttgtac ttgctgcatt tcgaactatt tccgtcattt taataaaaaa
aacactttta gAGAAGCGAT GAAGTAGATA AGGAGCTCAA AGCTCTCGTA AAGAGCAAAG ACAAGGATCT CTCAAAAGTC
AAGGCTTCGT GCAAAGACAC CATGgtgggt aacatcagtg atattaaaaa cttctcaaat gtgtttcagG AATGCATTAA
AGAGCTGAAA TGTAAACCAC TCGAAGAAGA TCACAAGGAG ACTCTCGACT TCTGTGGTCG TGCTCTTTTC CAGGACGAGT
TCAAGGATTG CAGCAAGAAA TTGCAGGCAT TGAAGAAAAG TGATAAGGAT GCTGCAAAGT GTTTGGAAGA TTTTAAATCA
GAGACAAAGA AGTCGACAAA GGACACTTGC ACATTTTTGA AGGGCTCGAA AGATTGTATC AAGGCTCATA TTAAGAAAGA
GTGCGGGGAT GATAAGTTGC AGGGATGGCA GAAGgtaagc ggatgatgtc ttagaaatcg gtttaggctt cggcgcaggc
cttaaactta gatgcttagg cttaggctta gagagcctca aactttcatt ttcatttcca ccaacttttc taaatttcta
tttttcagTT TGCTCAAGAC TCGTTCAAAG AGAACGAATG CGAGAAGGCA ATTGGAGCAA AATGGAACTA Acttcattaa
atgttcttcg atttcttgtt ccttgttaca gtaaccctgc aaaactgcaa taattaattt aaaagtcttt aaaattccgg
ttgagtaccc attcctccct
```

C

```
aaaaccgacg agaaaacaac ATGAAATCTA TTATCATTTT ATCAGTTTTC TTAGTTTGTT CCGCGTTGGC TGCCAGTTCT
GATAAGGATA CTGAATCAAA AGTGGCCAGA TATTTGAAAT CCTGTGGAGT TATCACTTCA TTGAAGACTA CTAAATGTGT
TAAGAGAAGC GATGAAGTAG ATAAGGAGCT CAAAGCTCTC GTAAAGAGCA AAGACAAGGA TCTCTCAAAA GTCAAGGCTT
CGTGCAAAGA CACCATGGAA TGCATTAAAG AGCTGAAATG TAAACCACTC GAAGAAGATC ACAAGGAGAC TCTCGACTTC
TGTGGTCGTG CTCTTTTCCA GGACGAGTTC AAGGATTGCA GCAAGAAATT GCAGGCATTG AAGAAAAGTG ATAAGGATGC
TGCAAAGTGT TTGGAAGATT TTAAATCAGA GACAAAGAAG TCGACAAAGG ACACTTGCAC ATTTTTGAAG GGCTCGAAAG
ATTGTATCAA GGCTCATATT AAGAAAGAGT GCGGGGATGA TAAGTTGCAG GGATGGCAGA AGTTTGCTCA AGACTCGTTC
AAAGAGAACG AATGCGAGAA GGCAATTGGA GCAAAATGGA ACTAActtca ttaaatgttc ttcgatttct tgttccttgt
tacagtaacc ctgcaaaact gcaataatta atttaaaagt ctttaaaatt ccggttgagt acccattcct ccct
```

Figure B7.2 Sequence annotation and gene predictions. The same gene as shown in Text Box Figure 7.1 is used here, but the orientation has been reversed to correspond to left to right reading of the sequences. (A) The genomic sequence for this region of the genome, with ORFs shown in capital letters. The ATG followed by an ORF is shown in red. (B) The second sequence highlights the predicted exons in the genomic sequence, with alternating exons in yellow and orange colors for clarity. The introns are in lower case and are not highlighted. The introns begin with GT and end with AG. The final exon ends with a TAA stop codon. (C) The third sequence is the cDNA sequence without the untranslated regions and with alternating yellow and orange exons.

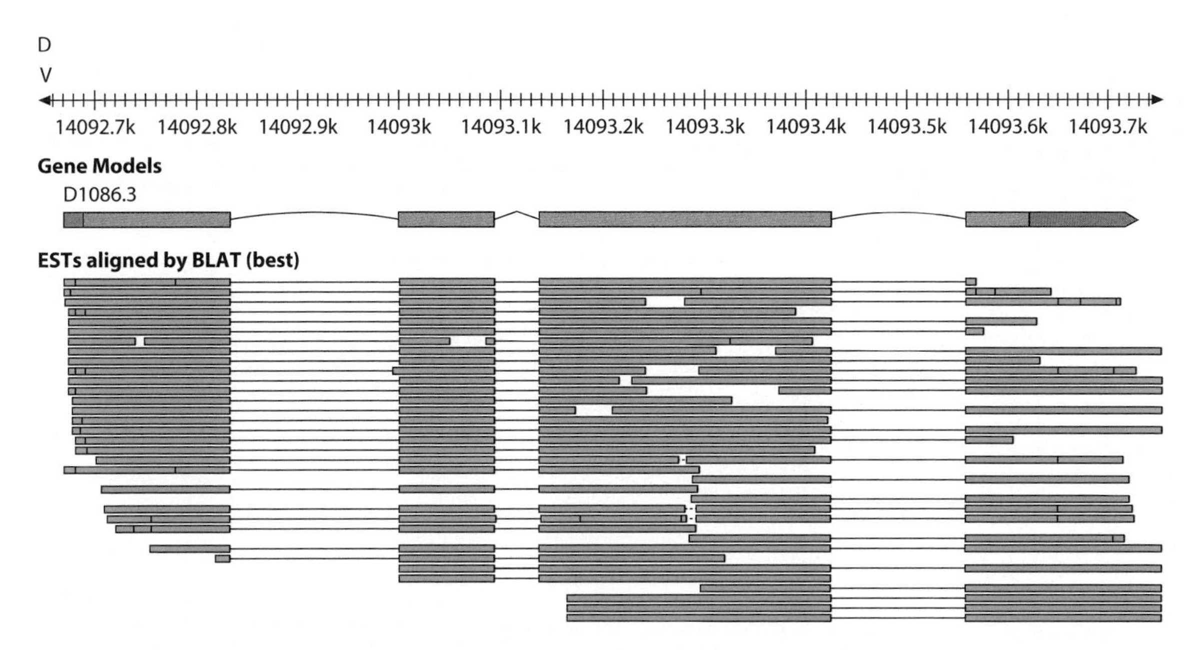

Figure B7.2 (Continued). (D) All of the ESTs associated with this gene. Although the gene is fairly small, none of the ESTs includes both the complete 5'-UTR and 3'-UTR. One EST differs from the others in having a few extra bases at the start of exon 2 and a few less at the end of exon 1; other than this EST, the splicing junctions are identical.

predicted to be exons *are* in fact exons. In about 75 percent of cases, the boundaries of the individual exon are also accurately predicted, and the exon itself is found perfectly.

The programs are good but not perfect at predicting intron/exon splices, and thus the specificity is lower than the sensitivity. That is, although it typically predicts the individual exons correctly, GENESCAN accurately predicts about half of mammalian gene structures, including splicing patterns. Given the number of exons and the size of introns in mammalian genes, this is an impressive achievement, and the numbers are much better for less complex genomes.

With only about half of the gene structures predicted correctly, however, an important refinement comes from ESTs and comparative genomics. Small exons can be overlooked or the exact splice boundary may be predicted incorrectly, which changes the predicted amino acid sequence. Therefore, an investigator working with an individual gene is still well advised to isolate a full-length cDNA whenever possible to confirm the prediction. With comparative genomics also used to inform the prediction, standard splicing patterns can usually be predicted accurately for evolutionarily conserved proteins.

The greatest current challenge in predicting gene structures comes from **alternative splicing**. The same primary transcript can be spliced different ways, and it is estimated that as many as two-thirds of mammalian transcripts may have more than one splicing pattern. *Ab initio* predictions use a set of common principles to predict gene structures, but few of the 'common principles' of alternative splicing are known.

Ab initio gene prediction is very powerful, but its strength relies on the set of gene structures used as the training set. That is, the statistical basis for the prediction is derived from the structures of known genes for that species (the training set), and the more genes that can be used to compile the statistical profile, the more accurate the prediction becomes. For this reason, *ab initio* approaches are inherently conservative in their predictions. It is also important that the training set of genes is drawn from the same species as, or a species closely related to, the genome being annotated, since many aspects of the structure of genes vary among species. With this proviso, *ab initio* approaches are currently the most powerful when used with species in which other experimental data are also available. As a practical matter, some of the most widely used programs such as Twinscan and the programs used in the Ensembl genome database combine *ab initio* predictions with comparative genomics and EST assembly to produce the most accurate gene predictions possible.

Further reading

Jones, S.J. (2006). Prediction of genomic functional elements. *Annual Review of Genomics and Human Genetics*, **7**, 315–37.

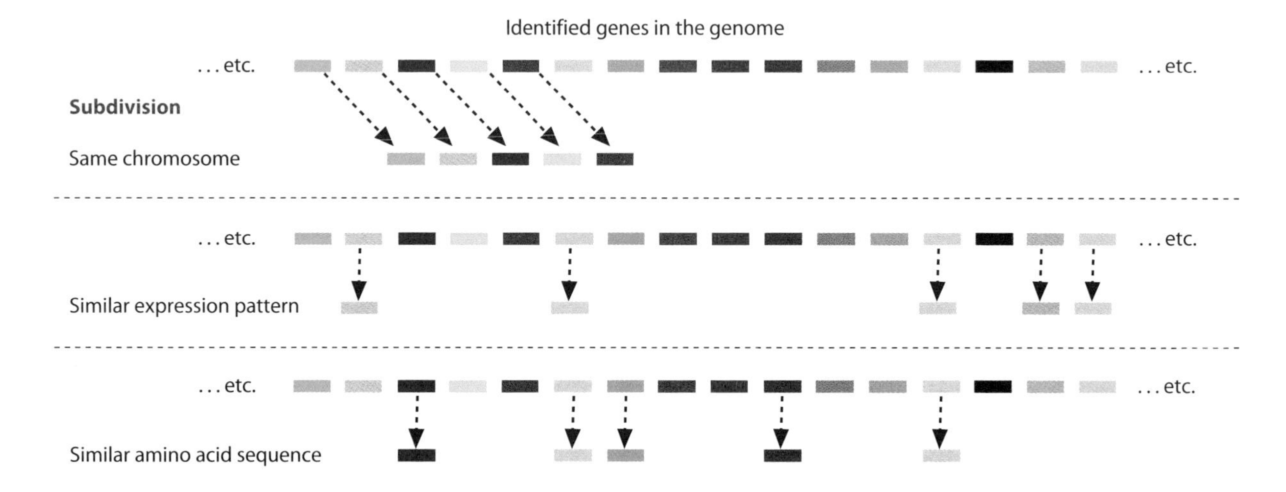

Figure 7.5 Some strategies for subdividing the genes in a genome-wide screen. Although the goal is that all genes will eventually be tested, genome-wide screens are often subdivided into groups of related genes for practical reasons. Three rationales for subdividing genes are shown here. The top line in each panel represents the genes in this genome, each shown in a different color. The order of the genes represents their physical order in the genome. The intensity of the colors indicates similarity in an expression pattern. The colors represent sequence similarity and inferred similarity in function. The first rationale is to test the genes on one particular chromosome or in one region of the genome, shown here as the first five genes. These five genes may have different expression patterns and no sequence similarity; the rationale for grouping them is simply location. A second rationale is to group the genes based on expression pattern, here shown as the five genes in pastel colors. These genes may all be expressed in the same tissue, for example. A third rationale is to group the genes based on amino acid sequence similarity, shown here as the five shades of blue. These genes may all encode zinc-finger proteins, for example.

provide information on gene expression that comprises a set of functional criteria for a genome-wide screen: the genes to be tested may be expressed in embryonic stem cells, in the germline during sexual maturation, or in particular cancer cell types. (These are the genes shown with similar expression patterns in the middle of Figure 7.5.) Many other examples of expression-based subsets have also been used, including genes whose expression is induced or repressed by drugs, hormones, or an environmental stimulus. The interplay between gene expression and mutation, which was described for traditional gene-by-gene cloning in Chapter 5, becomes all the more important in genome-wide approaches since the expression results confirm and extend the results of the mutant screen, and vice versa.

Another characteristic that has been used to test subsets of genes is **homology** or **postulated function** based on sequence comparisons. As genomes are sequenced or annotated, the sequences are searched for similarity with all other DNA sequences; for protein-coding regions, these tests are based on inferred amino acid sequences. Such homology searches are an important part of the annotation process to refine predictions about the functions of genes. One result of genome-wide homology comparisons is to provide a complete set of proteins in one family, such as all the G-protein-coupled receptors or all the serine proteases. Some gene families in multicellular organisms have dozens or hundreds of members, so the process of testing each one approaches being a genome-wide screen. The standards for inclusion among the genes to be tested can be even more loosely defined, such as ones encoding proteins with a predicted structure or domain—all C2H2 zinc-finger proteins for example, as shown in the bottom subdivision in Figure 7.5. Homology predictions are also combined with expression data to produce the subset of genes to be tested, such as all coiled-coil proteins expressed in the germline.

Although some of these procedures may not seem like genome-wide screens in that not all genes are being tested, it often does make sense to limit the scope of the screen. This is not so different from searching for mislaid car keys, in which it makes sense to focus the search on the locations where they are most likely to have been left. One important difference is that a search for lost keys has one very specific goal and will stop when that goal is achieved. Genome-wide screens, whether limited to certain subsets or not, have much wider goals, closer perhaps to producing an inventory of all objects whether lost or not, whether keys or not. All genome-wide screens begin somewhere but have the same ultimate destination: **to test mutations in all of the genes in the genome**.

7.3 Disrupting and perturbing genes

The second necessary part of a genome-wide mutant screen is to develop a method or methods of producing the mutant phenotypes. As noted in our definition of genome-wide screens, these methods rely on molecular tools that are developed from sequence information for the genes. Broadly speaking, two different approaches to producing genome-wide mutant phenotypes have been developed: gene disruptions and RNA interference. In *S. cerevisiae*, genes can be specifically targeted for disruption or deletion by homologous recombination, as described in Chapter 6. In the most extensive genome-wide screen that has been done, targeted deletions were screened for nearly all of the 6130 genes in the yeast genome. The second approach does not affect the structure of the gene itself but instead disrupts or perturbs its expression. This method, known as RNA interference (RNAi), has been widely applied to many organisms and has very broad applications and significance; in fact, the significance of RNAi was recognized by the award of the 2006 Nobel Prize for Physiology and Medicine to Andy Fire and Craig Mello. We begin by describing the gene disruption methods used in yeast.

Genome-wide mutant screens in yeast have been performed by targeted gene disruptions

More than 96 percent of the 6130 predicted genes in yeast have been disrupted by targeting insertions into the genome by homologous recombination. These strains are maintained as diploids, either homozygotes or heterozygotes; because 1159 genes are essential for growth, these are maintained as heterozygous diploids, whereas haploid deletion strains have been made for each of the 4757 non-essential genes. The deletion strain collections are commercially available and are widely used in yeast genetics.

The disruptions were made by inserting a kanamycin-resistance (kan^R) gene into each target gene, disrupting the coding region of the target gene with an easily selected marker. A different disruption cassette was made for each of the 6100 targeted genes. The process of targeted integration is detailed in Figures 7.6 and 7.7. Figure 7.6 shows the first step in the process: the use of PCR to amplify the kan^R gene, but with specially designed primers to produce an insertion cassette. The primers to amplify the kan^R gene were long—approximately 74 bp each—and included both gene-specific and gene-non-specific components. This can be illustrated by a careful examination of the right (or downstream) primer in Figure 7.6; the left (or upstream) primer has a similar structure in reverse. The first part of the primer corresponds to 18 nucleotides from the end of the kan^R gene and is used to amplify the selectable marker gene. The next part is a sequence of 20 nucleotides referred to as the **molecular bar code**. This sequence is different for each gene knockout. We will describe the molecular bar code in detail below, but for now we will focus on the general structure of the primer sequences.

Immediately downstream of the molecular bar code is a gene-non-specific sequence, which is the same for each of the 6100 genes. This sequence is not found in the yeast genome. This universal sequence allows a single primer pair to be used to amplify the molecular bar code regardless of the gene. Downstream of that is a gene-specific sequence of 18 nucleotides corresponding to a sequence in the gene to be disrupted. This sequence serves to target the insertion to a specific gene. Note again that the same structure is shown in reverse in the upstream primer sequence: the region of specific yeast homology is first, followed in order by a gene-non-specific sequence, the upstream molecular bar code sequence, and a portion of the kan^R gene. These two primers used together will amplify the kan^R gene with a long tail at each end as an insertion cassette, parts of which are specific to each gene in the yeast genome and parts of which are general to all 6100 insertion cassettes.

Molecular bar codes

The most unusual and cleverest aspect of this collection of insertion cassettes is the inclusion of the molecular bar code sequence, so this needs some further explanation. Additional information about how molecular bar codes have been used in other genome-wide screens

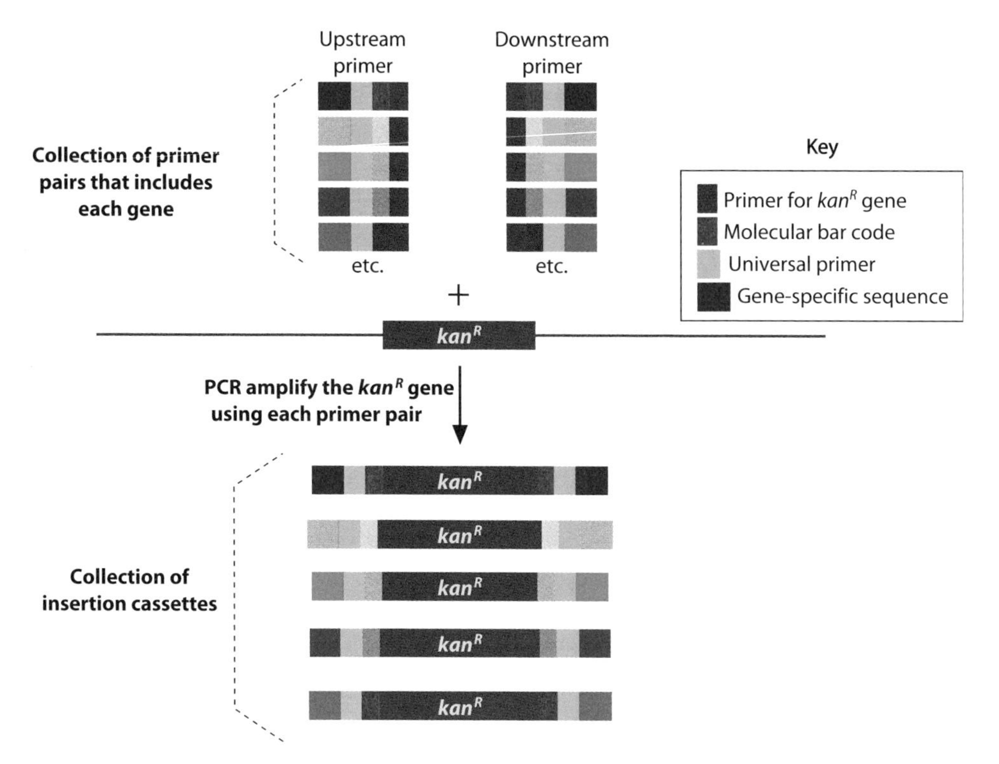

Figure 7.6 Schematic diagram of the PCR primers and insertion cassettes used in genome-wide gene disruption in yeast. A collection of PCR primer pairs is made, one for each gene in the genome. Five pairs of primers are shown at the top. Each primer is 70 bases long and has four sequence components, two of which are specific to that gene and two of which are common to all primers. The common components are shown in the same color in each primer, whereas the gene-specific components are shown as different shades of the same color. The color key for the primer components is shown at the top right. Each primer pair has sequences corresponding to the target gene in the genome, here shown in shades of blue. The gene-specific sequence is indexed with a specific defined sequence known as the molecular bar code, shown in shades of red. Note that each gene-specific sequence in blue is associated with one specific bar code in red. By knowing the bar code sequence, the investigator can determine which gene has been disrupted. The components that are common to all primer pairs are the sequences to amplify the *kan*^{R} gene (in purple) and the sequence used to amplify the molecular bar code (in green). By using an adjacent universal primer, all bar codes could be amplified regardless of their own sequence. Each primer pair is used to amplify the *kan*^{R} gene, resulting in a collection of insertion cassettes specific to each gene in the genome. Although five cassettes are shown here, the actual genome-wide screen had nearly 6000 different insertion cassettes, one for each gene in the yeast genome.

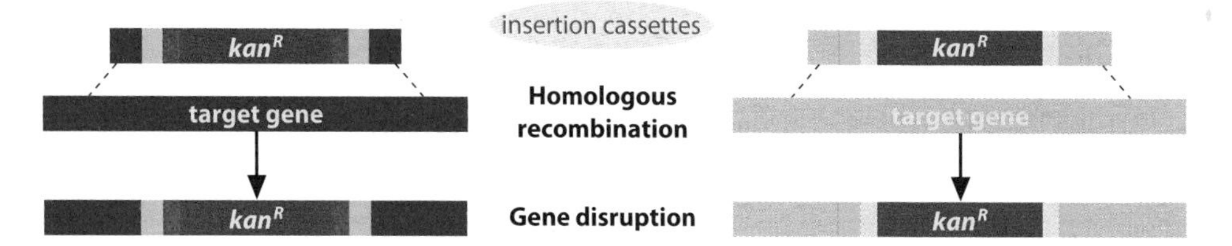

Figure 7.7 The insertion cassettes are used to disrupt the genes. Two of the insertion cassettes from Figure 7.6 are shown here, corresponding to two different yeast genes in two different yeast strains. The insertion cassettes are transformed at random into yeast cells. Homologous recombination between the ends of the insertion cassette and the target gene inserts the *kan*^{R} gene and the specific molecular bar code into each gene. Kanamycin is used to select for yeast cells that have an insertion, and the bar code will be used to tell the investigators which gene has been disrupted, as shown in Figure 7.8.

can be found in Text Box 7.2. In yeast gene disruptions, the molecular bar code is a sequence of approximately 20 nucleotides at each end that is not found in the yeast genome; the upstream and downstream sequences are different, and the bar codes are different for *each* yeast gene (i.e. there are 6100 different bar codes). The bar codes provide information to the investigator about which gene has been disrupted in each strain; it is a unique identifier for each gene in the same way that a bar code provides a unique identifier for each item in a retail store or each personnel badge and card key in a corporation.

It is instructive to consider what would happen if these molecular bar codes were omitted. A typical insertion cassette for a single yeast gene would include the *kan*^{R} gene flanked by sequences corresponding to the yeast gene targeted for disruption. In Figure 7.7, these are the blue elements. If this standard primer sequence were used, the *kan*^{R} gene would insert into the targeted gene by homologous recombination, as described for a typical yeast gene in Chapter 6. While this procedure works for a single gene, it has the following serious limitation when applied in a genome-wide screen in which 6000 different genes will be disrupted: How will the investigator know which gene has been disrupted in each yeast strain?

It is necessary to have a **gene-specific sequence** for each gene in the genome. The molecular bar code provides that information with a uniquely designed tag. The investigators designed the bar codes knowing which unique sequence has been assigned to and associated with each gene-specific homology region. When the gene homology regions target recombination to a specific gene, the recombination inserts not only the selectable *kan*^{R} gene but also the bar code that allows the specific gene to be tagged and identified in a large-scale assay.

The disruption procedure is similar to other yeast transformation experiments, and is shown in Figure 7.7. A library of insertion cassettes is created by PCR and then transformed into yeast cells under conditions under which most yeast cells will receive at most a single cassette. In the cell, homologous recombination between the insertion cassette and the target gene inserts the *kan*^{R} gene and the molecular bar code. As a result, each gene is individually disrupted and tagged in a different yeast cell. The cells are grown in the presence of kanamycin to select for transfomants that inserted the *kan*^{R} gene. The collection of transformed cells can then be tested under a wide range of growth conditions and subjected to other phenotype assays. For example, the cells can be grown in the absence of uracil to identify the genes needed for uracil biosynthesis. An example is shown in Figure 7.8. The mutant yeast cells are pooled and grown in the absence of uracil. The cells that do not need additional uracil (i.e. uracil prototrophs) will grow, whereas those that require uracil (uracil auxotrophs) will not grow.

The investigator now uses the molecular bar codes to determine which gene has been disrupted in each of the cultures, as shown in Figure 7.8. The bar code is amplified for each gene using the gene-non-specific primer and the results are displayed as DNA microarrays. Microarrays for genome-wide gene expression analysis are described in Chapter 8. For now, it is sufficient to imagine a grid in which an oligonucleotide for each of the bar codes is placed in a defined array. Because of the design, the investigator knows which bar code sequence is present at each row and column junction, as in a spreadsheet with 6000 entries. The amplified bar codes from each of the two cultures are hybridized to these defined arrays and the results are noted, as shown in Figure 7.8 for our five disruption mutants. Each spot corresponds to the bar code for a gene disruption, so the absence of hybridization indicates a failure of that cell to grow. The disrupted gene is inferred to be needed for growth in the absence of uracil. The microarray display of the bar codes allows the investigator to view the complete collection of 6000 gene disruption strains at one time without having to sequence any of the mutants to know which gene has been disrupted.

Our simple example with uracil biosynthesis greatly understates the power of this approach. A better demonstration is provided by some numbers. The yeast deletion collection has been cited in the scientific literature more than 600 times so far, with more than 100 different assay conditions. Some of these are described below when we discuss phenotypic assays. More than 5000 of the roughly 6000 genes have been found to have a mutant phenotype by at least one assay, and many additional genes have been found even for processes

BOX 7.2 Molecular bar codes

The project to disrupt every gene in the yeast genome required the use of a unique DNA sequence tag for each gene. Each gene is disrupted individually by using regions of homology to integrate the selectable marker, but the analysis of these disruption mutants is most efficiently done by looking at pools of mutant strains. The unique sequence tag, also known as a molecular bar code or DNA signature tag, allows the investigator to find the particular gene among the pool of mutants.

The pooling process can be illustrated by a thought experiment. Imagine a biological process in yeast that requires the products of 20 genes (e.g. a metabolic pathway). Yeast has a genome of about 6100 genes, and so the 20 genes that affect this pathway will be less than 0.5 percent of the gene disruptions—99.5 percent of the disruptions will not affect this pathway. Rather than individually screening through thousands of mutants that do not affect the pathway, the mutants can be pooled into groups of, say, 100 mutants. Now the process of screening the genome involves screening 60 pools of mutants rather than 6000 individual mutant strains. Even with 100 mutants in a pool, most of the pools will not have a mutation that affects our pathway, and so all of those pools with all of those mutants can be set aside. The number of pools that have a mutant of interest will be no more than 20 since there are 20 genes that affect our pathway (a number we do not know at the outset); it could be less than 20 if two or more mutants that affect the process end up in the same pool by chance. Now those pools that have a mutant of interest can themselves be subdivided and pooled, and each pool can be tested. By a regular cycle of testing and subdividing pools of mutants, the individual mutants can be identified with less effort than would be needed to test each mutant separately. However, even with this method of pooling mutants, several rounds of growth and plating are needed, and the process could take several weeks.

The bar codes provide an even more effective method of identifying individual genes within a pool that includes a mutant of interest. The original pool of mutants can be thought of as the **input** group. This entire group is then subjected to the phenotypic assay (e.g. growth under particular nutritional conditions) and the pool of strains that **passed** the test assay is collected. These strains define the output group—genes that **do not** affect the process of interest. For example, these are the mutants that were able to grow normally despite the lack of a nutrient; whatever gene is affected in each mutant, it is **not** needed for the process. By comparing the input group of mutants with the output group (in effect, subtracting the output results from the input group), the investigator can determine which genes are needed for the process. The molecular bar codes provide the information required for comparison to be made. Each gene is tagged with a bar code during the mutation process, and so the investigator compares the bar codes rather than trying to isolate individual mutants. The bar codes that are missing among the output group identify the genes of interest.

Molecular bar codes or DNA signature tags were first used for high-throughput analysis with pathogenic bacteria. Pathogenesis is often a difficult process to study in the laboratory because so many hosts must be tested. An early use of bar codes involved *Salmonella enterica* virulence in mice. Bar codes of about 40 nucleotides were incorporated into transposable elements, which were then inserted randomly throughout the *Salmonella* genome. The mutant bacteria were pooled and tested in mice; bacteria were collected from mice that became ill. These *Salmonella* mutants had retained the ability to become virulent. The virulent output pools were compared with the input pools using radioactive hybridization or a restriction digest in the universal sequence flanking each bar code to identify the genes that were not needed for virulence.

Detecting bar codes

A variety of methods have been used to **detect** the bar codes in the output pool. The key element is the sensitivity to detect the absence of individual bar codes in a large pool. Some investigators have used PCR amplification or restriction digests based on the universal sequence flanking each bar code. As techniques have developed, the bar codes have usually been detected by hybridization to a microarray, as was done in yeast. All of the bar codes from the input group are placed in a defined array pattern; bar codes in the output group are amplified by PCR and hybridization to the array. Those bar codes that do not show a hybridization signal identify the genes that are needed for the process.

Although they were first used in gene disruption screens, molecular bar codes have recently been applied in genome-wide screens in mammalian cells using RNAi and short interfering (siRNA). The strategy is essentially similar to what is done with the gene disruption screens: first creating a pool of 'mutagenic agents' tagged with a DNA signature bar code (in this case, a set of siRNA clones), secondly using these clones as an input group to produce mutant phenotypes, and thirdly using microarrays to compare the output group with the input group to detect genes with no effect and infer that mutants are needed for the process.

Further reading

Mazurkiewicz, P., Tang, C.M., Boone, C., and Holden, D.W. (2006). Signature-tagged mutagenesis: barcoding mutants for genome-wide screens. *Nature Reviews Genetics*, **7**, 929–39.

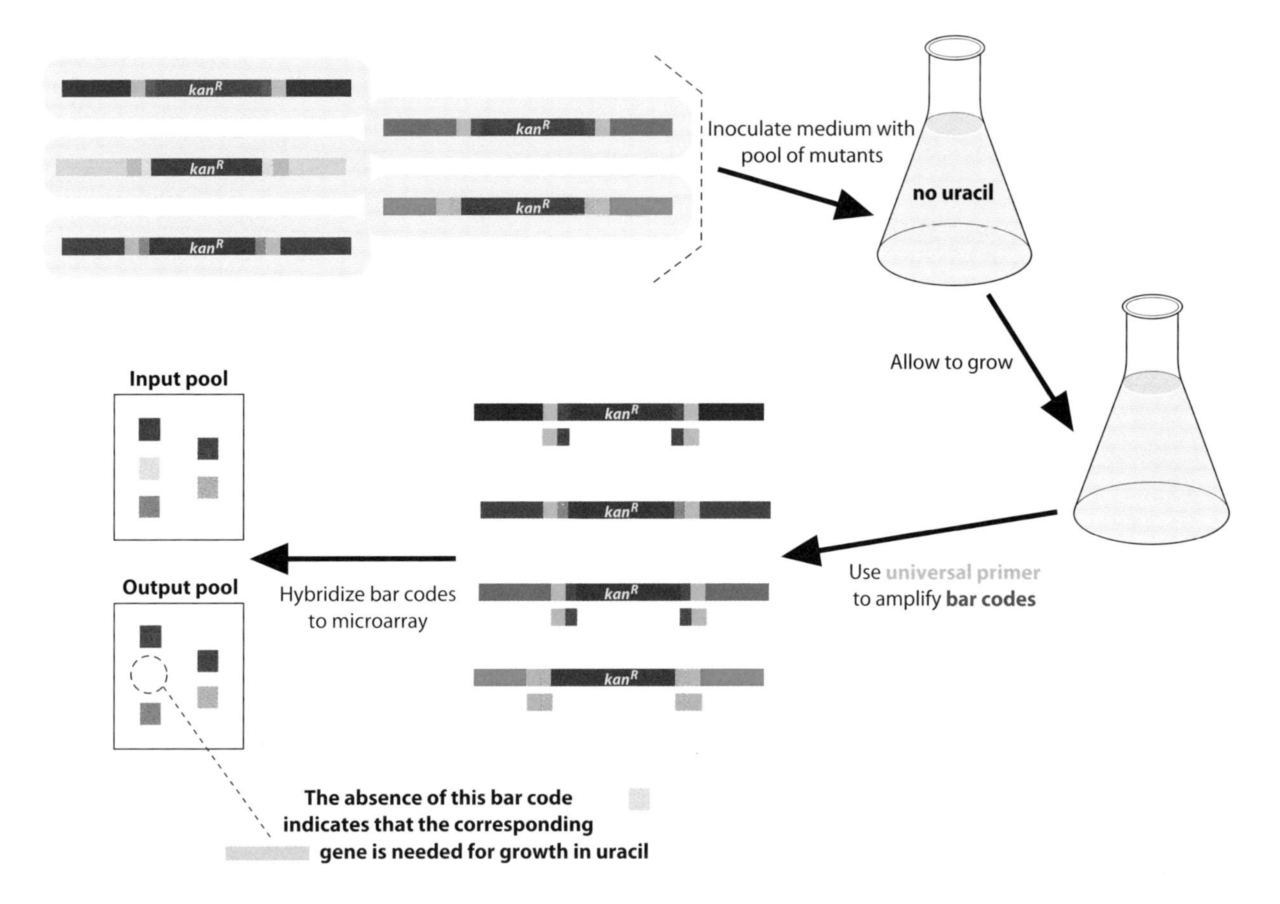

Figure 7.8 The use of the molecular bar code. Once disruptions are made for each gene, as shown in Figure 7.7, the mutants are pooled and tested for a mutant phenotype. Each yeast cell at the top left has a single gene disruption, and the pool of mutants is grown in a medium that lacks uracil. Once the culture has grown, the cells are removed and assayed by PCR. The universal primer will amplify each bar code in the culture, as shown. The collection of bar codes is hybridized to a microarray that contains all of the input bar codes in a defined pattern, here the same pattern as shown in the pool of mutants. The bar code for any gene that is needed for growth in the absence of uracil will not hybridize in the output array.

such as sporulation and DNA repair, which have been thoroughly studied by other procedures. Although the process seems completely removed from an individual investigator carefully streaking yeast cells on different culture plates, the intellectual basis and the fundamental question remain the same: **Find a mutant**.

7.4 RNAi and large-scale mutant analysis

One of the most unexpected and powerful methods of mutant analysis for genome-wide screens involves the use of double-stranded RNA (dsRNA) to interfere with gene expression, in a technique called **RNA interference (RNAi)**. In a typical RNAi experiment, double-stranded RNA corresponding to a portion of the transcript from a gene is introduced into the cell or organism. The dsRNA specifically blocks or reduces expression of that gene, thereby producing a mutant phenotype. An overview of the process of RNAi is shown in Figure 7.9. Let us consider an example: if dsRNA that corresponds in sequence to part of the mRNA encoding a myosin heavy-chain gene is introduced into *C. elegans*, the offspring of the injected worm are paralyzed, with a phenotype that resembles a structural change in the myosin heavy-chain gene. With some exceptions, the effect is not heritable so strictly speaking these are not mutants. That is, the paralyzed worms have an intact and unmutated gene for the myosin heavy chain and will thus produce offspring with normal

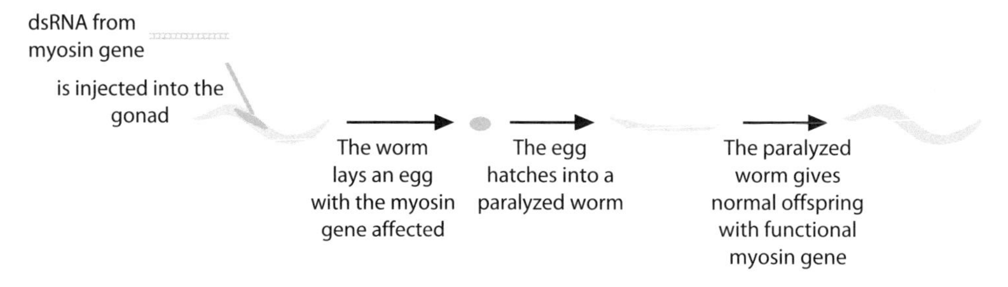

Figure 7.9 An overview of RNAi. The example here uses a muscle myosin gene from *C. elegans*. dsRNA corresponding to a portion of the coding region of the myosin gene is made *in vitro*. This dsRNA is injected directly into the worm gonad. The injected worm lays eggs, some of which have received the injected dsRNA. The affected egg hatches and the resulting worm is paralyzed, as if it had a mutation in the myosin gene. The gene itself is not structurally altered in the paralyzed worm, and so its offspring will not contain the dsRNA and will not show an RNAi phenotype.

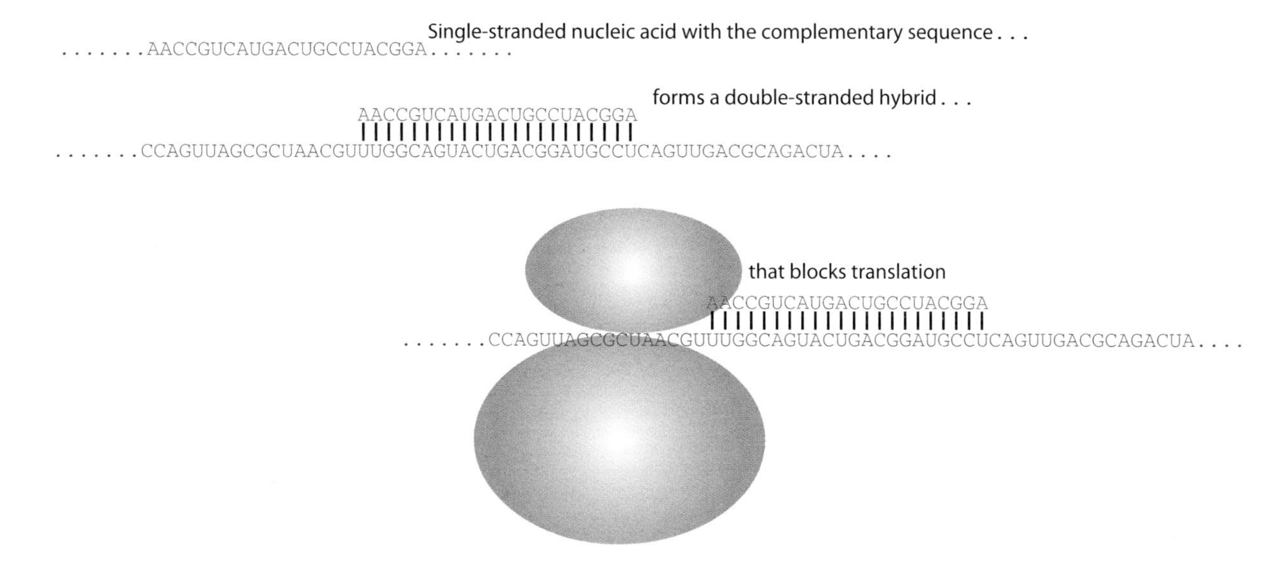

Figure 7.10 One postulated mechanism of antisense experiments. A single-stranded nucleic acid (RNA or DNA or a synthetic molecule) is made that is complementary in sequence to the target mRNA. The single-stranded molecule is shown in red, and the target mRNA is shown in blue. This single-stranded molecule is introduced into the cell where it is postulated to form a double-stranded hybrid with the mRNA. This double-stranded hybrid is then postulated to be blocked in translation, as shown at the bottom of the figure. Although antisense experiments of this type do block gene expression, the actual mechanism is rarely explored. It seems likely that many of these antisense experiments work by triggering RNA degradation via the RNAi pathway, as shown in Figure 7.13.

movement. In classical genetic terms, a non-heritable mutant phenotype induced by environmental agents is referred to as a **phenocopy**, and the phenotypes arising from RNAi are probably most accurately called phenocopies rather than mutants. However, the use of RNAi has spread quickly, with its practitioners calling these 'mutants', so it seems pedantic to insist on calling them phenocopies.

The antecedents of RNAi in other experiments

Although RNAi as an experimental method has a relatively brief history—the first RNAi experiments in animals were done less than 15 years ago—it is rooted in an older process known as antisense technology. Antisense techniques involved producing a single-stranded RNA or another modified nucleic acid that is complementary to the mRNA or sense strand. The single-stranded RNA was predicted to form a double-stranded hybrid with the mRNA and thereby block its translation. This hypothetical mode of action for antisense experiments is shown in Figure 7.10.

Many antisense experiments were successful and many more were attempted and not reported, mostly without investigation of the actual method by which the RNA molecule was exerting its effect. An important control for understanding the mechanism of antisense

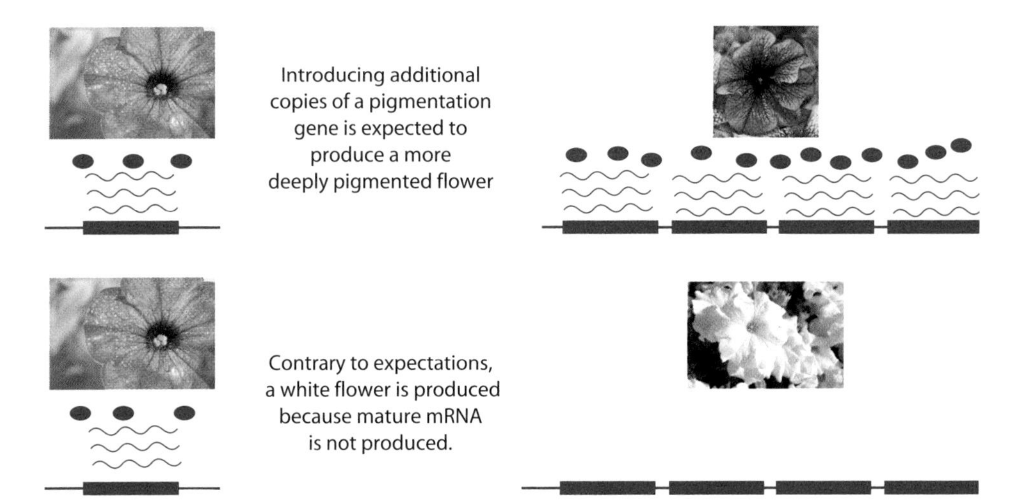

Figure 7.11 Post-transcriptional gene silencing in petunias. A petunia with light purple flowers is shown on the left. The flower is lightly pigmented because it has a hypomorphic mutation in the blue pigmentation gene shown beneath the flower; the gene is transcribed and protein is produced, but in reduced amounts compared with darkly pigmented petunias. Additional copies of the pigmentation gene were introduced into the plant. The top half of the figure summarizes the expected result of introducing additional copies of the gene, while the bottom half shows the actual result. Increasing the copy number of a hypomorphic mutation is expected to result in more pigmentation, a more nearly wild-type phenotype as described for an allelic series in Chapter 4. Contrary to expectations, gene expression was decreased rather than increased and a white flower resulted. Further investigation showed that the effect occurred post-transcriptionally by RNA degradation. Post-transcriptional gene silencing in petunias was one of the first demonstrations of the RNAi pathway.

effects was the introduction of a single-stranded sense RNA. If the antisense strand were blocking translation by forming a double-stranded hybrid with the mRNA, the sense strand should not produce an effect as it would be unable to form this hybrid. Although this control worked for some antisense experiments, it failed for others. Careful experiments in *C. elegans* with purified single- and double-stranded RNA revealed that the dsRNA hybrid formed *in vitro* had the greatest effect. This suggests that the silencing in some of the antisense experiments arose because a small amount of dsRNA had contaminated the sense or antisense RNA preparations. RNAi as an experimental technique was developed from these careful controls on antisense experiments.

Similar effects had first been observed in plants, particularly for flower color in petunias. Figure 7.11 is a diagram of the key experiment, done some years earlier. The petunia variety had purple flowers, the result of a hypomorphic mutation in the pigmentation gene. When multiple copies of the pigmentation gene were introduced as a transgene, the expression of the normal gene was unexpectedly blocked rather than increased. Further investigation of this effect revealed that it was mediated by RNA, and was also seen with RNA viruses that replicate through a dsRNA intermediate. Apparently, the plant has mechanisms to synthesize a dsRNA from a sense transcript. In plants, this effect is called **post-transcriptional gene silencing** (PTGS), but the general mechanism for gene silencing by dsRNA appears to be the same in plants and animals. Thus, although the experimental history of RNAi is quite recent, the evolutionary history of the response to foreign dsRNA predates the divergence of plants and animals. It has been widely postulated, with some supporting evidence, that the RNAi response is an ancient type of immune response, protecting eukaryotic organisms from RNA viruses and the effects of transposable elements, both of which involve dsRNA molecules.

In addition, it has recently been realized that small and non-coding RNAs are important naturally occurring regulators of gene expression. Such small regulatory RNA molecules have been found in both unicellular and multicellular eukaryotes, and a large and growing number of cellular processes involve **microRNAs** (miRNAs). Remarkably, the normal cellular response to these miRNAs involves many of the same biochemical steps exploited in RNAi experiments. The relationship between the dsRNA introduced in RNAi and some of the naturally occurring regulatory RNA molecules is described in Text Box 7.3. Curiously, RNAi is able to

BOX 7.3 Regulatory microRNAs

The discovery of RNA interference (RNAi) as an experimental technique is intimately connected with the discovery of another naturally occurring dsRNA species at about the same time in the same organism. These non-coding RNA molecules, which are a very diverse group, include hundreds of very short regulatory RNAs, which we refer to generally as **microRNAs** (miRNAs). Just as RNAi has emerged as a technical breakthrough in genetic analysis, regulatory miRNAs have become a conceptual breakthrough in thinking about gene regulation. miRNA-mediated regulation occurs when two complementary single-stranded RNAs encoded by two different genes hybridize *in vivo* to form a dsRNA molecule. That is, eukaryotes are now known to have a very large number of genes whose functional product is the RNA transcript rather than a protein. Humans are estimated to have more than a thousand different miRNA genes, but the actual number may prove to be higher. The story of regulatory miRNAs again begins with *C. elegans*, and the connection between miRNAs and RNAi is both mechanistic and personal. Not only do miRNAs and RNAi work by some of the same mechanisms, but the experiments were often done by the same laboratories.

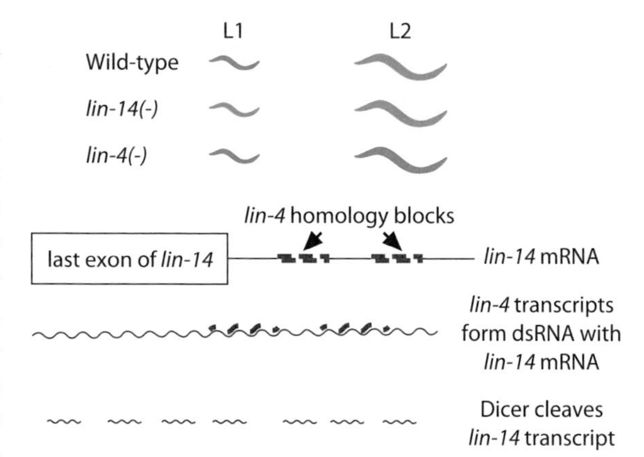

Figure B7.3 Regulation of *lin-14* by *lin-4* in *C. elegans*. In a wild-type worm, shown at the top, the L1 and L2 stages differ in size, cuticle, and other cellular features. These are represented schematically as blue for L1 stage worms and green for L2 stage worms. In a *lin-14* mutant, shown in the second line, the L1 stage is skipped and the worm takes on the L2 characteristics at hatching, although it is smaller. In a *lin-4* mutant, the L1 stage is reiterated. The method of regulation is shown in the drawing at the bottom. *lin-4* regulates *lin-14* by a post-transcriptional mechanism. The 3′-untranslated region of the *lin-14* mRNA has two blocks that are complementary in sequence to the *lin-4* miRNA. The *lin-4* RNA forms a dsRNA hybrid with *lin-14* mRNA, which targets it for degradation by the enzyme Dicer.

miRNAs regulate the *C. elegans* life cycle

C. elegans goes through four larval stages, termed L1–L4, before becoming an adult, as described in Chapter 2. Each larval stage is characterized by a distinct body cuticle, which is shed in a molt to produce the next stage. Mutations in two genes, termed *lin-4* and *lin-14*, affect these larval stages in an unusual way. This is illustrated in Text Box Figure 7.3. Loss of *lin-14* function results in a phenotype called precocious development. That is, at a chronological time and size when the worm should be an L1 larva, it skips the cuticle and other cellular features of the L1 stage and has many features of the L2 stage. A mutation in *lin-4* has the opposite phenotype in that a *lin-4* mutant reiterates the L1 cuticle and other features of the first larval stage even as it grows larger.

Experiments of the type to be described in Chapters 10 and 11 showed that *lin-4* turns off *lin-14* gene activity but did not reveal the mechanism by which *lin-4* regulates *lin-14*. When *lin-14* was cloned, the gene was found to encode a conventional if novel protein, although the mRNA transcript has an unusually long 3′-untranslated region (3′-UTR). The surprise came when *lin-4* was cloned. The gene is extremely small: it produces a transcript of 70 nucleotides that has no sustained open reading frame and could not be translated into a protein. The *lin-4* transcript is processed into a 22-nucleotide species, which is the functional molecule. As depicted in Text Box Figure 7.3, the 3′-UTR of the *lin-14* transcript has 14 copies of the complementary sequence of the *lin-4* RNA, arranged in two blocks of seven repeats. The *lin-4* RNA forms a double-stranded hybrid *in vivo* with the *lin-14* transcript and prevents its expression. In the wild-type, *lin-4* blocks the expression of *lin-14* at the end of the L1 stage, allowing the worm to molt to become an L2.

Although it may seem obvious that *lin-4* represses *lin-14* expression by preventing its translation, in the same way that antisense regulation was presumed to work, this is not the only (or perhaps even the most significant) means of regulation. Instead, the dsRNA hybrid formed by the *lin-4* miRNA and the 3′-UTR of the *lin-14* mRNA also targets the *lin-14* mRNA for degradation. Subsequent work, mostly done simultaneously with the investigation of RNAi, has shown that the degradation requires the enzyme Dicer and much of the same cellular machinery as involved in RNAi. RNAi is an experimental application of a naturally occurring mechanism of gene regulation mediated by miRNAs.

lin-4 was the first of a large number of regulatory miRNA genes to be found

A similar mechanism for regulating developmental timing in *C. elegans* involving other genes was found to work at later larval stages. For example, one important miRNA gene is called *let-7*. Although genes similar in sequence to *lin-4* have been found only in nematodes, genes related to *let-7* have been found in many other animals; at least nine orthologs of the *let-7* gene have been found in humans. Estimates of the number of miRNA genes in humans are constantly changing, but it is clear that there are hundreds if not thousands of such genes in our genome.

The miRNA genes can be grouped into families based on sequence relationships, much as genes encoding proteins can be grouped into families as discussed in Chapter 1. Some miRNAs such as *let-7* are evolutionarily ancient; humans have a *let-7* ortholog that is identical in sequence to the *let-7* gene in *C. elegans*. Other microRNAs are more recent in evolutionary origin, with some genes specific to one species. For example, the *lin-4* gene is nematode-specific, while there are 10 miRNA genes that are specific to primates and not found in other mammals.

With so many potential regulatory RNA molecules, the question immediately becomes: 'What do they regulate?' For most genes, we are only beginning to learn the targets. The *let-7* miRNA in *C. elegans* has been shown to have a wide variety of target genes, including many not involved in developmental timing. In fact, by scanning the *C. elegans* genome for sequences complementary to the *let-7* sequence, *let-7* is predicted to have at least 54 different direct regulatory targets, many of which have been experimentally confirmed. Thus, *let-7* is clearly a key developmental regulatory gene. Similar experiments have been done in other organisms. Among the many targets of the *let-7* gene family in worms and humans is the *ras* gene, whose expression has been implicated in a wide range of different human tumors. The Ras signal transduction pathway has been thoroughly studied in many organisms and for many different biological processes; in fact, it is found in nearly every molecular and cellular biology textbook. Yet the involvement of an RNA regulatory mechanism in this pathway was not suspected.

Computational approaches to predict target genes with other miRNA genes indicates that *let-7* is not unusual in having 54 direct targets—in fact, the number of target genes for different miRNAs in *C. elegans* ranges from two to 377, with a mode of 142 direct targets. This is an extraordinary number. One wonders if there are any genes that are not regulated by miRNAs!

miRNAs and the regulation of gene expression

At this point, it may be useful for us to compare regulation by miRNA genes with regulation by transcription factors. As a point of comparison, preliminary genome-wide studies with transcription factor proteins in yeast and worms find that a few transcription factors regulate dozens of genes but that the vast majority of transcription factors directly regulate the expression of only one or two genes. (We discuss some of these studies in Chapter 12.) In other words, based on the evidence acquired so far, a regulatory miRNA has perhaps 10 times as many direct target genes as a transcription factor protein. This is contrary to most of our models of gene regulation, so we will need to re-examine our thinking about the regulation of gene expression.

Thus, although the story of regulatory miRNAs began in *C. elegans*, it certainly does not end there; in fact, the end is not nearly in sight. miRNA molecules are being identified for dozens, if not hundreds, of different processes in nearly all multicellular organisms. Every month brings new reports of a miRNA that regulates some cellular or developmental process, and there is no indication that we have identified all or even most of the processes, the targets, or even the miRNA genes themselves. This is definitely one of the most rewarding and active areas of research in molecular genetics, and probably will continue to be for the foreseeable future.

Further reading

Many excellent review articles have been written about miRNA regulation. Here are two brief ones that provide a good overview and set of additional references.

Du, T. and Zamore, P.D. (2005). microPrimer: the biogenesis and function of microRNA. *Development*, **132**, 4645–52.

Alvarez-Garcia, I. and Miska, E.A. (2005). MicroRNA functions in animal development and human disease. *Development*, **132**, 4653–62.

exert an organism-wide effect in both plants and worms, suggesting that the dsRNA or the gene silencing effect is somehow passed between cells in these organisms. Although effects on the entire organism are observed in plants and worms, most of the RNAi screens in other organisms have been done on cells in culture. Whether on organisms or cells, the impact on genetic research has been profound nonetheless.

Despite its recent discovery, RNAi has a deep and expanding literature. It has proved to work in all multicellular plants and animals that have been thoroughly tested, although not for all genes or all tissues. Recall from Chapter 2 that an RNAi screen was used to identify genes involved in regeneration in planaria, an organism for which no mutations are known. Because the technique has such promise for altering gene expression, which could provide useful therapeutic agents, many experiments have been directed at understanding the mechanism of RNAi and improving its effects. Many features that are common to RNAi effects for most genes and most organisms have been identified, but differences in the details have also been found. Procedures that work well for many genes in one organism may not work for *all* genes in that organism, and may not work very well at all in another organism. In addition, although we write about the mechanism by which RNAi works, it is likely that there are many different mechanisms with some steps in common.

The effectiveness of RNAi for any particular gene is also likely to depend on the normal transcription pattern for that gene, which is often not known or not easily correlated with a specific effect. This is particularly true in describing genome-wide screens in which the goal is to test hundreds, thousands, or even tens of thousands of genes as rapidly as possible. The procedures are sometimes presented as if one common protocol fits all genes, but this is clearly not true. With those cautions, we will describe some of the protocols that are used and the mechanism by which RNAi is thought to work. We will describe the procedure and mechanism generally, and then note some of the specific differences that are appropriate in different organisms.

Introducing dsRNA into cells or organisms

The first requirement for conducting an RNAi screen is to be able to introduce the dsRNA into the cell or the organism. Several different methods are shown in Figure 7.12. The dsRNA can be transfected for cells grown in culture, as for mammalian cell lines or *Drosophila* cells. In *Arabidopsis* or in mammalian cells, the transfected molecule may be in a vector that has the dsRNA sequences in a hairpin configuration, as shown in Figure 7.12A. The dsRNA is cleaved and induces a transient effect in the cultured cells. For worms and flies, the dsRNA can be injected directly into the germline of either the adult mother (worms) or the developing embryo (flies). This method is shown for worms in Figure 7.12B. The effect of the dsRNA is then seen in their offspring. Injection is a laborious process, and only a few experimenters in worms have used this method for a genome-wide screen. More commonly, injection is used to test individual genes that have been identified by other means.

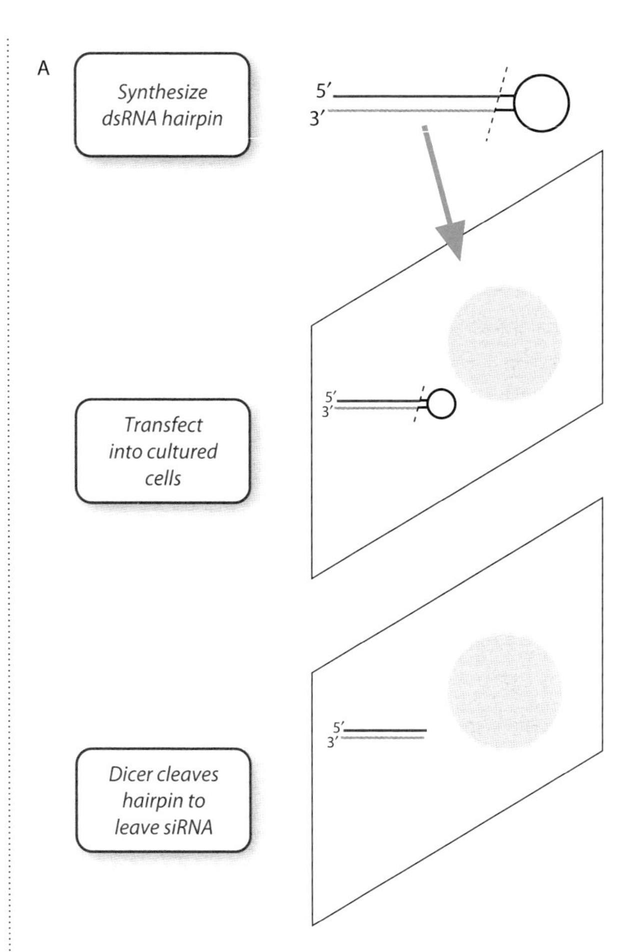

Figure 7.12 Introducing dsRNA for RNAi experiments. (A) dsRNA is transfected into mammalian cells as a hairpin. An RNA hairpin is synthesized *in vitro* with a stem of 22 nucleotides including a two-base overhang at the 3′ end. The loop is usually about 10 nucleotides long. The dashed line indicates the site of future cleavage by Dicer. The hairpin RNA is transfected into mammalian cells in culture, where it is cleaved by Dicer to make an siRNA. As described in Figure 7.13, the siRNA triggers RNAi. The RNAi effect in the cultured cells is transient. Other methods of introducing dsRNA into mammalian cells are described in Text Box 7.4 and shown in Figure 7.13C.

The most common method in *C. elegans* is to introduce dsRNA by feeding the worms on an appropriate bacterial strain. As noted in Chapter 2, worms in the laboratory eat *E. coli*. The feeding technique is shown in Figure 7.12C. Instead of dsRNA made *in vitro* by the investigator, the source of the dsRNA is the worms'

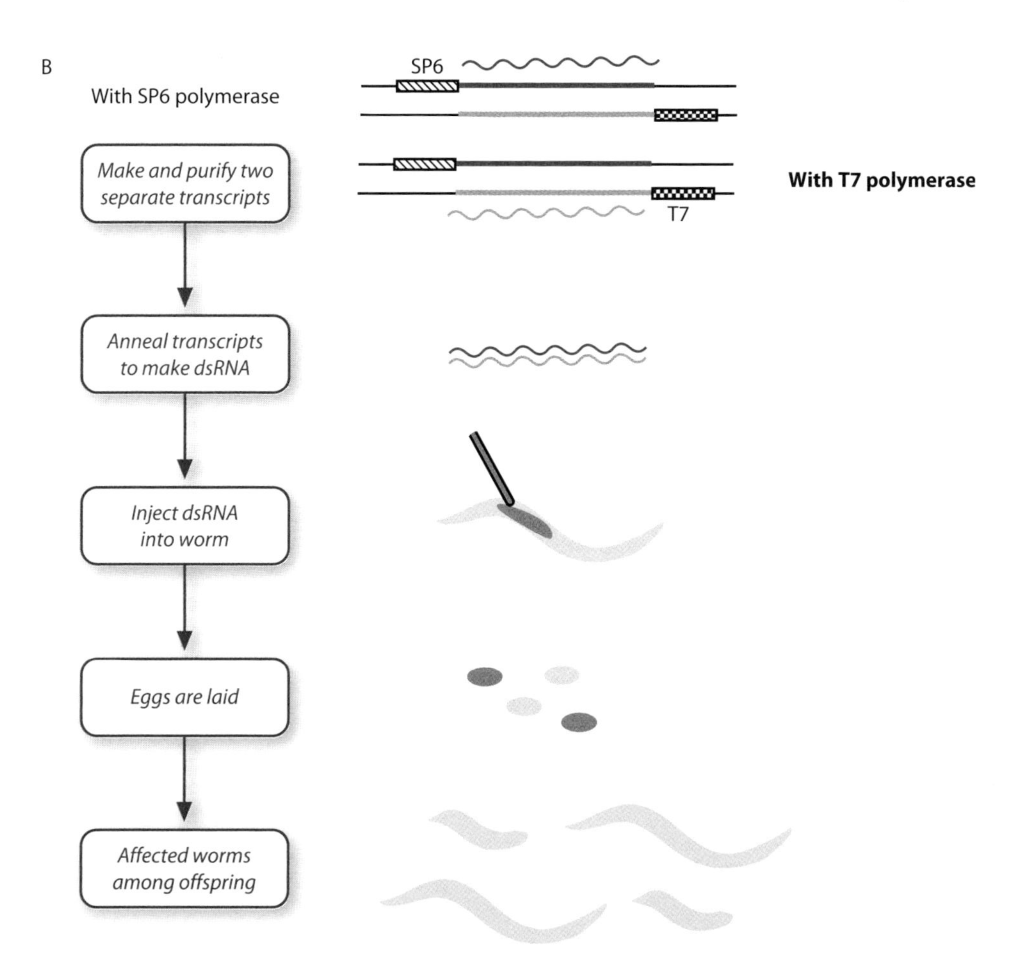

Figure 7.12 (Continued). (B) RNAi is performed by direct injection of the dsRNA. A cDNA from the gene of interest is cloned into a vector with different transcriptional promoters on each side of the cloning site. A vector with SP6 and T7 promoters is commercially available, but other promoters can be used. Two reactions are done *in vitro* in separate vials, one with SP6 polymerase added and the other with T7 polymerase added. Each reaction produces one single-stranded RNA molecule, which can be annealed *in vitro* to make dsRNA. The remainder of the procedure is similar to that shown in Figure 7.9. The dsRNA is injected into the animal, which lays eggs with the dsRNA. However, not all eggs will have the dsRNA but these cannot be detected until the offspring hatch. The eggs hatch, and some of the offspring show the RNAi phenotype, here a dumpy worm. Similar procedures are done by injecting directly into eggs in *Drosophila*.

food. Specially designed RNAi plasmids are available in which a cDNA insert is flanked by T7 promoters, as shown at the top of Figure 7.12C. These plasmids are transformed into an *E. coli* strain in which T7 RNA polymerase has been inserted into the chromosome under the control of a *lacZ* promoter region, so expression of the polymerase is induced by the addition of IPTG. Thus, upon IPTG induction of the bacteria, i.e. the worms' food, T7 RNA polymerase is expressed, which in turn results in the transcription of the plasmid insert. Since the T7 promoters flank the insert, each strand is transcribed and a dsRNA hybrid is made by the bacteria. As shown in Figure 7.12C, the worms eat the bacteria, and the dsRNA spreads from the gut throughout the organism.

Although the feeding procedure may seem strange, it works well and is extremely easy to do. Libraries with cDNA inserts from nearly all of the predicted worm genes are available, which makes a genome-wide screen no more technically demanding than growing bacteria and feeding them to worms. Many genome-wide screens have been done in *C. elegans* using RNAi by feeding. In direct comparisons involving the same genes in the same genetic strains, RNAi induced by injection is

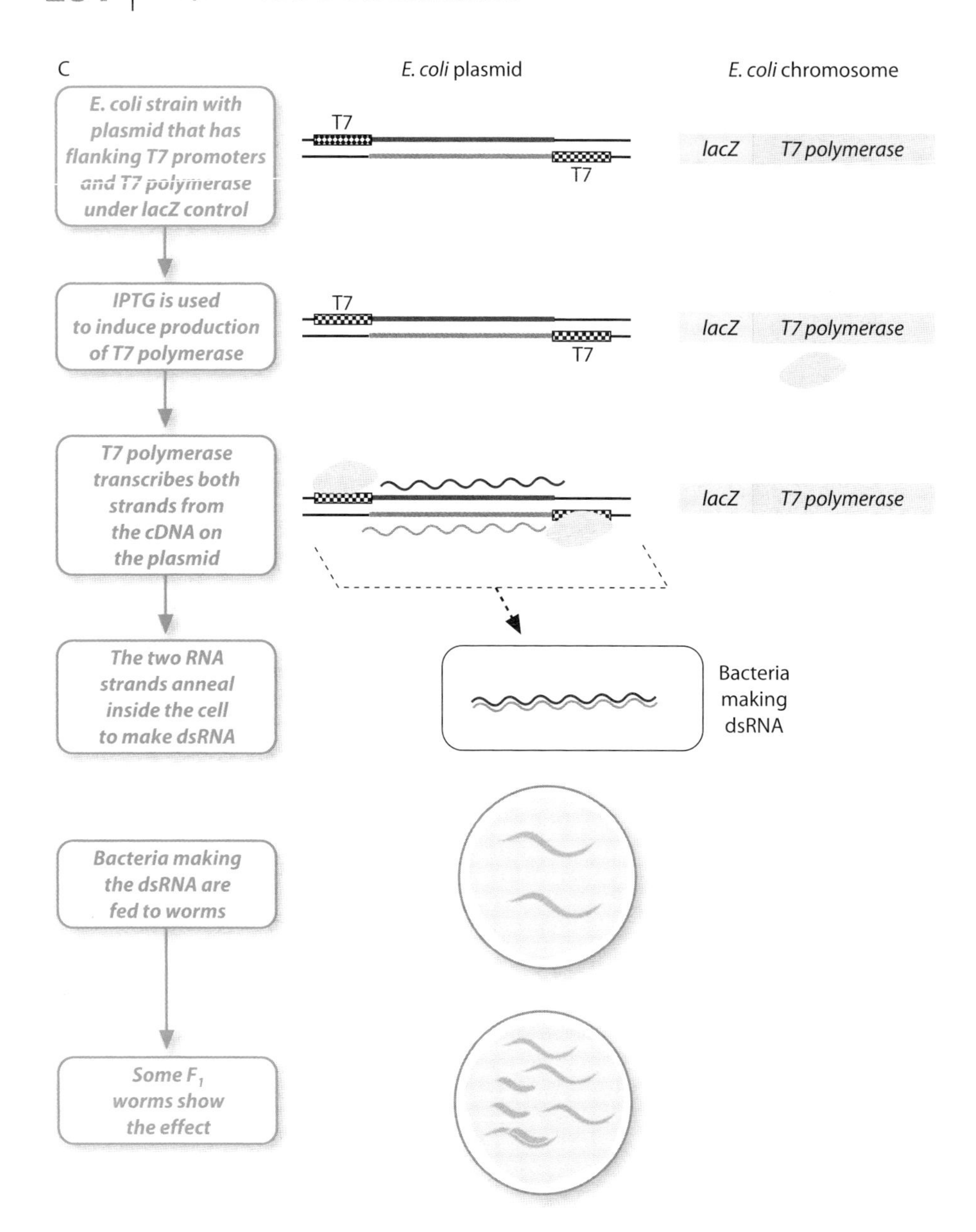

Figure 7.12 (Continued). (C) RNAi is performed by feeding. It will be recalled from Chapter 2 that the worm *C. elegans* eats *E. coli*, and molecules introduced into the gut have a systemic effect throughout the worm. The cDNA for the target gene is first cloned into a plasmid that has T7 promoters on each side. This plasmid is transformed into an *E. coli* strain that has a T7 polymerase gene under the control of the *lacZ* promoter integrated into the chromosome. The relevant genes on the plasmid (in the middle) and the chromosome (on the right) are shown. Both plasmid and chromosome have other selectable markers that are not shown for simplicity. The bacteria with the plasmid are induced by IPTG, which results in the production of T7 polymerase. T7 polymerase works on the plasmid promoters and transcribes both strands of the cDNA for the target gene. These two complementary RNA strands can anneal with each other inside the bacterial cell to make a dsRNA molecule. The bacteria making the dsRNA are spread onto plates, and worms are placed on the plates. As the worms eat the bacteria, they ingest the dsRNA, which then induces RNAi in the target gene. The worms reproduce, and some of their offspring exhibit the RNAi phenotype, here shown as a dumpy worm. A library of *E. coli* strains with each worm gene cloned into the plasmid vector is commercially available, and genome-wide screens in *C. elegans* are usually done using this feeding method.

considered more reliable than that induced by feeding, but feeding works very well as a first screen to find genes of potential interest.

Mechanism of RNAi

How then does the dsRNA produce its silencing effect? It may be helpful to compare RNAi with what investigators believed was happening in antisense experiments. Although an antisense strand that base pairs with its complement in an mRNA is the critical step, the principal mode of gene silencing by RNAi is probably not by blocking translation but rather by triggering mRNA degradation, at least in worms and plants. This mechanism of RNAi is shown in Figure 7.13A. In *C. elegans* or in flies, dsRNA that is 300 or more nucleotides long and corresponds to the gene is introduced; with such a length, many different parts of the transcript are represented. One strand is identical in sequence to the mRNA (the sense strand) whereas the other is the complementary sequence (the antisense strand). The long dsRNA is then cleaved into fragments by the enzyme Dicer. As its name implies, Dicer cuts the dsRNA into fragments of 19–24 nucleotides. Orthologs of Dicer have been found in all multicellular organisms for which genomic information is available, consistent with an evolutionarily conserved mechanism.

The RNA fragments produced by Dicer have a characteristic structure—a central dsRNA region of about 19 nucleotides with a two- or three-base overhang at each end (Figure 7.13B). These fragments are called short interfering RNAs (siRNAs), and their appearance in the process is common to all organisms. The sense strand of the siRNA is then degraded, leaving the antisense strand. In flies and mammals, the antisense strand becomes incorporated into a protein complex referred to as RISC (for RNA-induced silencing complex) where the siRNA targets this complex to the specific complementary mRNA. RISC degrades the corresponding mRNA without degrading the antisense strand of the siRNA, so the same complex can target and degrade many copies of an mRNA. Because the same RISC is used repeatedly, the mRNA from a gene can be completely degraded and no protein is produced. Not all of the mRNA population is degraded, so RNAi is best considered as knocking down gene expression rather than completely knocking it out.

An additional or slightly different step has been described for worms and plants, as shown in the left hand part of Figure 7.13A. Once the sense strand of the siRNA is degraded, the antisense strand forms a double-stranded hybrid with the mRNA. Such a double-stranded hybrid with its overhangs can serve as a primer for an RNA-dependent RNA polymerase, which makes multiple copies of long dsRNAs. These are cleaved (probably by Dicer) to make more siRNAs, which target the RISC to the complementary mRNA. The amplification step does not appear to occur in flies or mammals since no RNA-dependent RNA polymerase has been found. The long dsRNA that is introduced in flies and worms allows many different parts of the same mRNA transcript to be targeted for destruction, and the amplification step that occurs in plants (and worms) also accomplishes this purpose—many different parts of the mRNA become targeted for degradation, and little protein is produced from the target gene.

The procedure in mammals is rather different because we have a better immune system. In particular, a long dsRNA of this type triggers an antiviral interferon response, and interferons degrade dsRNAs non-specifically. Therefore, a long dsRNA does not yield siRNA, which makes it inappropriate for RNAi. A key insight in the development of RNAi for knocking down gene expression in mammalian cells came from the recognition that RNAi requires siRNAs with the structure shown in Figure 7.13B—a molecule that can be synthesized *in vitro* or *in vivo*. Such a synthetic siRNA is too short to trigger the interferon response. The synthetic siRNAs are incorporated into RISC, bypassing the need for Dicer, and gene expression can be blocked. One limitation of this procedure is that the short synthetic RNAs target a very specific region on the mRNA; different siRNAs corresponding to different parts of the same mRNA sometimes have different effects, some working very well and others not working at all. This target specificity is avoided in flies and worms because the long dsRNA is used, and in worms and plants because the siRNAs can prime the synthesis of a long dsRNA. Neither of these effects occurs in mammalian cells. Therefore, most investigators working with mammalian cells design multiple siRNAs for different regions of the same transcript.

siRNAs with slight mismatches to the target mRNA can also be effective in knocking down gene expression

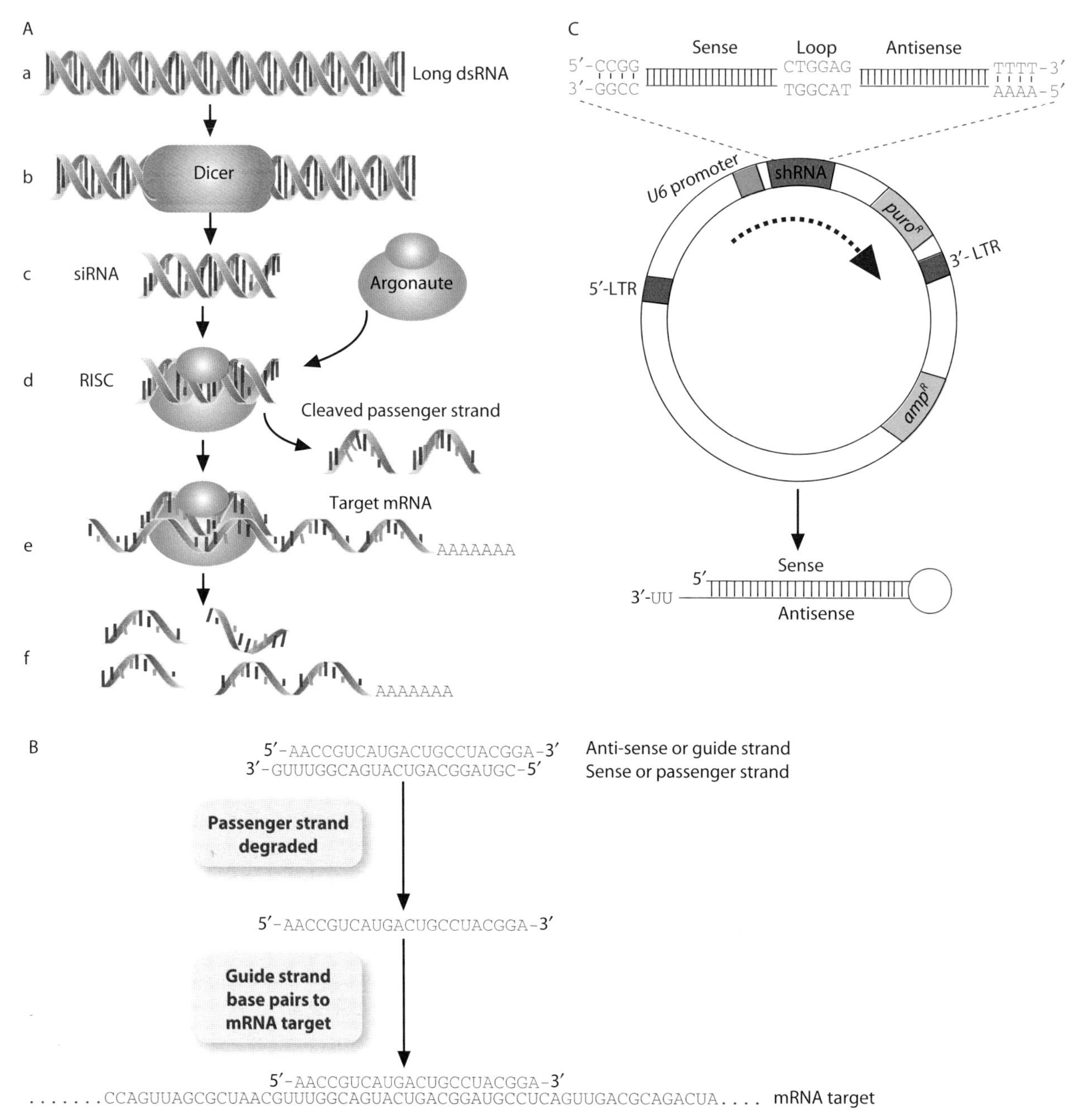

Figure 7.13 (A) The mechanism of RNA interference. A long dsRNA molecule is the target for an endoribonuclease known as Dicer. Dicer cuts the dsRNA into 22-nucleotide segments called short interfering RNAs (siRNAs). The structure of the siRNA is shown in more detail in panel B. The siRNA complexes with an unknown number of proteins including members of a family of proteins known as the Argonaute proteins. This produces the RNA-induced silencing complex (RISC). The active RISC unwinds the siRNA and degrades one of the two strands, the sense strand with the same sequence as the target mRNA; this leaves the antisense siRNA strand, complementary to the mRNA target. Each RISC apparently has a single siRNA and a nuclease (which is not Dicer). The remaining single-stranded RNA in the complex base pairs with the target mRNA sequence, which is then degraded by RISC. **(B) The structure of the siRNA.** Cleavage by Dicer produces a 22 bp/dsRNA product with a two-base overhang at each end. In a subsequent step in the RISC, one strand is degraded to leave the guide strand, which targets RISC to the mRNA by base pairing. **(C) One plasmid used for shRNA libraries in mammalian cells.** The sequence corresponding to the siRNA is shown at the top; it corresponds to the two arms flanking the loop. The loop sequence will be cleaved and is not important. The sequence corresponding to the short hairpin RNA (shRNA) is cloned into the expression vector plasmid in the middle of the drawing. In this particular library, the shRNA sequence is under the control of the promoter from the *U6* gene, which encodes an RNA involved in splicing, so the shRNA will be constitutively transcribed. Other promoters can also be used. Drug markers are included on the plasmid for growth in bacteria (ampicillin) and mammalian cells (puromycin). The dashed arrow indicates the direction of transcription, and viral long terminal repeats (LTRs) are involved in transcriptional regulation and termination. When the sequence corresponding to the shRNA is transcribed, the transcript forms the short hairpin shown at the bottom of the figure by intrastrand base pairing. Cleavage of the loop, as shown in Figure 7.12A, produces the siRNA involved in RNAi.

in mammalian cells, and probably in other organisms as well. It appears that translational blocks are more important than RNA degradation when there are mispaired bases in the dsRNA molecule, whereas Dicer and RNA degradation appear to be the major mechanism when siRNA makes a perfect match. The ability of siRNA to trigger a response even with a slight mismatch will become an important topic when the limitations of siRNAs are considered below.

Despite the variability in mechanism, RNAi is proving to be a powerful method of knocking down gene expression in mammalian cells, and different procedures have been tried to introduce siRNA. In addition to being transfected as a synthetic molecule, siRNAs can also be made from plasmid vectors that are introduced into the cells. This is usually a more effective way of doing genome-wide screens. Libraries representing tens of thousands of siRNA sequences have been constructed and are used in both academic and commercial laboratories. The libraries are also constructed with the insert as a longer sequence that has an inverted repeat structure. Such a structure is shown in Figure 7.13C. When this insert is transcribed, the presence of the repeats results in the formation of a hairpin or stem–loop RNA, providing the necessary dsRNA for effective RNAi. Many of these libraries place the siRNA sequence under the control of an inducible promoter, allowing the investigator to trigger RNAi with an environmental signal. Methods of introducing siRNAs into mammalian cells are described in more detail in Text Box 7.4.

Genome-wide screens based on RNAi are increasingly common in the genetics literature, and will certainly become even more common in the years ahead. RNAi is an inexpensive and relatively simple method of reducing gene expression, and is certainly less labor-intensive than genome-wide screens based on gene knockouts. It can also be used in many organisms for which gene knockouts are not feasible, and even for organisms for which little traditional genetics has been available, as described in Chapter 2 for planaria. The process requires none of the standard tools of traditional genetic screens, such as mutations, a genetic map, or balancer chromosomes. It is easy to see why RNAi now dominates the genetics literature, both for genome-wide screening and for analysis of single genes. Even if RNAi produces no new therapeutic agents, it has provided an easy and powerful means of silencing gene expression in diverse experimental systems, opening them up to the power of genetic analysis.

Limitations of RNAi

Despite its attractiveness for knocking down gene expression, RNAi has some limitations. First, the effects are variable for different genes and different organisms for reasons that are largely unknown. This limitation is one that is likely to become less important as more screens are done and investigators learn more about the mechanism of RNAi, although it may never go away entirely. In any method that attempts to perform a uniform assay on tens of thousands of diverse genes, some will prove to need individual attention. This variability is not too surprising to one who has done traditional genetic screens and realizes that even traditional mutants have variable expressivity, as described in Chapter 4.

Secondly, RNAi-induced phenotypes are probably best regarded as similar to genetic hypomorphs, as defined in Chapter 4, rather than null mutations. This may not be a limitation, because hypomorphs are often useful for constructing an allelic series, as described in Chapter 4. Nonetheless, it must be kept in mind that the RNAi phenotype may not be the same as the null mutant phenotype. With genome-wide screens involving thousands of genes, such tests are not feasible. However, as more is learned about the effectiveness of different siRNA sequences, it is not inconceivable that the equivalent of an allelic series will eventually be created by RNAi. In fact, with heat-inducible promoters for the dsRNA transcription, the equivalent of temperature-sensitive mutations has been made in *Drosophila*. This may be the first step towards intentionally producing hypomorphic mutations.

The problem of cross-reactivity

The third limitation is that non-specific or cross-reactive effects between a target gene and another gene have also been seen. Cross-reactivity or non-specific effects can arise because the siRNA is quite short, only 19–24 nucleotides. Figure 7.14 shows two mismatches between the siRNA and cross-reactive sequence. Notice how, in this figure, the molecules will base pair with each other despite the mismatches. Furthermore, some

BOX 7.4 RNAi in mammalian cells

The ability to use RNAi to knock down the expression of genes at will has immense promise for genetic analysis and drug therapy in mammalian cells. For example, imagine that the treatment for a rapidly growing or metastatic tumor is not chemotherapy or radiation but instead is the injection of a library of RNA molecules that will block cell proliferation or metastasis. Or that the treatment for a viral infection is to use dsRNA to block the replication or expression of the viral genome. It is far too early to know whether RNAi will ever become a realistic treatment, but many different laboratories and companies are investing resources in developing RNAi in mammalian cells. No one has yet done a true 'genome-wide' screen that tests all 30 000 or more mammalian genes, but numerous smaller-scale screens have been performed that test a subset of genes. Many different human and mouse RNAi libraries are commercially available, with more libraries and new approaches being developed constantly. A few of the approaches are described in this text box.

siRNA transfection

One of the first RNAi screens in mammals used chemically synthesized short interfering RNA (siRNA) molecules corresponding to more than 500 different human genes. This collection of siRNA molecules was transfected into HeLa tissue culture cells to look for genes that made the cells more or less sensitive to apoptosis. Since the siRNA collection included genes already known to be involved in apoptosis, the screen constituted a test of the method. One of the genes found by this method was a known apoptotic gene, but the method has several disadvantages for general use. First, and in addition to the labor and expense involved with synthesizing a large number of RNA molecules, the effects of the siRNAs are not long-lasting once the cells have been transfected. As the transfected cells divide, the siRNA becomes diluted, so the effects are transient. Consequently, large excesses or multiple transfections of the siRNA collection had to be used to see a consistent effect. This limits the utility of this method to genes whose transcripts are rapidly turned over. Furthermore, although HeLa cells and many other cultured cell lines can be readily transfected, many primary cell cultures cannot. Thus, the cells that would be the most important therapeutic targets are the cells least likely to be transfected. Transfecting synthetic siRNA is currently being used to test candidate genes identified by other methods, but other large-scale screens usually employ one of the alternative methods.

Plasmid-based screens

To avoid the problems of transient expression and dilution of the effective molecules, some other large collections use plasmids with short hairpins. These plasmids can either be transfected into cell lines or introduced with a viral vector. The cells have a plasmid that has a sequence with an inverted repeat (the hairpin) corresponding to the targeted genes. Often this sequence has been placed under the control of an RNA polymerase III promoter so that its transcription is controlled separately from the normal cellular genes, most of which are transcribed from a polymerase II promoter. When the sequence on the plasmid is transcribed, an RNA of approximately 50 nucleotides is generated that folds into a double-stranded hairpin with a stem of 21–23 paired bases and a loop of 6–10 bases, as shown diagrammatically in Figure 7.12A. The short hairpin RNA (shRNA) is cleaved by Dicer to produce the siRNA that becomes incorporated into the RNA-induced silencing complex (RISC) by which RNAi occurs.

A variation of this strategy is to insert a known miRNA gene on the plasmid and embed the shRNA sequence within that gene. The plasmid is introduced into the cells as before, and the miRNA gene is transcribed. The initial transcript is processed in the nucleus to a precursor miRNA of 70–90 nucleotides, as shown in Figure 7.13. This molecule is then exported to the cytoplasm where it is cleaved by Dicer to produce the siRNA that becomes incorporated into the RISC. In some direct comparisons, the collections that have the shRNA embedded into the miRNA gene have knocked down gene expression more effectively than the equivalent sequence cloned directly into the plasmid. Many commercially available shRNA collections are based on this approach.

Viral shRNA screens

Another strategy uses RNA viruses rather than plasmids. Many different classes of RNA virus can infect mammalian cells, replicate, and integrate their genome directly into the host genome. Thus, the shRNA genes can be incorporated into a viral genome *in vitro*, and the virus will transduce and integrate the shRNA molecules. The shRNA is stably inherited because the viral genome is integrated into the host genome, and expression can be monitored for many cell generations after the original infection. Viruses that have been used for such a collection of shRNA genes include retroviruses, adenoviruses, and lentiviruses. The expression of the shRNA gene can also be controlled by an inducible promoter because the shRNA can be incorporated as a sequence cassette at a cloning site. This provides the opportunity to turn the RNAi silencing effect off or on as needed for particular experiments.

With the potential importance of such therapeutic agents, other approaches are certainly being developed, and refinements of these basic strategies are forthcoming.

Further reading

Dillon, C.P., Sandy, P., Nencioni, A., Kissler, S., Rubinson, D.A., and Van Parijs, L. (2005). RNAi as an experimental and therapeutic tool to study and regulate physiological and disease processes. *Annual Review of Physiology*, **67**, 147–63.

Moffat, J. and Sabatini, D.M. (2006). Building mammalian signalling pathways with RNAi screens. *Nature Reviews Molecular Cell Biology*, **7**, 177–87.

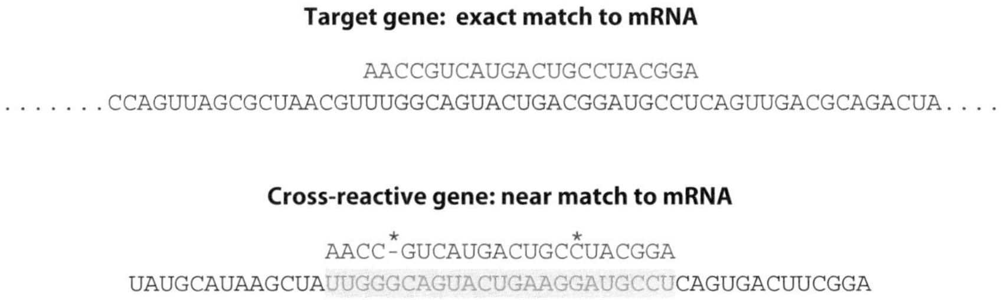

Target gene: exact match to mRNA

AACCGUCAUGACUGCCUACGGA
. . . . CCAGUUAGCGCUAACGUUUGGCAGUACUGACGGAUGCCUCAGUUGACGCAGACUA . . . .

Cross-reactive gene: near match to mRNA

*   *
AACC-GUCAUGACUGCCUACGGA
UAUGCAUAAGCUAUUGGGCAGUACUGAAGGAUGCCUCAGUGACUUCGGA

Figure 7.14 The cross-reactive effects of siRNA. The siRNA molecule is shown in red, with the target mRNA shown in blue. The siRNA has been synthesized to correspond in sequence to a portion of the target mRNA and will form an exact match to that sequence. Cross-reactive effects are seen when other unrelated mRNA molecules have a sequence that matches or nearly matches the siRNA sequence. Such a cross-reactive sequence is shown in black. In this hypothetical case, the cross-reactive sequence has one additional base and one mismatched base with the siRNA, as indicated by the red asterisks. Cross-reactive effects are known to occur by similar mechanisms, although it is not known how many or what type of mismatches trigger a cross-reactive effect.

experiments indicate that absolute identity may not be required for all positions in the siRNA so that the region of complementarity may be even shorter than 22 nucleotides. Although these dsRNA molecules with a mismatch may trigger an effect on translation rather than on RNA degradation, gene expression is still blocked. The tolerance for mismatches increases the probability that the same or a closely related sequence will be found at other sites in the genome, either another member of the same gene family or even a functionally unrelated gene. Thus, in complex genomes, transcripts with unrelated functions that happen to have short regions of sequence identity may be inadvertently targeted by the same siRNA.

This cross-reactive effect may be more widespread than is typically realized. One computational study used a 21-nucleotide sequence as the predicted size of the siRNA and found off-target similarity for more than 17 percent of the dsRNAs in the *Drosophila* database; when a 19-nucleotide sequence for the siRNA was used, more than 40 percent of the dsRNAs were predicted to have off-target similarity. This suggests that the use of siRNA must be viewed carefully. On the other hand, being aware of the frequency and the problems of mismatches can be compared with similar effects in any experiment in which short nucleotide sequences are used, such as in the design of oligonucleotide primers for PCR. Careful searches for potential off-target sequence similarity are routinely done to avoid this problem with PCR, and it seems likely that this will also become routine with siRNA.

In mammals, and perhaps other organisms, non-specific effects can also arise from physiological sources, in addition to this unexpected sequence similarity. Some evidence indicates that even such short sequences as found in siRNA can sometimes induce an interferon response in mammals that degrades RNA generally. In addition, it should be recalled that miRNA molecules similar to siRNAs are involved in a large number of physiological processes. Since RNAi takes advantage of these normal cellular processes, the machinery might be overwhelmed or shut down if too much dsRNA is used. As with many other aspects of RNAi experiments, these off-target effects will become easier to avoid as more is learned about the normal cellular response and better designs are incorporated.

7.5 Screening for mutant phenotypes

While most of the attention in genome-wide screens is focused on the techniques involved in disrupting genes or disturbing gene expression, screening the mutants that arise cannot be overlooked. As with a traditional genetic mutant hunt, the phenotypes that are recognized depend on the eyes and assays of the investigators. With genome-wide screens that test thousands or tens of thousands of genes, this becomes an important large-scale problem. Many possible phenotypes could be scored and, in many cases, the genome-wide screen is done with one or a few specific phenotypes in mind. For example, the genome-wide screen may be done by one set of investigators looking for genes that affect chromosome behavior in meiosis. Any mutant phenotype that does not affect meiosis may be noted but it is not pursued. This is similar in logic to traditional gene-by-gene genetic screens but is boosted by the power of the genome-wide screen.

Other genome-wide screens attempt to catalog as many possible phenotypes as can be observed. Such screens are similar in principle (if not in detail) to the screens for the first mutants in *C. elegans* or other model organisms. In these cases, the assays often test for very broad phenotypes such as growth and viability or fertility. The initial genome-wide screen of the yeast deletions tested growth under eight different conditions, with subsequent tests done for many other effects. The RNAi screens of the *C. elegans* genes recorded mutant phenotypes that can most readily be observed in worms—viability, fertility, general morphology, and movement. Subsequent experiments tested for many other effects, including life span and behavior.

In many ways, characterizing the phenotype of mutations generated by genome-wide screens is similar to characterizing mutations generated by gene-by-gene screens. Since all of the strains with gene disruptions or perturbations are maintained, they are available to be tested by another investigator with another idea for a screen. Thus, the same strains can be tested repeatedly. This is only slightly different from what has been done with traditional genetic screens, but the difference is very important. This difference is illustrated in Figure 7.15. In both traditional genetic screens and genome-wide screens, mutants are found and characterized;

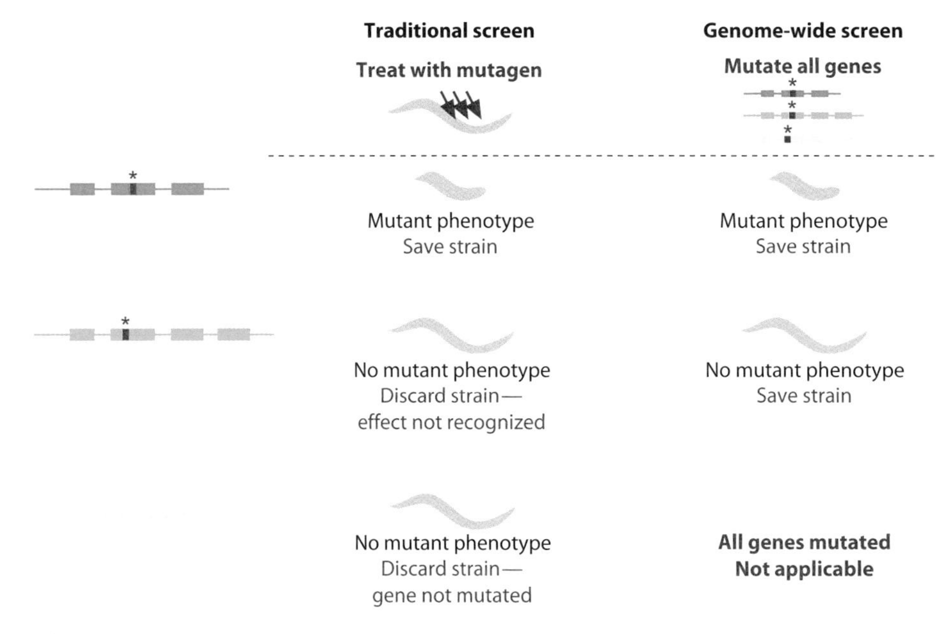

Figure 7.15 Comparison summary of traditional genetic screens with genome-wide screens. Three different genes are shown on the left. In a traditional screen, genes are mutated at random and mutant phenotypes are found. In a genome-wide screen, all genes are mutated by directed methods before looking for phenotypes. If the mutation causes a mutant phenotype, as shown for the gene in blue, the two approaches are the same. However, if the mutated gene does not give a mutant phenotype, as shown for the green gene, or if the gene is not mutated at all, as shown by the yellow gene, traditional methods miss the effect that genome-wide screens find.

these strains are saved to be available for other investigators. This is the gene in blue shown at the top in Figure 7.15. The difference between the two methods lies in the strains that *do not* show a mutant phenotype, the genes shown in green or in yellow in Figure 7.15. In traditional genetic screens, these are discarded. They may have phenotypes that are interesting or important to another investigator, but because the original investigator was not looking for that specific phenotype, the strain is not saved. Other investigators will need to produce and identify the mutants themselves. This is inevitable since no investigator can look for every possible phenotype and not every strain can be maintained if it is not known to have a mutation.

In genome-wide screens, all of the gene disruptions and perturbations are saved, even if the original investigator sees no mutant phenotype. Retention for the strain collection depends *not* on a *mutant phenotype* but rather on a *mutant gene*. Thus, other investigators need only screen the original collection of genes and not produce their own collection of mutations each time. This is a very important advantage, and is one of the reasons why many more phenotypes can be tested by genome-wide screens. Of course, for technical reasons, investigators may repeat the genome-wide screen; perhaps they do this because they are working with a more refined set of gene predictions or a better method of performing RNAi. Nonetheless, the collection of affected genes is saved and available to other investigators with other assays.

Because genome-wide screens involve examining thousands or tens of thousands of mutants, attempts are being made to score phenotypes by automated or high-throughput methods. For example, the phenotypic assay might involve expression of a reporter gene that can be detected in cell sorting or other automated methods. Mutants that affected expression of the reporter could be detected in a high-throughput assay. Automated methods to perform these high-throughput assays are also being developed. With the number of mutant strains that are generated by genome-wide screens, high-throughput methods are very attractive, and new assays will regularly be developed. However, whatever the method, the assay for the mutant phenotype will only be as refined as the observer, and many phenotypes will be missed in the initial screens. Because all strains exist for subsequent analysis, this is not usually a problem.

7.6 Confirming the effects

Reading the results of a genome-wide screen is roughly similar to reading a comprehensive encyclopedia—abundant amounts of information are available, but little of it can be assimilated at one time. The results of a genome-wide screen are in effect a database, and the precise roles of the individual genes usually need to be pursued separately by the traditional methods of genetic analysis described elsewhere in this book. The expanded flowchart in Figure 7.16 summarizes this analysis. One laborious method that is available in some model organisms is to compare the mutant phenotype from the genome-wide screen with that observed in a traditional single-gene screen, as previously described in Chapter 3, or with a reverse genetics method as described in Chapter 6. As with a traditional mutant phenotype, it may be possible to determine if the mutant is rescued or complemented by a wild-type version of the gene or if a related (possibly opposite) mutant phenotype is observed when the candidate gene is over-expressed. The mutant phenotype can be compared with the expression pattern of the gene by using some of the methods described in Chapters 8 and 9 to determine if these are consistent. Gene interactions can be assayed, as described in Chapters 10, 11, and 12.

Genome-wide mutant screens, as with traditional mutant screens, provide the material for analyzing almost any biological process. Compared with traditional genetic screens, genome-wide screens have the significant advantages that all genes can be tested in a statistically unbiased manner and that any candidate gene has already been cloned. Although the field of genetics has not quite moved to this situation, it is not hard to envision a time when all genetic screens for any biological process will be done as genome-wide screens. In fact, for some model organisms, that day is nearly upon us.

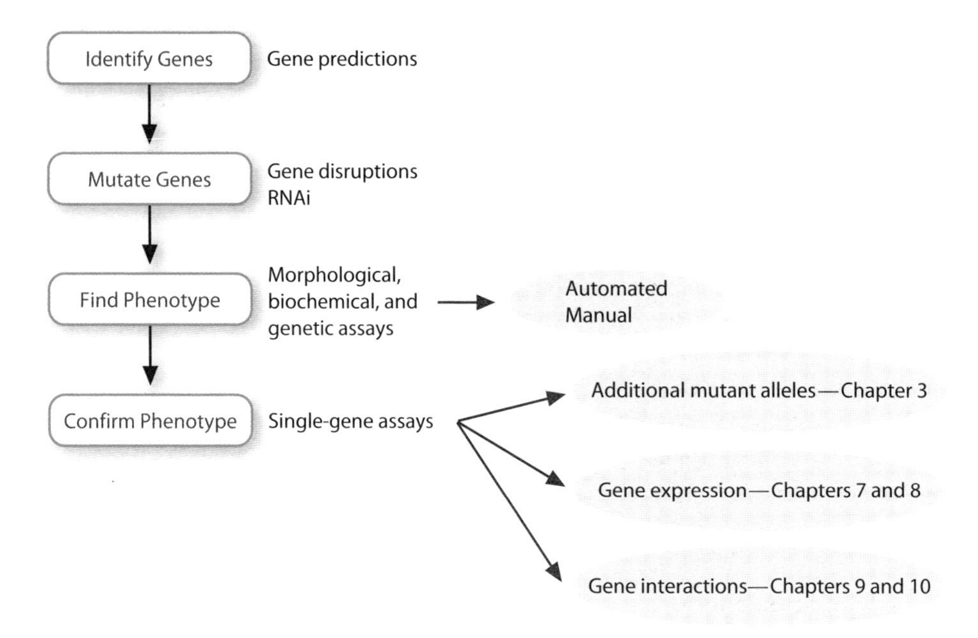

Figure 7.16 An expanded flowchart of genome-wide screens. Genes are identified or predicted within the genome. Mutants are created for each gene, using gene disruptions or RNAi, and each mutant strain is tested for its phenotype. Literally hundreds of different assay conditions have been used for different screens. Both automated high-throughput screens and manual screens have been done successfully. Once the gene has been found, the follow-up methods to confirm the phenotype are similar to those used for other genes. Additional mutant alleles can be found, as described in Chapter 3. Gene expression is tested by methods described in Chapters 8 and 9. The interactions of gene with other genes can be determined by methods described in Chapters 10, 11, and 12. Some or all of these follow-up methods will be important for any particular mutant recovered from a genome-wide screen.

7.7 Lessons from genome-wide screens

Every genome-wide screen yields its own lessons; the principal result is often the library of mutant strains itself. These screens can also be viewed as a whole rather than as a collection of individual effects, which may provide additional insights into global aspects of physiology, evolution, and genetics that are not readily available from other methods. For example, genome-wide screens recognized that few X-linked genes in *C. elegans* are involved in gametogenesis, a result that could not readily have been shown by traditional genetic screens.

It may seem premature or foolish to try to draw even more general conclusions or cross-species comparisons from the genome-wide mutant screens that have been done so far. After all, different methods have been used in each model organism, and various levels of 'genome-wide' analysis have been achieved. At a fundamental level, the genome-wide screens are not comparable. Among eukaryotes, deletions for nearly every predicted gene have been made only for *S. cerevisiae*. In other organisms, such as worms and flies, the phenotypes are primarily the result of RNAi screens. Thus, in comparing the results of these screens, we may be comparing null mutations with hypomorphic mutations. Despite these cautions about the nature of the mutant phenotypes and the number of genes, we will attempt to draw a few lessons from the results of genome-wide mutant screens in yeast, flies, and worms.

Lesson 1. New genes are found even for well-studied processes

The first and perhaps most important general lesson is that genome-wide screens work very well and find genes that have been missed or might have been missed by traditional genetic screens. At the beginning of the chapter, we cited one example from yeast involving new genes involved in DNA repair. The genome-wide screen identified three genes that had never been implicated in DNA repair before. Another example in yeast involves galactose metabolism, a pathway that has been thoroughly studied by traditional genetic methods. From the genome-wide selection in which each deletion was tested for its ability to grow in the absence of galactose, 10 genes were found that had escaped detection in traditional gene-by-gene mutant selections. Finding new genes that are involved in such familiar cellular processes may be sufficient justification for genome-wide screens. There is great satisfaction in knowing that all possible genes have been tested for an effect, and that no one critical gene remains to be found.

However, this is not the only justification—new phenotypes are also found for known genes. In the yeast genome-wide screens, additional mutant phenotypes were also found for genes that had been studied previously; for example, two genes known to affect the calcium-regulated phosphatase calcineurin were also found to affect the ability of yeast cells to grow under ionic stress. Similar examples can be found from nearly every study that uses genome-wide screens to investigate a biological process; new genes are found, and previously known genes are found to have unexpected effects.

Lesson 2. Many genes do not have a mutant phenotype

The second general lesson is that many genes do not appear to have a mutant phenotype. For geneticists who make their living from mutant phenotypes, this result is a bit humbling. For example, among the yeast deletion collection, only 18.7 percent of the deletions affected the ability of the yeast cells to grow in rich media. For RNAi experiments in worms, about half of the genes have recognizable mutant phenotypes; the remainder have no mutant phenotype under the RNAi conditions tested.

These results can be understood a number of ways, as summarized in Figure 7.17. The first explanation, which is particularly relevant for the screens based on RNAi, is that the gene in question is not affected by

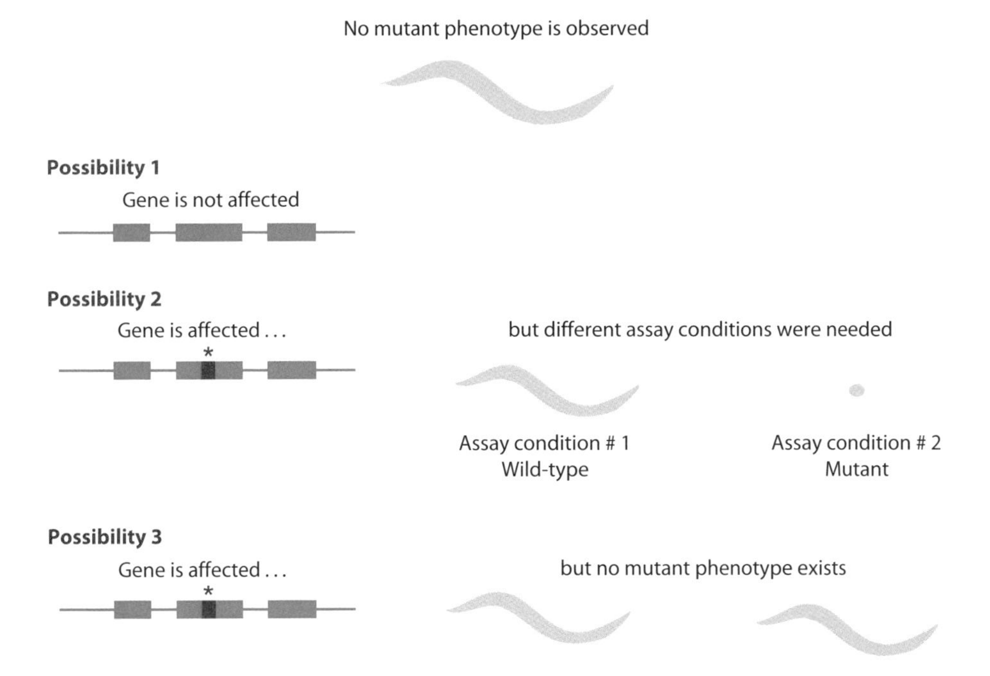

Figure 7.17 Three possible reasons why no mutant phenotype is observed for some genes in genome-wide screens. The first and least interesting possibility is that the procedure did not affect the gene. This is a particularly important problem for RNAi-based screens. The second possibility is that different assay conditions are needed to see the mutant phenotype, here seen as a worm that does not hatch from the egg. The third possibility is that mutations in the gene do not produce a mutant phenotype. These are discussed in more detail in Chapters 10, 11, and 12.

the procedure. This is shown at the top of Figure 7.17. Thus, failure to see a mutant phenotype may be the result of the failure of the procedure itself, at least as it relates to the idiosyncrasies of that particular gene. This possibility can be addressed by additional tests more specific to that gene or by comparing the RNAi results with the effects of deletions when available. Again, this explanation accounts for some of the failure to see a mutant phenotype. Even in yeast knockout strains, the knockout procedure is known occasionally to have produced duplicate copies of the wild-type gene elsewhere in the genome, and so the gene was not truly 'knocked out'. This explanation can be summarized as follows: **the gene might have a mutant phenotype but this procedure has not produced a mutant**.

A second possibility, shown in the middle of Figure 7.17, is that only a limited number of phenotypes and environmental conditions are tested in any one screen. When thousands of mutants are being scored, subtle defects can easily be missed. Thus, under one set of assay conditions the strain looks normal, whereas under another set the strain has a mutant phenotype. This explanation can be summarized as follows: **the gene has a mutant phenotype, but the investigators have not yet looked in the right place.** This certainly explains some of the results, although it is impossible to know if it explains all or even most of them. After all, there are always other conditions to test and other phenotypes to assay. In fact, with repeated screens under a wide range of genetic conditions, a phenotype has been found for about 5000 of the 6130 genes in yeast. As mentioned previously, one of the strengths of genome-wide screens is that the mutant strains are retained for further tests regardless of their original phenotype.

A third possibility is biologically the most intriguing. This is shown in the bottom part of Figure 7.17. Alterations or even deletions in some genes might simply have no effect on the organism that can be detected even in the most careful and thorough assay. For some genes, the mutant effects may only be apparent in combination with other mutations in the genome. That is, gene products interact with each other and compensate for each other such that the loss of one gene only shows an effect if other genes are also altered. The analysis of gene interactions is the topic for Chapters 10, 11, and 12, and these mutants with no mutant phenotype will be considered in more detail there. This explanation can be summarized as follows: **the gene has no mutant phenotype by itself.**

The results of genome-wide mutant screens are often combined with the results from other genome-wide screens, such as expression profiling as described in Chapter 8 and interaction networks as described in Chapter 12. Each of these methods brings the power of a traditional form of genetic analysis to a new and even more awesome level. The combination of these genome-wide methods is rapidly becoming the standard method of genetic analysis.

SUMMARY

Genome-wide screens attempt to identify phenotypes for every gene in the genome

The genome sequence and the list of annotated genes are sometimes compared to a parts list for a complicated machine. The parts list is certainly important as a first step to ensure that everything that will be needed is present. We know that we will not have to run to the store at the last minute to buy one more piece that we had overlooked. However, knowing how much sheet metal is needed and how many screws, bolts, and nuts are necessary to hold it together does not allow us to build an automobile. Likewise, knowing how many putative transcription factors, potential receptors, and so on there are does not tell us how a complex biological process works. However, with genome-wide screens, we have the opportunity to begin to understand even the most complex biological questions. We have all the parts, and we have tested them for their role in the biological question of interest. Genome-wide mutant screens combine the information from genome projects with the power of traditional genetic analysis in an attempt to infer a function for every gene in a genome. By itself, a genome-wide screen provides new information about even well-studied biological questions. But even more importantly, genome-wide screens allow us to answer questions more complicated than we yet know how to ask.

Chapter Capsule

Genome-wide screens

Genome-wide screens are quickly become the standard methods in those organisms whose genomes have been sequenced. The steps in a genome-wide mutant screen include:

- identifying the genes to be mutated
- disrupting or perturbing genes
- examining the mutant strains for phenotypic effects
- confirming the mutants that are found, usually by gene-specific methods.

Every gene in the yeast *S. cerevisiae* has been disrupted by homologous recombination. For multicellular organisms, RNAi has become the best method for perturbing genes.

Unit 3
Gene activity

The activity or the function of genes can be understood from a combination of molecular approaches examining gene expression and mutant phenotypes examining gene activity.

CHAPTER MAP

- **Unit 1 Genes and genomes**
- **Unit 2 Genes and mutants**
- **Unit 3 Gene activity**

8 Molecular analysis of gene expression

This chapter describes approaches to understanding the expression of a cloned gene or a set of cloned genes.

9 Analysis of gene activity using mutants

- **Unit 4 Gene interactions**

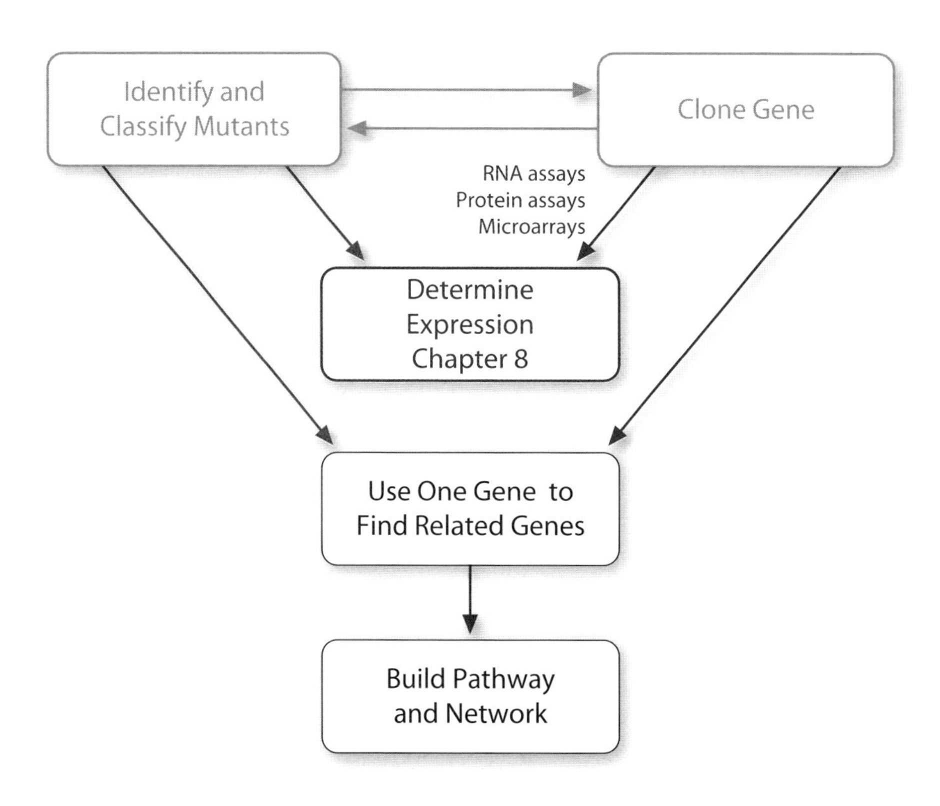

8

CHAPTER 8

Molecular analysis of gene expression

TOPIC SUMMARY

The expression pattern of a gene is critical to understanding its overall biological role. The molecular analysis of gene expression can be approached using methods from many different perspectives. These perspectives include:

- RNA or protein expression patterns
- Static 'snapshots' of gene expression patterns or patterns based on *in vivo* dynamics
- Expression of an individual gene or a genome-wide approach, such as a microarray that examines thousands of genes simultaneously.

Although every gene has its own characteristic pattern of expression, the same approaches can be applied to almost any gene that has been cloned.

IN BRIEF

A molecular understanding of the expression of a gene or a group of genes is a fundamental property in genetic analysis.

INTRODUCTION

Mendel recognized that genes affect some characteristics but not others—the gene affecting round or wrinkled peas acted separately from the gene determining green or yellow seed color, for example. Because we know that all of the cells in an organism have the same genetic constitution, with some interesting exceptions that we will not consider here, a logical inference is that genes are expressed in some tissues or structures but not in others. Thus, the characteristic qualities of each cell type, tissue, organ, developmental stage, or response to a change in environmental conditions reflect the expression of a subset of the genome.

The expression pattern of a gene is the outcome of a complex, highly regulated process. Some of the basic mechanisms common to the expression of all genes are well understood, but the processes determining specific expression patterns remain largely unknown. Common questions that arise in discussing gene expression include:

- What genes are expressed in a particular cell type or developmental stage?
- What genes are expressed in response to a particular stimulus?
- What determines which genes are turned on or off?
- What step in the expression process is the principal control point? Is it transcription, splicing, RNA silencing, translation, protein modification, protein stability, RNA or protein movement, or what?
- What is the significance of changes in gene expression that are not 'on' or 'off' but rather somewhere in between?
- How is the expression of this set of genes coordinated with other genes?

An investigator will want to ask all of these questions for any biological process, but the answers are different for each gene, cell type, developmental stage, and growth condition or treatment.

Instead of focusing on the *answers* to these and other questions, we will describe how the answers to these questions can be obtained. This is the *analysis* of gene expression. In this chapter, we will look at molecular approaches for analyzing gene expression—the types of experiments that can be performed when a gene has been cloned or a genome has been sequenced and annotated. The tools we will describe are assays for RNA expression and protein expression, either for a single gene or on a genomic scale. In the next chapter, we will consider how to analyze the activity of a gene using mutants, regardless of whether or not the gene has been cloned yet.

8.1 Molecular analysis of gene expression: an overview

Any study of gene expression begins with a careful definition of what the investigator hopes to learn. Gene expression can be studied using a variety of methods, but the method that is most appropriate in a given instance depends on what aspect of gene expression the investigator is immediately interested in. There are several ways of categorizing the methods used to analyze gene expression. Below, we describe some of the common classification schemes, which are summarized in Figure 8.1. The associated methods will be described in more detail throughout the chapter. A brief description of each of the methods is provided in Table 8.1 for those readers who may not be familiar with any of the terms.

The most fundamental way of separating out different approaches for studying gene expression is based on the molecular phenotype that is being examined—RNA or protein expression (shown in Figure 8.1 as the difference in the colors of the words). For example, some approaches focus on **RNA** expression by looking at the transcription pattern of a gene and the levels of mRNA. Other approaches examine the expression pattern of the **proteins**, using methods such as antibody-based techniques and translational reporter genes. This is the principal division used in Table 8.1 and Figure 8.1. While the expression pattern of most genes is determined primarily by their transcription pattern, there are many other possible points of control. As a result, a combination of RNA and protein expression methods is necessary for a full understanding of the regulation of a gene's expression.

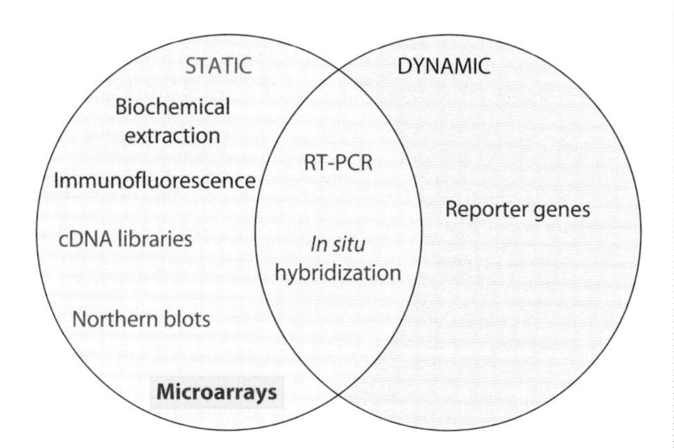

Figure 8.1 Methods of characterizing gene expression. Gene expression can be characterized by examining transcripts using the methods shown in red letters, or by examining proteins using the methods shown in blue letters. Expression patterns can also be seen as static snapshots at a particular time, as indicated by the pale blue circle, or as dynamic changes, as indicated by the pale pink circle. RT-PCR and *in situ* hybridization often combine both static and dynamic elements, as indicated by the area of overlap. Expression patterns can also be examined on a genome-wide basis, as indicated by the yellow box around microarrays.

Another way of dividing different approaches for studying gene expression is to separate the methods that take **snapshots** of the biology from those that detect **dynamic** changes. This is another way of saying that some approaches examine gene expression at a particular moment in time by using fixed specimens or extracted RNA or protein, whereas others can monitor the situation *in vivo*, allowing us to study the system in a continuously changing way. In Figure 8.1, these two approaches are represented as the two overlapping circles. For example, a biochemical extraction of proteins or the construction of a cDNA library depends on isolating proteins or RNA from a particular tissue or stage and thus shows a snapshot of the expression pattern of the gene at that time. In contrast, a dynamic method, such as the use of a reporter gene with GFP monitored *in vivo*, allows the investigator to follow the changes in gene expression directly, often as they are happening in live specimens. Any static method can approximate a dynamic method if the 'snapshots' are taken close enough together, just as the action of rapidly flipping a series of pictures can appear to animate them. The ability to see dynamic differences as they occur is often important for a full understanding of a process.

Finally, differences in gene expression can be assayed on an **individual gene** or **genome-wide** basis. Most methods for studying gene expression are used to analyze individual genes, but genome-sequencing projects have led to a growing interest in global gene expression studies. Genome-wide methods were introduced in Chapter 7 in the context of finding mutant phenotypes. Microarrays are the most common method of analyzing gene expression across the genome, but other techniques include large-scale RNA sequencing and proteomic approaches. These methods are

Table 8.1 Expression methods discussed in this chapter

RNA-based methods

Northern blots

RNA is isolated and separated on a denaturing gel. The RNA is transferred to a nylon filter or other medium. A labeled probe from the gene of interest is hybridized to the filter to detect the transcript or transcripts corresponding to the gene.

RT-PCR, or reverse transcription PCR

A crude extract is made from the tissue or stage of interest; the RNA may or may not be purified away from the DNA, proteins, and cellular debris. The RNA is reverse transcribed into cDNA, and the specific cDNA for a gene is amplified by PCR using primers that are specific to that gene.

In situ hybridization

A tissue section or an entire organism is fixed and mounted on a microscope slide. A labeled probe from the gene of interest is hybridized to the fixed specimen to display the location of the transcript.

Transcriptional reporter genes

The upstream regulatory region of a gene of interest is fused to the coding region of a gene encoding a reporter protein (a protein whose expression can easily be assayed and quantified, such as β-galactosidase from *E. coli* or GFP from jellyfish). The construct is then transformed into the organism, and the reporter protein expression is assayed in cell extracts or whole organisms. The reporter protein will be expressed in the pattern directed solely by the regulatory elements of the gene of interest, specifically revealing the transcription pattern of the gene.

Microarrays

Tens or hundreds of thousands of different DNAs corresponding to all or most of the transcripts in the genome are robotically spotted (two-channel arrays) or directly synthesized (one-channel arrays) on the surface of a support, such as a glass slide. RNA isolated from a sample is used as a template to make labeled cDNA or cRNA. The labeled cDNA or cRNA is hybridized to the array, and the pattern of hybridization is read by an optical reader. The fluorescent intensity at each spot on the grid provides a measure of the expression of the corresponding gene in the original sample. In a two-channel array, cDNAs from two different samples (usually a control and a treatment sample) are labeled using different fluorescent molecules and are simultaneously hybridized to the same array. The two different hybridization patterns are compared computationally to identify genes whose expression is different in the two RNA samples.

Large-scale RNA sequencing

Small RNAs or fragments of mRNAs isolated from a particular sample are cloned, and 17–23 bp of hundreds of thousands of clones are sequenced. The gene represented by each clone is determined by homology to the genome. The number of clones corresponding to each gene is a quantitative measure of RNA frequency in the original sample.

Protein-based methods

Western blots

Protein extracts are separated by denaturing or non-denaturing polyacrylamide gel electrophoresis (PAGE). The proteins are transferred from the gel to a membrane, which is then probed for the protein of interest using an antibody against the protein or an epitope tag. The antibody is conjugated to an enzyme that catalyzes the breakdown of a substrate, emitting light in the process. The light emission can be detected by exposing the membrane to an X-ray film.

Immunofluorescence

An antibody is made to the protein product of the gene under study; this is known as the primary antibody. The primary antibody is bound to fixed specimens of the tissue or stage of interest. In order to detect the presence of the primary antibody, a secondary antibody with a fluorescent label is used. The secondary antibody, available commercially, is directed against the part of the primary antibody molecule that is conserved in all antibodies of that species. (For example, anti-rabbit secondary antibodies will bind to any primary antibody produced in a rabbit.) The fluorescence pattern of the secondary antibody indicates the location of the primary antibody, which in turn indicates the location of the protein of interest. Strictly speaking, when a secondary antibody is used, the process is referred to as indirect immunofluorescence, but it is often referred to simply as immunofluorescence.

Translational reporter genes

As in transcriptional reporter genes, the protein-coding region of a reporter protein is expressed under the control of the upstream regulatory region of the gene of interest. In this case, the reporter protein is also fused in frame with the protein-coding region of the gene of interest. The construct is introduced back into the organism and expressed as a fusion protein. The expression pattern of the fusion protein (monitored by assaying for the reporter protein) should represent the pattern of stable protein expression of the gene of interest.

Proteomics

A genome-scale technique that separates the proteins from an extract on a two-dimensional gel. The proteins are separated by one property, such as isoelectric point, in the first dimension and by a second property, such as molecular weight, in the second dimension. The gel is stained, and the proteins of interest are re-isolated from the gel and identified by mass spectrophotometry. The complete set of proteins made from a genome is referred to as the proteome.

designated by the yellow box in Figure 8.1. Genomic techniques have greatly increased the throughput of studies and have provided a mountain of data. As no gene acts in isolation in an organism, the real value of genomic approaches lies in their ability to allow the researcher to view the expression of a gene in relation to those of all of the other genes in the genome, thus allowing a better understanding of the entire organism.

We have organized this chapter using two of these classification schemes. In the first half of the chapter, we will consider methods for studying individual genes, divided by their usefulness for uncovering RNA or protein expression patterns. In the second half, we will discuss global gene expression approaches and RNA techniques, followed by a brief description of some protein methods.

8.2 RNA expression analysis of individual genes

Because the expression pattern of most genes is determined at the level of transcription, many expression analyses of individual genes begin by looking at the RNA expression pattern. At one time, Northern blots were the only method for studying RNA expression. While they are still prevalent, other techniques have been developed that provide an improved level of quantitation or a more refined picture of the pattern of RNA expression. Here, we discuss four assays used to analyze RNA expression: Northern blotting, RT-PCR, *in situ* hybridization, and transcriptional reporter genes.

Northern blots are important for identifying the variety and size of transcripts expressed by a gene

Northern blotting was developed from the Southern blot technique used to look at DNA structures; however, Northern blotting uses RNA instead of DNA. The basic technique of a Northern blot has been employed for more than 25 years, so our discussion will be very brief. The procedure is shown diagrammatically in Figure 8.2.

1. RNA is purified from different tissues or developmental stages and run on a denaturing gel, which eliminates any secondary structure and separates the RNA population based on size. The RNA is sometimes called the 'target'.
2. The gel is blotted with a filter to transfer the target RNA to the filter.
3. The gene being investigated, or at least a portion of it, is labeled *in vitro* to make a specific DNA **probe**. Either a radioactive or a non-radioactive label can be used to make the probe for this gene. The probe will have the complementary sequence to a specific RNA on the filter.
4. The probe is hybridized to the RNAs on the filter, and the hybridization is seen as a band or bands, usually by exposure to film.
5. The same blot can often be used to examine the expression of multiple genes, one gene at a time, by chemically stripping or removing the first probe and then probing the same blot with a different probe.

Northern blots have a few significant advantages over other methods. The principal advantage is that *all* of the RNAs—full-length transcript, cleaved fragment, small RNA, and so on—with sequence hybridization to the probe and a sufficient level of expression will be detected. For instance, alternatively spliced products can be detected by using a probe complementary to any sequence that the transcripts have in common, as shown by the hypothetical example in Figure 8.2. The length of each of these RNA species can also be measured from the film by including an RNA size ladder on the gel. In fact, the size of a full-length transcript is often determined using Northern blots before it is known by other methods, which is particularly important for finding the lengths of the 5′- and 3′-untranslated regions. In addition, the transcription pattern can be observed by loading different lanes on the gel with RNA from separate tissues, developmental stages, or treatments. Our hypothetical example in Figure 8.2 shows a situation in which two tissues differ in their splicing pattern. This type of result is most readily seen with a Northern blot.

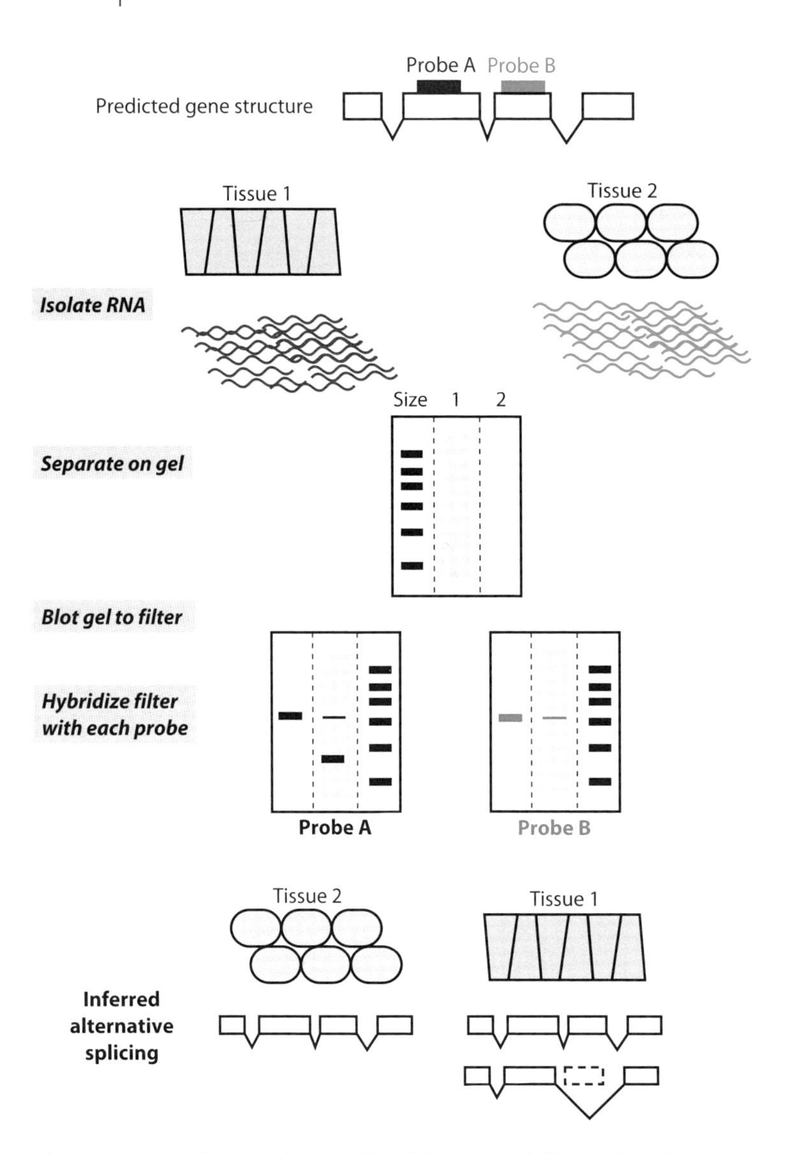

Figure 8.2 Transcript analysis by Northern blots. The goal of this Northern blot analysis is to examine the expression of the predicted gene, shown at the top, in the two tissues labeled Tissue 1 and Tissue 2. Two labeled probes are prepared from the predicted gene, probe A (in black) from exon 2 and probe B (in blue) from exon 3. RNA is isolated from each of the tissues and separated on a gel with suitable size markers. RNA from the gel is transferred to a nylon or nitrocellulose filter by blotting; the pattern of the filter is the reverse of the gel, as shown. The filter is hybridized with each probe. Probe A identifies a single transcript from Tissue 2 but two transcripts with different abundance from Tissue 1. The lower-abundance transcript from Tissue 1 is the same size as the single transcript from Tissue 2. Probe B identifies a single transcript from each tissue, the larger of the two from Tissue 1. The inference, as shown at the bottom, is that the gene is alternatively spliced in Tissue 1 to produce splice variants that include or exclude exon 3. Tissue 2 transcripts all include exon 3, and no alternative splicing is seen. Although the relative abundance of the transcripts is inferred from the strength of their signals, Northern blots are not truly quantitative. However, a probe from a conserved exon of the gene identifies all of the transcripts.

Despite their advantages, Northern blots have often been supplanted by other methods for many types of routine transcription analysis. A Northern blot requires much more RNA than most other methods; even when a large amount of RNA is used, many rare or rapidly turned over transcripts cannot be detected on a Northern blot. Thus, Northern blots are best suited to experiments in which getting a substantial amount of high-quality RNA is not too difficult and the transcript of the gene of interest is abundant. Many investigators also prefer to avoid Northern blotting for technical reasons. The gel used to separate the RNA employs hazardous materials (acrylamide, formamide, and formaldehyde) and RNA is also easily degraded. Because the entire process of Northern blotting involves working with RNA, great care must be taken throughout the process to avoid degradation. Furthermore, although Northern blots can provide snapshots of the transcription pattern of a gene, the cDNA cannot be cloned during the procedure or sequenced directly.

Until recently, Northern blot analysis was the standard in quantifying RNA expression quantitation. However, this method is not truly quantitative because the expression level of the gene of interest is only determined *relative* to that of another gene. Relative quantitation is done by re-probing the same blot with a second probe against a gene of constant expression that can act as an internal loading control. The expression of the loading control is used to normalize the expression of the transcript of interest. Other techniques, such as real-time RT-PCR, are much more quantitative.

RT-PCR is another widely used method of analyzing the transcription of individual genes, even when the transcripts are present at very low levels

For many applications, reverse transcription PCR (RT-PCR) is both faster and easier to perform than Northern blots. Conceptually, RT-PCR is a two-step process, although the two steps are often performed with a single reaction mix. The process is illustrated in Figure 8.3.

RNA is again purified from a particular stage, tissue, or treatment, although crude extracts can also be used. In the first step, the RNA is reverse transcribed to make cDNA using an enzyme called reverse transcriptase. Reverse transcriptase is an RNA-dependent DNA polymerase that polymerizes a new DNA strand from an RNA template but not a DNA template. To do this, the enzyme requires an oligonucleotide DNA primer that is complementary to the predicted RNA sequence, and the four deoxyribonucleotides in an appropriate buffer. Instead of using a gene-specific primer for the reverse transcription reaction, it is also possible to add either an oligo(dT) primer, which hybridizes to the poly(A) tail of every mRNA, or a mix

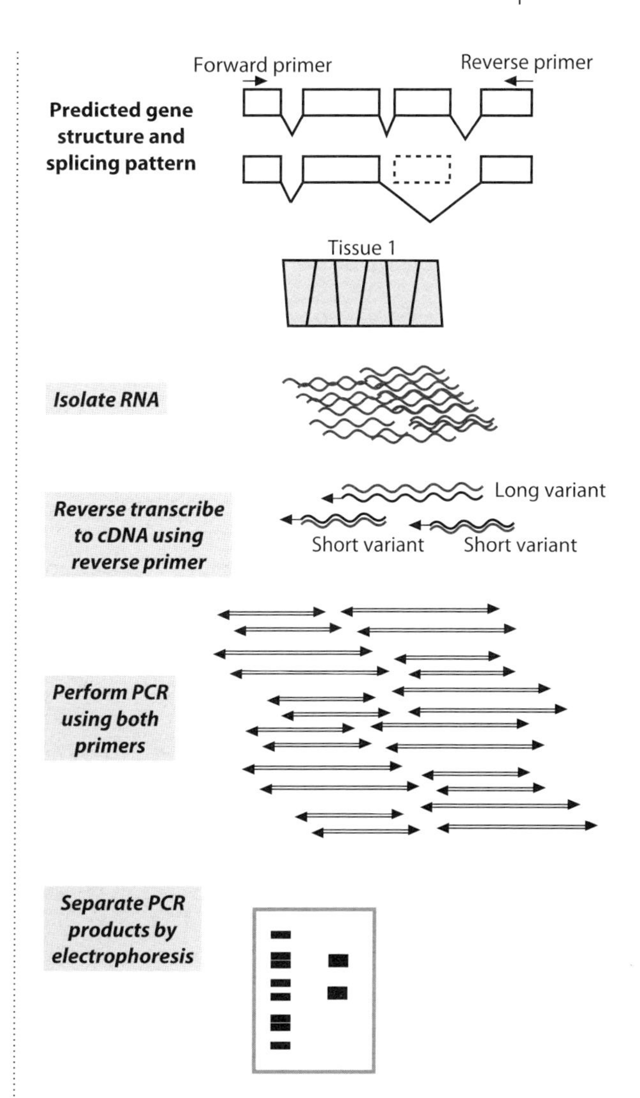

Figure 8.3 Transcript analysis by reverse transcription PCR (RT-PCR). The same gene structure and tissue are shown here as used in Figure 8.2. Two PCR primers are designed for the predicted gene, as shown at the top; in this case, primers near the 5′ and 3′ end of the gene were chosen. The RNA is isolated from Tissue 1 and reverse transcribed into cDNA. The gene-specific reverse primer from the 3′ end of the gene is used to prime the reverse transcription so that only this specific gene is made into cDNA. Oligo(dT) primers that bind to the poly(A) tail of mRNA or a collection of hexamer oligonucleotides with random sequence can also be used to prime reverse transcription, but these will make cDNA from all transcripts in the pool. Without further purification of the cDNA, both PCR primers are added with *Taq* or an equivalent DNA polymerase and PCR is performed. Note that both the short and the long variants are present but the relative abundance of each version is not preserved. The PCR products are resolved on a gel by electrophoresis. RT-PCR requires less RNA than a Northern blot and does not require labeled probes or blotting, which often makes it easier to perform. However, it will only detect sequences between the two primers, which must be carefully chosen.

of short random sequences, usually **random hexamers**, which can bind to any RNA with the complementary sequence. The choice of primer depends on the population of RNAs to be reverse transcribed; the latter two priming methods will allow for the reverse transcription of entire populations of RNAs rather than one specific species.

The primer hybridizes to the RNA and provides a 3′-OH to prime polymerization by reverse transcriptase, and the enzyme synthesizes a first-strand cDNA. In the second step of RT-PCR, two PCR primers are added that correspond to the sequence within the expected cDNA product; these are used to PCR amplify the cDNA product, which can then be detected by standard DNA gels. One of these primers (the forward primer) anneals to the cDNA, and the other can be the same primer used for the reverse transcription or can be a different reverse primer that will anneal to the reverse complement.

Because RT-PCR involves a much smaller amount of RNA and less handling of the RNA sample, it is typically easier to perform than a Northern blot. For these reasons, RT-PCR is usually the method of choice for studying the RNA expression of single genes when the transcripts are either expressed at very low levels or are rapidly turned over, and when working with very small tissue samples. Another significant advantage is that the PCR product can be used for cDNA cloning into a convenient vector. However, a few cautions are associated with RT-PCR. First, the primers need to be chosen carefully in case alternative splicing occurs; for example, a primer from exon 3 in Figure 8.3 would hybridize to only one of the transcripts. Secondly, unlike Northern blots, typical RT-PCR usually does not give an accurate picture of the amount of the transcript. RT-PCR is not truly quantitative because PCR amplification of the cDNA product is an exponential rather than a linear process (i.e. the amount of DNA product present increases exponentially with the number of rounds of amplification). Because accurate quantitation of the amount of transcript is often important, quantitative PCR methods have been developed and are in widespread use. Some of the more accurate and sensitive PCR-based methods for determining transcript abundance are described in Text Box 8.1. Of all of the methods used to study the RNA expression of single genes, these so-called real-time RT-PCR techniques are currently the most quantitative.

In situ hybridization detects transcript locations within the specimen

If RNA is isolated from different samples, such as different tissues, developmental stages, or stimulus treatments, either Northern blots or RT-PCR can be used to examine the expression pattern of the gene to which the RNA relates. Because these samples are comprised of a mix of cell types, tissues, or stages, these methods have a limited ability to discern the expression pattern for a specific transcription in a specific cell type or developmental stage. A more detailed approach is to detect the presence of the transcript directly in the cells of the organism or tissue using the process of *in situ* hybridization. This allows the precise localization of specific cells expressing a given RNA.

For *in situ* hybridization, a labeled probe that is complementary in sequence to the mRNA is made *in vitro*. The probe can be RNA produced by *in vitro* transcription, DNA produced by either PCR or *in vitro* nick translation, or an oligonucleotide. RNA probes are more sensitive (because RNA–RNA bonds are stronger than RNA–DNA bonds), but DNA probes are more specific and easier to prepare and use. Both radioactive and non-radioactive labels for probes are in widespread use. The common non-radioactive methods use digoxigenin (DIG) or biotin, among other labels. Detection of digoxigenin involves an anti-DIG antibody; biotin-labeled probes are detected using streptavidin. Both anti-DIG antibodies and streptavidin conjugated to different fluorescent labels are commercially available. The labeled probe is hybridized to fixed samples and detected using fluorescently labeled anti-DIG antibodies or streptavidin, and the location of the probe is analyzed. In some cases, *in situ* hybridization can detect as few as 20 copies of an mRNA in a cell.

In situ hybridization cannot be performed in all organisms, particularly those where it is hard to isolate and fix tissues with their transcripts intact, and is difficult in many others. However, for those in which it can be done, it is a very powerful technique. Figure 8.4 shows how *in situ* hybridization can be used to look at transcription patterns during development. The procedure does not require the isolation of RNA and does not require a guess, no matter how well informed, about the expression pattern in certain cells within a tissue. It reveals the expression pattern directly. By using

BOX 8.1 Quantitative PCR methods

The polymerase chain reaction (PCR) is the most widely used method in all molecular genetics for producing usable quantities of DNA or cDNA from a small amount of nucleic acid sample. For many purposes, the goal of using PCR-based techniques is to amplify the nucleic acid without attempting to be quantitative about the amount in the starting material or the product. However, sensitive quantitative methods are also important, particularly in expression assays, and reliable quantitative PCR (Q-PCR) assays have been developed. Because Q-PCR assays are conducted in real-time (as the amplification is occurring), the technique is also known as real-time PCR. For the most part, the terms quantitative PCR and real-time PCR are synonymous.

In order to understand the principles of real-time PCR, it is helpful to compare it with conventional PCR. Conventional PCR is usually assayed at its end point, when the reaction has been completed. The assay usually involves separating the products by agarose gel electrophoresis and determining their presence, size, and amount by ethidium bromide or similar staining in different samples. The reaction itself is not truly quantitative, and the detection assay is not very sensitive. To consider the detection method first, attempting to measure the concentration of a nucleic acid from the intensity of staining on the gel is usually an estimate, and even 10-fold differences in signal may be difficult to distinguish.

For quantitative assays, a more sensitive detection is required and an understanding of the kinetics of the reaction is important. When the amount of product is assayed at the end of each cycle, PCR follows rather complex kinetics, with a curve that can be divided into four phases, as shown in Text Box Figure 8.1. In the first phase, the amount of product is below the limit of detection, but the kinetics should be similar to the exponential phase. During the exponential phase (Phase 2 in Text Box Figure 8.1), all reaction components are new and abundant, so the amount of PCR product is doubling with each cycle. Replicate samples with the same amount of starting template show the least variability during this phase, so the exponential phase of the reaction is important for quantitation by real-time PCR. As the reaction components are depleted and degradation of the template and the product begins, the accumulation of the products becomes linear rather than exponential, as shown as Phase 3 in the figure. Replicate samples during the linear phase are highly variable in the amount of product, even when the same amount of template was present at the beginning. As the reactants are further depleted, the reaction enters the plateau phase (Phase 4), when no new

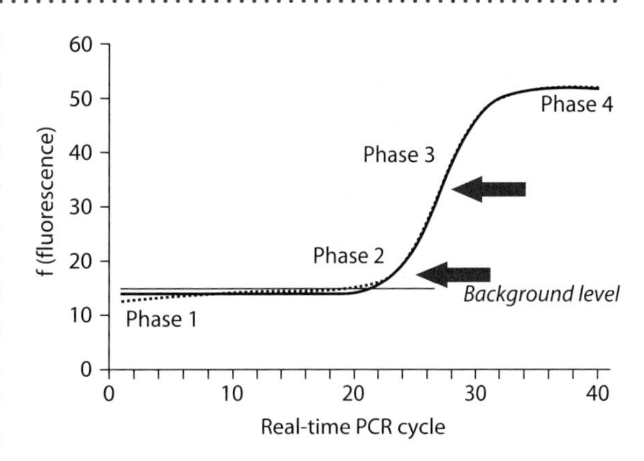

Figure B8.1 The kinetics of a typical PCR reaction. Quantitative or real-time PCR measures products during the exponential phase, the second phase in the diagram.

products are being made; this is the end point that is typically used for most experiments. The cycles at which the reaction switches from the exponential to the linear phase and from the linear to the plateau phase vary depending on the initial concentrations of primers, nucleotides, and template, as well as the temperatures used and the length of each cycle. In summary, the early cycles with exponential kinetics are the least variable, independent of these other factors.

Unlike conventional end point PCR, real-time PCR quantifies copy number using the exponential phase of the reaction. The key to real-time PCR is the ability to assay the accumulation of the products as they are being made. As a result, the amount of product can be measured at each cycle, and a graph can be constructed such that the cycles corresponding to the exponential phase can be determined for each sample. The higher the amount of starting template, the sooner the exponential phase begins (by cycle number). Quantification is determined by comparison with a dilution series from a control DNA sample. With sensitive assays, differences as low as twofold can be accurately measured.

Two different assay methods are in widespread use. The first of these uses SYBR Green, a fluorescent dye that binds to double-stranded DNA (dsDNA). SYBR Green is only weakly fluorescent on its own, but the fluorescence greatly increases upon binding dsDNA. Thus, as dsDNA is made during the extension phase of PCR, SYBR Green binds and the overall fluorescence increases. In real-time PCR using SYBR Green, the dye is included in the reaction mix, and the intensity of the signal is monitored at each cycle. Since the dye binds only to dsDNA, the intensity of the dye indicates the amount of the

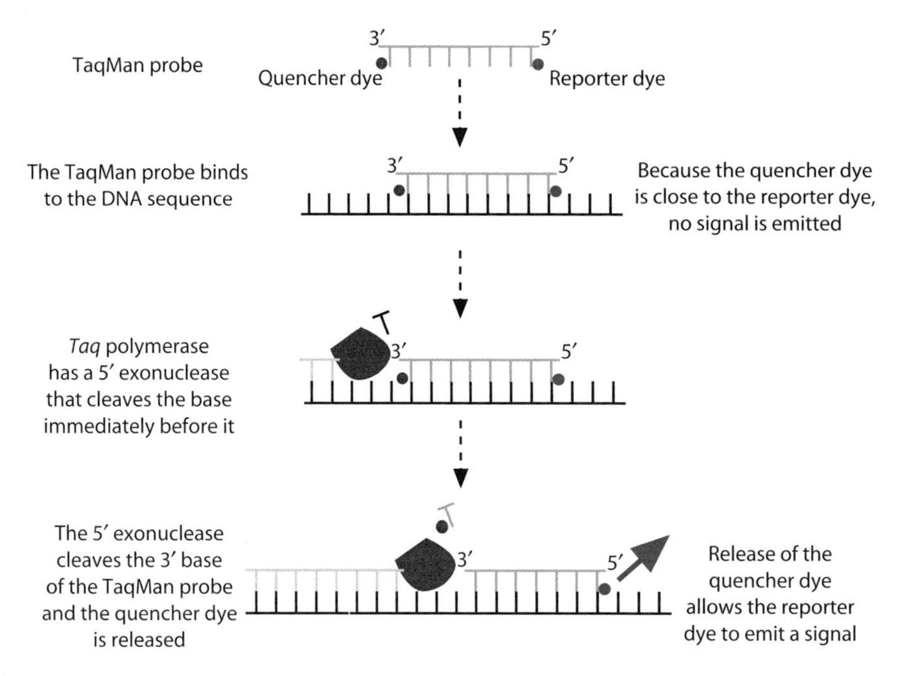

Figure B8.2 The TaqMan assay for quantitative or real-time PCR.

product. Of the two assay methods, SYBR Green is less expensive and requires fewer special reagents; however, it is often somewhat less sensitive.

The second real-time PCR method uses two primers to amplify a specific sequence in addition to a special oligonucleotide probe that binds to a sequence between the two primers. This probe is coupled to a different dye molecule at each end. In the widely used TaqMan system from Applied Biosystems, shown diagrammatically in Text Box Figure 8.2, a high-energy reporter dye is coupled to the 5′ end of the specific oligonucleotide probe and a low-energy quencher dye is coupled to the 3′ end of the same probe. Because the dyes are close together, the low-energy dye is able to quench the signal from the reporter dye by a property known as fluorescent resonant energy transfer (FRET). The oligonucleotide sequence of the reporter probe is complementary to a sequence found on the expected PCR product between the two PCR primers. The reporter probe hybridizes to the internal sequence, but the proximity of the two dyes prevents the emission of any signal from the reporter probe.

The *Taq* polymerase used in a TaqMan reaction has a 5′→3′ exonuclease activity in addition to its 5′→3′ polymerase function. If this *Taq* encounters an annealed sequence ahead of it as it synthesizes new products, the enzyme cleaves the nucleotides immediately, one by one. During the annealing step of each PCR cycle, the two gene-specific primers and the probe all bind to their complementary sequences—the probe to a region between the two primer-binding sites. As a result, during the extension phase of each cycle, the presence of the reporter probe will interrupt the amplification by *Taq*. Whenever the enzyme encounters the reporter probe oligonucleotide, the exonuclease cleaves the first nucleotide (the 5′ end with the reporter dye coupled to it), releasing the reporter from the probe. When this nucleotide is cleaved, the distance between the reporter and quencher dyes increases greatly, quenching no longer occurs, and the reporter dye emits a signal. Thus, the emission of the signal from the reporter dye records the synthesis of a new PCR product, and the progress of the reaction can be accurately and sensitively monitored.

The TaqMan system is reported to be more sensitive than SYBR Green methods, but each of them is far more sensitive and rapid than conventional end-point PCR. However, the special reporter probe of TaqMan must be made specifically for each PCR product, which makes it rather more expensive. Each of the real-time PCR systems requires a special thermocycler with an imager that can monitor the signal, but these machines are now standard laboratory equipment, and real-time PCR is quickly becoming a routine assay for studying DNA copy number as well as RNA expression (called real-time reverse transcription PCR, or real-time RT-PCR).

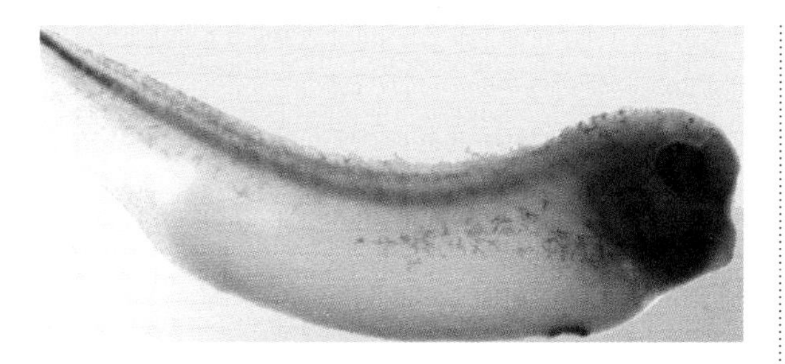

Figure 8.4 An RMA probe from the sonic *hedgehog* gene is hybindized to a *Xenopus* tedpole. The gene is transcribed in the head, the optic region, and throughout the ventral nerve cord.

in situ hybridization to examine the expression pattern of a gene in a whole population of cells or multiple developmental stages, a nearly dynamic picture of the expression pattern can be obtained. For example, if one isolates a group of *Drosophila* embryos, the embryos will almost certainly be at slightly different stages of development. By *in situ* hybridization of the same probe to the entire group, slight changes in the timing of RNA expression can be observed.

In situ hybridization has been particularly important in understanding *Drosophila* embryogenesis. In Case Study 3.1 in Chapter 3, we introduced the genetic screen for mutations affecting segmentation in the early embryo in *Drosophila*. One of the genes identified in this screen was the homeotic gene *bicoid*. Case Study 8.1 on *Drosophila* embryogenesis describes the use of *in situ* hybridization for analyzing the expression pattern of *bicoid*.

Reporter genes can be used specifically to monitor transcription

Northern blots, RT-PCR, and *in situ* hybridization all reveal the stable RNA expression pattern of a gene. However, none of these methods can identify what is controlling the expression pattern. A stable transcript is the end product of a series of biochemical processes affecting RNA—transcription, RNA splicing, post-transcriptional silencing, RNA turnover, and even RNA movement—all of which can be regulated to alter the final RNA expression pattern. For example, microRNAs (miRNAs), discussed in Chapter 7, regulate the expression of target genes via direct RNA cleavage and translation inhibition. For most genes, however, transcription is the most important point of control in regulating gene expression.

Another approach to studying gene expression is to use a **reporter gene**—a gene encoding a protein whose expression can easily be monitored. Reporter genes have a very long history in molecular genetic analysis and can be used specifically to study the transcription pattern (**transcriptional reporter genes**) or stable protein level (**translational reporter genes**) of another gene. A diagram comparing the structure of transcriptional and translational reporter genes is shown in Figure 8.5. In both cases, the coding region of the reporter gene is fused to the regulatory region of the gene of interest. The difference between them lies in the extent of the protein-coding region that is retained from the gene under study. Briefly, as shown in Figure 8.5A, a transcriptional reporter gene retains the regulatory region of a gene of interest but replaces all or most of the coding region with that of a reporter gene. A translational reporter gene retains both the regulatory and coding regions of the gene of interest, as shown in Figure 8.5B, but the coding sequence of the reporter gene is inserted in frame such that a fusion protein is encoded. In both cases, the reporter construct is then introduced into the organism, often as a stable transgenic line, and the expression pattern is determined by assaying for the reporter protein.

Transcriptional reporters allow the investigator specifically to examine the initiation of transcription, the first step in the regulation of gene expression, while translational reporters allow the investigator to view the location of the stable protein, the final output of all of the steps regulating gene expression. In contrast to the other methods we have discussed for studying RNA expression, transcriptional reporter genes can be used to analyze the transcriptional pattern determined solely by a gene's regulatory region, eliminating the effects of RNA splicing, silencing, turnover, and movement.

One of the many advantages of using reporter genes to monitor gene expression is that a stable genetic line with the reporter construct is produced. These reporter lines can be used for many other experiments in gene regulation. For example, a green fluorescent

Case Study 8.1

Molecular analysis of gene expression during *Drosophila* segmentation

In Case Study 3.1 in Chapter 3, we described how mutations affecting embryonic patterns and segmentation were isolated in *D. melanogaster*. This was followed by descriptions of how one of these genes, *patched*, was cloned in *Drosophila* and mice (Case Study 4.1 in Chapter 4) and how knockout mutations for *patched* were obtained in mice (Case Study 5.2 in Chapter 5). In this case study, we return to a few of the segmentation genes in *Drosophila* to describe some of the methods used to analyze gene expression. We will focus on two of the genes, *bicoid* (*bcd*) and *even-skipped* (*eve*), whose expression patterns were particularly important in understanding their functions during *Drosophila* embryogenesis. Our focus is not on precisely how the expression of genes is controlled—some aspects of that are not yet known. Instead, it is on how the expression pattern of these genes was studied.

The *bcd* gene shows translational control

In Chapter 3, we mentioned that some of the mutants affecting embryogenesis and segmentation exhibit a maternal effect. That is, if the female carries a wild-type allele of the gene, her offspring show normal embryonic development, even if they are genotypically mutant. Maternal-effect mutations occur because the oocyte has many gene products that direct early processes in embryogenesis. The gene product deposited in the oocyte by the mother could be in the form of sequestered RNA, which is then translated during embryogenesis, or in the form of protein. For most maternal-effect mutations, it is not known if the deposited gene product is RNA or protein.

Mutations in the *bcd* gene show such a maternal effect. A female that is a *bcd*/+ heterozygote produces normal embryos regardless of the embryonic genotype. However, a female that is *bcd*/*bcd* and has a normal phenotype will produce mutant embryos. The *bcd* mutant phenotype from homozygous mothers is particularly striking: the embryos lack structures characteristic of the anterior end. In their place, the embryo has posterior structures at both ends. Because the *Drosophila* egg is asymmetrical, it is possible to distinguish the presumptive anterior from the presumptive posterior regardless of the structures that are produced. In the language of developmental genetics, the anterior portion has been replaced by a **mirror-image duplication** of the posterior region. The mutant phenotype extends about a third of the way along the embryo, and the posterior two-thirds of a *bcd* mutant embryo look reasonably normal. Thus, the normal function of the wild-type *bcd*+ gene product is to direct the development of anterior structures, and its effect appears to be graded in severity from the most anterior end.

(The origin of the gene name *bicoid* requires a brief history lesson in *Drosophila* genetics. A different mutation with a 'two-tailed' phenotype had been described many years earlier; the gene was termed *bicaudal*, meaning two-tailed, which was abbreviated and spoken as *Bic*. When a second gene was found with a similar two-tailed mutant phenotype, it was named '*bic-oid*' because it was similar to *Bic*; *bicaudal*, the gene identified first, has been overshadowed by its namesake gene *bicoid*.)

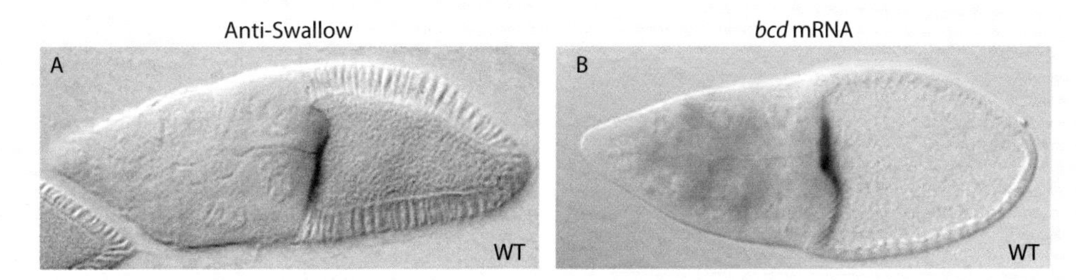

Figure C8.1 ***In situ* hybridization showing the location of the *bcd* mRNA.** An egg chamber from a normal female is shown, with the embryo forming to the right. The anterior of the embryo forms from the region immediately adjacent to the nurse cells of the egg chamber. In panel B, a biotin-labeled RNA probe from the *bcd* gene was hybridized to the developing egg chamber. The location of the biotin was detected using strepavidin. Note the dark blue band along the anterior margin, demonstrating that *bcd* mRNA is localized to this region. In panel A an antibody against the protein product of the *swallow* gene is used. The Swallow protein is localized to the same anterior margin of the embryo as the *bcd* mRNA, and is involved in tethering the *bcd* mRNA to this end.

How is *bicoid* involved in the specification of the anterior–posterior axis of the embryo? The answer immediately became clear when the gene was cloned and *in situ* hybridization experiments were performed. The result is shown in Case Study Figure 8.1, which depicts a *Drosophila* egg chamber with the developing embryo as the large cell at the right-hand side. The hybridization in this example was done with an RNA probe from the gene, but similar results are seen with DNA probes labeled in different ways. The *bcd* mRNA is deposited by the ovary into the developing embryos and is tethered to the extreme anterior margin of the embryo. This tethering process requires the protein products of the genes *swallow*, *staufen*, and *expurentia*. The Swallow protein is localized to exactly the same region of the embryo as the *bcd* mRNA, as shown by immunofluorescence imaging (Case Study Figure 8.1). In a *swallow*, *staufen*, *or expurentia* mutant, the *bcd* mRNA does not remain tethered at this end and normal anterior structures do not form.

The *bcd* mRNA is deposited by the mother in the embryo but is not translated for some time. This is important for understanding how *bcd* establishes the anterior–posterior axis. The early *Drosophila* embryo is a syncytium with all of the nuclei in a common cytoplasm; no cell membranes have been formed. Therefore, macromolecules are free to diffuse in this common cytoplasm and can affect any nucleus. Because the *bcd* mRNA is tethered, it does not diffuse throughout the embryo. However, as the mRNA is translated, the Bcd protein is free to diffuse. Case Study Figure 8.2 compares the localization of the *bcd* mRNA with the distribution of the Bcd protein. The protein appears in a gradient, with the highest concentration near the mRNA source at the anterior end.

The Bcd protein activates the transcription of gap genes in a concentration-dependent fashion

It will be recalled from Chapters 3 and 4 that positional specification in the *Drosophila* embryo occurs sequentially by genes with increasingly limited effects. The earliest-acting genes are the gap genes that specify broad regions of the embryo, with each gap mutant affecting a particular region. These are the genes whose expression is controlled by Bcd.

The Bcd protein is a transcriptional activator that has binding sites in the enhancer regions upstream of a number of other gap genes. These binding sites comprise nine bases

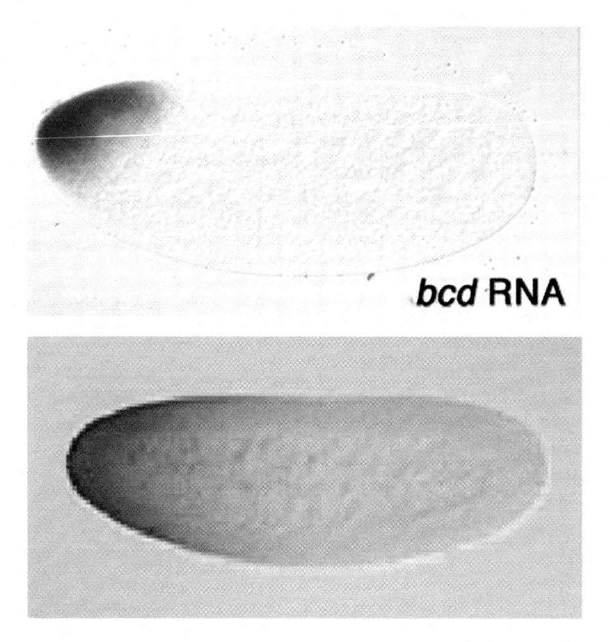

Figure C8.2 **Comparison of the localization of the *bcd* mRNA with the Bcd protein.** In the developing embryo, the *bcd* mRNA is localized to a tight region at the anterior end, seen by *in situ* hybridization in the upper panel as the region in dark blue. As the mRNA is translated, the Bcd protein diffuses away from this source, and spreads back along the embryo. This is shown by an immunofluorescence image using an antibody directed against the Bcd protein (lower panel). Note the highest concentration at the anterior edge and a deceasing gradient of protein over the anterior third of the embryo. The protein gradient is much more extensive than the RNA localization because the protein is free to diffuse while the RNA is tethered.

that differ slightly in sequence. The differences in sequence result in varying affinities for the Bcd protein, with some sequences comprising high-affinity sites and others comprising low-affinity sites. For example, the gap gene *hunchback* (*hb*) has three high-affinity sites and three low-affinity sites in a 300 bp region upstream of the gene. An Hb transcriptional reporter gene with *lacZ* fused to this 300 bp regulatory region was used to demonstrate the effects of Bcd on *hb* gene expression. By deleting or mutating each binding site separately, it was demonstrated that the six sites work additively or synergistically; if any site is mutated or lost, the expression of the Hb reporter gene is reduced.

This suggests how the gradient of Bcd protein expression sets up the pattern of expression for Hb and other gap genes. The Hb reporter provides the model, shown schematically in Case Study Figure 8.3. At the highest concentrations of Bcd protein, i.e. in those nuclei nearest the anterior end of the embryo, all six binding sites will be occupied and *hb* transcription is expected to be maximal. Nuclei progressively further from the anterior end of the embryo will be exposed to a progressively lower level of Bcd protein, resulting in lower *hb* expression. The differences in binding affinity provide some distinction between the sites, ensuring that the high-affinity sites will be occupied at the lowest concentrations of Bcd. Because Bcd expression is graded, the level of *hb* transcription depends on the occupancy of these Bcd-binding sites. As *hb* is transcribed, it is also translated, so that Hb protein is also expressed in a gradient. (*hb* also exhibits a maternal effect arising from protein sequestration by the mother. For our purposes in this case study, we will concentrate on the activation of transcription in the embryo.) The Hb protein gradient occurs in a slightly different but overlapping pattern with the Bcd gradient. If the Bcd protein is over-expressed,

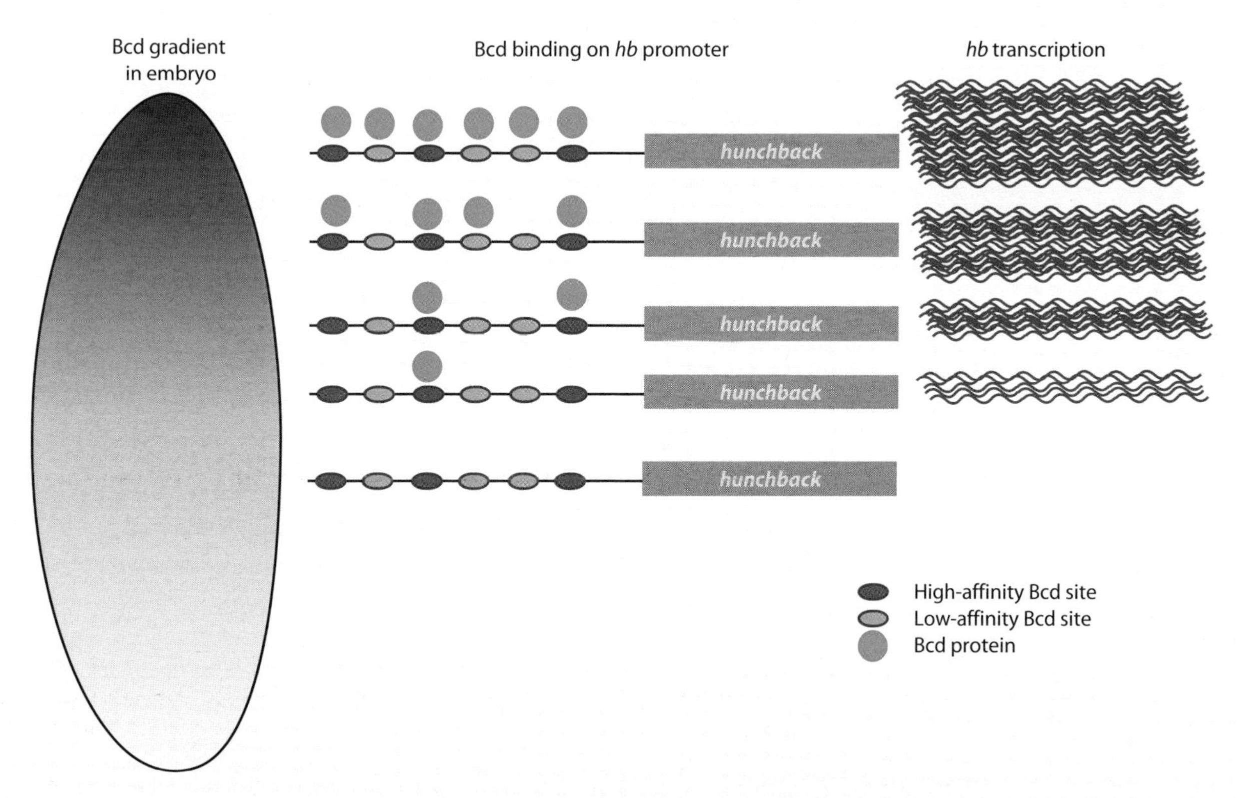

Figure C8.3 Bcd regulation of *hb* transcription. As illustrated on the left, Bcd protein is expressed in a gradient in the embryo, with the highest expression at the anterior (top). The Bcd protein binds to regulatory sequences upstream of the *hb* gene and activates its transcription. At the highest level of Bcd in nuclei near the anterior end of the embryo, all of the binding sites are occupied and *hb* is transcribed maximally. More posterior nuclei are exposed to a lower concentration of Bcd protein, resulting in progressively less transcription of *hb*. This schematic diagram is illustrative only, and the differences in the binding sites and the location in the embryo are simply to demonstrate the effect.

for instance by introducing additional copies of the gene, both the Bcd protein gradient and the Hb expression pattern are extended more posteriorly. Thus, different locations in the embryo will have somewhat different levels of Bcd protein and Hb protein.

The Hb protein is also a DNA-binding protein and activates the transcription of other gap genes. Experiments similar to those done with Bcd and Hb have also been performed with reporter constructs for other gap genes such as *Krüppel*, *knirps*, *caudal*, and *giant*. Although *hb* is not the only gap gene with Bcd-binding sites, transcriptional regulation of the *hb* gene is probably the simplest to explain. However, regulation of the other genes follows a similar strategy: the protein products of *hb*, *Krüppel*, *knirps*, and the other gap genes bind to specific regulatory sequences upstream of the other genes in a concentration-dependent fashion so that each gap protein is expressed in a specific region of the embryo. A complicated set of interactions between the regulatory sequences of one gene and the protein products of other genes refine the boundaries of expression of these proteins. Not all of the regulation is activation; for example, high levels of the Knirps protein repress the transcription of the *Krüppel* gene.

The gap genes regulate the expression of the pair-rule genes

Each of the gap genes encodes a DNA-binding protein that regulates the expression of other genes, of which the regulation of the pair-rule genes is the most impressive. Recall from Case Study 4.1 that mutants in the pair-rule genes affect alternating pairs of segments. This mutant phenotype was completely unexpected and illustrates how mutants can reveal an underlying cellular logic that is not obvious from other analysis. To be more precise, mutant phenotypes of the pair-rule genes are offset in register by half a segment with respect to the segment boundaries, affecting the posterior half of one segment, one complete segment, and the anterior half of the adjacent segment. The domains of action of the pair-rule genes are termed **para-segments**. This is analogous to the hems and seams in a pair of pants, each of which divides a pant leg in half. To an observer, the seam may be the more obvious division; the seam is analogous to the segment boundary. However, the strategy by which pants are assembled relies on the hems rather than the seams; the hem is analogous to the para-segment boundary, which becomes hidden as the embryo develops. The pair-rule genes reveal the presence of the para-segments.

Among the pair-rule genes, the regulation of *even-skipped* (*eve*) has been especially well investigated, with particularly dramatic results. As shown by the immunofluorescent image using antibodies against Eve (Figure 8.9), *eve* is expressed in seven stripes across the embryos. These seven stripes correspond to the regions that are deleted in embryos with *eve* mutations. This striped pattern allowed the detailed analysis of different portions of the regulatory region. The *eve* regulatory region is large and complicated for *Drosophila* genes, with regulatory sequences spread over a region of 15 kb both upstream of the gene and within the introns. As shown in Figure 8.9, expression of the *eve* gene in different stripes is regulated by specific enhancer regions. Because its regulatory region is organized in modules and expression is readily monitored by reporter gene constructs with different regulatory modules, the strategy by which *eve* expression is regulated is an outstanding example of the control of gene expression that transcends the role of *eve* in *Drosophila* embryogenesis.

The regulatory region of *eve* has been shown to have binding sites for the proteins encoded by *bcd*, *giant*, *Krüppel*, *knirps*, and *hb* as well as the pair-rule gene *runt*, and probably other proteins as well. Different regions of the embryo have different levels of each of these proteins, arising from the gradient pattern of expression of the gap proteins. In addition, the gap proteins may act as activators or repressors of transcription, and different sequences in the *eve* regulatory region have different affinities for these binding proteins. The combination of varying protein levels, protein activities, and binding affinities specifies unique regions in the embryo; as a result, *eve* is expressed in a series of stripes. This is shown diagrammatically in Case Study Figure 8.4. For example, the enhancer for the expression in stripe 2 has five binding sites for Bicoid, one for Hb, three for Krüppel, and three for Giant, with somewhat different affinities for their respective proteins. Bicoid and Hb activate transcription, whereas Krüppel and Giant repress transcription. There is a region in the *Drosophila* embryo at which the balance in the levels of these proteins precisely allows transcription of the *eve* gene. Slightly more posterior nuclei may be exposed to more Krüppel protein, for example, and so *eve* is not transcribed. Of course, the same regulatory region is found upstream of the *eve* gene in all nuclei, but this particular combination directs expression in one specific stripe. The refinement of the pattern into a stripe with precise boundaries comes from a complicated series of cross-regulation and autoregulation among the pair-rule genes, not all of which act transcriptionally.

The power of this striped image may be hard to imagine nearly 20 years after the publication of these original papers. However, if geneticists of a certain age are asked what motivated them to study developmental biology, the answer may well be 'stripes', i.e. the

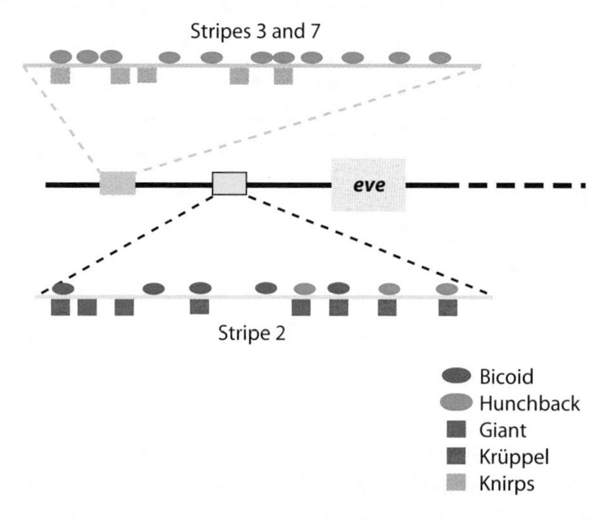

Figure C8.4 ***even-skipped*** **regulation by the gap genes.** The Eve protein is expressed in a pattern of seven stripes and is controlled principally by transcriptional regulation. Different enhancers control different stripes, as shown here for the enhancer for stripes 3 and 7 (controlled jointly) and the enhancer for stripe 2. Enhancers for the other stripes are located within introns of the *eve* gene and are not shown. Each of the regulatory proteins is expressed in specific regions of the developing embryo. Activating proteins are shown as ovals above the line, and repressor proteins are depicted as squares below the line. Thus, activation of transcription from the green enhancer will occur when Hb protein is present and will not depend on the presence of the Bcd protein. It will be shut off when the Knirps protein is low or absent. Activation of transcription from the yellow enhancer will occur when both Bcd and Hb are present but not when Giant and Krüppel proteins are made. Since Giant is expressed near the anterior and Krüppel in the middle of the embryo, *eve* transcription is limited to a region between these two repressors where both Bcd and Hb are present.

immunofluorescence pictures or the reporter gene constructs that showed the expression pattern of pair-rule genes such as *even-skipped.*

Further reading

There are many references and reviews that describe the genetic and molecular analysis of *Drosophila* embryogenesis. The following outstanding book is still available:

Lawrence, P.A. (1992). *The making of a fly: the genetics of animal design.* Blackwell, Oxford.

A widely cited review article is:

Ingham, P.W. (1988). The molecular genetics of embryonic pattern formation in *Drosophila. Nature,* **335**, 25–34.

Papers of note on *bicoid* include:

Driever, W. and Nüsslein-Volhard, C. (1989). The bicoid protein is a positive regulator of *hunchback* transcription in the early *Drosophila* embryo. *Nature,* **337**, 138–43.

Struhl, G., Struhl, K., and Macdonald, P.M. (1989). The gradient morphogen *bicoid* is a concentration-dependent transcriptional activator. *Cell,* **57**, 1259–73.

Papers of note on *even-skipped* include:

Harding, K., Rushlow, C., Doyle, H.J., Hoey, T., and Levine, M. (1986). Cross-regulatory interactions among pair-rule genes in Drosophila. *Science,* **233**, 953–9.

Macdonald, P.M., Ingham, P., and Struhl, G. (1986). Isolation, structure, and expression of *even-skipped*: a second pair-rule gene of Drosophila containing a homeo box. *Cell,* **47**, 721–34.

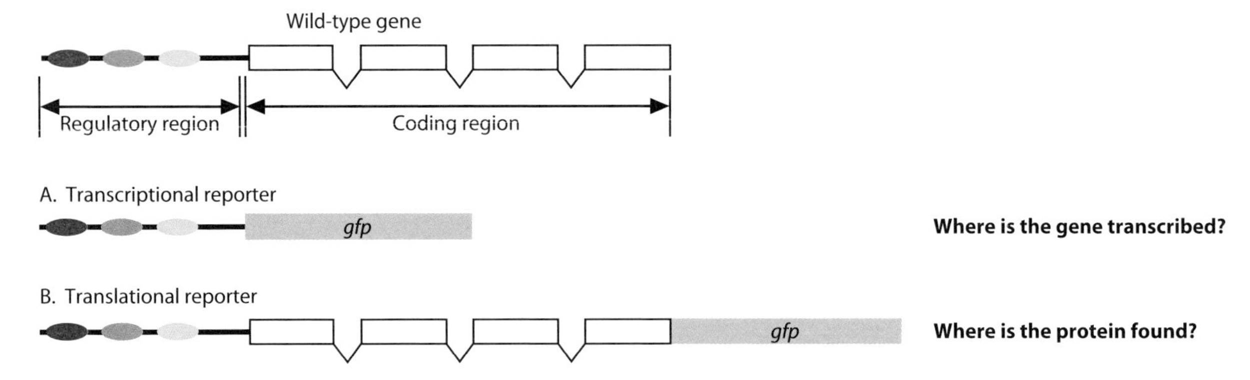

Figure 8.5 Diagram of reporter gene constructs. The hypothetical gene is shown at the top with a coding region of four exons and a regulatory region with three suspected elements shown in different colors. Two different types of reporter gene constructs are shown as fusions to the green fluorescence protein (*gfp*) gene; the reporter constructs address different questions about the expression of the gene. Panel A shows a transcriptional reporter in which all or nearly all of the coding region of the gene has been replaced by the coding region of *gfp*. The regulatory region has been retained. A transcriptional reporter is used to ask where the gene is transcribed. Panel B shows a translational reporter gene in which the coding region of *gfp* has been fused at the 3′ end of the last exon of the gene. This reporter gene has the entire coding region of the gene and will presumably be subject to the same post-transcriptional regulation and localization as the wild-type protein. A translational reporter is used to ask where the protein product is found.

protein (GFP) reporter gene that is expressed in the germline can be used to look for mutants that affect gene expression in the germline more easily than monitoring fertility or other germline phenotypes. We will return to this use of reporter genes in Chapters 10 and 11 when we discuss the analysis of genetic pathways.

Many of the basic principles for working with transcriptional and translational reporter genes are identical, so we will consider these together here. Because the methods used to make transgenic organisms have been discussed previously in Chapter 2, we will not consider them further now. Instead, we will focus on the choice of a reporter protein and the regulatory region of the gene of interest. A more detailed discussion of topics specific to translational reporter genes is reserved for the section on protein expression assays.

A good reporter protein is easy to detect and quantify

As mentioned above, the expression of a reporter protein must be easy to visualize, assay, and measure accurately. To meet these requirements, the expression of most common reporter proteins is examined using colorimetric assays. The strongest conclusions about expression differences between two different conditions or tissues can be made when the expression levels can be measured quantitatively. The reporter protein should also be stable so that it is not quickly degraded in the organism or during extraction. (The one caveat to this is that stable reporter proteins are not always advantageous when studying rapid decreases in gene expression because the reporter proteins will often persist longer than the endogenous expression.) Furthermore, in order to have the most sensitive assays, the model species cannot have a protein whose properties are similar to the reporter protein. The widely used reporter gene *lacZ* is not a good reporter gene in plants because there is a plant enzyme that can catalyze a similar reaction. The size of the reporter protein is also important. The reporter has to be large enough that it cannot diffuse out of the cell but not so large that it affects the activity of the protein being studied, at least in the case of translational reporter genes.

Although many different reporter proteins have been employed, three reporters are widely used in animal cells: β-galactosidase (β-gal) encoded by the *lacZ* gene of *E. coli*, luciferase from the firefly *Photinus pyralis*, and GFP from the jellyfish *Aequorea victoria*. Examples with each of these reporter genes are shown in Figure 8.6. Luciferase and GFP are also used in plants. However, plants have some endogenous β-gal activity, so the *E. coli uidA* gene, which encodes β-glucuronidase (GUS), is typically used instead of *lacZ*.

All of these reporter proteins are readily detected, even in single cells, by colorimetric assays that can be performed quantitatively by spectroscopically measuring the intensity of the color. *lacZ* and GUS reporters are particularly sensitive methods for studying cell-, tissue-,

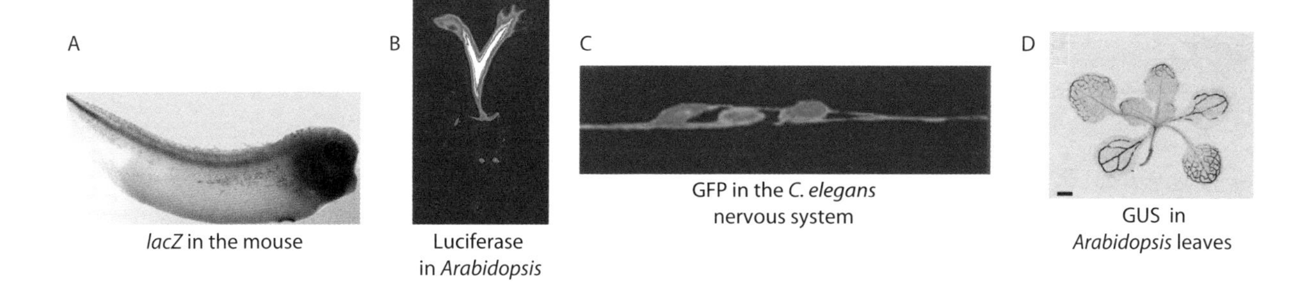

Figure 8.6 Examples of different reporter genes. (A) *lacZ* in the mouse embryo. (B) A luciferase reporter expressed throughout the plant in *Arabidopsis*. (C) A GFP reporter expressed in the nervous system of *C. elegans*. (D) A β-glucuronidase (GUS) reporter in the leaves of *Arabidopsis*.

and development-specific expression patterns in fixed specimens. For β-gal detection from *lacZ*, the specimen is fixed, and the substrate 5-bromo-4-chloro-3-indolyl-β-D-galactopyranoside (more commonly called X-gal) is added. β-Gal hydrolyzes X-gal to produce galactose and the deep-blue precipitate 4-chloro-3-brom-indigo, so the presence of the blue color indicates the activity of β-gal. Colorimetric substrates of β-gal other than X-gal are also commercially available, some of which allow more sensitive assays. GUS assays in plants work similarly, with the substrates being X-gluc and MUG.

Luciferase is the enzyme that makes fireflies glow or luminesce. A small amount of light is released when luciferase acts on its substrate luciferin. Because the light released from the metabolism of one luciferin molecule is small and transient, luciferase is primarily used as a reporter when working with isolated cells or cell lines because detection can happen quickly. In luciferase assays, the cells are lysed and combined with the substrate immediately before quantitation in a luminometer, which measures the amount of light emitted.

GFP has natural green fluorescence under blue light, so no additional substrate needs to be added for detection. This provides the important advantage that, unlike β-gal, GUS, and luciferase, GFP can be assayed in living cells. This makes GFP the reporter of choice for monitoring dynamic patterns of gene expression and its subcellular localization. In general, detection of GFP is rather less sensitive than detection of β-gal, although enhanced GFP with increased fluorescence is commercially available, so both β-gal and GFP reporters are widely used. In addition to its natural green fluorescence, the *gfp* gene has been modified to emit yellow, cyan, or red fluorescence; these modified versions were given the original names YFP, CFP, and RFP. These permutations of GFP allow the simultaneous labeling and tracing of multiple genes in the same cell or organism. As an example, the expression of four different fluorescent reporter genes in the *C. elegans* nervous system is shown in Figure 8.7.

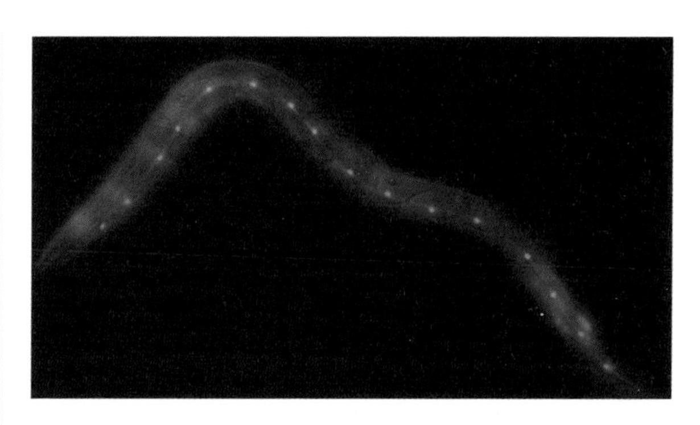

Figure 8.7 A *gfp* reporter gene expressed in the nuclei of seam cells of *C. elegans*.

Reporter constructs include most of the regulatory sequences of the gene of interest

In addition to having a protein product that is easy to assay, a transcriptional or translational reporter gene

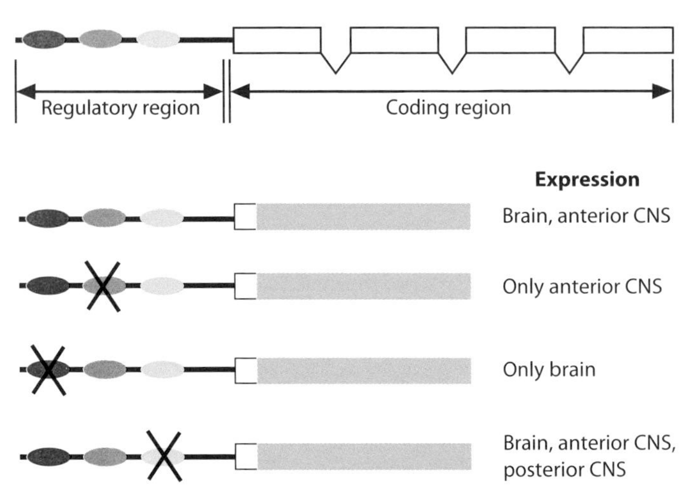

Figure 8.8 Testing different parts of the regulatory region using a transcriptional reporter. A hypothetical transcriptional reporter gene is expressed in the brain and the anterior central nervous system (CNS). A mutation that eliminates the putative regulatory element (the orange circle) results in loss of expression in the brain. An inference is that this regulatory element is needed for transcription in the brain. By a similar inference, the regulatory element in pink is hypothesized to be needed for transcription in the anterior CNS. Elimination of the yellow regulatory element results in a broader pattern of transcription, suggesting that this element is needed to repress transcription in the posterior CNS.

also requires the presence of the regulatory region of the gene of interest. As summarized in Chapter 1, the RNA expression pattern of a gene is transcriptionally determined by several regulatory elements—the upstream promoter, the 5′- and 3′-UTRs, enhancers, and other elements that may be found elsewhere associated with the gene. Enhancer sequences can be located at some considerable distance from the genes they regulate, which makes it difficult to include the entire regulatory region in a reporter construct. Likewise, with a transcriptional reporter, it is usually not possible to include regulatory elements that are found elsewhere in the gene. Thus, the most common transcriptional reporter gene fuses the region upstream of the core promoter, the core promoter, and the 5′-UTR to the reporter coding region.

The size of the core promoter varies among genes and among organisms, so the most conservative approach is to clone as much of the region upstream of the gene as possible. One way to verify that all of the key elements are present in a transcriptional reporter is to compare the expression pattern in a transgenic organism with the pattern seen with *in situ* hybridization. These expression patterns should coincide for most genes. (Notable exceptions include cases of RNA silencing, RNA turnover, and specific RNA localization.) Other approaches for verifying the relevance of the promoter region chosen are considered during the discussion of translational reporter genes.

Transcriptional reporter genes are useful for determining the importance of specific regulatory sequences for transcription

Transcriptional reporter genes are particularly good for identifying the roles of different sequences in the upstream regulatory region for gene expression. Figure 8.8 shows a hypothetical example of a neural gene with three suspected regulatory regions, two of which enhance expression and one of which represses expression. A particularly impressive real-life example involves the analysis of the regulatory region of the *even-skipped* (*eve*) gene in *Drosophila*. As seen in Figure 8.9A, the wild-type Eve protein is expressed in seven stripes in the *Drosophila* embryo, where it patterns segmentation. By deleting or replacing different portions of the upstream regulatory region in the *eve* transcriptional reporter gene, it has been possible to assign the control of expression in different stripes to different specific enhancer sequences upstream of the *eve* gene. As shown in Figure 8.9, if the enhancer responsible for expression in stripe 2 is the only regulatory element, *eve* is expressed only in this location. With the identification of the different enhancers, it has also been possible to determine which

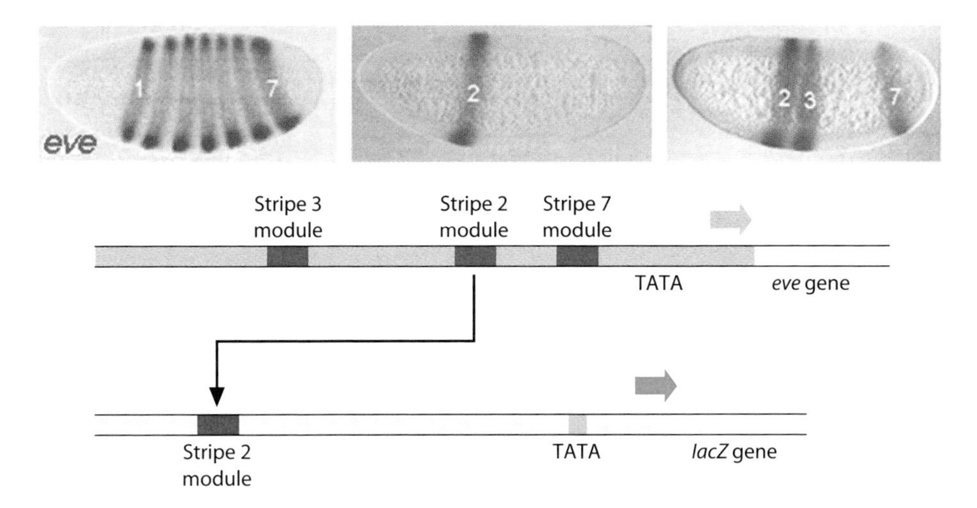

Figure 8.9 Regulatory regions of the *even-skipped* (*eve*) gene and protein in *Drosophila* embryos. The regulatory region of the *eve* gene was fused to *lacZ* and reintroduced into *Drosophila* females. The gene is expressed in seven distinct stripes in the embryo, numbered 1–7 from anterior on the left to posterior on the right. Fusion of the reporter protein to different portions of the regulatory region revealed the presence of enhancer for each stripe. For example, fusion with one particular module of the regulatory region results in the pattern of expression seen in the middle photograph; this module is inferred to be needed for expression in the second stripe, and thus is called the stripe 2 module. The results with the stripe 3 and stripe 7 modules are also shown. Regulation of *eve* expression has been one of the best examples of localizing particular expression control to specific enhancers.

of the other segmentation genes regulate *eve* via these various enhancers. This analysis is described in more detail in Case Study 8.1 on *Drosophila* development.

Enhancer traps are one application of transcriptional reporter genes

One special kind of transcriptional reporter is an enhancer trap. Enhancer traps are described in Chapter 5 as a means of cloning genes based on their expression pattern. Briefly, an enhancer trap uses a minimal promoter, consisting of a TATA box and a transcription initiation site, fused to a reporter gene. This construct is randomly inserted into the genome, and the expression pattern of the reporter gene depends on the enhancers and regulatory regions near the insertion site. The enhancer trap acts as a transcriptional reporter for the gene near the insertion site, and is under the regulation of *all* of its regulatory elements. Enhancer traps reflecting the expression of a particular gene can be identified by screening a population of transgenic organisms. Alternatively, a population of enhancer trap lines can be screened for transcription patterns of interest, thereby identifying new genes involved in a particular process.

8.3 Protein expression analysis of individual genes

The RNA expression pattern of a gene is usually a reliable indicator of the protein expression pattern since most gene expression is regulated by transcription. However, the expression of many genes is regulated by post-transcriptional controls. Thus, it is informative to have techniques to analyze the protein expression pattern more directly.

We consider three methods commonly used to analyze the expression pattern of a protein product from an individual gene: Western blotting, immunofluorescence, and translational reporter genes.

Antibodies can be used to monitor protein expression

Both Western blotting and immunofluorescence rely on antibodies to detect the presence of a protein. In Western blotting, antibodies are used to identify the presence of a protein in protein extracts. In contrast, immunofluorescence utilizes antibodies to study protein expression patterns in fixed but structurally intact tissue samples. Both techniques rely on the ability of an antibody to bind specifically to the protein of interest. The steps in generating and using antibodies against a protein of interest are shown in Figures 8.10 and 8.11 and can be summarized as follows.

1. The protein product of the gene, or at least a portion of the protein, is purified.
2. The purified protein is used as an antigen and injected into a mammal, typically a rabbit.
3. The rabbit raises antibodies against the protein, and those antibodies are isolated as antisera.
4. The purified antibodies are reacted with proteins on a Western blot or in a tissue sample or organism

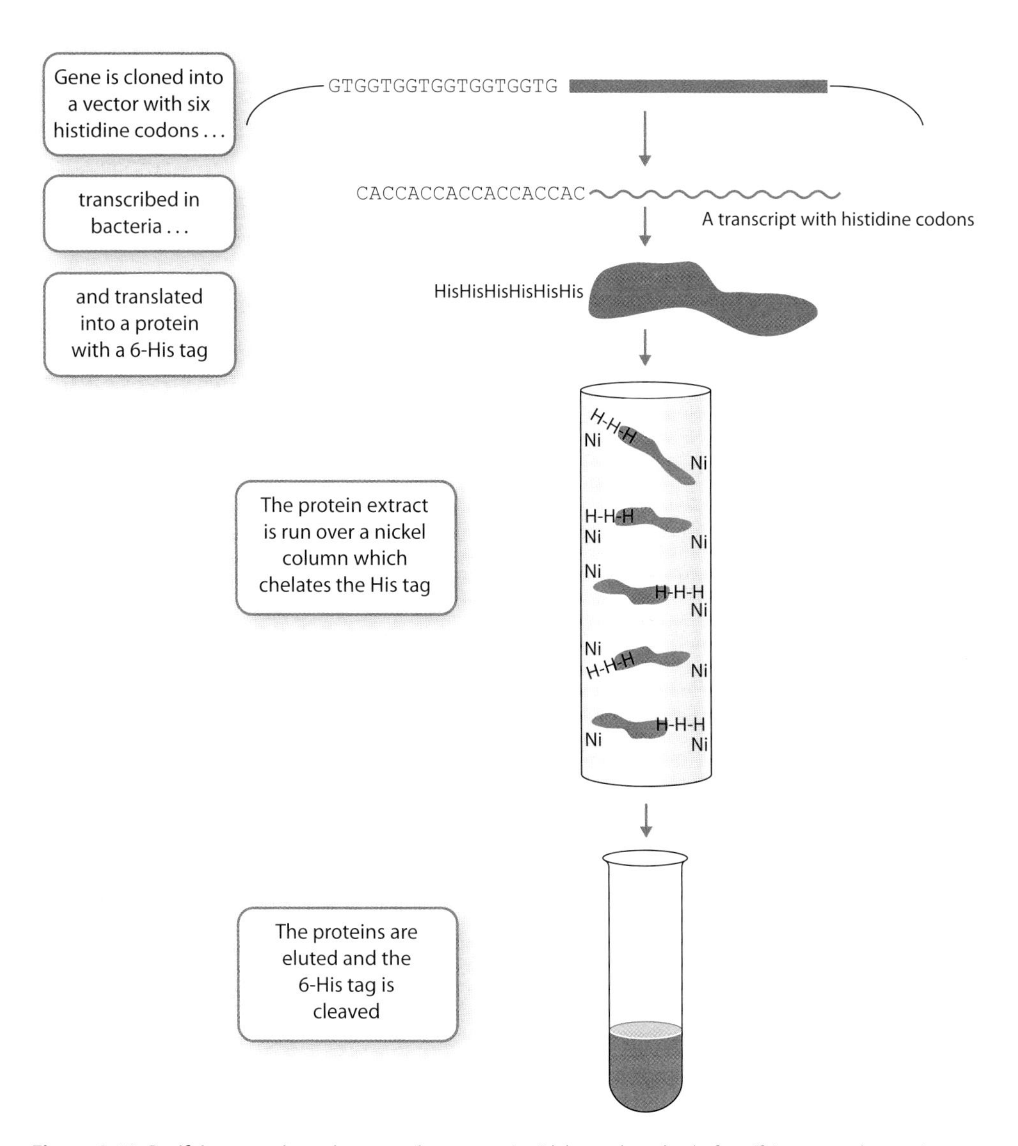

Figure 8.10 Purifying proteins using an epitope tag. A widely used method of purifying enough protein to make an antibody against it is to use an epitope tag such as the six-histidine tag shown here. The gene is cloned into a protein expression vector, which includes six histidine codons at one end—here the 5′end but either end can be used. When this is transcribed and translated in *E. coli*, a fusion protein is made that includes six histidines at the N terminus. A protein extract is prepared from the bacteria and passed over a nickel column (commercially available). The nickel in the column binds to the histidine tag and the fusion protein is retained while the other proteins in the extract pass through. The fusion protein is eluted from the column by a change in the buffer. Many expression vectors include a protease cleavage site that allows cleavage of the epitope tag.

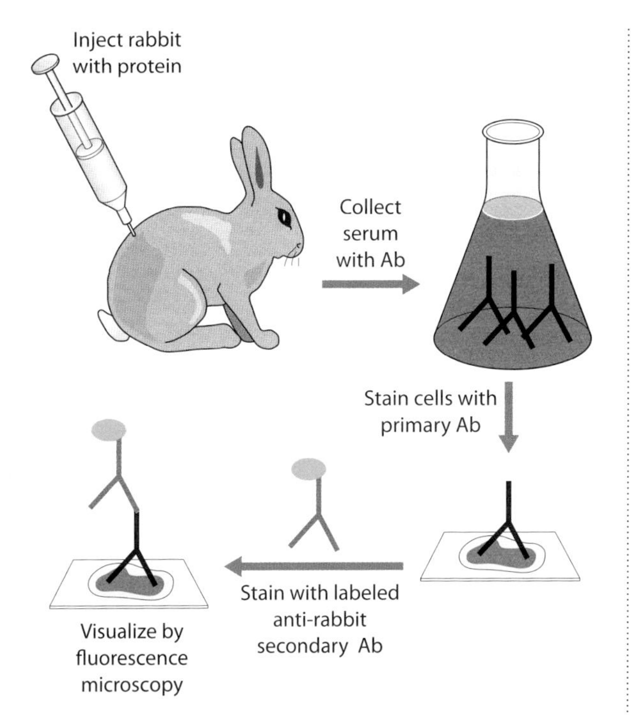

Figure 8.11 The procedure for indirect immunofluorescence. The protein that has been purified in Figure 8.10 is injected into a rabbit or other mammal, which mounts an immune response. The antibodies are termed the primary antibodies. The serum, which contains the primary antibodies against the protein, is collected and used to stain cells. In order to visualize the primary antibody, the cells are then stained with a secondary antibody, which reacts with any rabbit antibodies. The secondary antibodies, which are commercially available, have a fluorescent label, which can be observed under a microscope.

fixed for *in situ* hybridization experiments. The antibodies bind to the protein, indicating its location.

Each of these steps requires a little more explanation. The protein product that is used as an antigen is often purified from bacteria expressing a cDNA of the gene. The cDNA is first inserted into a translational expression vector that will make a fusion between the protein of interest and an **epitope tag**. Epitope tags are short amino acid sequences that are not found in the organism but can be used to purify and monitor the fusion proteins; they act as 'handles' with which the fusion protein can be 'grabbed'. Many different expression vectors that encode epitope tags are commercially available. In Figure 8.10, the gene has been cloned into a plasmid vector that has six consecutive histidine codons at one end (a 6-His tag); either the N terminus or the *C*-terminus of the protein may be tagged. Other epitope tags used for purification include glutathione *S*-tranferase (GST), hemagglutinin (HA), c-Myc, Flag, and even GFP itself.

The plasmid is transformed into *E. coli*, where the gene is transcribed and translated. The protein product will include the epitope tag at one terminus, as the means of purifying the fusion protein. In the case of a 6-His tag, a crude protein extract from the bacteria is passed over a nickel column, as shown diagrammatically in Figure 8.10. The histidine tag binds to the nickel and the protein of interest is retained on the affinity column, while the rest of the proteins pass through. The flow-through of the column is discarded. The column is washed, and the epitope-tagged protein is then eluted from the nickel column by washing with a higher salt concentration. The 6-His tag is then enzymatically cleaved from the purified fusion protein. In the case of the other epitotpe tags, antibodies against the tags are used in the affinity column rather than nickel, but the principle is the same.

Assuming that the original cDNA is full length, the protein purified by this technique will also be full length, and antibodies are raised against its entire sequence. As an alternative, the amino acid sequence of the protein can be studied to identify a short stretch predicted to act as a strong and specific immunogen. A peptide of this sequence is synthesized commercially and used as an antigen for the next step in the process.

The purified protein or peptide is injected into an experimental animal as an antigen to make a primary antibody. In Figure 8.11, the protein is injected into a rabbit to produce polyclonal antiserum. The antiserum contains the primary antibodies against the protein antigen, and antisera are collected at various times and surveyed for the production of antibodies. The antibodies are purified from the antisera using the purified protein as a target. The antibodies are then released and purified from the target proteins and used for many kinds of study, including Western blotting and immunofluorescence.

Western blotting uses antibodies to study protein expression in extracts

The development of Southern blotting not only led to the creation of Northern blotting for studying RNA expression but also to Western blotting for studying protein expression. (Southern blots were named for Edward Southern who developed the method; the other types of blot are nicknames derived from Southern.) Because antibodies are used to detect the presence of the protein, Western blots are also sometimes called **immunoblots**. In Western blotting, protein extracts are run on a native or denaturing polyacrylamide gel, the proteins are transferred from the gel to a membrane, and the membrane is probed with an antibody against the protein of interest.

The antibody forms a complex with the protein from the cells, but this complex itself cannot be seen directly. The presence of the bound antibody is usually detected using a secondary antibody that recognizes the first. The secondary antibody is conjugated to a fluorescent molecular or an enzyme whose presence can be measured by its catalytic activity. As shown in Figure 8.11, secondary antibodies are made in a different species by using the constant region of the primary antibody protein as an antigen. Because the constant region of the antibody protein is used as the antigen, a secondary antibody raised against one antibody from the species will react with all primary antibodies from that species. For example, the primary antibody may have been produced in a rabbit, and the secondary antibody may be an antibody from a goat that reacts with any rabbit primary antibody; the secondary antibody is referred to as a 'goat anti-rabbit' antibody. In the case of Western blotting, secondary antibodies are generally conjugated to alkaline phosphatase, horseradish peroxidase, or some other easily assayed label.

After the blot is treated with the secondary antibody, a substrate for the conjugated enzyme that will luminesce when metabolized is added. Immediately after addition of the substrate, the membrane is placed next to an X-ray film, and the light released by the reaction exposes the film to reveal bands corresponding to the presence of the protein of interest on the membrane. Western blotting has been streamlined by the creation of primary antibodies conjugated to a variety of labels, preventing the need for secondary antibodies. These labeled primary antibodies are more expensive but are available for detecting common proteins and epitope tags. Similar to Northern blots, Western blots can be stripped and re-probed with antibodies against other proteins. Re-probing with antibodies against a control protein can be used to determine the relative expression of the protein of interest.

Western blotting is useful for determining gross differences in protein expression between samples, whether they are from different genotypes, tissues, developmental stages, or treatments. Like Northern blots for RNA, however, Western blots are only quantitative relative to a control protein, and they do not provide fine details of the expression pattern. They also require the generation of antibodies against the protein of interest or the creation of transgenic organisms expressing epitope-tagged versions of the protein. Antibodies against all of the common epitope tags are available, and these can be used to detect the expression pattern of the protein of interest in a transgenic organism without generating protein-specific antibodies. While studying the endogenous protein expression pattern requires protein-specific antibodies, using transgenics expressing epitope-tagged translational reporters can approximate the native condition.

Immunofluorescence uses antibodies against the protein to monitor expression in fixed specimens

Another widely used method of monitoring protein expression of a specific gene involves a method known as immunofluorescence or, more accurately, **indirect immunofluorescence**. Like *in situ* hybridization with RNA, immunofluorescence studies gene expression using fixed samples. In this method, the organism or tissue is fixed by any of several procedures, and a primary antibody against a protein of interest is reacted with the fixed sample. The sample is then treated with a fluorescently conjugated secondary antibody. By monitoring the fluorescence of the secondary antibody with a fluorescence or confocal microscope, the location of the primary antibody, and thus the protein of interest, can be detected. Many different fluorescent conjugates are available, and combining multiple primary antibodies made in different species with differently fluorescently conjugated secondary antibodies can be used to study the expression of multiple proteins in one sample.

Indirect immunofluorescence is a standard and sensitive technique for analyzing the protein expression of individual genes. An example of the use of this technique is shown in Figure 8.12. Its principal limitation is the need for a high-quality primary antibody with low cross-reactivity to other proteins, but techniques and commercial facilities for producing antibodies are widespread. Because the samples are fixed, a single

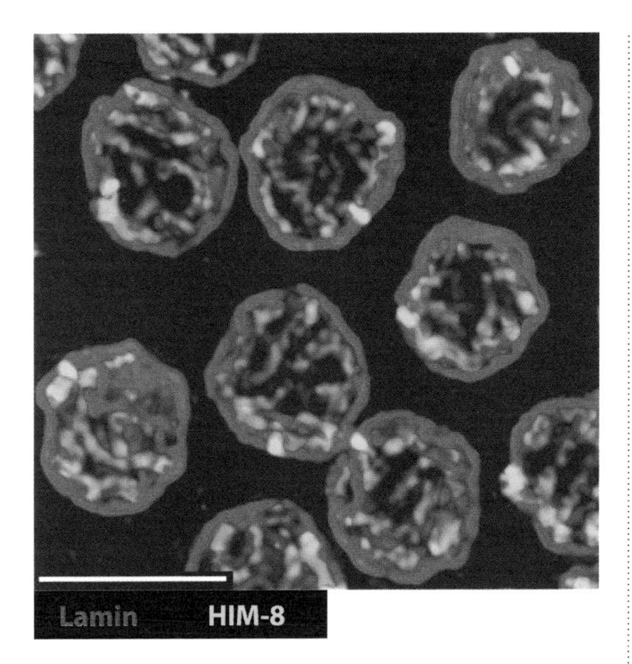

Figure 8.12 An example of immunofluorescence. Meiotic nuclei in the ovary of a *C. elegans* hermaphrodite are shown. The chromosomes are the blue rods, stained with the DNA dye DAPI. Two different antibodies are shown, with different fluorescent labels. One antibody is directed against the meiotic protein HIM-8, shown in yellow, and the other is directed against one of the lamin proteins in the nuclear envelope, shown in red.

immunofluorescence image does not provide much information about the dynamics of the expression pattern. However, a fairly complete expression pattern can be inferred by comparing samples from different tissues or developmental stages.

Translational reporter genes can be used to monitor protein expression

The last of the methods that we will describe for analyzing protein expression relies on the use of translational reporter genes, which were introduced earlier. Like transcriptional reporters, translational reporter genes use the expression of a reporter protein to study gene expression. In this case, the reporter is expressed wherever the *protein* of interest is expressed. Through comparisons with transcriptional reporter gene expression patterns, translational reporters can help to determine the influence of post-transcriptional regulation on gene expression. Translational reporter genes using GFP are particularly important for monitoring the subcellular localization and dynamics of proteins.

In addition to their role in analyzing the location of the wild-type protein, deletion and point mutations in the protein-coding region of translational reporters can identify residues and domains involved in its subcellular localization, stability, and movement. Our hypothetical example in Figure 8.13 shows a translational reporter

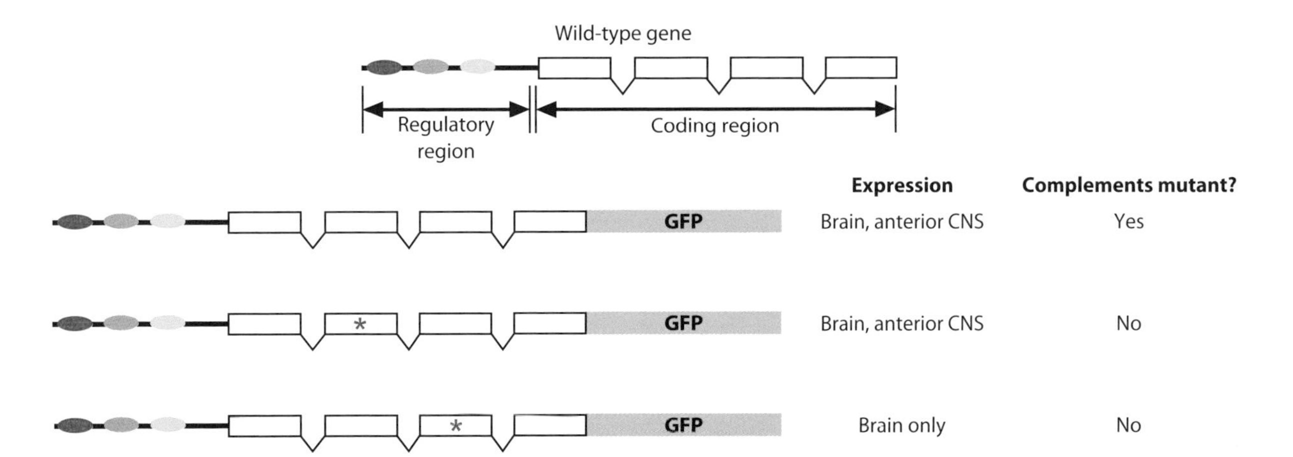

Figure 8.13 A translational reporter and protein localization. A hypothetical translational reporter with a GFP fusion is shown. A reporter gene with the full-length protein-coding region is expressed in the brain and anterior central nervous system (CNS). This gene with the full-length protein complements the defect in a null mutation for the gene, indicating that the reporter protein is functionally equivalent to the wild-type. A mutation in the second exon, shown by the red star, does not change the expression pattern of the protein; however, since this mutation no longer complements the null mutation, it must impair the normal function. A mutation in the third exon is only expressed in a subset of the tissues that are seen in the unmutated reporter gene, suggesting that the site in the third exon is needed for proper localization and expression in the anterior CNS. The mutation could alter a protein–protein interaction site or a residue involved in a post-translational modification, for example.

gene that is expressed in the brain and anterior central nervous system. The regulatory region and the entire coding region are fused to the reporter. Point mutations in exons 2 and 3 prevent the normal function of the protein, but the mutation in exon 3 also affects its localization.

For translational reporter genes, it is crucial to verify that the reporter protein is not changing the distribution, subcellular localization, or activity of the protein of interest. One way of ensuring that the translational reporter gene retains the normal function is to ask if the fusion protein is able to complement a null mutation in the gene. Complementation tests are described in Chapters 4 and 5. If the construct does not complement the mutation, the reporter protein may be altering the expression or function of the protein of interest. While complementation of the null mutant is ideal, additional information can also be gained from trying to understand the basis for partial or non-complementation.

8.4 RNA and protein expression patterns do not always coincide

In concluding this discussion of assays for the expression pattern of individual genes, we return to the broader topic of what is meant by 'gene expression'. As noted previously, gene expression is often used synonymously with transcription; even more precisely, it is used to mean the level of stable transcripts. The expression of many genes, perhaps most genes, is regulated primarily at the level of transcription, so it makes sense to think about determining the expression pattern in terms of controlling transcription. The connection between transcriptional regulation and gene expression is firmly fixed in the minds of most investigators.

While the RNA and protein expression patterns of a gene usually coincide, this is not always the case. In some instances, the two expression patterns may not overlap at all or may be only partially overlapping. These differences may be explained by translational inhibition by miRNAs, protein stability issues, or RNA or protein movement from the site of synthesis to a different location. Because these processes are sufficiently common, it should not be assumed that the RNA and protein expression patterns of a gene are identical. To illustrate this point, we will consider two examples of protein movement from *Arabidopsis*. Case Study 8.1 on the role of the *bicoid* gene in regulating *Drosophila* segmentation provides another example in which the RNA and protein have different patterns of expression.

Movement of the SHORT-ROOT protein in the roots of *Arabidopsis*

An example that illustrates the role of protein movement in gene expression is provided by root development in *Arabidopsis*. Roots are radially symmetrical structures with vascular or stele tissues (xylem and phloem) in the center, a layer of endodermis surrounding the stele, cortex beyond this, and an outer epidermal layer (Figure 8.14A). The xylem transports water and minerals from the roots to the shoot, and the phloem traffics nutrients throughout the plant. The endodermis protects the plant against many toxic or undesirable materials in the soil that try to enter through the xylem.

SHORT-ROOT (*SHR*) mutants have short determinant roots that lack an endodermis. Based on this phenotype, it is reasonable to expect that *SHR* is expressed in the endodermis; however, the *SHR* mRNA is not found here. As shown in Figure 8.14B, transcriptional and translational reporters revealed that the RNA and protein expression patterns of *SHR* are not identical. *SHR* is transcribed in the vasculature of the root, but the protein is found in both the stele cells and the endodermis. The protein is trafficked into the endodermis. This movement is required for SHR function, as shown by mutations in the SHR reporter gene. While a complete SHR translational reporter can complement a null mutation in *SHR*, a single amino acid change in the SHR-coding region generates a protein that retains normal

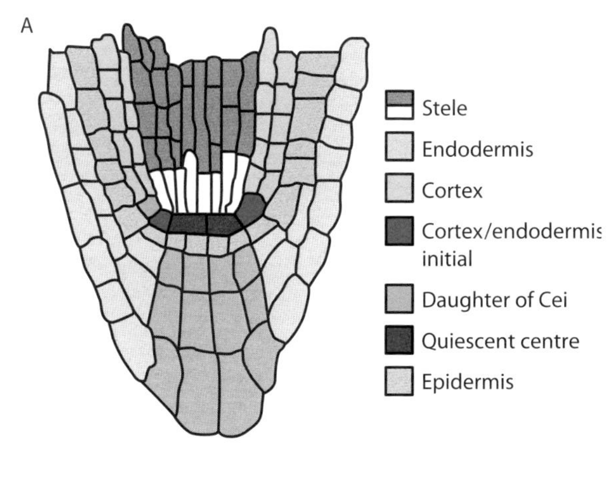

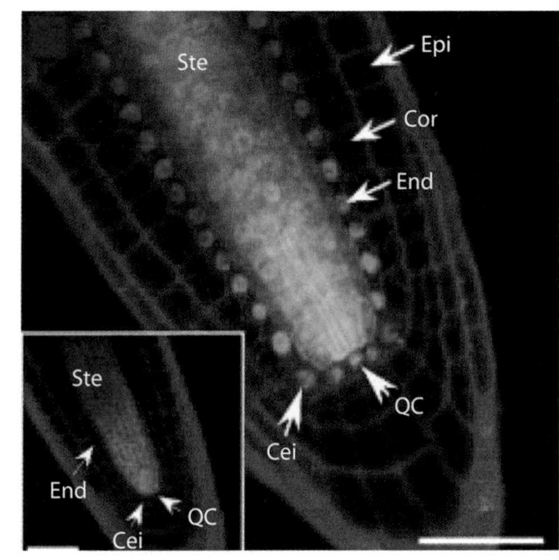

Figure 8.14 The site of transcription and the protein localization may differ. (A) The structure of a root in *Arabidopsis*. (B) The inset shows the transcriptional reporter. Note that the gene is specifically transcribed in the stele cells. The larger image shows a translational reporter with protein expression in both the stele and the endodermis.

biochemical activity but which prevents protein movement and cannot complement a mutant allele. Thus, the different RNA and protein expression patterns revealed a significant aspect of the biology that would have been missed by examining transcription alone.

KEY ARTICLES

Nakajima, K., Sena, G., Nawy, T., and Benfey, P.N. (2001). Intercellular movement of the putative transcription factor SHR in root patterning. *Nature*, **413**, 307–11.

Gallagher, K.L., Paquette, A.J., Nakajima, K., and Benfey, P.N. (2004). Mechanisms regulating SHORT-ROOT intercellular movement. *Current Biology*, **14**, 1847–51.

Protein movement controls the reproductive onset of *Arabidopsis*

Flowering provides a second example in which differences in RNA and protein localization play an important role. One of the key stimuli that induces flowering is day length. Many species, and sometimes even different ecological varieties, have an ideal day length for promoting flowering. In *Arabidopsis*, the optimal day length is 16 hours of light and 8 hours of darkness. Curiously, treating even a single leaf with the optimal photoperiod has the same effect as treating the entire plant (Figure 8.15). Similar results are seen when a leaf from a plant induced to flower is grafted onto an uninduced plant—the leaf induces the plant to flower. These studies suggested that an unknown substance moves from the leaves to the apex and initiates flowering. In the 1930s, this substance was named florigen because of its perceived ability to act as a hormone and induce flowering.

Many unsuccessful attempts were made over many decades to identify florigen using biochemical methods. It is only recently that work in *Arabidopsis* and rice has identified the protein FT as florigen. It has been known for some time that *FT* mRNA expression is induced in the phloem of *Arabidopsis* leaves in response to long day length conditions, as shown by the transcriptional fusion to GUS in Figure 8.16, and that null mutations in the *FT* gene delay flowering. However, the expression of the FT protein is found at such a low level that it becomes much more difficult to study. Because *FT* mRNA is expressed in leaves but the protein is required to induce flowering at the apex, it was hypothesized either that *FT* mRNA or FT protein was trafficked from the leaves to the shoot apex or that FT regulated the expression or activity of another gene that was trafficked.

To avoid the complications arising from the low level of FT protein expression, the *FT* gene was expressed under the control of a strong phloem-specific promoter. Under such conditions, the FT protein localizes to both the phloem and the apex. In addition, because the FT protein in rice is expressed at a higher level than the ortholog in *Arabidopsis*, investigators using a translational reporter with the native promoter could see the protein at both the transcription site in leaves and within the shoot apex. It is now

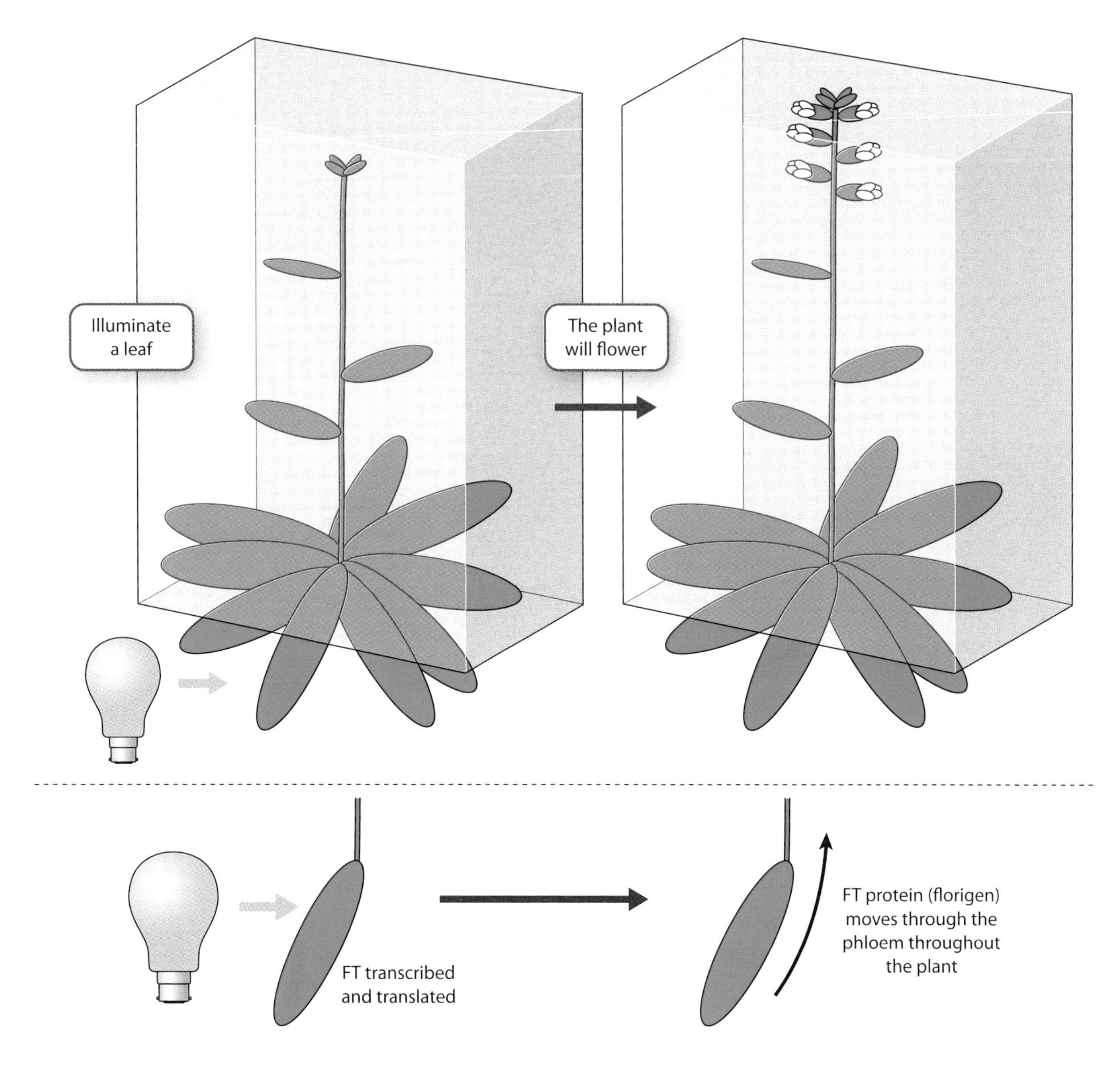

Figure 8.15 An experiment demonstrating florigen. When the plant is kept in the dark, it will not flower. Illuminating a leaf will induce flowering, as shown in the top half of the figure. The bottom half of the figure depicts the mechanism. When the leaf is illuminated, the *FT* gene is transcribed and the FT protein is made. The FT protein moves through the phloem and induces flowering.

believed that, to induce flowering, the FT protein moves from the leaves to the apex through the phloem via intercellular connections called plasmodesmata. Thus, like the example of SHR movement, different transcriptional and translational expression patterns can signal post-transcriptional regulation of gene expression.

KEY ARTICLES

Takada, S. and Goto, K. (2003). TERMINAL FLOWER2, an Arabidopsis homolog of HETEROCHROMATIN PROTEIN1, counteracts the activation of *FLOWERING LOCUS T* by CONSTANS in the vascular tissues of leaves to regulate flowering time. *Plant Cell*, **15**, 2856–65.

Corbesier, L., Vincent, C., Jang, S., *et al*. (2007). FT protein movement contributes to long-distance signaling in floral induction of *Arabidopsis*. *Science*, **316**, 1030–3.

Jaeger, K.E. and Wigge, P.A. (2007). FT protein acts as a long-range signal in *Arabidopsis*. *Current Biology*, **17**, 1050–4.

Tamaki, S., Matsuo, S., Wong, H.L., Yokoi, S., and Shimamoto, K. (2007). Hd3a protein is a mobile flowering signal in rice. *Science*, **316**, 1033–6.

The methods described so far in this chapter have been applied to thousands of genes in many different

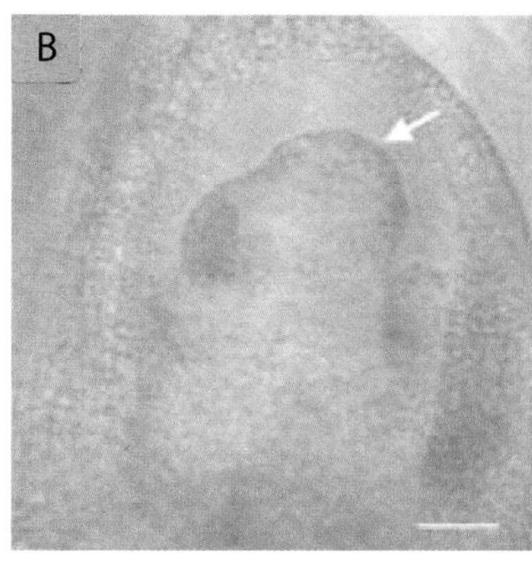

Figure 8.16 FT expression parallels the action of florigen. (A) The expression of an FT reporter gene construct in the leaf. (B) The protein is found in the shoot apex, where it induces flowering.

organisms. Although the techniques are not simple, they are standard; it is hard to publish a description of a gene or a mutation that does not include one or more of these experiments. The genes have been studied individually, much as the mutations in Chapters 3 and 4 were found one at a time. However, as described in Chapter 7 for mutations, the availability of complete sequences for some genomes has opened the door to looking simultaneously at all of the genes in one organism. We now consider these genome-wide approaches to analyzing gene expression.

8.5 Microarrays and genome-wide transcriptional analysis

The phenotype of a cell, tissue, organ, or organism is the outcome of the orchestrated expression of hundreds or thousands of genes. One of the major challenges is understanding just how this expression is orchestrated. Several methods are available for studying gene expression on this genome-wide scale. Specifically, microarrays and large-scale RNA sequencing are used to study RNA expression, and proteomics is used to examine protein expression.

Microarrays record the expression profile of many genes simultaneously

One of the first and most powerful applications of genome sequence information has been to assay gene expression using **microarrays**. Microarrays have been mentioned in Chapter 5 in the context of cloning genes based on expression patterns and in Chapter 7 in the context of identifying genes by molecular bar codes. In effect, a microarray places a diagnostic and specific sequence from each of many different genes onto a slide or support in an ordered pattern.

Microarrays allow the investigator to determine the transcription pattern, also called an **expression profile**, of an entire genome in a particular cell, tissue, or developmental stage or in response to a treatment. An expression profile tells us which genes are being transcribed in the same tissue or stage under a particular set of conditions. The collection of all of the transcribed genes in a sample or, more commonly, in a whole organism, is referred to as the **transcriptome**. Comparing the transcription patterns of two samples allows identification of genes that may be responsible for the biological differences between them.

In 1995, the first report of a microarray compared the expression of 45 genes in *Arabidopsis* in different tissues and wild-type plants and transgenics overexpressing a transcription factor. By 1997, a full genome yeast microarray with 6200 genes had been developed. Few techniques have been adopted as quickly as microarray analysis, and thousands of papers reporting research using microarrays are published annually. Microarrays of various levels of completeness are available for many different species.

Case Study 8.2

Microarrays and cancer cell expression profiles

Cancer involves changes in nearly all aspects of cellular metabolism and is one of the leading causes of death throughout the industrialized world. It is not too surprising that one of the earliest applications of the power of microarray analysis was to expression profiles of cancer cells. In this case study, we review some of the first successes obtained using microarray analysis in cancer cells.

Microarrays confirm and extend histological classifications

It is an oversimplification to refer to cancer as if it were one disease. Pathologists classify different types of tumors based on cell lineage, cellular phenotypes, and metastasis beyond the primary tissue focus. These histological phenotypes are reproducible and characteristic for different types of tumors, and are one of the principal factors in determining a course of therapy and prognosis. For microarray analysis to be useful in cancer studies, it must demonstrate that there are also reproducible differences in transcription patterns between different types of tumors and between tumor cells and normal cells that parallel the histological phenotypes. Furthermore, the transcription differences must be detectable regardless of the individual genetic differences among patients. If this is not true, then microarrays would not be as useful as histological analysis.

It is now clear that different tumors have different expression profiles, and these differences are reproducible. A tumor is typically derived from a particular cell lineage, so the expression of normal cells of that lineage is compared with that of tumor cells. For example, lymphomas and leukemias arise from hematopoetic cell lineages, so the tumor cells are paired with normal cells from the bone marrow to study the changes in expression profile. In an early study, the expression profiles of acute myelogenous leukemia (AML) and acute lymphocytic leukemia (ALL) were found to be different from each other but reproducibly similar in different AML and ALL tumors. These types of cancer are easily distinguished histologically, so the differences in the expression profiles were reassuring evidence that microarray analysis was able to classify the tumors in the same way that histological studies could. Similar results have also been seen for many other types of cancer, including lung, breast, and colon; tumor cells have a reproducible difference in expression from the normal cells, and the difference is characteristic for each type of tumor. In each case, about 5 percent of the genes are expressed differently in the tumor and the normal cells of the same lineage.

Of course, if microarrays were only giving the same results as histological studies, there would be little reason for pathologists and oncologists to work with microarrays; the hope is that microarrays will provide more detailed and more accurate information. Studies have also focused on differences between pre-cancerous lesions and tumor cells of the same type, and between tumor cells at different stages of progression. In nearly every case, the tumor cells at different stages from the same patient were more similar to each other than to tumors of the same type and stage from different patients. Thus, on a large scale, the expression profile of a tumor is highly reproducible. However, the profiles are not identical at different stages, so the differences could be used as molecular markers of different stages.

Different oncogenic agents produce different types of tumors

In addition to their use in tumor diagnosis and classification, microarrays are also being used to study the cause and origin of cancers. Since it is difficult to determine the specific cause of any human tumor, most of the studies have focused on tumor induction in mice or other mammals. In one study, six different known oncogenes were introduced as transgenics (as described in Chapter 6) and expressed in the mammary gland using a different specific promoter. Three of the oncogenes were known from previous studies to affect the *ras* signaling pathway; the mammary tumors derived from these three oncogenes have very similar expression profiles. However, the expression profile of these tumors was different from that of mammary tumors induced by the *myc* oncogene, which works via a different cellular pathway. Thus, although the mammary tumors appear histologically similar, the underlying expression pattern is different, reflecting differences in the mechanism of action by the oncogenic agent. It was also possible to correlate changes in the expression of particular downstream target genes with the class of oncogenic agent, suggesting that the expression profiles of these downstream targets may themselves be useful for classifying mammary tumors of unknown origin.

A similar conclusion about oncogenic agents and expression profiles has been drawn from ovarian tumors in humans. Most ovarian tumors are sporadic and of unknown genetic cause, but some arise from an inherited mutation in either *BRCA1* or *BRCA2*. Although the tumors are histologically similar, tumors arising from *BRCA1* mutations have a different expression profile from those arising from *BRCA2* mutations. In fact, the two profiles are quite distinct, and relatively few genes are affected by both. Interestingly, the expression pattern of different sporadic ovarian tumors could be classified into these two categories, either *BRCA1*-like or *BRCA2*-like expression. This result was somewhat unexpected, and suggests that one of the same two underlying genetic profiles is affected no matter what the causative agent of the ovarian tumor. Thus, microarray expression profile is extending the information obtained from histological analysis and finding new underlying principles of tumor formation.

Expression profiles may be useful in tumor progression and prognosis

Thus far, the evidence has suggested that expression profiling by microarrays can separate into different categories not only tumors based on histological differences but also histologically similar tumors. Pathologists use histological differences to determine types of therapy and the prognosis for the patient. Can microarray analysis be helpful here as well?

One study used melanomas, one of the most common forms of skin cancer. Different melanomas with the same histological phenotype can have very different invasive and metastatic properties. A microarray using only 7000 genes (perhaps a quarter of the human genome) identified distinct expression profiles between non-invasive and invasive melanomas. (An *in vitro* assay was used to distinguish invasive from non-invasive tumors in this study, but similar results have been seen with patients for this and other tumor types.) The profiles of these two categories were used to classify the *in vitro* response to different therapeutic agents. The categories of invasive or non-invasive expression profile precisely paralleled the response or non-response to a therapeutic vaccine. This is evidence that expression profiles may be useful in decisions about appropriate therapies.

Other studies have been able to correlate differences in progression and prognosis in lung and prostate cancer with differences in the expression profile. Based on tumors from patients in which the outcome of the tumor was known, different profiles were associated with differences in survival and relapse. Thus, the profile may be useful as a prognostic tool. Since the specific genes involved with each category are also known, it may be possible to use the expression profiles to focus on specific therapeutic targets or cellular activities.

Many other attempts to use expression profiles for prognosis have been made. One of the most thoroughly studied examples is breast cancer. Current therapies use the estrogen receptor (ER) as a diagnostic tool, dividing tumors into those that may respond to hormonal therapies (ER-positive tumors) and those that are unlikely to respond to them (ER-negative tumors). Microarray analysis used unsupervised clustering of expression differences of about 25 000 genes from different tumors; about 5000 genes were used to compile the expression profile differences. These profiles revealed that the ER status is one of the principal differences among tumor expression profiles, and ER-negative tumors cluster together in one branch of the dendrogram. However, among the ER-positive tumors, two subcategories that had not been previously recognized were seen. These two clusters were themselves quite different. Those ER-positive patients in one subcategory had a high rate of metastasis and an overall poor prognosis, whereas those in the other subcategory had much less metastasis and a better prognosis. Accurate separation between the categories could be done based on the expression profiles of fewer than 100 genes.

Summary and limitations

Although expression profiling of tumor cells has been widely performed—thousands of papers are published annually in this field—it is still too soon to know how valuable this approach will become. The major limitation appears to be the amount of tissue that is needed to prepare enough RNA for the microarray; histological analysis can be based on very few cells, whereas RNA expression analysis cannot. A related weakness is that the tissue sample may itself be heterogeneous, and tumor cells and normal cells of the same cell lineage, as well as cells from different lineages, may be included in the same sample.

On the other hand, microarray analysis has every promise to be a powerful tool in cancer biology. Reproducible differences in the expression profiles are seen when different tumors are compared. Furthermore, tumor types are detected by expression profiles that are not seen histologically, and these different profile types appear to be clinically significant for many kinds of cancer. Expression profiles have also shown that different types of oncogenic agent may produce different types of tumor, although these are not always histologically distinct. Microarrays have been able to define more precisely which genes change expression with oncogenesis, and which may be useful targets for future therapies. As costs come down and the procedures become simpler and more widely available, it may not be long before a trip to the doctor includes an expression profile as one more clinical indicator.

Further reading

Liu, E.T. (2003). Classification of cancers by expression profiling. *Current Opinion in Genetics and Development*, **13**, 97–103.

van t'Veer, L.J., Dai, H., van de Vijver, M.J., *et al.* (2002). Gene expression profiling predicts clinical outcome of breast cancer. *Nature*, **415**, 530–5.

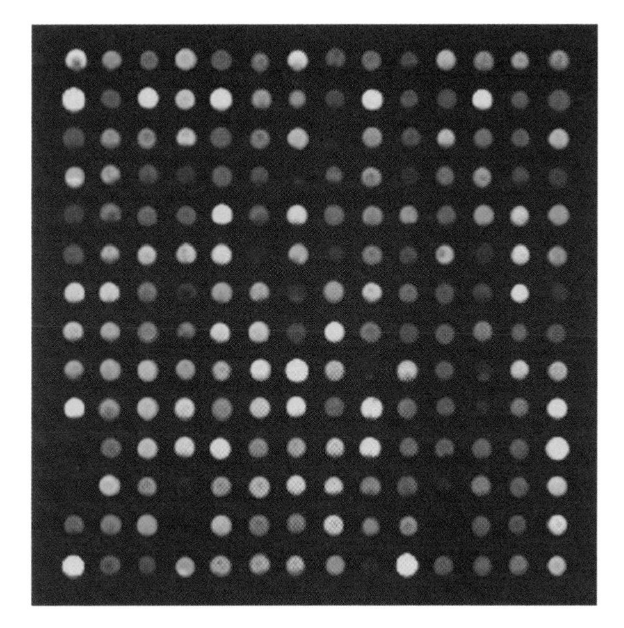

Figure 8.17 A two-channel microarray. RNA is isolated from each sample, made into either double-stranded cDNA or cRNA with a fluorescent label, and hybridized to the microarray slide. Genes transcribed at a higher level in one sample than in the other are either green or red, whereas those transcribed at similar levels are yellow. Genes not transcribed appear black.

The powerful attraction of microarrays does not come from seeing the expression of a single gene; any of the other methods described in this chapter could be used to do that. The attraction of microarrays is that *all* of the genes expressed in that sample are shown in one picture, with intensity differences representing different levels of expression for each gene in the genome. An example is shown in Figure 8.17. Suppose that the investigator is interested in genes that affect ovarian development in *Drosophila*. By looking at an expression profile of the ovary from young *Drosophila* females, all of the hundreds or thousands of genes expressed in the ovary are displayed. Furthermore, by reading the position of hybridization on the grid, the identity of each gene is known and each gene that is expressed in the ovary can easily be cloned.

Because the picture presents an expression pattern for the tissue or developmental stage, a relationship involving two or more genes can readily be detected. For example, if two genes of unknown function are expressed similarly, or co-expressed, it is reasonable to infer that they are functionally related. This principle can be extended to cross-species comparisons and to a range of other co-expression assays. Because so many microarrays have been performed and the expression data are deposited in online databases, it is now often possible to identify genes co-expressed with a gene of interest across thousands of arrays from hundreds of laboratories by examining archived data rather than by repeating the microarrays.

Microarrays are modified Northern blots that determine the transcription profiles of thousands of genes simultaneously

Microarrays are based on the ideas of Northern blotting, but, instead of examining the expression of single genes, they simultaneously measure the mRNA expression of up to tens of thousands of genes. The similarities and differences between Northern blots and microarrays are summarized in Figure 8.18. The experimental question for a Northern blot is: 'What tissues or stages express this particular gene?' The question for an expression profile compiled from a microarray is: 'What genes are expressed in this tissue or stage, and how does that compare with other tissues or stages?'

Of course, the difference between the experimental questions is only the beginning of the differences between the two techniques, but it may be helpful to think of them in parallel. As discussed earlier, with a Northern blot, target RNA is run on a denaturing gel, transferred and fixed to a membrane, and probed with a labeled DNA. In contrast, a microarray has thousands of different DNA probes distributed across its surface. Unlike a Northern blot, these probes are unlabeled. RNA is extracted from the sample of interest and reverse transcribed to make the first strand of a double-stranded cDNA. This cDNA is fluorescently labeled or used as a template for the synthesis of the second strand of cDNA that is transcribed *in vitro* into complementary RNA (cRNA), which is then labeled. cRNA has the same 5′→3′ sequence as cDNA, with U substituted for T. (cRNA is discussed further below.) The labeled cDNA or cRNA—the target in the case of microarrays—is hybridized to the array surface, and the amount of hybridization to each individual probe is measured by taking a digital image of the array. A gene transcribed at a very high level in the cell will have a much more intense signal than a gene transcribed at a low level. This makes a microarray exquisitely sensitive, and transcripts as rare as a few copies per cell can be detected reliably.

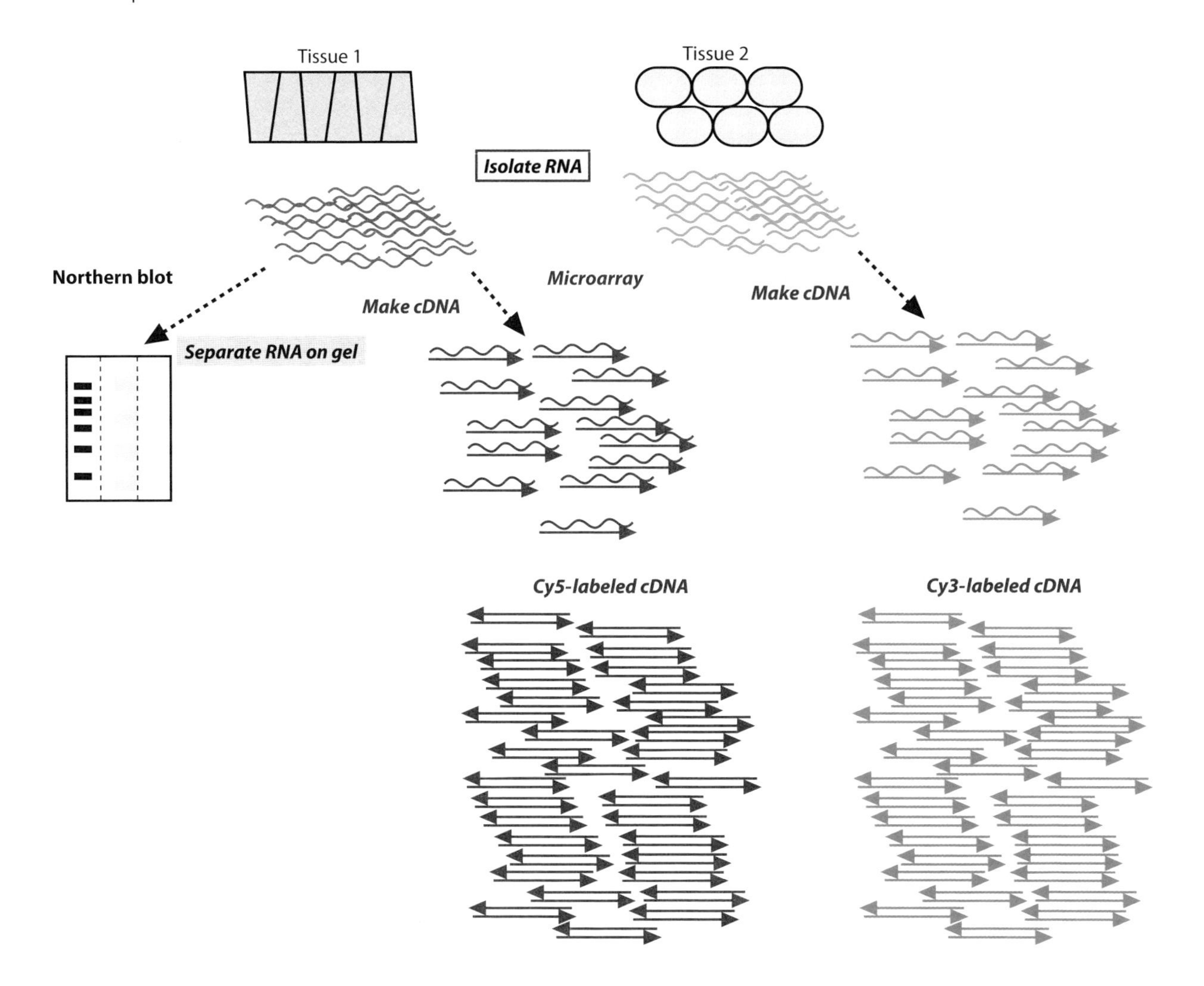

Figure 8.18 Microarrays and Northern blots: RNA treatment. Both microarrays and Northern blots begin by isolating RNA from different tissues. In Northern blots, as reproduced here on the left from Figure 8.3, the RNA is separated on a gel. For microarrays, as shown on the right, the RNA is reverse transcribed to make cDNA or cRNA. The cDNA is labeled with the fluorescent dye Cy5 or Cy3.

The principal task for the investigator wanting to conduct a microarray experiment is to isolate the RNA and produce the labeled target. The hybridization and detection of the signal are generally done in a special microarray facility because the cost of the necessary equipment is beyond the budget of most laboratories.

Two major microarray technologies are used today —two-channel arrays and one-channel arrays. The differences between the two types of array are summarized in Table 8.2. We will consider each of them separately and then discuss their advantages and disadvantages.

Two-channel microarrays are co-hybridized with two differently labeled targets

The first microarrays were two-channel arrays. They are made by spotting DNAs, either long (72–80-mer) oligonucleotides or, less commonly, full-length cDNAs, in a precisely defined pattern onto a glass slide similar in size to a standard microscopy slide. Each spot on the array has hundreds of thousands or millions of copies of a DNA sequence that ideally hybridizes to cDNA and cRNA corresponding to only one gene. Typically, robots are used to spot or 'print' the arrays, allowing for tens or hundreds of thousands of different **features** (spots) to be placed onto a single array at a spacing of less than 20 μm. The probes are then crosslinked to the slide using UV light. Despite the use of robotic technology, the shape and size of the features can vary significantly from spot to spot for the same array and from array to

Table 8.2 Comparison of one-channel and two-channel microarrays

Characteristic	One-channel microarray	Two-channel microarray
Source of the probes	Synthesized *in silico* on the array surface	Robotically spotted and then UV-crosslinked to the surface of the array
Size and type of probes	25 bp DNA oligonucleotides	72–80 bp DNA oligonucleotides or full-length cDNAs
Number of probes for each gene	More than 10; the set of probes for a given gene is called a probe set	One probe for each gene
Reproducibility and similarity of the features between and within arrays	High	Poor
Number of expression samples studied with each array	One	Two, but two arrays must be done for each pair (one where the control sample is labeled with Cy3 and the treatment with Cy5, and the other where the labels are swapped) to account for label bias
How is each target labeled?	Biotin	Cy3 or Cy5
How quantitative are the expression data?	Very—provide absolute expression values	Less—provide relative expression values
Expense of one microarray	High	Low

array for the same spot. This is due to slight variations in the printing pins, as well as the amounts of solution picked up and deposited by the pins each time.

Because of the irregularity of the spots, reliable and reproducible data are not achieved by hybridization with a single-labeled target. Therefore, the same array must be hybridized with two samples—a control and a treatment—each labeled with a different fluorescent label, the fluorophore Cy3 or Cy5. Thus, the expression of each gene is seen relative to other genes on the microarray. The gene expression profiles of more than two samples can be compared with one another by hybridizing one common sample, called the common reference, to each of the arrays with a different treatment sample for each slide. The common reference provides an internal standard to control for minor experimental variations in the labeling and hybridization methods.

The microarray in Figure 8.17 is a two-channel array. Notice the differences in the colors of each of the spots. The color of each spot signifies the relative expression of an individual gene between two different target samples. For example, assume that the first sample is prepared from the ovary of a female fruit fly at the beginning of oogenesis and labeled with Cy3 (false colored in green on the digital image by convention); the second sample is from the testes of a male at the beginning of spermatogenesis and is labeled with Cy5 (shown in red by convention). A microarray is co-hybridized with both of the samples, and the results are examined.

Many genes will show no expression in either sample because they are not expressed in these tissues at this time point; these spots on the grid will have no hybridization above background and will appear black. Other genes will represent functions common to both oogenesis and spermatogenesis and will be expressed in both samples; such genes might include those involved in DNA replication or general cellular housekeeping genes for example. In the two-color array, these features are yellow, a digital composite of the red and green signals. In most circumstances, the vast majority of expressed genes will be similarly expressed in both samples; thus, most of the non-black spots will appear yellow.

Still other genes will be uniquely expressed in only one of the two samples, representing ovary- or testis-specific genes. For instance, because only female fruit flies undergo meiotic recombination, dozens of genes whose products are needed for recombination will appear as oogenesis-specific; these will be green in our example. Genes that are expressed only during spermatogenesis will be red.

While investigators always hope to find genes that are specifically expressed in one sample and not at all in the other, most of the genes differentially expressed between two samples are expressed in both of them but at different levels. One of the principal strengths of microarrays is that they can be used to detect even small differences in the level of transcription. Consequently, a gene expression profile is not a series of on–off signals but rather a series of highly graded signal differences: the microarray pattern will have yellow, red, and green spots of varying intensities, as well as black spots for genes that are not expressed. The analysis of signal differences will be an important point in our discussion of the interpretation of microarray data.

One-channel microarrays are hybridized with one labeled target

In contrast to two-channel arrays, one-channel arrays are made by synthesizing short (25-mer) DNA oligonucleotides directly on the surface of the array. This technique was developed by the company Affymetrix. Rather than having only one probe per gene, these one-channel arrays use **probe sets**—more than 10 probes or features corresponding to different sequences within each gene, thereby increasing the accuracy of the expression data. These probes are shorter than those for oligonucleotide arrays, which could lead to cross-hybridization by related but different genes. Specificity for one gene is generally maintained by including probe sequences within the 3′-UTR, which are usually sufficiently divergent even for recently duplicated genes. Because Affymetrix named its microarrays GeneChips, microarrays also acquired the nickname 'chips'.

The synthesis of the probes *in silico* produces remarkably consistent features. The reproducibility of the spots means that each chip can be hybridized with nucleic acid from a single sample, and expression differences between samples are observed by comparing the results of different arrays. Accordingly, the expression pattern seen with a one-channel array is a true quantitative intensity, while expression values from two-channel arrays are relative to another sample.

The target RNA in one-channel arrays is biotin-labeled and detected with fluorescently conjugated streptavidin, a technique also used in the non-radioactive *in situ* hybridization protocols discussed earlier. To increase the hybridization signal, biotinylated anti-streptavidin antibodies can be added, followed by additional conjugated streptavidin. The net effect of this is that more fluorescent signal is produced from every hybridized molecule.

Comparison of one- and two-channel arrays

One-channel arrays have incredible reproducibility, dependability, and improved quantitation, so they are often preferred to two-channel arrays. Using one rather than two different fluorophores avoids a possible problem involving differences in the incorporation of the two fluorophores. Cy3 is more easily incorporated than Cy5 when making labeled targets, is much less photolabile, and fluoresces more strongly than Cy5. To deal with these issues, two-channel arrays are repeated with the labels on the control and treatment samples reversed. Any expression differences between the arrays are due to the different properties of the labels rather than differences between the samples. These technical issues are corrected during the data analysis phase.

Despite their advantages, one-channel arrays are much more expensive to produce and purchase. As a result, two-channel arrays provide a cheaper alternative for many laboratories and classrooms. Two-channel arrays with a customized subset of the genome are available for many model and non-model species. While the full power of microarrays is seen when using probes against all or most of the genes in a genome, smaller non-complete arrays have been made to study individual pathways, such as apoptosis or defense responses, or individual classes of genes, such as signal transduction components or transcription factors.

Technical advances have improved the range, sensitivity, stringency, and reproducibility of microarray experiments

The powerful attraction of microarrays has led to a number of techniques to improve the range of signals that can be detected in an expression profile, as well

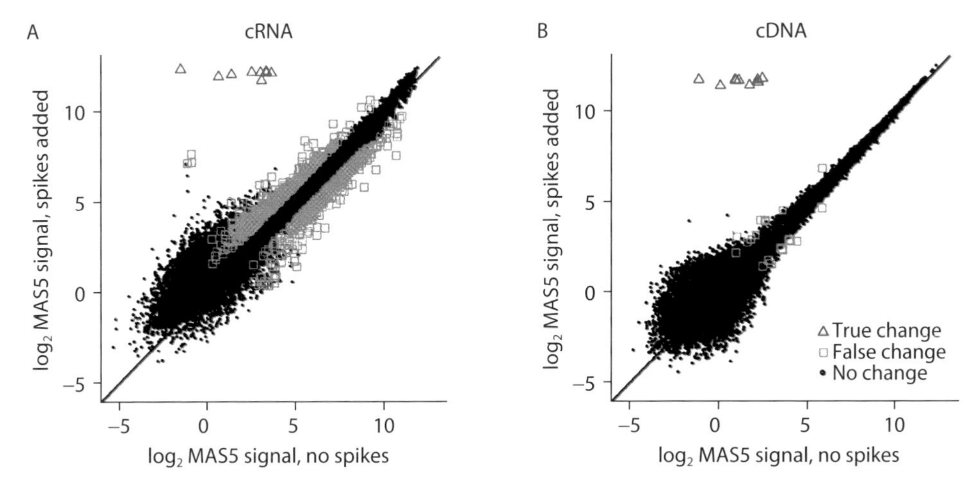

Figure 8.19 Comparison of the use of cDNA with the use of cRNA for hybridization in a microarray. The experiment is described in the text. The key results are shown in the blue boxes, which represent false positives, i.e. genes whose expression is the same in the two samples but appear to change. Many more false positives are seen with cRNA (panel A) than with cDNA (panel B).

as the reproducibility and stringency to reduce cross-hybridization between related but distinct sequences. Linear PCR amplification of the RNA or first-strand cDNA is a standard practice for all microarray experiments because it increases microarray sensitivity and allows the identification of many more genes whose expression changes. These amplification techniques were initially developed to allow smaller amounts of starting RNA to be used. With linear amplification, it is now possible to profile the global gene expression pattern of even a single cell.

Typically, labeled cRNA has been used with one-channel arrays instead of labeled cDNAs, in part because RNA–DNA hybrids are stronger than DNA–DNA hybrids. However, although RNA–DNA hybrids are stronger, DNA–DNA hybrids are more specific, and the use of cDNA instead of cRNA targets increases the specificity and lowers the false-positive rate. To explore this specificity, one study took RNA isolated from a human T-cell line and split it into two samples. One sample was spiked with high levels of nine RNAs corresponding to globin genes, which are not normally expressed in T cells. The other sample did not include the globin gene RNA. The spiked and non-spiked RNAs were then separately used to make both cRNA- and cDNA-labeled targets, which were hybridized to different microarrays. The results of the experiment are shown in Figure 8.19, where the expression value of every probe set in the non-spiked sample is plotted against the corresponding probe set for the spiked sample, one graph for each of the labeling techniques.

KEY ARTICLE

Eklund, A.C., Turner, L.R., Chen, P., *et al.* (2006). Replacing cRNA targets with cDNA reduces microarray cross-hybridization. *Nature Biotechnology*, **24**, 1071–3.

The only differences in the starting RNAs used for these arrays were the nine globin RNAs in the spiked samples. Therefore, only those nine globin genes should be differentially expressed when comparing the spiked and non-spiked samples. Any other difference is due to the difference in specificity between cDNA and cRNA. In addition to globin gene expression (shown by the red triangles in Figure 8.19), 791 other differences (the blue boxes in the image on the left) were seen when comparing the cRNA targets, while only 19 false positives (the blue boxes in the graph on the right) were seen for the cDNA samples, confirming the greater specificity of cDNA over cRNA. Similar results have been seen in related experiments, and so investigators are increasingly relying on cDNA targets with one-channel arrays.

8.6 Interpreting microarray data

The primary goals of a microarray experiment are to characterize the full genome transcription profile of two or more samples and to learn which genes are differentially expressed between them. With thousands or tens of thousands of features being compared in an experiment, the extraction and interpretation of microarray data is itself a sophisticated research field. Data interpretation involves several steps, which are briefly summarized below and in Figure 8.20.

1. Digital imaging of the microarray using a laser confocal microarray scanner. The resulting image constitutes the raw data from a microarray experiment.
2. Data extraction to convert the digital image of the fluorescence signal to numerical values, called the signal, for each spot.
3. Examination of the microarray image and the raw numerical values to determine the success and dependability of the target labeling and hybridization.
4. Normalization of the extracted data to remove experimental biases influencing the raw signal, thereby making the results within and between different arrays directly comparable.
5. Statistical testing to independently identify genes that are differentially expressed between samples and to cluster similarly expressed genes.
6. Functional analysis of the genes of interest. Assignment of functional annotations to these genes and identification of gene family, pathway, and subcellular localization categories that are over-represented or enriched among the genes of interest.

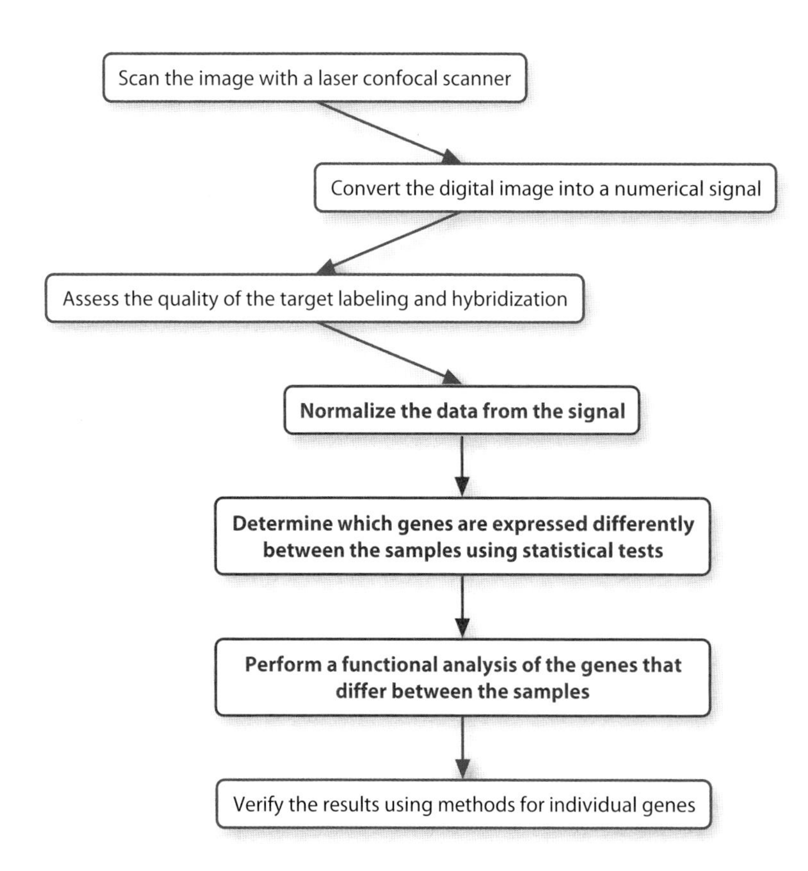

Figure 8.20 Flowchart for interpreting the results of a microarray experiment. The steps in red are described in more detail in the text.

7. Independent experimental validation of the results on a gene-by-gene basis using the RNA expression methods for individual genes discussed above.

Rather than going into detail about each of these steps, we will focus on three of the most important—normalization, statistical analysis, and functional analysis.

Normalization corrects the raw numerical expression values to remove experimental bias

The numerical values extracted from a microarray image are not directly usable measures of gene expression because their quantities are influenced by both biology, i.e. what is actually happening in the cell or organism, and experimental bias. The sources of experimental bias include differences in the amount of labeled target added to each array, the quality of each hybridization, the background level, the brightness of the scanned images, the properties of the two dyes in two-channel arrays (discussed above), printing effects for spotted arrays, and so on. The process of **normalization** takes the raw values extracted from the image and corrects them to remove within- and between-chip experimental biases such that the corrected values directly and solely reflect the biology, which is what the investigators are trying to understand. The final normalized data are then often transformed to a logarithmic scale so that expression changes between samples will be linear rather than exponential.

There are many different computational strategies for normalizing microarray data, but most of them include a background correction and data scaling. Correcting for background hybridization to the slide or support is done for all hybridization experiments, but data scaling requires further explanation. Data scaling is used to adjust the extracted expression values within and between arrays to normalize the signals from different features (spots). Microarray normalization works on the assumption that the total RNA expression and the mean spot intensity are similar for two chips because the target signal added at the start of the experiment is the same. Global mean normalization compares the total signal from two chips and corrects for chip-to-chip differences. Local mean normalization compares different regions on the same chip to correct for local differences in signal intensity. Because of these normalization procedures, the corrected expression values reflect the transcript levels of the original sample more accurately.

Statistical tests are applied to the normalized data to identify genes whose expression differs significantly between samples and clusters of co-expressed genes

Normalized microarray data can be used to examine transcriptional profiles directly and consider questions of differential expression between samples. What constitutes a statistically significant change in gene expression? Statistical analyses of the data, including *t*-tests and ANOVAs, are used to draw non-biased conclusions. The details of such tests are beyond the scope of this text, but in general, *t*-tests are used to identify significant differences between two groups, while ANOVAs can consider differences between more than two groups. The goal of these analyses is to identify genes differentially expressed between sets of samples, while reducing the number of false positives in the list and also limiting the number of false negatives.

These statistical tests require biological replicates, i.e. many independent samples representing the same treatment. While some conclusions can be drawn using only two replicates, the strongest statements require a minimum of three, and increasing the number to five or six greatly reduces the number of false negatives. *t*-tests and ANOVAs measure the variation between replicate samples and between conditions to determine the significance, or likelihood, of the difference being due to chance rather than being a true biological difference.

The outputs of statistical tests are *P*-values, which provide a numerical value (between 0 and 1) of the likelihood of the particular difference being due to chance. For many purposes, *P*-values less than or equal to 0.05 are considered significant in statistical tests, but the cutoff depends on the purpose. Consider the situation as it applies to microarrays. A *P*-value of 0.05 implies that there is a 5 percent chance that the difference in expression of a particular gene as revealed by microarray analysis is a false positive. That is, the samples appear to differ in expression, but the signal differences can be attributed to sampling variation rather than biological differences. In the case of a microarray experiment involving the simultaneous measurement of 20 000 genes, using a P-value of 0.05 might yield 1000 false positives. Because this number of false positives is not acceptable, the P-value cut-off is changed using a **multiple-testing correction** (MTC), which takes into account the fact that a large number of measurements are being made. There are many ways of doing this, but the most conservative way is to use the Bonferroni method, which sets a new P-value by taking the original P-value cut-off and dividing it by the total number of genes being considered. In the case of our example above, our new P-value cut-off is $0.05/20\,000 = 2.5 \times 10^{-6}$, a much more rigorous cut-off. This method is so rigorous that it often produces a higher rate of false negatives, i.e. some real differences in the expression signal are being missed because the test is too conservative. Other less strict MTC methods with fewer false negatives have been devised. The balance between the false-positive and false-negative rate is tricky, and is best dealt with by increasing the number of replicates, though the expense of microarray experiments has meant that two to five replicates are done.

Statistics can also be used to **cluster**, or group, genes based on similar expression patterns across a set of samples. Clustering genes is important for identifying co-expressed genes that are potentially involved in similar processes. Clustering can also be used to group different microarray samples based on the similarity of their global transcription profiles. Ideally, in a multiple condition microarray experiment, the replicates will be clustered together. This technique also has potential clinical applications because it can be used to compare the transcriptome of a tissue biopsy with samples that are known to be healthy or diseased.

There are many statistical approaches to clustering, but the easiest to visualize are hierarchical clustering techniques. These methods build a phylogenetic tree by measuring the expression differences between genes on an array or between different arrays. These trees are built by first identifying the most similar genes or samples and then extending the tree by identifying the next most similar genes or samples until all of the members are included.

Functional analysis of microarray data

While most of the interest in a microarray experiment is placed on the list of significantly upregulated and downregulated genes, such a list is not the final goal of most analyses. A microarray experiment strives to use whole-genome transcriptional information to understand the biology of the organism. Thus, we must go from a list of genes whose expression levels change to an analysis of the molecular functions of these genes and the effect of these changes on expression.

Although microarrays are available for many non-model species, the genomes of model organisms have been intensively studied and annotated to identify candidate genes and their possible functions. These functions are based on information from studies of individual genes using the genetic, molecular, cellular, and biochemical experiments discussed earlier in this chapter and throughout this book. Even in model organisms, many of the genes have not been genetically identified and studied, so this functional information is also based on homology to other genes of known function within the organism and in other species.

Using this functional information, all of the genes in the genome are placed in categories according to their gene families, pathway, and subcellular localization categories. One of the most common systems for categorizing genes uses gene ontology (GO) terms, which we introduced in Chapters 2 and 4. GO terms categorize genes in three different ways. Each gene is classified using terms for GO biological processes, GO cellular component, and GO molecular function. These categorize the gene based on the *process* or pathway the gene and its products regulate, the site of *subcellular localization* of the product, and the gene family and *biochemical function* of the product, respectively. As noted in Chapter 2, the genome databases for model organisms include the GO terms for genes with a postulated function.

A variety of computer programs and websites look for statistical enrichment of GO terms in a gene list. One impressive example of their power comes from studies of the gene expression of all of the cell types found in *Arabidopsis* roots. These studies used different GFP enhancer traps with cell-type-specific expression patterns in the root. The root cells were separated from one another using enzymes that digest their cell walls, and fluorescence-activated cell sorting (FACS) was used to isolate the cells expressing GFP. Microarrays were used to identify the transcription profile of each cell type, and the enrichment of specific GO terms was associated with different cell types. By combining these experiments with microarrays examining temporal expression patterns, the researchers have obtained one of the most complete transcriptional profiles of any plant or animal organ.

Microarrays can help identify previously unknown genes functionally involved in a biological process

In Chapter 5, we discussed the fact that the expression pattern provides a strategy for cloning the gene, even without knowing more about the gene's function. The expression pattern obtained from microarrays is a powerful complement to genetics and biochemistry for identifying new genes related to a particular process. One of many examples involves genes expressed in the germline in *C. elegans*. By comparing the expression pattern of normal worms with that of worms lacking a germline, a list of more than 100 genes expressed specifically in the germline was produced. The function of each of these genes was tested using RNAi, and several of these knockdowns had meiotic defects, suggesting their involvement in meiosis in *C. elegans*. Since the effects on meiosis were sometimes subtle changes in fertility or in the progeny in a later generation, the functions of these genes were missed in genome-wide RNAi screens. Furthermore, many of these genes do not have sequence similarity to genes known to be involved in meiosis in other organisms, so they would not have been identified by evolutionary conservation either. Their possible involvement in meiosis was identified using microarrays and then verified by functional tests of individual genes.

Expression profiles can provide biological markers for clinical and research applications

Expression profiles characterized using microarrays and other techniques have been very important for identifying biological markers distinguishing different cell types, stages of cell growth and division, tissues, developmental stages, and responses to environmental conditions. One particularly early application of expression profiles was to provide biological markers for cancer and other diseases. For most cancers, early detection greatly improves the prognosis for treatment. Unfortunately, early detection is often difficult because of the absence of a clear change in cellular morphology or the lack of a molecular marker that can easily be monitored during oncogenesis. Prostate cancer screening has benefited greatly from having an effective tumor marker—prostate-specific antigen (PSA). PSA is a protein that is produced by the prostate gland and circulates in the blood. Most men have very low levels of PSA. Elevated levels of PSA in the blood are an indication that more comprehensive tests for prostate cancer should be considered. PSA itself is not involved with prostate cancer; it is simply an indicator that a problem may exist. Although far from perfect, PSA levels are an effective and relatively simple test for the early detection of prostate cancer.

Very few known tumor markers work as well as PSA. Other tumor markers have a higher rate of false positives and, even more serious, a higher rate of false negatives, in which the cancer is present but undetected. Cancer is a complex cellular event involving expression changes of many genes, so it is not too surprising that it would be difficult to detect by monitoring the expression of a single gene. However, microarrays have considerable promise for both identifying possible new tumor markers and screening multigene expression changes in individual patients because of their capacity to detect changes in the expression profile of thousands of genes simultaneously. For example, the change in the expression profile of a set of 70 genes has been able to detect metastatic breast tumors. Using microarrays to measure the expression of a combination of dozens or even hundreds of genes may offer a much more sensitive assay for cancer detection in individual patients. In fact, a recent experiment identified a group of between five and ten genes whose expression profile appears to be an even more sensitive and reliable test for prostate cancer than PSA levels.

A similar application can be imagined for many other complex traits and diseases in humans or other organisms. It is not completely far-fetched to envision that a microarray expression profile may be useful in diagnosing the risk for heart disease, adult onset diabetes, and other traits associated with increased mortality. Perhaps an expression profile may some day become a routine diagnostic vital sign in the way that blood pressure, heart rate, electrolyte balance, and other vital signs are used now. Like other such physiological phenotypes, a microarray expression profile is presenting a complex and integrated body of information as one picture.

Although the use of expression profiles for clinical assays is still largely hypothetical or experimental, their utility for basic research is very well established. Microarrays have been used to identify markers for studying different biological processes. For example, the gene-line-specific genes identified in the *C. elegans* microarrays mentioned above could act as markers for germline tissue. Similarly to the PSA example above, misexpression of such markers can be a sign that something is not occurring normally. While many of these germline-specific genes do have a role in meiosis, markers may or may not directly function in the process or tissues being studied. However, this does not matter, because the usefulness of a marker is in its ability to help track the tissue, growth, development, or environmental response it is designating.

Even investigators working with individual genes rely on previous microarray data that has been collected for their organism. The importance of having expression profiles can be illustrated by remembering the characteristics of genes used for cloning presented in Chapter 5. One characteristic of a gene is its expression pattern, which a microarray provides. If an investigator working in one of the established model organisms has mapped a mutant phenotype to a region of a chromosome—the first of our characteristics of genes in Chapter 5—it makes sense to immediately use the genome database to look at the microarray data for all of the genes in that region to determine which are the most likely candidate genes for the phenotype. It is no wonder that relying on microarray expression profiles identified in other laboratories and recorded in genome databases has quickly become standard for nearly every molecular genetics laboratory.

Microarrays can provide answers for other questions about gene structure and expression

Because the power of microarrays is immense, more research applications for this technology are appearing regularly. Different questions can be asked by changing the probe sequences used on the microarray. For example, microarrays have been developed for identifying alternatively spliced transcripts (exon arrays) and previously unknown transcripts (tiling arrays). We mentioned previously that typical gene expression microarrays use probes that recognize the 3′-UTR of transcripts because this is generally the most gene-specific region in an mRNA. However, exon arrays have probes against each of the exons in every gene. A schematic diagram of the procedure is shown in Figure 8.21A. For transcripts that are not alternatively spliced, the level of hybridization should be similar to all of these probes. However, with alternatively spliced transcripts, the amount of hybridization to each exon probe will not always be the same, at least under some conditions. These arrays have been used to identify previously unknown alternative splicing events.

One important advance for identifying previously unknown genes employs a technique known as a **tiling array** (Figure 8.21B). Tiling arrays contain probes complementary to every region of the entire genome, both those predicted to encode genes and those not known or expected to contain genes, so they are not biased for known or predicted genes. Because DNA is double-stranded and mRNA can be transcribed from either DNA strand, a complete tiling array includes probes for both DNA strands. Often, all of the probes corresponding to one strand are placed on one chip, and all of the probes for the other strand are on a second chip. For example, the Affymetrix *Arabidopsis* tiling array consists of two chips, each having one 25 bp probe for every 35 bp of sequence of one strand of the genome. Using a tiling array like this one for expression studies has enabled the identification of previously unknown transcripts, as well as alternative splicing patterns.

Tiling arrays have also been used to study questions not directly related to transcript expression. In Chapter 12, we consider how tiling arrays are used to identify transcription factor binding sites across the genome. Using similar techniques, tiling arrays can be used to identify changes in genome methylation. Tiling arrays have also been hybridized to genomic DNA from a wild-type organism and a deletion mutant to identify the deleted region. Thus, we see that tiling arrays allow for a truly complete global view of gene expression.

SNP arrays are a variation of tiling arrays used to identify single-nucleotide polymorphisms (SNPs). These arrays are even denser than standard tiling arrays. For

A. Exon array

Probes on array: exons

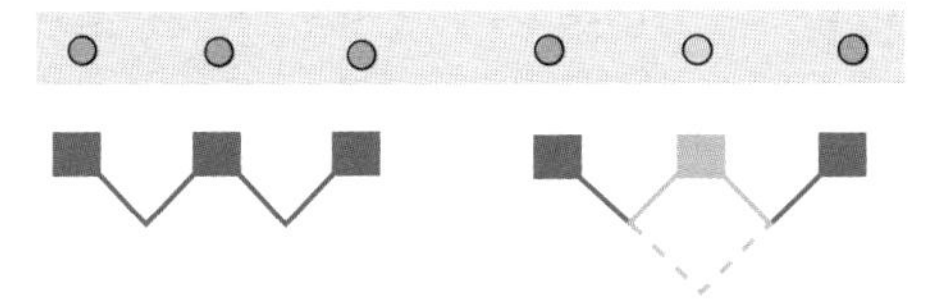

B. Tiling array

Probes on array: DNA oligonucleotides

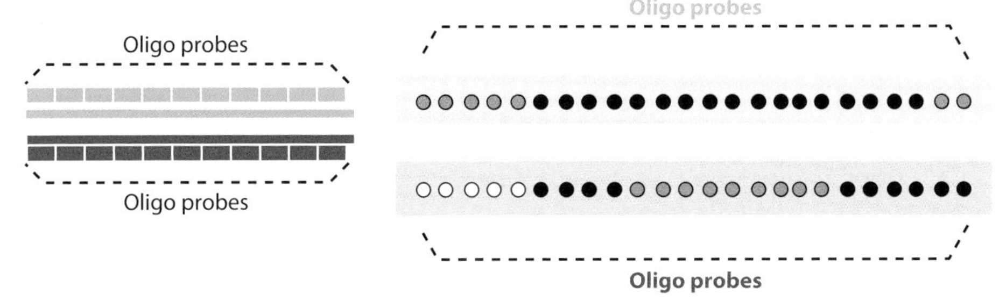

Figure 8.21 Other examples of the uses of microarrays. (A) An exon array. The array contains probe sets for known or predicted exons. Depicted here are six exons from two different genes; in actual experiments, the exon probes would not be next to each other. The gene on the left shows similar expression from all three exons, indicating that these are expressed at the same level. The gene on the right has reduced hybridization to the middle exon, indicating that this exon is missing in some transcripts and thus is alternatively spliced. (B) A tiling array. Two strands of DNA are shown, one in green and one in purple. Many small probes are made from each strand, shown here as broken lines. These probes are hybridized, with the hypothetical result shown. One interpretation of this pattern is that adjacent genes are transcribed from opposite strands in this portion of the genome. Since the oligonucleotide probes in a tiling array are about 35 bases long, an actual array would have thousands of oligonucleotides corresponding to any one gene rather than the few shown here.

every nucleotide on each strand of DNA in the genome, a SNP array has four probes, each with a 12 bp sequence perfectly complementary to the genome on either side of a variable base, i.e. A, C, G, or T. SNP arrays can be used to identify point mutations and copy-number differences responsible for a mutant phenotype and have become very influential in studies of natural variation within a species and association mapping, discussed in Chapter 5.

We have not described all of the ways in which microarrays are being used or the picture they provide of genome-wide transcription. The amount of data available from a microarray is astonishing, so it is not surprising that the data from microarrays are routinely re-analyzed or clustered in different ways to show the relationships among genes. It is almost certain that the potential applications of microarrays for genetic analysis are nowhere near completely explored.

8.7 Other genome-wide expression assays

While microarrays are definitely the most common technique for studying genome-wide expression patterns, other techniques are also used for both RNA expression and protein expression. We briefly summarize two other methods.

Large-scale RNA sequencing is an alternative to microarrays

As the name implies, **large-scale RNA sequencing** involves sequencing RNAs from a sample. RNA sequencing has been particularly important in the effort to identify and characterize small RNAs on a genome-wide scale. Modified microarray protocols such as tiling arrays are one option for finding small RNAs, but since microarrays rely on hybridization, transcripts present in very low abundance could be missed. The analysis of tiling array data also relies on the identification of **windows** of hybridization, which involves annealing to multiple consecutive probes in order to ensure specificity, something that cannot be done with small

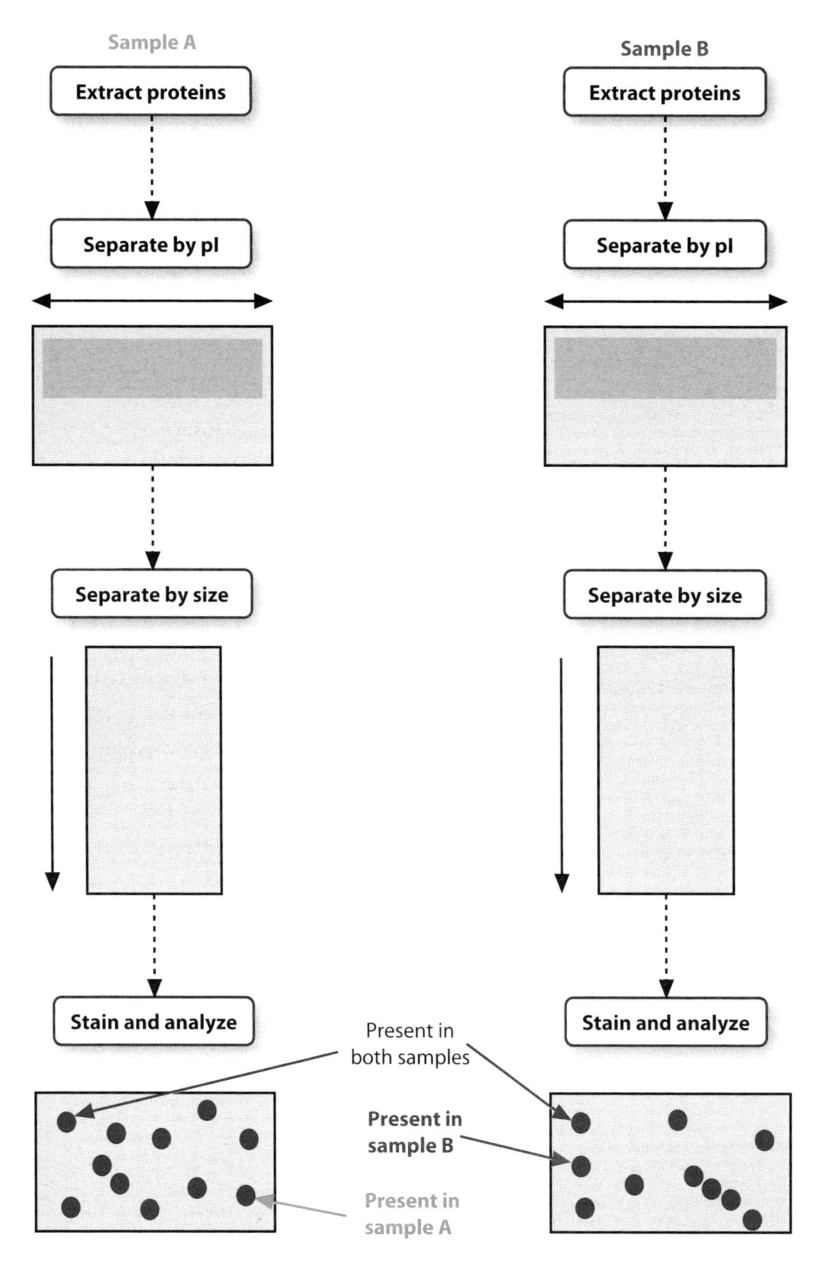

Figure 8.22 A two-dimensional gel used in analysis of the proteome. The goal of this analysis is to compare the proteins made by two different samples in the same way that microarrays are used to compare the RNA made from two different samples. Total proteins are isolated from each sample and run on a gel. Often the first gel separates the proteins based on the isoelectric point (pI). The gel is then rotated or the direction of the current is changed and the proteins are separated by a second property. Here the second dimension is the molecular weight. The gels are then stained and the protein pattern is analyzed. As suggested by the drawing here, some proteins will be found in both samples, whereas others will be found in only one of the samples. Thousands of proteins are likely to be different between the two samples.

RNAs. Several different methods, including massively parallel sequencing (MPS) and 454 Sequencing, have been developed to clone the small RNAs expressed in a particular sample and to sequence each of the individual clones. These large-scale or deep sequencing techniques involve the sequencing of millions of individual clones per sample. Such a technique is feasible because only short sequence reads are necessary and sequencing methods have become much cheaper.

Deep sequencing methods directly count individual RNA molecules, so the results are both sensitive and quantitative. These approaches have now been applied to sequencing mRNAs. Small mRNA fragments are cloned, and random clones are sequenced. The short sequences (17–23 bp) are used with the genome sequence to identify the gene of origin, so that the other tools of bioinformatics we have discussed can be applied.

Proteomics approaches can be used for global studies of protein expression

While most global gene expression methods explore the RNA component of gene expression, genome-wide protein analysis is also an important goal. After all, the functional product for most genes is a protein, so a functional analysis of gene activity should include the properties of the protein. Because the DNA component of an organism is called its genome and the RNA component is called the transcriptome, the set of proteins expressed by an organism has been dubbed the **proteome**.

Global gene expression studies at the protein level are much more difficult and labor-intensive than studies of RNA expression. The basis most proteomic techniques is to isolate all proteins made by a cell type and separate these on two-dimensional gels (Figure 8.22). In the first dimension, proteins are separated by their isoelectric point (pI), the pH at which the protein has no net charge. After separation is attained in the first dimension, the gel is turned through 90° and the proteins are separated in a second dimension using another of their properties, for example their molecular weight. The gel is stained with Coomassie blue, and the pattern of hundreds of protein spots on the gel is characteristic of the particular sample, similar to a transcription profile from a microarray.

Proteomic methods do not directly connect the protein product to a specific gene, as do the methods for studying the transcriptome. As a result, the individual proteins have to be isolated from the gel and sequenced by mass spectrometry. Because this is labor-intensive, this is generally done only for proteins differentially expressed between two samples of interest. Differential expression is observed by comparing the relative expression level of a spot with other spots on its own gel and then with the same spots on a second gel with protein extracts from a different samples. While technically more challenging than transcriptional techniques and requiring large amounts of protein extracts to visualize low-abundance proteins, proteomics allows a global view of the end product of gene expression, namely stable protein levels.

8.8 How much change in gene expression is biologically significant?

One of the underlying assumptions about the interpretation of gene expression data is that a change in the expression pattern of a gene or a group of genes is correlated with a change in the phenotype of the organism, tissue, or cell. This may seem so obvious that it needs no further exploration, but hidden behind this assumption is a question that is biologically challenging: How much change in gene expression (RNA or protein) is biologically meaningful and how much is biological noise? Biological noise is the variation in gene expression between biological replicates. The question takes on increased importance because real-time RT-PCR, microarrays, and large-scale RNA sequencing measure very small differences in expression. Furthermore, genomic techniques can produce long lists of differentially expressed genes that are significant based on statistical tests, despite the fact that the expression of some of these genes may be barely changing.

There are many examples in which small expression changes have major biological relevance. As noted in

Chapters 1 and 2, sex determination in flies (and worms) takes place early in embryo development and depends on twofold differences in the copy number of certain key genes on the X chromosome, which in turn lead to twofold differences in expression. Even smaller changes in gene expression levels would be found for some of the phenotypic differences between a Down syndrome child with trisomy 21 and an unaffected disomic 21; analysis has revealed that some phenotypes are sensitive to gene expression differences in the range of two versus three copies, a difference of 1.5-fold. In Chapter 4, we also noted examples of genes whose expression is particularly tightly regulated. These included several genes that are important in human cancers, such as *p53* and *Rb*. Thus, a small difference in gene expression is not merely a theoretical question. The difference could have significant clinical implications.

On the other hand, for many more genes, either small or large changes in gene expression seem to have no effect on the phenotype. As discussed in Chapter 3, most loss-of-function mutations are recessive, i.e. the phenotype affected by the gene is insensitive to a twofold difference in copy number. One caveat to this discussion is that, because of the intricate feedback patterns controlling gene expression, the expression of a single copy of a gene is not always exactly equal to half of the expression seen with two copies. Nonetheless, reducing gene expression by half does not appear to have much of an effect on the phenotype for most genes.

A similar conclusion is drawn from experiments that over-express a gene. Using transgenic organisms over-expressing or ectopically expressing genes, it has been shown that changing the expression of many genes by four-, five, or even tenfold has very little effect on phenotype. Many different investigators have over-expressed many different genes in the course of assaying gene expression. It is notable how few of these over-expressions actually affect the phenotype of the cell or the organism. For most genes, then, a three- or fourfold change in transcription may simply not make a difference to the biology of the organism. For these genes, microarrays may be attributing statistical significance to a change in expression that does not have biological relevance.

Therefore, the first answer to the question that opens this section is: 'It depends on the gene.' The gene-to-gene specificity is likely determined by the threshold required for normal and maximum activity of a particular gene and the normal expression level of a gene relative to this threshold. When working with a single gene, standard experiments can be performed to determine the sensitivity of the gene in question, but in a study involving global changes in gene expression, this answer is very unsatisfactory. If the goal of such an experiment is to determine an expression profile rather than identifying a collection of individual genes likely to affect a process, this answer is rather more acceptable. For example, if the goal is to determine the gene expression patterns that distinguish a metastatic tumor cell from a non-cancerous cell, it makes no difference if any of the individual genes is dose sensitive in its effects. What matters is that the profile is distinct.

The level of biological noise

A more satisfactory and complete answer to our question comes from studies that combine microarrays, quantitative RT-PCR, and reporter gene assays to measure biological noise affecting RNA and protein expression. We are using the term 'biological noise' here to refer to the stochastic variation in the level of gene expression that occurs under natural conditions. That is, if one could precisely measure the level of expression from the same gene under the same conditions in two identical cells, how much would the measurements differ? Biological noise arises from a complex and unknown combination of variations in all of the factors that affect gene expression: transcription initiation, the rate of mRNA synthesis and decay, initiation and rate of translation, protein turnover, and so on. In the most comprehensive studies to date, more than 4100 yeast strains expressing GFP translational fusions to different individual proteins were monitored using flow cytometry to measure the protein expression level, as summarized in Figure 8.23. The reporter genes were inserted at their usual site in the genome by homologous recombination, as described in Chapter 6, and thus were under the control of the regulatory region that the wild-type gene has at that site. The assay allowed fluorescence measurements of single cells from cultures with more than 50 000 cells grown to mid-log phase in either rich or minimal medum. More than 2500 genes comprising nearly 40 percent of

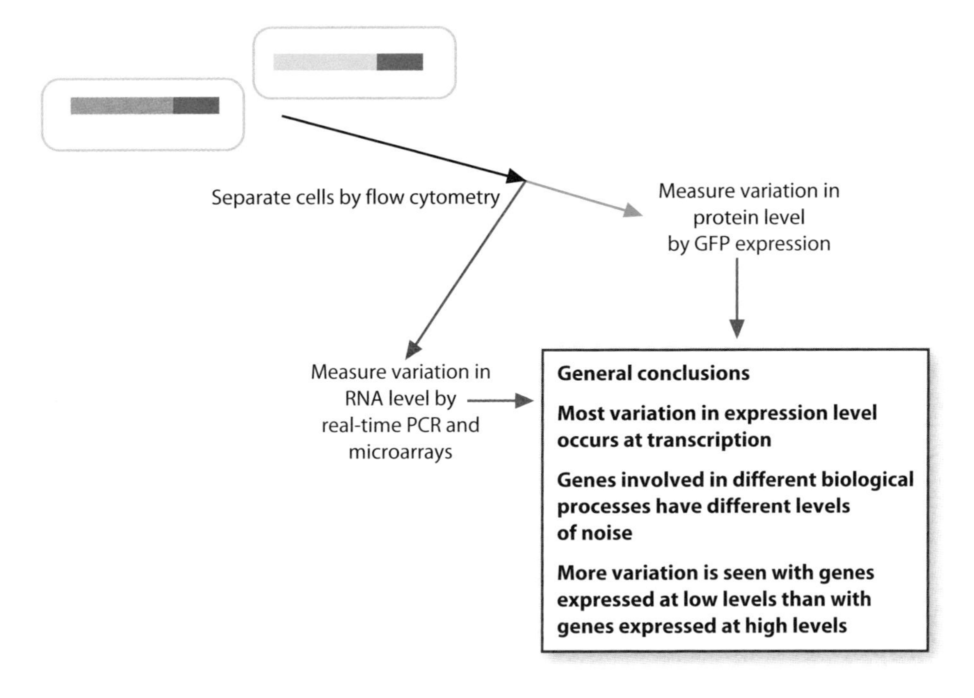

Figure 8.23 Measuring biological noise. Biological noise can be considered as the naturally occurring variation in the level of expression of a gene. In order to measure biological noise, yeast cells were made with GFP fusions to known genes; the reporter genes were integrated at their normal sites in the genome and were thus under the control of their native regulatory regions. Cells were separated by flow cytometry and the expression of GFP was measured in individual cells with the same reporter. The variation in RNA expression was measured using real-time PCR and microarrays. The variations in protein levels and RNA levels for each gene were compared.

the yeast genome could be detected and scored in the assay. These assays were followed by microarray and quantitative RT-PCR experiments to determine RNA expression patterns.

KEY ARTICLES

Newman, J.R., Ghaemmaghami, S., Ihmels, J., *et al.* (2006). Single-cell proteomic analysis of *S. cerevisiae* reveals the architecture of biological noise. *Nature*, **441**, 840–6.

Bar-Even, A., Paulsson, J., Maheshri, N., *et al.* (2006). Noise in protein expression scales with natural protein abundance. *Nature Genetics*, **38**, 636–43.

A few general conclusions emerge from the study. The first is that most of the variation in protein levels is also seen in mRNA levels. Thus, most of the regulation of gene expression occurs during transcription. On the other hand, twofold changes in mRNA levels were not seen as frequently as corresponding changes in protein levels. This indicates that most biological noise arises from the rate of mRNA synthesis and turnover, but that not all of this transcriptional noise is reflected in protein noise. Additional regulation occurs post-transcriptionally, and it seems likely that mechanisms have evolved to buffer the level of protein expression from random fluctuations in the level of transcription.

Secondly, the variation in protein expression is different for different biological processes. For example, proteins involved in chromatin remodeling, ATP synthesis, the Krebs cycle, environmental sensing, and stress response have higher than average noise, as if the cell can tolerate variation in these processes. On the other hand, proteins involved in translation and protein degradation exhibit lower than average noise, as if variations in these processes are less easily tolerated by the cell. In general, the amount of protein noise is inversely proportional to protein abundance, so low-abundance proteins have the greatest amount of variation in expression level. The same is true for mRNAs—mRNAs expressed at low levels in microarray studies have a much higher variance than those of greater abundance.

This inverse correlation between the amount of noise and mRNA abundance may reflect the fact that biological noise arises from inherent imprecision in the machinery of gene expression. The number of transcripts produced from a gene may vary somewhat because all biological processes have variability. For a gene expressed at high levels, a small change in the number of transcripts may not have much of an effect. On the other hand, for a gene expressed at low levels, a small change in the number of transcripts could change the expression by two- or threefold, either up or down. That is, while the *amount* of noise might arise from imprecision in gene expression, which should affect different transcripts similarly, the *impact* of the noise depends on the abundance of the normal transcript and thus on the individual gene.

Many technical questions arise during experiments of this type, and it is often difficult for investigators to correct for all sources of noise. As genomic experiments become even more sophisticated, our understanding of the normal variation in gene expression levels will certainly also become more sophisticated. A more complete understanding of the effect of biological noise on biological processes has not yet been developed, so the question posed at the beginning of this section is still not easy to answer. Or rather, perhaps the best answer remains: 'It depends on the gene.' To determine the impact of variation in expression levels for an individual gene, it helps to look at the phenotype of a mutant—the topic of the next chapter.

SUMMARY

Assays of gene expression

The expression pattern of a gene is one of its most fundamental biological properties. For most genes, the expression pattern arises from changes in transcription, although many important exceptions are known. For genes that have been cloned, both the transcription pattern and protein expression pattern can be determined by a variety of different well-developed assays. Some methods determine the stable level of mRNA or protein, the static level of expression; others are able to determine dynamic changes in the expression pattern. Although methods to determine the expression patterns for individual genes will continue to be important, microarrays and similar genome-wide assays are supplanting these for routine tests in well-studied model organisms.

Knowing the transcription or the protein pattern of a gene may seem to be the best possible assay for gene expression. However, these assays may not connect directly to the mutant phenotype that defines many genes. Expression assays based on mutant phenotypes were developed well in advance of the molecular assays, and still provide important and additional insights. These are discussed in Chapter 8. Molecular assays for gene expression are an essential component of genetic analysis. The methods and strategies to examine gene expression are a well-developed and mature technology, representing a foundational aspect of genetics.

Chapter Capsule

Molecular analysis of gene expression

The expression pattern of a gene is one of its fundamental functional characteristics, so it is among the primary things that an investigator will want to know. Assays for gene expression can focus on RNA or protein expression. The expression of most genes is regulated at the level of transcription, so RNA assays such as Northern blots, quantitative real-time RT-PCR, *in situ* hybridization, and transcriptional reporter genes provide detailed information about gene expression. Assays of protein expression include indirect immunofluorescence and translational reporter genes. Results from these experiments are compared with the results of assays of transcription to determine the importance of protein localization and other post-transcriptional regulation. Different assays can also provide a static picture of the stable pattern of expression, while others allow a dynamic picture of the changes in expression over time.

Increasingly, gene expression assays are being done on a genome-wide scale, using microarrays, large-scale RNA sequencing, and proteomics. These genomic approaches compare the expression patterns in different tissues or developmental stages, or under different conditions, to identify expression profiles and differentially expressed genes. Because many of these approaches for studying gene expression can detect very small differences in the level of transcription between samples, it is important both to determine the statistical significance of the changes and to consider how much variation in gene expression occurs naturally.

CHAPTER MAP

This chapter describes how mutant phenotypes are used to study the pattern of gene expression. The analysis begins with the mutant allele, as shown on the upper left of the figure, and uses conditional alleles and mosaic analysis to examine the expression pattern.

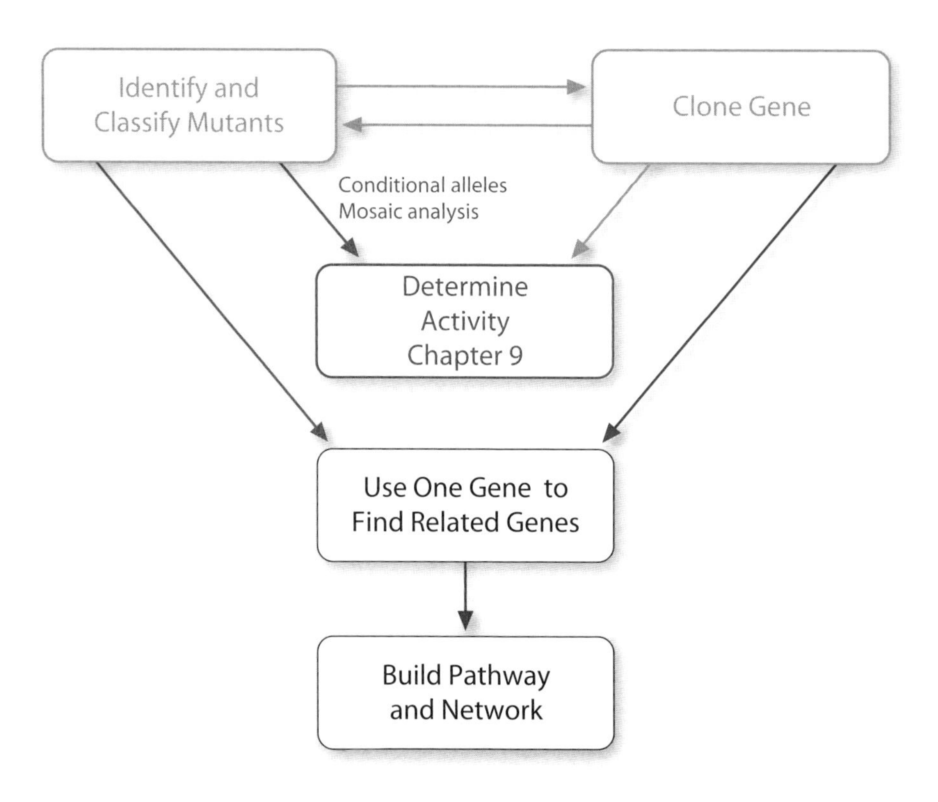

9

CHAPTER 9

Analysis of gene activity using mutants

TOPIC SUMMARY

The expression pattern of a gene is critical to understanding its overall biological role. For genes identified by mutations, the mutant phenotype itself can be used to analyze the activity of the gene. It is often helpful to distinguish the terminal phenotype of the mutant from the initial mutant phenotype, i.e. the phenotype exhibited during the early stages of an organism's development, a period of time that is sometimes called the phenocritical phase. With both conditional alleles of the gene (such as temperature-sensitive mutations) and temperature-shift experiments, the investigator can determine *when* the activity of the gene is needed. With mosaic analysis in which the wild-type function of the gene is removed in certain cells, the investigator can determine *where* the activity of the gene is needed. These are helpful in supplementing the molecular analysis of gene expression described in Chapter 8.

IN BRIEF

Mutant alleles themselves can be used to determine the time and location at which the activity of a gene is needed.

INTRODUCTION

Although molecular analysis of the type described in Chapter 8 is a familiar way of thinking of gene expression, much can also be learned from using mutant phenotypes. To be precise, an analysis based on mutant phenotypes should be referred to as the analysis of **gene activity** or gene action since mutant phenotypes alone cannot tell us about transcription or translation. However, by not making the distinction between the different molecular steps in gene expression, a genetic analysis based on mutant phenotypes can provide information on when and where the activity of the gene—whatever it may be in molecular terms—is actually needed.

This topic may seem a bit old-fashioned, and it certainly has a long history in genetic analysis, one that predates our knowledge of the molecular nature of genes. The richness of the tradition is one of the reasons why using mutants to understand gene activity is still important to understand. Another reason is that this type of approach is still used and is still useful, in part because it gives rather different information about the role of the gene than does gene expression.

In this chapter, we describe three approaches to the analysis of gene activity. In the first part of the chapter, we discuss some ideas related to understanding the mutant phenotype itself; many of these ideas connect directly to topics introduced in Chapters 3 and 4. The second topic involves the use of conditional mutations, particularly temperature-sensitive mutations. These are mutations that exhibit a mutant phenotype at one temperature and a wild-type phenotype at another, usually lower, temperature. Temperature-sensitive mutations have been used for understanding the *time* at which the activity of a gene is needed. Finally, we discuss a genetic approach known as mosaic analysis, which produces an animal with cells of two different genotypes. Mosaic analysis is most often used to determine the *tissues* in which the activity of the gene is needed.

Our task and our tools are depicted in Figure 9.1. The mutants themselves were found in Chapter 3 and classified in Chapter 4. Now it is time to begin to put them to work to help solve a biological question.

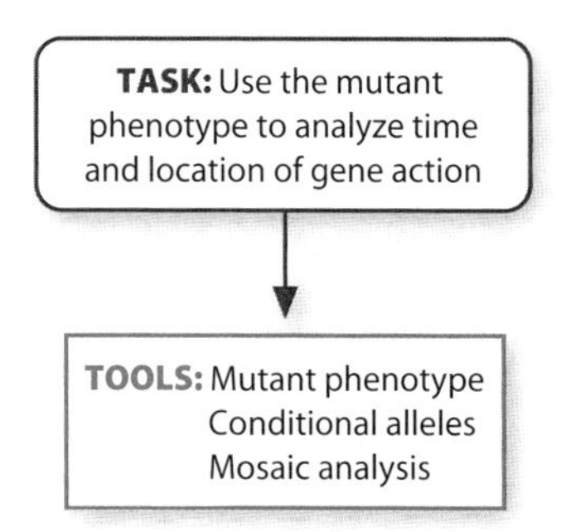

Figure 9.1 The task of analyzing gene activity using mutants involves the mutant phenotype, conditional alleles, and mosaic analysis.

9.1 Mutant analysis of gene activity: an overview

Tissues, organs, and cells differ in their function and morphology because different genes are active in different locations within the body. Morphological changes occur during the development of an organism because different genes are active at different times in the life of an organism. The concept of **gene activity** is different from gene expression. Part of the difference comes from the manner in which gene activity and gene expression are each assayed and what is needed to perform the assay. The analysis of **gene expression** uses the assays based on transcription and translation as described in Chapter 8; assays of gene expression use cloned genes for the starting material, and involve determining the location and time of RNA or protein.

Often the pattern of expression will also tell the investigator when and where the gene is active. That is, gene expression assays will tell us which cells transcribe the gene or make the protein, and it is reasonable to assume that the function of the gene is important in those cells. However, as illustrated in Figure 9.2, the time and location of a gene's *expression* may not be precisely the same as the time and location of when a gene is *active*. Gene activity is a functional assay—an assay that asks, 'Which cells or developmental stages require the normal function of this gene?' Figure 9.2A illustrates a situation when the gene is expressed (as shown in green) in the same cells in which it is active (shown in blue). Figures 9.2B and 9.2C illustrate situations when the locations of gene expression and gene activity are not the same. For example, as shown in Figure 9.2B, a protein may be *present* in a cell as revealed by immunofluorescence (therefore its gene has been expressed), but it is not *active*; perhaps it needs to be phosphorylated, proteolytically cleaved, or otherwise modified to become the active form. In such an example, the pattern of gene activity (in blue) is more limited or narrowly defined than the pattern of gene expression (in green). However, the opposite situation could prevail for some genes, as shown in Figure 9.2C. Perhaps the gene product is involved in extracellular signaling to other cells nearby or throughout the body. In this case, the gene might be transcribed and translated in one or a few locations, but the protein has effects throughout the tissue or body. Thus, in this situation, the pattern of

Gene expression | Gene activity

A. Gene expression pattern and gene activity pattern are the same

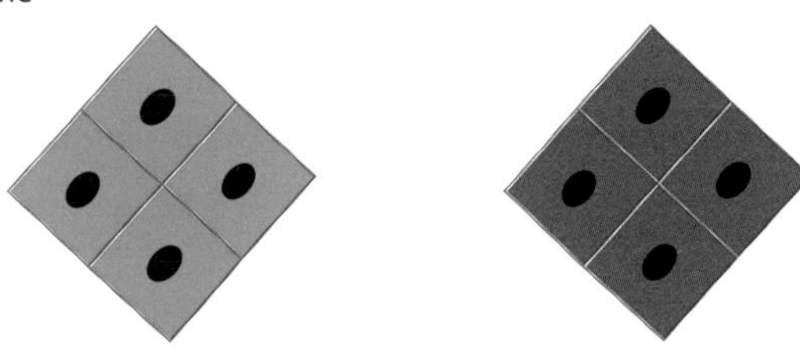

B. Gene expression pattern is broader than gene activity pattern

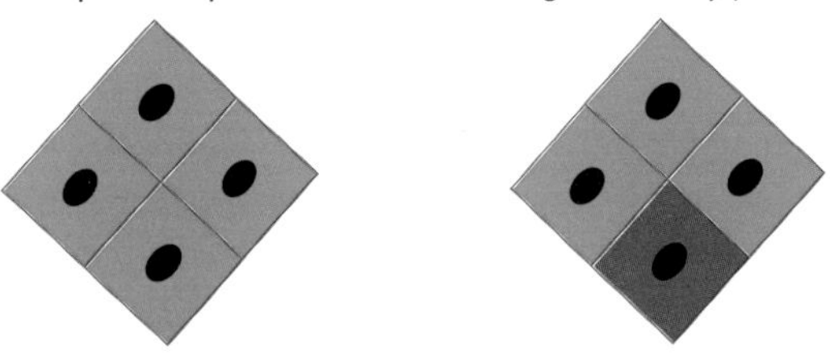

C. Gene activity pattern is broader than gene expression pattern

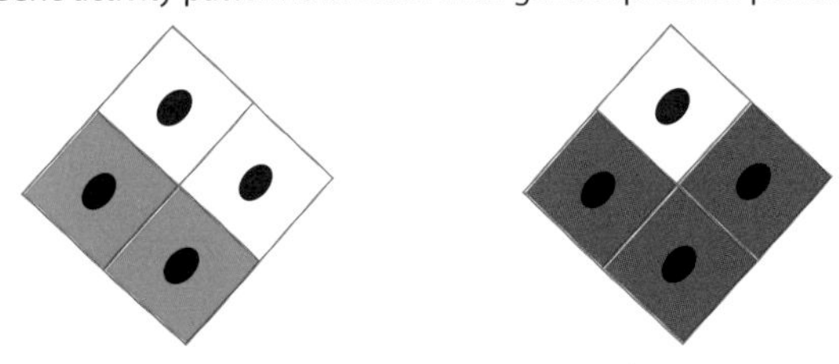

Figure 9.2 The difference between gene expression and gene activity. Four hypothetical cells are shown in each panel. A cell that expresses the gene, as determined from transcription assays or protein assays as described in Chapter 8, is shown in green in assays on the left. A cell in which the gene is inferred to be active, as described by assays described in this chapter, is shown in blue in assays on the right. A cell in which the gene is neither expressed nor active is white. The cells could be from the same tissue or from different tissues. (A) The cells that express the gene are the ones in which the gene is active. This is a common pattern in which gene expression and gene activity assays identify the same cells. (B) All the cells express the gene but gene activity is found in only one cell. This arises when the gene is only translated in some of the cells in which it is transcribed, or the protein is expressed in all cells but needs to be modified post-translationally to become active. The post-translational modification occurs in only one cell, so only that cell has gene activity. (C) Gene activity is found in cells in which the gene is not expressed. For example, perhaps the gene expression assay on the left measures transcription. The protein product is secreted to nearby cells that do not transcribe the protein; nonetheless the gene would be identified as being active in those cells.

gene activity as determined by functional assays such as mutant phenotypes is much broader than the pattern of gene expression. As we see here, genetic analysis is most informative when both the expression and activity are known because the pattern of a gene's expression may not be exactly the same as the pattern of its activity.

But what are the reagents and tools by which the pattern of gene activity can be studied? We have answered this question throughout the preceding chapters. The most informative reagent for studying the pattern of gene activity is the mutant allele. We have introduced this topic in Chapter 3 where the isolation of mutant alleles was described. In Chapter 4, we described how mutations can be classified as null or hypomorphic, and how an allelic series may reveal the diverse phenotypes of a gene. We return to these topics in this chapter, with a slightly different emphasis. We repeatedly emphasized the geneticist's mantra—find a mutant! But it is not enough to find a mutant; we must also use the mutant to study a biological process. So perhaps we should add a second line to the mantra—**Find a mutant! Examine the mutant!**

9.2 Interpreting mutant phenotypes

As we described in Chapters 3 and 4, most mutant phenotypes arise because a gene product is missing or reduced in amount. An allelic series of mutations (described in Chapter 4) allows the investigator to know if gene activity is completely eliminated, reduced, over-produced, ectopically produced, or altered in other ways. Generations of geneticists have thought and written about interpreting mutant phenotypes; the concepts of an allelic series can be directly traced to papers by the *Drosophila* geneticist H.J. Muller in the 1930s. It is informative that many of these geneticists did most of their work without knowing about the molecular nature of genes or the processes of gene expression. Upon re-reading with our understanding of genes and genomes, the ideas and interpretations in some of these papers do not age well. Others remain remarkably insightful.

A classic work on interpreting mutant phenotypes was written by the Swiss geneticist Ernst Hadorn in 1955; an English translation by Urusla Mittwoch entitled *Developmental Genetics and Lethal Factors* appeared in 1961. Many of the concepts that Hadorn introduced or discussed in terms of 'lethal factors'—what we would now call essential genes—set up a framework for thinking about mutant phenotypes. The book surveys a wide range of mutant phenotypes in many different animals (and a few plants), attempting to synthesize the effects in a coherent interpretation. We will briefly review some of these ideas using his terminology (although much of it has not been widely accepted). We will also put some of his ideas into modern terminology, with some contemporary examples.

Most mutant phenotypes are a complicated mixture of some cells and tissues that develop normally and others that develop abnormally. A traditional view is that mutations are **pleiotropic**, meaning that mutant effects are observed in more than one tissue or organ. With pleiotropic mutations, it is sometimes helpful to refer to a single isolated aspect of the mutant phenotype. Traditionally, each of the individual aspects of a complicated mutant phenotype is referred to as a phene. For example, the *patched* mutation in *Drosophila* affects para-segment boundaries during embryogenesis, eye development, and wing development, among other effects. The effect on wing development, which we will describe in some detail later in the chapter, is one of the mutant phenes that make up the complicated phenotype of the *patched* mutation; the effect on eye development is a different phene.

Lethal mutations arrest at particular developmental stages

Since Hadorn was addressing lethal effects, he focused attention on the time at which the gene is active, basing his interpretation on the mutant phenotypes. He referred to this as the phase of gene activity. He noted that lethal mutations in a variety of different species arrest development at or around certain transitions in the life cycle—hatching, molts, pupation, and so on. The stage of arrest is called the **effective lethal phase**. For many different mutations, the effective lethal phase corresponded to a 'boundary' in the life cycle or a

DEFINITION

phene: One aspect of a mutant phenotype is called a phene.

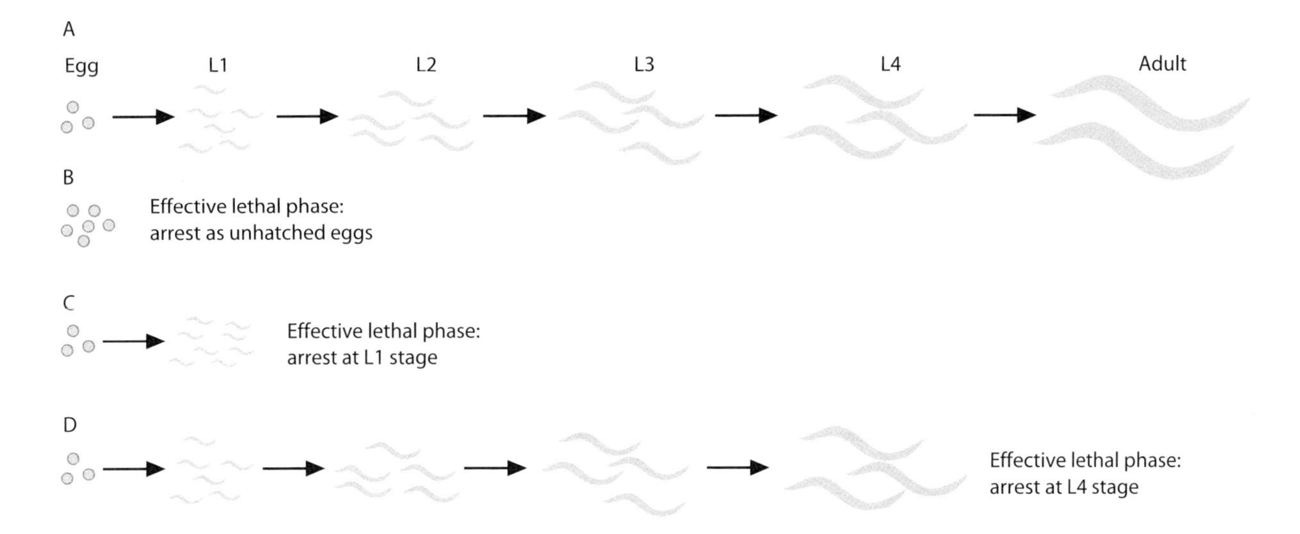

Figure 9.3 An effective lethal phase. (A) The normal life cycle of a worm, with an egg and four larval stages before the adult stage. The effective lethal stage or the terminal phenotype does not affect all stages equally. (B) Many lethal mutations in worms cause an arrest at the egg stage, so unhatched eggs accumulate on the plate. (C) Many lethal mutations allow the larvae to hatch from the egg but arrest during the L1 stage before the L2 molt. (D) Very few mutations arrest development during the L2 or L3 stage, but some will allow the worm to hatch and undergo the first three larval stages before arresting during the fourth larval stage.

transition between one stage of the life cycle and the next, so these were also referred to as **boundary mutations**. These concepts are shown diagrammatically in Figure 9.3, using the life cycle of *C. elegans*.

Many, many more mutations have been analyzed since Hadorn summarized lethal phenotypes in the late 1950s, but the concept of an effective lethal phase and a boundary effect is still useful. For example, the genome-wide RNAi screens in *C. elegans* found that the first larval stage molt is the effective lethal phase for many genes; the affected worms hatch from the egg but arrest development at the first larval stage. Interestingly, relatively few genes have the second or third molt as an effective lethal phase. Similar effects have been seen with lethal mutations in many other genetic screens in many organisms. There are particular stages of the life cycle when many different mutations arrest, and other stages when relatively few mutations arrest.

How should the effective lethal phase be understood? One possibility is that the effective lethal phase represents an observer bias with the human tendency to assign diverse things into the same few categories. That is, we see mutants arrested at particular stages because these stages are easy to observe. In reality, however, the mutant may arrest at subtly different points in a stage. Although an observer bias should not be completely discounted, a more fundamental biological interpretation is also likely. Certain life-cycle stages involve many complicated events: cells are migrating into different locations, a burst of gene expression is occurring, cells are dividing, and so on. A mutation that affects any part of these complicated events could result in an arrest at this stage. Effective lethal phases may represent physiologically critical stages in a life cycle. This is the interpretation that Hadorn favored.

We might be able to extend this concept of physiologically critical stages to thinking about the effects of mutations on different tissues and cells. For example, many pleiotropic mutations in animals exhibit an effect in the nervous system. Perhaps this is not surprising when we consider the physiological complexity of the nervous system compared with, say, the digestive system or the excretory system, particularly in invertebrates such as worms or flies. (This is not to imply that the digestive or excretory systems do not have their own level of complexity, but rather that the nervous system is more complicated in nearly all organisms.) Neurons extend lengthy cytoplasmic processes and secrete and receive signals from other cells, so a mutation that affects cell shape, secretion, or transport could exhibit effects in the nervous system. A gene that has general effects throughout the body could appear to be affecting the nervous system preferentially because proper function of the nervous system is a more demanding process. The effect on the nervous system of a particular mutation may indicate the underlying physiology of neurons rather than the neuron being the principal focus of gene activity.

The effective lethal phase refers to the **terminal phenotypic stage** in the phenotype. Even for mutations that are not lethal, the mutant phenotype is often described in terms of the terminal phenotype that is observed. For example, an adult fly has forked bristles, an adult worm cannot move properly, or an adult mouse has small eyes; all of these are the terminal phenotypic stages. However, often it is the phenotype observed at earlier times that is most instructive in understanding the activity of the gene. Hadorn used the term phenocritical phase to describe the time at which the mutant effects are seen for the first time. Consider, for example, a neurological mutation resulting in abnormal movement. The terminal phenotype may be that the adult animal moves poorly. However, the phenocritical phase—the phase when the earliest defects are detected—could be much earlier in development, perhaps when a particular subset of neurons fails to extend axons properly. The terminal phenotype is probably the outcome of a number of different failed or abnormal cellular processes, and it can be difficult to know which defects are the most informative in understanding the function of the gene. The earliest mutant effect, the phenocritcal phase, is often one that is particularly helpful in interpreting the terminal mutant phenotype.

Hadorn also recognized that different phenes that make up a complicated phenotype may exhibit different phenocritcal phases, i.e. the mutant effect may be seen in some tissues or organs earlier than it is seen in others. This difference in timing may be because the gene is active at different times in different tissues or because one tissue may need to develop normally for another tissue to develop via a signaling process. The distinction between the phenocritical phases in different tissues can be extremely helpful in understanding a mutant phenotype.

Let us consider an example—the *Sonic hedgehog* (*Shh*) mutation in the mouse. *Shh* encodes a secreted protein that is involved in many inductive interactions that pattern the mouse embryo; orthologs are found in many animals, including *Drosophila*. (In fact, as we will show genetically in Chapter 11, the *Drosophila* ortholog *hedgehog* is the ligand for the *patched* receptor.) Mutations in *Shh* are highly pleiotropic, causing defects in the development of the heart, limbs, eyes, ribs, skin, hair, and many other organs. The terminal phenotype is simply too complex to be analyzed fully. However, the earliest defects are in the establishment of structures along the dorsal midline of the embryo, including the notochord and the dorsal floor plate. The recognition that *Shh* affects the development of these structures, which then induce the formation of other organs, helps to explain the complexity of the mutant phenotype.

Mutations affect different cells, tissues, and organs

As we saw in Chapter 8, different genes are expressed in different parts of the body; a description of a mutant phenotype will usually include the effects on many different cells or tissues. This is such a fundamental point that only a little elaboration is needed. One of the key questions in observing the effects of a mutation in different parts of the body is whether the gene is active in that tissue or organ or if the mutant phene is the result of effects in other tissues or organs. Determining the phenocritical phase in different parts of the body can be helpful in answering this question. *Shh* mutants, as just discussed, are an example.

We can express the question about the focus of gene activity in a slightly different way: Does the gene exhibit a **cell autonomous** effect on the phenotype? That is, if it were possible to move genotypically mutant cells into a wild-type genetic background, would the cells still show the mutant phenotype? (This is a cellular example of the age-old nature versus nurture debate. Does the cell behave according to its own genotype or does the cell adapt to its new environment?) This type of experiment can be done by a number of genetic and physical means, which we describe later in the chapter. Figure 9.4 illustrates the concept of cell autonomy using a hypothetical example with bristle-producing cells. If the mutant cells surrounded by wild-type cells continue to exhibit a mutant phenotype, the mutation is said to be **cell autonomous**. If the genotypically mutant cells show the wild-type phenotype of their neighboring cells,

DEFINITION

phenocritical phase: The phenocritical phase is the stage at which the earliest mutant defects are observed.

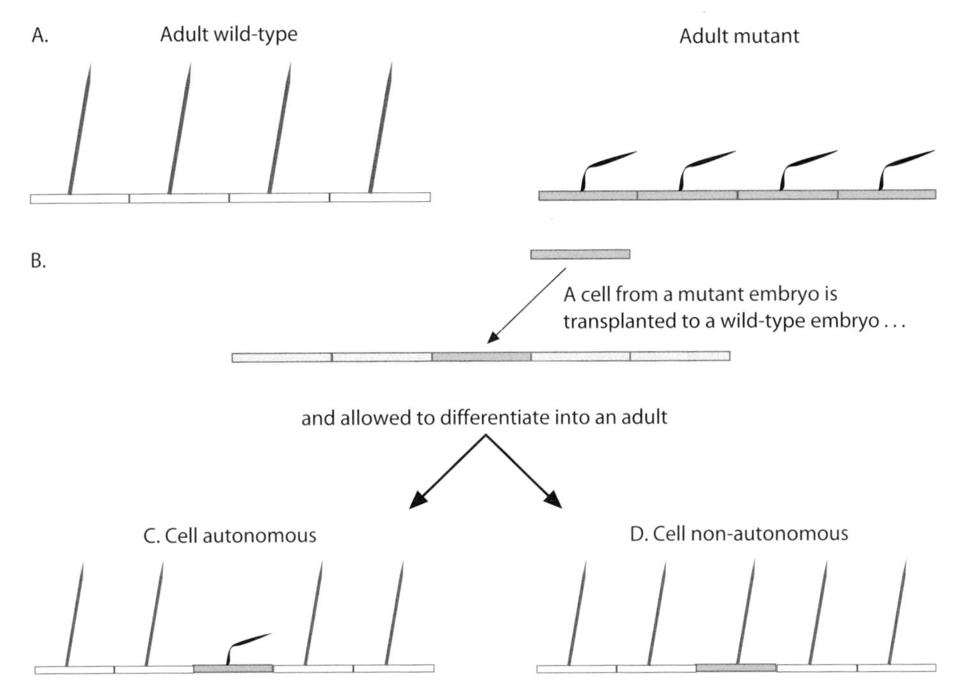

Figure 9.4 Cell autonomous and cell non-autonomous, as illustrated by a transplantation experiment involving bristles. (A) The bristle phenotype in adult animals. In this hypothetical example, each cell produces a bristle that is straight in the wild-type but bent in the mutant. The cells in the mutant are shown in green for clarity in subsequent panels. (B) A cell is removed from a mutant embryo and transplanted into a field of wild-type cells before bristle differentiation has occurred. The embryo is allowed to differentiate into an adult. Two possible results are shown in panels C and D. (C) The mutant cell produces a bent bristle. This defines the gene that produces the bent bristle as being cell autonomous: a cell with a mutant genotype shows the mutant phenotype regardless of surrounding cells. This normally identifies a gene product that works within a cell. (D) The genotypically mutant cell produces a phenotypically normal bristle. This means that the gene is cell non-autonomous, and often identifies a gene product that diffuses between cells.

the mutation is considered to be **cell non-autonomous**. To a first approximation, mutations that are cell autonomous define genes whose products act *within* a cell. Mutations that are cell non-autonomous define genes whose products are likely to be involved in a cell–cell signaling process, either through direct cell–cell contact or through a secreted signal molecule. The secreted ligand *Shh* is an example of a gene product that is cell non-autonomous in its action, whereas its membrane-bound receptor *patched* is principally cell autonomous in its effects.

A careful analysis of the mutant phenotype, particularly the phenocritical stage, is often very helpful in understanding the time when and the location where a gene is active. The remainder of the chapter describes genetic methods that provide additional means to determine either when a gene is active or where in the body it is active. We reiterate the important distinction made earlier between gene activity and gene expression. Gene activity is a general concept arising from inferences about gene function; gene expression examines the molecular explanations for the activity of the gene.

9.3 Conditional mutations and the time of gene activity

The concept of a conditional mutation was introduced in Chapter 3. However, we need to refine it here. In its broadest sense, **a conditional mutation is one whose mutant phenotype is observed only under certain environmental conditions**. Under other environmental conditions, the mutation has a wild-type phenotype.

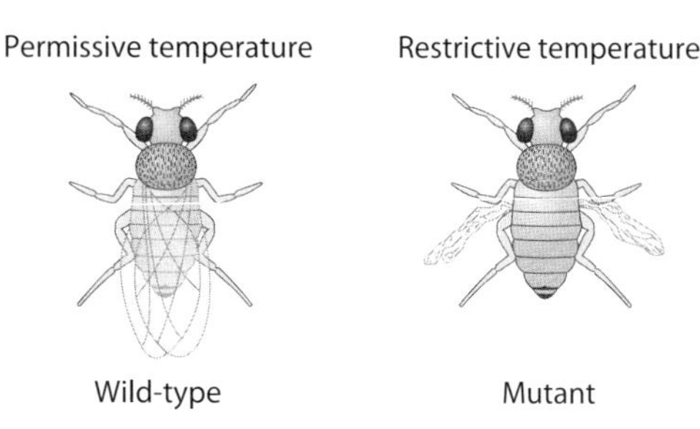

Figure 9.5 A conditional temperature-sensitive mutation. As described in Chapter 3, a temperature-sensitive mutation exhibits the mutant phenotype only at high temperature. As shown here for a temperature-sensitive mutation for the vestigial wings mutation in *Drosophila*, the genotypically mutant fly has normal wild-type wings at the low or permissive temperature. However, the genotypically mutant fly has vestigial mutant wings at the restrictive or high temperature.

According to this definition, many types of mutation are conditional, including auxotrophic and drug-resistant mutations.

For the purposes of analyzing gene activity, the most useful type of conditional mutation is **temperature sensitive**. A temperature-sensitive mutation is one that shows a mutant phenotype at one growth temperature but a wild-type phenotype at a different growth temperature. We illustrate this in Figure 9.5 with a mutant phenotype seen in the wing of *Drosophila*. Most temperature-sensitive mutations are heat sensitive—the mutant phenotype is seen at the high temperature and the wild-type phenotype is seen at the low temperature. A few temperature-sensitive mutations are cold sensitive and show the mutant phenotype at low temperature. However, since heat-sensitive mutations are by far the more common category, most investigators will use the term 'temperature sensitive' to refer to mutations that show a mutant phenotype at high temperature, and 'cold sensitive' to refer to mutations that show a mutant phenotype at low temperature.

A few other terms need to be defined. The temperature at which the mutant phenotype is observed is called the **restrictive temperature**. This reflects the fact that temperature-sensitive mutations are used most extensively in working with lethal mutations, and this temperature restricts the organism's growth. The temperature at which the wild-type phenotype is seen is called the **permissive temperature**, since it permits normal growth of the organism.

Furthermore, we define restrictive and permissive temperature relative to the normal growth conditions of the organism. For example, *C. elegans* is normally grown at about 20°C in the laboratory. Thus, the restrictive temperature for temperature-sensitive mutations in worms is about 25–26°C, near the top of the temperature range for wild-type worms. The permissive temperature for a worm temperature-sensitive mutation is usually 15°C. *S. cerevisiae* is normally grown at 30°C, so the restrictive temperature is 36–37°C and the permissive temperature is 22–23°C. In most cases, the molecular or biochemical explanation for temperature sensitivity is not known. However, the simplest and most likely explanation is that a missense mutation has made the protein less stable; at elevated temperature, the less stable mutant protein unfolds while the wild-type protein maintains its normal conformation. An important and interesting caution against this simple interpretation comes from the role of chaperone proteins, as discussed in Text Box 9.1.

Temperature-shift experiments can determine the approximate time of gene activity

In Chapter 3, we described temperature-sensitive mutations as a means of working with lethal mutations; at the permissive temperature, the mutant cells or organism develop normally, so an otherwise lethal mutation can be maintained in culture as a homozygote. In addition to maintaining lethal mutations, temperature-sensitive mutations have also been widely used to estimate the *time* of gene activity. The method involves doing a **temperature shift**, i.e. growing the organism at the permissive temperature for one part of its life cycle and at the restrictive temperature for another part. Both a **shift-up** from the permissive temperature to the restrictive temperature and a **shift-down** from the restrictive temperature to the permissive temperature provide informative data. Typically, only one shift is done for any set of experiments, so that, once shifted to a growth temperature, the organism remains at that temperature.

We can illustrate with a simple hypothetical example from *C. elegans*, beginning with the shift-up experiments. The example is shown diagrammatically in Figure 9.6; the temperature-sensitive mutant phenotype is that the worm is dumpy in shape. A worm that is a homozygous mutant for a temperature-sensitive dumpy mutation

BOX 9.1 HSP90 buffers against environmental and genetic changes

Although we discuss temperature-shift experiments in terms of one particular gene and its protein product, any of the other proteins being expressed in the organism could also be destabilized by higher temperatures. Yet many or even most proteins remain stable and do not unfold under environmental changes. Many proteins retain function and wild-type organisms develop normally at higher temperatures because of the action of a class of proteins called **heat shock proteins** (HSPs). HSPs are found in both prokaryotes and eukaryotes, and fall into several different protein families. All of them have roles as **molecular chaperones**—proteins that help other proteins fold properly. HSPs are important for buffering an organism against environmental changes, helping proteins fold into their active conformations even under non-optimal growth conditions. The expression of many HSPs is induced by heat and other stresses such as exposure to heavy metal ions. Other HSPs are constitutively expressed rather than heat induced and have more general roles in protein folding.

The heat shock protein HSP90 is important for stabilizing unstable proteins under both non-stress and stress conditions. HSP90 activity is required for all organisms. *Drosophila* has one HSP90 gene and flies that are homozygous mutant in the HSP90 gene die. Other organisms have several copies of the HSP90 gene in their genome. HSP90 is constitutively expressed but can be further induced by stress. Under non-stress conditions, HSP90 recognizes a small number of inherently unstable proteins involved in signal transduction pathways. HSP90 helps to stabilize these proteins until they are phosphorylated or otherwise modified by an activated signaling pathway. Interestingly, the ability of HSP90 to recognize these proteins is not based on a specific amino acid sequence in the proteins. Instead, HSP90 is hypothesized to recognize and bind to unstable proteins more generally. Under stress conditions, such as high temperatures, many more proteins become unstable; HSP90 binds to these unstable proteins, enabling them to carry out their functions even under non-ideal situations.

Work by Susan Lindquist and her colleagues suggests that this ability of HSP90 to stabilize unstable proteins is important for buffering not only against environmental change but also against genetic variation. Her group examined *hsp90* loss-of-function phenotypes in *Drosophila* (by genetic mutation of the single HSP90 gene or by pharmacological inhibition of HSP90 function) and *Arabidopsis* (by pharmacological inhibition because there are seven *hsp90* genes in the *Arabidopsis* genome). In both organisms, a significant number of unexpected mutant phenotypes are observed when HSP90 function is knocked out. These mutant phenotypes are genetic in origin rather than environmentally induced phenocopies, and vary in the type and percentage depending on the genetic background of the organism tested. Many different structures and processes are affected. Developmental defects in *Drosophila* include deformed, duplicated, or transformed abdominal, bristle, eye, haltere, leg, thorax, and wing phenotypes. For *Arabidopsis*, the mutant effects included abnormal cotyledon, root, hypocotyl, and leaf phenotypes. In each organism, multiple mutant phenotypes are often observed in the same individual. Interestingly, weaker effects on many of these same phenotypes are observed under high-temperature stress, even in a wild-type HSP90 background. This suggests that these are naturally temperature-dependent processes, and that HSP90 helps to buffer the processes against the effects of environmental change.

Although the results were observed with *environmental* changes, the generality of the effect leads to an intriguing hypothesis about *genetic* change. HSP90 also binds to and stabilizes many proteins that have acquired missense mutations, allowing these mutated proteins to retain some residual function. Thus, mutations can accumulate in an organism without strong negative selection because the effects of the mutation on the protein folding are buffered by HSP90. Under stress, these proteins become even less stable, gene activity is reduced, and some of the genetic variation is revealed as phenotypic variation.

These results pose an interesting question about temperature-sensitive mutations. If HSP90 buffers protein folding against temperature changes, why do temperature-sensitive mutations exist? There are several possible explanations. It could be that temperature-sensitive mutations alter protein function rather than protein stability, such that HSP90 does not recognize them. Another possibility is that temperature-sensitive mutations may be so destabilizing to the protein that HSP90 cannot fully suppress their effects.

Further reading

Queitsch, C., Sangster, T.A., and Lindquist, S. (2002). Hsp90 as a capacitor of phenotypic variation. *Nature*, **417**, 618–24.

Rutherford, S.L. and Lindquist, S. (1998). Hsp90 as a capacitor for morphological evolution. *Nature*, **396**, 336–42.

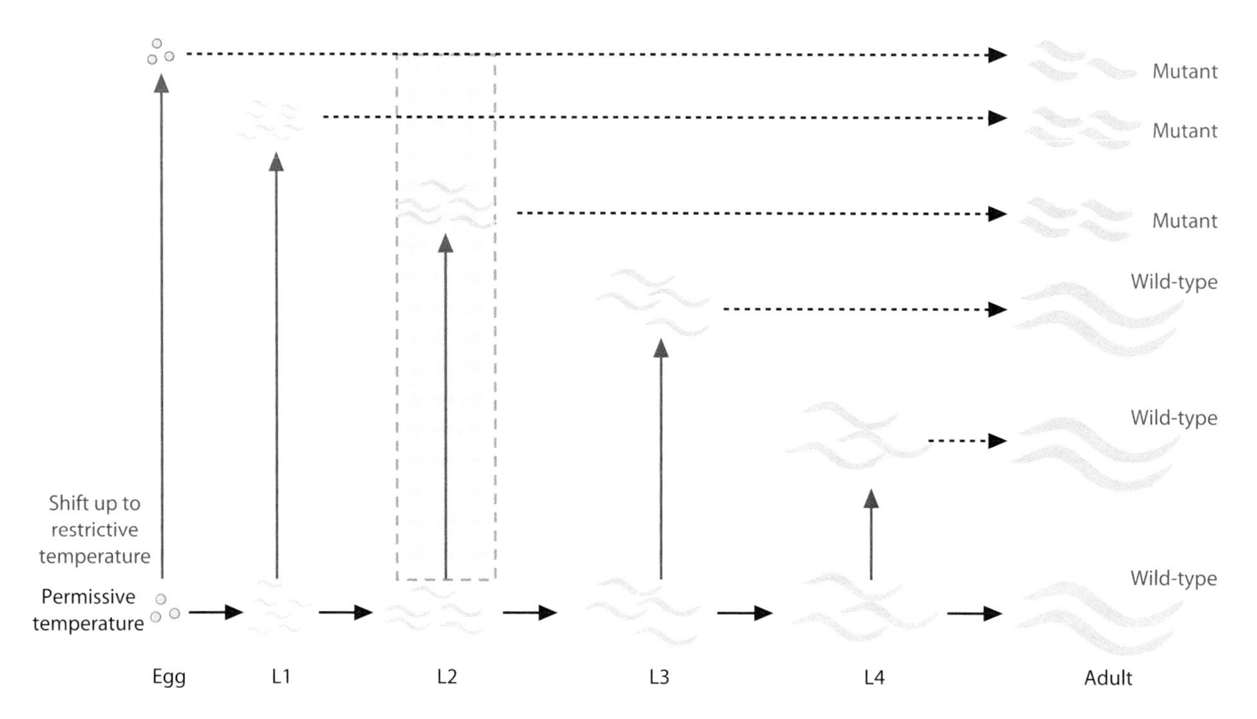

Figure 9.6 Shift-up experiments. The goal of these experiments is to determine when a gene is active. All of the worms are homozygous for a temperature-sensitive dumpy mutation. A population of worms is grown at the permissive temperature. At various times during the life cycle, some of the worms are shifted up to the restrictive temperature, as indicated by the upward-pointing red arrows. Once shifted up, the worms are maintained at the restrictive temperature and the phenotype is observed in adults. The diagram illustrates that worms shifted up at the egg stage or at the L1 or L2 stage become dumpy mutant adults. Worms shifted up at the L3 or L4 stage become wild-type adults. The inference is that the gene activity is needed after the egg and L1 stages and before the L3 and L4 stages, i.e. during the L2 stage, as shown by the green-boxed area.

(abbreviated *ts*, so the mutant is *ts/ts*) is allowed to lay eggs at the permissive temperature. The eggs will develop and hatch into larvae normally because the worm is at the permissive temperature where the mutant phenotype is not observed. At various times, some of the eggs and larvae are shifted up to the restrictive temperature and the phenotype of the worms is observed. If the worms are shifted to the restrictive temperature *before* the time when the gene is active, the *mutant* phenotype will be seen and the worms are dumpy. This is because the function of the gene is impaired at the restrictive temperature, and the function of the gene must have been needed during the time when the worm is at the restrictive temperature. On the other hand, if the worms are shifted to the restrictive temperature *after* the time when the gene is active, the *wild-type* phenotype will be seen. This is because the function of the gene has occurred normally at the permissive temperature before the shift-up.

Imagine that our hypothetical gene is active only during the second larval stage, as shown in Figure 9.6. If the worms are shifted to the restrictive temperature during embryogenesis in the egg or at the first larval stage, the worm will experience the second larval stage at the restrictive temperature when the gene will not be able to function. Mutant worms will arise. If the worms are shifted to the restrictive temperature at the third or fourth larval stage, the worms will experience the second larval stage at the permissive temperature when the gene will function normally. The worms will be wild-type in morphology. By doing shift-up experiments at different times of the life cycle, the time at which the gene is active can be determined.

We can now perform the same type of analysis with the same temperature-sensitive mutant using shift-down experiments. This experiment is illustrated in Figure 9.7. The *ts/ts* mutant worm is allowed to lay its eggs at the restrictive temperature; the eggs and larvae are shifted to the permissive temperature at different times after egg-laying. Worms that are shifted to the permissive temperature during embryogenesis or the first larval stage will go through the second larval stage at the permissive temperature, and therefore will be wild-type. The gene that affects the morphology functions normally. Worms that are shifted to the

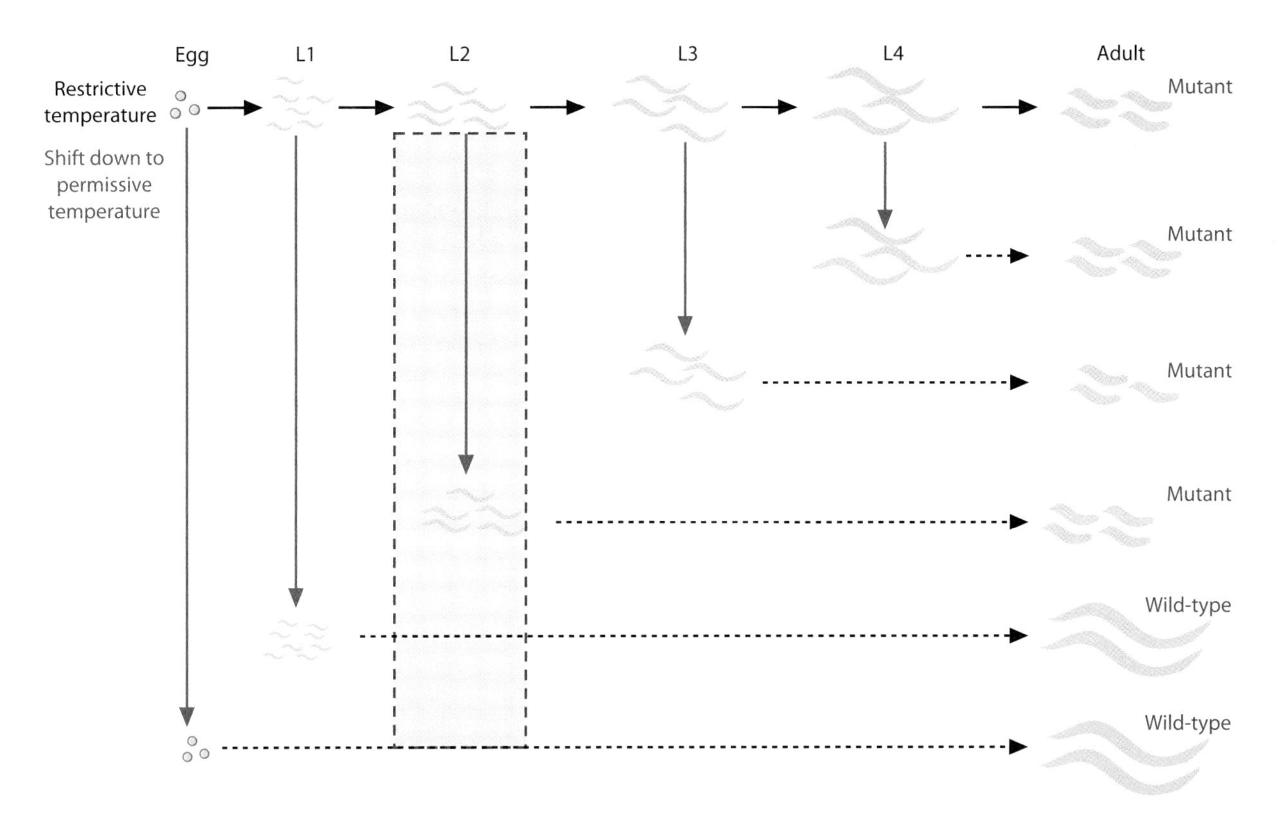

Figure 9.7 Shift-down experiments. As with the shift-up experiments in Figure 9.6, the goal of shift-down experiments is to determine the time when the gene is active. A population of worms that is homozygous for a temperature-sensitive dumpy mutation is grown at the restrictive temperature. At different stages of the life cycle, worms are shifted down to the permissive temperature, as indicated by the downward-pointing blue arrows. Once shifted down, the worms are maintained at the permissive temperature and the adult phenotype is observed. The figure illustrates that worms shifted down before the L2 stage become phenotypically wild-type adults, whereas worms shifted during or after the L2 stage become phenotypically mutant adults. The inference is that the gene is active after the L1 stage and during the L2 stage, as shown in the green box.

permissive temperature at the third or fourth larval stages have already experienced the critical second larval stage at the restrictive temperature and therefore will be mutant because the gene (or its product) was unable to function normally.

The reciprocal shift experiments (shift-up and shift-down) have each identified the second larval stage as the stage of gene activity. The results of the shift-up and shift-down experiments are summarized in Figure 9.8. The second larval stage has been variously termed the **t-crit** for the critical time, the **stage of execution**, or more commonly, the **temperature-sensitive period** (TSP). If a population of the temperature-sensitive mutants can be synchronized sufficiently in development, it may be possible to define the TSP even more narrowly by shifting the worms up or down at specific times during the second larval stage. However, in practice, it is often more useful to think of the TSP broadly or as an approximate time of gene activity rather than attempting to pinpoint it very precisely. Let us now consider the analysis of cell-cycle mutations in the yeast *S. cerevisiae*, an illustration of one of the most successful applications of temperature-shift experiments.

The cell cycle in yeast was analyzed using temperature-sensitive mutations

The analysis of the yeast cell cycle is one of the venerated examples of the uses of temperature-sensitive mutants. In the decades since Hartwell and his students began this analysis, the eukaryotic cell cycle has become a fundamental and familiar topic in molecular and cell biology courses, and it dominates our thinking about cancer cell therapy. Modern students might learn the cell cycle as a cascade of phosphorylation steps, i.e. as a biochemical pathway, and might not realize that it all began with temperature-sensitive mutations in yeast. For this and other work on the cell cycle, Hartwell shared the Nobel Prize in Physiology and Medicine in 2001. In an

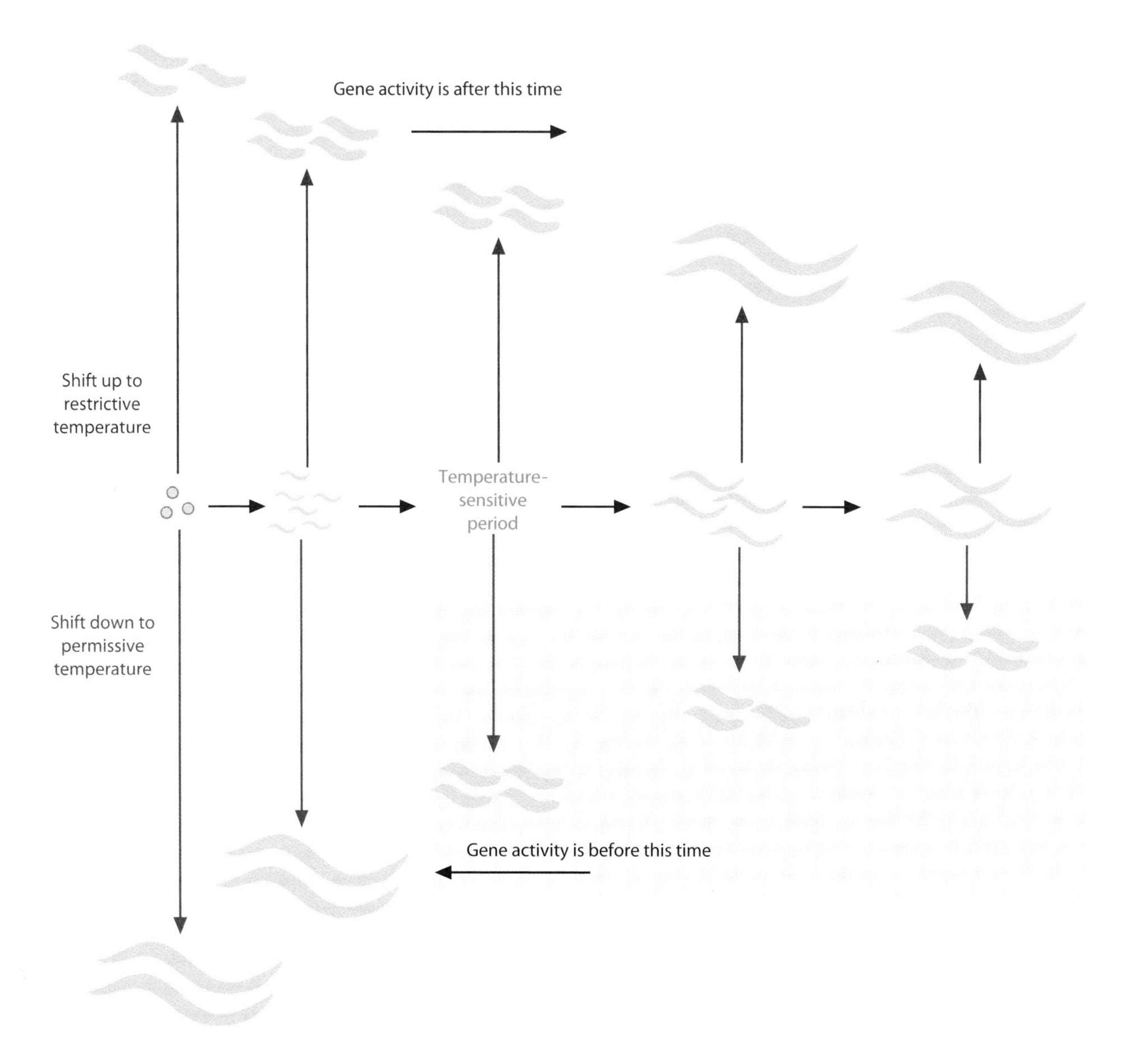

Figure 9.8 The temperature-sensitive period. The results shown in Figures 9.6 and 9.7 are summarized here. If the temperature-sensitive period is experienced at the restrictive temperature, the mutant phenotype will be observed. Shifts up to the restrictive temperature before the L2 stage result in a mutant phenotype, so the activity of the gene is required during or after this stage. Shifts down to the permissive temperature after the L2 stage result in a mutant phenotype, so the activity of the gene is required before or during this stage. The region of overlap is shown in green as the temperature-sensitive period.

autobiographical sketch, Hartwell mentions that his thinking about the yeast mutants was influenced by his contact with Bob Edgar, who pioneered the use of temperature-sensitive mutations in his analysis of the bacteriophage T4 assembly.

The yeast *S. cerevisiae* is called budding yeast because a cell divides by producing a bud (Figure 9.9). Budding is correlated with DNA replication, and the size of the bud is indicative of the stage of the cell division cycle. There are a number of steps in the cell cycle that are morphologically recognizable, such as bud formation, spindle pole body duplication, microfilament ring formation, and others. Other steps such as DNA replication were recognizable by biochemical assays. Hartwell referred to these recognizable steps as **landmark events**. He and his students isolated 150 mutations that grew normally at the permissive temperature of 22°C but that arrested cell division at the restrictive temperature of

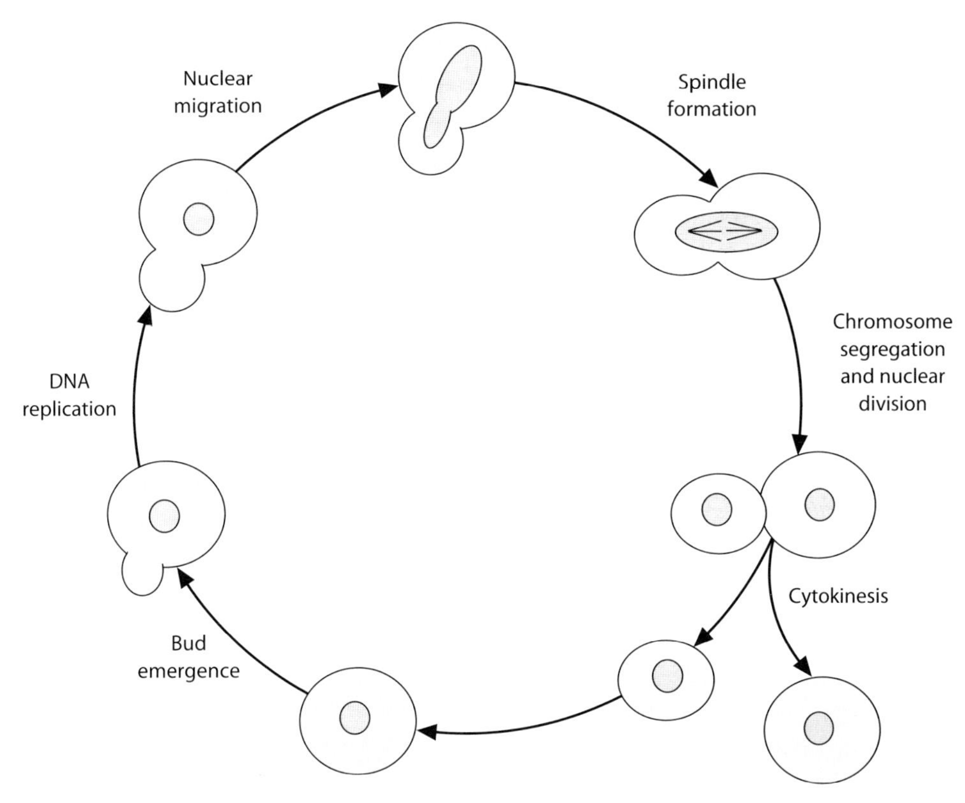

Figure 9.9 The cell cycle of the budding yeast *S. cerevisiae*. Some of the key morphological events in the cell cycle are shown. Temperature-sensitive mutations that arrested the yeast cell cycle at different points, identifying the cell division cycle (*cdc*) genes, were used to understand the genetic, molecular, and biochemical regulation of the cell cycle of nearly all eukaryotic cells. A shift to the restrictive temperature before the critical stage results in a cell-cycle arrest with a diagnostic phenotype. A shift to the restrictive temperature after the critical stage allows the cell to complete one cell cycle before arresting at the next cycle.

36°C; by complementation tests as described in Chapter 3, these mutations defined 35 genes. (Many more mutations and genes have been identified since the initial studies.) The genes were termed *c*ell *d*ivision *c*ycle defective, abbreviated *cdc1*, *cdc2*, etc.

The mutations had discrete terminal arrest phenotypes, recognized by their failure to execute one of the landmark events. For example, the mutation shown diagrammatically in Figure 9.10 results in cells that arrest with buds forming but no DNA replication. These landmark events are comparable to Hadorn's effective lethal phase. The first mutant defect that could be observed in a given mutant was termed its **diagnostic landmark**, comparable to Hadorn's phenocritical phase.

When a culture of growing yeast cells was shifted to the restrictive temperature, the cells arrested at the same point in the cell division cycle, even though they were at different stages in the cell cycle when shifted. With temperature shifts at different times in the cell cycle, a

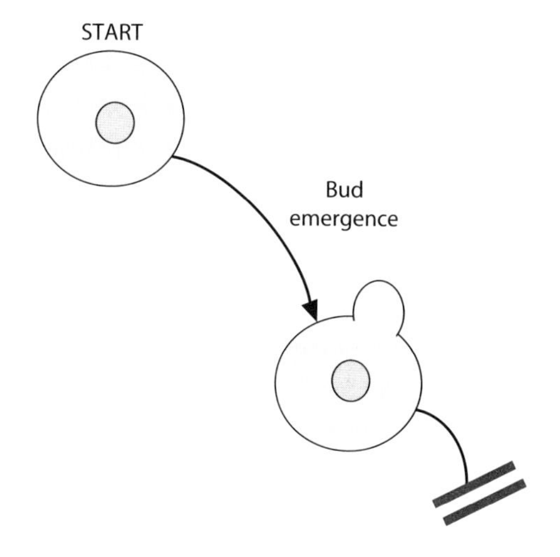

Figure 9.10 The phenotype of a cell division cycle (*cdc*) mutant. The drawing indicates a mutant that forms a bud but cannot initiate DNA replication. The cells arrest at this stage until shifted to the permissive temperature.

temperature-sensitive period for each mutation could be established; if a cell was shifted after this time, it completed its current cell cycle and did not arrest until the next cell cycle. Hartwell referred to this temperature-sensitive period as the **execution point**—the time in the cell cycle when the cell is no longer sensitive to shift-up experiments. He made the point that the execution point, and by analogy the temperature-sensitive period, could be allele specific rather than characteristic for a gene. This is probably because different alleles have differing amounts of residual gene activity after the shift to the restrictive temperature. As a result, a mutant allele with more residual activity after the shift up will be able to carry out more of the cell-cycle stages than one with little residual activity.

Because the cell cycle has an observable sequence of morphological events, it was possible to use the diagnostic landmarks to place different genes in order of gene activity. For example, since spindle pole body emergence occurs before the chitin ring forms to separate the budding cells, mutants whose diagnostic landmark is spindle pole body emergence must affect an earlier stage in the cell division cycle than mutants whose diagnostic landmark is chitin ring formation. It was also possible to use the temperature shifts to determine which events in the cell cycle occur independently; that is, a mutant that fails to carry out one landmark event may still be able to carry out other landmark events normally because the events are regulated separately. For example, bud emergence can still occur even if chitin ring formation is blocked and vice versa, indicating that these processes are independently regulated. Therefore, although yeast cell division appears to be a cycle, the cycle branches into two or more separable pathways at certain points. This will become an important topic in Chapter 11 when the construction and interpretation of genetic pathways are described.

Although we have presented relatively simple examples here, the interpretation of temperature-sensitive periods can be complicated. Some of the ways that temperature-sensitive periods can be interpreted are discussed in Case Study 9.1 on dauer larvae formation in *C. elegans*. Because this process has also been analyzed by molecular analysis, it is possible to reinterpret the temperature-sensitive period in light of gene expression results. One of the interesting reasons learned from studying these genes is that the temperature-sensitive periods correspond to the activity of the genes rather to their expression; that is, the temperature-sensitive period accurately found the time that the *process* of dauer larva formation occurs and not the time when individual genes are expressed as part of this process.

Case Study 9.1

Analyzing temperature-sensitive periods

Temperature-sensitive (*ts*) mutations have been used for decades to estimate the *time* of gene activity. In this case study, we discuss one example of a biological process in which *ts* mutations were an important part of the analysis. The genes involved in this study have subsequently been cloned and the biological process has been analyzed in molecular detail. This allows us the opportunity to compare what was learned from genetic analysis before the genes were cloned with what has been learned from the cloned genes. The biological process that we will discuss is the formation of dauer larvae during the life cycle of *C. elegans*.

The dauer larva is an alternative stage in the nematode life cycle

In the laboratory, *C. elegans* grows through four larval stages, termed L1 through L4, as described in Chapter 2. Each of these stages is punctuated by a molt in which the cuticle is shed and replaced. Under laboratory conditions at room temperature, the entire life cycle takes about 4 days, as summarized in Case Study Figure 9.1. However, when the worm is starved, crowded, or otherwise stressed, it can enter a larval stage known as the dauer larva as an alternative to the L3 stage; in this variation of the life cycle, the worm molts directly from L2 into the dauer larva, with a distinct cuticle and morphology (Case Study Figure 9.2). When dauer larvae (also known as dauers) are moved to fresh food on uncrowded plates, they molt to become L4 larvae and resume the life cycle.

Mutants that affected dauer formation were identified and sorted into a number of genes known as the *daf* (for *da*uer *f*ormation) genes. Null mutant alleles in some genes fail to form dauers even under stress conditions; these are known as dauer-defective genes and will not be discussed further. Null mutations in other genes always form dauers even under non-stress conditions; these are known as the dauer-constitutive (*daf-c*) genes and will be the subject of this case study.

Shift-up experiments suggest that the activities of the *daf-c* genes are needed well before dauer formation itself occurs

The *daf-c* mutations arrest before sexual maturation so these mutations are effectively sterile or lethal, although with a distinctive phenotype. Many of the *daf-c* genes were defined by temperature-sensitive mutations so that the strains could be maintained as homozygotes, a strategy for maintaining mutant strains introduced in Chapter 3. A typical example is a temperature-sensitive allele of *daf-7*. At the permissive temperature of 15°C, the *daf-7* homozygote grows and reproduces normally. When grown at the restrictive temperature of 25°C, the *daf-7* homozygote forms dauer larvae even when food is present. A similar phenotype is observed for mutations in at least five other *daf-c* genes. Each of these six genes, including *daf-7*, was subjected to temperature-shift experiments to determine when the gene is active. It is worth noting that the *daf-c* mutations imply an interesting logic for

DEFINITION

dauer-constitutive: Mutants that always form dauer larvae, regardless of the environmental conditions.

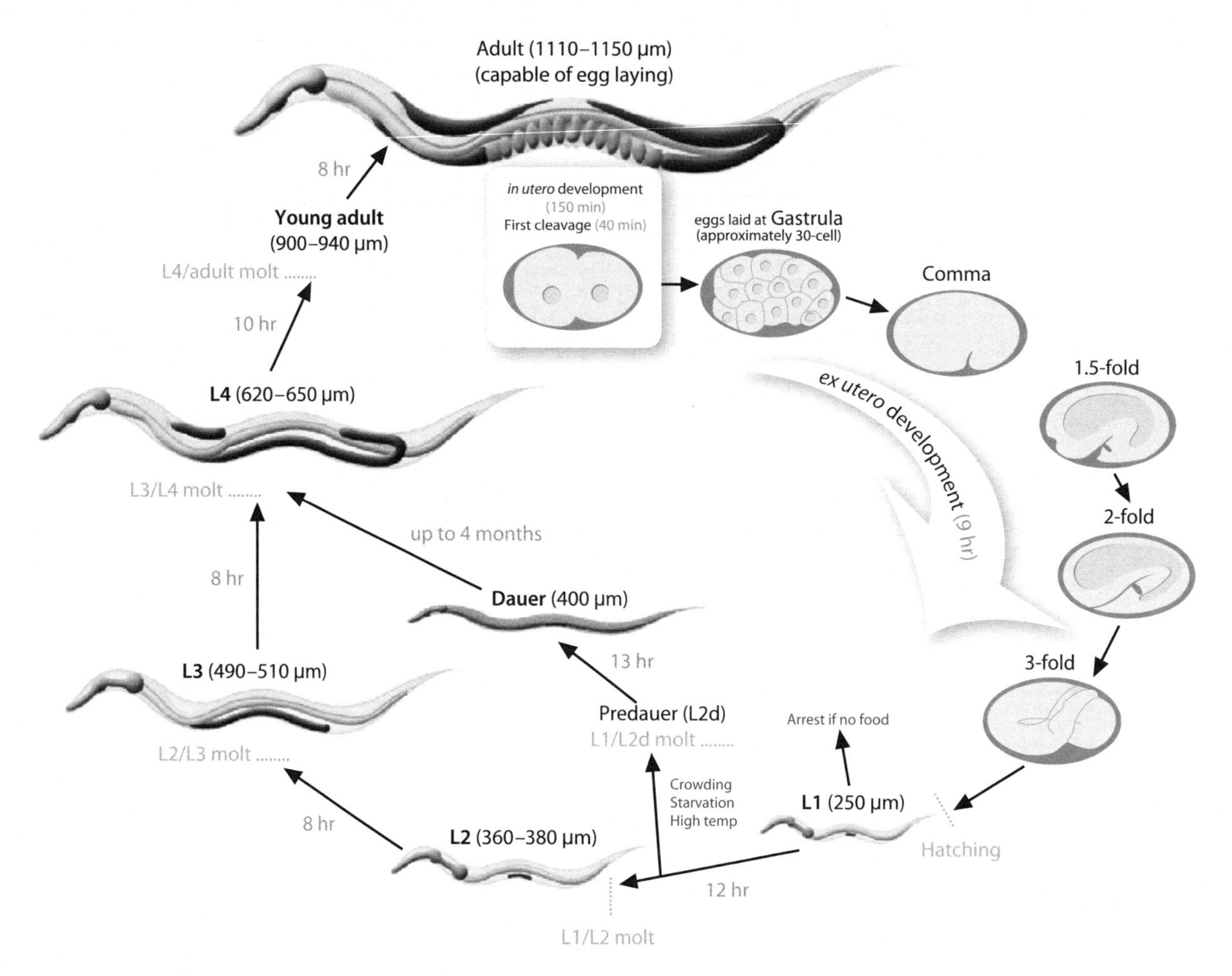

Figure C9.1 The life cycle of *C. elegans*, showing the dauer larva stage. The times are the approximate duration of each stage at 20°C. Under conditions of crowding, starvation, or high temperature during the late L1 stage, the worm will form a distinctive dauer larva rather than the L3 stage. The dauer larva molts to become an L4 stage larva, bypassing the L3 stage altogether. The time of commitment to dauer larva formation was determined by temperature-shift experiments.

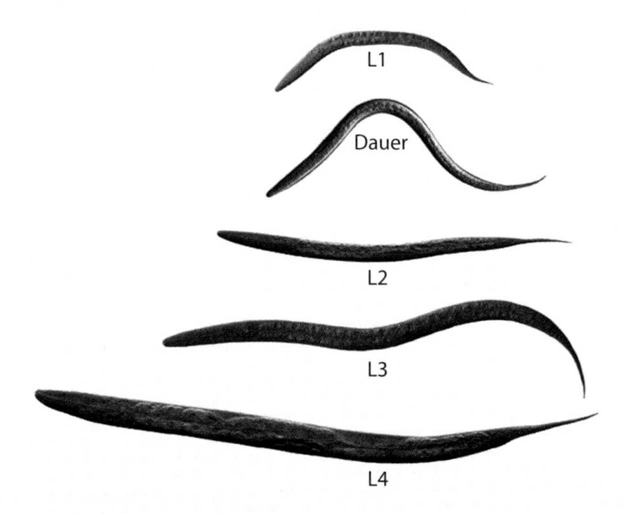

Figure C9.2 The morphology of a dauer larva. Compared with an L3 larval worm, which is the equivalent age, the dauer larva is thinner and shorter. It is also a darker color, which does not show up in the figure.

dauer formation. The mutations always form dauers, so the wild-type activities for these genes must be involved in *preventing* dauer formation under non-stress conditions.

The first complication in the temperature-shift experiments is that the life cycle itself is dependent on temperature, as is true for many animals. Wild-type worms raised at the restrictive (warmer) temperature mature faster than worms raised at the permissive (cooler) temperature. As a result, simple chronological time (e.g. 24 hours after hatching) could not be used for the temperature shifts. For a culture of worms at 15°C, the first molt from L1 to L2 occurs at about 24 hours; for a culture at 25°C, the first molt occurs at about 13 hours, and the worm is already in the middle of the L3 stage by 24 hours. Thus, in order to standardize the shift-up and shift-down experiments, the timing of the shifts was related to developmental events such as the molts themselves, and worms were shifted up or down based on their larval stage rather than chronological time.

In the temperature-shift experiments, populations of larval worms that hatched at the same time were maintained at one growth temperature until a particular molt had occurred, and then shifted to the other temperature and assayed by counting the percentage of dauer larvae in the total population. The slope of the transition between mostly dauer larvae and mostly wild-type was characteristic for each mutant surveyed. These are shown for some of the mutations in Case Study Figure 9.3. For example, for *daf-7*, the transition from mostly wild-type to mostly dauers in the shift-up experiments occurred over approximately 6 hours,

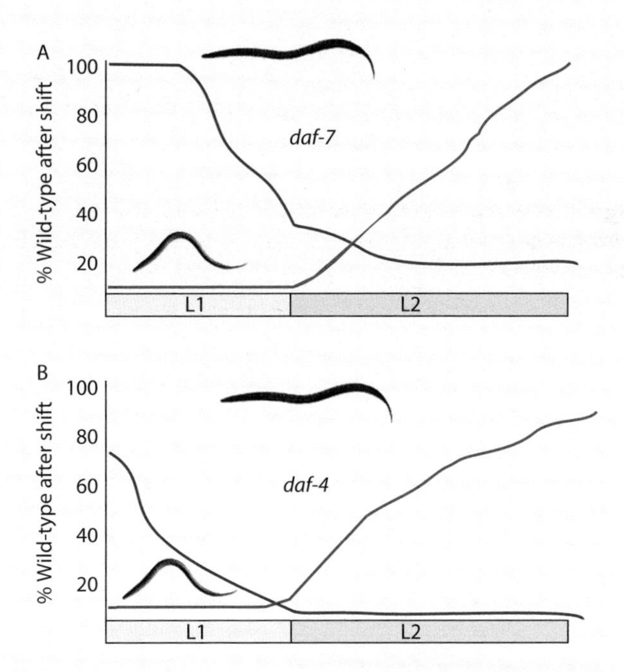

Figure C9.3 Temperature-shift experiments for (A) *daf-7* and (B) *daf-4*. The *x*-axis depicts the developmental stage of the worm, corrected for the temperature dependence of the life cycle. As described in the text, a population of worms is grown at one temperature and then shifted to the other temperature at particular stages. The percentage of worms that developed as wild-type is plotted on the *y*-axis. Shift-down experiments are plotted in blue and shift-up experiments are plotted in red. For example, if *daf-7* mutant worms are shifted up as newly hatched L1s, only a few percent develop as wild-type worms and most become dauers. Conversely, if *daf-7* mutant worms are shifted down as newly molted L2s, about 30 percent develop as wild-type and about 70 percent are dauers. Not all *daf-4* worms become dauers at the restrictive temperature; we return to this observation in Chapter 11.

whereas for *daf-4*, the transition occurred much more gradually over a period of about 25 hours. The physiological reason for this difference is not clear from the studies. Temperature-sensitive alleles are likely to be hypomorphic rather than null, as discussed in Chapters 3 and 4, so the difference in the slope of the transition curve could simply represent the amount of residual gene activity in each mutant; alternatively, the differences in transition may represent the sensitivity of the gene product to temperature changes, or that the gene activity is needed over a broader period of time. Other explanations are also possible.

Nonetheless, by taking the midpoint of each curve—the time at which half of the population becomes dauers—it was possible to construct a temperature-sensitive period (TSP) for each mutant. Consider the data for *daf-7*, as shown in Case Study Figure 9.3, beginning with the shift-up experiments plotted in red. If worms are shifted up before the L1–L2 molt, they become dauer larvae; if they are shifted up after the L1–L2 molt, they do not form dauer larvae and develop as wild-type. Thus, in a simple interpretation, the TSP for *daf-7* activity *begins* before the L1–L2 molt in the middle of the first larval stage, well before the dauer larva itself actually forms after the L2 stage. Similar results were observed for several of the other mutants—shift-up experiments after the L1–L2 molt did not result in dauer formation, although the worm is homozygous for a *daf-c* mutation. The notable exception was the *daf-14* gene in which the TSP is much earlier; shift-up experiments conducted even a few hours after hatching were largely ineffective in producing dauer larvae, suggesting that the activity of the gene is needed much earlier than the other *daf-c* genes.

Shift-down experiments give rather complicated results for several genes

One way of interpreting the shift-up experiments is that the activity of the genes *begins* to be needed during the first larval stage or, in the case of *daf-14*, shortly after hatching. The appearance of the mutant phenotype following a shift-up indicates that the activity of the gene is needed after the time of the shift. By the same reasoning, the **shift-down** experiments should define the times when the activity of the genes *ends*. A shift-down *before* this time results in a wild-type phenotype, whereas a shift-down *after* this time gives a mutant phenotype—the wild-type gene activity is being supplied too late to 'rescue' the mutant defect.

This interpretation fits well with the results for some of the genes. Consider the shift-down experiments with *daf-7* again (plotted in blue in Case Study Figure 9.3). The worms are raised at the restrictive temperature and then shifted down to the permissive temperature. If the period of gene activity has already occurred while the worm is at the restrictive temperature, the temperature shift-down will result in only mutant worms; this begins to be observed for *daf-7* at approximately the time of the L1–L2 molt. Thus, the TSP, as indicated by the appearance of mutant phenotypes in temperature-shift experiments, is slightly before or around the L1–L2 molt.

However, this interpretation does not work so easily for some of the other *daf-c* mutations. Consider the results for *daf-2* as shown in Case Study Figure 9.4. The shift-up experiments indicate that mutant phenotypes are no longer seen when worms are shifted after the L1–L2 molt. In the shift-down experiments, shifts after the early L2 stage do not rescue the mutant phenotype. In a standard interpretation, the beginning of the TSP is defined as the midpoint of the shift-up curve and the end of the TSP is the midpoint of the shift-down curve. However, for *daf-2* and some other *daf-c* genes, the 'beginning' of

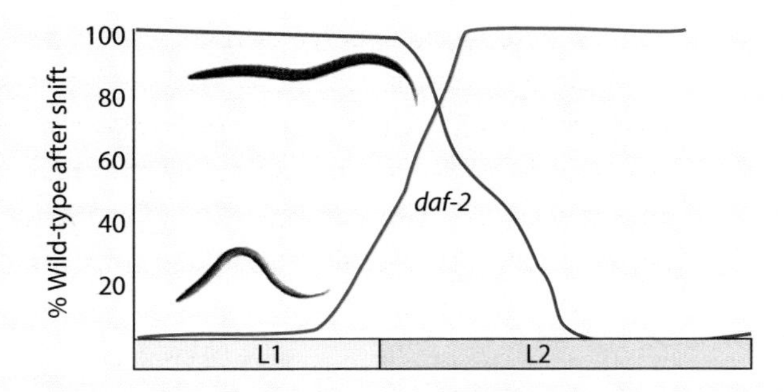

Figure C9.4 Temperature shift experiments for *daf-2*. The experiments were done as described in Case Study Figure 9.3 and the results are depicted similarly. Shift-down experiments are plotted in blue, and shift-up experiments are plotted in red. Note that the midpoint of the shift-up curve occurs slightly before the midpoint of the shift-down curve, the reverse of the situation in Case Study Figure 9.3.

the TSP occurs after the 'end' because the shift-up midpoint is later than the shift-down midpoint. Since it is clearly impossible to have a process end before it begins, how should these data be viewed?

Many different explanations have been advanced for these 'reversed' TSPs, but the most likely is that looking only at the midpoints of the temperature-shift curves oversimplifies what is occurring. A better interpretation is that these shifts define the time when the gene activity is *needed* for this biological process without attempting to be too precise about the beginning or end points. In other words, to appropriate the term used by Hartwell in his analysis of the cell cycle in yeast, the TSP is an **execution point**, a time when the function of the gene is needed. This interpretation does not require over-analyzing the data or making untested assumptions about the nature of the temperature-sensitive defect or the activity of the gene and the process it affects. By using this interpretation, the functions of all of the *daf-c* genes (except *daf-14*) are needed around the time of the molt from the L1 stage to the L2 stage. The TSP represents a band of time when the function of the gene is needed.

Temperature-shift analysis has some inherent limitations

Because temperature-shift experiments are among the simplest to perform—they involve moving cultures of the animal from one temperature to another—they provide an attractive way of approximating the time of activity for a gene. There is extensive, albeit older, literature on the analysis of temperature-shift experiments, including the use of temperature pulses, attempts to distinguish mutations that affect the assembly of macromolecular complexes from mutations in the synthesis of the components, and other intellectual efforts. As seen with the TSPs in the preceding section, temperature-shift experiments have some inherent limitations that affect these interpretations.

First, the life cycle itself must be adjusted for temperature dependence, as was done carefully in the analysis of dauer formation. It is much easier to use chronological time than developmental landmarks, but the appearance of certain developmental or morphological events provides the best timing method. As was done with the molts in *C. elegans*, Hartwell used morphological landmarks for his cell-cycle analysis to avoid the problem that yeast cells grow at different rates at different temperatures.

Secondly, as noted by Hartwell but often overlooked, the TSP is best regarded as characteristic for an *allele* rather than for a *gene*. We know that different mutations can have somewhat different phenotypes based on the amount of residual gene activity in the mutant; this is the basis for constructing an allelic series as described in Chapter 4. The same understanding should be applied to temperature-sensitive mutations. A shift-up for some alleles may greatly reduce or eliminate the activity of the gene whereas a shift-up for other alleles may only slightly reduce the activity of gene. Such a difference, which would be unknown to the investigators as the temperature shifts are being done, could give rise to different TSPs for the same gene. In the case of the dauer mutations that have been tested, different alleles of the genes have similar TSPs, suggesting that this was not a significant problem in the analysis.

Thirdly, since the life cycle itself is temperature dependent, many other developmental processes are also temperature dependent. If a given process is temperature sensitive, *all* mutant alleles of a gene required for that process will be temperature sensitive regardless of their molecular lesions. The TSP for such genes may be extremely broad, reflecting all the times when the function of the gene is needed.

Finally, in most cases, little follow-up analysis has been done to compare a temperature-shift analysis with other types of data in order to validate the conclusion from the temperature-sensitive mutations. Although we present this as an analysis of the activity of individual genes, it can also be thought of as the time when the biological process affected by this gene is regulated.

Temperature-shift analysis of the *daf-c* genes can be compared with the molecular analysis of these genes

The dauer formation pathway has become one of the best-studied developmental pathways in worms, in part because many of the *daf* genes affect other processes including aging and life span. (We discuss this pathway again in Chapter 11.) This subsequent analysis makes it possible to compare the thorough temperature-shift analysis done in 1981 with the contemporary knowledge of these genes derived from molecular biology and genomics.

Several of the *daf-c* mutations proved to be alleles of genes in a transforming growth factor (TGF)-β signaling pathway. *daf-4* and *daf-1* encode the subunits of the TGF-β receptor, with *daf-4* being a type II receptor that interacts with the ligand directly and *daf-1* being a type I receptor that interacts with *daf-4*. Reporter gene constructs with these genes, using GFP as described in Chapter 8, have been used to show that gene expression begins during the middle of embryogenesis, or much *earlier* than the TSP for any of the mutants in the dauer formation pathway. Once expressed, the receptors are present throughout the life cycle, including the larval stages. The receptors are expressed in many neurons in the head, including those involved in dauer formation.

daf-7 encodes a member of the TGF-β superfamily of proteins, likely to be the ligand that binds to the receptor encoded by *daf-4* and *daf-1*. This is shown diagrammatically in Case Study Figure 9.5A. As shown by a GFP reporter construct, *daf-7* expression begins at about the middle of the L1 stage and continues throughout the subsequent larval stages. Importantly, *daf-7* is only expressed in the presence of abundant food when dauers are not expected to form. The expression of *daf-7* is limited to those neurons that are known from cell ablation experiments to be involved in dauer formation. In keeping with its role as a key regulator of the dauer pathway, *daf-7* is not expressed in starved worms.

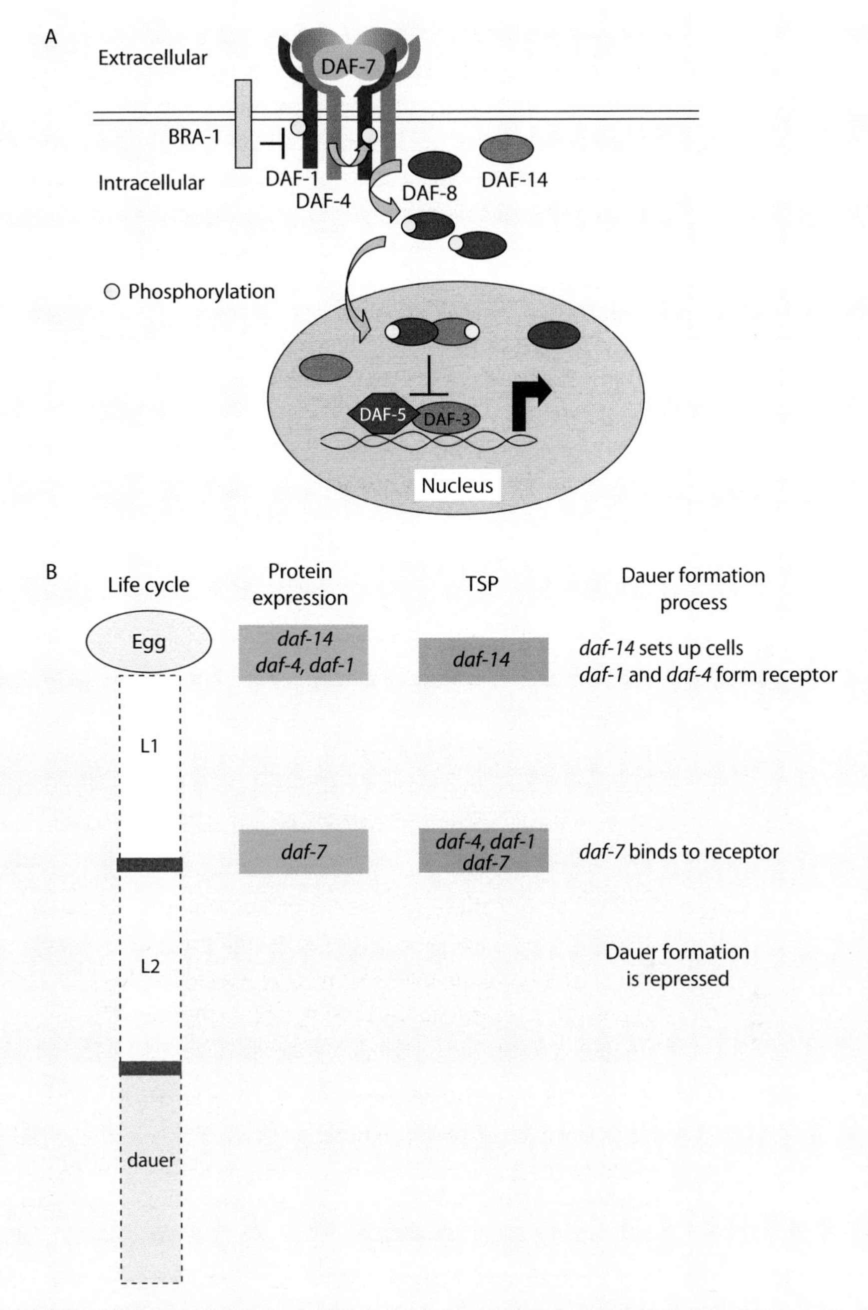

Figure C9.5 The dauer formation pathway. Panel A shows the dauer formation gene products as components of a TGF-β signal transduction pathway: *daf-14* encodes a cytoplasmic protein involved in signal transduction, and *daf-1* and *daf-4* encode the components of the TGF-β receptor. The ligand for the receptor is encoded by *daf-7*, and binding of DAF-7 to the receptor triggers the pathway. The results of the protein expression and temperature-shift experiments are compared in panel B. Stages in the life cycle are on the left, with the molts indicated by red bars. Based on the protein expression results with GFP reporter gene constructs, shown as green boxes, *daf-14*, *daf-4*, and *daf-1* are expressed during embryogenesis, whereas *daf-7* is only expressed in the late L1 stage, and only in well-fed worms. Based on the TSPs, as shown by the blue boxes, *daf-14* activity is required during embryogenesis, whereas *daf-4*, *daf-1*, and *daf-7* activities are needed at the end of the L1 stage. The inferences from these data are shown on the right. During embryogenesis, *daf-14* sets up the cells that will be involved in dauer formation; the receptor formed by *daf-1* and *daf-4* is synthesized but is not active. The *daf-7* ligand is produced during the L1 stage and binds to the receptor, which represses dauer formation.

Most of the other *daf-c* genes encode proteins corresponding to known or suspected signaling molecules in the TGF-β pathway downstream of the receptor, as shown in Case Study Figure 9.5A. These proteins are present in many other cells in addition to neurons involved in dauer formation, possibly because this pathway regulates other processes in the worm. *daf-14* encodes an atypical member of the SMAD protein family that is expressed in many of the cells and tissues are that undergo morphological changes during dauer formation. It may be involved in signal transduction in these tissues.

A simple and interesting interpretation of the temperature-shift, molecular, and gene expression data is that wild-type *daf-7+* expression is stimulated by the presence of food shortly after the time of hatching. The DAF-7 protein binds to the receptor complex formed by DAF-4 and DAF-1, and the signal is transduced via the other *daf-c* genes to *shut down* dauer formation. The absence of *daf-7* expression or a mutation in any of the other *daf-c* genes blocks the pathway and prevents the shut-down of dauer formation, so that dauers form constitutively.

An interpretation of temperature-sensitive periods

Based on the interpretation above, how should we understand the TSPs? The relationship between the TSP and the protein expression results is shown diagrammatically in Case Study Figure 9.5B. We begin by discussing *daf-14* whose TSP is much earlier than the other genes. The exact role of *daf-14* in dauer formation is not clear, but its expression in the tissues involved in dauer formation suggests that it sets up these tissues to respond to the dauer signal. Thus, one interpretation is that its expression is earlier than the others because the other gene products require DAF-14 to establish the appropriate tissues for them to be functional.

The most compelling result is that the TSP for *daf-7* corresponds very well to the time when the DAF-7 protein is being produced, as shown by the GFP reporter. The genes that are expressed earlier than predicted from their TSP, such as *daf-1* and *daf-4*, encode molecules that are part of the signaling process regulated by *daf-7*. Thus, these proteins are present but apparently not active in dauer formation until *daf-7* is active. The dependence of their activity on *daf-7* explains why these genes have TSPs that are approximately the same as the one for *daf-7*. Only when the ligand binds to the receptor will the pathway become functional; the activity of all of the genes in the pathway reflects the activity of the DAF-7 ligand, which triggers the signal transduction. Thus, the TSP for all of these genes accurately captures the time of gene activity for dauer formation, i.e. the time of activity for the *biological process*, which corresponds to the activity of its key regulatory gene, *daf-7*.

Further reading

Swanson, M.M. and Riddle, D.R. (1981). Critical periods in the development of the *Caenorhabditis elegans* dauer larva. *Developmental Biology*, **84**, 27–40.

Information on individual genes is available from Wormbase.

9.4 Mosaic analysis and the location of gene activity

Genes are active at particular times and in particular locations in the organism. It is often possible to infer something about both the time and location of gene activity from an analysis of the mutant phenotype, particularly by examining the phenocritical phase when the mutant defect is first evident. Because many mutations affect the development and function of more than one structure, finding the phene with the earliest observable defect may help to explain the other defects.

However, the analysis of mutant phenotypes alone allows only a limited perspective on where the gene is active, and additional methods are often used. One of the most powerful ways of analyzing the location of gene activity is to make an organism with two different genotypes. In such an organism, any individual cell has a *particular* genotype, but different cells have one of two *different* genotypes. This immediately raises two questions. First, how does the investigator make an organism with cells of different genotypes? Secondly, how does the phenotype of such an organism inform us about the location of gene activity?

Organisms with cells of different genotypes can be constructed by a number of different methods. We will begin by describing a physical method involving the micromanipulation of cells. Although these physical methods can be technically challenging, they are the easiest methods to explain. We will then use those organisms to explain how this tells the investigator about the location of gene activity. Finally, we will describe other genetic and molecular methods that are used to make organisms with different genotypes.

Chimeras are produced by physical manipulation of cells

Physical manipulation of individual cells probably began shortly after cells were identified. Certainly microscopists working in the late nineteenth and early twentieth centuries had already become adept at physically manipulating cells. Often this involved killing individual cells with a needle—the intellectual precursor to our modern methods using laser ablation—or separating cells from their neighbors and observing the effects. Once cells were separated from a developing organism, it was only a short experimental step to try to re-aggregate them into a different organism and watch that organism develop. For many organisms, including molluscs and most vertebrates, embryos with re-aggregated cells could develop for some time, or even grow into an adult animal. Since the re-aggregated cells did not have to come from the same individual—in fact, not even from the same species in some cases—the developing embryo had cells of different genotypes. An organism produced by physical manipulation of cells is referred to as a **chimera**. The key to this definition of a chimera is that it is produced by a physical method, such as re-aggregating cells. In a subsequent section, we will discuss **mosaic analysis**, in which individuals with cells of two different genotypes are made by genetic means.

Chimeric embryos have already been introduced in Chapter 6 in the description of transgenic mice, as illustrated in Figure 9.11. Since the focus of Chapter 6 was manipulating *genes* and reintroducing them into the mouse rather than manipulating the *cells* of the mouse, we did not discuss how these chimeras could be used to illustrate gene activity. Nonetheless, some aspects of the interpretation for gene activity were implied in our description. The production of chimeric embryos does not require the use of embyomic stem cells since cells of any two genotypes could be re-aggregated and introduced into a female. These cells can differ at many different loci, not only the gene that has been engineered *in vitro*, and these other genes can also be monitored for gene activity. We will consider the activity of five different hypothetical genes using such a chimera.

The first of the genes we will consider for chimera analysis involves coat color. In the production of chimeric embryos, it is standard practice to use cells from mice of two different colors—a black mouse and a white mouse for example. These cells are re-aggregated, go through many rounds of mitosis, and eventually differentiate into epidermal cells (as well as many other cell types). Those epidermal cells that came from the white mouse will produce white fur, whereas cells coming from the black mouse will produce black fur. This will be seen as patches of black and white on the chimeric mouse. On close examination, any individual cell will

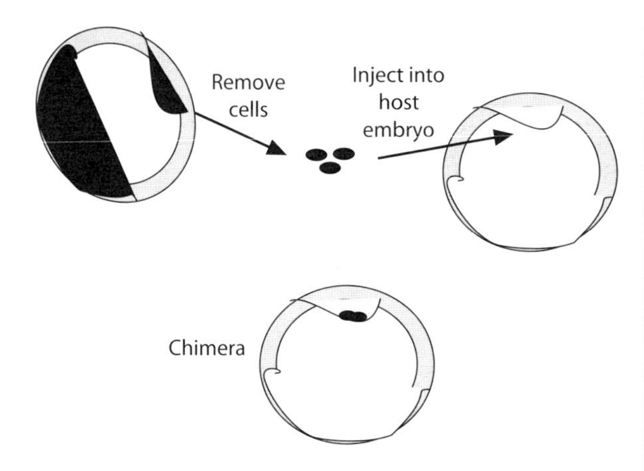

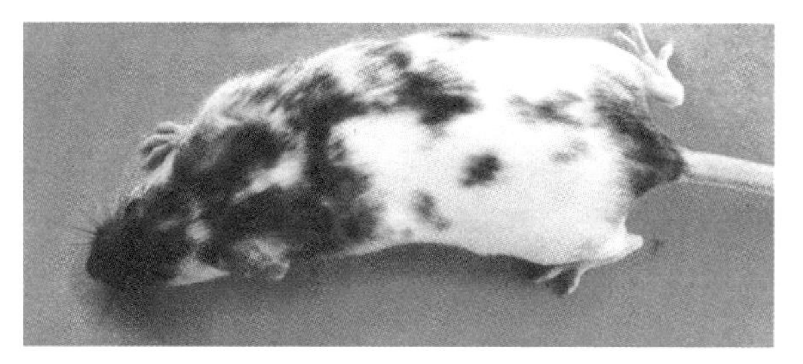

Figure 9.11 A chimeric mouse. A chimera is made by physically aggregating cells of two different genotypes and phenotypes. The top left of the figure depicts a blastocyst-stage embryo from a strain with black fur. Cells are removed and injected into the blastocyst from a strain with white fur, as shown at the top right. The chimeric embryo has cells of both types in the inner cell mass, which will become the adult mouse. An example of a chimeric mouse arising from this type of procedure is shown at the bottom, with the characteristic white and black fur.

produce either a white hair or a black hair. (The colors are seen as patches or stripes in the embryo because the melanocytes that give rise to hair and pigmentation do not migrate extensively throughout the body.) The coat color gene is **cell autonomous**, i.e. the gene product acts within the cell that produces it and does not diffuse to nearby cells or throughout the body. Because coat color is cell autonomous, it makes a useful phenotypic marker to determine which cells came from which parent.

In the description of the construction and analysis of chimeric and mosaic embryos (to be defined in a subsequent section), cell autonomous genes will play an important role. Many genes act within the cell in which they are produced—transcription factors, intracellular signaling molecules, chromosomal proteins, and many more. The important aspect of these genes for our purpose is not their function. Instead, the important aspect is that these genes have a mutant phenotype that can easily be scored within a single cell. These form the genetic markers that make it possible to monitor the activity of other genes in the same cell, which may or may not act autonomously.

Cell autonomous markers are used to analyze the activity of other genes

Now we turn to some of the other phenotypes that may be observed in chimeric mice. Suppose that the white mouse has a mutation in another gene whose activity we want to monitor. The white cells carry the mutation for this other gene, whether or not a mutant phenotype is always observed. This point is very important: the cell autonomous marker tells the investigator which cells have a mutant *genotype* regardless of the *phenotype* of that cell. Now, imagine that the investigator examines the phenotype of those white cells and determines that they are developing abnormally; that is, in the individual mouse being studied, cells are also showing a mutant phenotype for the gene of interest. The logical inference is that the gene must be active for that cell or cell type to develop and function normally. If that gene is genotypically mutant, the cell is phenotypically mutant. **Thus, the presence of a mutant phenotype demonstrates that those cells require the normal activity for that gene.**

As the investigator examines other white cells in the chimeric mouse, he/she finds other white cells that develop and function normally, i.e. other than being white (an innocuous phenotype itself), the cell is wild-type in appearance and function. However, because the cell is white, it must still be genotypically mutant for the gene being studied. The inference is that the cells or cell types that develop and function normally do not require the activity of the gene. The gene is not expressed in those cells or its expression is not needed for normal cellular function. Thus, the chimeric embryo demonstrates which cells and tissues require normal activity for the gene and, just as important, which cells and tissues do not require normal activity for the gene. **The presence of a wild-type phenotype from cells that are genotypically mutant implies that those cells do not require the activity of the gene of interest.**

Other results may also be observed for some genes. We illustrate this diagrammatically in Figure 9.12. Cells

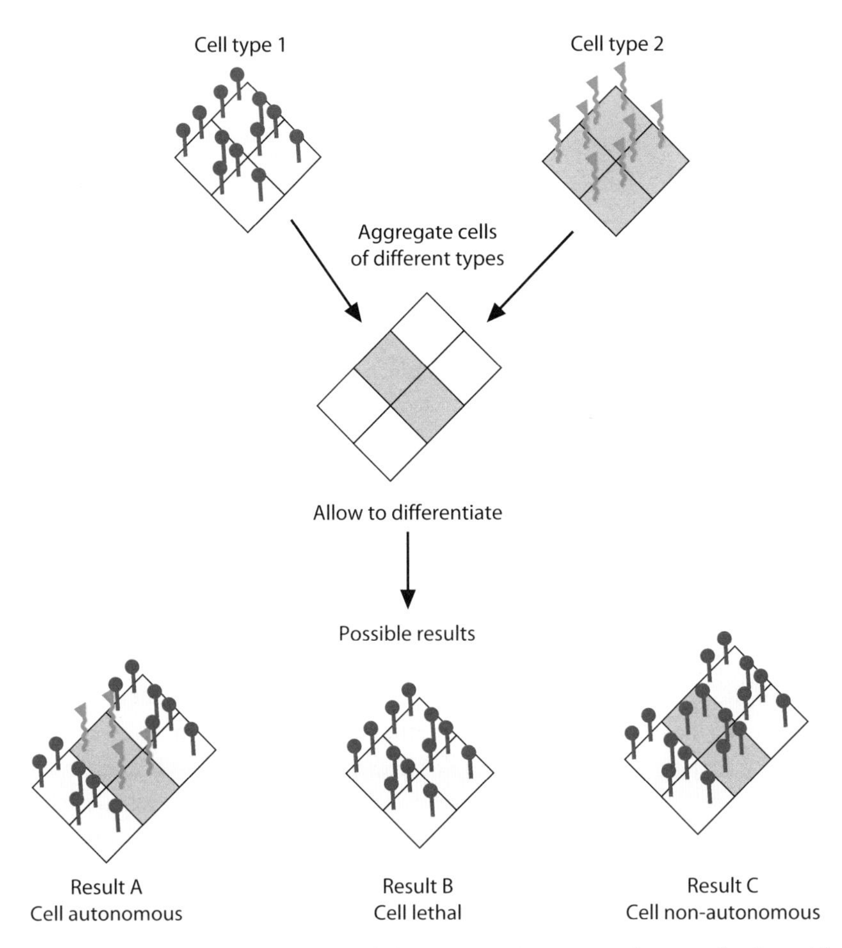

Figure 9.12 Interpreting the results of chimeric experiments. In a chimera, cells of two different genotypes are aggregated. In this diagram, the phenotype being tested is the blue or green activity on the cell surface; the cells are shown in different colors to indicate the marker being used to follow the cell types. The blue and green activities could be different activities arising from different alleles of the same gene. The yellow cells have the blue activity for the gene of interest, while the purple cells have the green activity. Yellow and purple cells are aggregated and then allowed to differentiate. One of three possible results could arise, as shown at the bottom of the figure. In result A, all of the cells with the yellow marker have the blue activity, and all of the cells with the purple marker have the green activity; that is, the cells have the phenotypic activity corresponding to their own genotype, so the activity is cell autonomous. This type of result is observed with gene products that function within a cell. In result B, none of the purple cells appear upon differentiation, and only yellow cells with the blue activity are observed. The inference is that the green activity is cell lethal, so that cells that have the green function cannot differentiate and survive in this tissue type. In result C, the cells with the yellow marker show the blue activity, but in addition the cells with the purple marker show the blue activity. That is, the blue activity is not cell autonomous and some cells do not express the phenotype of their own genotype. A cell non-autonomous result such as this is often observed with gene products that diffuse between cells.

of different types (e.g. with two different genotypes) are aggregated, and the aggregated cells are allowed to differentiate. In Figure 9.12, cell type 1 is shown in yellow and has the gene activity represented by the blue icons, whereas cell type 2 is shown in purple and has the green icon to represent the gene activity. Notice that the yellow and purple coloration of the cells represents an observable marker phenotype; in a chimeric mouse, this could be the black or white fur. The gene activity icon (either blue or green) represents the phenotype that is to be tested for its autonomy. Different results could occur once the aggregated cells differentiate, as shown in Figure 9.12. If the activity of the gene in question is cell autonomous, the marker phenotype (yellow or purple) will correspond to the activity phenotype (blue or green), as shown in result A. Another

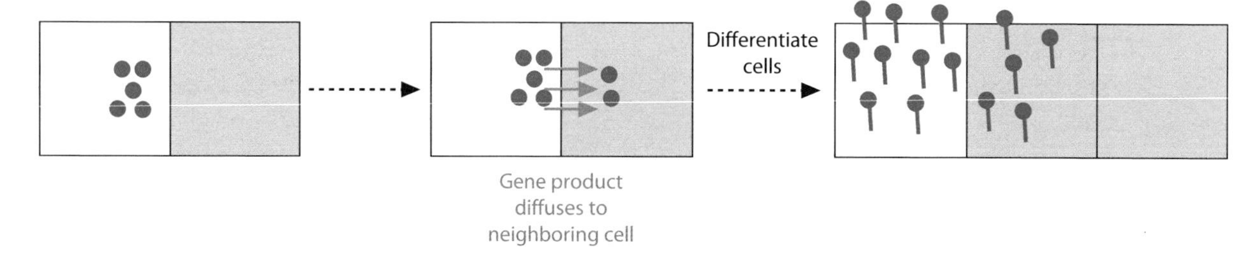

Figure 9.13 Cell non-autonomous differentiation. Two cell types are shown, one marked by a yellow color and the other by a purple color. The yellow cells produce a gene product, here indicated as blue balls, which diffuses to nearby cells. The purple cells themselves do not make this gene product, possibly because they have a mutation in the gene. Because the gene product diffuses from the yellow cells, both cell types produce the blue activity, regardless of their own genotype. Therefore, the blue activity is cell non-autonomous. Purple cells that are not adjacent to the wild-type yellow cells do not produce the gene product. Such a result suggests the distance over which the gene product or signal can spread.

possibility is that the activity of the gene is essential for cell survival upon differentiation, as shown in result B. Thus, although the aggregate had cells of each marker type, the cells with the green activity do not differentiate and no purple cells are observed in the final tissue. As shown, it is the blue activity that is apparently needed for cell survival. Thus, the chimera has identified a mutation that is **cell lethal**—a gene whose activity is essential for the survival of cells of a particular type.

A third possibility is that the gene activity is not cell autonomous, shown as result C in Figure 9.12. In this case, the cells with the purple marker do not differentiate according to their own activity but instead show the blue activity of the neighboring cells. The lack of correspondence between the phenotype of the marker and the gene activity phenotype is characteristic of a gene activity that is cell non-autonomous.

Genes whose activity is not cell autonomous are potentially the most complex to interpret using this analysis. In this case, the phenotype of the cells does not agree with their genotype but may depend on the genotype and phenotype of neighboring cells. This is illustrated as result C in Figure 9.12. An interpretation for **cell non-autonomous** gene activity is that the gene product signals to other cells, diffuses between cells, or works throughout the organ or body.

For genes that are cell non-autonomous, further analysis of the pattern of wild-type and mutant cells is sometimes helpful, as shown in Figure 9.13. Suppose that a gene product diffuses or signals to other cells. In Figure 9.13, the purple cells do not produce the gene product themselves, but the neighboring yellow cells do and the gene product diffuses into the purple cells. Thus, the purple cells in the figure that lie immediately adjacent to yellow cells with a wild-type phenotype also have a wild-type phenotype. The inference is that the proximity to the genotypically wild-type yellow cells is producing the wild-type phenotype. This may imply that, in the normal organism, cells of one type, in this case marked by the yellow color, signal the development of their neighboring cells, in this case marked by the purple color. Although the recipient cells lack the normal activity, they develop normally because the signal from the yellow cell type is present. The investigator can also examine the phenotype of cells that are not immediately adjacent to the wild-type cells to determine how far the normal phenotype extends among the genotypically mutant cells. This can suggest how far the signal diffuses in normal development.

Patterns of gene activity for cell non-autonomous mutations should be interpreted with caution. Cells exchange many signals with each other, and the process is often reciprocal rather than one way. Nonetheless, a lack of correspondence between the cell autonomous markers and normal cellular function usually indicates that the gene of interest is not cell autonomous. The inference is that the gene product does not act exclusively in the cells where the gene is active and is likely to be involved in signaling between cells.

Mosaic organisms are made by several different genetic techniques

We have described how the phenotype of chimeric embryos can be used to make inferences about gene activity in order to illustrate the principles involved in this type of analysis. We now turn to the genetic methods that also analyze gene activity according to these principles, using cells with two different genotypes that exhibit two different phenotypes. An organism in which cells of two different genotypes have been produced by genetic methods is referred to as a **mosaic**.

The genetic methods used to make organisms with cells of two different genotypes are rather more difficult to understand than the physical methods that simply aggregate cells. However, the analysis of gene activity uses the same inferences regardless of how the organisms are made. All of the methods use a cell autonomous marker that gives a mutant phenotype that is easily observed in single cells. However, the cell autonomous marker is not the gene of interest; it is simply the means used to follow the gene of interest.

All, or nearly all, of the genetic techniques used to make mosaics are performed beginning with heterozygous individuals. Since most mutations are recessive, the heterozygous individual has a wild-type phenotype; however, if the wild-type allele is removed in some cells, the cell will have a mutant genotype. If possible, the mutant allele should be a null allele so that no possible gene activity is generated from the mutant. Two general methods are used to make mosaic organisms. In *Drosophila* and especially in *C. elegans*, mosaics are made by the loss of a chromosome or a chromosome fragment. Mosaics can also be made in *Drosophila* by inducing recombination in somatic cells, a process known as mitotic recombination. Mosaics produced by loss of a chromosome fragment in *C. elegans* will be described first.

Mosaic organisms can be made by the loss of a chromosome fragment in *C. elegans*

As first discussed in Chapter 2, chromosomes in *C. elegans* have an unusual feature that is important in producing mosaics. In worm cells undergoing mitosis, the microtubules attach along the length of the chromosome rather than at one localized centromere. This property means that any chromosome fragment can exist free of attachment to other chromosomes and will be inherited semi-stably during cell division. That is, a worm may be homozygous for a recessive mutation *a/a* and also have a free duplication fragment of that region of the chromosome that carries the wild-type allele *a+*. (Nearly all free duplications in *C. elegans* carry only wild-type alleles because of the genetic screens used in their isolation.) Such a worm is a **segmental trisomic**—it has three copies of a particular region of one chromosome—and because it carries the wild-type *a+* allele on the duplication, it has a wild-type phenotype. This situation is illustrated in Figure 9.14. However, the duplication will occasionally be lost during mitosis: some of the cells will be *a+/a/a* with a wild-type allele and other cells will be *a/a* and have no wild-type allele. Consequently, the worm will be **mosaic**. If a cell autonomous marker that allows the investigator to determine which cells have retained or lost the duplication is also present, the locus of gene activity can be determined.

Let us illustrate this with a simple example. Imagine that an investigator has found a recessive mutation that results in paralysis when homozygous; we will denote this mutation *m*, so the paralyzed worm is *m/m*. Paralysis could arise from defects in a gene that acts either in body muscles or in neurons that connect to the body wall muscles. (For simplicity in the example, we will consider only these two possible origins of the paralysis.) In order to determine where this gene acts, the investigator makes a segmental trisomic that is homozygous for the mutant *m* allele on its normal chromosome and which also carries a free duplication of the same region of the chromosome with the *m+* allele. This experiment is illustrated in Figure 9.15. Such a worm moves normally because it carries the dominant wild-type *m+* allele. Also included on the duplication is the wild-type allele for a cell autonomous marker denoted as *a+*, while the recessive mutant allele *a* of the marker gene is homozygous on the normal chromosome. We will assume for simplicity that *a* is linked to *m* so that the same free duplication has wild-type alleles of both genes. Different cell autonomous markers have been used in worm mosaics; a mutation in a gene called *ncl-1* results in a misshapen nucleolus that is easily recognized by Nomarski light microscopy, and has been widely used. The genotype of the worm is *Dp(m+ ncl-1+)/m ncl-1/m ncl-1*; that is, it has a free duplication with the *m+* and the *ncl-1+* allele linked to each other, and the *m* and the *ncl-1* recessive mutant alleles linked on the chromosome.

This worm is allowed to self-fertilize to produce hundreds of offspring, some paralyzed and some moving normally. The neurons and muscle cells of the paralyzed worms are examined for the presence or absence of the cell autonomous marker *ncl-1*. Many of

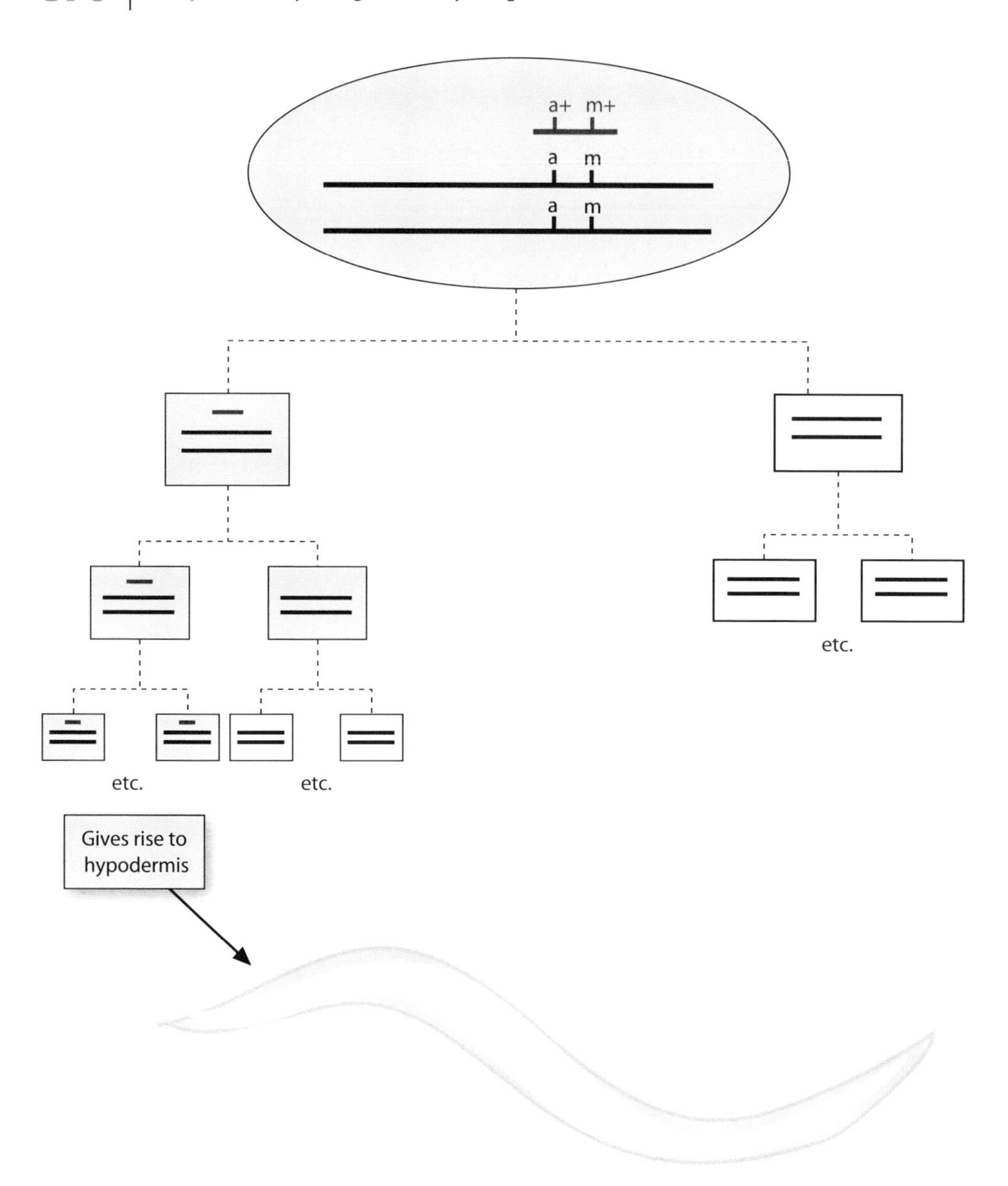

Figure 9.14 Generating mosaic animals in *C. elegans*. In worms, mosaics are made by the somatic loss of a chromosome fragment. The single-celled embryo at the top of the figure is a partial trisomic, homozygous on the chromosomes for the marker *a* and the mutation *m*. The chromosome fragment in red includes the wild-type allele for both of these genes, and so the cell has a red phenotype. As the embryo divides, the chromosome fragment is lost in some cell lineages but not in others; the cells that retain the fragment are red, whereas those that lack the fragment are white and mutant. In reality, the cells that lose the chromosome fragment will be identified because they are mutant for the marker *a*. In this example, the fragment that will give rise to the hypodermis or skin cells of the worm is retained in the cells. The other cells are homozygous mutant for the marker and the mutation of interest.

the paralyzed worms will lack the duplication in all of their cells, which is uninformative. However, some of them will be mosaics, having retained the duplication in some cells and lost it in others, which can be determined by the cell autonomous marker.

Now the question is: In these paralyzed mosaic worms, has the duplication been lost in the neurons or has it been lost in the muscles? This is illustrated in Figure 9.15. For our example, suppose that the paralyzed mosaic worms have lost the duplication in the muscles but have retained it in the neurons. This result indicates that the presence of the wild-type allele *m+* in the neurons is not sufficient to produce normal movement, but that its absence in the muscle results in paralysis. The *m* gene is inferred to act in the muscles. A similar examination is done for the worms that move normally.

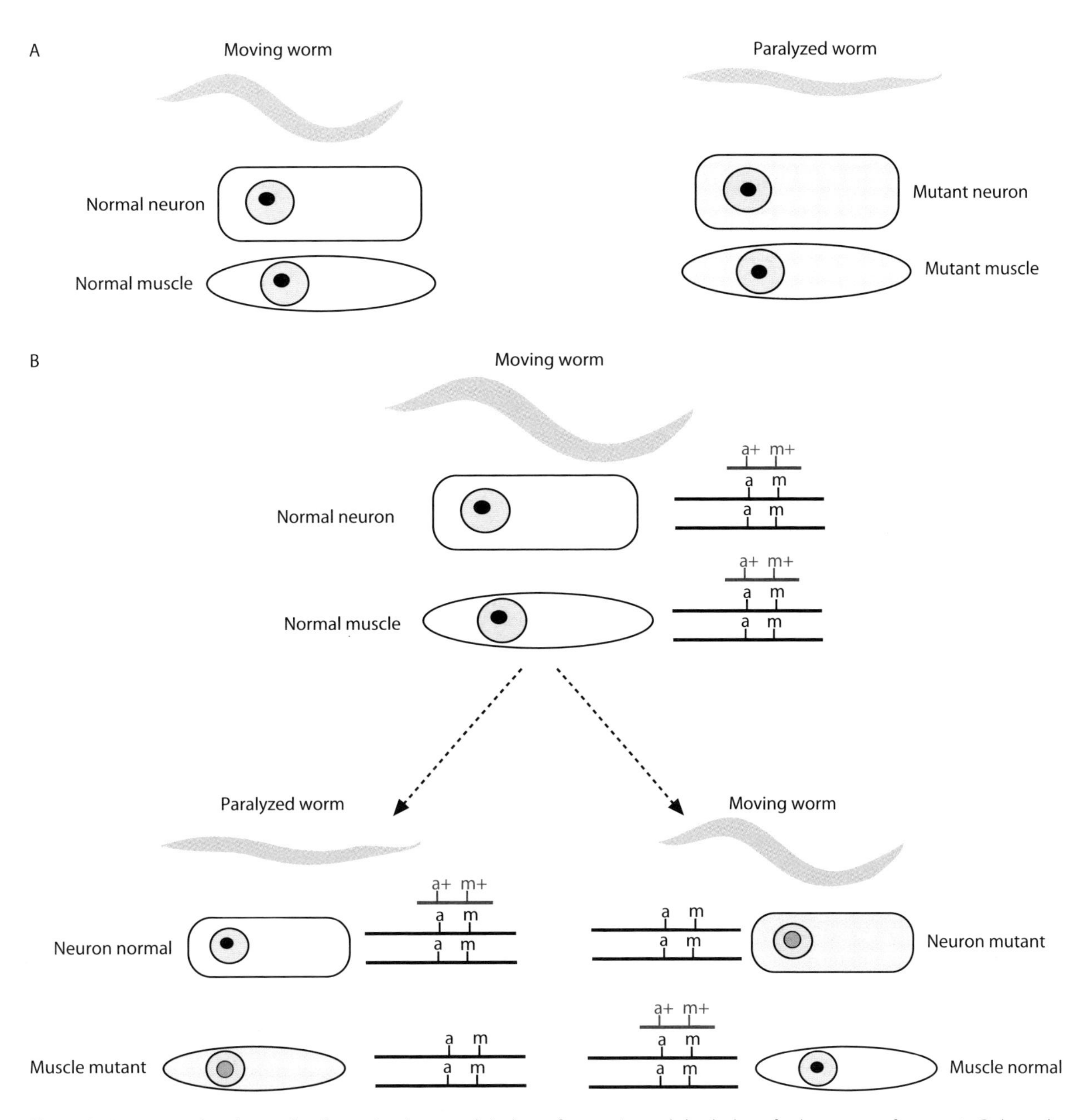

Figure 9.15 Interpreting the results of mosaics. An example is shown for mosaics made by the loss of a chromosome fragment in *C. elegans* by a procedure similar to that shown in Figure 9.14. The gene of interest, shown in panel A, is needed for movement so that mutants for the gene are paralyzed. The gene could act in neurons or muscles, and the experiment is designed to determine which tissue is the focus of action. In the drawing, muscles and neurons are distinguished by the different shape of the cell; the genotypically mutant cells are shown in green for clarity. Panel B shows a moving worm that has the wild-type alleles for both the gene of interest, *m*, and the linked genetic marker, *a*. The cell marker affects the structure of the nucleolus, as shown in the mutant cells in panel B; in nucleolar mutant worms, the nucleolus is larger and more diffuse. Cells with this phenotype will also lack the wild-type allele for the gene affecting movement. A population of worms carrying the chromosome fragment with the wild-type alleles is allowed to reproduce, and some of the paralyzed (or partially paralyzed) or moving offspring are examined. As shown at the bottom of the figure, the paralyzed worms have normal neurons but mutant muscles. The worms that move normally have mutant neurons but normal muscles. Because the loss of the chromosomal fragment in muscles correlates with paralysis, the gene must be needed in muscle cells but not in neurons.

Some of them will be mosaics, as recognized by the cell autonomous marker. Have the mosaic worms that move normally retained the duplication in the muscles or in the neurons? If our inference about the gene activity is correct, some of the worms with normal movement will have lost the duplication in the neurons, but all of them will have retained it in the muscles.

Although slightly simplified, this example is not very different from a standard mosaic analysis in *C. elegans*. The semi-stable mitotic inheritance of free duplications makes it possible to produce mosaic worms, and the cell autonomous marker provides the necessary means to monitor duplication presence or loss. The analysis of genetic mosaics in *C. elegans* depends on knowledge of the cell lineages that give rise to adult tissues. The duplication fragment is lost at some frequency at each cell division; it is possible to infer when the duplication was lost during development by observing which cells have the duplication and which lack it.

Among the first examples to be analyzed by mosaic analysis in *C. elegans* were genes that acted in environmental sensing. Environmental sensing such as response to particular chemicals or uptake of the dye FITC depends on the sense organs, whose structure is illustrated in Figure 9.16. The sense organ in the head of the worm consists of neurons known as amphids and two non-neuronal cells called the socket cell and the sheath cell. Some of these mutants have morphological defects in the structure of the sense organ, and mosaic analysis was used to determine if the genes acted in the neurons or the non-neuronal cells. Two genes particularly illustrate the utility of this approach. Mosaic analysis with a free chromosomal duplication was done with *osm-1* mutations, which cannot take up the dye. If the amphid cells lacked the wild-type allele of *osm-1+*, the sense organ could not take up the dye; in contrast, if the non-neuronal cells lacked the wild-type allele of *osm-1*, dye uptake still occurred normally. Thus, *osm-1* acts within the neurons since the mutant phenotype was seen when only the amphids were mutant. A similar analysis of the gene *daf-6* found that it works within the socket or sheath cells. These examples were the proof that mosaics could be generated using free duplications, and mosaic worms have been made for many other genes. For instance, in Chapter 11 we show that *her-1*, one of the key genes involved in sex determination, is likely to encode a secreted ligand. This result was

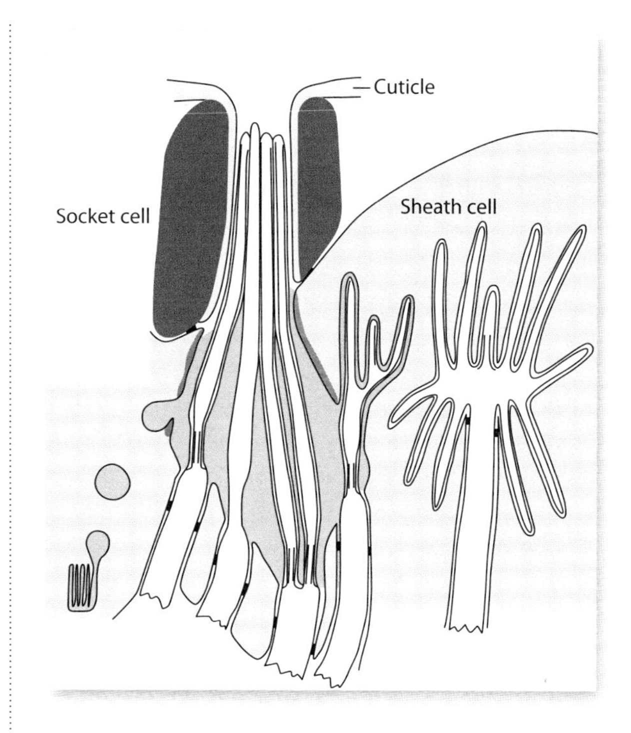

Figure 9.16 The structure of the amphid sensillum in *C. elegans*. This diagram of an electron micrograph shows the socket cell underneath the cuticle, with the sheath cell wrapping around the sensory neurons with exposed endings. The amphid is the principal sense organ in the head and is involved in dauer larva formation, chemotaxis, and most other sensory behaviors.

supported by mosaic analysis showing that the mutant effects of *her-1* are not cell autonomous.

Although most investigators working with mosaics in *C. elegans* have used chromosome fragments, it is also possible to use microinjected DNA to make mosaics. Recall from Chapter 2 that transgenic worms are made by simply injecting DNA into the gonad of a worm. The injected DNA forms a multicopy array that is semi-stable during mitosis, similar to the behavior of a duplication fragment. An injected array can apparently be attached along the length to microtubules and segregated like a miniature chromosome fragment, only with defined sequences. This is particularly useful because the cloned *ncl-1+* gene, a *gfp*-tagged gene, or another cloned cell autonomous marker can be co-injected and will be co-segregated, even if the cell autonomous marker is not linked to the gene of interest in the worm genome. With chromosomal duplications and injected DNA multicopy arrays, the investigator has to find an appropriate level of segregation and loss. If the duplication is extremely stable, mosaic animals are rare. On the other hand, if the duplication or injected array is lost too frequently, only the first few cell divisions can be analyzed and the strain is very difficult to maintain as a heterozygote. Often finding the appropriate rate of loss is a matter of trial and error.

Mosaics based on X chromosome loss can be made in *Drosophila*

The use of free duplications to make mosaics in worms depends on the unusual properties of *C. elegans* chromosomes. No other well-studied model organism has semi-stable segregation of free chromosome duplications; in organisms with conventional centromeres, a duplication including the centromere is segregated normally, whereas a duplication lacking the centromere is lost immediately. However, a similar principle has been used in *Drosophila*, except by employing complete chromosomes rather than chromosome fragments. This use preceded the application of chromosome duplications in worms by several decades.

D. melanogaster has four chromosomes—the X chromosome, two large autosomes, and a small fourth chromosome. Loss of one of the large autosomes is lethal; in fact, deletions of more than a few dozen genes are inviable. Thus, loss of one of the autosomes cannot be used for mosaic analysis. *Drosophila* can tolerate loss and monosomy for the fourth chromosome, so this could be a method to make mosaics for genes on that chromosome. However, the fourth chromosome has only about 75 genes, so this is clearly not a generally useful strategy. However, mosaics based on the loss of the X chromosome have proved to be a useful tool for genetic analysis.

Sex determination in *Drosophila* depends on the number of X chromosomes rather than on the presence or absence of a Y chromosome. Unlike mammals, the Y chromosome is not needed for male sex determination. An organism or a cell that is X0 (i.e. with one X chromosome and no other sex chromosome) has male development, whereas an organism or cell that is XXY or XXX shows female development. This is shown in Figure 9.17. A fly can live and develop as a sex chromosome mosaic, with X0 cells and XX or XXY cells. Such a fly will have a mixture of X0 male cells and XX or XXY female cells, a phenotype known as a **gynandromorph**.

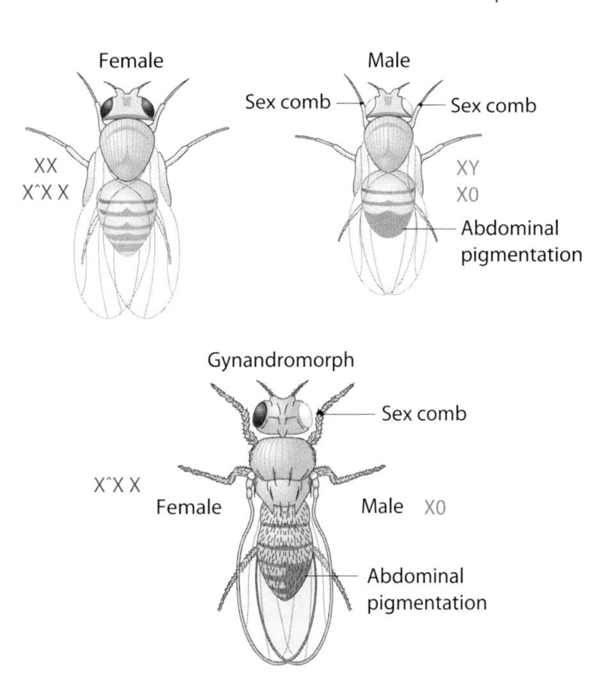

Figure 9.17 Gynandromorphs in *D. melanogaster*. A normal female and male are shown at the top of the figure. The female could be XX or X^XX with two attached X chromosomes and a regular X chromosome. The male could be either XY or X0. An X^XX/X0 gynandromorph is shown below, with the male side on the right and the female side on the left. Notice the male characteristic sex combs on the forelegs and the abdominal pigmentation. This gynandromorph also carried recessive alleles for the X-linked genes white eyes and vestigial wings, which appear as hemizygous on the X0 side. Most gynandromorphs are not as neatly bilaterally symmetric as this diagram suggests.

Gynandromorphs arise when the female fly has a particular type of chromosome rearrangement called an attached X chromosome; such rearrangements are also known as **compound chromosomes**. The *Drosophila* X chromosome has its centromere near one terminus. Muller and other early geneticists used X-irradiation to fuse two X chromosomes at their centromeres, resulting in one large fusion chromosome with a single centromere. The mitotic segregation of such a chromosome is shown in Figure 9.18. These two X chromosomes, symbolized X^X, will tend to segregate together (although recombination between the two arms breaks the fusion chromosome into two halves at some frequency). Thus, a fly with an attached X^X is a female because it has two X chromosomes, albeit fused into one unit, and will produce oocytes that have the attached X^X. The offspring that inherit this attached X^X will be a female regardless of the sperm that fertilizes this ovum, either X^XY or X^XX. We will

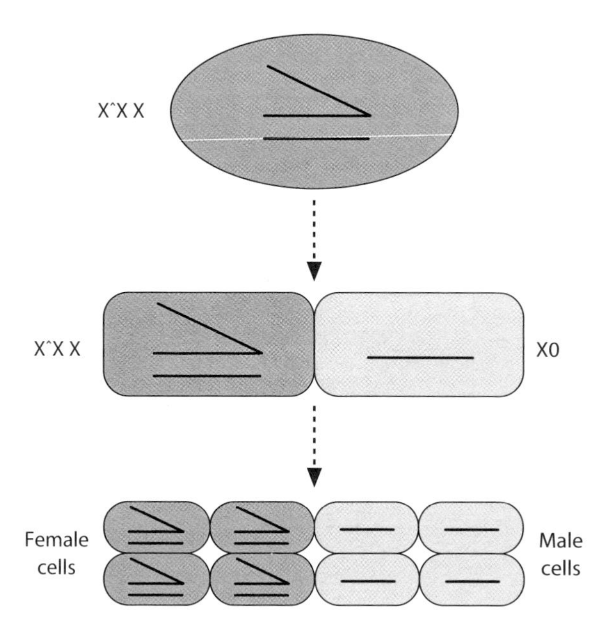

Figure 9.18 Generating mosaics in *Drosophila* by attached X chromosomes. An embryo with a pair of attached X chromosomes and a third unattached X chromosome is shown at the top. The embryo is female, as shown in red. As the cells divide mitotically, the attached X chromosome is retained in some cells, shown on the left, and lost in other cells, shown on the right. When the attached X chromosome is lost, the cell differentiates male structures rather than female structures.

consider the X^XX case, as illustrated in Figure 9.18. As the embryo begins mitosis, the X^X is lost mitotically in some cells to produce X0 cells, with a mixture of XX and X^XX cells elsewhere in the organism. The X0 cells will follow male development, whereas the XX and X^XX cells will follow female development. **As a result, the fly is mosaic for sex determination itself as well as for any recessive X-linked mutations.**

Gynandromorphs have been particularly useful in a mosaic analysis of sex determination itself. The first important observation is that individual cells follow either male or female development, with none that are intersexual. In other words, sex determination itself is cell autonomous. This has been interpreted to mean that sex determination in flies occurs within cells and that the key signals do not diffuse between cells or throughout the organism. Note that a similar experiment in mammals, if it were possible, would not give this result because mammalian sex determination relies on hormones. Although *Drosophila* has hormonal signals for many developmental processes, sex determination is not one of them. In fact, as discussed in Chapters 1 and 2, subsequent genetic and molecular analysis of sex determination in *Drosophila* has shown that the early steps in sex determination depend on alternative splicing of RNA transcripts and a transcription factor known as *doublesex*. In other words, sex determination depends on gene products that act within a cell, just as gynandromorph analysis indicated.

In addition to their utility in mosaic analysis of sex determination, gynandromorphs can also be used to perform mosaic analysis of X-linked genes. The procedure is similar to what is done with chromosome fragments in *C. elegans*. A fly heterozygous for an X-linked gene of interest is made with an attached X chromosome; that is the female is X^XX, in which the attached X has the wild-type allele and the single X has the recessive mutant allele. As the attached X is lost in particular cells, the locus of activity for the gene can be observed. Since the sexual morphology of the cell is itself a cell autonomous trait, no other marker is actually needed in these experiments, although most investigators include another recessive X-linked marker gene for clarity. The gynandromorphy shown in Figure 9.17 includes the recessive X-linked markers white eyes and vestigial wings.

Most mosaic analysis in *Drosophila* relies on somatic crossing-over

Although gynandromorphs have been a useful and appropriate means of carrying out a mosaic analysis for some genes in *Drosophila*, a very different method of generating mosaics is much more widely used. This method involves somatic crossing-over, also known as **mitotic recombination**. The procedure is shown diagrammatically in Figure 9.19 for the *patched* (*ptc*) gene.

As with other types of mosaic production, the fly (typically at the larval or pupal stage) at the beginning is heterozygous for the gene of interest and one or more cell autonomous markers. A typical configuration is shown in Figure 9.19, which has two cell autonomous markers on the same homolog as the mutation of interest and a different cell autonomous marker on the other homolog. In Figure 9.19, the gene of interest is *ptc*, and the cell autonomous markers linked to *ptc* are two bristle markers known as *straw* (*stw*) and *pawn* (*pwn*). On the other homolog is a different cell autonomous marker known as *shavenoid* (*sha*). Although the fly

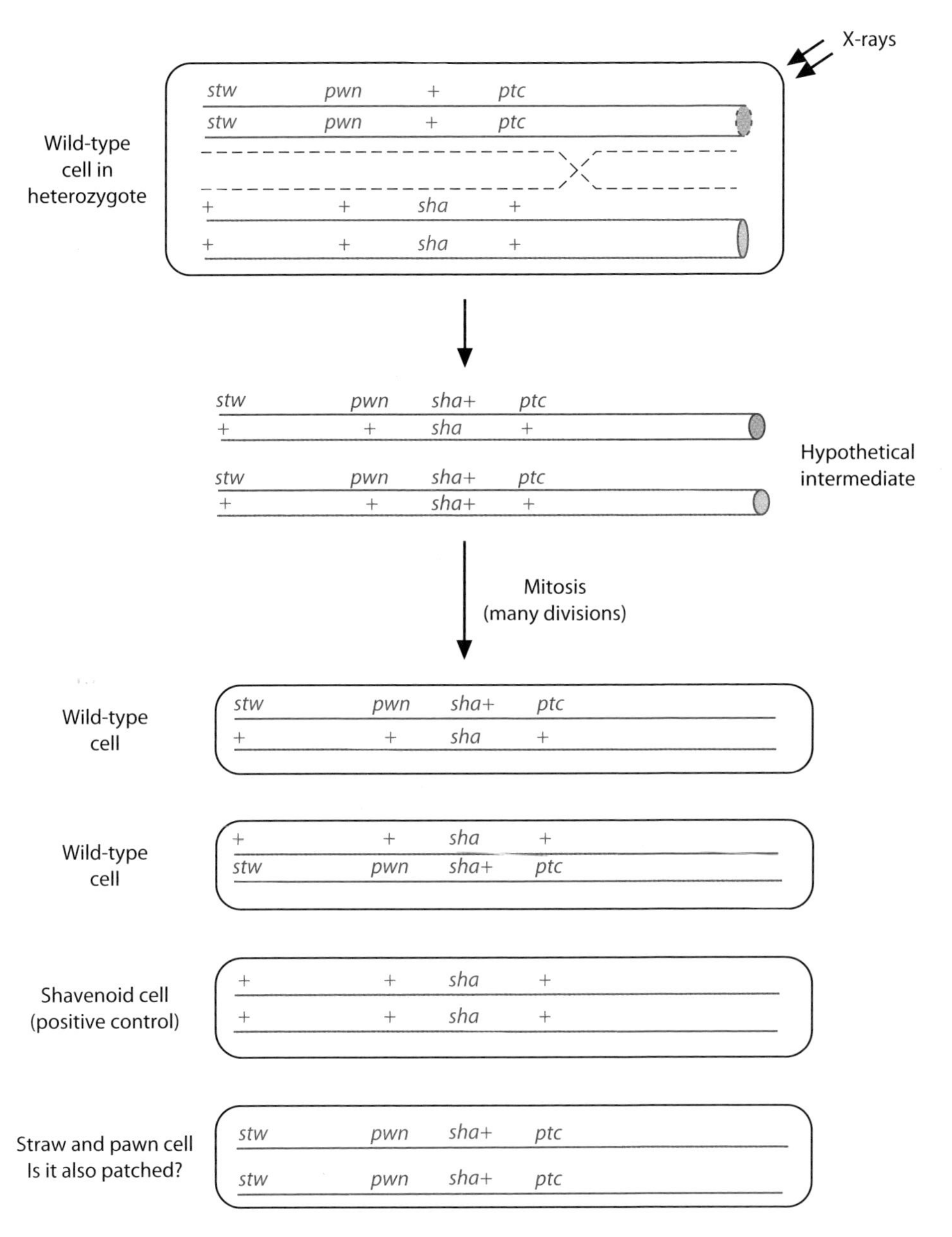

Figure 9.19 Generating mosaics by somatic crossing-over in *Drosophila*. A wild-type fly that is heterozygous for the gene of interest and cell markers is subjected to low-level X-irradiation. The gene of interest in this example is *ptc*, shown in red on the sister chromatids at the top. On the same chromosome as *ptc* are the cell markers straw bristles (*stw*) and pawn bristles (*pwn*), which will allow the homozygous *ptc/ptc* cells to be identified in the wing no matter what phenotype these cells display. On the other chromosome is the cell marker shavenoid bristles (*sha*) in blue. This marker serves as a positive control. The X-rays produce chromosome breaks, and as a result crossovers occur by an unknown mechanism. The hypothetical intermediate is shown for clarity, but this may never form. Notice that at the four-strand stage of the cell cycle, sister chromatids of different genotypes are shown attached to the same centromere because of the somatic crossover. When these sister chromatids separate at mitosis, four possible mitotic products occur, as shown at the bottom of the figure. The first two mitotic descendant cells are heterozygous for all genes and will be wild-type, like the unirradiated parent. The third mitotic descendant will be homozygous for the shavenoid marker but homozygous wild-type for the other genes. These cells will have a shavenoid phenotype. The occurrence and frequency of these cells indicate that somatic crossing-over has occurred. The experimental result is shown in the cell at the bottom. This cell is homozygous for the markers *stw* and *pwn* and is necessarily also homozygous mutant for *ptc*. The phenotype of this cell indicates the role of *ptc+* in that cell or cell type. If the cell develops normally, *ptc+* is not needed. If the cell fails to develop at all, the *ptc* mutation is cell lethal and no straw and pawn cells will be seen. If the cell has abnormal development, *ptc+* is required within the cell.

itself is wild-type, somatic crossing-over will be used to produce cells that are homozygous for each of the cell autonomous markers.

Somatic crossing-over is induced by low-level irradiation with X-rays. The irradiation induces double-stranded DNA breaks, and chromatids are exchanged in the repair of these breaks; the exact mechanism is not known, although it is almost certainly related to the processes that organisms use to repair double-stranded breaks that occur during normal growth. As the irradiated cell divides by mitosis, some of its descendants will become homozygous for the cell autonomous marker linked to the mutation of interest and others will become homozygous for the cell autonomous marker on the other homolog. That is, the resulting fly will have clones of cells homozygous for each marker against an overall background of wild-type development, as shown in Figure 9.19. As the somatic recombination occurs between the genes and the centromere, as shown in Figure 9.19, both types of marked cells will occur. In the diagram, one set of marked cells will be straw and pawn in phenotype, whereas the other will have a shavenoid phenotype. These clones of cells marked with the different recessive markers, known as **twin spots**, are diagnostic that somatic crossing-over has occurred. This mosaic analysis of *patched* expression in the wing is described in more detail in Case Study 9.2.

Because the entire larva or pupa has been irradiated, twin spots could be found in any part of the body. Their size and shape will depend on the time when somatic crossing-over occurred and on the orientation and number of the subsequent cell divisions; larger clones of cells arise from earlier irradiations and more cell divisions. Of course, populations of pupae or larvae are irradiated, not single flies, so some of the flies will have no twin spots and some may have died from the irradiation. In others, the twin spots will be located in parts of the fly that are not of particular interest to the investigator. For example, if the investigator is interested in the role of a gene in wing development, twin spots on the head or the abdomen may be of minimal interest. Nonetheless, as the investigator sorts through the population of irradiated flies, some will have twin spots in the part of the fly under study. With these, the analysis is done much as we have already described for other procedures.

As before, we can ask which cells require the activity of the wild-type allele of the gene in order to show a normal phenotype. Case Study 9.2 describes how this was done for the *patched* gene to show that the activity of the gene differs in the anterior and posterior halves of the wing. Suppose that the gene of interest is on the same chromosome as the cell autonomous marker *forked* bristles, and opposite the marker *yellow* body, as shown in Figure 9.20. With twin spots of forked and yellow cells, the investigator looks at the forked patches to determine if their phenotype is normal or mutant for the gene of interest; they *must* be genotypically mutant for the gene of interest because they exhibit the *forked* marker phenotype. If the forked spots have a mutant phenotype for the gene of interest but nearby yellow clones are normal, the gene must be active in that region of the fly. If, on the other hand, the forked spots have a wild-type phenotype (other than having forked bristles), the gene must not be required for the normal development and differentiation of those cells. This is precisely the same analysis that we used in describing chimeras; the only difference is that these mosaics have been made by somatic crossing-over rather than by physically aggregating cells.

Twin spots are also used to determine if the gene activity is cell autonomous or cell non-autonomous. If the forked spots in our example have a mutant phenotype but the nearby cells are wild-type, the gene of interest is likely to act within the cells. On the other hand, if the forked spots do not precisely correspond to the mutant phenotype, the gene product may be diffusing between cells. For example, suppose that forked cells bordering wild-type cells or yellow cells have a wild-type phenotype. One inference is that the normal gene product is being made in the wild-type or yellow cells and is diffusing or signaling the nearby cells; these cells develop normally because they are receiving the wild-type gene product from their neighbors. Conversely, if the wild-type or yellow cells that border the forked patches show a mutant phenotype, the inference is that the forked cells are failing to send a signal to their neighboring cells.

Twin spots also provide an important internal control for mutations that are cell lethal. For example, the presence of yellow patches of cells indicates that somatic crossing-over has occurred; forked cells must have been produced because they are the reciprocal

Case Study 9.2

Mosaic analysis of *patched* expression in the wing

In describing how mosaic analysis is performed and used in *Drosophila*, it is appropriate to consider a mosaic analysis of the *patched* (*ptc*) gene. The *ptc* gene was introduced in Case Study 4.1 in Chapter 4 as one of the mutations in *Drosophila* that affects the polarity of embryonic segments. We also described how the gene was cloned from *Drosophila* and from vertebrates and mice (in Chapter 5), and how *patched* gene disruptions were made in mice (in Chapter 6). Genetic interactions involving the *ptc* gene in *Drosophila* are used in Chapter 11 as an example of how phenotypes are used to infer a genetic pathway.

In describing the isolation and characterization of the *Drosophila ptc* gene, we noted that some hypomorphic alleles of the gene are viable but have defects in adult wing formation, as well as in structures in the eye and antenna and elsewhere in the adult. The wings develop from a pair of imaginal disks that are present in third-instar larvae; imaginal disks are described in Chapter 2 as pockets of larval cells that form adult structures. The wing forms when the disk cells essentially balloon outwards and then flatten. Case Study Figure 9.6 compares the location of some structures in the adult wing disk with the location of their primordia in the disk.

The anterior–posterior compartment boundary

The most important feature of wing development is the presence of a boundary shown by a red dotted line in Case Study Figure 9.6. This imaginary line, called the anterior–posterior compartment boundary, separates the anterior part of the wing blade from the posterior

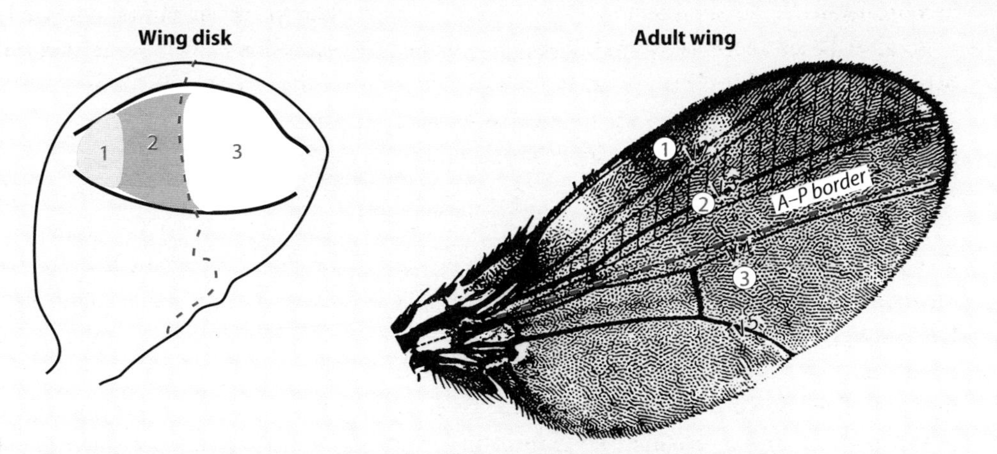

Figure C9.6 The wing disk and the adult wing blade in *Drosophila*. The wing disk, on the left, is a pocket of cells in the larval stages that will form the adult wing. A larva has a pair of wing disks. The dotted red line is the anterior–posterior (A–P) compartment boundary, detected by mosaic analysis and the phenotypes of some mutants. The shaded areas numbered 1, 2, and 3, will become the wing blade. Both the disk and the wing have many structures, such as bristles and hairs on the cuticle, that are not shown. During wing formation, this region of the disk expands outward and then flattens in the dorsal–ventral direction to make a blade. The regions numbered 1, 2, and 3 are shown in the corresponding regions of the disk and wing. (Redrawn from Phillips *et al.* 1990.)

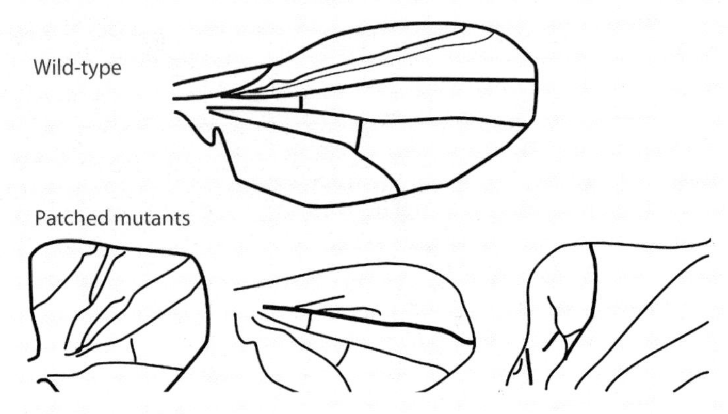

Figure C9.7 Effects of *ptc* mutations on wing development. The wings in *ptc* mutants and mutant combinations have abnormal shapes and wing vein patterns. (Redrawn from Phillips *et al.* 1990.)

part. The boundary does not correspond to obvious morphological structures in either the disk or the adult wing, although it runs near the fourth wing vein. Its presence was determined from mutant phenotypes and from mosaic analysis with genes that pattern the structures in the wing. For example, mitotic clones induced by somatic crossing-over at the larval stage do not cross this boundary; in fact, they often form a straight line along this imaginary line rather than the irregular edges of most mitotic clones. In addition, mutations that affect wing development, including *ptc* mutations, typically affect only one of the compartments and have no effect on the other compartment.

The viable *ptc* mutations exhibit a range of defects in wing development, mostly notably in the overall shape of the wing and alterations in the pattern of wing veins, as shown in Case Study Figure 9.7. Analysis by *in situ* hybridization and GFP reporter genes (using the methods described in Chapter 8) shows that the *ptc* gene is expressed strongly along the anterior–posterior boundary in the adult wing, with a lower level of expression throughout the anterior compartment. A mosaic analysis of *ptc* further confirms that the activity of the gene differs depending on the region of the wing being considered.

Mosaic analysis of the *ptc* gene

In order to examine the cellular effect of *ptc*, somatic crossing-over was induced by irradiating third-instar larvae by the method described in the chapter. The genotype of the heterozygote used to produce the mosaic flies is shown in Figure 9.19. From that figure, note that the presence of *shavenoid* cells indicates that somatic crossing-over has occurred and that a twin spot of genotypically *ptc* mutant cells is expected to be found nearby. The presence of *pawn* and *straw* cells indicates cells that are genotypically mutant for *ptc*. The *ptc* mutations in the mosaic analysis were among the lethal alleles that affect segmentation in the embryo, and are probably null alleles of the gene. The use of a null allele is important so that the gene is no longer making any functional product after the mosaic clone is induced. (Some genes continue to have residual functional activity even after the clone has been induced. This phenomenon, known as **perdurance**, has been attributed to the stability of a protein product that lingers even though no new gene product is being made.)

These mosaic clones were particularly well marked, using the cell autonomous markers *straw* bristles (*stw*), *pawn* bristles (*pwn*), and *shavenoid* bristles (*sha*). By having cell autonomous markers on each of the homologous chromosomes, twin spots could be induced and monitored. Cells that are *stw* and *pwn* in phenotype are examined for normal wing vein

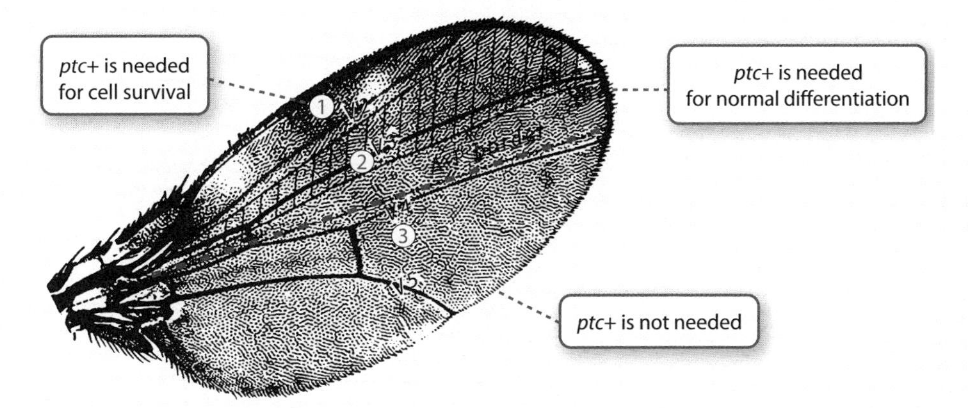

Figure C9.8 Effects of *ptc* on different regions in the wing. Three different results were observed for *ptc* somatic clones, depending on the region of the wing in which the clone arose. In region 1, along the anterior margin of the wing, *ptc*+ is needed for cell survival. In region 2, *ptc*+ is needed for normal wing vein pattern and differentiation but is not needed for cell survival. In region 3, the posterior compartment, *ptc*+ is not needed at all. (Redrawn from Phillips *et al.* 1990.)

pattern as well as other wing structures. These cells are genotypically homozygous for the *ptc* mutation since *ptc* is found on the same homolog. The interpretation of these mosaics is as follows: if the *stw* and *pwn* cells have a normal wing vein pattern, the normal function of the *ptc* gene is not needed in those cells.

The results of the mosaic analysis depended on which region of the wing was being examined. Three different results were observed, as summarized in Case Study Figure 9.8.

- A total of 36 clones was found in the posterior compartment of the wing, all of which had normal wing venation. Since this region of the wing had normal development despite the absence of *ptc* gene activity in these cells, the activity of the *ptc* gene must be dispensable for this region. This is region 3 in the diagram.
- Sixty different clones were found along the anterior–posterior border (region 2 in the diagram). More than two-thirds of these clones showed defects in wing venation in this region. Thus, in this region of the wing, *ptc* gene activity is needed for normal wing venation.
- Along the most anterior region of the wing in region 1, very few clones marked with *stw* and *pwn* were found. The apparent absence of *ptc* mutant clones demonstrates the importance of using *shavenoid* as a marker to find twin spots. That is, *sha* clones *were* found in this region, indicating that somatic crossing-over had occurred and that cells mutant for *ptc* had been produced. However, these *ptc* mutant cells apparently do not proliferate into clones of cells, suggesting that normal *ptc* activity is needed for cell survival in this region of the wing.

The effects of the *patched* gene in *Drosophila* (and in other organisms as well) are complex, and the expression of the gene is dynamic. Mosaic analysis reflects some of the complexity in gene function, and demonstrates the need for careful controls and analysis. Molecular analysis of the gene (i.e. analysis of the structure of the Ptc protein and its expression) has demonstrated a role of the gene in cell–cell interactions and signal transduction. Although the specific function of the *patched* gene could not be inferred from mosaic analysis, the results indicate that the gene acts within cells, and suggest its role in cell signaling.

Further reading

A reference that covers both classical and more recent methods of performing mosaic analysis in *Drosophila* is:

Blair, S.S. (2003). Genetic mosaic techniques for studying *Drosophila* development. *Development*, **130**, 5065–72.

The mosaic analysis of *patched* was done by:

Phillips, R.G., Roberts, I.J., Ingham, P.W., and Whittle, J.R. (1990). The *Drosophila* segment polarity gene *patched* is involved in a position-signalling mechanism in imaginal discs. *Development*, **110**, 105–14.

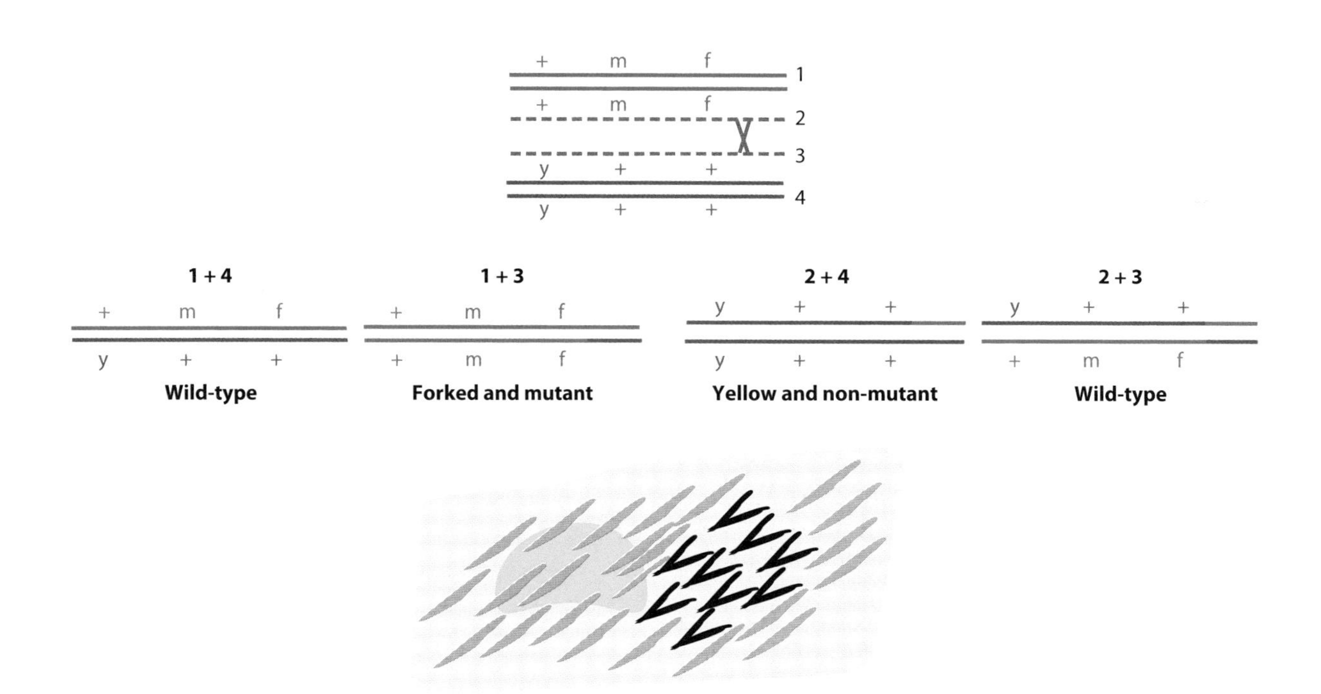

Figure 9.20 Twin spots from somatic crossing-over. Somatic crossing-over is induced in a fly that is heterozygous for the cell autonomous markers forked bristles (*f*) and yellow cuticle (*y*). The mutation of interest *m* is on the same homolog as forked and between the two markers. The somatic crossover occurs to the left of forked, as shown. Upon cell division, the four chromatids (numbered) segregate to the daughter cells as shown. The somatic clone of yellow cells has a nearby twin-spot somatic clone that is forked, as shown in the diagram at the bottom.

product of the yellow cells. The presence of yellow cells with no nearby forked cells is evidence that the gene of interest is needed for cell survival; the forked cells form but do not divide or grow.

Somatic crossing-over can also be used to estimate the time of gene activity

We have just described mosaic analysis as a means of determining the **location** of gene activity. Although this is the principal use of this technique, mosaics generated by somatic crossing-over can also be used to estimate *when* a gene is active. In methods that rely on chromosome loss, such as duplications in *C. elegans* and the X chromosome in gynandromorphs in *D. melanogaster*, the investigator cannot control the time at which the mosaic is produced; the chromosome or chromosome fragment is lost randomly, typically very early in

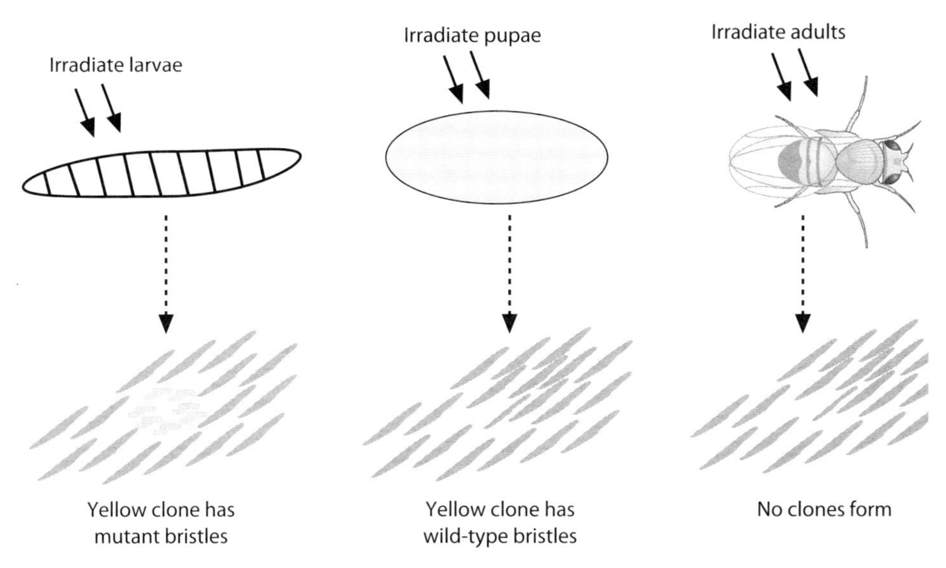

Figure 9.21 Using somatic crossing-over to determine time of gene activity. The gene of interest affects bristle formation in the adult fly. The yellow gene is used as the cell marker in this hypothetical experiment. If somatic crossing-over is induced during larval stages, as shown on the left, the yellow clones have mutated bristles. This indicates that the genes specifying yellow cuticle and mutant bristles act after this time. If somatic crossing-over is induced during the pupa stage, as shown in the middle, the yellow clones have normal wild-type bristles. This indicates that the gene for yellow cuticle acts after this time but that bristle formation has already occurred. Finally, if adult flies are irradiated, neither yellow clones nor mutant bristle clones are observed because both genes have already acted.

development. However, by using somatic crossing-over to produce mosaics, the investigator can exercise some control over when the mosaic is produced. That is, by irradiating the fly larvae or pupae at different times, it is possible to produce mosaics at different stages of development.

Like temperature-shift experiments, the time of gene activity can also be estimated from these experiments, as shown in Figure 9.21. Suppose that a gene acts during the last larval stages to affect bristle formation. Mosaics induced before this time remove the wild-type gene activity and form mutant bristles. Mosaics induced after the time of gene activity (e.g. by irradiating pupae or adults) do not have mutant bristles because the time of gene activity has already occurred.

One difference between using temperature-shift experiments and somatic crossing-over to estimate the time of gene activity is that temperature-shift experiments may have residual gene activity. When the temperature-sensitive mutant is shifted (e.g. shifted up), the protein product may not be inactivated immediately and the gene activity may persist for an undetermined period of time. The gene may continue to be transcribed and some functional protein may still be made at the restrictive temperature. Somatic crossing-over removes the wild-type copy of the gene so there will be no continuing gene activity once the clones have been induced. Of course, pre-existing functional proteins are not removed by the irradiation, but protein turnover will eventually remove these and no new functional protein will be synthesized.

Mosaic analysis provides other information about gene activity and normal development

Mosaic analysis, as done by any of these procedures, can also be used to estimate how many cells contribute to an adult structure as well as to infer the developmental relationship between cells and tissues. Once the mosaic cells are produced, they will continue to divide and display the phenotype. Thus, mosaicism that occurs early will result in larger patches of cells than mosaicism that occurs later. By estimating the fraction of a structure that is found in marked patches, it is possible to infer how many cells were present at the time mosaicism was induced. Consider a simple example of mosaicism that is introduced when only four cells are present. Based on the frequency at which mosaics are produced by somatic crossing-over, it is reasonable to assume that only one of these cells (at most) will be mosaic. The resulting patches will comprise about a quarter of the wing, assuming that all cells produce a similar number of mitotic descendants.

Mosaicism can also be used to determine when cells are sent on different developmental paths. If patches of cells cross morphological boundaries and affect more than one structure in the adult, the boundary between the structures has not been established at the time of mosaicism. We can illustrate this with an example from the *Drosophila* wing (Figure 9.22). For our purpose in this example, the wing can be considered in three main parts: the top (or ventral) side of the wing blade, the bottom (or dorsal) side of the wing blade, and the notum at the base of the wing blade. Mosaics induced early might have a patch of cells that spreads over the notum and both sides of the wing blade, as shown at the top. The inference is that, at the time mosaicism was induced, a cell has not yet become committed to any one of these structures but still contributes to descendants in all three. Somatic crossing-over induced slightly later might produce clones that affect either side of the wing blade but not the notum; at an even later stage, the clone might contribute to only one of these three regions. The inference is that mitotic descendants of a mosaic cell produced at this stage contribute to only the wing blade or only one side of the wing blade. Many mosaic patches need to be examined to make such inferences, but these have been helpful in understanding developing patterns. One of the most important was the recognition that the wing blade itself is divided into anterior and posterior compartments, as described in Case Study 9.2. The compartment boundary between anterior and posterior is not an observable anatomical feature. Its presence was inferred because mosaic patches of cells did not cross this boundary; in fact, they formed a straight edge along it, as indicated in the drawing at the bottom of Figure 9.22.

Somatic crossing-over has been a powerful method of producing mosaics in *Drosophila*. Mitotic crossing-over can be induced for any region of the genome, so it is not limited to the X chromosome or a region that has a free duplication. Furthermore, with a multitude of bristles, hairs, and pigmentation patterns to examine, many different cell autonomous markers are available for nearly every region of *Drosophila* chromosomes. Mosaic analysis by somatic crossing-over is one of the primary analytical tools for *Drosophila* geneticists.

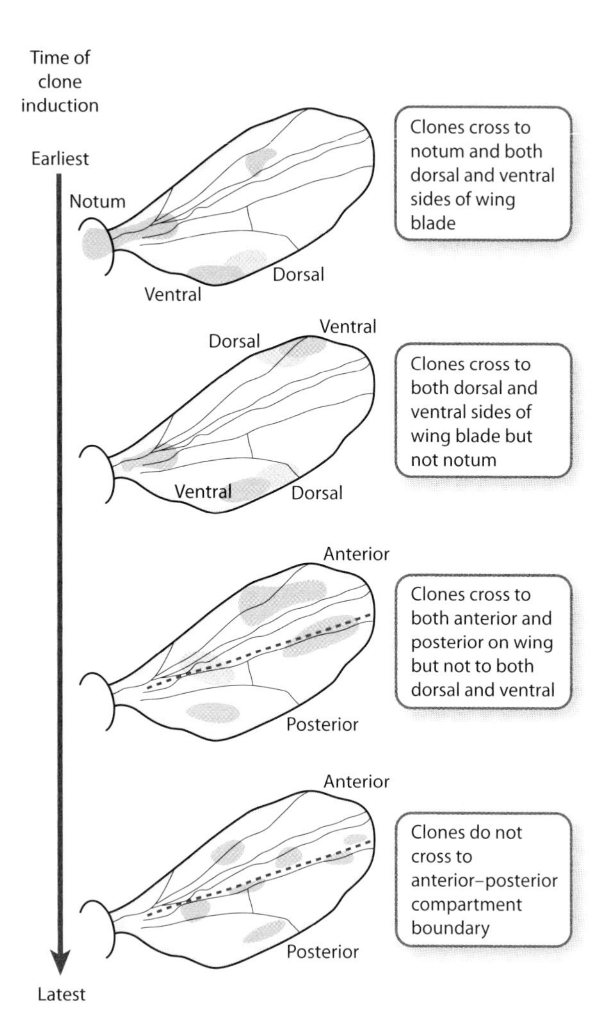

Figure 9.22 Using somatic crossing-over to show developmental restrictions. Somatic crossing-over can be done at different times and the cell clones will affect different structures. When induced during the early larval stages, mitotic cell clones in the wing can cross from the notum to the wing blade and to both ventral and dorsal sides. Clones on the dorsal side are indicated by the lighter shade of blue. With later induction, the clones will not cross to the notum and the wing blade, and will not affect both dorsal and ventral sides. Clones induced even later will affect only the anterior or the posterior part of the wing blade, with a compartment boundary separating them, indicated by the dotted red line.

Mosaicism can also be induced in transgenic organisms

We began this section by discussing transgenic mice, which are chimeric for cells of two different genotypes. To conclude it, we refer back to another technique used in transgenic mice and other transgenic organisms. In Chapter 6, the use of the Cre-*lox* system of targeted recombination was described. (The Flp–*FRT* system of targeted recombination, also introduced in Chapter 6, can be employed in a similar way.) It was noted that the Cre recombinase can be placed under the control of a stage-specific or tissue-specific regulatory region, so that Cre is made only in particular cells or at particular times. The gene of interest is 'floxed' or flanked by the *loxP* recombination sites, as shown in Figure 9.23. When Cre expression occurs, the floxed gene is deleted or altered by site-specific recombination. Thus, a mosaic organism is produced in which some cells have the wild-type (though floxed) gene and other cells have a deleted or altered gene.

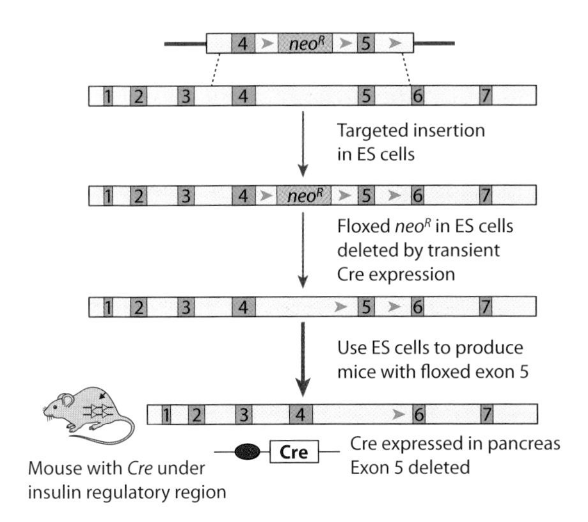

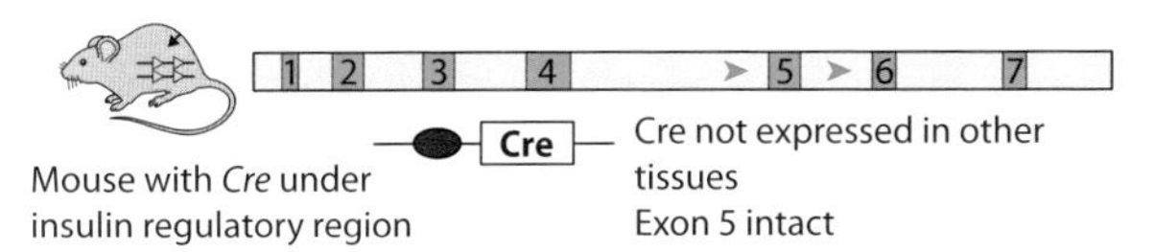

Figure 9.23 Mosaic analysis in the mouse using tissue-specific Cre expression. This is the same figure as Figure 6.14 illustrating genetic analysis in the pancreas. Exon 5 of the gene of interest is shown flanked by the *loxP* sites, or floxed. A transgenic mouse is made in which Cre recombinase is placed under the control of the insulin regulatory region. As a result, Cre is only expressed in the pancreas and exon 5 is only deleted in the pancreas. The mouse is a mosaic with a mutant gene in the pancreas and the normal gene in other tissues.

The organism can be analyzed by the same logic that has been used for other types of mosaic analysis. If the floxed gene of interest is not required in the cells where Cre expression is induced, the cells will develop normally; just a mosaic patch reveals which cells do and do not require the activity of the gene. It is also possible to determine if the floxed gene works cell autonomously, i.e. within the cell where it is produced, or non-autonomously, diffusing or signaling between cells. In other words, although the techniques to make mosaics have changed, the underlying analysis remains the same. The techniques have simply made mosaic analysis available in other model organisms.

SUMMARY

Mutant analysis of gene activity

Mutant alleles themselves are helpful indicators of when and where a gene is active. The phenocritical phase, the time or tissues when or where the mutant effects are first observed, often demonstrates which cells and stages are affected. Conditional alleles, particularly temperature-sensitive alleles, have been used in many different organisms to analyze the time at which a gene acts. The experimental technique with temperature-sensitive alleles involves shifting mutants between different growth temperatures and thus could not be simpler. Mosaic analysis, carried out by producing organisms with cells of different genotypes, can be used to determine the location at which a gene is active. A variety of methods has been used to produce mosaic organisms, but the analysis of the mosaic phenotype is the same no matter which procedure was used. Most of these powerful methods have been available for many years, and genetic analysis has used all of them successfully.

In the modern genomic age, gene activity is usually thought of in terms of molecular analysis of transcription and translation. There is no doubt that these are the most precise measures of gene expression, so every investigator will probably want to use them for any gene of interest. However, the powerful genetic methods of conditional mutations and mosaic analysis should not be ignored. These provide a functional assay of gene activity that extends our understanding based on gene expression assays.

Chapter Capsule

Mutant analysis of gene activity

Genes are active in different cells at many different times. By using mutations and some well-established methods, an investigator can infer when and where the function of a gene is needed. This can be done without knowing anything about how the activity of the gene is regulated, although knowledge of the expression pattern is critical to complement understanding.

Temperature-sensitive mutations are useful for understanding *when* a gene is active. By shifting the temperature-sensitive mutant between the permissive and restrictive temperatures at different times, a temperature-sensitive period can be determined. The temperature-sensitive period is inferred to be the time when gene activity is needed.

Mosaic analysis of mutant alleles is useful for understanding *where* a gene is active. Mosaic analysis employs different methods of removing the wild-type allele from particular cells in a heterozygote so that homozygous mutant cells are produced. These cells can be analyzed to determine if:

- the gene functions cell autonomously or non-autonomously
- the activity of the gene is essential in certain cells
- some cells do not require the activity of the gene.

Unit 4
Gene interactions

Genes carry out their functions by interacting with other genes and their products. A biological process can be understood by the pathways and networks of interacting gene products.

CHAPTER MAP

- ▶ **Unit 1 Genes and genomes**
- ▶ **Unit 2 Genes and mutants**
- ▶ **Unit 3 Gene activity**
- ▼ **Unit 4 Gene interactions**

10 Using one gene to find more genes

In this chapter, both mutants and cloned genes are used to identify additional genes involved in the same biological process.

11 Epistasis and genetic pathways

12 Pathways, networks, and systems

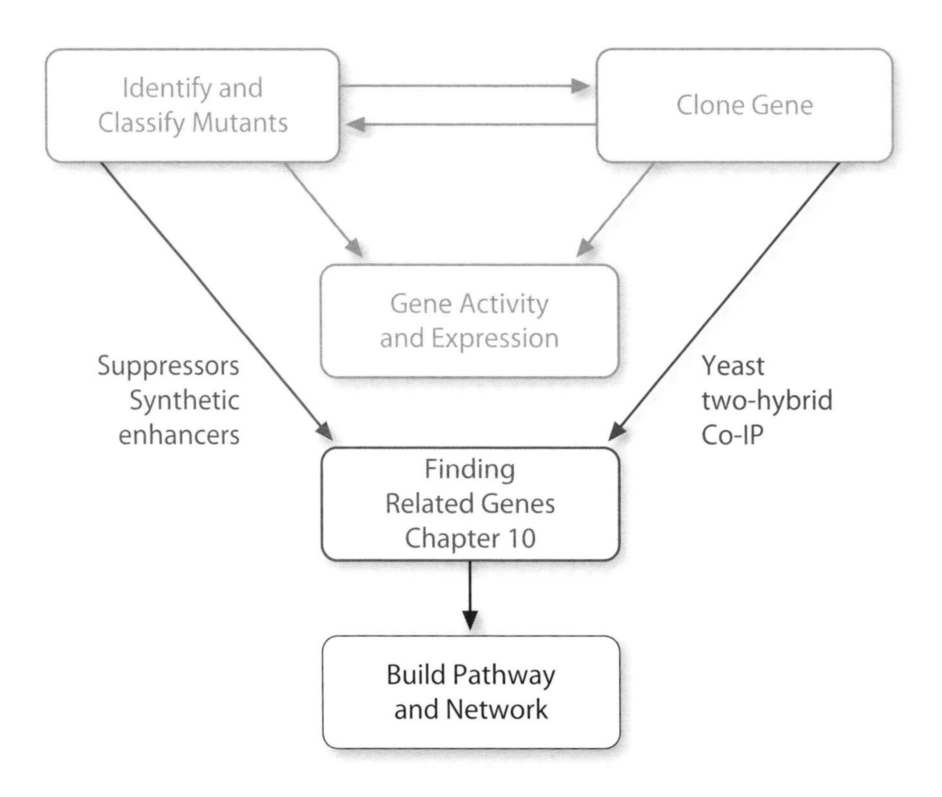

10

CHAPTER 10

Using one gene to find more genes

TOPIC SUMMARY

If the investigator begins with a mutation in a particular gene, genes involved in the same cellular or physiological process can be identified by the following:

- **Suppression**—mutations in a second gene that overcome or compensate for the defect in the original mutant.
- **High-copy suppression**—over-expression of the wild-type function of a second gene that overcomes the defect in the original mutant.
- **Synthetic enhancement**—mutations in a second gene that exacerbate the defect of the original mutant.
- **Non-allelic non-complementation**—recessive mutations in a second gene that fail to complement a recessive mutation in the original gene.

Suppressor mutations can affect either the function of the gene or the lesion of the specific mutation allele. Suppressors that affect the specific allele but not the function of the gene are informational suppressors. Suppressors that affect the function of the gene without regard to the type of mutant allele can bypass the original function or be epistatic to the original mutation. Suppressors that affect both the function of the gene and the specific mutant allele are interactional suppressors.

If the investigator begins with a **cloned wild-type gene**, proteins that interact with the original gene product can be identified by:

- yeast two-hybrid selection
- co-immunoprecipitation of protein complexes and other biochemical methods.

Both experimental methods are useful.

- Genetic methods find functionally related genes on the basis of the effects of additional mutations on the phenotype. The methods yield a greater number of potentially interacting genes than the molecular methods but do not necessarily indicate direct physical interactions between gene products. Some effects on phenotype will be due to mutations in upstream or downstream components of the pathway being investigated or in components of a related pathway. The genetic approach has the virtue of simplicity as the the initial experiments are easy to do.
- The molecular methods use expression of the wild-type cloned gene to find other proteins that physically interact with its gene product. These methods yield fewer potential candidates than the genetic methods, but indicate direct and immediate interactions.

IN BRIEF

One gene can be used as a tool to find other genes involved in the same biological process.

INTRODUCTION

Finding a gene with an interesting phenotype is an early step in the genetic analysis of the biological process in which it participates. In this chapter, we see how we can use any gene as a tool to find other genes that affect the same biological process or pathway. The methods for doing this use two contrasting but complementary approaches (Figure 10.1). The conventional genetic approach starts from the mutant phenotype of the gene, while the molecular genetic approach starts from the cloned gene and the protein it produces. The genetic and molecular approaches are both powerful, and they often identify the same genes. However, because they use different starting materials, the results from a combined approach are usually much more complete than with either alone.

We will first describe the methods that rely on the mutant phenotype: suppression, synthetic enhancement, and non-allelic non-complementation. We then briefly discuss two applications that begin with the cloned gene: the yeast two-hybrid screen and co-immunoprecipitation. By the end of the chapter, we will have described techniques that, when combined, have the potential to identify all of the genes that are involved in a biological process.

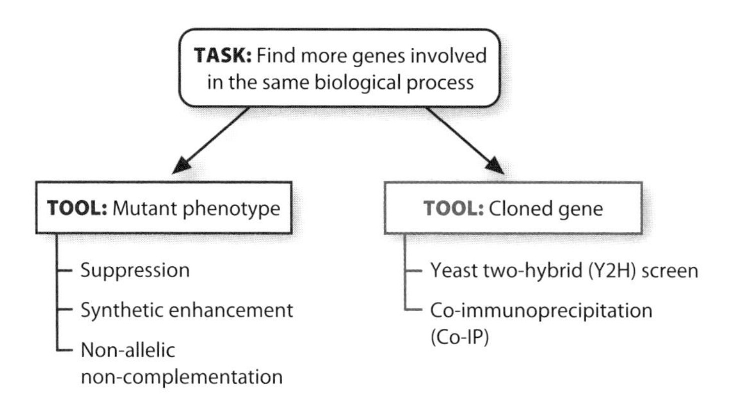

Figure 10.1 The task for Chapter 10 and the tools we will develop. The task is to use one gene as a starting point to find other genes involved in the same biological process. The tools can begin with either the mutant phenotype of the gene or the cloned gene. Those that begin with the mutant phenotype include suppression, synthetic enhancement, and non-allelic non-complementation. Those that begin with the cloned gene include yeast two-hybrid screens and co-immunoprecipitation.

10.1 Using one gene to find more genes involved in the same biological process: overview

So far in this book, we have discussed ways of finding a mutation that affects a given biological process or pathway, and how to clone and analyze that gene. Having found one such gene, we will often want to find more genes that affect the same process, especially when little is known about how the process operates.

One obvious way of finding functionally related genes is simply to repeat the mutagenesis procedure employed previously, searching for more mutants that have similar phenotypes. This is the approach used in large-scale mutation screens, as described in Chapter 3. We might also look for genes whose DNA sequence or expression pattern is similar to our original gene, as described in Chapter 5. These approaches have been very successful, as illustrated by the large-scale saturation screens for mutations involved in embryonic development in *Drosophila* described in Chapters 3 and 4. Many other examples could also be described.

In principle, it might seem as though it would be possible to find all of the genes involved in a process by mutation screening or genomic screens if one searches long enough. However, this approach will never provide all of the functionally related genes, even if continued to the exhaustion of the investigator. As discussed in Chapter 3, there are limitations inherent in mutation screens, powerful as they are. The genome-based approaches described in Chapter 6 have fewer limitations, but can only find all of the genes involved in a process if the genome has been completely sequenced and thoroughly annotated. Even then, recognizing and testing all of the candidate genes identified by a genomic screen is a formidable task.

This chapter describes the other methods that can be used to find functionally related genes that do not suffer from the limitations of saturation and genomic screens. The premise of these methods is that one identified mutant phenotype or the physical gene underlying it are used as the starting points, and it is not necessary to repeatedly begin anew with the wild-type organism or its genome. The methods described in this chapter are not limited by the requirement for similarities in mutant phenotypes, expression patterns, or sequence, and thus many more candidate genes can be found. The most significant feature of these methods is that they do not depend on a prior detailed understanding of the process being investigated. Functionally related genes are found because they interact with the first gene. The nature of that interaction then enables us to understand the relationship of the individual genes to the biological process that all of them affect.

In practice, there is a close relationship between the tasks of finding functionally related genes and of ordering these genes into a pathway or network, and these procedures are usually done at the same time. However, to make the explanation clearer, we will describe the techniques for finding functionally related genes in this chapter and the techniques used to place the genes and their products into cellular or physiological pathways in Chapter 11 and 12. The close connection between these chapters will soon become apparent.

10.2 Using a mutant phenotype as a tool to find related genes

We begin this chapter by describing the genetic methods of suppression, synthetic enhancement, and non-allelic non-complementation. These approaches require nothing more to begin with than one well-characterized mutant phenotype. We will see how these methods have been used to investigate muscle formation and movement, vulva development, and neurotransmitter function and release in worms, and axonal guidance in *Drosophila*. Detailed descriptions of how these methods were applied to investigate spindle morphogenesis in budding yeast are given in Case Study 10.1.

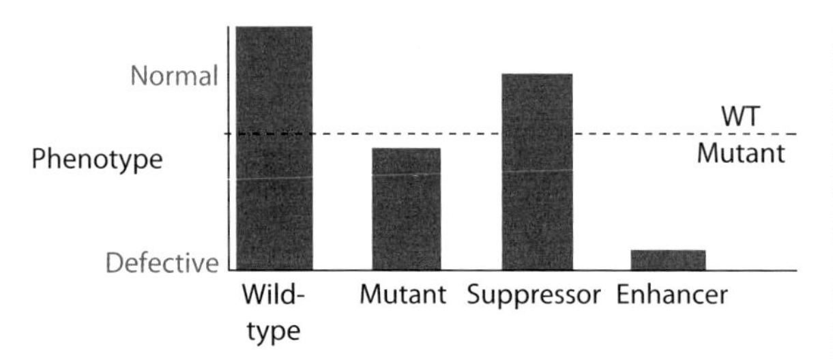

Figure 10.2 A graphical depiction of suppressors and enhancers. The height of the bars indicates the amount of gene activity as inferred from the phenotype of the different strains. A wild-type (WT) strain has a normal phenotype. A mutant strain may have some activity of the gene but the level of activity is below the inferred threshold needed for a wild-type phenotype. A suppressor mutation raises the level of gene activity in the mutant above this inferred threshold, and the strain has a more normal phenotype. An enhancer mutation lowers the level of gene activity in the mutant, and the strain has a more severely defective phenotype.

Suppressors and enhancers modify the phenotype of another mutation

Once a mutation has been found that affects the biological process of interest, a common and powerful strategy to find functionally related genes is to search for mutations that modify the original mutant phenotype. The approach can be summarized as follows, with the possible results shown diagrammatically in Figure 10.2. A homozygous mutant designated *a/a* is reated with a mutagen, and the offspring in a subsequent generation are examined. These will include a wide variety of double mutants, all of which are still homozygous for the original *a/a* mutation. Among these are double mutants that carry a second mutation that modifies the phenotype of the *a/a* mutant strain. Because they affect the same mutant phenotype, these modifying mutations identify candidate genes that affect the same biological process as the original mutant gene. Different modifying mutations can affect the mutant phenotype in different ways, so the double mutants will also often have different phenotypes from each other. We are using homozygous diploids in this example, but the analysis in organisms usually studied in the haploid state, such as yeast, is conceptually similar.

The modifying mutations are considered to fall into two major categories (Figure 10.2). The first category comprises those that make the mutant phenotype less severe and more like wild-type. These mutations are called **suppressors**. This can be illustrated with an example. Suppose that the original mutation (*a/a*) renders a yeast cell unable to divide at low temperatures. A suppressor mutation (*b*) in a second gene overcomes that defect and allows the cell to grow normally at low temperatures. The ability to restore the wild-type function is the defining characteristic of a suppressor mutation. It should be noted that the suppressor mutation makes the double-mutant phenotype more similar to wild-type although it still may not be completely like wild-type in all characteristics. In the notation of genotypes, the *a/a* strain has a mutant phenotype, whereas the *a/a;b/b* double-mutant strain has a more nearly normal phenotype. When this situation is observed, the *b* mutation is said to be a suppressor of the *a* mutation.

The second category of modifying mutations comprises those that make the mutant phenotype more severe. These mutations are generically called **enhancers**. It is an unfortunate confusion that the name 'enhancer' is used both for these modifying mutations and for regulatory sequences that specifically stimulate the expression of a gene. To avoid confusion, we will refer to these modifying mutations as **synthetic enhancers**. Again, consider an example in which the original mutation renders the yeast cell unable to divide at low temperatures but capable of dividing at high temperatures. In other words, as defined in Chapters 4 and 9, the original mutation (*a/a*) is cold sensitive. A synthetic enhancer mutation (*x*) in the original cold-sensitive yeast cell could make it unable to divide at any temperature. This more severe mutant phenotype is the defining characteristic of a synthetic enhancer. In the notation of genotypes, the *a/a* strain has a mutant phenotype, but the *a/a;x/x* double-mutant strain has a more severely mutant phenotype. When this situation is observed, the *x* mutation is said to be a synthetic enhancer of the *a* mutation.

DEFINITIONS

suppressor: A mutation that abolishes or greatly reduces the phenotypic defect caused by another mutation.

synthetic enhancer: A mutation that increases the severity of the phenotypic defect caused by another mutation.

Suppressors and synthetic enhancers are two of the most useful tools in our genetics toolbox. They have been used for decades in genetic analysis to find genes with related functions and to determine the nature of the relationships between different genes. Until recently, suppressors were more widely used than synthetic enhancers. This is in part because suppressor mutations have proved to be very successful in identifying functionally related genes, whereas enhancer mutations were thought to be less likely to identify a specific functional relationship. To use an analogy familiar to homeowners with plumbing problems, many different actions can make a problem worse (enhancers) but relatively few are able to fix it (suppressors). Fixing the plumbing problem (analogous to a suppressor mutation) requires that the nature of the problem is understood and addressed. Turning a leaky pipe into a flooded basement (analogous to an enhancer) does not require such specific interaction.

In the past few years as genome-wide screens for mutations have become more widely used (see Chapter 6), synthetic enhancers have been found to be extremely useful. It is true that, like the leaky pipe and the flooded basement, many mutations can result in a more severe mutant phenotype when combined with another mutation, and can do so even if the mutations have unrelated functions. However, the key to the successful exploitation of synthetic enhancers lies in identifying those modifying mutations that are specific to the function of the original mutation rather than simply making some general physiological problem worse. Our greater knowledge of genomes and greater ability to manipulate genes allows us to determine these specific interactions more readily, so the real power of synthetic enhancers has emerged. Rather than replacing these two classical genetic tools, genomics has served to illustrate how useful and powerful suppressors and synthetic enhancers can be.

Suppressor and enhancer gene nomenclature can be extremely confusing

One of the difficulties in reading the literature on suppressors and synthetic enhancers arises from the ways these genes are named. The nomenclature varies between different model organisms and can be confusing even to experienced geneticists. The important thing to keep in mind is that, in general, genes are named for their **mutant** phenotype rather than their wild-type function. For example, the *white* gene in *Drosophila* is named because the mutant phenotype is white eyes. The wild-type *w+* gene produces red eyes. A similar general rule of nomenclature applies to naming suppressors and enhancers. The mutant phenotype of a suppressor gene is, by definition, suppression of something else. Therefore, given a suppressor gene name such as *sup-5+* in *C. elegans* or *SUP45* in budding yeast, it will be the mutant version, *sup-5* or *sup45*, that acts as the suppressor. The function of the wild-type *sup-5+* or *SUP45* gene may have nothing to do with suppression.

But do the names *sup-5* or *sup45* tell us anything about which mutation is being suppressed, or about the wild-type function of these genes? In the case of yeast and worms, the answer is usually no, although, as in most cases of genetic nomenclature, exceptions exist. The numbers 5 and 45 in the gene names have nothing to do with the mutation being suppressed or the wild-type function of the gene, at least for yeast and worms. Suppressor gene names in *Drosophila* follow different rules, and often do include the name of the mutation being suppressed. From the gene name *su(Hw)* in *Drosophila*, we can infer that a *mutation* in the gene *su(Hw)* acts to suppress the defect caused by a *mutation* in the gene *Hw* (*Hairy wing*). The wild-type function of the *su(Hw)* gene is not to suppress either the mutant or the wild-type version of *Hairy wing*. In fact, in this example, it turns out that the wild-type function *su(Hw)* has nothing to do with the wild-type function of *Hw*.

A similar convention is used for naming synthetic enhancers, and the genes are again named for their mutant phenotype, which is enhancing the mutant phenotype of another gene. With synthetic enhancers, the name of the mutant phenotype being enhanced is often included in the gene name. This is probably to show that the interaction is specific to that mutant phenotype or gene, and that the enhancer mutation is not simply making many mutations more severe by a non-specific interaction. For example, from the gene name *E(spl)*, we infer that a mutation in the gene *Enhancer of split* acts as a synthetic enhancer of the *split* mutation in *Drosophila*. The wild-type function of the *E(spl)* is not to enhance the wild-type function of *split*; it is the mutants that show the interaction, and the genes are named for the mutant phenotype.

Table 10.1 Genes discussed in this chapter

Organism	Wild-type gene	Homozygous mutant phenotype	What does this example illustrate?
C. elegans	*sup-5+*	Suppresses nonsense mutations in many genes	Informational suppressors
C. elegans	*unc-54+*	Mutant is paralyzed	Beginning mutation for many of the suppressor screens
C. elegans	*myo-3+*	Suppresses *unc-54* alleles	Bypass suppressors
C. elegans	*ace-1+* *ace-2+* *ace-3+*	Reduced levels of acetycholine esterase A mutant phenotype is seen only when more than one gene is mutant	Synthetic enhancement by paralogous genes
C. elegans	*lin-8+* *lin-9+*	No mutant phenotype alone No mutant phenotype alone Multiple vulvae in the double-mutant homozygote	Synthetic enhancement by functional redundancy
Drosophila	*slit+*	Axonal guidance defective	Non-allelic non-complementation with *robo* mutations
Drosophila	*robo+*	Axonal guidance defective	Non-allelic non-complementation with *slit* mutations
C. elegans	*unc-13+*	Paralyzed	Non-allelic non-complementation with *unc-64* mutations
C. elegans	*unc-64+*	Paralyzed	Non-allelic non-complementation with *unc-13* mutations

This type of nomenclature is not very informative and can be difficult to follow. Many suppressor or synthetic enhancer genes have mutant phenotypes of their own, separate from their modifying phenotype, and this other phenotype is often the name given to the genes. The modifying effect on other mutants may not be observed until later in the analysis. In this chapter, we have mainly used as examples genes that do not use the SUP or *su* type of gene name. The different genes that will be discussed in this chapter are listed in Table 10.1.

10.3 Suppressor mutations: more similar to wild-type

As with many other types of genetic analysis, the use of suppressors began with *Drosophila* when, in 1920, A.H. Sturtevant recognized that suppressor analysis could find genes with related functions. Suppressor analysis was subsequently more fully developed as a powerful technique in bacterial and phage genetics. The reason for this is that large populations of double mutants need to be examined because suppressors are often rare. Thus, suppressors are more widely used in yeast than in worms or flies, and more widely used in worms and flies than in mice, reflecting the ease with which thousands or tens of thousands of double-mutant organisms can be examined. In yeast, worms, and flies, suppressor analysis has proven to be an extremely powerful method of finding genes that modify or interact with the original gene. As we shall see in our examples, suppressor mutations often find genes that might not have been so easy to find by simply doing a more extensive search for more mutations of the same type as the original one. Recall that the defining characteristic of a suppressor mutation is to modify the phenotype of another mutation. Therefore, even if the suppressor mutation has a different phenotype from the original mutation or has no other mutant phenotype on its own, it can be found by virtue of its interaction with the original gene. This reflects one of the general principles of mutant analysis stressed throughout this book. With a suppressor, we do not need to know what phenotype to look for among the double mutants—if it suppresses the original mutation, the suppressor probably defines an interacting gene that is worth studying. The suppression phenotype also reflects some aspects of the underlying biology that may not be apparent to the investigator by another means.

Suppressors are particular easy to find if the original mutation has a pronounced and severe phenotype so that correction of that defect is more obvious. Notable biological examples that have given up their secrets to suppressor analysis include many different cells or organisms that cannot grow or reproduce under certain growth conditions: worms that cannot move, flies that do not move towards light, yeast cells that cannot repair particular kinds of DNA damage, plants that cannot flower, and many more. In fact, when the original mutation has an easily recognized phenotype and it is possible to screen many mutagenized offspring, suppressor analysis has proven to be so useful that it is almost compulsory if one wants to perform a thorough investigation of the phenotype and the biological process.

Figure 10.3 shows an example of finding suppressors in yeast. Yeast cells with a particular mutation in the β-tubulin gene do not divide at low temperature; they do not complete nuclear division and no growth is seen at the restrictive temperature. These mutant yeast cells were treated with a mutagen and replated, and their progeny were screened for cells that could grow at the restrictive temperature. Several different suppressor mutations were found and are discussed in further detail in Case Study 10.1.

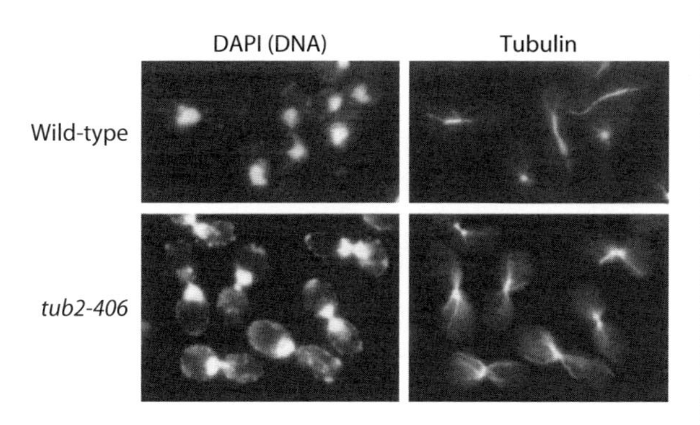

Figure 10.3 The phenotype of the yeast β-tubulin mutation *tub2-406*. Wild-type budding is shown in the top two panels, with the *tub2-406* mutant phenotype at the bottom. The DAPI stain in the left-hand panels indicates the effect on chromosome segregation, and the α-tubulin antibody staining in the right-hand panels shows the mitotic spindle. Notice from the DAPI-stained cells that the *tub2* mutant arrests with its nuclear DNA in the neck of the budding cell and does not complete nuclear division, as the wild-type does. In addition, microtubules in the *tub2* mutant do not assemble into normal-length spindles, as shown in the α-tubulin-stained cells. Because the *tub2-406* mutation prevents nuclear division at the restrictive temperature, suppressors of this phenotype could be found by looking for normal division.

Another example of the identification of suppressors is provided by *C. elegans*. In this case, recessive mutations in the gene *unc-54* result in paralysis. The paralyzed *unc-54/unc-54* worms were mutagenized and suppressors were identified by looking for moving worms on the growth plates. These will be double mutants of the genotype *unc-54/unc-54;sup/sup* in which an additional mutation, the suppressor (abbreviated *sup*), has reversed the effects of the *unc-54* mutation and allowed the worm to move. Although we have shown the suppressor as a homozygote, the suppressor mutation may be dominant or recessive just like any other mutation. The assumption of this screen is that the second mutation is identifying another gene that affects movement. The combination of the mutation in *unc-54* and this second mutation *somehow* restores movement to paralyzed *unc-54/unc-54* worms. This 'somehow' in the preceding sentence is the crux of suppressor analysis—once a suppressor is found, we still want to know how it acts to restore the function to the original mutation.

Although finding suppressors seems uncomplicated, particularly when the original mutation phenotype is easy to observe, the follow-up genetic characterization is usually more demanding. The main complication is that the suppressor's phenotype is initially defined only in relation to its effect on another mutant. Therefore, all strains used for the initial outcrossing, mapping, and maintenance of the suppressor mutation must also carry the original mutation. (Outcrossing and mapping are described in Chapter 3.) Thus, one of the first steps in working with a suppressor is to determine whether it has a mutant phenotype of its own, independent of its suppressor property. If it does, this separate mutant phenotype can be used instead of the suppression phenotype and simplifies the routine crosses of genetics. When a suppressor mutation has its own mutant phenotype, that phenotype is often used to name the gene so that the suppression may not be indicated in the nomenclature. In this chapter, we will encounter examples of both types of gene names, such as *C. elegans sup-5* on the one hand and *myo-3* on the other (see Table 10.1). Although the gene names do not reflect it, both of these genes were discovered because of the ability of mutations in them to suppress the mutant phenotypes of other genes.

Suppressor mutations can be either intragenic or extragenic

The principal motivation for most suppressor screens is to find additional genes that affect the same biological process as the original gene. Thus, we want to determine quickly whether the suppressor does in fact define a different gene or whether it is another mutation in the original gene, such as a reversion. This is usually done by determining whether the suppressor maps to a different genetic location. Mapping is often done concurrently with the crosses aimed at determining whether the suppressor has a mutant phenotype of its own. Text Box 10.1 provides a brief overview of the procedure for mapping suppressor mutations.

Suppressors that map into the same gene as the original mutation are called **intragenic suppressors**. They include true revertants that affect the mutated base or amino acid, and pseudo-revertants that affect a different base or amino acid in the same gene. In addition, many dominant hypermorphic or antimorphic alleles of a gene that over-express or alter the function of the gene can be suppressed by recessive null alleles of the same gene that eliminate the gene function. (Chapter 3 defines hypermorphic and antimorphic alleles more fully). Although this is interesting and important, particularly in the intensive study of a single gene, our goal is to find mutations in different functionally related genes. Since these intragenic events do not define a new gene, they are not of interest for our present purpose. We are interested in the genes known as **extragenic suppressors**.

Extragenic suppressors fall into three main functional classes

Once it has been determined that a mutation maps to a different location and acts as an extragenic suppressor, the first goal of a suppressor screen has been accomplished: we have found a new gene. However, the analysis of suppression is only beginning. We now want to know *how* the suppressor mutation restores the original function. The answer will help to determine whether the suppressor mutation does in fact define a gene with a related function to that of the original gene. Many decades of genetics can be summarized in the flowchart of suppressor analysis presented in Figure 10.4. In order to understand an extragenic suppressor's mechanism of action, two fundamental questions have to be answered.

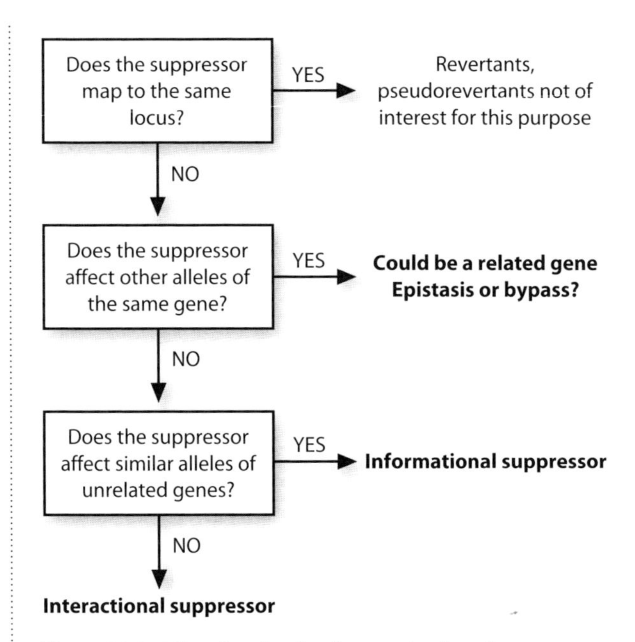

Figure 10.4 A flowchart for the characterization of suppressors. The arrows indicate logical directions and may not indicate the order in which the experiments are done. The suppressor is mapped, tested with other alleles of the same affected gene, and tested against similar alleles (e.g. null alleles, splicing defects, and so on) of genes with unrelated functions.

- Does the suppressor mutation also affect other mutant alleles of the original gene? In other words, is the suppressor allele specific?
- Does the suppressor mutation also affect other genes in addition to the original gene? In other words, is the suppressor gene specific?

In answering these questions, it is important to realize that the suppressor was identified because it modified the phenotype of a particular mutant allele. The specific mutant allele used in the original screen is actually what is being suppressed. This mutant allele could represent any one of many different molecular lesions that

BOX 10.1 General strategy for mapping a suppressor

Suppressor mutations can be notoriously difficult to map since the complete phenotype of the suppressor is only seen in the presence of the mutation being suppressed. The strategy for mapping a suppressor varies with the organism and the available genetic tools, but an overview is given here. Let us call the original mutation used to find the suppressor *a* so the mutant strain is *a/a*. The suppressor will be designated *sup*. The suppressed strain with a wild-type phenotype has the genotype *a/a;sup/sup*.

The suppressor will be mapped with respect to the marker *m*. The marker *m* could be a morphological or nutritional mutant or it could be a molecular polymorphism such as a SNP. Molecular polymorphisms have the potential advantage that linkage to many different sites can easily be monitored in the same recombinants, and most linkage mapping of the type described here is currently done using a collection of SNPs.

Initial mating

a/a;sup/sup X a/a;m/m

The suppressed strain is mated with *a/a* mutants that are also carrying the mapping marker *m*. The *a* mutation must be present in the mapping strain since the only known phenotype of the suppressor is its ability to suppress *a*.

F_1 generation

a/a;sup/+;m/+

The F_1 progeny will be homozygous for the a mutation since each parent was homozygous for *a*. It will be heterozygous for both the suppressor and the linkage marker. This cross will show whether the suppressor is dominant or recessive. If the F_1 progeny are identical to *a/a*, the suppressor is recessive. If the F_1 progeny are wild-type, the suppressor is dominant. We will assume that the suppressor is recessive; the progeny ratios in the F_2 generation will be different for a dominant suppressor and the overall process is somewhat easier.

F_2 generation

a/a;sup/sup;m/+ etc.

The F_1 offspring can be mated among themselves to produce an F_2 generation. Other crosses such as a backcross to the suppressed parent could also be used. The F_2 generation will have a variety of different phenotypes since both the suppressor and the marker mutation are segregating. A quarter of the progeny will have the genotype *a/a;sup/sup*, i.e. the same phenotype of the original suppressed parent, and these are the progeny to be examined. They can be recognized unambiguously because all other offspring will have the *a/a* mutant phenotype.

The suppressed *a/a* F_2 offspring are then examined for the presence or absence of the *m* marker phenotype. The results are shown in Text Box Figure 10.1. If the suppressor is unlinked to the marker being tested, a quarter of these offspring will be *m/m*, half will be *m/+*, and a quarter will be *+/+*; that is, the suppressor has segregated independently of the marker mutation. The independent assortment of *sup* and *m* shows that these genes are unlinked. On the other hand, if *sup* and *m* are linked, most of the *a/a;sup/sup* will not have the *m* marker.

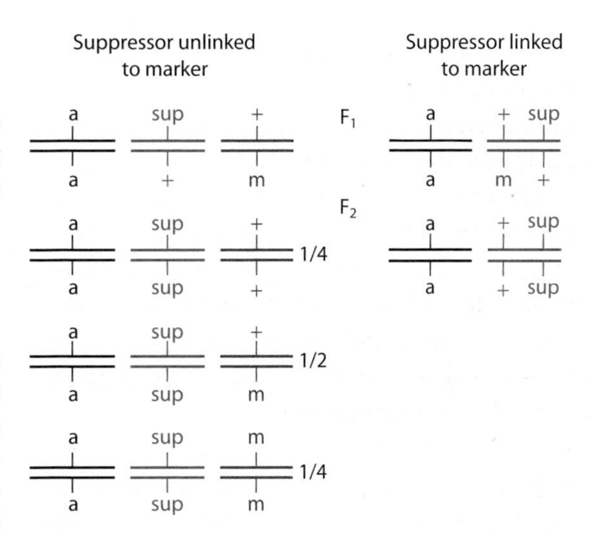

Figure B10.1 Mapping a suppressor with respect to the marker *m*. The F_1 offspring will have the genotype *a/a;sup/+; m/+*. The configuration of the chromosomes if the suppressor and the marker are unlinked is shown on the left; the configuration if the suppressor and the marker are linked is shown on the right. The F_2 offspring that are homozygous for the suppressor are shown. Note that, for an unlinked marker, the suppressor and the marker segregate independently. For a linked marker, most of the suppressed *a/a;sup/sup* progeny will not have the marker in question, although the exact frequency depends on the map distance between the marker and the suppressor.

could be found for that gene. Therefore, a suppressor mutation could either be modifying the effect of a particular mutant *allele* by correcting its specific molecular defect, or it could be modifying the general effect of the mutated *gene* by compensating for the overall loss of its function.

We will illustrate this with the hypothetical situation shown in Figure 10.5. (Although our example is

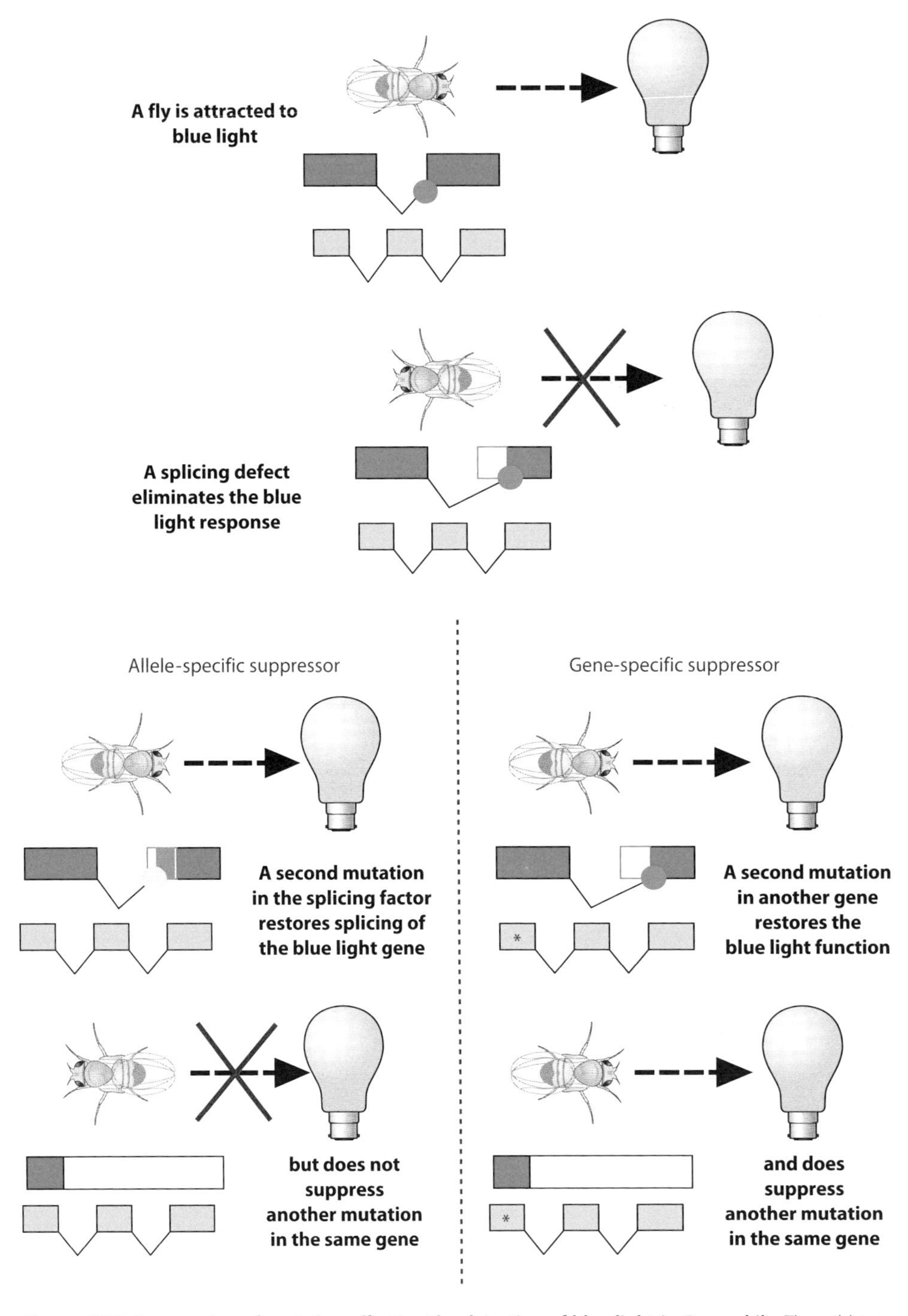

Figure 10.5 Suppression of mutation affecting the detection of blue light in *Drosophila*. The wild-type fly is attracted to blue light, as shown at the top. A mutant that alters the splicing of a blue light detection gene (as shown in the second figure) eliminates the response to blue light. By finding flies with a restored response, suppressor mutations are identified. An allele-specific suppressor, as shown on the left at the bottom, restores the splicing defect of the blue-light-sensing gene. This suppressor will not suppress another mutation in the gene. A gene-specific suppressor, as shown on the right at the bottom, affects another gene, indicated as a yellow gene. This type of suppressor will affect many types of alleles of the blue-light-sensing gene.

hypothetical, a screen similar to this one found genes involved in the RAS signal transduction pathway in *Drosophila*.) Let us return to the original example of mutant analysis that was introduced in Chapter 1. *Drosophila*, like other insects, will fly towards a light source. In order to understand this process, we find a mutant fly that, unlike wild-type fruit flies, does not move in the direction of blue light. The mutant can still move and it is still attracted to white light, for example, so this is not some general inability to fly or detect light. We infer from this phenotype that the mutant gene impairs the ability to detect specifically blue light and we want to use the mutant fly to find other genes involved in light detection. The mutant flies are themselves mutagenized and their offspring are exposed to blue light. A double mutant that can move towards blue light is found; the suppressor mutation maps to a different location so it is an extragenic suppressor. In other words, a new gene that may affect the ability to detect light has been identified. How do we characterize it further and understand its method of suppression?

The inability of our original mutant fly to move to blue light was due to one specific molecular lesion in one specific gene. Let us assume that the original mutation is due to a defect in the splicing of the mRNA from a blue light detector gene, i.e. the molecular lesion is a splicing defect. Our extragenic suppressor could then represent a mutation in a gene involved in the splicing process itself, thereby fixing the specific defect (splicing) rather than the function of the gene (blue light detection). What should we expect when we analyze this type of suppressor? Think again about the flowchart in Figure 10.4. What else is this suppressor expected to affect? If it is truly a defect in splicing itself, the suppressor may be able to modify and suppress similar splicing-defect mutations in other genes. On the other hand, it will have no effect on mutant alleles of the blue light gene that involve molecular lesions other than splicing, such as nonsense mutations, missense mutations, or deletions. Thus, this type of suppressor is **allele specific** for the defect in the blue light detector gene, but it is **gene non-specific** as it can also suppress splicing mutations in other genes. This type of suppressor is referred to as an **informational suppressor** (see Figure 10.4). The suppressor does not identify a gene that affects light detection at all, but instead finds a gene involved in a fundamental process in the cell.

However, this is only one type of suppressor. What other types of suppressor might be found? A suppressor mutation could correct the blue light defect by affecting another gene involved in light detection. For example, it might modify such a gene so that the modified gene product now reacts to blue light. Exactly how this suppressor works is unimportant for now; the important point is that it restores the biological function lost in the original mutant. Because it affects the light-sensing function, this suppressor could also modify the phenotype of many different mutant alleles of the blue-light-detecting gene, including missense alleles, nonsense alleles, and deletions, as well as the splicing defect in the original mutation. Thus, this suppressor is allele non-specific in its ability to modify the phenotype of many molecular lesions in this gene. However, it is unlikely to affect any mutations in genes that are not involved in light detection. Thus, it is gene specific to the light-detecting pathways. We will refer to these as gene-specific but allele-non-specific suppressors (see Figure 10.4). Although the terms allele specific and gene specific are widely used in suppressor analysis, they do not tell the whole story They could be defined in a more informative way: 'allele-specific' suppressors modify the molecular lesion that caused the missing function; 'gene specific' suppressors affect the function itself. Although the term 'gene-specific' is often used to distinguish these two types of suppressors, the term '**function specific**' might be a more accurate description of the effect. A gene-specific suppressor will often suppress mutations in several different genes if all of the genes have related functions. The key to distinguishing allele-specific effects from gene-specific or function-specific effects is to determine if many different alleles of the same gene are suppressed, and then to determine if alleles of functionally unrelated and functionally related genes are also suppressed.

Finally, a suppressor mutation can be both allele specific and gene specific. These are the most specific suppressors in that they suppress one mutant allele of only one gene and thus affect both the molecular lesion of the allele and the function of the gene. This type of suppressor is known as an **interactional suppressor** (see Figure 10.4).

The two tests involved in the suppressor analysis are to determine if the suppressor modifies the phenotype of other mutant alleles of the same gene and if it modifies the phenotype of mutant alleles of functionally unrelated genes. These two tests will determine the type of suppressor even when the molecular lesion of the original mutation is unknown. In general, if suppression is being used to identify additional genes with related functions, interactional suppressors and gene-specific suppressors are the most useful because they affect the function of the gene, whereas informational suppressors will be the least informative because they simply affect the defect in the mutation. We will consider each possibility in turn, beginning with the interactional suppressors.

Interactional suppressors are specific to both the gene and the allele

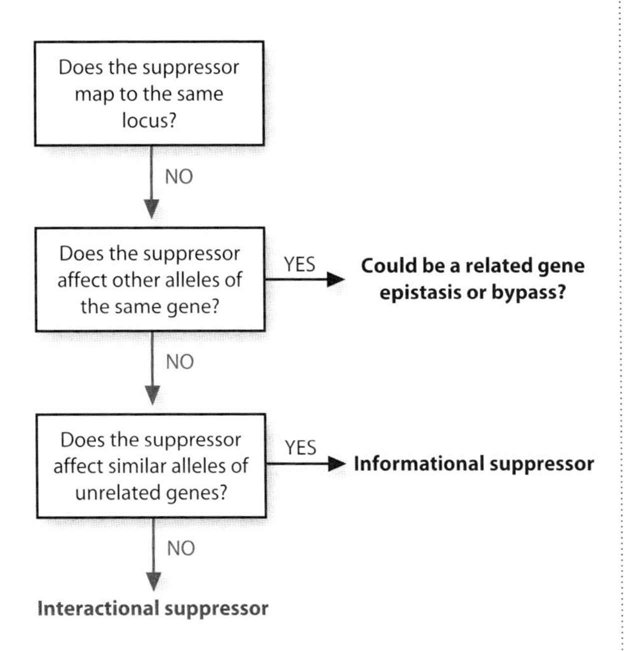

A few suppressors affect only the original mutation used and only the original gene tested. A suppressor that is specific to one or a very few alleles of only one gene indicates that suppression is affecting *both* the defect in the original mutant allele *and* the function of the original mutant gene. Although rare, these are among the most informative of all classes of suppressor. Nearly all instances of such highly specific suppression that have been studied indicate a direct physical interaction between the product of the original gene and the product of the suppressor; this is equally true whether the products of the two genes are RNA or protein. This is why these mutations are called **interactional suppressors.**

To appreciate how interactional suppressors work, and why they are both allele specific and gene specific, we will use an example of a ligand binding to its receptor. Although our example is hypothetical, similar examples are known in both bacteria and yeast. A schematic diagram of a ligand binding to its receptor is shown in Figure 10.6. In the first mutation shown, the mutant phenotype is the result of a change in the conformation of a specific amino acid in the ligand so that the mutant ligand can no longer bind to the receptor. A second mutation, this time in the receptor, restores the interaction by altering an amino acid at the binding site, thereby enabling the mutant ligand to bind. The second mutation in the receptor is a suppressor: the double mutant functions better than either single mutant by itself. The specificity between the two mutations is the key to recognizing interactional suppressors. Other mutations that alter the function of the ligand or change its conformation in other ways will not be suppressed since these altered ligands will still be unable to bind to the mutated receptor. Thus, this suppressor is allele specific. In addition, the suppressor will not suppress mutations in genes other than that for the ligand; thus, the suppression is gene specific.

Interactional suppressors are an important tool in the analysis of a biological process because they not only identify genes with a related function but also help to elucidate the type of relationship between the genes. Interactional suppressors provide information about direct physical interactions and the sites or residues that are involved in that interaction that other types of suppressor cannot provide. Thus, they are highly sought after, and suppressor screens can be modified to enrich for them. For this, it is essential to use as the original mutation a missense allele that is known or thought to be at the interaction site of the gene product, as only this type of allele is expected to be suppressed by an interactional suppressor. Case Study 10.1, Part 1, illustrates how interactional suppressors affecting a gene involved in yeast spindle morphogenesis were sought and discovered. It can be seen that the investigators began with a temperature-sensitive mutation that affected one, but not all, aspects of the

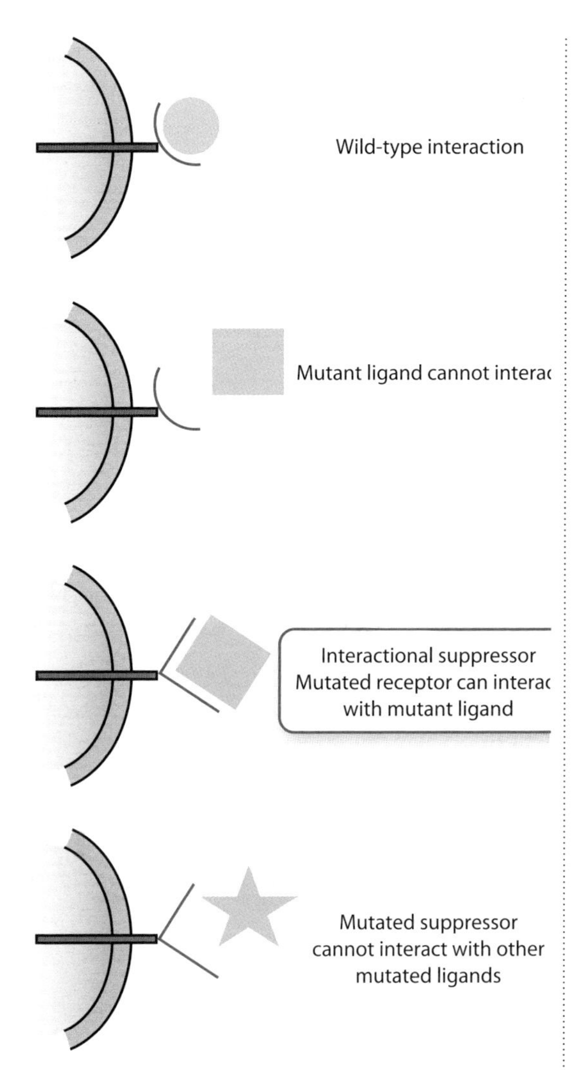

Figure 10.6 Interactional suppressors are specific to both the mutant allele and the gene. A ligand (green) binds to a receptor (red) on the surface of the cell. A missense mutation in the gene encoding the ligand results in an alteration of the shape of the protein, which reduces or eliminates the function of the gene because it can no longer interact with its receptor. A suppressor mutation in the gene encoding the receptor may be able to correct the defect and restore the interaction with the mutated ligand. The suppression is allele specific and will not suppress mutations in the ligand gene with other defects, as shown in the bottom panel.

mutant phenotype of the gene, thereby greatly increasing the likelihood of finding the desired interactional suppressors. Because they are specific to both the function of the gene and the defect in the mutant allele, interactional suppressors are often rare; in order to find and use them, it must be possible to screen large numbers of mutants easily.

Informational suppressors affect the molecular lesion in the specific allele but not the function of the gene

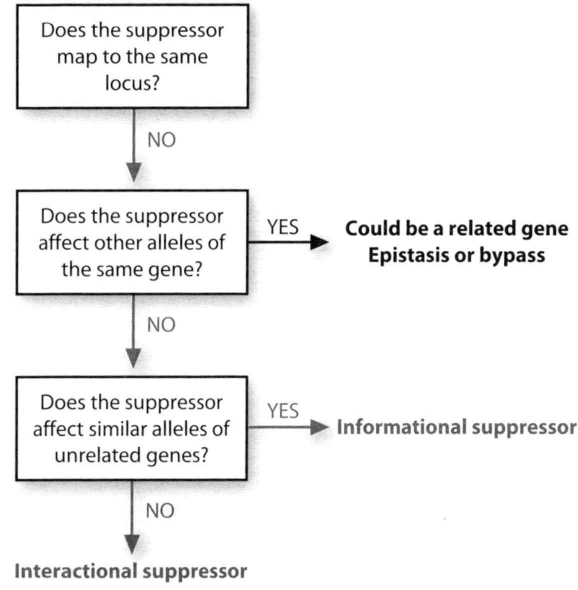

Some suppressors will affect only particular mutant alleles, but they may do so for different genes, even when the genes have unrelated functions. These are known as **informational suppressors.** As we see from Figure 10.4, informational suppressors are allele-specific suppressors that affect only the molecular defect of the allele but have nothing to do with the function of the gene itself. They are called informational suppressors because they often affect the general transcription and translational machinery, and thus the information flow from gene to protein. When we want to find genes with functions related to the original gene, informational suppressors are less helpful than interactional and gene-specific suppressors.

Informational suppressors were first identified in bacteria and were subsequently found in yeast, *Arabidopsis*, worms, and flies. An example was introduced earlier as part of our general description of finding suppressors. The *sup-5* gene in *C. elegans* was identified by its ability to suppress a mutant allele of the muscle-specific gene *unc-54*. Subsequent work showed that *sup-5* affected only a few particular null alleles of *unc-54*, suggesting that *sup-5* mutations are suppressing some specific molecular defect in the *unc-54* null allele rather than

Case Study 10.1

Part 1: Genetic analysis of spindle morphogenesis in budding yeast

Interactional suppressors identified both known and previously unsuspected interacting proteins

The use of suppressors and synthetic enhancers to find genes with related functions has a rich history in genetic analysis. In this case study, we focus on one biological process, the morphogenesis of the mitotic spindle in budding yeast (*S. cerevisiae*), starting with a mutation in a gene for β-tubulin. Several properties of this system make it a good example. First, nearly all of the tools described in this chapter were used in the experiments. Secondly, cytological experiments on spindle morphogenesis in yeast and other organisms provided an extensive background by which to interpret the genetic analysis. Thirdly, more detailed molecular and biochemical experiments confirmed and extended the conclusions from the genetic analysis. Thus, we have an independent method of determining whether the genetic analysis using suppressors and enhancers is working as expected. An overview of the yeast life cycle and some of the genetic and molecular methods that will be mentioned are given in Chapter 2.

In dividing *S. cerevisiae*, a cytoplasmic bud emerges early in the cell cycle and grows during cell-cycle progression. During the S phase, the cell nucleus migrates to the neck region connecting the mother cell to the bud. Mitosis partitions the chromosomes into the two cells (which are now nearly the same size), and cytokinesis completes the division (Case Study Figure 10.1).

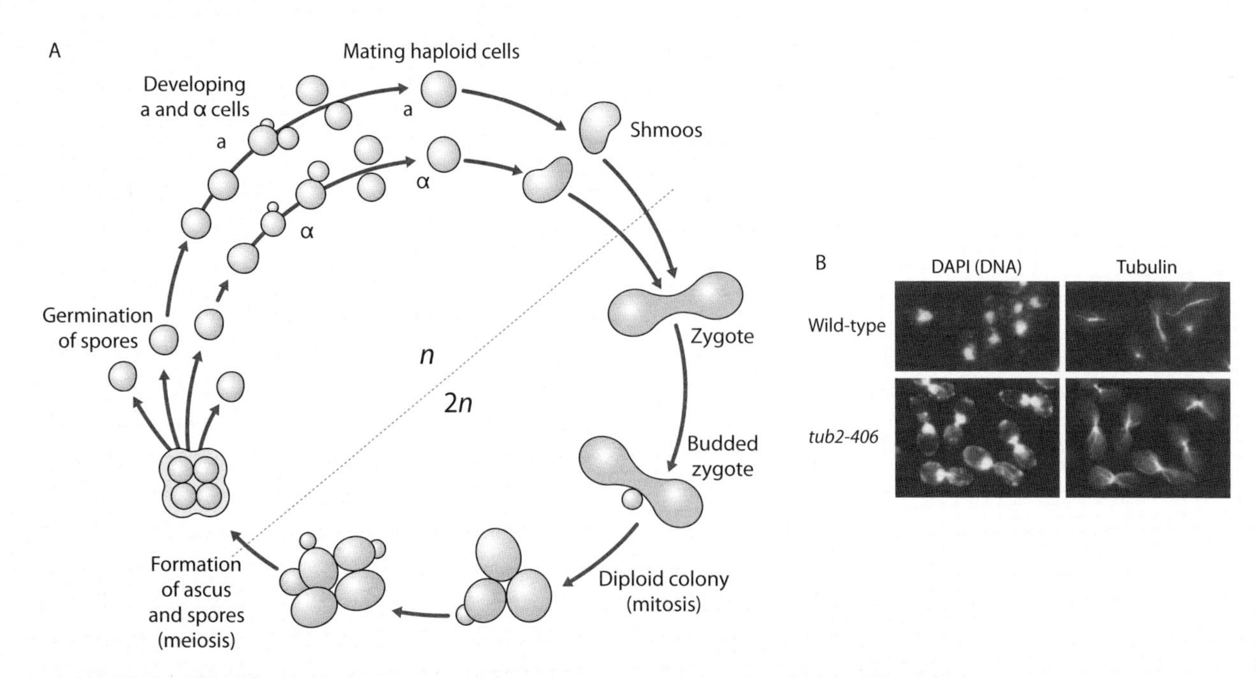

Figure C10.1 The cell division cycle of the budding yeast *S. cerevisiae* is shown diagrammatically (panel A) and by fluorescence microscopy (panel B). A bud begins to form and the nucleus migrates to the neck of the mother cell and bud. The mitotic spindle is established and the chromosomes are segregated to the two cells. The cells in panel B were stained with an anti-tubulin antibody to show the orientation of the spindle and with the DNA dye DAPI to show the chromosomes.

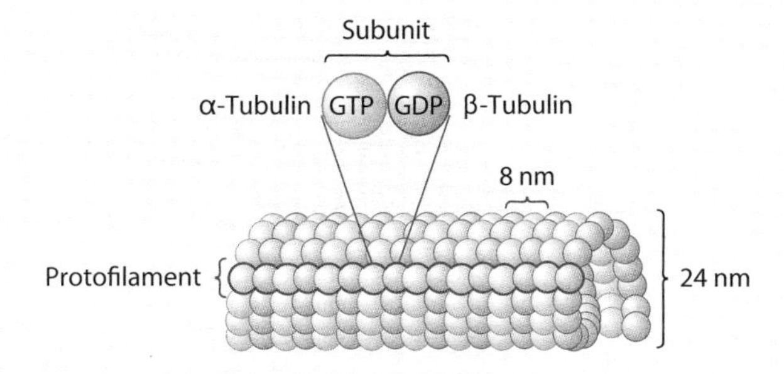

Figure C10.2 **Drawing of a microtubule.** Note the alternating structure of the α- and β-tubulin subunits and the close physical contact between the two tubulin subunits.

As in other eukaryotic cells, chromosome separation depends on the function of the mitotic spindle. The spindle is formed of microtubules, which are heterodimeric polymers of two different but paralogous proteins, α-tubulin and β-tubulin (Case Study Figure 10.2). Associated with the microtubules in the spindle are many other different proteins, known generally as microtubule-associated proteins, which enable it to carry out its function of segregating sister chromatids during mitosis. Because the amino acid sequences of α- and β-tubulin are highly conserved across evolution, vertebrate tubulin genes were used to carry out the molecular cloning of the tubulin genes from yeast. (Cloning based on molecular homology is discussed in Chapter 5.) This molecular analysis showed that yeast has two structural genes for α-tubulin (*TUB1* and *TUB3*) and one structural gene for β-tubulin (*TUB2*).

A cold-sensitive mutation of *TUB2* provided the basis for suppression experiments

The roles of each tubulin gene were studied by genetic analysis. Reverse genetic approaches were used with the cloned genes to disrupt the chromosomal copy (as described in Chapter 6). In addition, forward genetic approaches beginning with wild-type yeast cells identified new mutations in each of these genes. Gene disruptions revealed that both the α-tubulin gene *TUB1* and the β-tubulin gene *TUB2* are essential for cell growth and division under normal laboratory conditions, whereas *TUB3* is not. Because the deletion mutations are lethal, it was necessary to use conditional mutations of *TUB1* and *TUB2* in the further analysis of yeast cell division. (The use of conditional mutations is described in Chapter 4.)

Of particular interest for our purposes here are the cold-sensitive mutations that were found in *TUB2*. At the permissive temperature of 28°C, these mutant cells grow and divide normally. At the restrictive (low) temperature of 14°C, the *tub2* mutations accumulate as cells with large buds with a single undivided nucleus and no mitotic spindle (Case Study Figure 10.3). Not all of the *tub2* cold-sensitive mutations have identical phenotypes, possibly because each mutation affects a different amino acid in β-tubulin. Some of the *tub2* mutations affect both the spindle and the cytoplasmic functions of microtubules so that the nucleus does not move to the neck of the budding cell or divide. However, one particular cold-sensitive allele of *TUB2* (*tub2-406*) arrests with large buds and undivided chromosomes at the region of the bud neck (Case Study Figure 10.3). The nucleus is found in the bud neck, indicating that the cytoplasmic microtubules needed for movement of the nucleus during budding appear to function normally in this mutant. However, the presence of non-separated chromatids indicates that this mutant has a specific defect in spindle microtubules. Thus, the *tub2-406* mutation separates two functions of β-tubulin

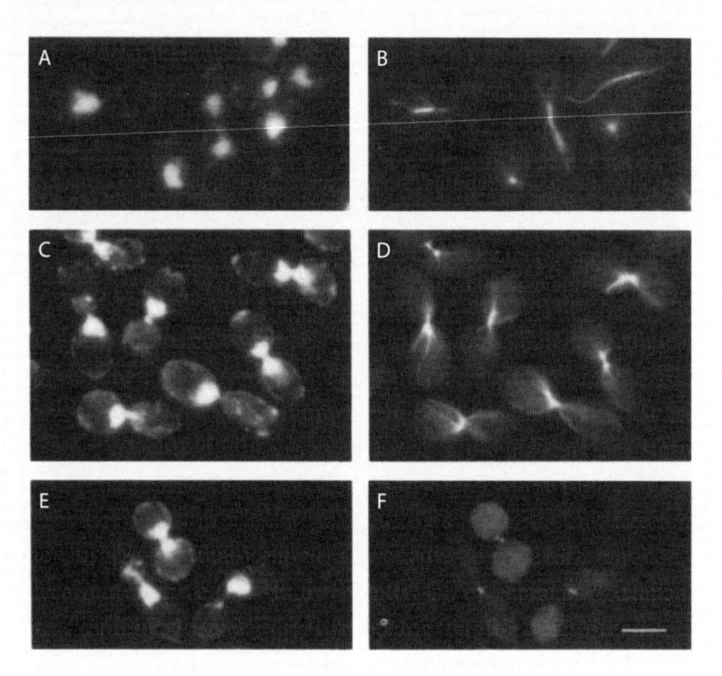

Figure C10.3 The phenotype of the mutant *tub2-406* at the permissive temperature (panels A and B) and the restrictive temperature (panels C–F). Panels A, C, and E are DAPI-stained cells and show the chromosome behavior. Panels B, D, and F are stained with an anti-α-tubulin antibody and show the mitotic spindle. Notice that the nucleus arrests at the neck of the bud at the restrictive temperature (panels C and E) and that the spindle does not assemble normally (panels D and F). (Reproduced from Pasqualone and Huffaker 1994.)

and affects only spindle structure. Therefore, it was used as the original mutation to search for suppressor mutations.

Suppressors of *tub2-406* were found by looking for cell division at low temperature

Because *tub2-406* is cold sensitive and affects only one of the several phenotypes of a *tub2* null allele, it seems very likely that *tub2-406* represents a missense mutation. In fact, subsequent molecular analysis has revealed that *tub2-406* is a missense mutation in which methionine is substituted for valine at amino acid 100. This cold-sensitive growth defect was used to find genes that affect spindle morphology and function by looking for suppressors of *tub2-406*. The genetic selection involved mutagenizing *tub2-406* mutant cells and looking for yeast colonies that grew and divided normally at low temperature. Such suppressors could be gene specific for many β-tubulin mutations and thus identify genes that affect the functions of the β-tubulin protein. Alternatively, the suppressors might be allele specific and affect only the specific defect on the *tub2-406* lesion.

Both categories of suppressor were identified. Eleven different suppressor mutations of *tub2-406* were found and mapped: five mapped to the *TUB2* locus itself and were presumed to be intragenic suppressors, two mapped to the *TUB1* locus, and four mutations mapped to another locus, subsequently named *STU1*. These results are summarized in Case Study Figure 10.4. Notice how closely these suppressors fit our flowchart in Figure 10.5 of the main text in that they identified α-tubulin (a protein known to interact with β-tubulin) and another previously unknown gene that may have a function related to β-tubulin.

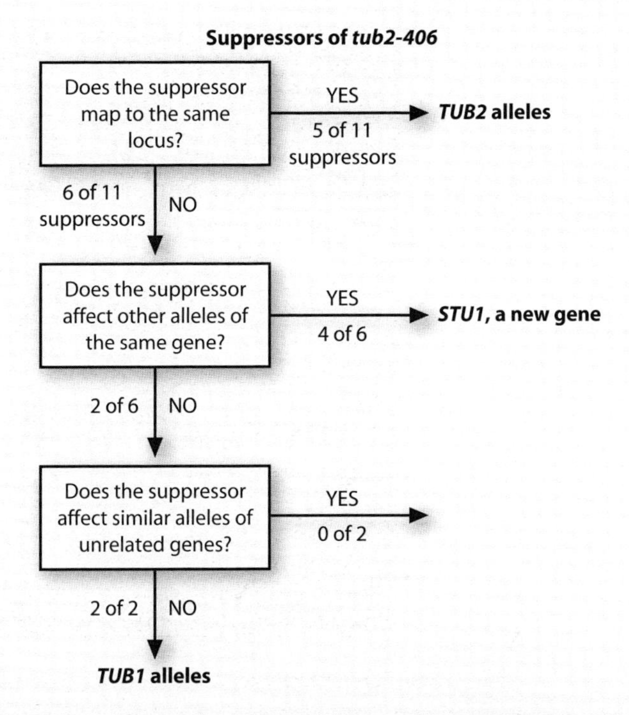

Figure C10.4 Flowchart showing the results of the genetic screen for suppressors of the cold-sensitive mutation *tub2-406*. Compare this flowchart with Figure 10.5 in the main text. Eleven suppressors of *tub2-406* were identified. Five map to the same locus as *tub2* mutations and are classified as intragenic suppressors. Two of the remaining six suppressors were specific to both the gene and the allele, and four were specific to the gene but not the allele. (No informational suppressors were found.) The four allele-non-specific suppressors identify a new gene known as *STU1*. The two interactional suppressors are alleles of the α-tubulin gene *TUB1*.

The next step in the flowchart is to ask if the suppressors affect other alleles of *tub2* to determine whether they are gene specific or allele specific. Different suppressors gave different results in this test. The *stu1* mutations did in fact suppress several alleles of *tub2* in addition to *tub2-406*, so we will deal with them in Part 2 of this case study. In contrast, the *tub1* mutations isolated as suppressors of *tub2-406* did not suppress other alleles of *tub2*, including null alleles of *tub*. In fact, other *tub1* mutations besides these suppressors do not suppress *tub2-406* or other *tub2* mutations. This immediately indicated that these *tub1* suppresssor mutations represent allele-specific or interactional suppressors of *tub2-406*. These mutations may identify residues in α-tubulin that specifically contact the amino acid of β-tubulin that is affected in *tub2-406*.

The inference is that the *tub2-406* mutation disrupted a specific interaction with α-tubulin at this site. The suppressors are postulated to be mutations in α-tubulin (*tub1*) that restore the interaction with the mutated β-tubulin but cannot restore the function lost in other mutations in β-tubulin. In support of this inference for interactional suppression, the *tub1* alleles that suppressed *tub2-406* map to a specific location on α-tubulin that appears to be needed for its interaction with β-tubulin in spindle formation (Case Study Figure 10.5). Thus, allele-specific, gene-specific suppressors were used to map the region of interaction between α-tubulin and β-tubulin in microtubules. Further tests to confirm this genetic inference will be described in Part 5 of this case study.

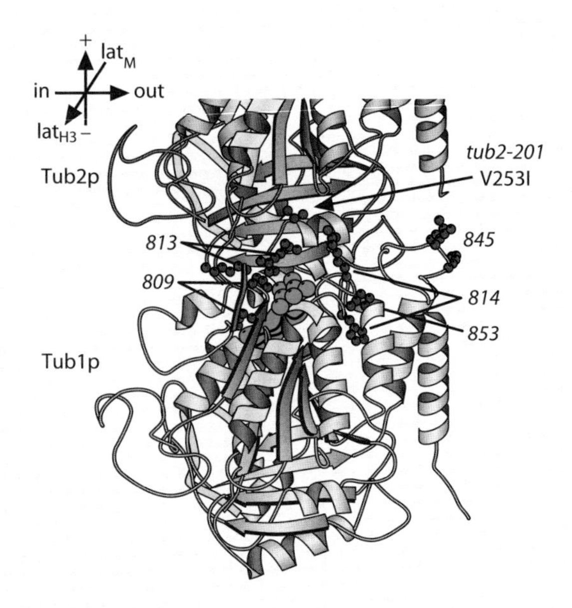

Figure C10.5 The location of the interactional suppressors as projected onto the three-dimensional structure of the tubulin subunit. The interactional suppressors of *tub2* and *tub1* are predominantly located at the interface between the two proteins. (Reproduced from Richards *et al.* 2002).

Further reading

Huffaker, T.C., Hoyt, M.A., and Botstein, D. (1987). Genetic analysis of the yeast cytoskeleton. *Annual Review of Genetics*, **21**, 259–85.

Pasqualone, D. and Huffaker, T.C. (1994). *STU1*, a suppressor of a β-tubulin mutation, encodes a novel and essential component of the yeast mitotic spindle. *Journal of Cell Biology*, **127**, 1973–84.

Richards, K.L., Anders, K.R., Nogales, E., Schwartz, K., Downing, K.H., and Botstein, D. (2000). Structure–function relationships in yeast tubulins. *Molecular Biology of the Cell*, **11**, 1887–1903.

the muscle defect of *unc-54* mutations in general. More significantly, *sup-5* also suppresses some mutant null alleles of many other genes, including ones that have nothing to do with muscle function. The inference is that, since the genes being suppressed have no common function, suppression must be related to the molecular nature of the alleles.

Once the genes and the mutations were sequenced, it became evident that the mutant alleles suppressed by *sup-5* are nonsense mutations resulting in a premature UAG stop codon. It may be recalled that UAG is a stop codon because no naturally occurring tRNA molecule has the complementary CUA anticodon (codons and anticodons have reverse complementary sequences). *sup-5* turned out to be a mutation in a tRNA gene, and results in the amino acid tryptophan being inserted at the UAG amber stop codon, thus suppressing the premature termination of translation, as shown diagrammatically in Figure 10.7. Thus, *sup-5* is an example of a common type of informational suppressor—mutations in tRNA genes. It may be helpful here to recall how the genes were named. The normal function of the *sup-5+* gene is to encode a tRNA that inserts tryptophan at UGG codons; it has the complementary CCA as its anticodon. The mutated gene now inserts tryptophan at UAG stop codons, and the gene is correspondingly named *sup-5* for this mutant phenotype of suppressor rather than its wild-type phenotype.

Mutations in tRNA genes are probably the best-studied group of informational suppressors, particularly in prokaryotes in which they were instrumental in understanding the mechanism of translation. However, informational suppressors have also been identified as mutations in the genes for ribosomal proteins and

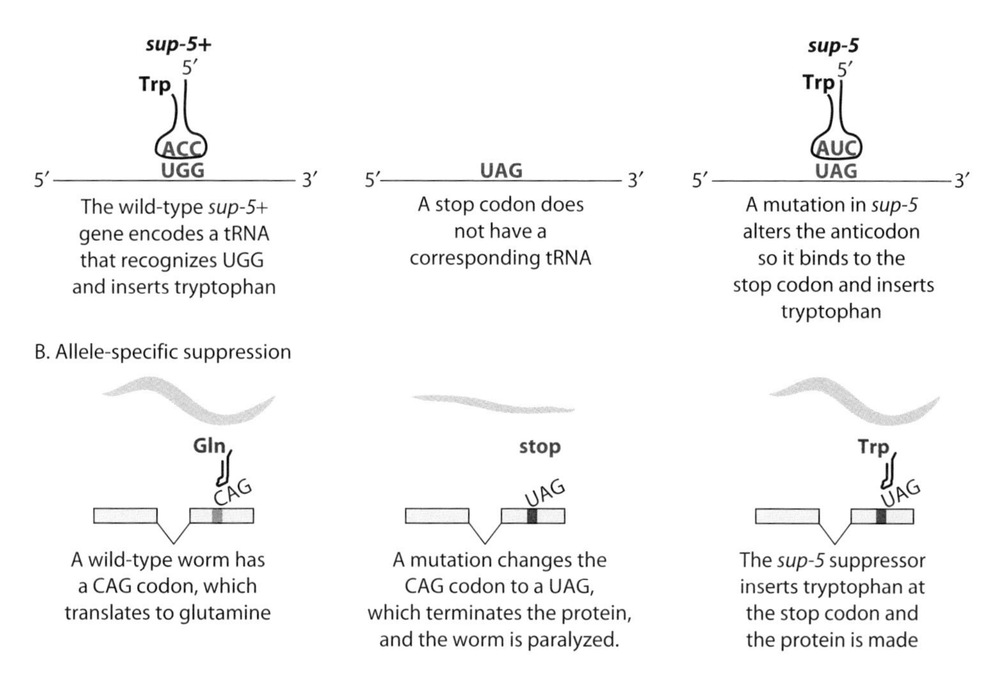

Figure 10.7 Informational suppression involving tRNA genes. An example with the *sup-5* gene in *C. elegans* is illustrated, but suppressor tRNAs similar to this have also been studied in many bacteria and in yeast. Panel A shows suppression in terms of the tRNA gene. The wild-type *sup-5+* recognizes UGG codons and inserts a tryptophan. If the UGG codon is mutated to UAG, the mRNA cannot be translated since no tRNA gene naturally recognizes UAG. This generates a stop codon. However, a compensatory mutation in the *sup-5* tRNA can suppress this stop codon if it recognizes UAG and inserts an amino acid. Although this is shown as inserting the same amino acid as in the wild-type protein, this may not be the case. Panel B shows suppression in terms of the worm and the allele. In the wild-type, the gene has a CAG codon at this site and a tRNA gene recognizes this. A mutation replaces the CAG with UAG. As a result, the protein is not made and the worm is paralyzed. The suppressor *sup-5* tRNA inserts a tryptophan at this stop codon, allowing the muscle protein to be made and movement to be restored. The protein in the suppressed worm has a missense replacement of glutamine with tryptophan, so movement may not be completely normal. Suppressor tRNAs will also act on the naturally occurring stop codons, generating read-through at the C terminus for some proteins. As a result of reading through natural stop codons, many suppressor mutant strains grow poorly.

rRNAs, translation elongation factors, components of mRNA processing, splicing, and many other genes involved in transcription or translation. Again, an example was used in our earlier discussion of suppressor nomenclature. The original *Hairy wing* (*Hw*) mutant allele in *Drosophila* described earlier arose from the insertion of a transposable element known as *gypsy* into the *Hw+* gene; inactivation of the gene results in the Hairy wing phenotype. The suppressor mutation *su(Hw)* suppresses the defect by affecting *gypsy* transposable elements, and thus acts as a suppressor for many other mutations involving *gypsy*. It has nothing to do with the function of the *Hairy wing* gene itself, and a wide variety of other genes have mutant alleles that can be suppressed by *su(Hw)*.

Informational suppressors have been important in gaining understanding of the fidelity of transcription and translation but, because they suppress the molecular lesion in a particular allele and not the functional defect in the gene, they are not very useful for finding additional genes that affect the same biological process. This is not a failure of the suppressor screens, as they are picking up interacting genes. It is simply that we already know that any gene needs to interact physically with RNA polymerases and general transcription factors, for example, while its mRNA needs to interact with tRNA molecules, ribosomes, and the rest of the translational machinery. Identifying mutations in these genes as suppressors gives no additional insight into the biological process influenced by the target gene.

Gene-specific suppressors affect many different alleles of a gene, but no or only a few other genes

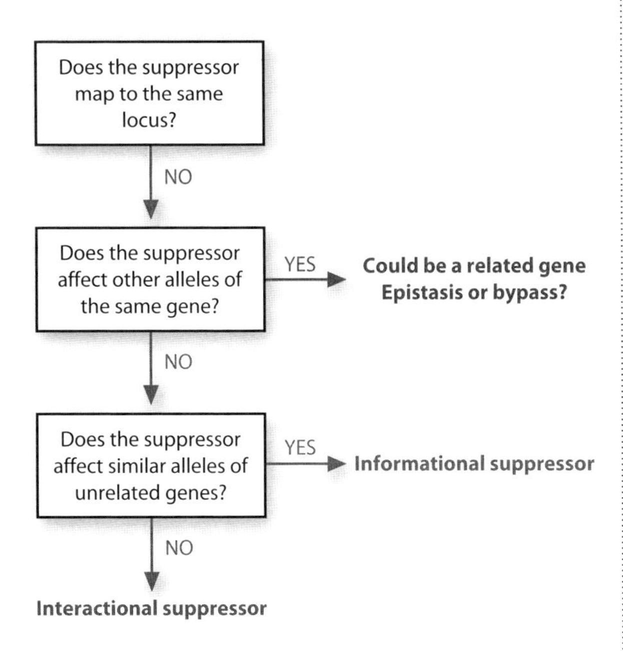

We now turn to the third class of suppressor mutation—those that suppress many mutant alleles of the original gene but are specific to this gene in their effect. These suppressors are the class found most frequently in a suppressor screen. They usually identify other genes that affect the same biological process—exactly the result we are looking for. However, in contrast to interactional suppressors, gene-specific suppressors often identify functionally related genes that do not interact directly.

Broadly speaking, there are two ways in which a suppressor of this type can work. It can affect a gene in a parallel or related pathway, such that the original defect is bypassed, or it can correct the defect by affecting another gene in the same pathway. These two categories are termed bypass suppressors and epistatic suppressors, respectively (see Figure 10.4). The differences in these mechanisms of suppression are illustrated in Figure 10.8. As will be discussed in more detail in Chapters 11 and 12, a cell or organism often has parallel pathways to accomplish a particular cellular function, and each of these pathways has several molecular steps. Each step is controlled by the wild-type allele of a gene so that a mutation in that gene knocks out that step. Suppression might arise in either of two ways. In the first step, a mutation that over-expresses or ectopically expresses the alternative pathway may be able to restore wild-type function because it *bypasses* the original mutation; hence this is called bypass suppression.

In the second step, the original mutation and the suppressor mutation are in the *same* pathway. For example, the original mutation could be a dominant hypermorphic mutation. A suppressor mutation arises whenever a gene downstream in the pathway is knocked out. Alternatively, the original mutation could be a null allele, but a hypermorphic mutation that activates or over-expresses a downstream gene in the same pathway can restore wild-type function. These are examples of suppression by epistasis. Epistasis will be described in detail in Chapter 11, so we will defer a more detailed discussion of this type of suppression until then. Often the only way to determine if the suppressor affects another pathway or another step in the same pathway is by a more extensive epistatic or molecular analysis. Fortunately, as will be seen in Chapter 11, such an epistatic analysis is usually not very difficult to perform.

The difference between bypass suppression and epistatic suppression can be illustrated by our analogy of a homeowner with leaky pipes. The homeowner may be able to alter another pipe or valve to divert the water into other pipes away from the leak; this is a bypass, so the altered valve is analogous to a bypass suppressor. Alternatively, the homeowner may be able to fix the leaky pipe by altering a pipe upstream or downstream of the leak in the same conduit, perhaps by reducing the water flow in an upstream pipe so that the leak is not so severe. This is analogous to epistasis, and the altered pipe in the same conduit is analogous to an epistatic suppressor.

Bypass suppressors that work by affecting a different pathway are known in many different organisms and for many different processes. More complicated examples will be encountered in Chapters 11 and 12, so we will introduce a relatively simple example here. The *C. elegans unc-54* gene encodes a myosin heavy-chain isoform, so mutations in the gene cause paralysis. One suppressor that restores movement to *unc-54* worms is a dominant mutation in another myosin heavy-chain gene known as *myo-3*. The mechanism of suppression was discovered when *myo-3* was cloned and analyzed. The suppressor mutations are dominant mutant alleles of *myo-3* that cause a stable over-expression of the encoded myosin heavy chain that is usually a minor component of the body-wall myosin. In other words, the suppressor mutations in *myo-3* were hypermorphic

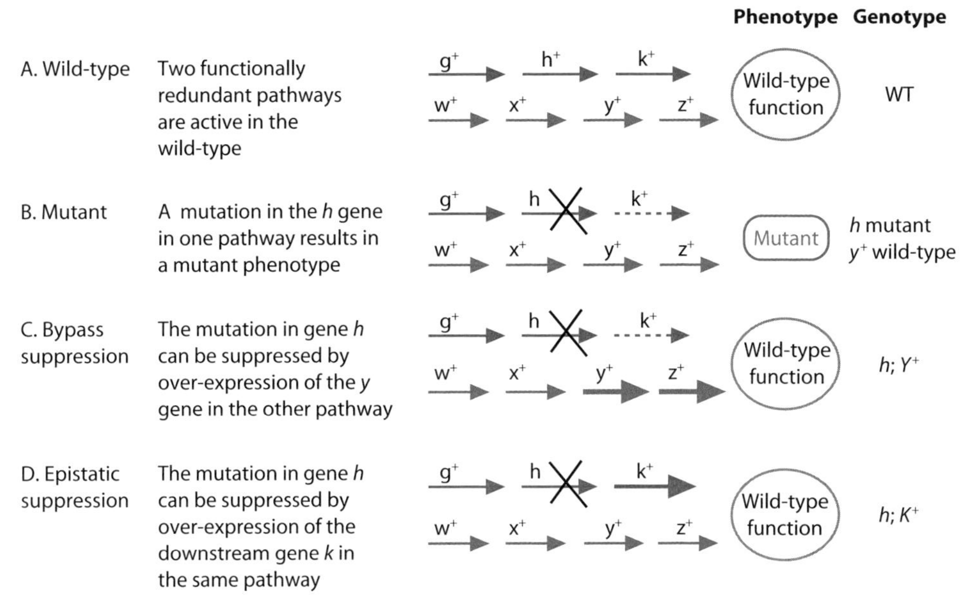

Figure 10.8 Comparison of suppression by bypass mutations and suppression by epistasis, as illustrated by hypothetical pathways. (A) As shown here, the two pathways have partially redundant functions in the wild-type, although the actual gene products are different and both pathways are needed for a wild-type phenotype. (B) A mutation that knocks out gene *h* in one pathway results in a mutant phenotype. Note that the mutation in gene *h* also reduces the activity of the downstream gene in its pathway (gene *k*), although gene *k* is still wild-type. The mutation in gene *h* can be suppressed by other mutations by either of two methods, as illustrated in panels C and D. (C) In bypass suppression, over-expression of a gene in the other pathway (shown by the heavy arrow in the step affected by gene *y*) results in enough gene product to restore function. The over-expression of gene *Y* (indicated as a capital) may also result in the over-expression of genes downstream of *y* in that pathway but have no effect on genes in the other pathway. (D) In suppression by epistasis, the suppressor mutation lies in the same pathway as the original gene. In the example shown, a mutation that results in the over-expression of gene *k*, a downstream gene in the same pathway, could restore normal activity and function of that pathway. Epistasis is considered more fully in the next chapter.

in the terminology of Chapter 3. Thus, *unc-54* is suppressed by *myo-3* hypermorphic mutations because the defect is bypassed by over-expression of a closely related protein that compensates for the lost function. In the context of Figure 10.8, the *unc-54* mutation knocked out the top pathway so the worm is paralyzed. The dominant mutation in *myo-3* is represented by the over-expression of the alternative pathway, which produces enough myosin to restore movement.

Bypass suppressors were among the early evidence that genomes contain redundant information. In the case of *myo-3* and *unc-54*, the bypass occurs because the worm has paralogous copies of myosin genes with partially redundant functions. We now know from the many eukaryotic genomes that have been sequenced that paralogs and gene families such as the myosin genes are a common phenomenon, as discussed in Chapter 1. Given the abundance of gene families, bypass suppression that activates another member of a gene family is expected to be rather common. Although it may not be easy to distinguish it genetically from other forms of suppression, molecular and genomic information will usually provide the insights to recognize a bypass suppressor.

However, not all bypass suppression is due to duplicate genes. Occasionally, bypass suppression indicates that the organism has redundant information or redundant pathways to a downstream phenotype. This is not due to similar genes but to alternative pathways that are partially redundant. In these situations, bypass suppression is often related to synthetic enhancement, so we will consider these two situations together later in the chapter. We will also return to this in Chapters 11 and 12, particularly as we consider complex networks of interacting genes. Another example of functional suppression involving spindle morphology in yeast is discussed in Case Study 10.1, Part 2. Although the mechanism of suppression here is not unmistakably due to either bypass or epistasis, the example illustrates how suppression studies identify genes with related functions.

Case Study 10.1

Part 2: Genetic analysis of spindle morphogenesis in budding yeast

Functional suppression of *tub2-406* by the *stu1* mutations identified a new gene involved in spindle formation and function

As we saw in Part 1 of this case study, the cold-sensitive *tub2-406* mutation identified a site in β-tubulin that is needed for spindle morphogenesis but not for cytoplasmic functions of microtubules. Suppressors of *tub2-406* might then identity other genes involved in spindle morphogenesis, either via a direct or an indirect interaction with β-tubulin. Four suppressor mutations of *tub2-406* mapped to the same gene and failed to complement each other for suppression. This gene was named *STU1* for 'suppressor of tubulin'. The *STU1* gene was cloned by complementation: a plasmid with the wild-type *STU1* gene was identified by its ability to complement the suppression in *stu1;tub2-406* strains and thus display the cold-sensitive arrest phenotype of *tub2-406*. (Cloning genes by complementation is described in Chapter 5.) This approach to cloning the *STU1* gene is shown in Case Study Figure 10.6. Molecular analysis of the complementing plasmid showed that the *STU1* gene, as defined by these suppressor mutations, was a novel protein whose function could not be directly inferred from similarity to other proteins.

The goal of the suppressor screen is to find other genes involved in spindle morphogenesis. The first gene identified by these suppressors encoded an α-tubulin (as described in Part 1 of this case study). The second gene identified is *STU1*. Did the suppressor screen work as expected, and is the *STU1* gene also involved in spindle morphogenesis and function? Several lines of genetic evidence indicated that *STU1* is needed for spindle formation.

First, both the tested *stu1* mutations suppressed not only *tub2-406* but also other alleles of *tub2* with similar spindle defects. This indicates that suppression of β-tubulin function is not allele-specific to *tub2-406* but is related to the function of β-tubulin in spindle

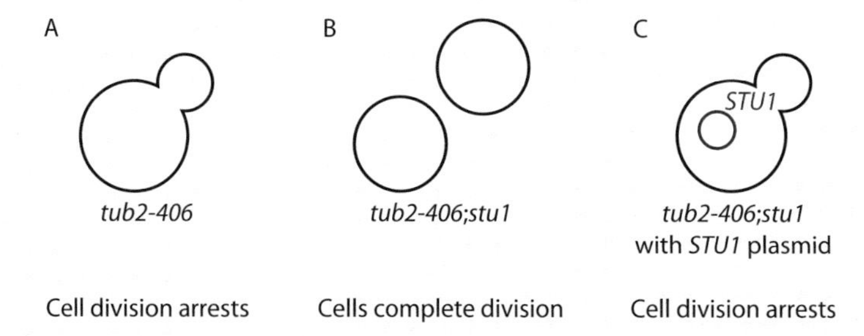

Figure C10.6 Cloning *STU1* by complementation of the suppression. The *tub2-406* mutant arrests cell division with a bud not separated from the mother cell, as shown in panel A. The *stu1* suppressor allows the cells to complete division (panel B). These *tub2;stu1* double-mutant cells were transformed with a library of plasmids to identify the *STU1* gene. The presence of the wild-type *STU1* gene on a plasmid complements the *stu1* mutant defect, and thereby results in cells that resemble the original *tub2-406* mutant, as shown in panel C. The *tub2-406* mutation is cold sensitive, and so these experiments were done at the restrictive temperature of 14°C.

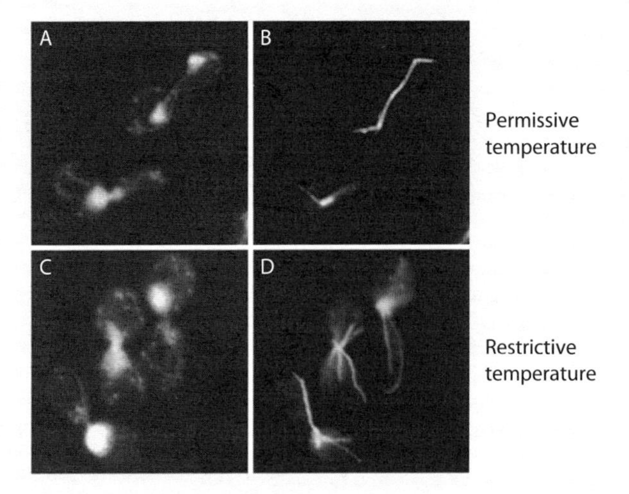

Figure C10.7 The phenotype of *stu1* mutant cells. DNA is visualized by DAPI staining (panels A and C) and the spindle by an antibody directed against α-tubulin (panels B and D). The cells in A and B are grown at the permissive temperature and show a wild-type phenotype. The cells in C and D were grown at a restrictive temperature and show the *stu1* mutant phenotype. Note the nuclear arrest at the bud neck (in panel C) and the abnormal spindle formation (in panel D). This phenotype suggests that STU1 has a function related to TUB2. (Reproduced from Pasqualone and Huffaker 1994.)

morphogenesis. Even more convincing, the *stu1* mutations did not suppress alleles of *tub2* with defects in cytoplasmic microtubules, indicating that *STU1* is not needed for these functions of β-tubulin. Thus, from the suppression analysis itself, it seemed likely that *STU1* defines a gene needed for spindle morphogenesis.

Secondly, the function of *STU1* was tested more directly by a disruption of the gene. (The procedure to make gene disruptions is described in Chapter 6.) These *stu1* deletion mutations, like the *stu1* point mutations isolated as *tub2-406* suppressor alleles, had spindle defects similar to those seen in *tub2-406*, i.e. many of the *stu1* mutant cells arrest with large buds and an undivided nucleus located at the neck. The phenotype of a *stu1* mutant is shown in Case Study Figure 10.7. Notice how similar the phenotype of a *stu1* mutation is to the phenotype of the *tub2-406* mutation shown in Case Study Figure 10.3. Thus, the suppressor gene had a mutant phenotype of its own, which resembled the mutant phenotype of the mutation being suppressed. This provides additional evidence that STU1 is functionally related to the spindle morphogenesis function of TUB2.

Thirdly, and finally, an epitope-tagged version of STU1 was produced and shown to complement the *stu1* gene disruptions. In other words, the only functional STU1 protein in the cell is the epitope-tagged protein, and the cell divides normally. By using an antibody directed against the epitope tag, the localization of the STU1 protein was determined by immunofluorescence. The tagged STU1 protein was found to localize specifically to the spindle. This directly confirmed the inferences from the genetic analysis that the STU1 protein is involved in spindle formation.

How well did the *tub2* suppressor screen work in identifying other genes involved in spindle morphogenesis? The suppression screen not only helped to define the region of α-tubulin and β-tubulin that interact in spindle microtubules (see Part 1 of this case study) but also identified *STU1*, a previously unknown gene required for normal spindle function. Thus both of the genes identified as extragenic suppressors of *tub2-406* are themselves involved in spindle morphogenesis.

Further reading

Huffaker, T.C., Hoyt, M.A., and Botstein, D. (1987). Genetic analysis of the yeast cytoskeleton. *Annual Review of Genetics*, **21**, 259–85.

Pasqualone, D. and Huffaker, T.C. (1994). *STU1*, a suppressor of a β-tubulin mutation, encodes a novel and essential component of the yeast mitotic spindle. *Journal of Cell Biology*, **127**, 1973–84.

Yin, H.L., You, D., Pasqualone, D., Kopski, K.M., and Huffaker, T.C. (2002). Stu1p is physically associated with β-tubulin and is required for structural integrity of the mitotic spindle. *Molecular Biology of the Cell*, **13**, 1881–92.

High-copy suppression involves the use of a wild-type cloned gene

In the examples of suppression that we have described so far, a *mutation* in one gene alleviates the defect of a *mutation* in another gene. However, this is not the only type of suppression that can be observed. Another type of suppression arises when over-expression of the *wild-type* copy of one gene ameliorates the effect of a mutation in a different gene. In this case, the wild-type allele of a gene is acting as an extragenic suppressor. (Note that this is a different situation from the rescue of a mutation by introducing a wild-type copy of the same gene, as described in Chapter 3.) This type of suppression has been applied particularly in yeast, where it is easy to over-produce a wild-type gene product by using a high-copy-number YEp plasmid, as discussed in Chapter 2. This technique is called **high-copy suppression**. High-copy suppression can be considered analogous to the presence of a hypermorphic mutation, as previously described for the muscle genes in worms. The suppressor gene itself is not mutated, simply over-expressed because it is present on a plasmid at high copy number.

Although high-copy-number YEp plasmids are widely used, there are other methods of over-producing a wild-type gene product. These include using a constitutive promoter or an inducible promoter. Regardless of the mechanism, the conceptual mechanism underlying a high-copy suppression screen is similar to other types of suppression. For example, suppose that the original mutation prevents yeast cell division. A library of wild-type genes on high-copy-number plasmids is transformed into the mutant strain. Colonies that grow and divide normally are selected for and thereby suppress the mutant phenotype. The plasmid from these suppressor colonies is isolated and the gene is analyzed.

High-copy suppression has many advantages over conventional suppressor studies. The primary advantage is that the genes being screened for suppression have already been cloned. This avoids one of the most challenging steps in conventional suppressor analysis in other organisms, namely mapping and cloning the suppressor gene. The plasmid with the cloned gene can also be readily transformed into other genetic backgrounds, simplifying the manipulations to test other alleles and other genes for suppression. Compared with the methods of suppression already described, high-copy suppression is relatively easy to perform in eukaryotes, especially yeast, and is more directly connected to the molecular analysis of the genes involved.

These rescuing plasmids could, in theory, be functioning by any of a number of methods, not all of which involve suppression (Figure 10.9). The most obvious alternative to suppression is that the plasmid contains the wild-type version of the mutant gene itself, so that wild-type function is restored by complementation. Although such a result is useful for ensuring that one has cloned the right gene, as discussed as one approach to cloning genes in Chapter 5, complementation is of no interest for identifying other genes that affect the same process.

However, many other types of interaction are possible with these high-copy suppressors, and several examples can demonstrate the power of high-copy suppression. One common mechanism is that the high-copy suppressor encodes a functionally related gene. Over-expression of this gene provides enough functional product to overcome the loss of the first gene. This situation is analogous to bypass suppression with hypermorphic mutations, as described above with the *myo-3* and *unc-54* genes in *C. elegans*. An example of high-copy suppression involving two α-tubulin genes in yeast is described in Case Study 10.1,

Case Study 10.1

Part 3: Genetic analysis of spindle morphogenesis in budding yeast

High-copy suppression helps to define the relationship between the α-tubulin genes

Let us return to the role of the two α-tubulin genes. As discussed above, there are two genes in budding yeast, *TUB1* and *TUB3*. Sequence analysis of the two genes for α-tubulin—*TUB1* and *TUB3*—shows that they are paralogous but not identical. However, the two genes behave differently in genetic analysis. Yeast cells in which *TUB1* is disrupted cannot grow and divide, indicating that *TUB1* encodes an essential function. In contrast, *tub3* mutants grow and divide normally. How then can two slightly different versions of the same protein, α-tubulin, have mutations with such different phenotypes?

Two possible explanations come to mind. First, it may be that the *TUB3* isoform of α-tubulin is not involved in growth and cell division under normal laboratory conditions, i.e. *TUB3* is functionally distinct from *TUB1* despite their sequence similarity. Secondly, it may be that the two proteins have very similar functions and the difference in their mutant phenotype reflects a difference in the level of expression. Under the second hypothesis, mutations that knock out *TUB3* are wild-type in phenotype because *TUB1* provides enough α-tubulin to compensate, whereas mutations that knock out *TUB1* cannot be compensated by the normal and low level of expression from *TUB3*.

These two hypotheses can be tested by varying the dose of the genes. Under the first hypothesis, changing the dose will have no effect since the genes are functionally distinct. Under the second hypothesis, increasing the dose of *TUB3* should be able to suppress a *tub1* mutant. A high-copy suppressor analysis (described in Chapter 2) was used to test these hypotheses. Plasmids were made carrying the wild-type copy of either *TUB1* or *TUB3*, and grown to high copy number in different yeast mutant strains. The growth defect in *tub1* mutants can be complemented by a plasmid that carries the wild-type *TUB1*, as expected. In addition, a *tub1* mutant strain with a wild-type copy of the *TUB3* gene cannot grow if *TUB3* is at normal (low) copy number. These are the two controls that indicate that the yeast cells with the plasmids are behaving as predicted. The experiment involves the yeast cells with *TUB3* on a high-copy-number plasmid. In this case, *tub1* mutations are suppressed and the cells grow. This dosage relationship is shown in Case Study Figure 10.8. Thus, it appears that over-expression of wild-type *TUB3* using a high-copy-number plasmid can provide enough α-tubulin for normal growth even when *tub1* is absent. This high-copy suppression indicates that either gene can provide functional α-tubulin for normal cell growth if expressed at a high enough level. This implies that the difference seen in the mutant phenotype for the two genes is a difference of expression level.

The significance of the different expression levels of *TUB1* and *TUB3* will become clearer in Part 4 of this case study.

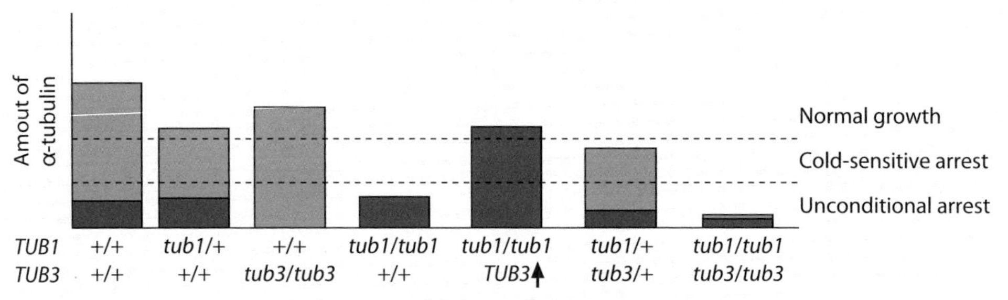

Figure C10.8 The dosage relationship of the two α-tubulin genes, *TUB1* and *TUB3*. The green boxes represent the amount of α-tubulin produced from *TUB1*, and the blue boxes represent the amount produced from *TUB3*. The genotype of each tubulin gene is shown below the boxes. Wild-type cells make more α-tubulin from the *TUB1* gene than from the *TUB3* gene. For simplicity, only two different mutant thresholds are illustrated. Cells with an amount of α-tubulin above the upper threshold have normal growth. This result is seen in wild-type cells, in heterozygous *tub1* strains that have normal *TUB3* function, or in *tub3* mutants that have normal *TUB1* function. A *tub1* homozygote arrests during mitosis (the exact phenotype depends on the allele), but this is suppressed when *TUB3* is over-expressed on a high-copy-number plasmid as indicated by the *TUB3*↑ symbol. *tub3* mutations also show unlinked non-complementation with *tub1* mutations, and the *tub1;tub3* double mutant is lethal under conditions when the corresponding *tub1* mutant will grow.

Further reading

Huffaker, T.C., Hoyt, M.A., and Botstein, D. (1987). Genetic analysis of the yeast cytoskeleton. *Annual Review of Genetics*, **21**, 259–85.

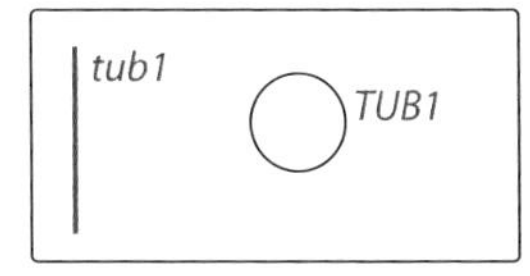

Complementation—the plasmid is providing the wild-type allele of the mutant gene

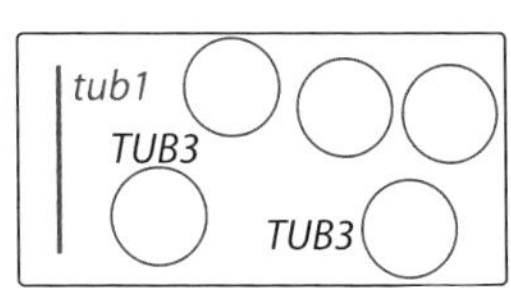

High-copy suppression—the plasmid carries the wild-type allele of a related gene
Over-expression of this related gene suppresses the mutant phenotype

Figure 10.9 Comparison of complementation with high-copy suppression, using the yeast α-tubulin genes *TUB1* and *TUB3* as an example. The *tub1* mutant strain is transformed with a library containing wild-type copies of yeast genes. Cells that can divide and grow normally are selected. These will include cells that received a wild-type copy of the *TUB1* gene, which complements the mutant defect. Complementation will be seen with *TUB1* in both low- and high-copy-number strains. Other cells will have a high copy number of the related α-tubulin gene *TUB3*, since over-expression of *TUB3* provides enough α-tubulin to suppress the *tub1* defect. Note that suppression in this case is due to over-expression of the wild-type copy of the *TUB3* gene rather than an effect of a second mutation.

Part 3. As with the *C. elegans* example, the nature of the suppression in the yeast case is clear from the molecular analysis of the genes involved, and the recognition that the suppressor gene is functionally very similar to the gene it suppresses and encodes a similar product that can directly substitute for the mutant product. However, it is important to realize that it is the *over-expression* of the related gene that confers suppression. The endogenous wild-type copy or copies of the suppressor gene that are present in the wild-type genome have not been sufficient to cause suppression on their own, possibly because they may not be expressed in the relevant cell type or at the required time during the organism's life.

High-copy suppression can be interpreted in the same way as other types of suppression, with the recognition that the suppression arises from over-production and not under-production of a gene product. Thus, high-copy suppressors can often be understood as analogous to dominant or hypermorphic alleles of the suppressor gene rather than recessive loss-of-function alleles.

One more strategy for suppressor analysis is to begin with a dominant gain-of-function mutation and to find recessive suppressors (this is the opposite of the set-up for high-copy suppression that begins with a recessive mutation).

10.4 Synthetic enhancers: modifying mutations that make a mutant phenotype more severe

We began this chapter by comparing a cell or organism with a mutation to a homeowner with a leaky pipe. In our analogy, the homeowner could modify the leak in two ways—by repairing the pipe or by making the leak more severe. Repairing the pipe is analogous to genetic suppression. Now we turn to the other possible modification—making the leak more severe. In genetic terms, these modifying mutations are known as synthetic enhancers. Although the homeowner wants to avoid synthetic enhancement, recent work has shown that these are among the most useful, although understudied, types of modifying mutations for understanding a biological process.

Synthetic enhancers are mutations that exacerbate the effect of the original mutation

Many different modifying mutations can occur that make a mutant phenotype more severe. Although such mutations can simply be called enhancers, for the reasons described at the beginning of the chapter we will call them synthetic enhancers. We call them 'synthetic' because the effect of the mutant enhancer is seen only when it occurs in combination with a mutation in a different gene; the word 'synthesis' means the combination of two or more things into a new thing. Until recently, enhancement has not received as much attention as suppression, and has not been as widely used as a genetic tool. In part, this arises from the point made at the beginning of the chapter—apparent enhancement may arise because two mutations, each of which is mildly deleterious to the organism, combine in a non-specific fashion to make the organism even less viable or even dead. If I am already in a bad mood because I have a cold, the lack of a parking space close to my building can put me in a really foul temper. This type of non-specific enhancement is rarely interesting or informative.

Synthetic enhancement includes two different but related effects on mutant phenotypes. The first effect is that a mutant phenotype arising from one gene is made more severe by a mutation in a second gene. Our analogy with the leaky pipe is applicable here: a defect is observed, and the second mutation makes it worse. For example, imagine that a mutation results in a phenotype that is 60 percent penetrant. The definition of penetrance was given in Chapter 4; in this context, it means that, among a population of genotypically mutant organisms, 60 percent of them display a mutant phenotype and 40 percent have a wild-type phenotype. A synthetic enhancer of this mutant phenotype will result in a higher penetrance, so that perhaps 100 percent of the genotypically mutant organisms are phenotypically mutant. We will use this effect in describing synthetic enhancement of the dauer formation phenotype in *C. elegans*.

Another and potentially more interesting situation arises when the mutation in one gene results in no obvious mutant phenotype. The mutant phenotype is only observed when a second specific gene is also mutated. These are also known as synthetic enhancers. We will use this effect in describing genes affecting vulva formation in *C. elegans*. In many cases, the mutation in the first gene results in no mutant phenotype, whereas the mutation in the second gene results in a lethal phenotype. In an example like this, the term synthetic lethality is used. In all of these manifestations, the interest in synthetic enhancement lies in the specificity of the interaction. In terms of genotypes, both the *a*/*a* mutant and the *b*/*b* mutant have a wild-type phenotype. However, the double mutant *a*/*a*;*b*/*b* has a mutant phenotype. In this case, the *b* mutation is a synthetic enhancer of the *a* mutation (and the *a* mutation is a synthetic enhancer of the *b* mutation). In these cases, enhancement does not arise from the combination of two different and quite unrelated deleterious effects. For example, perhaps mutations in any gene other than the *b* gene do not give a synthetic phenotype in combination with mutations in the *a* gene. As we shall see, synthetic enhancement is the result of a functional relationship between the two genes.

Like bypass suppression, synthetic enhancement is often evidence that genomes have redundant information. Synthetic enhancers can involve two different types of redundant information, as will be discussed in more detail in Chapter 12. The redundant information can indicate the presence of paralogous genes with very similar or duplicate functions in the genome. Alternatively, the redundant information may not involve paralogous genes with similar functions but, rather, redundant or duplicate pathways that produce the same function. The ability to make targeted mutations, particularly in yeast and mice, has awakened new interest in synthetic interactions, which are currently in wide use.

We need to make a clarification in the nomenclature and gene notation before beginning the discussion of synthetic enhancement. In the cases that are described, all other mutations are recessive and the strains are homozygous mutant diploids; in the case of yeast, the strains could also be haploids. We will also describe a different but related interaction that occurs among heterozygotes in a later section; this is known as non-allelic non-complementation.

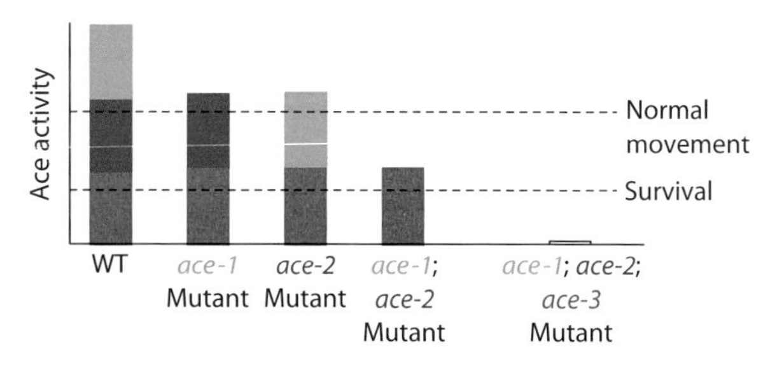

Figure 10.10 Synthetic enhancement illustrated by the acetylcholine esterase (*ace*) genes in *C. elegans*. Each of the three genes *ace-1*, *ace-2*, and *ace-3* makes a different isoform of the enzyme, as shown by the different colored bars. Threshold levels for different phenotypes are inferred from the results with the mutants. In worms homozygous mutant for either *ace-1* or *ace-2*, the amount of acetylcholine esterase is reduced. However, the residual amount produced from the two remaining genes is enough to ensure normal movement. A double mutant, which reduces the amount of the enzyme further, leaving only the *ace-3* gene active, has impaired movement but is still alive. The triple mutant, which completely lacks the enzyme, does not survive. The presence of three genes was inferred from the results with synthetic enhancement.

Synthetic enhancement can involve duplicate or paralogous genes

Our first example illustrates how synthetic enhancers can indicate the presence of functionally redundant paralogues, even if such genes have not been previously detected. The genes involved in this example encode the enzyme acetylcholine esterase in *C. elegans*. Acetylcholine esterase breaks down the neurotransmitter acetylcholine, and can be detected by a relatively simple colorimetric assay in worms. Mutations were found by finding worms with reduced histochemical staining for the enzyme. Worms homozygous for a null allele of the gene for acetylcholine esterase (*ace-1/ace-1*) have reduced levels of the enzyme, but have no obvious defect in either movement or behavior and look completely like wild-type. Given the widespread importance of acetylcholine as a neurotransmitter, the lack of a mutant phenotype was initially puzzling.

To address this, the *ace-1* mutant strain was itself mutagenized; among the offspring were some that had a distinctive uncoordinated movement. Standard mapping experiments similar to those described for suppressors showed that the uncoordinated mutant phenotype was due to a second unlinked recessive mutation in a hitherto unknown gene, which was then called *ace-2*. This situation is shown graphically in Figure 10.10. The uncoordinated worm was a double mutant, homozygous for the *ace-1* mutation and an *ace-2* mutation. Colorimetric assays for the enzyme indicated a further reduction of acetylcholine esterase in these uncoordinated *ace-1/ace-1;ace-2/ace-2* double mutants. Both *ace-1/ace-1* and *ace-2/ace-2* single mutants have a wild-type phenotype but the double mutant has the uncoordinated mutant phenotype.

Although it was not certain until both genes were cloned, the reduction of acetylcholine esterase in each of the *ace* gene mutations suggests that *ace-2* also encodes acetylcholine esterase. The interpretation of this interaction is that normal movement requires a particular threshold level of acetylcholine esterase, and wild-type worms make an excess of the enzyme well above this threshold. Even if one of the *ace* genes is knocked out, the level of the enzyme is above the threshold and the worm moves normally. Only when both *ace* genes are knocked out does the enzyme level fall below the movement threshold so that the double-mutant worm is uncoordinated.

However, by colorimetric assays it was found that even the double mutant has some residual acetylcholine esterase. This suggests the presence of a third *ace* gene. The *ace-1;ace-2* double mutant was mutagenized to search for this third gene. This time, synthetic lethal mutations were found; these are mutations that resulted in lethality only when both *ace-1* and *ace-2* were also homozygous mutant. These studies identified the third acetylcholine esterase paralog, *ace-3*. As shown in Figure 10.10, this suggested the presence of another mutant phenotype threshold, this time one needed for viability.

Mutants that knock out only one of *ace-1, ace-2*, or *ace-3* have a wild-type phenotype. The inference is that, even if the isoform encoded by one *ace* gene is not being made, enough of the enzyme is made from the other two wild-type genes that the worm is normal. However, each knockout of an additional *ace* gene results in a more severe mutant phenotype. If two genes are knocked out, the worm is uncoordinated. If all three genes are knocked out, the worm is dead. The number of acetylcholine esterase genes was not known at the time the genetic analysis was done. The presence of three genes with redundant functions was inferred from the synthetic enhancement and confirmed when the genome was sequenced many years later. By beginning with one mutation that had no obvious mutant phenotype, it was possible by synthetic enhancement to find the other genes that had related functions. Both ways that synthetic enhancement can manifest itself were observed in this example since no mutant phenotype is observed unless two different genes are mutated; the mutant phenotype is then further enhanced by the mutation in the third gene.

Synthetic enhancement and bypass suppression can be two indications of the same effect

As pointed out in Chapter 1, genomes have paralogous genes for many gene families. If the paralogs have retained duplicate functions, as the *ace* genes in *C. elegans* appear to do, knocking out one of them may not cause a difference in the phenotype. Earlier in the chapter we used the existence of paralogous genes to explain the action of some bypass suppressors. How then can the same phenomenon—the existence of genes with duplicate or highly similar functions—account for two apparently opposite genetic effects? One explanation is shown diagrammatically in Figure 10.11. The keys lie in the nature of the mutations for each gene and in the threshold at which a mutant phenotype is observed. For synthetic enhancement to occur, as shown in Figure 10.11A, a knockout of the first gene has *no effect* on the phenotype because the product

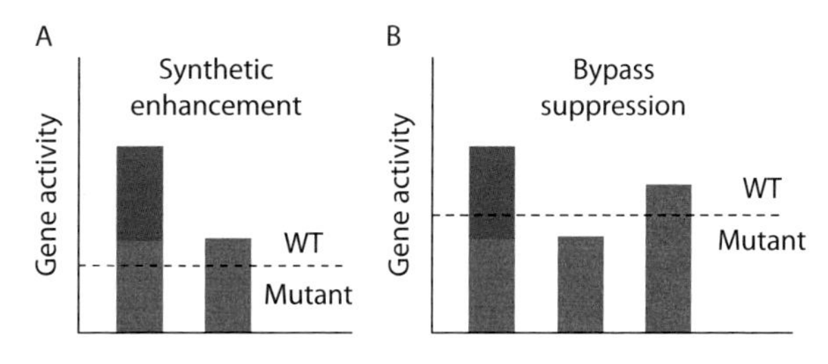

Figure 10.11 Graphical comparison of synthetic enhancement with bypass suppression. Although the two phenomena seem to differ, they are often closely related; the difference lies in the amount of the gene product needed for normal function. A case with two genes with redundant functions is illustrated. The activity of the two genes is illustrated by the red and blue bars. Elimination of the first gene (the blue bar) results in a mutant phenotype in the case of some genes (panel B) but not others (panel A). Synthetic enhancement is observed when a second gene is also eliminated and a mutant phenotype results (panel A). Bypass suppression is observed when the second gene is over-expressed, giving an amount of gene function that exceeds the threshold.

of the second gene is sufficient for activity above the critical threshold. Thus, a mutant phenotype will only be observed if the second gene is also mutated.

In the example of bypass suppression in Figure 10.11B, the critical threshold of gene activity needed is higher or the functions of the paralogous genes are not completely equivalent. Thus, a knockout of the first gene results in a *mutant* phenotype. However, this mutant phenotype can be suppressed by over-expression of the second gene. The investigator will observe this as bypass suppression. In each example, the two mutated genes have related functions. The differences between these two situations lie in two characteristics: the critical threshold of gene activity needed to observe a mutant phenotype, and the nature of the mutation in the second gene. In each case, we have been able to use a mutation in one gene as our tool to find other genes with related functions.

Synthetic enhancement can involve parallel biological pathways

Another example from *C. elegans* illustrates that synthetic enhancement need not be limited to one gene. In fact, entire pathways of related genes can be found by synthetic enhancement screens. In Chapter 2, we described genes that affect vulva formation in worms. Some of the mutations resulted in a phenotype referred to as multivulva or Muv, in which the mutant worms

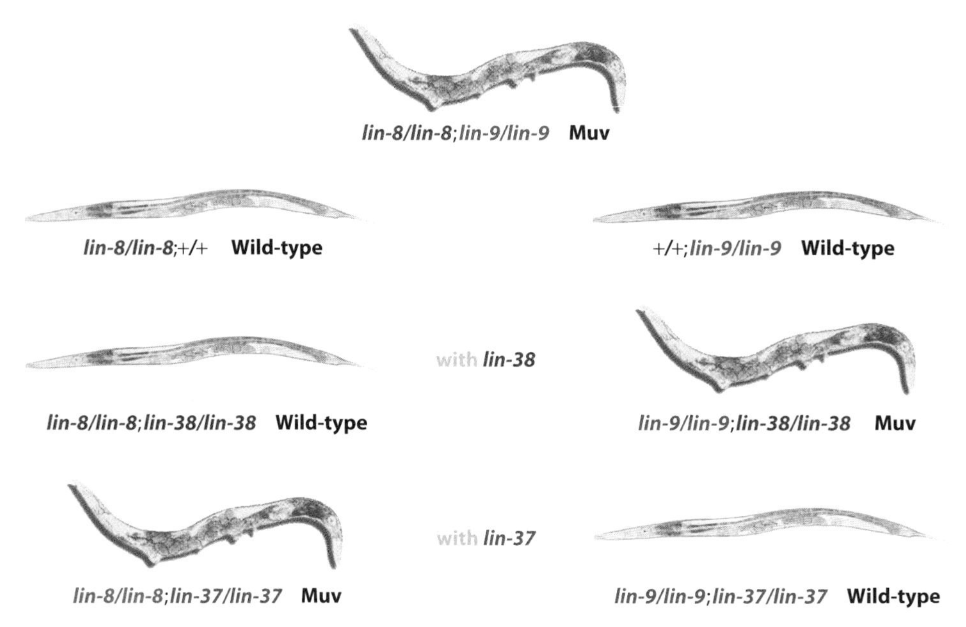

Figure 10.12 Synthetic enhancement, as illustrated by genes affecting vulva lineages in *C. elegans*. A double mutant that is homozygous mutant for *lin-8* and *lin-9* exhibits a multivulva (Muv) phenotype. A worm that is homozygous mutant for *lin-8* or *lin-9* alone exhibits a normal phenotype. The genes function in redundant pathways, and both pathways must be mutant before a mutant phenotype is seen. This is illustrated by mutations in two other genes. The *lin-38* gene works in the same pathway as *lin-8*, so a *lin-8;lin-38* double mutant is wild-type. However, *lin-38* and *lin-9* are synthetic enhancers of each other, so a *lin-9;lin-38* double mutant has a Muv phenotype. Similarly, *lin-37* works in the same pathway as *lin-9*, so does not enhance *lin-9* mutations but does act as a synthetic enhancer of *lin-8* and is also a synthetic enhancer of *lin-38*. This is shown graphically in Figure 10.13.

have multiple partial vulvae. This phenotype is shown in Figure 10.12. Upon analysis, one Muv mutant strain proved to be a double mutant homozygous for two unlinked recessive mutations, subsequently placed into the genes *lin-8* and *lin-9*. It was fortuitous that this double mutant was observed since the *lin-8/lin-8* and *lin-9/lin-9* single-mutant strains have a wild-type phenotype, as shown in the second line of Figure 10.12. (In fact, each gene may well have been mutated numerous times previously in the Muv mutant screen but, because the mutant phenotype is only seen in double mutants, the mutations were overlooked.) Both mutations needed to be homozygous in order to observe the Muv phenotype. Thus, *lin-8* and *lin-9* are synthetic enhancers of one another, suggesting that they affect duplicate functions.

If this were the only effect of the two genes, the mechanism of synthetic enhancement between *lin-8* and *lin-9* could be similar to the preceding example of the *ace* genes. However, a difference became apparent when each of the single-mutant strains was used in a direct screen for additional synthetic enhancers: *lin-8/lin-8* worms, which have a wild-type phenotype, were mutagenized and Muv offspring were recovered. Likewise, *lin-9/lin-9* worms were mutagenized and Muv offspring were recovered. These Muv offspring are double mutants with the phenotype arising from synthetic enhancement. If *lin-8* and *lin-9* were simply functionally redundant paralogs, the synthetic enhancers of *lin-8* would include alleles of *lin-9* and possibly a few other functionally related genes. Furthermore, if *lin-8* and *lin-9* were paralogs, the genes that were synthetic enhancers of *lin-8* would also be expected to be synthetic enhancers of *lin-9*. This is what was observed with the functionally equivalent *ace* genes —each mutation acts as a synthetic enhancer of the others.

However, this was not what was observed for *lin-8* and *lin-9*. When *lin-9/lin-9* worms were mutagenized and synthetic enhancers were recovered, several genes in addition to *lin-8* acted as synthetic enhancers of *lin-9*. In addition, several genes in addition to *lin-9* acted as synthetic enhancers of *lin-8* in the analogous experiment. Most significantly, the mutants that enhanced *lin-8* did not enhance *lin-9* and did not enhance each other; likewise, genes that were synthetic enhancers of *lin-9* did not enhance *lin-8* and did not enhance each other. Two examples, for the genes *lin-38* and *lin-37*, are shown in Figure 10.12. Mutations in *lin-38* acted as synthetic enhancers of *lin-9* but not as synthetic enhancers of *lin-8*. Likewise, mutations in *lin-37* acted as synthetic enhancers of *lin-8* but not as synthetic enhancers of *lin-9*. The synthetic enhancers fell into two distinct classes—the *lin-8* synthetic enhancers and the *lin-9* synthetic enhancers.

Thus, the newly identified genes were not paralogs of either *lin-8* or *lin-9*, and they were not paralogs of one another. Instead, the two sets of synthetic enhancers were defining two functionally redundant pathways, both involved in vulval development. The mutant Muv phenotype from these pathways is only observed when both pathways are mutated; if either pathway is functional, the worm is wild-type. The two classes of genes, called class A and class B, now comprise at least eight different genes. A Muv mutant phenotype is observed for these genes only when the worm is doubly mutant for both a class A mutation and a class B mutation. *lin-8* and *lin-38* are members of A class, whereas *lin-9* and *lin-37* are members of B class.

Molecular analysis of these genes has shown that, unlike the cases above, the two classes do not encode duplicate genes. In fact, genes in class B appear to affect cell-cycle progression in G_1 and include the *C. elegans* ortholog of the human *Rb* gene, a cell-cycle-related gene that is mutated in human retinoblastoma. Because of its relevance to human cancer, the role of the *Rb* gene has been extremely thoroughly analyzed. However, none of the class A genes encodes a protein with recognizable similarity to any other proteins in other organisms, let alone ones that interact with *Rb*. Clearly the two pathways have related functions. However, the functional redundancy of the pathways, inferred from the synthetic enhancement, remains to be solved.

As more genomes are being sequenced and analyzed, there has been renewed interest in synthetic enhancers and in functionally redundant pathways. Figure 10.13 shows the general logic for synthetic enhancers and functionally redundant pathways, a concept that we consider in much more detail in Chapter 12. In particular, targeted gene knockouts in yeast and mice, as

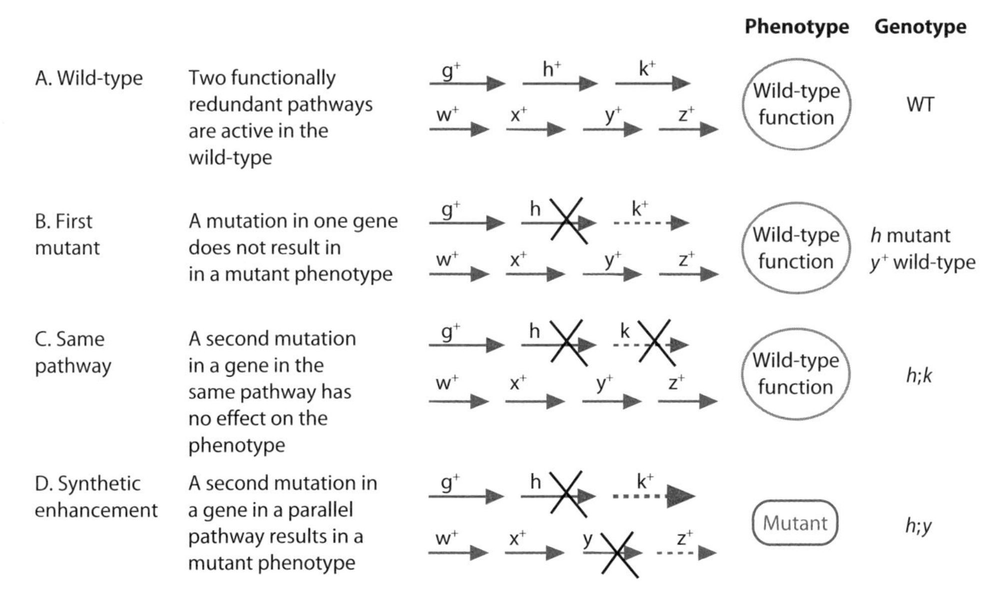

Figure 10.13 Graphical illustration of synthetic enhancement. (A) Two hypothetical pathways with partially redundant functions are active in the wild-type. Either pathway is needed for wild-type function. (B) A mutation that eliminates a step in one pathway (gene *h* in the red pathway) will not produce a mutant phenotype because the other pathway is still active. Notice the difference between the wild-type phenotype and the mutant phenotype for a similar pathway shown in Figure 10.8. (C) A second mutation that eliminates another step in the same pathway such as gene *k* will not act as a synthetic enhancer of the gene *h* mutation. (D) Conversely, any mutation that eliminates the second pathway, such as the one in gene *y*, is capable of synthetic enhancement of a mutation in the first pathway. Compare this figure with Figure 10.8.

described in Chapter 6, often result in no mutant phenotype. For example, targeted deletions have been made for each open reading frame in yeast, and the resulting strains have been tested under a number of growth conditions. Only about 20–25 percent of genes are essential for growth under laboratory conditions, with the remaining 75–80 percent of the knockouts giving no mutant phenotype. However, most of the genes with no mutant phenotype alone have a synthetic lethal phenotype in combination with another mutation. This will be discussed in more detail in Chapter 12.

It is humbling for a geneticist to realize that much of what we have learned from Mendelian genetics has come from a minority of the genes in the genome, namely those that have an easily recognized mutant phenotype. Potential synthetic enhancer mutations are probably found in every mutagenesis screen but, because there is no mutant phenotype, the mutation has been overlooked. The interaction between *lin-8* and *lin-9* in *C. elegans* was found because a fortuitous double mutant arose in the screen and exhibited a mutant phenotype. The interaction between the *ace* genes was known because a colorimetric assay was available. Synthetic enhancers are easier to find when the first gene has a mutant phenotype and the second mutation makes it more severe. On the other hand, this enhancement is more likely to be non-specific or to involve genes whose functions are not closely related, as in the example of the cold and the parking space used earlier. The synthetic enhancers in which no mutant phenotype is observed from the first gene are much more difficult to find but are probably much more likely to identify genes with related functions.

The findings described in these experiments fulfill the goal of a synthetic enhancement screen—to use mutations in one gene to find functionally related genes. Thus, synthetic enhancement screens can be a very useful method of finding other genes affecting the same biological process. A genome-wide analysis of synthetic enhancers is just beginning. Many of the characteristics of suppressors, such as mutations that modify only one or a few alleles, have not yet been explored with synthetic enhancers. For example, most synthetic enhancement has focused on null alleles, but the possibility of allele-specific interactions for one or both genes should not be overlooked. It seems very likely that we have only begun to explore the potential information that can be gleaned from synthetic enhancers. We will return to this topic in Chapter 12 when we discuss biological networks.

Non-allelic non-complementation is a type of synthetic enhancement that occurs when both mutations are heterozygous

Synthetic enhancement as described so far involves organisms that are mutated for two different genes, neither of which has a mutant phenotype on its own. In the cases we have described, each gene has to be a homozygous mutant for an effect to be seen, i.e. all of the cases have involved recessive mutations and the interaction requires that the mutations be homozygous. Another form of synthetic enhancement can occur in heterozygotes. This is most often found accidentally during a particular type of mutagenesis scheme.

Suppose that an investigator has a mutation and wants to find more alleles of the same gene. A standard method is a non-complementation screen to identify mutations that fail to complement the original mutation, as described in Chapter 4. The expected and most common result is that these newly arisen mutations are additional alleles of the same gene. Occasionally, however, mutations in a second unlinked gene will fail to complement the first mutation, a phenomenon known by the unwieldy names of **non-allelic non-complementation** or unlinked non-complementation.

The effect is more easily understood by looking at the genotypes. In non-allelic non-complementation, both the *a*/+ heterozygote and the *b*/+ heterozygote are wild-type because each mutant allele is recessive to wild-type. The *a*/*a* and the *b*/*b* homozygotes display a mutant phenotype, often a very similar mutant phenotype. Although the mutations are recessive, the *a*/+;*b*/+ double heterozygote has a mutant phenotype. (For comparison and review, with synthetic enhancement the *a*/*a* genotype has a wild-type phenotype, whereas the *a*/*a*;*b*/*b* genotype has a mutant phenotype.) This is not a rare phenomenon, and examples have been reported in many different organisms for many different processes. One particular example is described in Text Box 10.2. Frequently, the non-complementing genes have been shown to encode products that interact physically, which suggests that non-allelic non-complementation might be particularly helpful in

BOX 10.2 Non-allelic non-complementation

As described in Chapter 4, a complementation test is used to determine if two mutations are alleles of the same gene or if they lie in two different genes. The test consists of constructing an organism that is heterozygous for each of the two alleles and examining its phenotype. We will denote this doubly heterozygous organism as *a*/+;*b*/+, in which *a* and *b* are the mutations being tested for allelism. The only prerequisite for performing a complementation test is that the alleles in question have to be recessive. Complementation tests are very widely used both to determine allelism and to find additional alleles of an existing gene by an F_1 screen, as described in Chapter 4. In an F_1 screen for new alleles of the *a* gene, mutagenized organisms are mated to *a*/*a* mutants, and the F_1 offspring are examined for those with *a* mutant phenotype. Since these new mutations fail to complement the original *a* mutation, these would be considered alleles of the *a* gene.

To be considered alleles, the non-complementing mutations must also map to the same locus on the chromosome. Yet as investigators have done F_1 screens, it is not unusual to find new mutations that fail to complement the original mutation but which map to a different location in the genome. These cannot be alleles, yet the mutations fail to complement each other. The name given to this phenomenon is non-allelic non-complementation or unlinked non-complementation. When these non-complementing but unlinked genes have been analyzed further, many if not most of them define genes whose products interact with each other. The interaction is functional and need not be a direct physical contact, similar to both suppression and synthetic enhancement. Thus, non-allelic non-complementation is another genetic method for finding interacting genes.

One well-studied example in *C. elegans* involves genes involved in neurotransmitter release. One such gene is *unc-13*. The *unc-13* gene encodes at least five different isoforms of a protein that is conserved in many (and perhaps most) other animals, including *Drosophila* and mammals. Mutations in the gene are recessive, and homozygous *unc-13/unc-13* worms are often paralyzed, depending on the allele used. The gene product regulates neurotransmitter release by its interaction with syntaxin.

Syntaxin is one of the proteins of the SNARE core complex, which mediates fusion between synaptic vesicles and the plasma membrane in animals. However, mutations in the gene *syn-1* have a wild-type phenotype, possibly because there are multiple syntaxin paralogs in worms. Although *unc-13* mutations are recessive and *syn-1* mutations appear to have no mutant phenotype, the mutations fail to complement each other. A worm that is *unc-13*/+;*syn-1*/+ has uncoordinated movement. This failure to complement is found with some alleles of *unc-13* but not with others, and the alleles that fail to complement do not fall into a simple allelic series.

The inference is that the proteins work together in mediating vesicle fusion to the plasma membrane. A mutation in one of the genes does not have an effect but the complex or the process is now more sensitive to a mutation in the other gene. The allele specificity of the effect could indicate that the interaction between the proteins is a direct physical contact. Although this inference is supported by other evidence, there are not enough examples to be sure that allele-specific non-complementation indicates a direct interaction in all cases.

analyzing protein complexes. Even if a physical interaction does not occur, non-allelic non-complementation is an effective, and somewhat under-utilized, approach to finding genes that affect the same biological process.

Two general models have been proposed to explain non-allelic non-complementation. First, the underlying biological process may be sensitive to *dosage* of the two products, such that a simultaneous reduction in both proteins produces a mutant phenotype. This type of interaction is shown in Figure 10.14. One example of this type of genetic interaction is seen between two genes involved in neuronal migration in *Drosophila*. During neuronal migration, the Slit protein binds to the receptor encoded by the *robo* gene (Figure 10.15). Mutant homozygotes for either gene have axons that cross the ventral midline inappropriately, resulting in a mutant phenotype. Mutations in either *slit* or *robo* are recessive, so each heterozygote is normal. However, flies heterozygous for both *slit* and *robo* have a mutant phenotype, with axons that cross the ventral midline. The interaction is not allele specific, as many different alleles of each gene fail to complement each other. This suggests that the interaction may depend on a dosage effect of the two gene products.

As we learned in Chapter 4, a standard method of evaluating a dosage effect is to use an allelic series, with

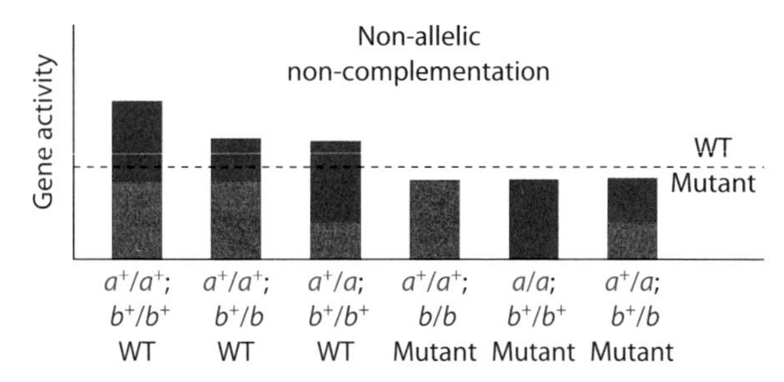

Figure 10.14 Graphical illustration of non-allelic non-complementation. Two genes with related functions (red and blue) are shown, with the threshold for normal activity represented by the dotted line. Each mutant allele reduces the function of that gene by half. However, each single heterozygote still has enough residual activity for normal function. The mutant phenotype arises when one gene is homozygous mutant, or when both genes are heterozygous. The appearance of a mutant phenotype in this latter double heterozygous case defines non-allelic non-complementation.

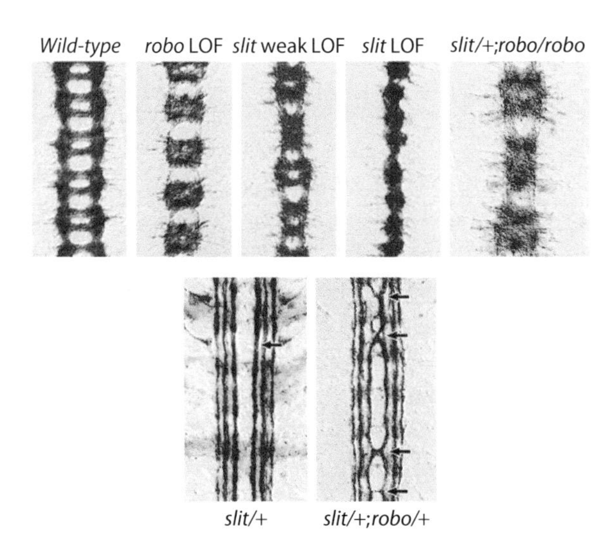

Figure 10.15 Non-allelic non-complementation illustrated by the interaction of *robo* and *slit* at the dorsal midline in the *Drosophila* central nervous system. The embryonic nervous system is stained with a monoclonal antibody that recognizes all commissural and longitudinal axons. Top row: In wild-type embryos, axons cross the midline in a regular pattern. In *robo* mutant embryos, the longitudinal axons are reduced and the commissural axons cross and recross the midline. A similar phenotype is seen in *slit* hypomorphic alleles and the *slit* null allele; in the *slit* null allele, the longitudinal axons do not form, and the axon scaffold is greatly compressed. A similar compression is seen in a *robo* homozygous and *slit* heterozygote; compare the width of the scaffold in the second picture and the fifth picture. This is a synthetic enhancement of *robo* by *slit* in *slit/+;robo/robo*. Bottom row: A *slit/+* heterozygote is normal, as is a *robo/+* heterozygote (not shown). However, a *slit/+;robo/+* double heterozygote has axons that cross the midline abnormally. The two genes exhibit non-allelic non-complementation. LOF, loss of function. (After Kidd, T., Bland, K.S., and Goodman, C.S. (1999), *Cell*, **96**, 785–94.)

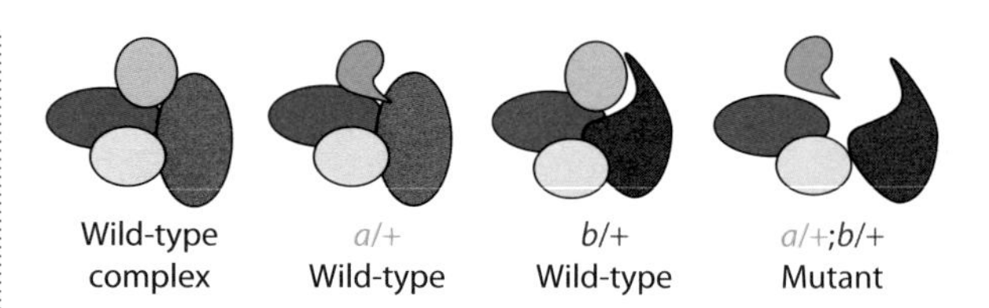

Figure 10.16 The 'sensitized product' model for non-allelic non-complementation. The wild-type proteins function as part of a multi-protein complex. A recessive mutation in one gene, shown here as an alteration of the green component, is wild-type because the complex still forms. Similarly, a recessive mutation in a second gene, shown here as an alteration in the purple component, also has a wild-type phenotype. In the double heterozygote *a/+;b+*, the complex cannot form normally and a mutant phenotype arises.

mutant alleles that have different residual levels of activity for each gene. This approach has been applied to the interaction between *slit* and *robo*. Many alleles of *slit* have been found, and it is possible to rank the strength of the alleles on the basis of the penetrance of the *slit* mutant defect. Different alleles in the series were tested for non-complementation with *robo*. The most severe loss-of-function mutants in *slit* homozygotes are also the most severe mutations for failing to complement *robo* in double heterozygotes. In fact, the allele order derived from the phenotype in *slit* homozygotes is similar to the allele order derived from the phenotype in *slit/+;robo/+* double heterozygotes. This is consistent with non-allelic non-complementation arising from a dosage effect of interacting proteins.

A second explanation is that non-allelic non-complementation may indicate a *sensitized product* in a macromolecular complex (Figure 10.16). In this model, an altered product of the first gene has no visible phenotype in a heterozygote, but renders the entire complex sensitive to a second defect. The second defect could come from either a second mutation in the same gene (i.e. a homozygote) or from a mutation in an interacting gene (i.e. two heterozygotes and hence non-allelic non-complementaton). An interaction with the sensitized product is not necessarily expected to follow an allelic dosage series, but will depend on the particular defect in each allele.

This explanation has been used to explain the interaction between two genes involved in neurotransmitter release. In *C. elegans*, a recessive mutation in *unc-64* fails to complement a recessive mutation in the unlinked gene *unc-13*. The proteins encoded by both *unc-64* and *unc-13* are involved in neurotransmitter release at the synapse. They are part of a larger complex of interacting proteins: a mutation in one gene, while having no mutant phenotype on its own, makes the complex more sensitive to mutations to other proteins in the complex. In contrast to *robo* and *slit*, some alleles of *unc-13* and *unc-64* do not show non-allelic non-complementation. In addition, the interactions do not follow an allelic series. This result suggests that the interaction between *unc-13* and *unc-64* is not simply a dosage effect. Instead, it has been inferred that the mutation in one gene (*unc-13*) alters the complex sufficiently for it to be more sensitive to mutations in other genes (e.g. *unc-64*) affecting the same process.

As with synthetic enhancement, non-allelic non-complementation has usually been observed between genes involved in the same or closely related processes. Although it is difficult to know how commonly this type of interaction occurs, non-allelic non-complementation has proved to be a useful and informative property of the two genes when it has been encountered. Examples of synthetic enhancement and non-allelic non-complementation in yeast mitotic spindle morphogenesis are described in Case Study 10.1, Part 4.

10.5 Summary: finding related genes using mutant phenotypes

Suppression, synthetic enhancement, and non-allelic non-complementation find genes that are related in function to an original mutated gene. These genetic approaches require only that the investigator has a mutation as the starting material and a method of generating more mutations. Suppressor mutations identify mutations that overcome the effect of the original mutation, often by affecting another gene in the same biological pathway. Synthetic enhancer mutations identify mutations that make the effect of the original mutation more severe, often by affecting another gene in an alternative or parallel biological pathway. Whereas the literature on suppressor mutations is rich and deep, most of the literature on synthetic enhancer mutations is much more recent. Synthetic enhancers provide a method of uncovering and analyzing redundant information in the genome, whether in the form of paralogous genes with duplicate functions or in the form of parallel biological processes. The genetic methods based on mutant phenotypes can often be used to provide a broad-brush picture of a biological process. The details of the process require methods based on cloned genes, which we address next.

Case Study 10.1

Part 4: Genetic analysis of spindle morphogenesis in budding yeast

Synthetic enhancement among *STU1*, *TUB2*, *TUB3*, and *TUB1* reveals complex interactions in spindle morphogenesis

In the previous parts of this case study we have shown how suppressor analysis has been used to identify potential regions of interaction between α-tubulin and β-tubulin, and to find a new gene involved in spindle morphogenesis. Part 3 shows how a high-copy suppressor experiment revealed that the two α-tubulin genes are functionally equivalent despite different mutant phenotypes. In this part, we describe experiments that reveal a more complex set of interactions between these genes involved in spindle morphogenesis. The experiments involve varying the wild-type or mutant dosage of two genes simultaneously, either in double heterozygotes or in homozygous double mutants. In each example, a more severe mutant phenotype is produced in double mutants than in single mutants. The interaction seen in heterozygotes (i.e. a dominant interaction) is an example of non-allelic non-complementation, and the interaction seen in homozygotes (i.e. a recessive interaction) is synthetic enhancement, as described in the main text.

Non-allelic non-complementation involving *stu1* and *tub1*

Recall from Parts 1 and 2 of this case study that certain specific *tub1* mutations and all *stu1* mutations suppress the cold-sensitive *tub2-406* mutation in β-tubulin, and restore spindle microtubule function in a *tub2-406* mutant at restrictive temperature. The *TUB1* gene encodes α-tubulin, and *STU1* encodes a novel protein that localizes to spindle microtubules and affects spindle function. This raises the question: How do the *tub1* mutations and the *stu1* mutations affect each other? The gene interactions discussed here are summarized in Case Study Tables 10.1 and 10.2.

Table C10.1 Gene interactions in spindle morphogenesis: non-allelic non-complementation.

tub gene	Interacting gene	Phenotype	Interpretation
tub2-406/tub2-406	*STU1/stu1*	Cold-sensitive arrest	*tub2* mutant phenotype
tub2–406/tub2-406	*stu1/stu1*	Growth at low temperature	Suppression of *tub2-406*
tub2-406/tub2-406	*TUB1/tub1-108*	Cold-sensitive arrest	*tub2* mutant phenotype
tub2-406/tub2–406	*tub1-108/tub1-108*	Growth at low temperature	Suppression of *tub2-406*
tub2-406/tub2-406	*TUB1/tub1-108; STU1/stu1*	Growth at low temperature	Non-allelic non-complementation
TUB2/tub2 (cs)	*TUB1/TUB1*	Normal growth	*tub2* mutation is recessive
TUB2/tub2 (cs)	*TUB1/tub1-1*	Cold-sensitive arrest	Non-allelic non-complementation

Table C10.2 Gene interactions in spindle morphogenesis: synthetic enhancement

tub gene	Interacting gene	Phenotype	Interpretation
TUB1/tub1	*TUB3/TUB3*	Normal growth	*tub1* mutation is recessive
TUB1/tub1	*TUB3/tub3*	Cold-sensitive arrest	Non-allelic non-complementation
tub1 (cs)/*tub1* (cs)	*TUB3/tub3*	Cold-sensitive arrest	*tub1(cs)* mutant phenotype
tub1 (cs)/*tub1* (cs)	*tub3/tub3*	Arrest at all temperatures	Synthetic enhancement
tub2 (cs)/*tub2* (cs)	*STU1/STU1*	Normal growth at 30°C	Conditional *tub2* mutant phenotype
TUB2/TUB2	*stu1/stu1*	Normal growth at 30°C	Conditional *stu1* mutant phenotype
tub2 (cs)/*tub2* (cs)	*stu1/stu1*	Cellular arrest at 30°C	Synthetic enhancement

Each suppressor mutation is itself recessive, so that *SUP/sup;tub2-406/tub2-406* cells do not grow and divide at low temperature (*SUP/sup* refers to a heterozygote between one of the suppressor mutations and wild-type). For example, *tub1-108* is one of the allele-specific suppressors of *tub2-406.* A *TUB1/tub1-108;tub2-406/tub2-406* strain is cold-sensitive and shows the spindle morphology defect characteristic of *tub2-406* mutants. Likewise, a *STU1/stu1;tub2-406/tub2-406* strain also has the mutant phenotype of *tub2-406.* In contrast to the results with heterozygotes in one of the suppressor genes, a *STU1/stu1;TUB1/tub1-108;tub2-406;tub2-406* grows and divides at low temperature; that is, a strain that is heterozygous for both suppressors simultaneously has the *tub2-406* mutation suppressed. To express this result somewhat differently, although *stu1* and *tub1* define distinct genes, mutations in *stu1* fail to complement *tub1-108* in their ability to suppress *tub2.* Although each of the suppressors is recessive, a strain that is heterozygous for both of them suppresses *tub2-406.* In other words, the mutations *tub1-108* and *stu1* show non-allelic non-complementation for their ability to suppress *tub2-406.* (Yeast geneticists often refer to this as unlinked non-complementation.)

The non-allelic non-complementation suggests that *tub1* and *stu1* may also interact genetically with each other since both show genetic interactions with *tub2.* The interaction between *tub1* and *stu1* may or may not be a direct physical interaction between the two proteins. In fact, experiments that will be described in Part 5 of this case study indicate that the two proteins probably do not directly contact each other. As described in the main text, this type of non-allelic non-complementation could be an example of a sensitized product. From this model, we infer that the mutation in *stu1*, although recessive, somehow affects the structure of the mitotic spindle. This altered structure, although functional, is then further compromised by the mutation in *tub1.* The spindle with both mutations cannot assemble properly. Either mutation alone can be tolerated as long as the wild-type protein is also present, but the combination of both mutations is defective.

Non-allelic non-complementation involving *tub1* and *tub2*

Non-allelic non-complementation is also observed between certain α-tubulin and β-tubulin mutations. A genetic screen was designed to find mutations that failed to complement cold-sensitive mutations in *tub2* but that were not themselves alleles of *TUB2.* One of the mutations found in this screen was an allele of the α-tubulin gene *TUB1*, designated *tub1-1.* This is also an example of non-allelic non-complementation. Both the *tub1-1* mutation and the *tub2* mutation are recessive. However, a strain that is heterozygous for both *tub1-1* and *tub2* has a cold-sensitive mutant phenotype similar to that seen in *tub2* (Case Study Table 10.1). In this example, interactional suppressors, molecular assays, and cytological assays have all shown that these two proteins directly contact each other in the mitotic spindle. The non-allelic non-complementation between mutations in the two genes provides one more piece of evidence for this interaction. Just as significantly, the molecular and cytological evidence justifies our interpretation that non-allelic non-complementation often involves proteins with direct physical interactions.

Non-allelic non-complementation and synthetic enhancement involving the two α-tubulin genes

We have been discussing the genetic interactions between the two different tubulin genes *TUB1* and *TUB2* and the microtubule-associated protein encoded by *STU1*. Another interaction was also observed between the two different α-tubulin genes *TUB1* and *TUB3*. As described above, mutations in *tub1* are recessive, and mutations in *tub3* have no mutant phenotype alone. Experiments using high-copy suppressors (see Part 3 of this case study) have indicated that the difference between these two phenotypes arises from a difference in the expression level of each protein.

Just as interactions with over-expression of the genes was described in Part 3, we can also test the effect of gene dosage by *lowering* the dose of each gene using heterozygotes. A strain that is heterozygous for both mutations (*TUB1/tub1;TUB3/tub3*) arrests cell division at low temperatures. Thus, the two α-tubulin genes also exhibit non-allelic non-complementation. Given what we know about the effects of gene dosage from high-copy suppressors, we infer from this that the non-allelic non-complementation is also indicative of a dosage effect. This is further illustrated by results with homozygous double-mutant cells. A *tub1;tub3* double-mutant strain arrests cell division under growth conditions when either *tub1* or *tub3* single mutants can grow. In other words, the two α-tubulin genes *TUB1* and *TUB3* display both dominant and recessive interactions, i.e. they display both non-allelic non-complementation and synthetic enhancement. These results are shown in Case Study Table 2.

It is worth thinking about these examples in more depth. As discussed in the main text, two different models have been postulated to explain non-allelic non-complementation. The interaction between *TUB1* and *TUB3* is a clear example of gene dosage effects with two closely related genes. *Over-expressing* wild-type *TUB3* on a high-copy plasmid *suppresses* the mutant defect in a *tub1* mutation. *Reducing expression* of *TUB3 enhances* the phenotype of *tub1* mutations. In heterozygotes for the two genes, this is seen as unlinked non-complementation. In homozygotes for the two genes, this dosage effect is seen as synthetic lethality. These effects are expressed graphically in Case Study Figure 10.8 (in Part 3 of this case study).

The interactions between *STU1* and *TUB1* and between *TUB1* and *TUB2* cannot be so easily explained as a gene-dosage effect arising from differences in gene expression level. Several observations indicate that a different explanation is needed.

1. The proteins encoded by these genes are not closely related, so it is unlikely that we are dealing with functional redundancy.
2. More significantly, the effects are seen with specific mutant alleles of *tub1* and *tub2* and not with gene knockouts. An allele-specific effect of this type is not consistent with an

effect of gene dosage when the effect should become more obvious as the gene function is reduced.

3. Over-expression of the wild-type function of one of the genes does not act as a suppressor of mutations in the other gene. Thus, a sensitized interaction model, as described in the main text, more easily explains these effects. In such a model, the first mutation renders the macromolecular complex (in this case, the mitotic spindle) more susceptible to a second mutation. This is not merely a matter of dosage; rather, it is due to the more specific molecular defect in the alleles being used.

All of the genetic interactions that we have discussed in this chapter were used in the analysis of the mitotic spindle in yeast. The cellular process under study—microtubule assembly into a mitotic structure—has been studied by a variety of other methods in addition to traditional genetics. In Part 5 of this case study we will discuss some of these other methods that have given a detailed look at this structure and that serve to justify our confidence in the genetic analysis.

Further reading

Huffaker, T.C., Hoyt, M.A., and Botstein, D. (1987). Genetic analysis of the yeast cytoskeleton. *Annual Review of Genetics*, **21**, 259–85.

Schatz, P.J., Solomon, F., and Botstein, D. (1988). Isolation and characterization of conditional lethal mutations in the *TUB1* α-tubulin gene of the yeast *Saccharomyces cerevisiae*. *Genetics*, **120**, 681–95.

Stearns, T. and Botstein, D. (1988). Unlinked non-complementation: isolation of new conditional lethal mutations in each of the tubulin genes of *Saccharomyces cerevisiae*. *Genetics*, **119**, 249–60.

10.6 Using a cloned gene to find genes that affect the same biological process

One of the primary advantages of the genetic analysis methods described in the previous part of the chapter is that prior knowledge of the gene products or their functions is not required. However, a complete analysis of the interactions needs to include much more molecular detail on what the proteins are and which ones interact directly. That is, while the genetic approaches can identify *functional* interactions, molecular methods are usually needed to identify direct *physical* interactions between the gene products.

Protein–protein interactions are central to every cellular process. In fact, it is hard to think of a protein that does not carry out its function by interacting with other proteins, or of cellular processes that do not depend on protein–protein interactions. In this section, we briefly describe two widely used methods to look directly at protein–protein interactions. These methods rely on having cloned one or more of the genes of interest and require some additional molecular tools. Although these methods have been used in many different organisms, they have been particularly useful in situations where mutants were not available. The methods will be described here as supplemental to the genetic methods, and we will not attempt an exhaustive review. The first method to be described, the yeast two-hybrid assay, uses a transcriptional activation procedure to detect binary protein interactions. The second method, co-immunoprecipitation, relies on the specificity of antibody–antigen interactions to precipitate a macromolecular complex. Many other biochemical methods have been developed to analyze protein complexes, but these are beyond the scope of this book.

Yeast two-hybrid assays use a genetic approach to discover protein–protein interactions

Protein interactions in macromolecular complexes have been analyzed for decades by biochemical methods such as cross-linking and complex purification. The technical challenges of these biochemical experiments led to the development of simpler alternatives to examine protein–protein interactions. Of these, the most widely used is the **yeast two-hybrid (Y2H) assay**, a genetic method using transcriptional activation as the assay. The method uses the selection power of a yeast genetic screen to examine protein interactions. Because it is fundamentally a genetic screen, the Y2H assay allows the screening and detection of large libraries of cloned genes, and can be scaled up to examine the entire genome. Y2H selections are also useful to detect transient interactions or other interactions that might not be stable during a purification scheme. Because the method is widely used, there are many variations and modifications for specific purposes. Our description will cover only the most general version.

The basis of the yeast two-hybrid screen is shown in Figure 10.17 and depends on the biochemical properties of the transcriptional activator protein GAL4. GAL4, like many transcription factors, has two separate domains that are capable of acting independently of each other. The DNA-binding domain (abbreviated DB) is needed for sequence-specific binding. The activation domain (abbreviated AD) interacts with the general transcription machinery to activate transcription. Each of these domains is cloned separately into a yeast centromere plasmid that is maintained at low copy number in the yeast cell. The two plasmids also include antibiotic-resistance markers to allow them to be grown in bacteria and a nutritional marker to allow their analysis and maintenance in yeast. (This type of centromere plasmid is described in Chapter 2.) In Figure 10.17, the plasmid with the DNA-binding domain includes the *LEU2* gene whereas the plasmid with the activation domain includes the TRP1 gene. These are the two plasmids of the two-hybrid assay.

Suppose that the investigator has a gene of interest, Gene A, and wants to find proteins that interact directly with the protein product of Gene A. The gene is cloned in frame to the sequence for the GAL4-DB domain, so that when this plasmid is transcribed and translated, a fusion protein that has GAL4-DB fused to Protein A will be made. This plasmid is called the **bait plasmid**. A second plasmid has the transcriptional activation domain of GAL4 fused in frame with the gene to be tested. This is called the **prey**. A more powerful and widely used version has a library of prey plasmids in which an individual gene is cloned with the AD domain, and many different genes are represented as prey in the library.

The two plasmids are transformed into a yeast cell that has a GAL4-binding site upstream of some easily selected marker. In Figure 10.17, the selectable marker is the *HIS3* gene. This gene is only transcribed, and the cell is only able to grow in the absence of histidine, if the GAL4-binding site is activated. Each plasmid produces its bait or prey fusion protein. The bait protein can bind to the GAL4-binding site, but it lacks the ability to interact with the transcriptional machinery and activate transcription. On the other hand, the prey protein can activate transcription but it cannot bind to the GAL4-binding site. However, if the bait and the prey physically interact with each other, the two different domains of the GAL4 protein are brought together, and transcription of the *HIS3* gene under GAL4 control is activated. If the bait and prey interact with each other, the cell can grow in the absence of added histidine; if the bait and prey do not interact with each other, the cell cannot grow in the absence of added histidine. Thus, there is a powerful and easy selection for protein–protein interactions.

In this simple form, Y2H assays provide a relatively easy method for testing postulated interactions between two proteins. It is also possible to map the portion of each protein needed for the interaction to occur. This is described in Case Study 10.1, Part 5, for the interactions between proteins in the spindle in yeast. Y2H assays are extremely powerful, and a well-designed prey cDNA library allows thousands of possible interactions to be screened at once. The method does have certain limitations, in terms of both failing to detect some interactions (false negatives) and finding interactions that are not physiologically meaningful (false positives).

Among the reasons that a Y2H screen may fail to detect an interaction that occurs physiologically are the following.

- The fusion protein with the bait or with the prey does not adopt the same conformation as is found in the cell.

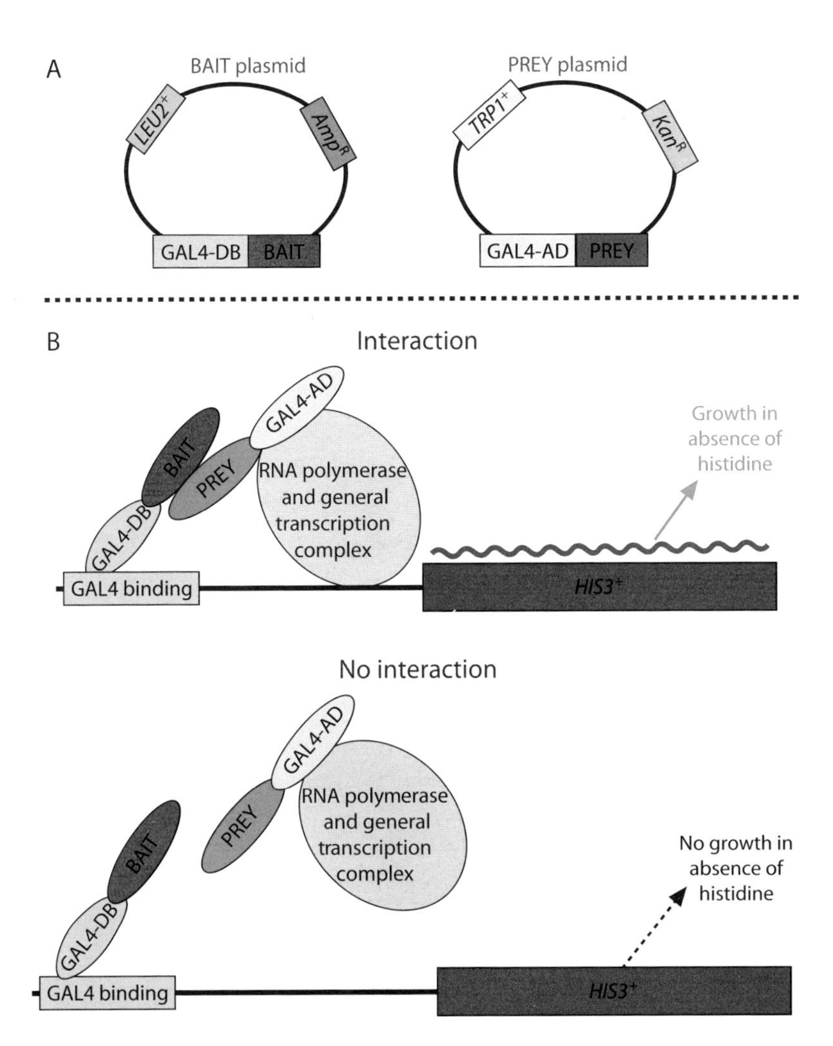

Figure 10.17 The yeast two-hybrid (Y2H) system for detecting protein–protein interactions. The Y2H screen relies on the domain structure of the transcriptional activator in yeast, GAL4. The GAL4 DNA-binding domain (DB) binds a specific and defined nucleotide sequence. The GAL4 transcriptional activation domain (AD) interacts with the transcriptional machinery. A gene-specific transcriptional activator is found when produced GAL4-DB is associated with GAL4-AD. In panel A, each domain is cloned into a yeast centromere plasmid with a selectable drug marker (ampicillin or kanamycin resistance in the figure), which allows selection in *E. coli*, and a nutritional marker (*LEU2* or *TRP1* in the figure), which allows selection in yeast. The gene of interest, known as the bait, is cloned in frame with GAL-DB. Genes to be tested, known as the prey, are cloned in frame with GAL-AD. Each prey plasmid has a different gene to be tested, so a library of prey genes is constructed. The two plasmids are transformed together into host yeast cells. The assay in yeast is shown in panel B. The presence of the selectable auxotrophic markers (*TRP1*+ and *LEU2*+ in the figure) ensures that only yeast cells with both plasmids will grow on minimal medium. The host yeast cells have the GAL4-binding site inserted upstream of one or more reporter genes. The figure illustrates the *HIS3* gene under the control of the GAL4 promoter, but *LacZ* or *URA3* are also useful reporters. Auxotrophic markers such as *HIS3* and *URA3* are selectable, whereas *LacZ* expression can be assayed quantitatively. The yeast cell will express the DB–bait fusion protein and the AD–prey fusion protein. If the bait and prey proteins interact with each other, the two domains of the GAL4 protein are brought together and transcription of the reporter gene is activated. Thus, in the figure, the yeast cells would be plated on media that lack tryptophan (to select for the prey plasmid), leucine (to select for the bait plasmid), and histidine (to select for the interaction). Only those cells that grow in the absence of histidine have an interaction between the bait and prey.

- A Y2H screen detects only binary interactions. If molecules Y and Z have to interact with each other before the Y–Z complex interacts with X, the interaction will probably be missed.
- A Y2H screen detects interactions between unmodified proteins. If the proteins only interact when one or both is phosphorylated or acetylated (for example), the interactions will be missed.

Case Study 10.1

Part 5: Genetic analysis of spindle morphogenesis in budding yeast

Genetic analysis has been confirmed and extended by biochemical and molecular assays

Confirming interactional suppressors

Throughout this case study we have described how suppressors and synthetic enhancers are used to identify new genes whose functions are related to the original gene—in this example, the yeast β-tubulin gene *TUB2*. Certain mutations in the α-tubulin gene *TUB1* are interactional suppressors of *tub2* missense mutations, and we implied that such suppressors define sites of interaction between the α- and β-tubulin proteins. However, it is fair to ask: Has anyone tested this inference to see if these mutations do in fact lie at the sites of interaction between the two proteins?

The assembly of tubulin into a mitotic spindle provides an example that allows us to answer this question, at least in part. The three-dimensional structure of bovine tubulin has been solved and, since the amino acid sequences of the tubulin subunits are evolutionarily conserved, this structure provides a reasonable model for the structure of yeast tubulin. Thus, the location of each mutation could be determined, not only in terms of its location in the gene and polypeptide chain, but also in terms of its location in the tubulin heterodimer. In principle, this would allow us to see if the *tub2* missense mutations used in the suppressor studies lie in a region of β-tubulin that interacts with α-tubulin.

To make the investigation even more powerful, many additional mutations in *tub1* and *tub2* were made by a procedure known as alanine-scanning mutagenesis. In this procedure, clusters of charged amino acids are changed to alanine using *in vitro* mutagenesis of the cloned *TUB1* or *TUB2* gene. The mutated gene is then reintroduced into the yeast cell as the only source of α- or β-tubulin, and the mutant phenotype of the cell is characterized. Alanine-scanning mutagenesis allows a more rapid and thorough investigation of different regions of a protein than single amino acid substitutions. In particular, charged residues that are replaced with alanine are often on the surface of the protein and thus more likely to affect interactions with other proteins. From the phenotypes of these mutations, different interactions could be tentatively assigned to regions of the proteins, which could then be compared with the expected location on the three-dimensional structure.

The results of these experiments for α- and β-tubulin provided striking confirmation of the value of the suppressor experiments. The cold-sensitive *tub2* mutations suppressed by specific mutations in *tub1* do in fact alter amino acids that are predicted to lie on the surface of β-tubulin, either in the region of interface between the two proteins or on the lateral surfaces of β-tubulin. These are sites on the tubulin protein that other proteins involved in spindle assembly might contact. Therefore, by comparing what we know about the three-dimensional structure of the tubulin heterodimer with what we have inferred from the suppressor analysis, we find that the genetic suppressor analysis gives a good approximation of the regions of interaction. Because this comparison of molecular and biochemical analysis works so well in an example when both are known, we have greater confidence that interactional suppression between other proteins identifies good candidates for amino acids that lie at the regions of interaction, even when the structures of these proteins are not known.

The role of the *STU1* gene

We have also described how suppressor analysis with *tub2* identified *STU1* as another gene involved in spindle morphogenesis. Further investigation has shown just how useful the suppressor analysis was in this case. Having been identified in *S. cerevisiae*, *STU1* orthologs have now been found in many other organisms, including humans. These proteins are also involved in spindle functions. In addition, a number of different *stu1* mutations have been studied in yeast, including several temperature-sensitive alleles. These *stu1* mutations have their own mutant phenotype independent of their ability to suppress a *tub2* missense allele. The phenotype of these *stu1* mutations, like that of the original *stu1* mutation identified as a *tub2* suppressor, displayed clear defects in the mitotic spindle.

One of the temperature-sensitive alleles of *stu1* has been used for its own suppressor analysis. The suppressors of this *stu1* mutation included mutations of *tub2*. That is, *stu1* mutations act as suppressors of a cold-sensitive *tub2* mutation, and, in turn, certain *tub2* mutations can suppress certain *stu1* mutations. Again, this suppressor profile is suggestive of a specific contact and interaction between the two proteins.

Further investigations of these suppressors suggest an even more complex picture. The *tub2-406* mutation that began this case study affects a site on the lateral surface of the β-tubulin protein. This is in a region of the tubulin heterodimer that is exposed and that may well interact with the Stu1p protein or another microtubule-associated protein. As described earlier, this location is consistent with suppression arising from direct interaction between the two proteins. On the other hand, the *tub2* mutations that acted as *stu1* suppressors did not map to the same region of the β-tubulin protein, suggesting that suppression in this case is not due to specific interactions or contacts. Another explanation is that these *tub2* mutations affect the folded structure of β-tubulin, and the altered structure of the protein changed its interaction with Stu1p. This shows one of the limitations of suppressor analysis. Because it is based on a functional relationship between the two proteins, suppressors only suggest an interaction between the two genes or proteins. In order to prove that a direct biochemical interaction exists, a molecular analysis is required.

Molecular analysis of the Stu1p protein

The genetic analysis indicating that Stu1p is a component of microtubules that interacts specifically with β-tubulin was confirmed from two types of molecular experiments, similar to those described in the main text. The first of these was a biochemical extraction experiment. In cellular fractions, Stu1p is found physically associated with microtubules during co-sedimentation in a centrifugation assay, consistent with the genetic results showing that Stu1p is important for spindle formation. Using the cloned *STU1* gene, the investigators made a series of truncated or mutated Stu1p proteins. Similar sedimentation experiments with truncations of the *STU1* gene allowed them to determine the region of Stu1p that is needed for its association with microtubules. The identification of the microtubule-binding region was important in the next set of experiments.

The association with microtubules by biochemical co-sedimentation assay did not indicate whether Stu1p interacts with α-tubulin, β-tubulin, or both. To determine what directly interacts with Stu1p, the region of Stu1p involved in microtubule association (as determined in the truncation experiments) was used in a yeast two-hybrid (Y2H) assay, as described in the main text. The microtubule-binding region of Stu1p was cloned into a Y2H vector in which it was fused to the GAL4 activation domain. Plasmids with either *TUB1* or *TUB2* fused to the GAL4 DNA-binding domain were transformed into the cells, and a Y2H assay was performed with each gene. The assay clearly showed that the microtubule-binding region of Stu1p interacts with β-tubulin and not with α-tubulin. Again, this is precisely what is expected from the suppressor analysis of the genes—suppressors of Stu1p were found in *TUB2* but not in *TUB1*.

However, this was not the end of the experiments. The region of β-tubulin that interacts with Stu1p was identified by repeating the Y2H assay with the *tub2* mutations induced by alanine-scanning mutagenesis. The assumption is that if the interacting region of *tub2* has been mutated, no interaction will be seen with the binding region of Stu1p. Such mutated versions of *tub2* that no longer interacted with Stu1p were found. The interacting region of *tub2* as identified by this assay was compared with the three-dimensional structure of tubulin, as before. This revealed that the putative interface between β-tubulin and Stu1p is a patch of amino acids on the lateral surface of β-tubulin, on the same face of the β-tubulin protein as the *tub2-406* mutation used in the original suppression studies. Therefore, the physical interaction detected by the Y2H assay defined precisely the same part of the proteins as was defined by the suppressor analysis. (It should be noted that other alanine-scanning mutations in *tub2* failed to interact with Stu1p in the Y2H assay. However, since these mutations also failed to interact with several other microtubule-associated proteins, it is reasonable to conclude that they are affecting the Stu1p interaction via some more general property of β-tubulin, such as its folded structure.) Once again, the striking similarity of results from the Y2H screen and the suppression analysis gives us confidence that the suppressor analysis is providing meaningful information about the protein–protein interactions, even when the structures are not known.

Summary

The mitotic spindle in yeast allows the analysis of an unusually rich set of interactions by both genetic and molecular assays. Because the mutant phenotype is a cell division arrest that can be scored easily, nearly all of the types of genetic interaction screens described in the chapter were exploited in this analysis. This included multiple types of suppressor analysis, synthetic enhancement, and non-allelic non-complementation. Furthermore, because tubulin is a protein whose biochemical properties are characterized and whose three-dimensional structure has been solved, it is possible to combine the results of the genetic analysis with biochemical analysis and molecular analysis, including Y2H assays. The combined results provided a far more complete picture than either approach could have done alone. A summary of the results will indicate how the experiments confirmed and extended each other.

Here then is a partial summary of what we learned from the genetic interactions and were able to confirm from the molecular interactions.

- **The α-tubulin genes *TUB1* and *TUB3* are functionally equivalent** The reason that *tub3* mutations can grow whereas *tub1* mutant cells cannot is explained by a difference of dosage of the two proteins. Increasing the dosage of *TUB3* on a high-copy plasmid suppresses *tub1* mutations, indicating that the two proteins are likely to be functionally equivalent. Synthetic enhancement and non-allelic non-complementation between *TUB1* and *TUB3* mutations provide further support for the conclusion that these genes differ in level of expression but not in function.
- **α-Tubulin and β-tubulin interact** This interaction was certainly known long before the genetic assays addressed it but our knowledge of the interaction did provide an important confirmation of the genetic assays. Mutations at the site of the interaction on β-tubulin can be at least partially suppressed by mutations on the corresponding interface on α-tubulin. The most important conclusion from these experiments is that the genetic assays are a reliable indicator of direct physical interactions.
- **Stu1p is a protein that is associated with β-tubulin in the mitotic spindle and is needed for normal spindle morphogenesis** The gene was identified by suppression analysis. Its function was inferred not only from the suppression profile but also from the analysis of *stu1* mutations and from the localization of Stu1p. Further molecular analysis confirmed that the interaction occurred with β-tubulin and not with α-tubulin, as the suppressor studies had indicated. Likewise, it was possible to map the region of the Stu1p protein and β-tubulin responsible for this interaction by biochemical and Y2H assays. This confirmed in part, but not entirely, inferences made from suppressor analysis.

This single biological example demonstrates how genetic analysis based on a simple mutant phenotype can find additional genes that affect the same function. In addition, it provides insights into how the genes and proteins interact. Therefore, it is not surprising that suppressors and synthetic enhancers are among the most informative tools that the geneticist has to work with. Chapters 11 and 12 will describe how this type of analysis has been expanded even further, to investigate genetic pathways and networks of functionally related genes.

Further reading

Pasqualone, D. and Huffaker, T.C. (1994). *STU1*, a suppressor of a β-tubulin mutation, encodes a novel and essential component of the yeast mitotic spindle. *Journal of Cell Biology*, **127**, 1973–84.

Reijo, R.A., Cooper, E.M., Beagle, G.J., and Huffaker, T.C. (1994). Systematic mutational analysis of the yeast β-tubulin gene. *Molecular Biology of the Cell*, **5**, 29–43.

Richards, K.L., Anders, K.R., Nogales, E., Schwartz, K., Downing, K.H., and Botstein, D. (2000). Structure–function relationships in yeast tubulins. *Molecular Biology of the Cell*, **11**, 1887–1903.

Yin, H.L., You, D., Pasqualone, D., Kopski, K.M., and Huffaker, T.C. (2002). Stu1p is physically associated with β-tubulin and is required for structural integrity of the mitotic spindle. *Molecular Biology of the Cell*, **13**, 1881–92.

Modifications of the basic Y2H assay have reduced these potential shortcomings and allowed more interactions to be detected. A more serious caveat to the Y2H method comes from the potential for false positives, i.e. proteins that do not interact physiologically may interact when expressed in this form in a yeast cell. For example, two proteins that interact in a Y2H assay may be expressed at different times or in different cells under physiological conditions. One way of avoiding some of these false positives is to compare interaction results with expression data to determine if the interaction is physiologically feasible. The suspected interactions can also be tested by other methods such as co-immunoprecipitation, as described below.

Despite their limitations, Y2H screens are widely used and extremely informative. Part of their utility comes from their relative simplicity, particularly as yeast strains with multiple reporter genes under GAL4 control have been developed. Furthermore, no protein purification or molecular reagents are required other than cDNA clones for the genes of interest and a good cDNA library. Unlike purification procedures, one protocol can generally be established for many different interactions. This is an enormous advantage, particularly when considering different organisms. The same procedure for detecting interactions is used whether the proteins are derived from mammals, higher plants, or invertebrates. Another significant advantage is that further investigations of the interaction, such as mapping important amino acids or regions of the proteins, can be done by routine manipulations of the cDNA clones rather than the proteins.

The greatest advantages of Y2H screens arise when using a cDNA library as the prey. In this case, a library is made in which each cDNA is fused to the activation domain. If the cDNA library is sufficiently complete, essentially all of the genes in a genome can be screened for their ability to interact with a bait protein. The initial selection requires only a few plates of yeast medium and a few days. Depending on the quality of the cDNA library and the number of colonies scored, one can be confident that even interactions among minor components or components with low affinity have been found. Just as significant, the gene responsible for each of these interactions is already cloned. For these reasons, Y2H selection has become the most widely used method for looking for interacting proteins. In fact, it is being extended to look at all of the protein–protein interactions in an organism, a topic we will return to in Chapter 12.

A yeast two-hybrid assay was used to examine brassinosteroid signaling in *Arabidopsis*

Y2H screens have been very important for identifying components of the brassinosteroid signal transduction pathway in *Arabidopsis*. Remember from Chapter 2 that brassinosteroids are plant hormones with a variety of developmental roles, many due to their positive effects on cell elongation. The primary brassinosteroid receptor (BR) is BRI1, a leucine-rich repeat receptor-like kinase. Null mutants of *BRI1* have a dwarf phenotype and are insensitive to exogenous brassinosteroids. In an attempt to understand the function of BRI1 better, Y2H screens for proteins interacting with the intracellular kinase domain of BRI1 have been done. (The kinase domain was used as the bait rather than the full-length protein because the researchers wanted to identify the direct phosphorylation substrates of BRI1.) These Y2H screens have identified one positive and two negative regulators of BR signaling, all of which are likely substrates of BRI1.

The interacting protein that acts as a positive regulator of BR signaling was another leucine-rich repeat receptor named BAK1 (BRI1-ASSOCIATED KINASE). BAK1 interacts with the full-length BRI1 receptor both in Y2H assays and *in vivo*. This interaction is strengthened *in vivo* after BRI1 binds brassinosteroids. While BAK1 does not itself act as a brassinosteroid receptor, its activity is required for normal brassinosteroid signaling. Null mutants in *BAK1* have a very weak brassinosteroid-insensitive phenotype, and a kinase-inactive version of *BAK1* acts as a dominant-negative mutation, abolishing all BR signaling. Interestingly, the *BAK1* gene was also identified as a suppressor of *bri* mutations by two different suppressor screens, demonstrating the utility of having approaches that identify both physical and genetic interactions.

One of the proteins identified as a negative regulator of brassinosteroid signaling, BKI1 (BRI1 KINASE INHIBITOR), clarified how BAK1 acts as a positive regulator. In contrast to the results with BAK1, overexpression of *BKI1* in plants results in dwarfism and a 100-fold reduction in brassinosteroid sensitivity. BKI1

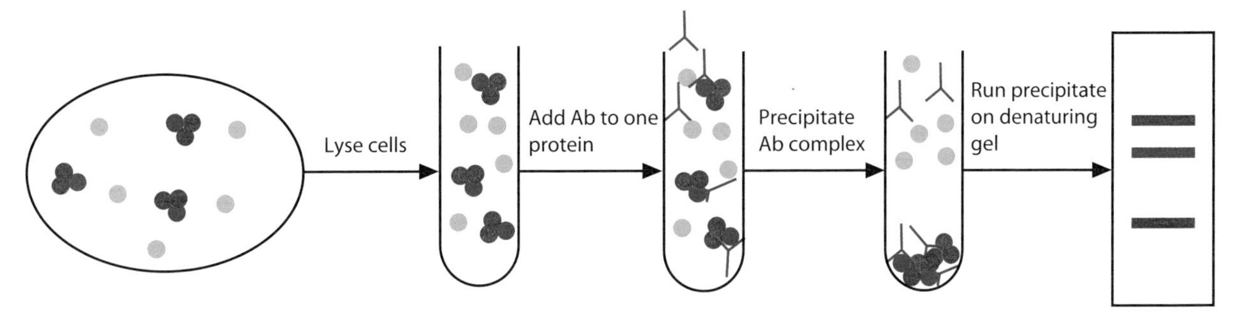

Figure 10.18 Co-immunoprecipitation to detect protein–protein interactions. Proteins within the cell are represented by small colored balls. Co-IP allows the detection of proteins that interact with a protein of interest (in this figure, the blue protein). The blue protein is found in a multimeric complex with the red and purple proteins, but not with the green protein. The cell is lysed by a procedure that preserves the macromolecular complexes. An antibody that is specific to the blue protein is added to the lysate, and the antigen–antibody complex is precipitated. The components of the precipitate are then separated on a denaturing gel. Although the antibody was directed only against the blue protein, the purple and the red proteins are also seen on the gel because of their interaction with the blue protein. The green protein, which is not part of the stable complex, is not found. Identification and characterization of the interacting components depends on analysis of amino acids and proteins, rather than cloned genes.

is localized to the plasma membrane, like BRI1 and BAK1. In the absence of brassinosteroids, BKI1 binds to the kinase domain of BRI1, blocking the interaction between BRI1 and BAK1. Upon brassinosteroid treatment, BKI1 dissociates from the plasma membrane, allowing BAK1 to interact with BRI1 and elicit downstream signaling events. BKI1 acts as a brassinosteroid signaling inhibitor by preventing the interaction of BRI1 and BAK1 in the absence of the hormone.

It is quite common to use only a single domain, such as the BRI1 kinase domain, or another portion of a protein as the bait in a Y2H experiment rather than the full-length protein. Often such bait constructs are used when the protein–protein interactions might only occur *in vivo* after a conformational change in one of the proteins. When only a domain is used as a bait rather than the full-length protein, it is even more imperative to demonstrate that the interaction also occurs *in vivo* since this approach can lead to more false positives. The interacting portions of the two proteins can be further narrowed down using various deletion constructs in pairwise Y2H binding experiments, demonstrating that Y2H assays can be used as both a screen and a specific binding assay.

KEY ARTICLES

Li, J., Wen, J., Lease, K.A., Doke, J.T., Tax, F.E., and Walker, J.C. (2002). BAK1, an *Arabidopsis* LRR receptor-like protein kinase, interacts with BRI1 and modulates brassinosteroid signaling. *Cell*, **110**, 213–22.

Nam, K.H. and Li, J. (2002). BRI1/BAK1, a receptor kinase pair mediating brassinosteroid signaling. *Cell*, **110**, 203–12.

Wang, X. and Chory, J. (2006). Brassinosteroids regulate dissociation of BKI1, a negative regulator of BRI1 signaling, from the plasma membrane. *Science*, **313**, 1118–22.

Co-immunoprecipitation is a physiological standard for protein–protein interactions

Our emphasis on gene-based tools to detect interacting proteins should not be interpreted as an underestimation of the value of biochemical methods based on the proteins themselves. Many sophisticated biochemical methods of looking at protein–protein interactions have been developed. Here, we will briefly describe one widely used method, namely **co-immunoprecipitation** (co-IP). The basis of the procedure is that antibody directed against one protein in a macromolecular complex will co-precipitate the other proteins in that complex, as shown schematically in Figure 10.18.

In order to perform immunoprecipitation, the protein of interest is purified and used as an antigen to produce an antibody. (Antibody production in general was described in Chapter 8 when we discussed immunofluorescence for detecting protein expression.) Cells or tissues that express the protein of interest are lysed and the macromolecules are extracted under gentle conditions that preserve the macromolecular interactions. In some protocols, the proteins in the cells are reversibly cross-linked prior to extraction. The cell extract is incubated with the antibody against the protein of interest, which then precipitates not only the original protein but also other proteins with which it has stable interactions. The precipitated complex is separated on a denaturing gel, and the individual proteins are analyzed.

Co-IP has its advantages and limitations. One limitation is that, compared with the other methods that we have described, it is more technically challenging. No standard set of incubation and extraction conditions works for all protein complexes or all organisms. The method also depends on the quality of the antibody used to detect the complex. Because different antibodies are needed for each protein, the procedure cannot readily be scaled up to look at genome-wide interactions, as can be done with Y2H assays. On the other hand, a key advantage is that co-IP experiments can find previously unknown interactions, involving even minor components. For example, co-IP experiments have shown that at least 55 proteins are needed for accurate transcriptional initiation and elongation in eukaryotes. Its most fundamental strength is that the method approximates physiological conditions and gives few false positives—proteins that can be co-immunoprecipitated probably do form a complex *in vivo*. For this reason, co-IP experiments are often used as the standard for confirming protein–protein interactions initially detected by other methods such as interactional suppressors or Y2H assays. For example, the Y2H interactions between BRI1 and BAK1 and between BRI1 and BKI1 described above were also demonstrated by co-IP to occur *in vivo*.

The power of protein interactions: an example from plant pathology

A combination of Y2H and co-IP experiments in *Arabidopsis* led to one of the largest recent paradigm shifts in the field of plant pathology. Plants are subject to infection and disease caused by many bacteria, fungi, viruses, and nematodes. For decades, research in many different crop species had noted a 'gene-for-gene relationship' between so-called resistance genes (*R* genes) in the host and avirulence genes (*avr* genes) in their pathogens. It had been proposed that the protein product of a specific *R* gene recognizes a single Avr protein from one pathogen via direct protein–protein interactions. According to this model, pathogen recognition by the plant then results in resistance.

Genetic experiments in *Arabidopsis* and various crop species supported such a model. A pathogen with a recessive mutation in an *avr* gene might no longer be recognized, and thus can overcome the resistance conferred by the *R* gene. Likewise, a plant with loss-of-function mutations in an *R* gene can lose resistance to a specific pathogen because the mutations are unable to recognize the pathogen.

The simplicity of this model is very appealing, and it makes the testable prediction that direct protein–protein interactions should occur between the specific R proteins and specific Avr proteins. In fact, very few examples of direct protein–protein interactions between R/Avr protein pairs could be demonstrated. Mutants in *Arabidopsis* and the protein interaction assays described here allowed a direct test of the model.

Arabidopsis defends itself against AvrB-expressing bacterial strains of *Pseudomonas syringae* using the R protein RPM1. The *P. syringae* pathogen infects plants by secreting AvrB proteins into the *Arabidopsis* cytoplasm, so it was thought that the cytosolic RPM1 protein might interact with and disable the Avr proteins. However, no direct interaction between AvrB and RPM1 could be found. Investigators performed a Y2H experiment using AvrB as the bait with an *Arabidopsis* cDNA library as the prey in order to identify the plant proteins directly binding AvrB. Although RPM1 was not found, one of the prey proteins called RIN4 was identified by the AvrB bait.

By using transgenic plants expressing an epitope-tagged version of RPM1, it was found that RPM1 and RIN4 interact *in vivo* in the absence of pathogens. In these experiments, RIN4 and its associated proteins were immunoprecipitated with an antibody against RIN4 and separated on a polyacrylamide gel. Using an antibody against its epitope tag, RPM1 was identified as one of these RIN4-interacting proteins. Similar experiments were done using an epitope-tagged *avrB* transgene under an inducible promoter, and showed that AvrB and RIN4 also interact *in vivo*. These immunoprecipitation results indicate that RIN4 interacts directly with both AvrB and RPM1, which do not appear to interact with each other.

Further co-IP experiments showed that another *P. syringae* Avr protein (AvrRpt2) and its R protein (RPS2) also interacted with RIN4, indicating that RIN4 is not specific for one pathogen and resistance gene but may play a more general role. The following model, named the guard hypothesis, has been proposed. In the absence of a pathogen, RIN4 is found in a protein complex with one of the R proteins. Upon infection with *P. syringae*, the Avr proteins are expressed; these proteins target and destabilize RIN4. The destabilization of RIN4 in turn signals the R proteins that an infection has occurred, and the plant defense pathways are engaged. Thus, rather than recognizing the pathogens themselves, R proteins detect the presence of an infection by guarding the proteins commonly targeted by the pathogen. Rather than preventing the destruction of these targets, these R genes recognize the damage and increase the defense response. Additional examples of this mechanism involving different guarded proteins have since been uncovered in *Arabidopsis*, supporting the validity of the model.

This example displays some of the strengths and weaknesses of the co-IP technique. The Y2H experiments identified RIN4 as a potential interacting protein. The co-IP experiments verified the Y2H results *in vivo*. The protein interaction assays were able to test and disprove a model that could not be ruled out by genetic data. The results prompted a reinterpretation of a fundamental aspect of plant pathology. On the other hand, co-IP requires antibodies against the proteins of interest or the expression of epitope-tagged versions of the proteins. As a result, co-IP is more challenging to use for identifying new protein–protein interactions than Y2H assays. In addition, once a protein complex has been precipitated, analysis of the other proteins is still more difficult than analysis of the genes encoding them, although advances in proteomics are making this more routine.

KEY ARTICLES

Mackey, D., Holt, B.F., Wiig, A., and Dangl, J.L. (2002). RIN4 interacts with *Pseudomonas syringae* type III effector molecules and is required for RPM1-mediated resistance in *Arabidopsis*. *Cell*, **108**, 743–54.

Mackey, D., Belkhadir, Y., Alonso, J.M., Ecker, J.R., and Dangl, J.L. (2003). *Arabidopsis* RIN4 is a target of the type III virulence effector AvrRpt2 and modulates RPS2-mediated resistance. *Cell*, **112**, 379–89.

SUMMARY

Finding more genes

Almost every biological process is affected or controlled by many interacting genes and their protein products. In previous chapters, we have described methods of finding and characterizing genes that affect a biological process. In these methods, we began with a wild-type organism. Now, in this chapter, we have added new tools to our genetic toolbox that rely on a previously identified gene, in the form of a mutant phenotype or a cloned gene or both. Two of the methods, suppression and synthetic enhancement, require only that the investigator has a mutation in the gene of interest. The gene of interest does not need to be cloned to use these methods, although that is certainly an advantage. The other two methods, co-IP and Y2H screens, require that the gene of interest be cloned and, in the case of co-IP, that an epitope-tagged protein or specific antibodies have been made against its protein. These latter two methods do not require that mutant alleles exist for the gene, although again that is certainly an advantage. In practice, an investigator will probably use several of these methods to find genes related to the gene of interest because no single method provides all of the information necessary or even all of the other related genes that might be identified. The comparative strengths and limitations of the four methods are summarized in the chapter capsule, and are described below.

The methods that begin with a mutant phenotype—suppression and synthetic enhancement—have the great advantage that the initial screen is usually easy to do. In addition, they require the least prior knowledge about the biological process and can be used to find genes involved in the same process, even if the interaction between them is indirect. Thus, they give the largest number of interacting genes. On the other hand, because these are methods based on mutant phenotypes, the interaction between the genes is occasionally too indirect to be helpful or unambiguous. Furthermore, the genetic methods in and of themselves do not result in cloned genes or molecular knowledge of genes and proteins or the biological process.

In contrast, the Y2H screen works only with cloned genes and yields cloned genes. It is relatively simple to perform and, with a good library of prey clones, it can be scaled up to identify all of the proteins and genes capable of a direct interaction. The protocols have been worked out in depth and can be applied to any organism regardless of the process under investigation. The disadvantages of Y2H screens are that certain types of interaction might be missed. Furthermore, it does not give information about the nature of the interaction, so the investigator does not know if the interacting protein activates or represses the bait protein. The most significant disadvantage is that some of the interactions may not be physiologically meaningful and must be confirmed by other assays. Nonetheless, the follow-up methods are feasible.

The most widely used follow-up method is co-IP. In addition to being useful for confirming interactions from Y2H or other screens, co-IP is itself useful for detecting interacting proteins. It has the advantage of being the most likely to identify interactions that are physiologically meaningful. It can also provide the most detailed biochemical information about these interactions. It comes the closest of all of the methods to examining dimeric and multimeric interactions in their native state, complete with the post-translational modifications. On the other hand, co-IP remains the most difficult of these methods to perform, and requires the most prior knowledge and additional reagents, particularly high-specificity antibodies. In addition, since it identifies interacting proteins, the corresponding genes still have to be identified and cloned, and the analysis of proteins is still less routine than the analysis of DNA.

All of these methods have a role, and all are widely used. Think of the genetic and molecular analysis of a biological process in terms of solving a jigsaw puzzle, with each gene or protein being represented by a puzzle piece. The gene that the investigator starts with can be thought of as a piece that depicts part of a particular scene and has a particular shape. Suppose that it is a piece that depicts part of the roof of a barn. In this analogy, the molecular and biochemical methods correspond to finding an interlocking puzzle piece with the right shape regardless of what scene the piece depicts. If the two pieces fit together in all particulars, we can have great confidence that they sit next to each other in the puzzle. In contrast, the genetic methods look only at the scene that is depicted and not at the shape of the piece. Again, this can be an advantage. The genetic methods are likely to yield far more genes than the molecular methods—in our puzzle analogy, there are more pieces that depict part of the farmer's barn than there are pieces that fit together with our individual piece. Some of those pieces are next to each other in the puzzle but other pieces are far apart. Thus, in order to complete the puzzle, we have to use both the scene that is depicted and the shapes of the individual pieces. Similarly, in order to analyze a biological process, the investigator will use both genetic and molecular methods.

Having used these methods, the investigator will have identified a substantial number of the genes that affect the process—in principle, every gene or protein that affects the process. In the course of finding these genes, the investigator has seen glimpses of the underlying logic of the process itself. To use our jigsaw puzzle analogy one last time, as we have collected the pieces, we begin to see the picture on the top of the puzzle box. In the next two chapters, we will describe how these methods can be used together to yield an even more complete picture of the biological process. We have the mutants, we have the clones—let us use them to solve some biological problems.

Chapter Capsule

Finding related genes

Using the mutant phenotypes of the original mutation: suppressors and synthetic enhancers

Method

Mutagenize a strain with the gene of interest. Look for offspring with a more wild-type phenotype (suppressors) or a more severely mutant phenotype (synthetic enhancers).

Advantages

- Easy to do.
- Requires little or no prior knowledge or insights about the process.

- Easily scaled up.
- Can find genes in the same process, even if the interaction is indirect. Highest yield of any of the methods.
- Can be used together with epistasis to make a pathway, and the types of relationship are known.

Disadvantages
- The interacting genes still have to be cloned.
- The follow-up genetics (such as testing other alleles and other genes) can be difficult or time-consuming.
- The interaction may be too indirect to be helpful.
- Interpretation can be ambiguous.

Using yeast genetics and the original gene (molecular genetics method): yeast two-hybrid screens

Method
Transcriptional activation in yeast is used to detect protein–protein interactions. If the bait (gene of interest) and prey (gene to be tested) proteins interact with each other, two domains of the GAL4 transcription factor are brought together and the downstream selectable gene is transcribed.

Advantages
- It relies on a selection, so tens of thousands of possible interactions can be screened at once. It is the easiest method to scale up to genome analysis.
- Because yeast has a shorter generation time than most other organisms, it is faster than most other methods.
- Direct protein–protein interactions are detected.
- Has a standard protocol that works for most proteins for all organisms so many follow-up procedures and modifications have been tested.
- Not very demanding technically.
- The interacting genes are cloned as a direct result of the procedure.

Disadvantages
- Interactions are tested under non-physiological conditions. Some interactions that are found are not physiologically relevant.
- Only binary interactions are found. Multimeric interactions are missed.

- Proteins with post-translational modifications might be missed.
- Requires having a cloned gene and a good prey library to screen.
- Does not yield mutant phenotypes, so cannot tell about the nature of interaction (either activating or repressing).
- Gives information about the direct interactions but not about the overall process or the pathway.

Using antibodies against the original protein (biochemical method): co-immunoprecipitation

Method

Extract protein complexes from wild-type tissue or cells. Use an antibody to the protein of interest to isolate (precipitate) that protein from the extract. Other proteins that directly interact with the protein of interest will be co-precipitated because of their biochemical interactions with that protein.

Advantages

- Most physiological of all of the molecular methods.
- Gives the most detailed information of all of the methods including:
  (a) stoichiometry of the components
  (b) strength of the interaction
  (c) Post-translational modifications of proteins.
- Can detect multimeric complexes.
- Ensures direct interaction of the proteins (and genes).

Disadvantages

- Most technically demanding of all of the methods.
- Requires the most prior knowledge and reagents.
- No 'standard protocol' exists so the procedure has to be adapted for each complex and each organism.
- Genes still have to be cloned, and the techniques for working with protein sequences are not as easy as those for working with DNA sequences.
- Does not yield mutant phenotypes.
- Gives information about the direct interactions but not about the overall process or pathway.

CHAPTER MAP

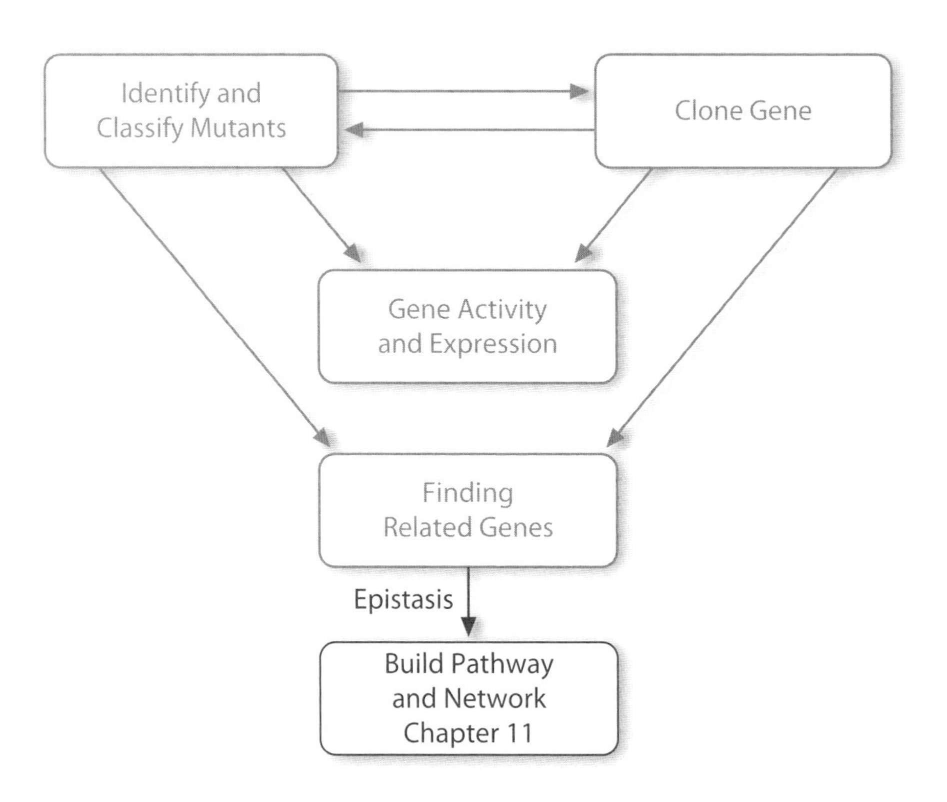
Identify and Classify Mutants
Clone Gene
Gene Activity and Expression
Finding Related Genes
Epistasis
Build Pathway and Network Chapter 11

11

CHAPTER 11

Epistasis and genetic pathways

TOPIC SUMMARY

Genes or their products modify the activity of other genes so that a phenotype is the outcome of genes interacting with each other in a pathway or a network. Genetic methods such as suppression, synthetic enhancement, and epistasis can be used to infer a logical sequence in which genes function in a pathway. Each of these genetic methods compares the phenotype of an organism that is mutant in two different genes with the phenotype of single mutants. The phenotype of the double mutant is used to infer which gene is downstream of the other in the pathway.

IN BRIEF

The pathway by which two genes interact can often be inferred from a double mutant that lacks the function of both genes.

INTRODUCTION

Genetics has prospered by the analysis of single genes with a simple mutant phenotype. One need only think of the pioneers of the field: Mendel with round and wrinkled peas, Garrod and inbound errors of metabolism, Morgan with a white-eyed male fly, and so on. There is a long and ever-expanding list of key discoveries that focus on individual genes. Even with the wave of new approaches made possible by genomics, one of the crucial stages in genetic analysis is characterization of the role of an individual gene.

However, genetic analysis of a biological process will almost never stop with the analysis of only one gene. Many different genes contribute to the biological process, and the investigator will want to identify and characterize all of them. Two methods of finding more genes have been described in previous chapters—mutant hunts that focus on a specific set of phenotypes (Chapter 3) and genome-wide screens (Chapter 6). Although it is rarely easy to accomplish, it is realistic to think of using these methods to identify all of the genes that affect a biological process.

Characterizing individual genes is only part of the process of genetic analysis, however. A biological process can be thought of as a pathway with the genes and proteins comprising individual components or steps. The pathway may be long or short, straight or branched, but genetics can be used to reveal its underlying logical structure. Genes and proteins interact with other genes and proteins. In fact, it could be argued that every identified gene or protein leads to a larger question involving other genes and proteins. If the protein is a kinase, we immediately wonder about its substrate; if it is a DNA-binding protein, we wonder about its regulatory targets. Even when we know all the parts of the pathway, we still need to know their interactions in order to understand a biological question.

In this chapter and the next, we will consider several different methods that expand genetics beyond the analysis of individual genes and proteins to the analysis of several genes interacting in a biological process. The next chapter will focus principally on genomic methods to infer biological networks. This chapter builds on genetic approaches, primarily using mutant phenotypes, to infer the pathways by which genes affect a biological process. Our task is to build the genes we have identified by mutant phenotypes into a pathway; our tools are suppression, synthetic enhancement, and epistasis (Figure 11.1).

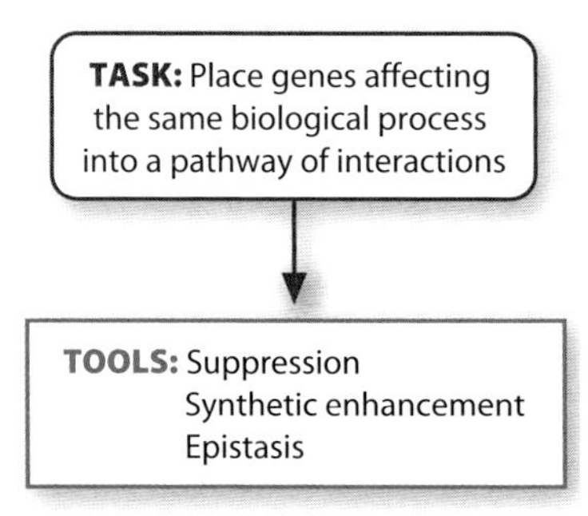

Figure 11.1 Our task and tools.

11.1 Epistasis and genetic pathways: an overview

Genes and proteins work together with other genes and proteins to carry out a biological process. In the preceding chapters of this book we have described how to find and characterize individual genes, even how to find all of the genes that affect a biological process. In Chapter 10, we described how one gene could be used as a starting point to find additional genes involved in the same process. In this chapter we will describe how to expand genetics beyond the analysis of individual genes and proteins to the analysis of several genes interacting together.

The genes and their products can be thought of as individual musicians. Just as it is possible to catalog each of the genes, it is also possible to isolate each musician in an orchestra and listen to his/her part separately. That is enough to describe some remarkable music. But musicians in an orchestra only occasionally play unaccompanied solos, just as genes and proteins in a cell only occasionally perform a function alone. If analysis of a single gene is similar to listening to a Partita for Solo Violin, analysis of many interacting genes may be comparable to listening to one of the Brandenburg Concertos—no matter how remarkable the activity of that one gene, it becomes all the more amazing and intricate when placed in the context of the activity of other genes.

In learning how genes (or, more often, their protein products) interact with one another, we turn from the analysis of single mutants to the analysis of double- or even triple-mutant strains. That is, rather than examine the phenotype of a cell or organism that has a mutation in one gene, the investigator examines the phenotype of a cell or organism that has mutations in several different genes—most commonly, in two different genes. To oversimplify slightly, the double mutant may have one of four possible phenotypes with respect to each of the single mutants. We summarize these possibilities in Figure 11.2. First, the double mutant may exhibit both mutant phenotypes. Mendel encountered such an example in using peas that were doubly mutant for round vs. wrinkled and for yellow vs. green. We learned long ago that these traits are **inherited** independently of one another, taught as Mendel's Law of Independent Segregation. But a simple but overlooked lesson from that same experiment is that these genes *act* independently of one another—the color of the pea has no effect on the shape of the pea.

Gene interactions

a/a;b+/b+ Mutant phenotype A

a+/a+;b/b Mutant phenotype B

a/a;b/b What is the phenotype of the double mutant? What effect does the mutation in gene *b* have on the phenotype of *a*?

Possibility 1. The double mutant shows both mutant phenotype A and mutant phenotype B.	**Inference:** Gene *a* and gene *b* act *independently.*
Possibility 2. The double mutant shows a wild-type phenotype.	**Inference:** Gene *b* is a *suppressor* of gene *a*.
Possibility 3. The double mutant shows a more severe mutant phenotype than A.	**Inference:** Gene *b* is a *synthetic enhancer* of gene *a*.
Possibility 4. The double mutant shows the same mutant phenotype as gene *b*.	**Inference:** Gene *b* is a *epistatic* to gene *a*.

Figure 11.2 Gene interactions that could be observed in a double mutant. Suppressors and synthetic enhancers are discussed in Chapter 10, while this chapter focuses on epistasis in which only one mutant phenotype is observed. Note that this refers to the phenotype and not to the inheritance of the two genes. Genes may be inherited independently because they are on different chromosomes but do not provide independent phenotypes.

What if the two genes do not function independently of one another? In this instance, the phenotype of one mutant will affect the phenotype of the other mutant. (Notice that we are describing phenotypes and not inheritance.) There are three possible double-mutant phenotypes as shown in Figure 11.2.

1. The double mutant may exhibit a phenotype that is wild-type or more like wild-type, exhibiting neither of the phenotypes of the individual mutations. This type of interaction, is known as **suppression**.
2. The double mutant may have a more severe mutant phenotype than either of the two single mutants. This type of interaction is known as **synthetic enhancement**. Both suppression and synthetic enhancement are discussed extensively in Chapter 10.
3. Only one of the two mutant phenotypes is seen and the other one is obscured or masked. This type of interaction is known as **epistasis**, and is the principal topic of this chapter.

The logic of epistasis requires close attention

Before describing epistasis and the analysis of genetic pathways, we need to make a few important points about the logic that will be used. First, recall that the normal function of the genes are being inferred from their *mutant* phenotype, i.e. from the absence of function of a gene or the absence of its protein product. In nearly all cases, the alleles used for the genes being studied are recessive null alleles; the occasional exceptions will be clearly indicated. One example that will be used is sex determination in *C. elegans*. As we see later, null alleles in the gene *tra-1* result in both 1X and 2X animals developing a male morphology. Thus, the role of the *tra-1+* gene is to produce females (or hermaphrodites) since its absence *prevents* female development.

Secondly, as discussed in Chapter 9, in working with mutant phenotypes, we are making inferences about *gene activity* rather than *gene expression*. When a gene is described as being ON or OFF, the molecular biology of gene expression is being oversimplified or temporarily ignored. A gene is inferred to be ON if the organism requires its function for a wild-type phenotype. This identification is based on mutant phenotypes and not usually on a molecular analysis. In writing that gene *A* turns OFF gene *B*, the mechanism by which this happens is often not known and, at least as a starting point, not necessary to be known. It might mean that the protein product of gene *A* blocks the transcription of gene *B*, or that the protein product of gene *A* directs the splicing of the pre-mRNA of gene *B* to an unproductive isoform, or that it blocks the translation of the mRNA from gene *B*. The statement might mean that the protein product of gene *A* modifies the protein product of gene *B* by phosphorylation or dephosphorylation, and thereby inactivates its function, or that it prevents the protein product of *B* from being localized to its proper place in the cell, or that it causes the degradation of protein B. As we can see from these examples, the protein product of gene *A* might have any of a host of effects that prevent gene *B* from carrying out its normal function, including effects that we are not clever enough to have thought of yet. (As a side note, the regulatory role of a microRNA was first identified by using epistasis of a genetic pathway many years before microRNAs were recognized as the functional molecule.)

Thirdly, and most significantly, just because we are often inferring the relationship between the two genes from an analysis of their mutant phenotypes, we cannot assume that gene *A* and gene *B* or their products physically interact—they might not interact at all. One of the great strengths of genetic analysis of single genes is that one need not know anything about the gene product in order to perform informed experiments that allow insightful conclusions. It is also true of the analysis of gene interactions—one need not know the molecular nature of the interaction to do experiments that analyze the interaction. Of course, if the molecular nature of the interaction is never investigated, some key insights will be missed. As with many aspects of genetic analysis, the mutants can tell us where to look, but they may not tell us what we will find.

The logic of using the interactions between mutant phenotypes to infer the wild-type functions of the genes can become intricate. A form of epistasis can also be done using a mutation in one gene and the wild-type phenotype for another gene. This analysis is usually a little easier to grasp and it sets the stage for the analysis using two mutant phenotypes. We will begin with that approach.

11.2 Combining mutants and molecular expression assays

Let us consider an analysis of the genetic interaction between two genes that uses one mutant phenotype and one wild-type phenotype. In this particular type of experiment, the investigator has a mutation in one of the two genes, and a molecular reagent such as an antibody or a reporter gene for the expression of the other (potentially) interacting gene. The molecular reagent can be drawn from any of the expression assays described in Chapter 8, including microarrays. We introduced the *patched* gene in *Drosophila* in Case Study 3.1 in Chapter 3, and we have used the genetic analysis of this gene in several previous chapters. Although much of our prior discussion of *patched* (*ptc*) centered on its role in the segmentation of the embryo, this gene also has a later role in wing development, as noted in Chapter 4 when we discussed some hypomorphic alleles of *ptc* and in Chapter 9 when the mosaic analysis of *ptc* function in the wing was used. We return to the role of *ptc* in the wing in this example.

As noted in Chapter 2, the wings and other limbs in adult flies are derived from particular pockets of cells set aside during embryonic development, which are known as imaginal disks. The cells of the imaginal wing disks are subdivided into two groups, an anterior compartment and a posterior compartment. As described in the case study of mosaic analysis (Case Study 9.2 in Chapter 9), mutations in the *ptc* affect the development of the anterior compartment, with little or no effect on the posterior compartment. For example, mutations in *patched* result in the loss of specific structures in the anterior wing compartment and the overgrowth of other anterior structures. Antibodies to the Ptc protein were used to show that *ptc* is expressed throughout the anterior compartment, but most strongly in the cells close to the compartment boundary.

A mutation in *ptc* changes the expression pattern of other proteins expressed in the wing

Another gene involved in imaginal disk development is decapentaplegic (*dpp*), which is needed for outgrowth of the disks into normal-sized limbs; in *dpp* mutants, the wings are quite small or even absent. Antibodies to

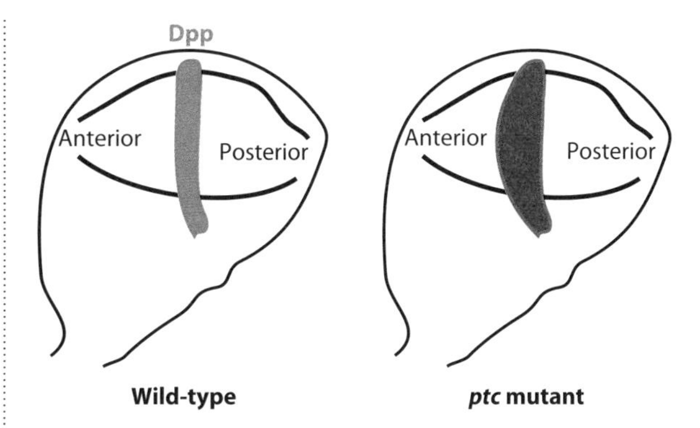

Figure 11.3 The effect of *ptc* mutations on Dpp protein localization. In a wild-type *Drosophila* wing imaginal disk, the Dpp protein is expressed in a band across the middle of the wing. In a *patched* mutant (*ptc*), the expression of the Dpp protein expands into the anterior compartment to include cells that do not express Dpp in the wild-type.

the Dpp protein showed that expression was confined to the cells on the anterior side of the compartment boundary, and it is not expressed throughout the anterior compartment. This is shown diagrammatically in Figure 11.3. Thus, both the Dpp and Ptc proteins are present in some of the same cells at about the same time, although Ptc is expressed in a broader region than Dpp. In light of the similar expression pattern, it is logical to ask if the expression of one gene affects the expression of the other.

The combination of the antibodies to detect the presence of each protein and the mutations in each gene allowed investigators to answer a fairly simple question: Does the *ptc* gene control the expression of the Dpp protein? (The results of the analysis are slightly simplified here for clarity; we are more interested in the method used than the information the method yielded.) To answer this question, the Dpp expression pattern was examined in a *ptc* mutant fly and compared with the wild-type pattern. The results are shown in Figure 11.3. In a *ptc* mutant, the expression of Dpp protein expands over a broader region of the anterior compartment than it does in a *ptc+* fly, and cells that do *not* express Dpp in a wild-type fly *do* express Dpp in a *ptc* mutant. The inference is that the expression of Dpp depends on *ptc* expression; in particular, *ptc+* turns off *dpp+*. Even if the fly is mosaic for *ptc* and *ptc+* by methods described in Chapter 9, this effect is seen: Dpp is expressed in *ptc* mutant cellular clones, whereas it is not expressed in

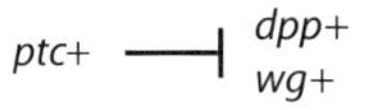

ptc+ ⊣ *dpp+* *wg+*

Figure 11.4 A symbolic representation of the relationship shown in Figure 11.3. The normal role of the *patched* gene (*ptc+*) is to repress or turn OFF the expression of the genes *dpp* and *wg*. This symbolism refers to the logical relationship without implying a direct interaction or a mechanism.

ptc+ cells. By a similar analysis, *ptc+* was also found to turn off the gene *wingless* (*wg+*). In contrast, the expression pattern of Ptc, as detected by antibody staining, is the same in wild-type and in *dpp* or *wg* mutant flies, so the localization of the Ptc protein does not depend on the activity of the *dpp* and *wg* genes.

The logical pathway describing these interactions is depicted as shown in Figure 11.4. Particularly note the symbol used for turning OFF the activity of the other gene. We are inferring the activity of the wild-type *ptc+* gene from the effects of the *ptc* loss-of-function mutation. We interpret Figure 11.4 as indicating that, in normal flies, the activities of the *dpp+* and *wg+* genes function downstream of the *ptc+* gene, and that the role of the *ptc+* gene is to turn off the *wg+* and *dpp+* genes. To express this relationship another way, the wild-type *ptc+* gene negatively regulates the expression of Dpp and Wg.

Although this seems like a simple result, it is worth a bit more thought. Recall from our discussion in Chapters 3 and 9 that the mutant phenotype can be used to infer the gene function but it does not provide details about that function. We have inferred that *ptc+* turns off Dpp and Wg. How exactly does *ptc* turn off the expression of *dpp* and *wg* in those anterior cells? We do not know from this experiment. The experiment does not tell us if the effect is on transcription, translation, or some other molecular process for the *dpp* and *wg* genes—if the *ptc* gene product interacts directly with the *dpp* DNA sequence, the mRNA splicing pattern, or the Dpp protein itself—or even that the same molecular process is involved in its regulation of both genes. Just as significantly, the experiment does not tell us if the interaction between *ptc* and *dpp* is direct or if other genes are involved. In fact, many other genes could be involved in this process. The experiment simply describes a functional relationship between the normal activity of the *ptc+* gene and the localization of the Dpp protein. Put another way, the interaction described this way gives us the *logic* by which *ptc* and *dpp* interact but it does not tell us the *mechanism*.

However, the logic for many complicated processes is often enough to provide useful information. Consider an example from daily life. When I need to stop my car suddenly, it is sufficient to know that if I push down hard on the large center pedal on the floor, the car will stop. I do not need to know the mechanism by which this pedal exerts its effect, only that it happens. While it might be helpful to understand hydraulics, brake pads, and rotors, this will not change the simple logic—step down here and the car stops. So it is with this simple example of gene interaction: we infer that *ptc+* turns off Dpp because Dpp is expressed in *more* cells in the *absence* of *ptc* activity.

The combination of genetic mutations with molecular reagents of this type is a frequently used analytical method in genetics, with hundreds or perhaps thousands of examples in the literature. The approach does not have a common name, although some investigators have referred to it as **molecular epistasis** to illustrate a connection with the genetic approach we describe next. By using mutant alleles of one gene and antibodies raised against the wild-type protein of the second gene, we can determine if one gene affects the expression and localization of another gene product. This does not require antibodies, and any molecular reagent can be used. Similar experiments are done using *in situ* hybridization to mRNA and by reporter gene constructs with *lacZ* or *gfp* instead of the antibody. Microarray experiments that compare the wild-type expression pattern with the expression profile in a mutant are another example of this logical approach.

In one sense, molecular epistasis experiments are simply obtaining a more detailed analysis of the mutant phenotype of one of the genes, *ptc* in our original example. Conceptually, the expression pattern of the target gene (*dpp* or *wg* in our example) is a phenotype of *ptc* in the same way that bristle number, survival, antibiotic sensitivity, amino acid prototrophy, or sexual morphology are the phenotypes of other genes. However, because the mutant phenotype of one gene is defined in terms of the expression pattern of another gene or genes, this approach allows us to build a **pathway** through which the genes or their proteins work in a way that morphological phenotypes alone do not.

In our example, we placed *ptc* upstream of *dpp* and *wg* in a functional pathway and showed that *ptc* is a negative regulator of these two proteins. This approach can be extended to include many other genes and proteins to construct a longer or more detailed logical pathway. The strength of this method is that the investigator needs to have only two components—a mutation in one gene and a molecular reagent to monitor the expression of the other gene—and no prior knowledge of functions of these genes.

A thoughtful consideration of this approach will also recognize its limitations. Although we can say that *ptc+* represses the *dpp+* and *wg+* genes, we have no idea whether the interaction between *ptc+* and these downstream genes is direct or indirect. We have no idea how many other genes might be involved and what they may be doing, and we have no idea of the mechanism by which the upstream gene is affecting the expression of the downstream gene. Of course, this method also requires that one or both of the interacting genes has been cloned, and that molecular reagents to monitor its expression (such as antibodies, fusion proteins, or *in situ* hybridization probes) are available. Those reagents may not always be available for one of the genes, at least not at the beginning of the analysis. Despite these limitations, we can begin to build a pathway or a network by which genes interact to determine a more complicated phenotype.

By understanding the rationale behind using molecular epistasis we can now consider some other genetic methods that do not rely on molecular reagents or cloned genes, and thus may be even simpler to perform. These genetic methods use only the mutant phenotypes to infer the pathway of interactions among genes without knowledge of any of the molecules involved. This is one of their advantages—the interpretation of the interaction does not require knowledge about the molecules involved. All that is required is the ability to construct and analyze a double-mutant strain.

11.3 Epistasis and genetic pathways

At the beginning of the chapter, we discussed the fact that a double mutant could have one of four possible mutant phenotypes with respect to the corresponding single mutants. One type of gene interaction between two genes mentioned earlier is **epistasis**. Epistasis refers to the ability of the activity of one gene to mask the activity of another gene. Since gene activity can be inferred from observing mutant phenotypes, epistasis often refers to the ability of mutations in one gene to mask the effects of mutations in a second gene. That is, in a double mutant, only one of the two mutant phenotypes is observed; the other is not seen.

An example used in many textbooks involves genes for coat color in mammals, as shown in Figure 11.5. In the conventional textbook problem, there are two genes represented by *C* and *B*. In order for color to be observed, the *C* allele must be present: *cc* animals are white. At the *B* locus, black (*B*) is dominant to brown (*b*), so that the genotypes *BB* and *Bb* are black and the genotype *bb* is brown. Double-mutant animals that are genotypically *cc;BB*, *cc;Bb*, and *cc;bb* are all phenotypically white because the C allele is needed to express color. We say that a mutation in the *C* gene to *cc* is epistatic to the *B* gene: *C* masks the effects of *B*. (In old terminology that is rarely used, the *B* gene is said to be hypostatic to *cc* because it is the gene whose effect is masked.) If the *C* allele is not present, the genotype at the *B* locus has no effect on the phenotype. Another way of stating this result is that the *C* allele is needed for the *B* locus genotype to be expressed in its phenotype; in *cc* animals, the genotype at the *B* locus is irrelevant to the final phenotype.

Let us consider this result in light of our previous discussion using expression pattern as our phenotype. The *C* allele is needed to turn ON expression of the *B* locus, because in *cc* mutants, the *B* locus is functionally OFF. As discussed previously, we do not know the molecular mechanism by which the *C* allele affects the expression or function of the *B* locus or if the interaction between the two loci is direct or indirect. However, we have been able to infer a simple genetic logic: the dominant *C* allele turns ON the function of the *B* gene, as symbolized in the pathway in Figure 11.5.

Epistasis is widely used to place the genes affecting the same biological process into a logical pathway or a framework of interactions. The 'technique' for epistasis is simple: the investigator makes a double mutant between two genes and examines its phenotype. The

Figure 11.5 Epistasis involving coat colors. In this familiar example, the wild-type *C* allele must be present for coloration, and the *B* gene determines black vs. brown. The possible coat color genotypes for each phenotype are listed below each picture, and the epistasis pathway inferred from these phenotypes is shown at the bottom of the figure. The double-mutant animal is the *cc;bb* white rabbit.

relative simplicity of making and analyzing double mutants makes epistasis a powerful and attractive genetic tool. Epistasis was also used to understand the logic of some important biological pathways long before molecular or biochemical information was available. To cite two examples other than the ones we will use in this chapter, both the Ras/Raf signal transduction pathway and the apoptosis pathway were first analyzed using epistasis with mutants in flies or worms.

However, like any powerful tool, epistasis needs to be used with caution. One of the first cautions is that the method describes a logical, not a biochemical, pathway. It is tempting to think of epistatic pathways in terms of metabolic pathways or biosynthetic pathways, but this is not the best way to approach epistasis analysis. A significant strength of the genetic approach is that epistasis experiments do not require any prior molecular or biochemical information about the genes or the process that they affect. For this reason, epistasis experiments are often an early step in the genetic analysis of a biological question. Epistasis provides a logical framework inferred from the phenotype of double mutants. It does not require having all of the steps in the pathway or knowing what any of those functions are. But having the logical pathway then allows the investigator to design new experiments to fill in the molecular and biochemical mechanisms by which the process occurs.

Although there are many published examples where epistasis has been used to infer a pathway of gene interactions, few authors have provided detailed descriptions of the logic that they used to produce the pathway. In the remainder of this chapter, we will take apart some of these experiments and show how epistasis was used in the construction of a few particularly well-studied genetic pathways, attempting to reconstruct the logic of the analysis.

General amino acid control in budding yeast is regulated by two types of gene

Budding yeast has an interesting regulation of amino acid biosynthesis known as general amino acid control. When starved for one amino acid, such as histidine, the yeast cell activates (or more formally, it de-represses) the transcription of genes encoding biosynthetic enzymes for many of its amino acids, as shown diagrammatically in Figure 11.6. The cell begins to transcribe the genes involved in biosynthesis not only for histidine but also for arginine, tryptophan, and lysine. It is easy to imagine how such a regulatory system might be advantageous in

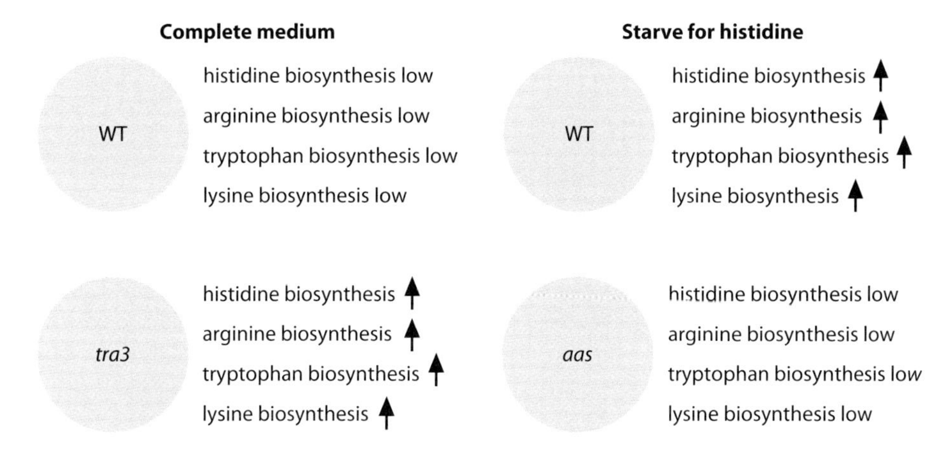

Figure 11.6 The general amino acid response in yeast. When wild-type yeast are grown in complete medium, amino acid biosynthesis is low, as shown in the cell at the top left. When the cell is starved of one amino acid, biosynthesis of several different amino acids increases, as shown at the top right. Two different categories of mutation are known to affect this response. In *tra3* mutants, on the bottom left, general amino acid biosynthesis is elevated even when the cells are grown in complete medium. In *aas* mutants, on the bottom right, amino acid biosynthesis is not elevated even when the cell is starved. Three different *AAS* genes are known: *AAS1*, *AAS2*, and *AAS3*.

nature since yeast cells would rarely be starved of only one amino acid without also being starved of the others; in the absence of one amino acid, the cell begins to synthesize many amino acids for itself.

KEY ARTICLE

Hinnebusch, A.G. and Fink, G.R. (1983). Positive regulation in the general amino acid control of *Saccharomyces cerevisiae*. *Proceedings of the National Academy of Sciences of the USA*, **80**, 5374–8.

Mutations that affect general amino acid control fall into two classes, summarized in Figure 11.6 as *tra3* and *aas* genes. Recessive mutations in TRA3 constitutitively activate general amino acid control. In a *tra3* mutant, amino acid biosynthetic genes are transcribed regardless of the growth medium. This mutant phenotype indicates that the wild-type *TRA3* gene acts as a negative regulator of the process, and its normal function is to turn OFF general amino acid control. In contrast, mutations in the *AAS* genes cannot activate general amino acid control under starvation conditions. Three *AAS* genes have been found, called *AAS1*, *AAS2*, and *AAS3*. These phenotypes are shown diagrammatically in Figure 11.7. Hence, the *AAS* genes are positive regulators of the general amino acid control and must somehow turn ON the process. Thus, the *TRA3* gene and the *AAS* genes have opposite mutant phenotypes and affect general amino acid control in opposite functional

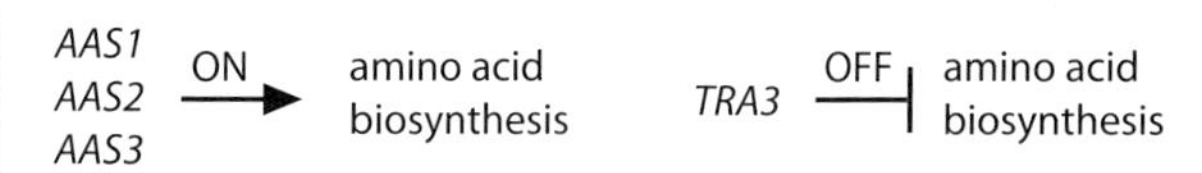

Figure 11.7 Diagrammatic representation of the genes involved in the general amino acid response. The three *AAS* genes are needed to turn the response on, and the *TRA3* gene is needed to turn the response off.

ways—the wild-type *AAS* genes turn the process ON, and the *TRA3* gene turns the process OFF.

In order to understand the pathway by which these genes regulate general amino acid process, double-mutant strains were made that carried the *tra3* mutation and one of the *aas* mutations. The *tra3;aas3* double mutant shows the phenotype of the *aas3* gene and cannot activate general amino acid control even under starvation conditions. In other words, the mutation in *aas3* is epistatic to the mutation in *tra3*—the mutation in *aas3 masks* the effect of the mutation in *tra3*.

Mutations in the other two other *aas* genes, *aas1* and *aas2*, have a similar mutant phenotype to *aas3* and do not activate general amino acid control. However, the phenotype of a double-mutant strain between one of these genes and *tra3* is completely different from the *tra3;aas3* double mutant. Both an *aas1;tra3* double mutant and an *aas2;tra3* double mutant have the phenotype of a *tra3* mutant and do not repress general amino acid control. Therefore, a mutation in the *tra3* gene is epistatic to mutations in *aas1* or *aas2* genes.

Thus, we cannot conclude that either the TRA phenotype or the AAS phenotype is inherently epistatic. Rather, the epistasis depends upon which pair of genes is being tested.

Let us think about these results as a logical pathway of interactions. When starved of amino acids, the AAS^+ genes are turned ON and the $TRA3^+$ gene is turned OFF. Although we know that the genes have opposite mutant phenotypes and hence functions, we do not know which gene is the upstream regulator of the other. We consider both possibilities in Figure 11.8. Suppose that the normal role of one of the AAS^+ genes is to turn off $TRA3^+$, as shown in Figure 11.8A. If both the *AAS* gene and the *TRA3* gene that it represses are knocked out by mutation, what is the expected phenotype? In the diagram, the *AAS* gene is shown upstream of the *TRA3* gene. The AAS gene is mutated so it will not be able to turn OFF the $TRA3^+$ gene. However, the activity of *TRA3* is itself knocked out because the gene is mutant, so the *tra3* gene will not be active. In other words, if both genes are knocked out by mutation, the *tra3* mutation is expected to be epistatic to the *aas* mutation in a gene upstream of it. This is the exact situation that we observe with *aas1;tra3* and *aas2;tra3*. Once their target gene *TRA3* is eliminated, the role of the *aas1* and *aas2* genes becomes irrelevant.

We can draw a pathway that illustrates this (Figure 11.8B). Although the pathway is inferred from the mutant phenotypes, it is drawn with wild-type gene functions. The function of the $AAS1^+$ and $AAS2^+$ genes is to turn OFF $TRA3^+$ when the yeast cell is starved of amino acids. When either AAS1 or *AAS2* is knocked out, $TRA3^+$ is not turned off and the yeast cell does not respond to amino acid starvation.

The double mutant between *tra3* and *aas3* shows an *aas3* mutant phenotype, so an *aas3* mutaton is epistatic to a *tra3* mutation. Although the phenotype is different for this pair of genes, the logic used previously to place them in a pathway still applies. If both *tra3* and *aas3* are knocked out, the cell has the phenotype of *aas3*. This implies that if *aas3* is mutant, the activity of *tra3* is irrelevant. Thus, as we reasoned before, $AAS3^+$ must act downstream of $TRA3^+$ in the pathway. Again, we can illustrate this with the pathway in Figure 11.8C, showing that $TRA3^+$ turns OFF $AAS3^+$. In fact, we can put the two pathways together and infer one longer pathway including all three genes: $AAS1^+$ and $AAS2^+$ turn OFF $TRA3^+$, and $TRA3^+$ turns OFF $AAS3^+$, as shown in Figure 11.8D.

Why do we not make a double mutant between two of the *aas* genes to test this pathway or to infer the relationship between the *aas1* and *aas2* genes? The reason is that this double mutant will not be informative. Neither of these mutants can activate transcription of the biosynthetic genes, so the double mutant will not activate transcription of the biosynthetic genes either. Since this is the phenotype of each of the single mutants, we cannot determine which phenotype is being observed in the double mutant. Thus, we have to leave the relationship between *aas1* and *aas2* unresolved, at least for now.

Now let us put this logical pathway back into the context of general amino acid control. When starved of an amino acid, the yeast cell activates $AAS1^+$ and $AAS2^+$. These genes repress $TRA3^+$. The repression of $TRA3^+$ results in $AAS3^+$ being activated, which somehow turns on amino acid biosynthesis. When the cells are grown in rich medium where amino acids are plentiful, *AAS1* and *AAS2* are not expressed. Therefore, *TRA3* is expressed (or not repressed). The expression of *TRA3* represses *AAS3*, and general amino acid biosynthesis is shut down.

This pathway illustrates many of the important points about epistatic analysis in a relatively simple system of control. For example, although we have inferred the order in which each of the genes functions, nothing about the molecular mechanism has been revealed. Many other genes could be involved in this process, but we have not included them in the pathway. We will now reinforce many of these points with another complex process in an attempt to reach some general principles for constructing and interpreting gene interaction pathways.

Sex determination in *C. elegans* involves both positive and negative genetic interactions

Differences in sexual morphology are one of the most familiar and fundamental biological phenomena. As discussed in Chapter 2, *C. elegans* has two sexes, males and hermaphrodites (Figure 11.9). The genetic control of sex determination in worms is one of the best examples of the use of epistasis to analyze a complex biological question.

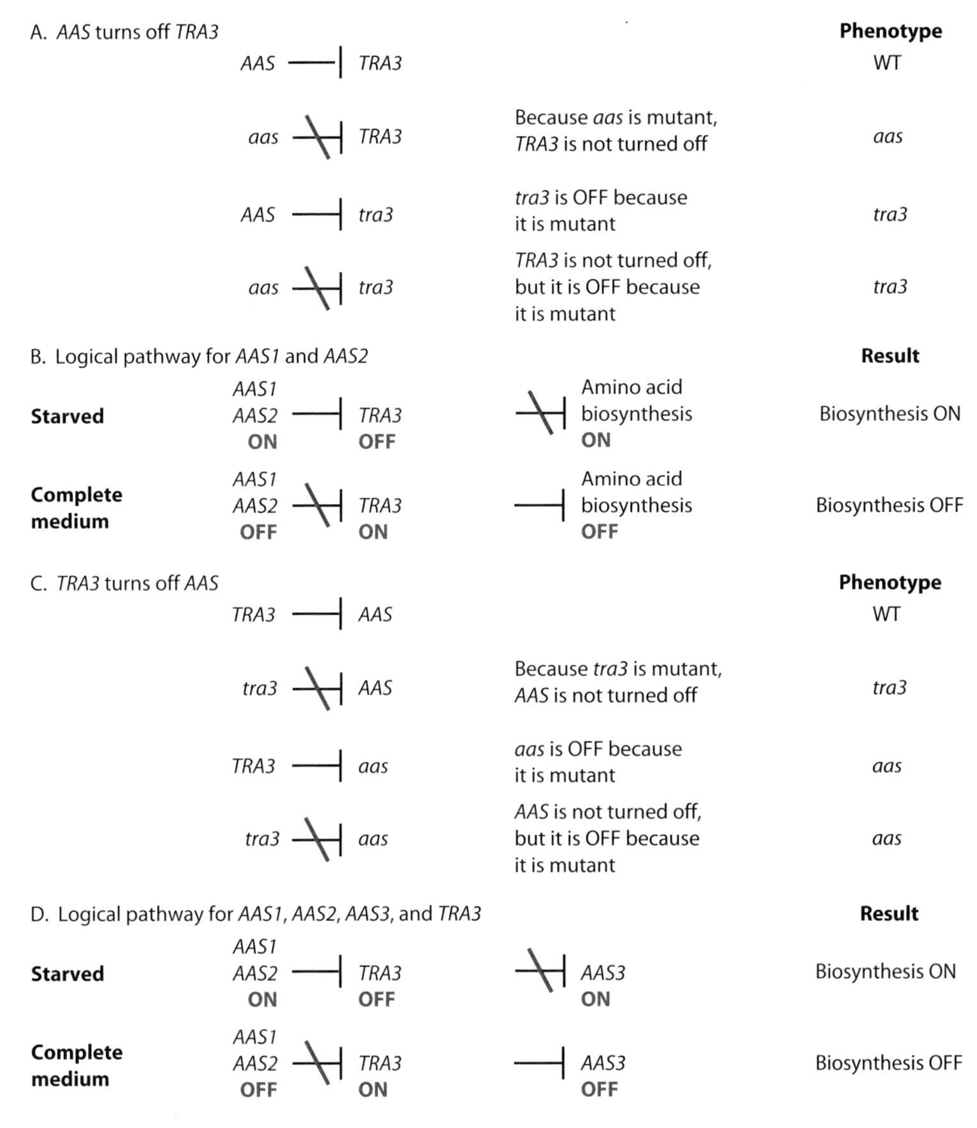

Figure 11.8 Logical analysis of the general amino acid response. Panel A presents a stepwise analysis of the expected mutant phenotypes when an *AAS* gene turns off *TRA3*. The top line presents the wild-type situation. The second and third diagrams are the two single mutants. If the *aas* gene is mutant, *TRA3* is not turned off, resulting in an *aas* mutant phenotype. Likewise, if the *AAS* gene is wild-type but the *tra3* gene is mutant, a *tra3* mutant phenotype is seen. The fourth diagram shows the result when both genes are mutant. The *aas* gene does not turn OFF the *tra3* gene because the *aas* gene is mutant. However, the *tra3* is OFF regardless because it is also mutant. Thus, the double mutant will have the same phenotype as the *tra3* mutation and the *aas* gene is irrelevant. The *tra3* mutation is epistatic to the aas mutation. This is the result observed with *aas1* and *aas2*. Panel B places the analysis from panel A into a logical pathway, showing that *AAS1* and *AAS2* have the phenotype expected of genes upstream of *TRA3*. When the cell is starved of amino acids, the *AAS1* and *AAS2* genes are ON. These genes turn OFF *TRA3*, so amino acid biosynthesis is not repressed and biosynthesis occurs. In complete medium, the *AAS1* and *AAS2* genes are OFF. Since they are not active, *TRA3* is not turned off, and the activity of the *TRA3* gene turns OFF amino acid biosynthesis. Panel C presents a stepwise analysis of the situation when *TRA3* turns OFF the *AAS* gene in the wild-type. As in panel A, the second and third diagrams are the single-mutant phenotypes for *tra3* and *aas*. The fourth diagram shows the double mutant. If the *tra3* gene is mutant, it cannot turn OFF the *aas* gene. However, the *aas* gene is OFF because it is mutant, and the *tra3* phenotype is irrelevant. In this situation, the *aas* mutant phenotype is epistatic to the *tra3* mutant phenotype. As described in the text, *aas3* mutants are epistatic to *tra3* mutations, placing *AAS3* downstream of *TRA3*. Panel D combines the analysis from the previous panels in one logical pathway. Under starvation conditions, *AAS1* and *AAS2* are ON, which turns OFF *TRA3*. Since *TRA3* is OFF, *AAS3* is not repressed and it is ON, so amino acid biosynthesis is ON. In complete medium, *AAS1* and *AAS2* are OFF. As a result, *TRA3* is not repressed and is ON, which turns OFF *AAS3*. Since *AAS3* is OFF, amino acid biosynthesis does not occur.

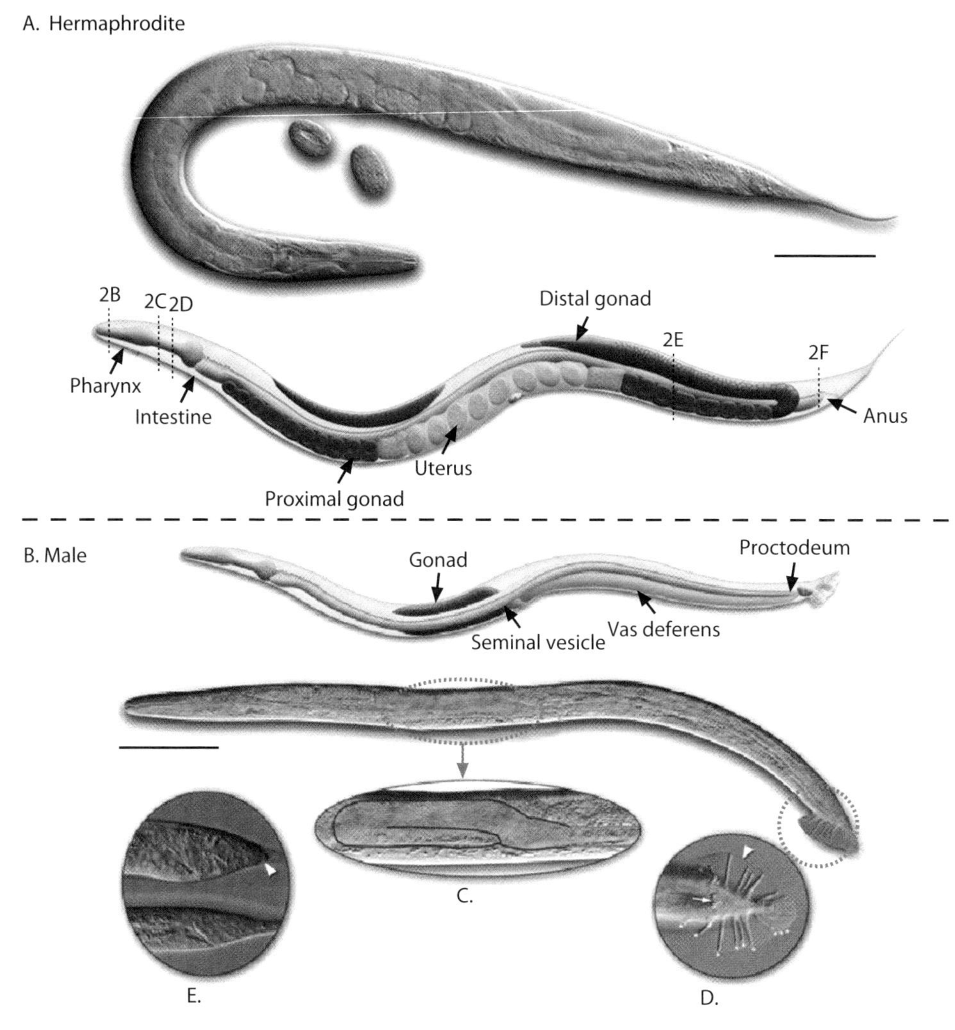

Figure 11.9 (A) Hermaphrodite and (B) male worms. The head is to the left, and the tail is to the right. Hermaphrodites are larger, and have a gonad with two lobes centered on a vulva at the midpoint of the animal and a plain tapered tail. Males have a gonad with a single lobe that exits in the posterior (C). The male tail is elaborate, with the fan and rays shown at the bottom of the figure (D), while the hermaphrodite tail is a simple spike (E). Hermaphrodites have two X chromosomes, whereas males have one X chromosome. Bar, 0.1 mm.

KEY ARTICLE

Hodgkin, J. (1990). Sex determination compared in *Drosophila* and *Caenorhabditis*. *Nature*, **344**, 721–8.

The primary signal for sex determination in normal diploid *C. elegans* is the number of X chromosomes: males have one and hermaphrodites have two. The hermaphrodite is essentially a somatic female that makes some sperm, which it can use for internal self-fertilization. Our example will concentrate on the somatic tissues; the regulation of sex in the germline (itself an interesting topic) will be ignored. The somatic differences between males and hermaphrodites are apparent from Figure 11.9. Males have an elaborate tail that is used in mating, whereas the hermaphrodite tail is morphologically simple. Hermaphrodites have a vulva and associated structures for egg-laying that males lack. There are also differences in size and behavior, and in the shape of the gonad. In fact, if one examines the cell lineages, approximately 40 percent of post-embryonic cell lineages show evidence of sexual dimorphism. Thus, males and hermaphrodite sex determination can be distinguished at the gross morphological level as well as by examining individual cells or lineages.

Mutations in a number of different genes that affect somatic sex determination have been identified, and

A. Mutant phenotypes

	1X	2X
tra-1	male	male
her-1	female	female
tra-2	male	male

B. Inferred wild-type role

	1X	2X
tra-1+	OFF	ON
her-1+	ON	OFF
tra-2+	OFF	ON

Figure 11.10 Sex determination genes in *C. elegans*. In wild-type, worms with one X chromosome are male and worms with two X chromosomes are female (or hermaphrodite). The mutant phenotypes of three genes in 1X and 2X animals are shown in panel A. In *tra-1* or *tra-2* mutants, both 1X animals and 2X animals develop as males. In *her-1* mutants, both 1X animals and 2X animals develop as females. The inferred wild-type roles of these genes are shown in panel B. In 1X animals (normally males), the *tra-1+* and *tra-2+* genes must be OFF and the *her-1+* gene must be ON. In 2X animals (normally female), the *her-1+* gene must be OFF and the *tra-1+* and *tra-2+* must be ON. We refer to the 2X animals as females based on somatic development and morphological similarity to females in related nematodes, although they are self-fertilizing hermaphrodites in *C. elegans*. These can be considered as females that make some sperm. Sex determination in the germline is not considered in this example.

epistasis among these genes has established the genetic pathway of sex determination. Our discussion will focus on three of the genes, but many more genes have been found by using suppression and synthetic enhancement as described in Chapter 10. The three genes that we will discuss are, in order, *tra-1*, *her-1*, and *tra-2*. All of these genes are defined by mutations that result in sex reversal of one sex, with little effect on the somatic sex differentiation of the other sex. Recessive mutant alleles of the *tra-1* gene result in both 1X and 2X animals developing as males. Thus, the wild-type function of *tra-1+* is apparently dispensable in 1X animals—they become males even in the absence of *tra-1* function—but is needed for 2X animals to become hermaphrodites. We can summarize with the chart in Figure 11.10: in an animal with one X chromosome, *tra-1+* activity is OFF, whereas in an animal with two X chromosomes, *tra-1+* activity is ON.

Negative pathways involve mutations with opposite phenotypes

In contrast to the *tra-1* mutant phenotype, recessive mutations in the *her-1* gene result in both 1X and 2X animals developing as hermaphrodites. This implies that the wild-type function of *her-1+* is needed for male development in 1X animals but not hermaphrodite development in 2X animals. (Recall from Chapter 3 that genes are named for their mutant phenotypes. A *her-1* mutation results in hermaphrodite development, so the gene is named for this phenotype. The wild-type function of the *her-1+* gene is then needed for males.) Because the hermaphrodite has somatic sexual characteristics similar to females in related species and we are ignoring the germline, we will refer to this phenotype as female. From our summary chart in Figure 11.10, *her-1+* activity is ON in an animal with one X chromosome and OFF in an animal with two X chromosomes. Therefore, as indicated in the chart, the *tra-1+* gene must be ON when *her-1+* is OFF and *tra-1+* must be OFF when *her-1+* is ON. In a logical sense, the role of one of the genes is to turn off the activity of the other. Note again that we are describing a functional role for the genes, and the molecular biology of this process is being set aside for now.

Which gene turns OFF the activity of the other? This question can be answered by making a double-mutant strain and examining the epistatic interaction between the two genes. We will scrutinize the logic of this procedure in detail, beginning with the flowchart in Figure 11.11. The step-by-step analysis is as follows.

Step 1. Determine if the pathway involves positive or negative interactions. Genetic interactions can result in either both genes being turned ON or one gene being turned OFF when the other is ON. These interactions are interpreted differently, and so it is imperative to infer the correct type of interaction pathway between the two genes. One useful way of determining if the genes are in a positive or negative relationship is to compare their mutant phenotypes. In a positive relationship, when gene *A* is ON, gene *B* is also ON. We recognize this type of relationship because the mutants have similar phenotypes. In our example with general amino acid control described previously, the *aas* mutations are in

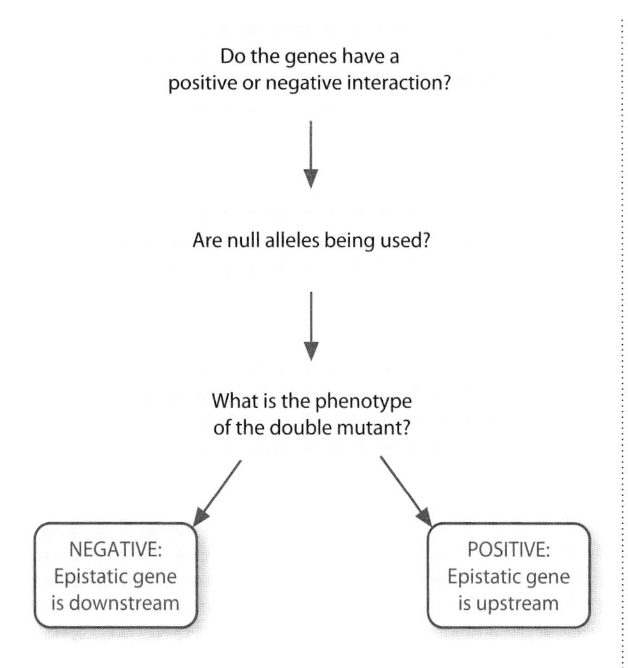

Figure 11.11 Flowchart for the analysis of epistasis, as developed in the text.

a positive relationship with each other because the mutants have the same or similar phenotypes. In contrast, in a negative relationship, the genes have opposite and distinct mutant phenotypes. Thus, in this type of interaction, when gene *A* is ON, gene *B* is OFF and vice versa.

Let us return to our example in which we see that, in a normal male, *tra-1+* will be OFF whereas *her-1+* will be ON. In a normal hermaphrodite, *tra-1+* will be ON and *her-1+* will be OFF. Therefore, *tra-1* and *her-1* must be in a negative relationship. This pattern of interaction is the easiest one to analyze by epistasis, so we will describe it first.

Step 2. Determine the types of allele. For most purposes, it is important that null mutations are used for each gene. (Different types of alleles are described in Chapter 4.) Because a null allele eliminates the function of the gene, the double mutant represents the situation when each gene is OFF. Null alleles are the one type of mutation whose activity is unambiguous—there is none.

Step 3. Compare the phenotype of the double mutant with the phenotype of each of the single mutants. The third step is to make the double mutant and compare its phenotype with each single mutant. The simplicity of this experiment should not hide its subtlety. The interpretation of the phenotype of the double mutant is ***the*** crucial step in doing epistasis experiments and the one that is most frequently debated by collaborators and competitors. Because interpretation of the double-mutant phenotype is the crucial and sometimes most subjective result, the genetic relationship inferred from epistasis needs to be tested by other means, such as additional genetic or molecular tests. It is important to examine as many different aspects of the phenotypes as possible to confirm this interpretation.

Since interpretation of the phenotype is the crucial step, let us use a non-genetic analogy for interpreting epistasis in a negative pathway. In a car, the brake pedal and the tires are in a negative relationship: when the brake is active, the tires are not moving, and when the tires are moving, the brake is not active. Defective brakes result in a moving car, whereas flat tires result in a stationary car. Those are the 'mutant phenotypes' of brakes and tires. Now consider a 'double-mutant' car in which the brakes are defective and the tires are flat. The observed phenotype is that the double-mutant car is stationary; in other words, it has the phenotype of the tire defect. To use the genetic term, mutant tires is epistatic to mutant brakes. To express this relationship another way, if the car is stationary because the tires are flat, the functional or non-functional brakes make no difference to the phenotype of the car. Because the mutant tire phenotype is epistatic to the mutant brake phenotype, we can infer that tires function downstream of the brakes. (Fortunately, this logical inference agrees with even a rudimentary knowledge of how an automobile functions.)

Now let us use the same logic in our biological example. Based on the mutant phenotypes, *tra-1* and *her-1* are in a negative relationship. All of the mutant phenotypes that we have described are those of null mutations, so the first two steps in our flowchart are completed. We turn to the third step, the analysis of the double-mutant phenotype.

There are two possible negative interactions between *tra-1+* and *her-1+*. Either *tra-1+* turns OFF *her-1+* or *her-1+* turns OFF *tra-1+*. These are shown

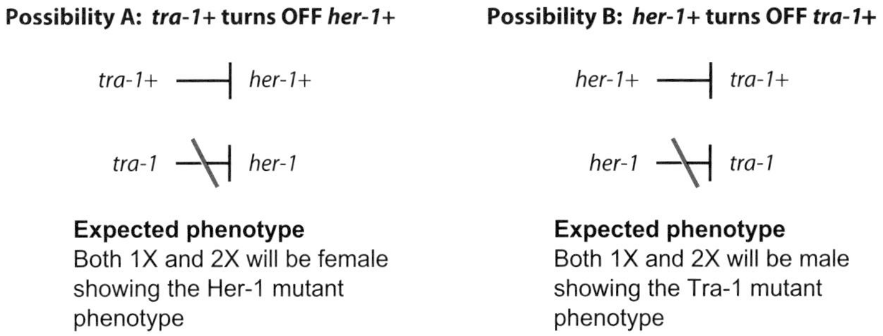

Possibility A: *tra-1+* turns OFF *her-1+*

tra-1+ —| her-1+

tra-1 —/| her-1

Expected phenotype
Both 1X and 2X will be female showing the Her-1 mutant phenotype

Possibility B: *her-1+* turns OFF *tra-1+*

her-1+ —| tra-1+

her-1 —/| tra-1

Expected phenotype
Both 1X and 2X will be male showing the Tra-1 mutant phenotype

Figure 11.12 Possible interactions between *tra-1* and *her-1*. The two genes are in a negative relationship since one must be ON when the other one is OFF. In possibility A, *tra-1+* turns OFF *her-1+* in the wild-type. In the double mutant, *tra-1* does not turn OFF *her-1*, but *her-1* is inactive because it is mutant. Thus, the double mutant will have a Her-1 mutant phenotype. In possibility B, *her-1+* turns OFF *tra-1+*. The double mutant will have a Tra-1 mutant phenotype.

diagrammically in Figure 11.12. Suppose that the biological pathway is that *tra-1+* turns OFF *her-1+* (possibility A in Figure 11.12). In the double mutant, both genes are OFF. Then when *tra-1* is OFF because it is mutant, *her-1+* would normally be ON. However, in the *tra-1;her-1* double mutant, the mutation in *her-1* means that it is also OFF. Thus, if the role of *tra-1+* is to turn OFF *her-1+*, the double mutant should look like the *her-1* single mutant—the mutation that knocks out the activity of *her-1* renders the state of *tra-1+* irrelevant.

Now let us consider possibility B in Figure 11.12 where the proposed pathway is that *her-1+* turns OFF *tra-1+*. In this case, the double mutant with both genes OFF should look like a *tra-1* mutant worm. Since *tra-1* is mutant and thus OFF, it makes no difference whether *her-1* is active or inactive. From this example, the one with general amino acid control, and the broken-down car, we can draw a general conclusion about genes in a negative pathway:

In a negative pathway, the phenotype of the *downstream* gene is epistatic to the phenotype of the upstream gene.

Having stated the analysis, what is the actual result with worms? The double-mutant worm carries null alleles of both *tra-1* and *her-1*. In the double mutant, both 1X and 2X animals are males, exhibiting the *tra-1* mutant phenotype. Since the *tra-1;her-1* double mutant has the same phenotype as the *tra-1* single mutant, the *tra-1* mutant phenotype is epistatic. We infer that, in the normal pathway, *tra-1* is the downstream gene and *her-1+* turns OFF *tra-1+* (possibility B in Figure 11.12).

It is important to remind ourselves what this result does and does not mean. Most significantly, it does not mean that the *her-1+* gene and the *tra-1+* gene or their products *directly* affect one another. Epistasis describes a genetic interaction that results in a mutant phenotype. Many possible molecular relationships involving the two genes are consistent with the same result from epistasis. It is possible that the *her-1+* gene product is a direct repressor of *tra-1* transcription; it is also possible that the *her-1+* gene product directly interacts with the *tra-1+* protein, somehow blocking its function. However, it is also possible that the two genes are expressed at different times, in different tissues, or, in this case, in different sexes. As a result, there may be no direct molecular interaction between the genes or their products. In fact, any number of genes could act in the pathway between *her-1+* and *tra-1+*. Epistasis by itself cannot distinguish among these and many other possibilities.

On the other hand, since we are comparing phenotypes, epistasis does not require that we have a molecular or biochemical understanding of the genes, their products, or the pathway they affect. If we want to control the movement of our car, we do not need to know how many mechanical steps occur between the brake pedal and the tires.

Positive pathways involve mutations with similar phenotypes

We have now established a portion of the logical pathway for somatic sex determination, namely that *her-1+* turns OFF *tra-1+*. Epistatic interactions involving other

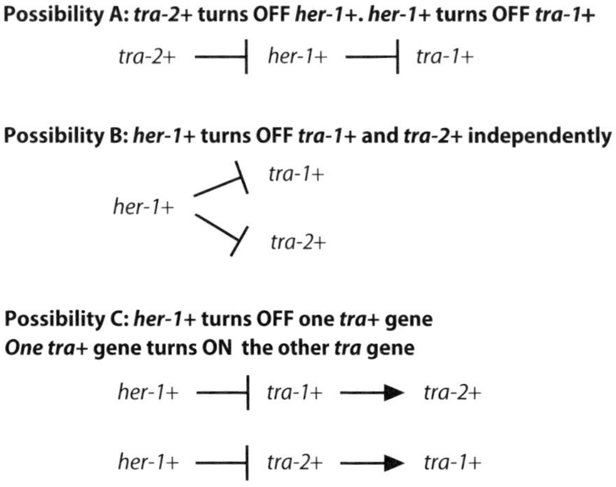

Possibility A: *tra-2+* turns OFF *her-1+*. *her-1+* turns OFF *tra-1+*

tra-2+ ⊣ *her-1+* ⊣ *tra-1+*

Possibility B: *her-1+* turns OFF *tra-1+* and *tra-2+* independently

her-1+ ⊣ *tra-1+*
her-1+ ⊣ *tra-2+*

Possibility C: *her-1+* turns OFF one *tra+* gene
***One tra+* gene turns ON the other *tra* gene**

her-1+ ⊣ *tra-1+* → *tra-2+*

her-1+ ⊣ *tra-2+* → *tra-1+*

Figure 11.13 Possible interactions among *her-1*, *tra-1*, and *tra-2*. The two *tra+* genes are in a positive or parallel relationship with each other and in a negative relationship with *her-1+*. From results in Figure 11.12, *her-1+* is upstream of *tra-1+*. This leaves three possibilities for *tra-2+*. In possibility A, *tra-2+* turns OFF *her-1+* and *her-1+* turns OFF *tra-1+*. In possibility B, *her-1+* turns OFF *tra-1+* and *tra-2+* in separate pathways and the *tra* genes act independently. In possibility C, *her-1+* turns OFF one of the *tra+* genes and the *tra+* genes turn each other ON. Either of the *tra+* genes could be upstream of the other.

genes have allowed investigators to infer the entire sex determination pathway. We will limit ourselves to considering only one such interaction here because it demonstrates an important extension of the use of epistasis.

The *tra-2* gene was identified with a mutant phenotype similar to *tra-1*: recessive null alleles of *tra-2* result in a male phenotype in both 1X and 2X animals, as summarized in Figure 11.10. Thus, *tra-2* must also be in a negative relationship with *her-1*. Numerous different pathways could explain this relationship. These possibilities are shown in Figure 11.13. For example, *tra-2* could turn OFF *her-1*, which turns OFF *tra-1* (possibility A in Figure 11.13). Alternatively, *her-1* could turn OFF both *tra-1* and *tra-2*, but those genes are independent of each other (shown as possibility B in Figure 11.13). Finally, *her-1* might turn OFF one of the *tra* genes, which turns ON the other *tra* gene. In this hypothesis, either of the two *tra* genes might be the more upstream gene and either the most downstream gene (possibility C in Figure 11.13).

Since *tra-2* and *her-1* are in a negative relationship, some of these models can be tested using the *tra-2;her-1* double-mutant animal. Based on the general principle worked out in the preceding section on negative interactions, if such an animal resembles *her-1* mutants, then *her-1* is epistatic to *tra-2* and must be the downstream gene of the two—the first of the three alternatives. However, if the *tra-2* mutant phenotype is epistatic to *her-1*, the first alternative is ruled out and either of the other two models must apply. The experiment has been done. When the double mutant between *tra-2* and *her-1* is made, it resembles a *tra-2* mutant, and the 2X animals show male somatic development. This epistasis places *tra-2+* downstream of *her-1+* in the pathway, thereby eliminating the first possibility.

How then can we distinguish between the other two possibilities in pathways B and C in Figure 11.13? We said earlier that epistasis works best when the mutations have opposite phenotypes, as in all of the preceding examples. The *tra-1* and *tra-2* mutant phenotypes are similar to each other in that 1X and 2X animals are males. As a result, the *tra-1;tra-2* double mutant will also almost certainly be male, which is uninformative. The mutations do not have opposite phenotypes; that is, *tra-1* and *tra-2* are in a *positive* relationship so that when one of the genes is ON, the other gene is also ON. In fact, both genes must be active in 2X animals and would be responsible for female development. In previous examples, we have not worked with a positive relationship and the genes were not ordered with respect to each other using epistasis. How can such an experiment be done?

There are two approaches for performing epistasis for genes in a positive relationship in which the mutant phenotypes are similar.

1. **Using subtle differences between the two phenotypes.** The phenotypes of the mutants in each gene may be similar but not identical in all respects. If this is true, the phenotype of the double mutant can be compared with each of the single mutants, particularly as it applies to these more subtle aspects. It is particularly important to work with null alleles in this case since slight differences could also be due to hypomorphic alleles for one or both genes.
2. **Using a dominant allele.** It may be possible to use a hypermorphic mutation for one of the two genes. As noted in Chapter 4, dominant gain-of-function or hypermorphic mutations often have a mutant phenotype that is the opposite phenotype from the recessive null mutation for the gene. Such a

difference could allow an unambiguous recognition of epistasis between the two genes. In this case, it is particularly important to understand the nature of the hypermorphic allele. If the dominant mutation introduces a novel phenotype or acts in some other way than a hypermorph, the results of the epistasis analysis could again be misleading.

Let us now consider each of these approaches in more detail.

Using subtle differences in the mutant phenotype

Either of the two methods listed above could be used to examine the relationship between *tra-1* and *tra-2*. In order for the first approach to work, subtle differences in the mutant phenotypes of the two different *tra* genes are important. The recessive mutant phenotypes of these two genes are similar in that 2X animals are male. However, there are numerous and easily observed differences between *tra-1* and *tra-2* mutants in 2X animals. Most noticeably, *tra-2* mutants are noticeably different from normal males and Tra-1 mutant males whereas *tra-1* mutants closely resemble normal males in all their somatic tissues. For example, *tra-2* mutants do not exhibit male mating behavior and the male tail is not as fully formed as in *tra-1* males or normal males. Here is a situation when it is important to know that one is working with null alleles. Some hypomorphic *tra-1* mutant alleles resemble a null *tra-2* allele and do not have the normal-looking male tail; an analysis with such a *tra-1* hypomorphic mutant would be difficult to interpret. However, these phenotypes can be readily distinguished with null alleles of each gene. It is also helpful to have multiple phenotypic differences between *tra-1* and *tra-2* mutants to examine in order to confirm that the interaction between the two genes can be generalized to different tissues and aspects of the phenotypes.

When the double mutant between *tra-1* and *tra-2* is made and analyzed, it has the phenotype of a *tra-2* mutation; the Tra mutant males have the less complete transformation of the tail that is characteristic of Tra-2 mutants and do not exhibit male-like mating behavior. Thus, in this interaction, *tra-2* is epistatic to *tra-1*. But what does this tell us about which gene is downstream in the pathway? Our first principle for interpreting epistasis was derived for a negative pathway, and placed the epistatic gene downstream. However, the relationship between *tra-1* and *tra-2* is positive, not negative. Each gene must be ON to produce female development. What principle governs epistasis for positive pathways? We will state the conclusion before developing the explanation. In positive pathways, the interpretation of epistasis is reversed and the epistatic gene lies upstream.

In a positive pathway, the phenotype of the *upstream* gene is epistatic to the phenotype of the downstream gene.

Illustrating positive pathways using non-biological examples

Positive pathways are familiar to us from many other aspects of our life, so we can use another non-biological analogy for illustration. My desk lamp and the building electrical circuit-breakers are in a positive pathway, with the circuit-breakers upstream of the lamp. If a circuit-breaker is tripped, all electrical appliances on that circuit go out, including my desk lamp. However, if only my desk lamp goes out and other appliances on that circuit in the building are working, the problem lies in my lamp and not in the building circuit-breaker. Knowing the pathway, I can interpret the 'phenotypes'.

Suppose that my desk lamp suddenly goes out and my room is dark. I can deduce the source of the problem (i.e. the source of the 'mutation' in the pathway) by looking at 'subtle differences' in the phenotype of the two defects. In this case, the subtle difference is other electrical appliances. If other electrical appliances in the building are still functional, then the darkness has the characteristics of a knockout of the desk lamp, the equivalent of the downstream gene. On the other hand, if other electrical appliances are also not working, then the darkness has the characteristics of a knockout of the circuit-breaker, the equivalent of the upstream gene.

Suppose that I turn off my desk lamp—so it is mutant and OFF—and I can move the circuit-breaker to the OFF position as well. This is the equivalent of making a double mutant in the pathway. What is the expected phenotype of this double knockout? In addition to the

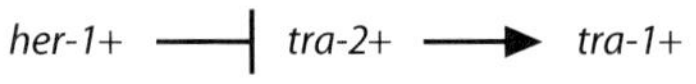

her-1+ ⊣ *tra-2+* → *tra-1+*

Figure 11.14 The relationship among *her-1+*, *tra-2+*, and *tra-1+*, as determined by epistasis. Double mutants placed *her-1+* upstream of both *tra-2+* and *tra-1+*. A double mutant between *tra-2* and *tra-1* showed the mutant phenotype of Tra-2 in subtle characteristics. Thus, if *tra-2* is inactivated, the activity of *tra-1* is irrelevant. This places *tra-2+* upstream of *tra-1+*.

desk lamp, other electrical appliances will also be OFF. By using subtle differences in the phenotype of my darkened room, I can infer the pathway. The double mutant has the mutant phenotype of the *upstream* switch, the circuit-breaker.

Returning to somatic sex determination

The epistasis experiments with *tra-1* and *tra-2* allowed the construction of a genetic pathway involving these three genes. This pathway is shown in Figure 11.14. The *her-1;tra-1* and *her-1;tra-2* double mutants showed the mutant phenotype of the Tra mutants and developed as males. This information placed the *tra-1+* and *tra-2+* genes downstream of *her-1+* in this negative step of the pathway. We previously noted that the *tra-1;tra-2* double mutant has the phenotype of *tra-2* mutants, placing *tra-1+* downstream of *tra-2+* in a positive step in the pathway. Therefore, the pathway can be summarized as follows: the wild-type function of *her-1+* is to turn OFF *tra-2+*. The wild-type function of *tra-2+* is then to turn ON *tra-1+*. This does not happen when *her-1+* has turned OFF *tra-2+*.

Using a dominant allele to confirm the results

Because subtle differences in the *tra-1* and *tra-2* mutant phenotypes could also arise from pleiotropic differences between the two genes, it is useful to have another method to infer the same information. We stated above that a second way of analyzing the epistatic relationship between two genes in a positive pathway is to use a dominant gain-of-function or hypermorphic mutation. In fact, such a hypermorphic mutation has been used to confirm the pathway. A dominant allele of *tra-2* has the opposite phenotype from recessive loss of function alleles. *tra-2* dominant mutations (abbreviated *tra-2*(dom)) do not affect somatic sexual development in 2X animals, but cause 1X animals to develop as females.

It is possible to make a double mutant between *tra-2*(dom) and *tra-1* and examine the phenotype. Since the phenotypes of the two mutations are opposite, the interpretation is also reversed. That is, although the wild-type *tra+* genes are in a positive relationship, the dominant mutation in *tra-2* with the opposite phenotype allows the investigator to interpret this like a negative relationship. In this particular case, the mutant phenotype resembles a *tra-1* mutant, the downstream gene in the pathway as constructed above.

This agrees with our inference that *tra-1* is downstream of *tra-2*. Again, this epistatic relationship can be illustrated by our simple example of the building electrical circuit-breakers and my desk lamp. Suppose that I force the circuit-breaker into the ON position and use electrical tape to hold it in place. I have produced, in effect, a dominant gain-of-function mutation in that part of the pathway (and an electrical hazard that, in true professorial fashion, we will ignore). Now if the desk lamp goes out (the 'loss-of-function' phenotype), the problem can only lie at that step in the pathway because we know that the upstream step is constitutively active.

This explains wild-type sex determination, at least as it relates to these three genes, as shown diagrammatically in Figure 11.15. Look at this figure and notice what we have been able to infer about the complex process of sex determination using only the careful analysis of double-mutant phenotypes. In a 1X animal, *her-1+* is normally ON. Because it is ON, it turns OFF *tra-2*, which is needed for hermaphrodite development. Since *tra-2* is OFF, *tra-1* is also OFF, and male development ensues. In a 2X animal, *her-1* is normally OFF. Because it is OFF, *tra-2* is ON, which

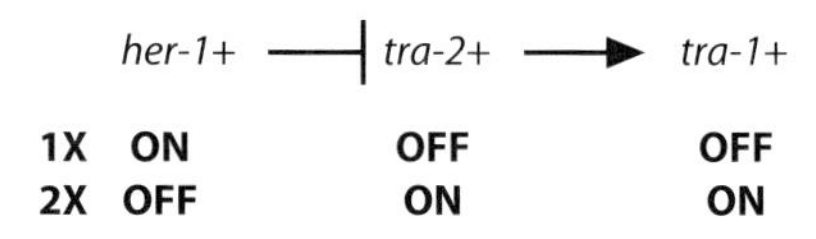

Figure 11.15 Sex determination in *C. elegans*. The pathway was determined by the phenotypes of double mutants. In 1X animals (males), *her-1+* is ON, which then turns OFF *tra-2+*; thus, *tra-1+* is not activated and male development ensues. In 2X animals (females), *her-1+* is OFF so *tra-2+* is ON. This activates *tra-1+*, and female development ensues.

turns on *tra-1*, and normal hermaphrodite or female development ensues.

The pathway inferred from epistasis informs further experiments to understand sex determination

This pathway has implications for additional experiments and their interpretation. For example, the *her-1* gene product and the *tra-1* gene product are unlikely to interact with each other directly because they do not encode consecutive steps in the pathway. Indeed, searching for such a direct interaction would probably be a waste of time and resources. Further analysis has supported this inference. Each of these three genes was cloned by one of the procedures described in Chapter 5: *tra-1* and *tra-2* by positional cloning, and *her-1* by a transposable element insertion. Molecular analysis of these genes suggests that *her-1* probably encodes a secreted protein that could function as a ligand, *tra-1* encodes a transcription factor, and *tra-2* encodes a membrane-spanning protein. Epistasis suggests that the *her-1* and *tra-2* gene products could interact with each other, as they are consecutive steps in the pathway, so a direct physical interaction between a possible HER-1 ligand and a possible TRA-2 receptor is a reasonable hypothesis. However, there is no evidence from epistasis that these genes or their products directly interact, so a direct molecular test of that hypothesis would be needed. (In fact, a direct physical interaction between HER-1 and TRA-2 has been demonstrated by a yeast two-hybrid assay and a co-immunoprecipitation assay, as described in Chapter 10.)

We need to return to another possible result in a double mutant that did not arise in this pathway. What if the two *tra* genes work in different pathways of male development and are independent in their effects? In this case, the effects of both mutations would be observed in the double mutant, so that it had characteristics of each of the two single mutants. Because most biological processes involve branched or parallel pathways rather than the linear stepwise pathway seen with somatic sex determination, double mutants often have intermediate phenotypes. Epistasis is the least ambiguous when the double mutant clearly has the phenotype of one of the single mutants. The clear epistasis of *tra-1* over *tra-2*(dom) helps to rule out the possibility that the genes work in separate pathways. We will return to this example later in the chapter.

Epistatic analysis has known limitations

Based on the description presented so far, epistasis may seem like an ideal tool for the analysis of complex biological problems. First, the process is fairly easy to perform. Making a double mutant is usually a simple technique, requiring few resources and little energy from the investigator. Secondly, the interpretation of the double-mutant phenotype also seems straightforward. There are two simple principles that make the analysis of the double mutant easy. In addition, the procedure does not require detailed knowledge of the physiology or biochemistry of the process or of the molecular biology of the genes or their products. Futhermore, epistasis has a history of success in solving complicated biological pathways. Most common cellular signaling pathways, including the Ras pathway, apoptosis, the Wnt pathway, and others, were deduced at least in part using epistasis in worms or flies.

However, epistasis simply presents a logical model of gene functions without providing information on the molecular biology of the process. In some situations, a faulty interpretation of epistasis can be misleading. It is worth considering how an investigator might be misled.

First and most commonly, the phenotype of one gene is not completely epistatic to the other gene, as described above. As a result, the double mutant shows some combination of both single mutant phenotypes. This presents complications for the investigator. The most conservative interpretation of this result is that the two genes under analysis act in parallel pathways that together contribute to the phenotype. An example of such an analysis involving the formation of dauer larvae in *C. elegans* will be discussed in the following section.

Secondly, the alleles may not be null. This may arise because that the investigator does not have true null alleles to work with, or because the investigator has chosen to work with non-null alleles (for reasons of viability and fertility, for instance.) In this case, the mutant phenotypes are difficult to interpret. They might indicate genuine epistasis between two genes with slightly different roles (such as *tra-1* and *tra-2*) or they might be indicative of partial phenotypes caused by non-null alleles.

Certainly, working with dominant hypermorphic alleles to analyze a positive pathway needs to be approached with caution. Not all dominant alleles are hypermorphic (Chapter 4), and not all 'hypermorphic' alleles have a simple over-production phenotype. The over-production may result in expression at specific times or in tissues that do not usually express the gene. Over-expression can result in unpredictable relationships among gene products whose stoichiometry is sensitive to dosage. And, like hypomorphs, hypermorphs fall onto a continuum of gene activity, which is often difficult to predict.

In addition, it is important to analyze the phenotypes carefully using multiple assays, including morphological, cellular, and molecular examination, if possible. Misleading or ambiguous interpretations may be detected when investigators attempt to use the proposed pathway to interpret the relationships of additional genes, only to find unpredicted results. The presence of internal inconsistencies is often a signal that some aspect of the epistatic pathway has been interpreted incorrectly. Epistasis is often a good 'first approach' that will suggest numerous additional experiments. It is important to be thorough in these follow-up experiments, and to make sure that the results are internally consistent before placing too much confidence in the inferred pathway. The examples above not only made use of all of the possible double mutants, but also used other alleles (such as hypermorphs or the genetic equivalent) to confirm the pathway. Helpful as they are, epistatic pathways should rarely be understood as the complete solution to the relationship among a set of genes. Case Study 11.1 completes our description of the *ptc* gene in *Drosophila*. Epistatic interactions between *hh* and *ptc* and between *smo* and *ptc* showed that Ptc is likely to be the receptor for the Hh ligand; molecular and biochemical analysis has confirmed this, as described in the case study.

11.4 A complex pathway involving dauer larva formation in *C. elegans*

The final example for this chapter will focus on a more complex branched pathway. The example draws together the use of suppression, epistasis, and synthetic enhancement, the gene interactions described in this chapter and the previous one. The analysis of this pathway, or more appropriately these pathways, led to unexpected insights into a number of fundamental processes, including mammalian longevity.

The process of dauer larva formation was described in Case Study 9.1 in the context of analyzing temperature-sensitive periods. As a brief review, under stress conditions such as starvation or crowding, *C. elegans* will form a distinctive developmental stage called the dauer larva as an alternative to the larval stage L3. Molecular and biochemical analysis has shown that the signal that triggers dauer larva formation is a pheromone secreted by all worms; a high concentration of the pheromone indicates crowding. A worm can stay as a dauer larva for months until conditions are favorable, at which time it molts to the L4 stage.

Dauer formation has been studied as a behavioral-response pathway, and recessive mutations in more than 30 genes have been identified that have some defect in dauer formation (see online access provided through Wormbase). Nearly all of the mutations (at least all those we will consider) are recessive, and null alleles are known for all of the genes we will consider. As mentioned in Case Study 9.1, the genes, called *daf* for *da*uer *f*ormation defective, fall into two broad classes: Daf-d (defective) and Daf-c (constitutive). Daf-d (defective) mutants cannot form dauer larvae even under stress conditions. An example of such a gene is *daf-3*. The wild-type function of the *daf-3+* allele is then needed for dauers to form.

KEY ARTICLE

Riddle, D.L. and Albert, P.S. (1997). Genetic and environmental regulation of dauer larva development. In: *C. elegans II* (eds D.L. Riddle, T. Blumenthal, B.J. Meyer, and J.R. Priess), Chapter 26. Cold Spring Harbor Laboratory Press, Cold Spring Harbor, NY.

Case Study 11.1

Epistasis and the *patched* pathway

This case study concludes our description of the *patched* (*ptc*) gene in *Drosophila*. The analysis of this gene has been a recurring theme throughout the book. We have previously described how mutations in this gene were identified and characterized (Chapters 3 and 4) and how the gene was cloned from both *Drosophila* and other animals (Chapter 5). A gene disruption of *Ptch* in mice is described in Case Study 6.1, and mosaic analysis of *ptc* in the wing of *Drosophila* is described in Case Study 9.2.

Epistasis was also used to infer the signaling pathway that includes the genes *hedgehog* (*hh*), *patched* (*ptc*), and *smoothened* (*smo*), a pathway that has implications for developmental patterning in nearly all animals and for cancer in humans. All three of these genes are in the segment polarity category of *Drosophila* genes, and all have proved to be evolutionarily conserved from flies to humans, with one ortholog of *smoothened* and several orthologs of *hedgehog* and *patched* in vertebrates. Thus, the pathway by which these three genes interact is of interest not only because of the biological significance but also because of the possible medical importance.

The effect of the mutations on downstream target genes

Each of the mutants has a distinctive morphological phenotype in *Drosophila* embryos and in adults that could have been used to perform an analysis of epistasis. The most informative phenotype is the effect of these genes on the expression pattern of *wingless* (*wg*) and decapentaplegic (*dpp*) in the wing disk. The data are described earlier in the chapter and summarized in Figure 11.3. Briefly, in a *ptc* mutant animal, the expression of the normal Dpp and Wg proteins extends to a more anterior part of the wing than it does in flies that are wild-type for *ptc+*. Thus, the wild-type function of *ptc+* includes restricting *wg+* and *dpp+* expression in the anterior part of the wing. We can summarize this by saying that, in the wing of a normal fly, *ptc+* turns OFF *wg+* and *dpp+*, as summarized in Case Study Figure 11.1. As in all such epistasis experiments, the molecular mechanism by which this regulation occurs is not known and not needed for an interpretation of the interaction pathway.

Similar experiments were also done to determine the effect of *hh* and *smo* mutations on Wg and Dpp protein expression and localization. In contrast to the results with *ptc* mutants, neither Wg nor Dpp is expressed in a *hh* mutant or in a *smo* mutant. Thus, the

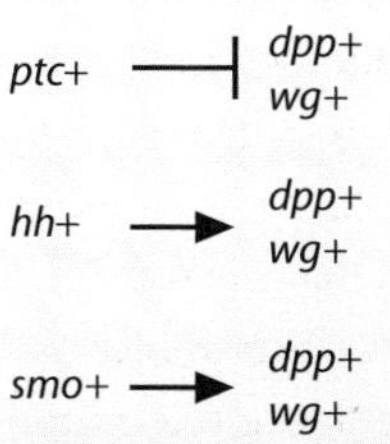

Figure C11.1 The normal role of *ptc+*, *hh+*, and *smo+* in the expression of *dpp+* and *wg+*. This has been inferred from changes in expression pattern using antibodies and reporter genes for *dpp* and *wg*.

Mutant	dpp+ wg+
ptc	High
hh	Off
smo	Off
ptc;hh	High
ptc;smo	Off
hh (constitutive)	High
hh (constitutive);smo	Off

Figure C11.2 The mutant phenotype of the single and double mutants.

wild-type function of *hh+* and *smo+* must be to turn ON Wg and Dpp proteins, as summarized in Case Study Figure 11.2. The expression of Wg and Dpp becomes a convenient phenotype for analyzing the epistatic interaction among *ptc*, *hh*, and *smo* mutants. Based on the results with Wg and Dpp expression and using our nomenclature above, *ptc* and *hh* are in a negative relationship, *ptc* and *smo* are in a negative relationship, and *hh* and *smo* are in a positive relationship. That is, when *ptc+* is ON, *hh+* and *smo+* must both be OFF; and when *ptc+* is OFF, *hh+* and *smo+* must both be ON. These relationships are summarized in Case Study Figure 11.2.

Epistasis involving negative pathways

Double-mutant combinations were made between *hh* and *ptc* and between *smo* and *ptc*. Each of these pairs of genes is in a negative relationship so, as discussed in the chapter, the phenotype of the double mutant is the phenotype of the downstream genes. In the *hh;ptc* double mutant, Wg and Dpp expression is high. In other words, the *hh;ptc* double mutant has the phenotype of a *ptc* single mutant, and *ptc* is epistatic to *hh*. Based on the rule for a negative relationship we derived earlier, we place the epistatic gene downstream, and we can infer that *hh+* turns OFF *ptc+* in wild-type flies, as depicted in Case Study Figure 11.3A.

smo and *ptc* are also in a negative relationship, so the experiment can be done and interpreted just as it was for *hh* and *ptc*. The double mutant between *smo* and *ptc* has the phenotype of *smoothened*, and Dpp and Wg are not expressed in the wing. Thus, *smo* is epistatic to *ptc* and we infer that *ptc+* turns OFF *smo+* using the same principle for negative pathways as before. This is shown diagrammatically in Case Study Figure 11.3B.

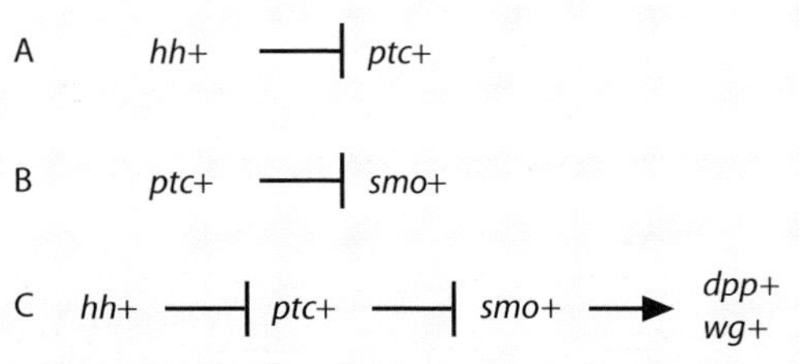

A *hh+* ⊣ *ptc+*

B *ptc+* ⊣ *smo+*

C *hh+* ⊣ *ptc+* ⊣ *smo+* → *dpp+* *wg+*

Figure C11.3 The interactions among *hh+*, *ptc+*, and *smo+*, as determined from the epistasis seen in double mutants.

Combining these results, the epistasis model for this pathway suggests that *hh+* turns OFF *ptc+* and *ptc+* turns OFF *smo+*. Then *smo+* turns ON *wg+* and *dpp+*, as shown in Case Study Figure 11.3C.

The use of an *hh* hypermorphic mutation

An additional experiment helped to confirm this model by directly investigating the relationship between *smo* and *hh*. Since these genes are in a positive pathway, the double mutant in which both genes are knocked out is expected to have the phenotype of either of the single mutants, and Wg and Dpp will not be expressed. Such a result will not be informative. To avoid this limitation, a different type of *hh* mutant phenotype was used. The wild-type *hh+* gene was cloned under a constitutive promoter and expressed in flies by procedures similar to those described in Chapter 2. The over-expression resulted in the equivalent of a *hh* hypermorphic mutation, and Wg and Dpp expression is high in this '*hh* constitutive'. In other words, the *hh* constitutive resembles the *ptc* null allele. Since this phenotype is the opposite of the *smo* mutant phenotype, the *hh* constitutive mutation could be combined with a null allele of *smo* to determine the epistatic relationship between these two alleles. Neither Wg nor Dpp is expressed in this double mutant, so it has the *smo* mutant phenotype. An interpretation of this result is that even when the *hh* gene is constitutively active, the *smo* mutant phenotype is epistatic. As in the analogy with the circuit-breaker and the desk lamp used in the chapter, the gene that is epistatic is downstream, indicating that *smo+* is downstream of *hh+*, confirming the pathway constructed with the null alleles of the three genes and shown in Case Study Figure 11.3C. *hh+* and *smo+* are in a positive relationship with each other as a result of two negative interactions involving *ptc+*. With *ptc* as the intermediate between them in the pathway, *hh* and *smo* are not necessarily expected to interact directly with each other.

The importance of the epistasis results became obvious when each of these three genes was cloned. *hh+* encodes a secreted protein, which is expected to function as a ligand; this is consistent with its upstream position in the pathway. However, both *smo+* and *ptc+* encode transmembrane proteins. Based simply on the molecular analysis and in the absence of epistatic analysis, it would be difficult to predict which of these two genes is the receptor for *hh*. However, because *ptc* acts in the pathway between *hh* and *smo*, the most reasonable genetic model is that *ptc+* is the receptor for the *hh+* ligand.

Biochemical confirmation of the epistasis pathway

Biochemical analysis using co-immunoprecipitation experiments of the type described in Chapter 10 has confirmed this relationship among *hh*, *ptc*, and *smo*. Vertebrate orthologs of each of these three genes were made as fusion proteins with epitope tags (described in Chapter 8) and each pair of proteins was expressed jointly in cells, as shown in Case Study Figure 11.4. Antibodies to the epitope tags were used to immunoprecipitate the complexed proteins from cells. When the Hh protein ortholog was precipitated using its epitope tag, the Ptc protein was co-precipitated, as shown in Case Study Figure 11.4A. This is consistent with a physical interaction between Hh and Ptc in the cell, and consistent with the inference from epistasis that placed Ptc downstream of Hh. On the other hand, the vertebrate version of Smo does not co-precipitate with any of the vertebrate Hh orthologs, as depicted in Case Study Figure 11.4B. Thus, Smo does not physically interact with Hh

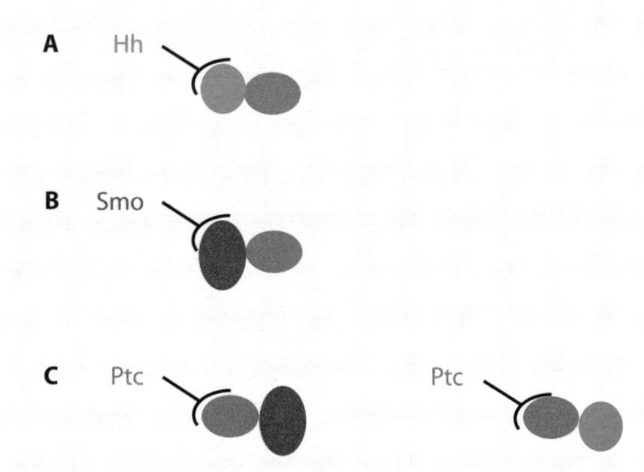

Figure C11.4 Co-immunoprecipitation (co-IP) experiments show the interactions among Hh, Ptc, and Smo proteins. Epitope-tagged versions were made of each of the three proteins, shown here in different colors: Hedgehog in green, Smoothened in red, and Patched in blue. Each pair of proteins was co-expressed in frog oocytes, which do not normally express these proteins. The epitope tags were used for co-IP. (A) The epitope-tag version of Hh precipitated both Hh and Ptc, indicating that these two proteins interact in the cell. It did not precipitate Smo in analogous experiments. (B) The epitope tagged version of Smo precipitated both Smo and Ptc but did not precipitate Hh. (C) The epitope-tagged version of Ptc could precipitate not only Ptc but also Smo or Hh, in separate experiments. This indicates that Ptc can interact with either Hh or Smo, whereas Hh and Smo do not interact with each other, providing additional support for the epistatic pathway shown in Figure C11.3.

and is probably not the receptor for the Hh orthologs. Conversely, antibodies against the vertebrate Ptc protein can immunoprecipitate the Hh orthologs in a complex with Ptc or precipitate Smo in a complex with Ptc (Case Study Figure 11.4C). Thus, Ptc, the middle gene of the three-gene pathway, physically interacts with both Hh and Smo. This suggests that Ptc is in a complex with Smo in the membrane. The binding of one of the Hh ligands to the Ptc protein may then release Smo from the Ptc–Smo complex, and it is able to turn on the expression of the downstream genes Dpp and Wg. The entire pathway is shown in Case Study Figure 11.5.

The concordance between the epistasis pathway derived from genetic interactions in *Drosophila* and the physical interactions demonstrated with vertebrate proteins re-enforces the power of each approach.

Further reading

The literature on the interactions among Hh, Ptc, and Smo is extensive. The following are some of the papers most relevant to this case study.

Alcedo, J., Ayzenzon, M., Von Ohlen, T., Noll, M., and Hooper, J.E. (1996). The Drosophila *smoothened* gene encodes a seven-pass membrane protein, a putative receptor for the hedgehog signal. *Cell*, **86**, 221–32.

Basler, K. and Struhl, G. (1994). Compartment boundaries and the control of *Drosophila* limb pattern by *hedgehog* protein. *Nature*, **368**, 208–14.

Capdevilla, J., Estrada, M.P., Sanchez-Herrero, E., and Guerrero, I. (1994). The *Drosophila* segment polarity gene *patched* interacts with *decapentaplegic* in wing development. *EMBO Journal*, **13**, 71–82.

Chen, Y. and Struhl, G. (1996). Dual roles for Patched in sequestering and transducing Hedgehog. *Cell*, **87**, 553–63.

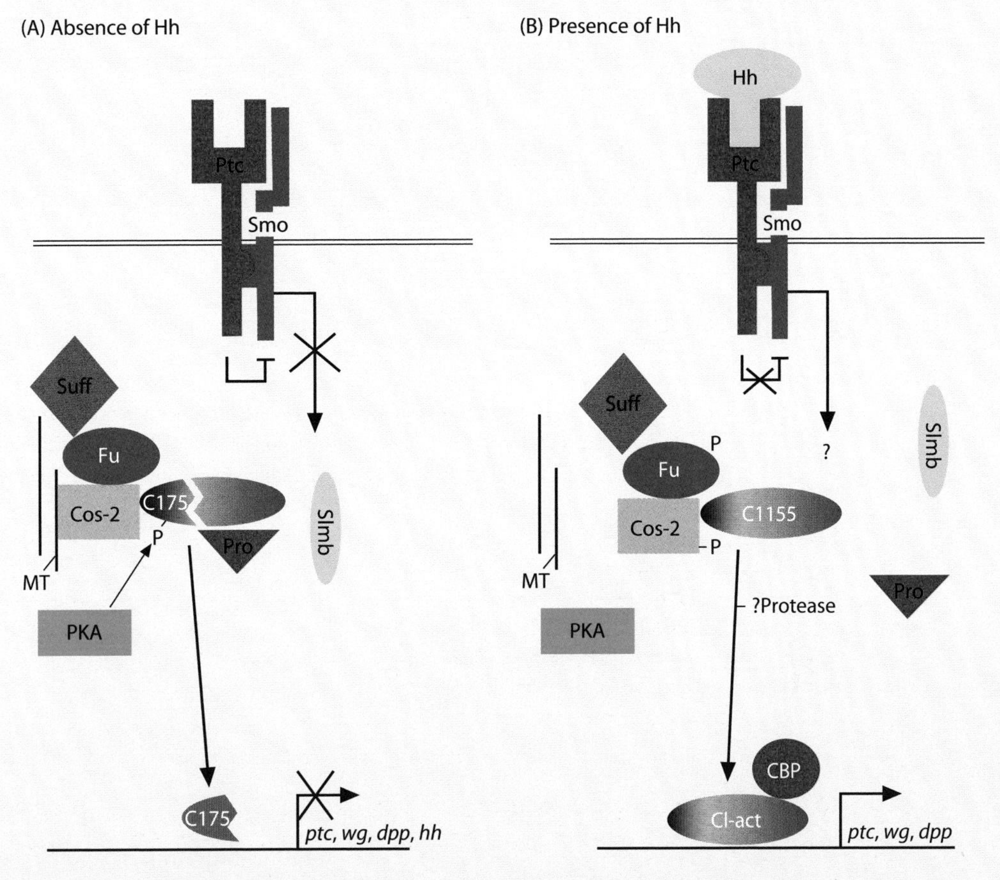

Figure C11.5 The Hh signaling pathway in *Drosophila*. Many genes act between *smo* and *wg* and *dpp*, as shown in this diagram. Both are transcribed by the transcription factor Cubitus interruptus (Ci) shown in the figure as C175 and CI. The vertebrate ortholog of Ci is the oncogene *Gli*, and most of the other genes in this pathway have one or more vertebrate orthologs.

Hidalgo, A. and Ingham, P. (1990). Cell patterning in the *Drosophila* segment: spatial regulation of the segment polarity gene *patched*. *Development*, **110**, 291–301.

Martinez Arias, A., Baker, N.E., and Ingham, P.W. (1988). Role of segment polarity genes in the definition and maintenance of cell states in the *Drosophila* embryo. *Development*, **103**, 157–70.

Schuske, K., Hooper, J.E., and Scott, M.P. (1994). *patched* overexpression causes loss of *wingless* expression in *Drosophila* embryos. *Developmental Biology*, **164**, 300–11.

van der Heuvel, M. and Ingham, P.W. (1996). 'Smoothening' the path for hedgehogs. *Trends in Cell Biology*, **6**, 451–3.

Gene	Mutant	Wild-type function
daf-3	No dauers	Needed for dauer formation
daf-7	Constitutive dauers	Needed to block dauers
daf-11	Constitutive dauers	Needed to block dauers

daf-c+ ⊣ *daf-d+*

daf-d+ ⊣ *daf-c+*

Figure 11.16 The dauer-constitutive and dauer-defective genes in *C. elegans*. In this and subsequent figures, the dauer-defective genes are shown in blue and the dauer-constitutive genes are shown in red. Mutations in a dauer-defective or *daf-d* gene such as *daf-3* result in a failure to form dauers even under stress conditions. The wild-type function of a *daf-d* gene is needed for dauer formation. Mutations in a dauer-constitutive or *daf-c* gene such as *daf-7* and *daf-11* result in dauer formation even under non-stress conditions. The wild-type function of *daf-c* genes is to block or prevent dauer formation. *daf-d* and *daf-c* genes must be in a negative relationship, and one must turn OFF the other. For any pair of *daf-d* and *daf-c* genes, the wild-type allele of the *daf-c* gene could turn off the wild-type *daf-d* gene; or the wild-type *daf-d* gene could turn off the wild-type *daf-c* gene, as shown at the bottom of the figure.

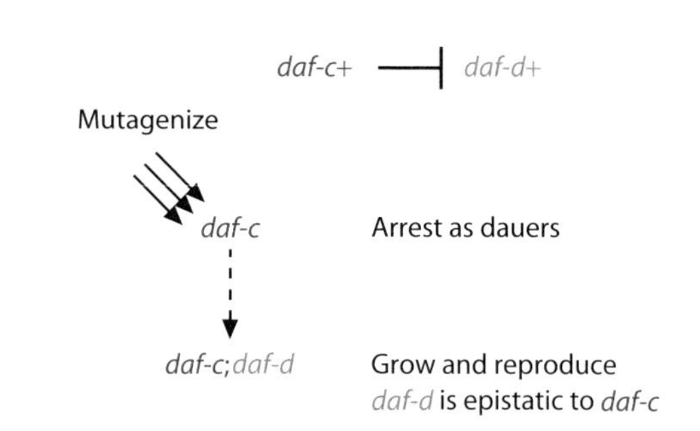

Figure 11.17 Finding *daf-d* genes as suppressors of a *daf-c* gene. In a negative pathway such as the one shown at the top, the double mutant has the phenotype of the downstream gene. In this case, a mutation in a downstream *daf-d* gene will act as a suppressor of a *daf-c* gene and the double mutant will not form dauers. A *daf-c* mutation is mutagenized and allowed to reproduce at the permissive temperature for two generation before the population is shifted to the restrictive temperature where the *daf-c* phenotype is observed. If a mutation has been induced in a downstream *daf-d* gene, the double mutant will grow and reproduce, whereas the *daf-c* single mutant arrests development. Such a screen was used to find *daf-d* gene downstream of different *daf-c* mutations.

In contrast to the *daf-d* genes, *daf-c* (constitutive) mutants *always* form dauers regardless of conditions. The worms arrest at this stage and never mature to the next larval stage or adulthood. Examples of *daf-c* mutants include the genes *daf-7* and *daf-11*. Both genes are defined by mutations that are temperature-sensitive, so that, at the restrictive temperature, these mutants form dauers regardless of the conditions. The wild-type alleles of *daf-7+* and *daf-11+* are required to prevent dauers from forming during good conditions. The opposing phenotypes seen with *daf-3* and with *daf-7* and *daf-11* mutants suggest that the *daf-3* gene is in a negative pathway with *daf-7* and *daf-11*. These relationships are summarized in Figure 11.16.

Suppression analysis was used to find additional genes affecting dauer formation

Recall from Chapter 10 that one way of finding additional genes affecting the same biological process is to look for suppressor mutations. A suppressor is a mutation that suppresses or overcomes the effect of another mutation in another gene. The phenotype of the *daf-c* mutants such as *daf-7* and *daf-11* easily lends itself to the identification of suppressors to find additional genes affecting dauer formation. A worm that arrests as a dauer larvae (a Daf-c phenotype) does not mature and lay eggs at the restrictive temperature. One can mutagenize the Daf-c worms and look for suppressors that allow the Daf-c worms to mature and give offspring, as shown in Figure 11.17.

What kind of mutations might be identified as suppressors of a *daf-c* mutant? One type is expected to be mutations in *daf-d* genes downstream of the *daf-c* gene. *daf-c* and *daf-d* mutations are in a negative relationship, with opposite mutant phenotypes, as shown in Figure 11.17. In other words, a *daf-d* mutation that is downstream of the *daf-c* mutant is epistatic and the *daf-d* mutant phenotype is seen. Such a *daf-d* mutation allows the *daf-c* mutant to mature and reproduce—the *daf-d* mutation acts as suppressor of an upstream *daf-c* mutation.

By finding *daf-d* mutations as suppressors of different *daf-c* mutations, it was possible to infer an epistatic pathway (Figure 11.18A). For example, *daf-7* dauer-constitutive mutations are suppressed by mutations in

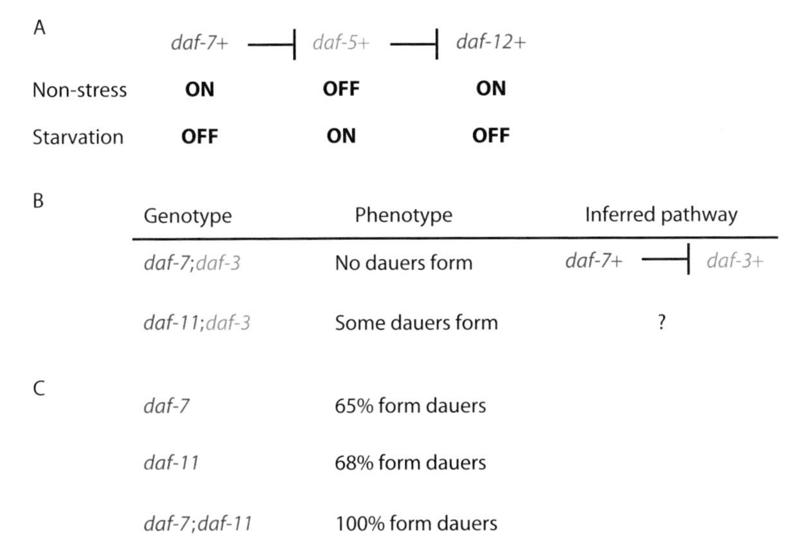

Figure 11.18 Interactions among dauer genes. (A) A simple pathway is inferred based on the phenotypes of double mutants involving *daf-7* and *daf-5*, and *daf-12* and *daf-5*. (B) *daf-3* is downstream of *daf-7* in a simple pathway, but a simple interaction cannot readily be inferred from the interactions between *daf-11* and *daf-3*. The *daf-11;daf-3* double mutant does not show clear epistasis between the two mutations. (C) Synthetic enhancement between *daf-7* and *daf-11* is used to infer the presence of two parallel pathways. Neither *daf-7* mutations nor *daf-11* mutations are completely penetrant, but the double mutant has 100 percent dauers at the restrictive temperature.

the *daf-d* gene *daf-5*. This means that *daf-5* is downstream of *daf-7* in the epistatic pathway. On the other hand, *daf-c* mutations in the gene *daf-12* are epistatic to *daf-5*, so *daf-12* must be downstream of *daf-5* in the pathway. This gives the simple pathway that *daf-7+* turns OFF *daf-5+*, which in turn turns OFF *daf-12+*; when *daf-7+* is active, *daf-5+* is turned OFF, so *daf-12+* is ON, as shown in Figure 11.18A. Many *daf-d* mutations were found by these suppressor experiments because it is easier to assay the ability to suppress a *daf-c* mutation than to assay the failure to form dauers.

Interactions among the dauer mutations are not always simple to interpret

As more mutations were identified and epistasis between the mutations was performed, some results were difficult to interpret. Some of the subtleties of epistasis analysis can be illustrated with three genes. These subtleties provide some helpful insights into the power and limitations of epistasis, suppressors, and enhancers.

One complicating feature is that the *daf-c* mutants are not 100 percent penetrant, even in null alleles, so there are always some worms that are homozygous mutant in genotype but do not form dauers. (This was noted for *daf-4* in Case Study Figure 9.5.) This means that a quantitative assay based on a population of worms rather than the phenotype of a single worm is essential to recognize the phenotypes of each of the mutants. A *daf-c* mutant may result in 70 percent dauer formation, whereas a *daf-d* suppressor may give only 10 percent dauers under the same conditions in the same mutant strain. Yet in some mutant combinations, epistasis works very clearly. For example, a double mutant between *daf-7* (constitutive shown in red) and *daf-3* (defective shown in blue) is defective for dauer formation, the phenotype of *daf-3*, as summarized in Figure 11.18B. No (or almost no) dauer larvae are formed by the double mutant. Since *daf-3* is completely epistatic to *daf-7*, the two genes are inferred to be in a linear pathway with *daf-3* downstream of *daf-7*. This part resembles the negative pathways we have considered previously.

In contrast, a double mutant between *daf-3* (*daf-d*) and *daf-11* (*daf-c*) is only partially dauer defective. In a population of *daf-3;daf-11* double mutants, most of the worms were dauer defective (the Daf-3 mutant phenotype) but some were capable of forming dauers (the Daf-11 mutant phenotype). This result presents a complication in thinking about how *daf-3* and *daf-11* are related to each other. The relationship between *daf-3* and *daf-7* is relatively easy to infer because *daf-3* is completely epistatic to *daf-7*, and no double-mutant worms form dauers. However, the relationship between *daf-3* and *daf-11* is more difficult to interpret because the phenotype of the double mutant is not clearly either dauer defective or dauer constitutive.

A key insight in the analysis of dauer formation is that the phenotype of a *daf-3;daf-11* double mutant can be understood as a mix or an intermediate of the two mutant phenotypes. We noted earlier that the most conservation interpretation of an intermediate phenotype is that the two genes act independently. If that interpretation is true here, there could be two separate parallel pathways to dauer formation, with *daf-3* affecting one of the pathways and *daf-11* affecting the other. The double mutant between the two pathways exhibits characteristics of both in the same way that a double mutant between any two independent genes shows two phenotypes. However, since clear epistasis is observed between *daf-3* and *daf-7*, they are hypothesized to be in the same pathway but different from the pathway with *daf-11*.

Synthetic enhancers support the presence of two pathways

The presence of parallel pathways affecting the same biological process was introduced in Chapter 10 in the discussion of the multivulvae mutations in *C. elegans*. The idea was developed that the presence of two pathways with functional redundancy can be inferred from synthetic enhancement. Synthetic enhancement refers to the ability of one mutation to enhance or make more severe the mutant phenotype of another gene.

The presence of two pathways in dauer formation is supported by synthetic enhancer experiments. As noted above and summarized in Figure 11.18C, neither of the Daf-c mutants, *daf-11* or *daf-7*, is 100 percent penetrant, and some genotypically mutant worms are phenotypically wild-type. The presence of non-dauer worms could be an indication of the functional redundancy of the two pathways; although one pathway is mutant, the other pathway is still functional.

If two pathways are present, the expectation is that knocking out both of them simultaneously will result in synthetic enhancement. The quantitative phenotypes of the Daf-C mutations can be used to detect such an interaction. Consistent with the hypothesis of parallel pathways, the *daf-11;daf-7* double mutant is completely penetrant, and all worms form dauers, as summarized in Figure 11.18C. Thus, the two mutations are synthetic enhancers of each other, supporting the hypothesis that *daf-11* and *daf-7* are in two different pathways leading to dauer formation. These pathways are shown in Figure 11.19. The *daf-7* pathway includes *daf-3*, and *daf-3* is downstream of *daf-7*, as noted above. The *daf-11* pathway does not include *daf-3* as a downstream gene.

However, this is only an inference based on the mutant phenotypes, suppression, and synthetic enhancement by these three genes. As noted above, more than 30 genes affect dauer formation. We can make predictions about what should be observed when those other genes are tested in double-mutant combinations with *daf-3*, *daf-7*, and *daf-11*, and among themselves. The genes in one pathway are expected to exhibit clear-cut suppression (by epistasis) with either *daf-3* or *daf-7*. They are not expected to be synthetic enhancers of the genes

Figure 11.19 The pathway for dauer formation in *C. elegans*, as inferred from genetic interactions involving double mutants. As before, genes in blue are needed for dauer formation so that mutations cannot form dauers. Genes in red are needed to prevent dauer formation, and mutations constitutively form dauers. Genes stacked at the same position cannot be ordered using gene interactions alone. The pathway begins and ends with a single pathway, but has parallel pathways in the middle. For negative interactions, the downstream gene is epistatic. For positive interactions, the upstream gene is epistatic. The existence of parallel pathways was inferred by synthetic enhancement.

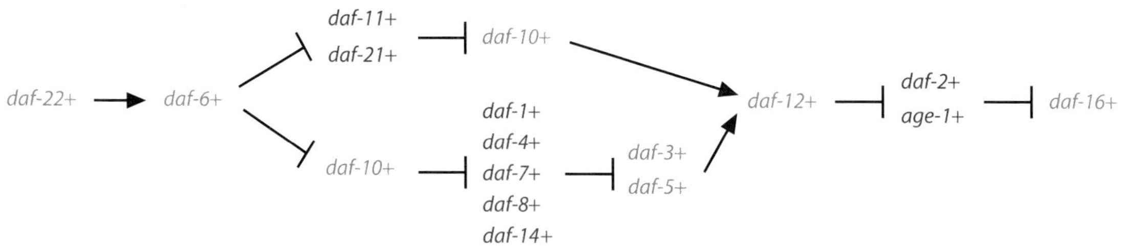

in the same pathway, in particular of *daf-7*. Genes in the other pathway are expected to exhibit clear-cut suppression by epistasis with *daf-11*. Genes in different pathways are expected to give a mixed phenotype when *daf-c* and *daf-d* mutations are combined, but *daf-c* mutations in two different pathways are expected to exhibit synthetic enhancement.

These predictions have been tested for many of the different genes, and the presence of two pathways leading to dauer formation, as shown in Figure 11.19, is consistent with all of the experiments. For example, mutations in *daf-5* (defective) suppress the constitutive phenotype of *daf-7* and *daf-1*, but give a mixed phenotype (as does *daf-3*) when tested with the constitutive mutants *daf-11* and *daf-21*. This places *daf-5* downstream of *daf-1* and *daf-7* but in the same pathway as these genes and *daf-3*. Mutations in *daf-21* act as synthetic enhancers of the constitutive mutations in *daf-1*, *daf-4*, and *daf-7* (among other genes) but not of mutations in *daf-11*. This places *daf-21* in a different pathway from *daf-1*, *daf-4*, and *daf-7*.

The two pathways share some common steps

The results with several of the genes are particularly informative in thinking about the presence of two pathways. Notably, as shown in Figure 11.19, some genetic steps are common to both the pathways and others are unique to one pathway. Genes common to both pathways exhibit epistatic interactions with genes that do not show interactions with each other. For example, the *daf-d* mutations *daf-22* and *daf-6* are suppressed by *daf-c* mutations in either of the pathways; any *daf-c* mutation is epistatic to mutations in *daf-22* or *daf-6*. This indicates that these two genes are the most upstream in each pathway, and that the two pathways must diverge after these two genes.

These two genes were themselves placed in order by a type of molecular epistasis experiment. Recall that dauer formation is triggered by a particular pheromone. Addition of dauer pheromone overcomes the defect (or suppresses, to use the genetic term) in *daf-22* mutants but not the defect in *daf-6* mutants. This implies that *daf-22* mutants are still able to transmit the pheromone signal, whereas *daf-6* mutants are defective in signal transmission. Thus, *daf-22* is inferred to be upstream of *daf-6* and also of the production of the pheromone; the gene may possibly be involved in pheromone production. The signal is transmitted via *daf-6* to either of the downstream genes *daf-10* or *daf-11* and then via either of the two pathways.

The pathways reconnect downstream, and the final signaling steps are in common again. Mutations in the *daf-d* gene *daf-12* mutations are epistatic to *daf-c* mutants in both pathways. This suggests that *daf-12* acts downstream of both pathways, possibly to integrate the signal generated from each one. Also downstream of both pathways—in fact, downstream of *daf-12*—are the genes *daf-2*, *age-1*, and *daf-16*. In fact, *daf-16+* encodes a transcription factor, so presumably its target genes are the same regardless of the branch of the dauer formation pathway. Daf-16 probably has many transcriptional targets, including some of the genes involved in the morphological changes between wild-type worms and dauer larvae. The inducing signal at the most upstream end of the branched pathway is the same and the responding signal at the most downstream end is the same; the functional redundancy occurs between the branches in the middle of the pathway.

This careful analysis reveals the power and limitations of this type of genetic analysis. The power is that the functional roles and relationships of more than 15 genes in complicated and interlocked signaling pathways have been described. In addition, some of the dauer mutations were used to find additional genes, some of which had previously been recognized for their role in other processes (such as thermotaxis or osmotic pressure avoidance) but not for their role in dauer formation. Because of the thoroughness of the screens in finding the new genes, it is likely that most or all of the genes involved in this aspect of dauer formation have been found.

On the other hand, the role of some genes is not conclusive even when the epistasis results are clear; notice that *daf-10* is placed in both branches of the pathway but upstream in the *daf-7* and *daf-3* branch and downstream in the *daf-11* branch. (A molecular analysis of the *daf-10* gene reveals that it encodes a transport protein, so its role is common to both pathways but at different points in the pathway.) It is also worth noting how carefully this analysis was done, and that some aspects were only possible because of the nature of the mutant phenotype and the versatility of working with *C. elegans*.

The synthetic enhancement experiments were possible because even null alleles are not 100 percent penetrant, which could be recognized because large numbers of worms could be grown and examined. The phenotypes had to be carefully and, in this case, quantitatively analyzed. In fact, other phenes of the mutations, such as cell morphology and dye uptake that are not discussed here, have also been useful in determining the relationships. Finally, many different double-mutant combinations have been used to confirm the resulting pathway, which can only be done in an easily manipulated genetic model organism.

The dauer pathway will be discussed once more, in Case Study 12.1. The thorough genetic analysis of the gene interactions laid the foundation for an impressive integration of many of the contemporary approaches in genetics, genomics, and systems biology.

11.5 The pathways unveiled

The descriptions we have given of the genetic analysis of the sex determination and dauer formation pathways have been short on molecular information. The exclusion of molecular details has been intentional so that the strengths and weaknesses of purely genetic screens can be seen more easily. None of the screens was done exactly as described here. Some of the details of the mutant phenotypes have been simplified (possibly oversimplified), complications such as temperature sensitivity and viability have occasionally been ignored, and some of the important genes have been omitted from our discussion. In addition, the genetic analysis of these genes often occurred at the same time as their molecular cloning, so the genetic and molecular analyses were usually part of the same description, and the results from molecular biology experiments were important in interpreting some of the genetic results. The description of how the pathways were deduced from genetic data should not be taken as a description of what each of the pathways actually does. For that, we return to these pathways and include just a bit of the molecular information that has been found from cloning the genes.

Although the HER-1 and TRA-2 proteins are not particularly closely related to proteins in organisms other than *C. elegans*, the target gene *TRA-1* is a transcription factor with orthologs in most eukaryotes, including yeast, flies, humans, and *Arabidopsis*. Its human orthologs are the GLI gene family, oncogenes associated with gliomas. The pathway by which GLI expression is regulated in humans does not appear to involve the same upstream genes as *tra-1* regulation in worms, so that remains to be elucidated.

The *hedgehog*, *patched*, and *smoothened* pathway in *Drosophila* is an evolutionarily conserved and crucial signaling pathway in the development of most, if not all, animals. The ligand is encoded by a *hedgehog* single gene in *Drosophila* but a family of related genes in vertebrates. Of these, the best studied is the gene *Sonic hedgehog* (*SHH*), whose developmental roles in mammals might be broadly summarized as 'the morphogen for everything' including the gut, the limbs and digits, the neural tube and the brain, the eye, the heart, the notochord, even left–right asymmetry. Given its broad role in development, it is not too surprising that people with inherited mutations in *SHH* have severe developmental problems, including holoprosencephaly, cleft lip and palate, and many more. Most such unfortunate individuals die *in utero*. Somatic mutations in SHH, *patched* (*PTCH*), and *smoothened* (*SMO*) have also been associated with basal cell carcinomas, medulloblastomas, rhabdomyosarcomas, and other human tumors. *SHH* and *SMO* are proto-oncogenes, and the *PTCH* gene family act as tumor suppressors. Thus, the epistatic pathway inferred from *Drosophila* wing development and embryo segmentation has given us fundamental insights into human development and some of the most common forms of human cancer.

The dauer pathway in *C. elegans* also has interesting implications for human biology. As noted in Case Study 9.1, the branch of the pathway that includes *daf-3* and *daf-7* is a TGF-β signaling pathway, one of the fundamental signaling pathways in eukaryotes. In fact, *daf-7* encodes a ligand in the TGF-β family and is distantly related to *dpp* in *Drosophila* (the gene used as a downstream marker in the *hh*, *ptc*, *smo* pathway). Many of the other genes in this branch of the dauer formation pathway are related to known human oncogenes. On the other hand, the branch of the pathway that includes *daf-11* and *daf-21* regulates cyclic GMP signaling, another fundamental biochemical property in all eukaryotes; *daf-11* is expressed in the neurons in the snout that have been implicated in dauer formation.

Perhaps most interesting are the genes downstream of both signaling pathways, which, by genetic logic, are responsible for integrating the different signals for dauer formation. *daf-2* encodes an insulin receptor, whereas *daf-16* encodes a transcription factor of the forkhead family, often found downstream in the insulin pathway. Orthologs of both genes are known in many other organisms, including humans. Mutations in the human ortholog of *daf-16*, FOX01A, are associated with the tumor rhabdomyosarcoma. Just as interesting, each of these mutants in the insulin branch of the dauer formation pathway has an unusually long lifespan in *C. elegans*. Mutations in the insulin signaling pathway are associated with longevity in flies and mice as well, and are being investigated for effects on aging and life span in humans.

No one would have had the foresight or the audacity to begin the analysis of these genes because they expected to gain insights into heart development, skin cancer, or longevity. All of those insights came later, when the molecular analysis of these genes found their orthologs in other organisms, including humans. The use of suppressors, enhancers, and epistasis gives us the expected knowledge about gene interaction in yeast, worms, and flies, and the unexpected knowledge into key processes in humans. Such insights about these complex processes would have been hard to obtain without the ability to examine gene interactions in model organisms. Because we can do the experiments in model organisms without prior knowledge of the process, we may not know what process we are actually modeling. But we do not have to know that—the genes can tell us.

Chapter Capsule

Epistasis and genetic pathways

Every biological process is the outcome of genes or gene products working together in pathways and networks. These pathways may be linear or branched, and may involve many genes or few genes. Gene interactions based on the phenotype of double mutants can be used to infer a logical order of gene function in the pathway.

In a ***negative*** pathway in which the wild-type function of one gene is to turn OFF the other gene, mutations in the two genes have opposite phenotypes. The double mutant has the mutant phenotype of the ***downstream*** gene in the pathway.

In a ***positive*** pathway in which the wild-type function of one gene is to turn ON the other gene, mutations in the two genes have similar phenotypes. The double mutant has the mutant phenotype of the ***upstream*** gene in the pathway.

When there are ***parallel*** or functionally redundant pathways for the same biological process, a double mutant with mutations from two different pathways may have a mixed mutant phenotype that is a ***composite*** of the two mutants. Alternatively, the double mutant may show ***synthetic enhancement*** of the one or both of the single mutations, resulting in more severe mutant phenotype than either single mutant alone.

CHAPTER MAP

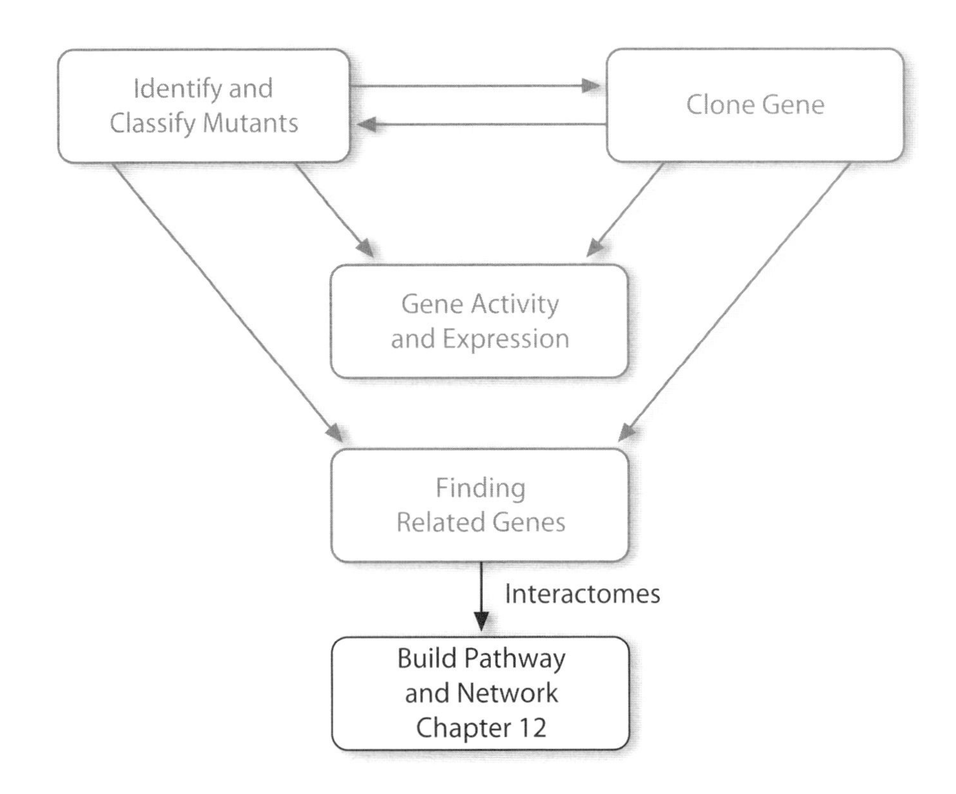
Identify and Classify Mutants
Clone Gene
Gene Activity and Expression
Finding Related Genes
Interactomes
Build Pathway and Network Chapter 12

12

CHAPTER 12

Pathways, networks, and systems

TOPIC SUMMARY

An organism can be thought of as a system made up of networks of interacting components. Two types of networks are transcriptional regulation arising from transcriptional factor proteins binding to DNA sequences, and protein–protein interactions. Attempts are being made to map each of these networks completely, at least in simple organisms or limited tissue types.

Each of these networks of interactions is dynamic: it changes over the life of a cell, the life cycle of an organism, and the evolutionary history of the species. The tools of genetics are being combined with computational and mathematical approaches in new ways in an attempt to describe all of the interactions that occur within an organism and to understand this system of interactions.

IN BRIEF

Genetic, genomic, and computational approaches are being combined to analyze biological questions by regarding the organism as a complex network of interacting components.

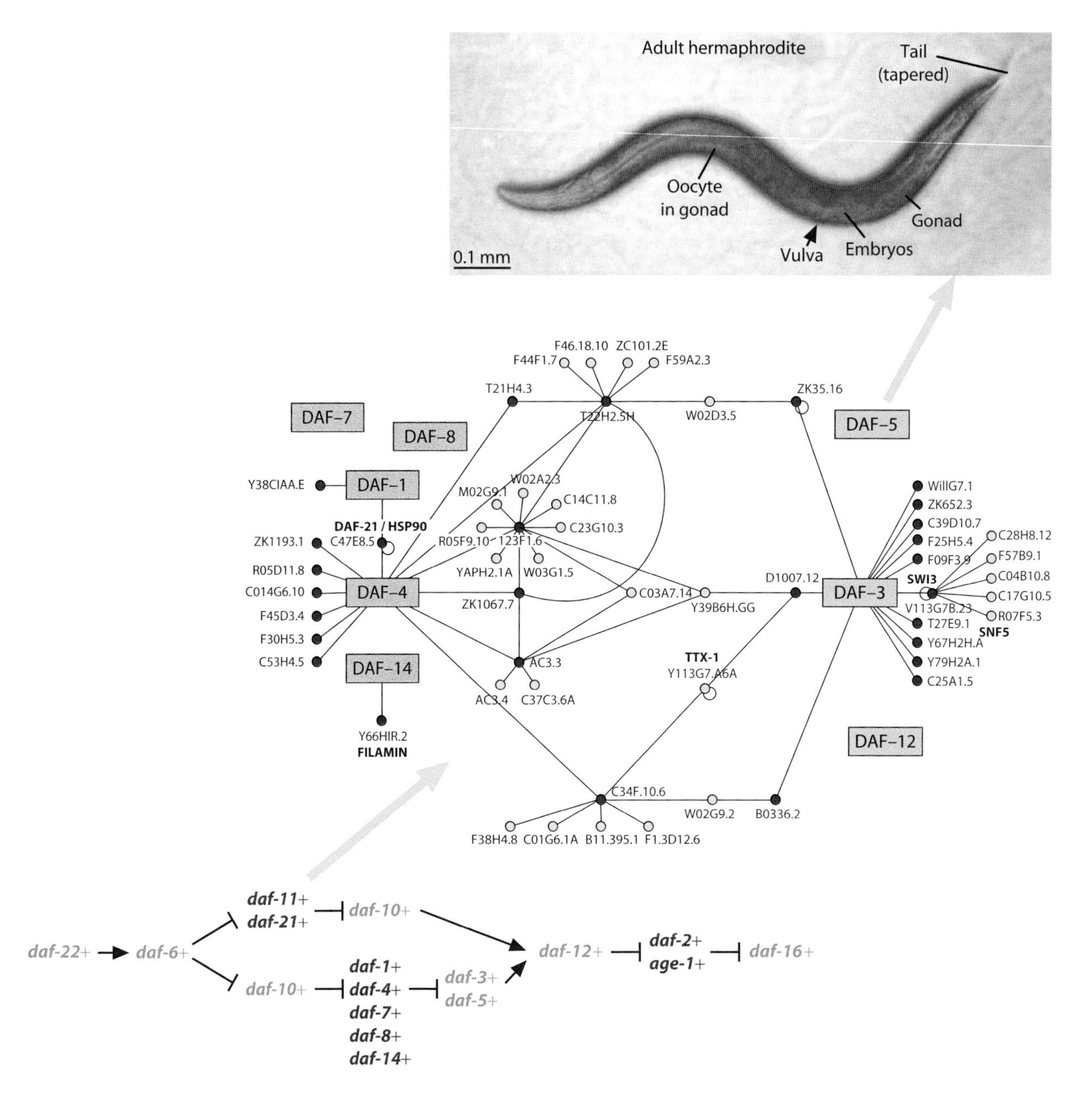

Figure 12.1 Integrating a pathway with a network with a system. The lower left corner shows the pathway for dauer formation in *C. elegans*. This pathway has also been investigated as a network of interacting components, as shown in the center figure. Understanding how to synthesize the network into worm biology, as shown at the top right, will be a major challenge for the years ahead.

INTRODUCTION

Genomics has made it feasible to identify and analyze the function of every gene in an organism. From the analysis of the function of individual genes, it is also possible to analyze the interactions among genes. As discussed in Chapters 10 and 11, suppression, synthetic enhancement, and epistasis (i.e. the set of genetic interactions) help to describe functional relationships among genes, allowing geneticists to construct pathways by which the genes function. Yet even this approach, powerful as it has proved to be, is limited. Within a cell or organism, genes and their products form a complex web of interactions. Some of these interactions can be described by simple linear motifs such as the pathways in Chapter 11. Other interactions cannot be described quite so simply yet are still crucial to normal biological functions. Rather than thinking of genetic interactions in a pathway, it is sometimes more useful to imagine them as a web or network of interacting parts. How can a network of interacting gene products be analyzed?

A new type of analytical strategy for elucidating these complex networks combines the genetic, genomic, and molecular tools of the preceding chapters with computational and mathematical approaches used in other complex systems such as airline flight patterns, the World Wide Web, and social interactions. Each of these complex networks comprises a system of interconnected components, with each component functioning independently and yet also performing as part of a more complicated unit. The system consists of the parts, their interactions, and properties that emerge as a result of the interactions. The same logic may be true of biological systems: each gene or gene product has its individual function, which can be analyzed by the tools of genetics and molecular biology, but it also works together with other genes and gene products to produce a functional organism (summarized in Figure 12.1). The biological process or the organism is thought of as a **system**, and the analytical strategy used to study them is being called **systems biology**.

Systems biology depends on all of the tools of genetics that we have described—mutations, cloned genes, microarrays, and other molecular assays of gene expression. But it attempts to move beyond individual genes to understand an entire process. Two complementary questions lie at the core of a systems approach to genetic analysis.

- Can the properties of a biological system be inferred and understood from the functions and interactions of individual genes?
- Can the activity and functions of individual genes be inferred from understanding the properties of the biological system?

Although we cannot yet answer these questions, the framework for thinking about them promises some new insights into genetics and biological systems.

We summarize a systems approach to genetic analysis in Figure 12.2. The tools that are used in this chapter include all those that we have discussed in previous chapters—mutations, cloned genes, genomics, RNAi, microarrays, and so on. The task is to use the products of these tools to reassemble a biological system.

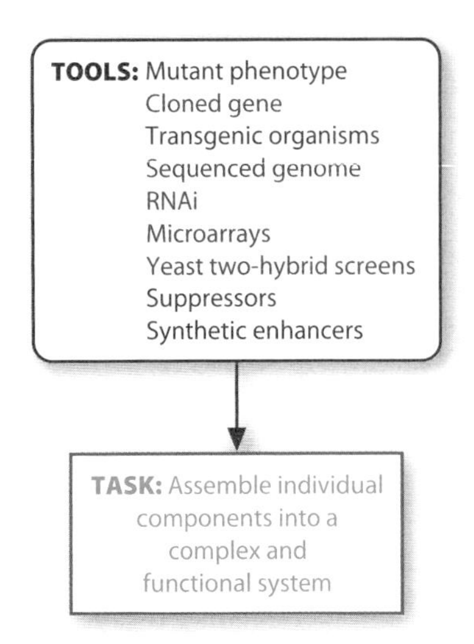

Figure 12.2 The tools and the task for Chapter 12. All of the tools that have been developed in the preceding chapters are brought together for our final task, which is to reassemble the results from individual tools into a complex system. The tools shown in red are based on mutant phenotypes; the tools shown in blue require cloned genes.

12.1 Pathways, networks and systems: an overview

The phenotypes that are most familiar to both geneticists and students of genetics arise from mutations in one gene or a few genes, with each gene having a significant impact on the phenotype. The phenotypes associated with such a single gene typically fall into a few discrete classes: red or white eyes, round or wrinkled peas, and so on. When intermediate mutant phenotypes are observed, they can often be explained as part of an allelic series, as discussed in Chapters 3 and 4. However, these discrete phenotypic classes do not easily describe much of the genetic variation that we observe in nature. Compared with the genetic variation in natural populations, model organisms under laboratory conditions present experimentally tractable but artificial examples.

For example, in natural populations allelic variation occurs at many different genes simultaneously, and homozygosity for the same allele is the exception rather than the rule. In model organisms, by contrast, inbreeding has produced lines that are homozygous at nearly all loci, and the effects of only one or a few genes are analyzed at one time. In a model organism, phenotypic differences arising from variation for other genes in the genome are minimized since the mutations are induced on a genotype that is wild-type and identical for all other genes. We illustrate this difference in Figure 12.3 by comparing the genetic variability in a wild fruit fly with a laboratory strain mutant in a single gene for wing formation. Laboratory strains are the same at essentially all loci except for the one affecting wing formation. In contrast, the wild fly has genetic variability throughout its genome. Because only one gene is varied at a time in the laboratory, interaction between alleles of different genes is limited to what is introduced and controlled by the investigator.

Furthermore, in model organisms, the mutant phenotypes are observed under defined and controlled environmental conditions in the laboratory. Thus interactions between the genotype and the environment are uniform for all mutant individuals. This is not the case in natural populations where environmental conditions vary widely; interactions between the genotype and the environment are important elements of genetic variability and natural selection.

Geneticists who work with natural populations, including those studying human populations, have worked with complex interactions between genes and between the genotype and environment for many decades. Sophisticated mathematical methods have

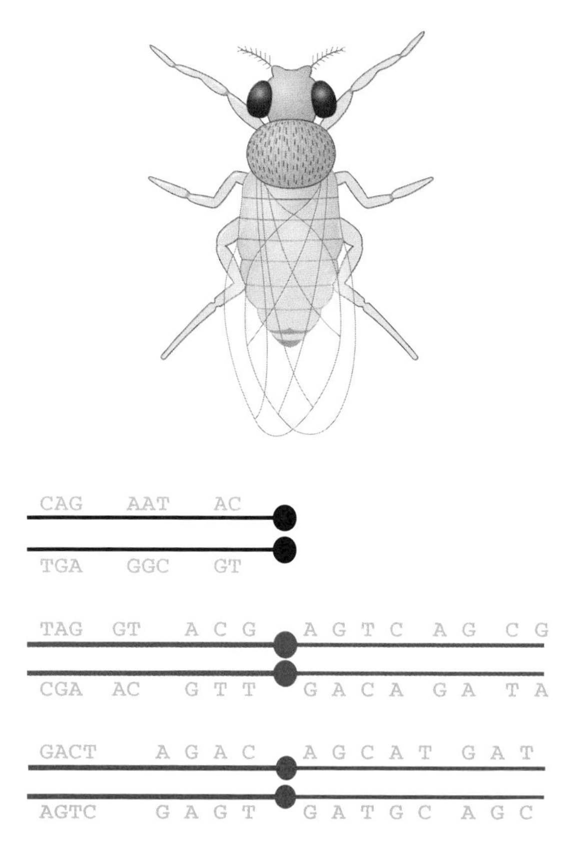

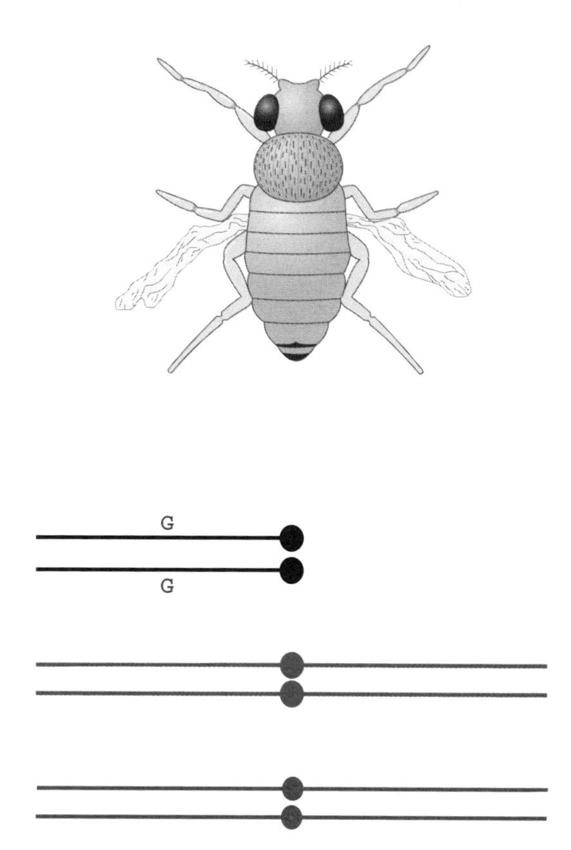

Figure 12.3 Natural populations and laboratory populations differ in their genetic variability. The three main chromosomes of *D. melanogaster* are shown beneath the drawings of two flies, with each homolog represented. Differences between the fly and laboratory wild-type fly are illustrated by the base changes at different sites. The fly on the left represents one from a natural population. As illustrated in the drawing, the fly is heterozygous or polymorphic for sequence differences throughout the genome; a wild fly differs from a laboratory wild-type fly at thousands of polymorphic sites. The fly on the right represents a laboratory strain that is a homozygous mutant for a vestigial wing phenotype. In general, the only sequence difference is the one that contributes to the mutant phenotype.

been developed to evaluate the relative contributions of genotypic and environmental variation, and the interaction between genes, and between genes and the environment. In many cases, it has not been possible to verify the results of these models experimentally. However, the availability of genome sequence information is beginning to change this situation, so that new and enhanced approaches to studying genetic interactions and gene–environment interactions are being used. Some of these new approaches that appear to hold particular promise are described in this chapter. However, we note at the outset that many of these approaches are themselves too new to have established their value. Our principal focus will be on using these approaches to analyze the effects of interactions between different genes; we return to the possible interactions between genotype and the environment in a more speculative section at the end.

Systems biology is made possible by genomics

The central tenet of these new approaches to genetic analysis is that the organism, cell, or biological process under study is treated as an organized system; this field has come to be called **systems biology**. The exact definition of 'systems biology' is somewhat fluid among its practitioners. For our purposes, a system has a large number of components, each of which has its individual characteristics and functions; furthermore, these components interact with each other both directly and indirectly to produce a functional entity. Systems biology includes both genetic and biochemical approaches, but we will limit our discussion of a systems approach to genetic analysis.

Systems biology is feasible because of genome sequencing and genome-wide screens, as described in Chapters 7 and 8. In a well-annotated genome, the genes and gene products—a most important set of components—have been cataloged. Furthermore, many of the characteristics and functions of these individual components can be deduced or demonstrated using methods and strategies described in the preceding chapters. In a system, these individual components interact with one another, just as genes and gene products interact in the cell. In Chapters 10 and 11, we described genetic and molecular methods to identify and analyze these interactions. Therefore, in principle, we have both elements needed for a systems approach: components of known function and a catalog of interactions between the components.

In thinking about an organism as a biological system, one point of exploration is the extent to which the properties of the overall system can be inferred from understanding the properties of the individual components, and which properties of the system are **emergent** rather than being inferred from the parts. One useful contrast between this and what has been done previously is to recall that the powerful experimental approaches of molecular biology are largely **reductionist**, i.e. these approaches reduce a complex biological system into its parts. Systems biology is largely **synthetic**; in systems biology, the parts identified by the reductionist approaches of molecular biology are then intellectually reassembled into a working system.

Systems biology may be best understood by analogies with other familiar systems. For example, the traffic pattern in a particular locality is a system. The number and movement of individual cars comprise the component parts. These components interact with each other via a series of highways. A direct interaction between the cars occurs when one car immediately follows another along the same highway or stops for another car at an intersection. Each car functions independently of the others, but the movement (or function) of one car also directly affects the movement of another. An indirect interaction in this highway system arises when the number of cars on one highway is particularly heavy, so that some cars are diverted to another highway. This also slows down the movement of cars on the other highway; although no car on one highway directly interacts with one on the other highway, the overall function of the traffic pattern is affected. By knowing the number and movement of cars in one part of the traffic system, a good traffic engineer can infer the impact on another part of the system.

This analogy illustrates several features of systems biology. For an individual commuter, the properties of the entire system provide useful knowledge insofar as they affect single components—his/her car and the route to work. Hearing that there is heavy traffic on one road, the savvy commuter chooses an alternative route. In such an example, the properties of the system as a whole can also be used to predict the effects on individual components. In addition, the properties of the highway system are dynamic rather than static; the direct and indirect interactions between the cars are different at rush hour and in the middle of the night. A traffic engineer can predict the system-wide effects of closing a particular road at a given time or of building an additional connecting highway.

Systems biology is not yet as sophisticated as traffic engineering in being able to infer the properties of the system from the individual components or in being able to predict the effects on the system of changing one component. In particular, the analysis of indirect interactions and dynamic changes in biological systems is in its infancy. For that reason, we will concentrate on procedures used to determine and analyze the *direct* interactions among biological components, and turn to the indirect and dynamic properties at the end of the chapter. We will consider two types of direct interaction between biological components that are known to regulate the overall properties of the organism and are being studied on a genome-wide scale:

- the interactions of transcription factor proteins and their target DNA-binding sequences
- the interactions between proteins.

These interactions are shown diagrammatically in Figure 12.4. Figure 12.4A shows the transcriptional regulation of a hypothetical gene, first as a conventional diagram with the proteins binding to upstream sequences and secondly as a network of interactions. Figure 12.4B shows a protein complex, first as a multimeric complex and secondly as a network diagram of interacting components. In each case, we will discuss how the interactions are determined experimentally and computationally, often using tools and approaches that have been developed in previous chapters. We

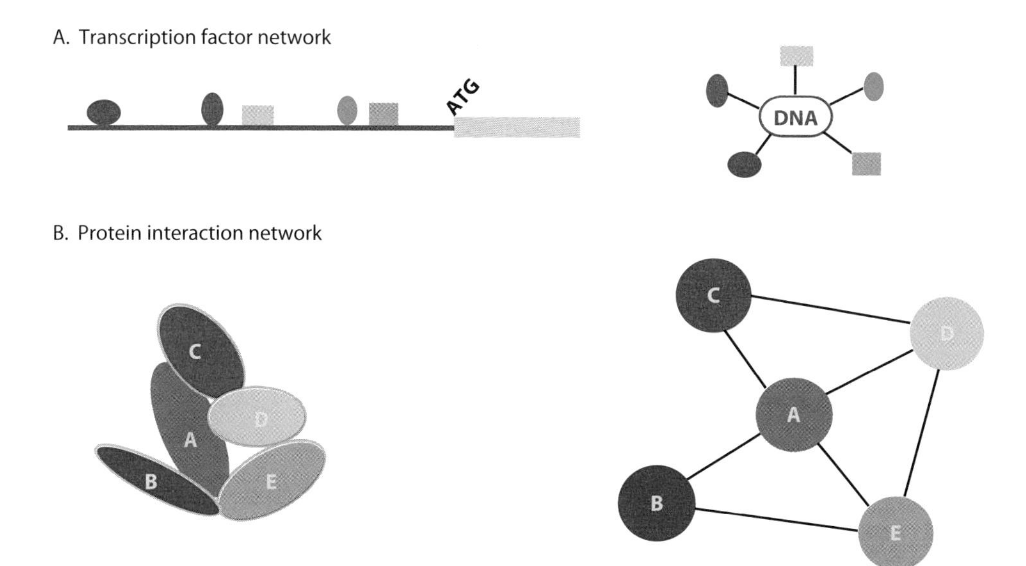

Figure 12.4 Examples of two biological interaction networks. (A) The regulatory region of a typical gene. Five different transcription factors bind to this region of DNA. This type of interaction is often represented by the drawing on the left. The network drawing on the right shows the same five transcription factors binding to the same DNA regulatory region, but in a different format. (B) A protein complex. The drawing on the left shows the complex with five members and illustrates their contacts with each other. The same interactions are shown as a network diagram on the right.

will also consider the methods used to test the functional significance of the interactions. The interactions together can be thought of as comprising a **network**. For our purpose, the system will have one or more networks that depict the interactions. Different network plots can display different types of interactions, many of which are dynamic. We begin with some background on the mathematical analysis of networks and a definition of a few of the key terms.

KEY ARTICLE

Zhu, X., Gerstein, M., and Snyder, M. (2007). Getting connected: analysis and principles of biological networks. *Genes & Development*, **21**, 1010–24.

12.2 Properties of networks: background and definitions

While it would be possible to catalog the various interactions that occur between components in a network, interactions are not usually depicted in such a table because there are so many different interacting components. Instead, the interactions are usually plotted as a graph. The difference in how a table and a graph are used to show the data will be familiar to travelers. We use a timetable, i.e. a table of connections, to determine if a certain train goes from Station A to Station B and when the next train arrives. We use a route map, i.e. a graph of connections, to see the best plan for a more complex itinerary. In a graph, the stations are depicted as a **node** or a vertex, and the travel between them as connections or **links**.

For biological networks of the type we are describing, a node is one of the macromolecules and a link is an interaction between them. As shown in Figure 12.4, for transcriptional networks, the nodes are transcription factors and regulatory sequences; the links depict binding of a transcription factor to that regulatory sequence. For a network of protein interactions, the nodes are proteins and the links are direct physical interactions between two of them. In the example shown in Figure 12.5A, the protein interacts directly with 10 other

A. First-degree interactions

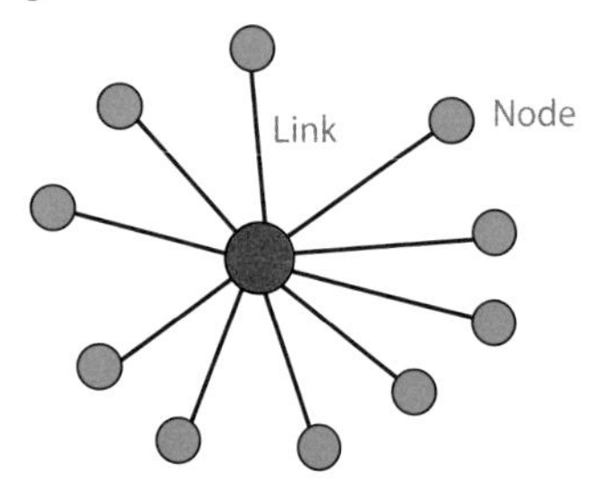

B. Interactions among the interactors

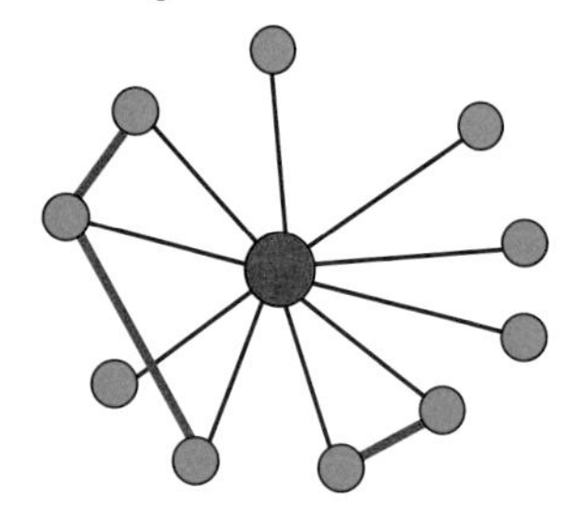

C. Second-degree interactions and a network

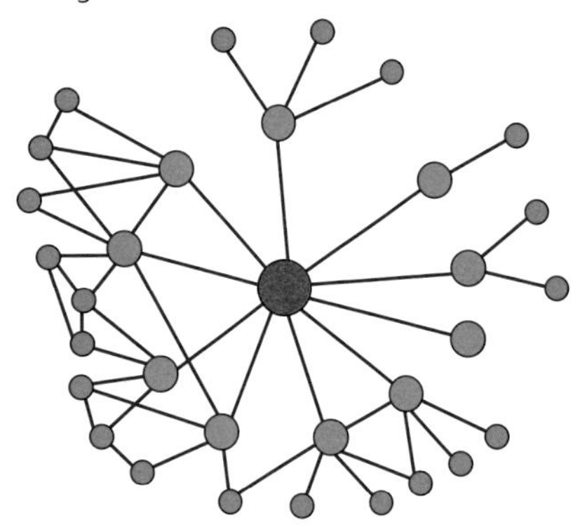

Figure 12.5 A simple biological network. A hypothetical protein interaction network is shown. Each protein is represented by a node and each interaction or link between the proteins is represented by a line connecting two nodes. (A) The protein shown in red at the center of the graph interacts directly with 10 other proteins, shown in green. These green proteins are its first-degree interactions. (B) The first-degree interactions (the green proteins) interact with each other as shown in red, suggesting some functional relationships among these proteins. (C) The first-degree interactors also interact with a number of other proteins, shown in blue. The blue nodes are second-degree interactions with respect to the red protein.

proteins. These 10 proteins are termed the **first-degree interactions**—the interactions that directly connect two nodes.

KEY ARTICLE

Barabási, A.L. and Oltvai, Z.N. (2004). Network biology: understanding the cell's functional organization. *Nature Reviews Genetics*, **5**, 101–14.

The graph can be expanded beyond first-degree interactions because each of the first-degree nodes has its own set of links in addition to the one described in the original network. Some of the first-degree interactors will also interact with each other, creating an interconnected web of interactions (Figure 12.5B). The first-degree interactors will also identify additional components of the network, which may be genes or proteins that function in the same biological process. These situations are shown in Figure 12.5C. These additional interactions are plotted in a similar way, resulting in a set of second-degree interactions. These second-degree interactions themselves do not interact directly with the original protein of interest, but they interact with one or more of its interacting proteins. In short, they can be thought of as 'friends of friends'. As each interaction is plotted, a network of connected nodes arises.

For example, consider what happens when a new subscriber signs onto a social network site such as Facebook or MySpace. The new person finds his friends who are also subscribers and invites them to be Facebook friends. A small network forms, composed of first-degree interactions between the new subscriber and his friends. Each friend has his/her own group of Facebook friends as well. Some of the new subscriber's friends will also be friends with each other, so they are already connected in the social network. Others will be friends of the first-degree interactor but are not known to the new subscriber. Some of these people will invite the new subscriber to become their friend as well, so the subscriber's network becomes more complex.

The network can be used to infer the functions of its components

In other parts of this book we have described three different methods of inferring the function of a gene: mutations that knock out or change its function, sequence similarity to other genes or proteins whose function is known, and expression patterns. A fourth possible method is to analyze its interactions. Analysis of the interactions can be done by looking at an individual node in a network and noting all its interactions. Most fundamentally, such a network allows us to infer the function of genes and proteins whose function was not known from other methods. This arises from the 'friends of friends' aspect of a network. Thus, networks

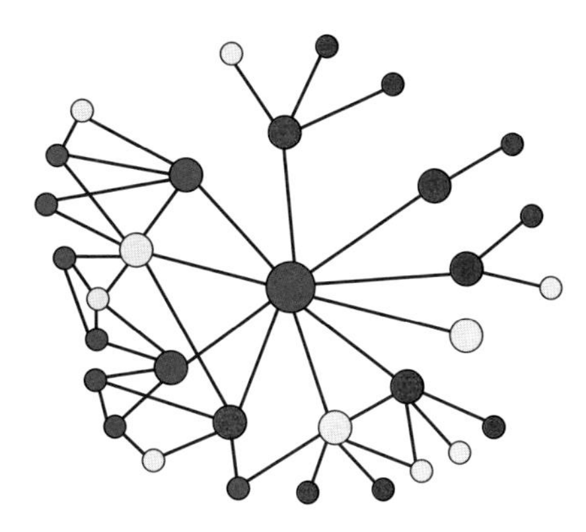

Figure 12.6 Interactions suggest biological functions. In this hypothetical graph of protein interactions, the biological function of the protein shown in red is unknown. The proteins shown in blue have been shown to be involved in chromosome structure. The proteins shown in gray have an unknown function, whereas those in pink have known functions other than chromosome structure. Since the red protein has numerous first- and second-degree interactions with proteins involved in chromosome structure, a reasonable hypothesis is that the red protein is also involved in chromosome structure.

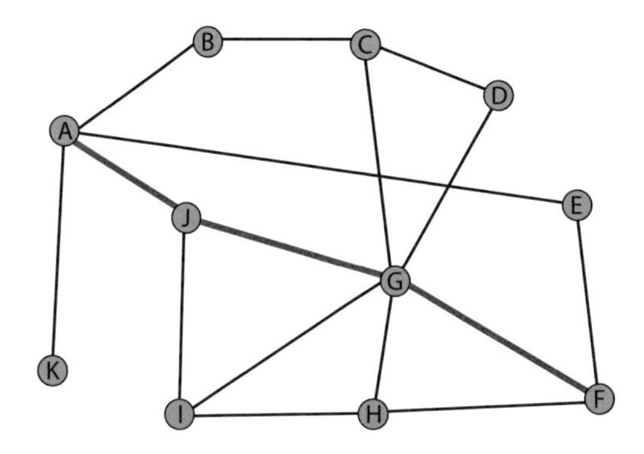

Figure 12.7 A path through a network. This simple network diagram with 11 nodes will be used to illustrate several features of networks. The heavy red line shows a path connecting node A with node F. The length of the path is 3. A path of length 2 goes from node A to node E to node F; there are many paths of length longer than 3 that also connect A and F.

provide another means of inferring the function of a gene. For example, imagine a protein of unknown function whose first- and second-degree interactions include a number of proteins known to be involved in the organization of chromosome structure. This is the hypothetical protein shown in Figure 12.6. A reasonable hypothesis is that the unknown protein is also involved in the chromosome structure. More specific examples are discussed below.

In addition to being a valuable method of inferring the function of an individual gene or protein, the interactions can be analyzed by studying the properties of the entire network. The plot of a biological interaction network comprises a graph whose properties can be studied by the well-developed field of **graph theory**. (An entire subfield of computer science is devoted to the best way to plot and represent such graphs, but this goes beyond our discussion here.) Although each component is a node or a vertex in the network and each interaction is a link or an edge, the individual nodes or links are not the subject of graph theory. Instead, graph theory is concerned with the characteristics or **topology** of the entire graph. Some authors have referred to this as molecular cartography—a map of the macromolecular connections that occur within a system. It is worth recalling the two questions that framed our introduction to systems biology. The network graph can give insights into the roles of individual nodes—proteins or genes in our case. The topology of the entire network graph can help to synthesize the properties of these individual components into a functional system. We will only discuss a few of the features of these graphs, but this field is a rich area for much more analysis.

Paths

In networks, the link that connects two nodes is known as a **path**. In the simple network shown in Figure 12.7, one path between node A and node F is shown in red. A useful way of describing the complex graph of the biological interactions is to consider the **path length**, or the *number* of links that connect any two nodes. In discussing the entire graph, one can describe the minimal path length, the average path length, or the longest path; each of these parameters might be useful in some contexts.

(Finding the minimal path length is the goal of 'The Kevin Bacon Game', in which players compete to find the shortest number of shared movie roles (links) that connect any other actor to Kevin Bacon. In this network of about 800 000 actors, the average 'Bacon number', or the number of links to Kevin Bacon, is slightly less than 3. In a similar vein, many mathematicians can tell you their Erdos number, the path of publications and collaborations that link them with the mathematician Paul Erdos, one of the pioneers of graph theory.)

The minimal path length between two nodes provides information about how easily or rapidly a change in one node affects another. In an example we will develop more fully later in the chapter, genes in yeast involved in responding to a change in environmental conditions have short path lengths. The average path length can be used to indicate the extent of interconnections among the nodes and the size of the network.

In most biological situations, the description of the graph has been cast in terms of the *average* path length. Biological networks are said to be **small world**, meaning that the average path length connecting two nodes in the network is short. We can define 'short' more precisely by computing the number of connections in the network. If the average path length is proportional to the logarithm of the number of connections, a network is defined as being small world. The World Wide Web, social networks, and most other familiar networks fit the definition of small world. Many biological networks are described as **ultra-small world**, meaning that the average path length is very short. A more technical description of ultra-small networks is that the average path length in biological networks is proportional to the logarithm of the logarithm (i.e. log–log) of the size of the network. The average path length between nodes in a biological network is a very small number; typically, less than four interactions connect any two nodes. An ultra-small network has an important property for genetics, namely that a change in one part of the network will quickly affect many other parts of the network. One consequence of very short path lengths is that a cell can process information quickly.

In attempting to describe a protein interaction network in a cell, analogies to other networks like the Kevin Bacon game and airline flight patterns abound. Some of these analogies are likely to be useful, whereas others will almost certainly be misleading. One challenge of such a young field is to know which examples are the most applicable. For instance, some analyses of biological networks indicate that the average path length between any two nodes is highly similar in different species. Has there been selective pressure to maintain a particular path length in a protein network? Or is this an incidental consequence of other properties of the network? At the moment, investigators do not know and arguments exist on both sides of the question.

The degree distribution and hubs

Every node in a network has some number of connections to other nodes, which varies for different nodes. One of the most important properties used to describe a network is the number of links per node, which is known as the **degree** or connectivity. The degree is usually represented by the parameter k. In any given network, k has a particular distribution called the **degree distribution**, i.e. the way in which the degree varies between nodes. Figure 12.8 shows a simple network, tabulating the number of links for each node. For example, although there is an average of three links per node, some nodes have three links, some have four, some have five, and so on. The degree distribution is plotted as a histogram with variation around the average degree; in complex biological networks, this histogram has thousands of data points.

A key objective of studying networks is to determine the mathematical function that describes the degree distribution. This function can then be used to describe the overall properties of the network and to calculate the probability that a node has exactly n links, or more or less than n links. This will allow the investigator to distinguish nodes that are highly connected to others in the network from those that are more loosely connected, or to determine if all nodes have approximately the same degree. Most of the original theoretical work on networks assumed that connections between the nodes formed at random. It has been shown that in such a random network, the degree distribution follows a Poisson distribution. (Many important genetic events can be described by a Poisson distribution. For example, in their fluctuation test, Delbrück and Luria showed that the occurrence of a mutation follows a Poisson distribution, thereby demonstrating that mutation is a random event.) This type of network is shown in Figure 12.9A. The histogram of the degree distribution for this network is shown in Figure 12.9B. A key feature of a Poisson distribution is that values that are far from the mean are quite rare, as shown in Figure 12.9C. In other words, if the average number of links per node is three, it is very unlikely that a node will have 10 or more links—to be precise, the probability is less than 0.004.

Biological networks cannot be described by assuming a random number of connections between nodes

In an analysis of a biological network, one of the key questions is to determine if the degree distribution can

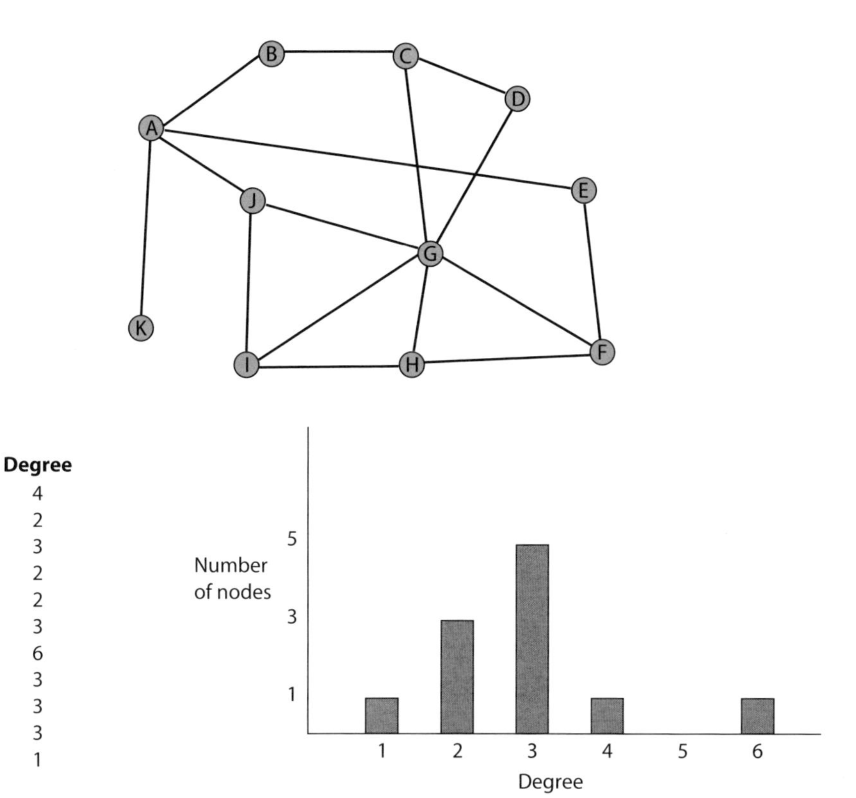

Node	Degree
A	4
B	2
C	3
D	2
E	2
F	3
G	6
H	3
I	3
J	3
K	1

Figure 12.8 The degree distribution of a network. The same network as shown in Figure 12.7 is shown here. Beneath the graph on the left is a tabulation of the degree for each node in the graph. This is plotted as a histogram on the lower right. Such a histogram corresponds to the degree distribution. Biological networks may have thousands of nodes and tens of thousands of links.

be described by a Poisson distribution; in statistical terms, a random network is the null hypothesis that is rejected or accepted. The degree distribution for many biological networks, including most of those we will discuss, *cannot* be described by a Poisson distribution. Instead, a variety of research has shown that biological networks, as well as many other familiar networks such as the World Wide Web and most social networks, can be described by a different degree distribution, referred to as a **power-law distribution**.

One type of power-law distribution is shown in Figure 12.10. In a power-law distribution, the degree distribution is inversely proportional to the degree parameter k raised to some power (called the degree exponent and symbolized by γ). Expressed mathematically, a degree distribution described by a power law is proportional to $k^{-\gamma}$.

The importance of the degree exponent for describing the properties of the network may not be intuitive, so a comparison with random networks will help. For the network of protein–protein interactions in the yeast cell, γ is 1.86. If γ is greater than 3, the degree distribution for the network approximates what is seen for a random network. A hypothetical network, in which the degree distribution is 1.9, is shown in Figure 12.10A. The histogram of the degree distribution is shown in Figure 12.10B. Notice in comparing the power-law distribution in Figure 12.10 with the Poisson or random distribution in Figure 12.9 that the histograms have a different shape and the curves that plot the degree distribution are also quite different.

One property of a power-law distribution with undeniable significance for biological networks is the occurrence of nodes with many links and nodes with very few links. A node that has many connections is referred to as a hub. Both the network in Figure 12.9 and the one in Figure 12.10 have 19 nodes. In the random network in Figure 12.9, only one node has more than four links, and that one has five links. Likewise, only one node has a single link. However, as seen in Figures 12.10A and

DEFINITION

hub: A node with a large number of links is defined as a hub.

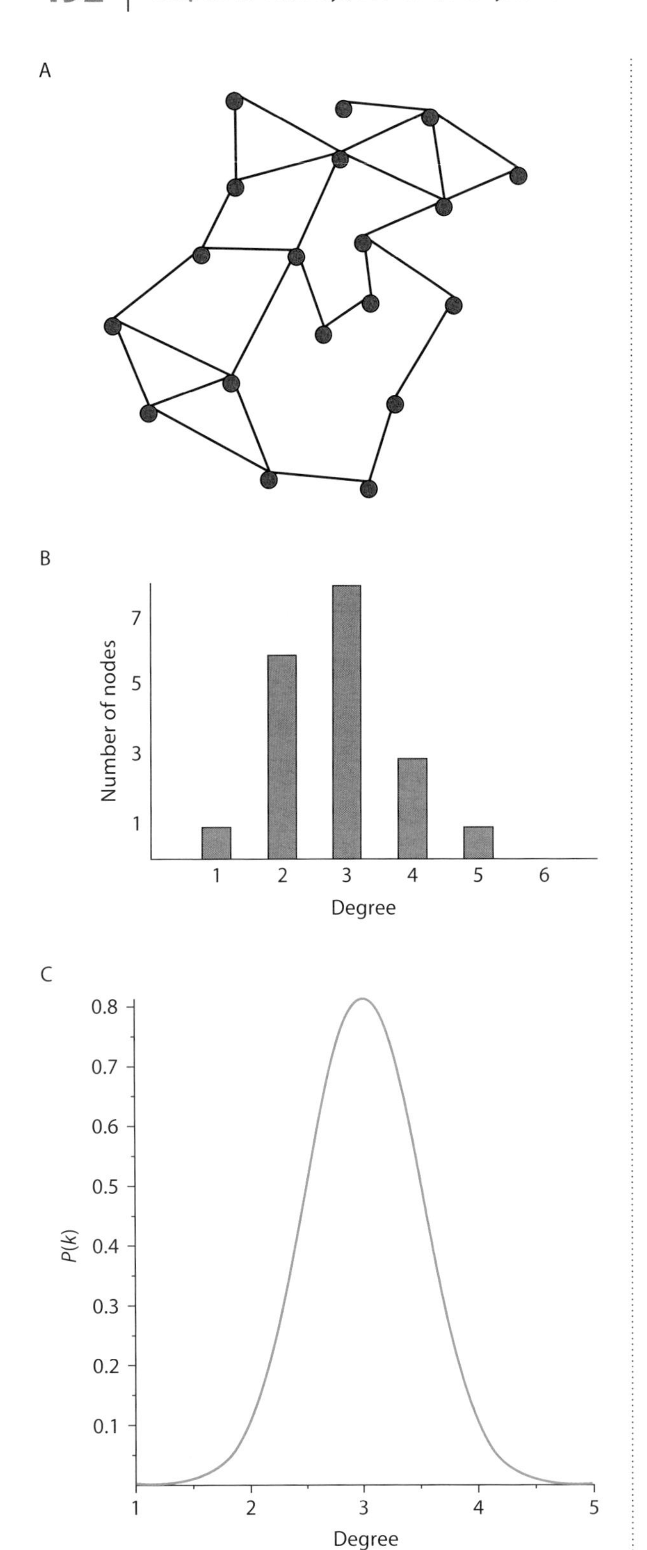

Figure 12.9 A random network. (A) A network of 19 nodes in which the links occur at random. (B) A histogram of the degree distribution for these nodes. Notice that the distribution centers around an average degree of 2.74. (C) As the number of nodes and links increases, the degree distribution $P(k)$ follows a Poisson distribution. Nodes with a large number of links are expected to be rare.

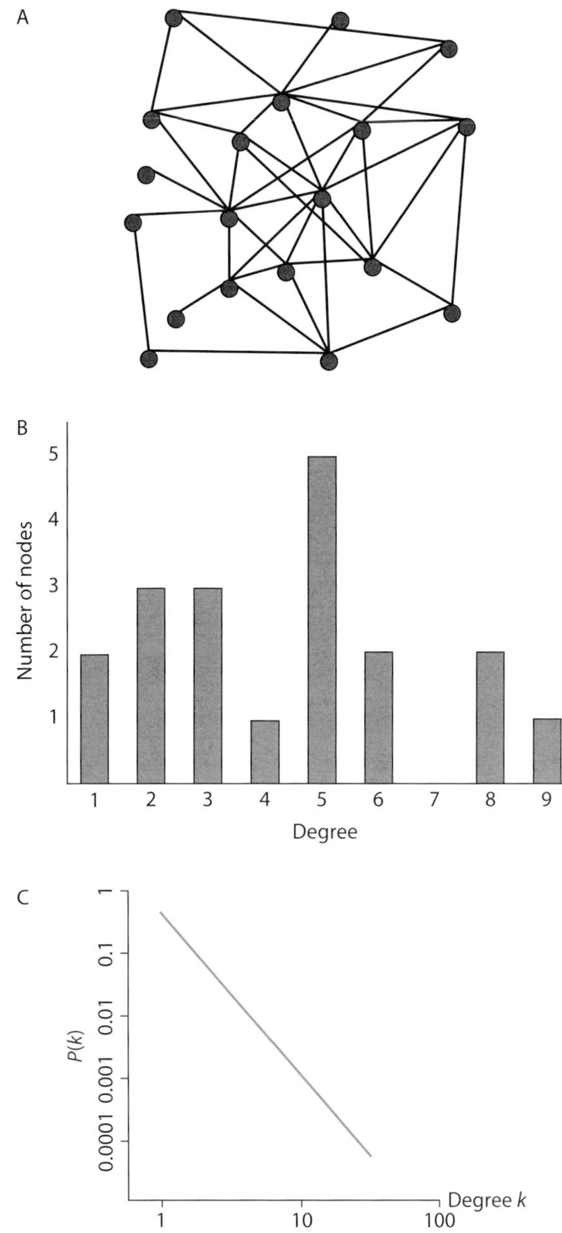

Figure 12.10 A network described by a power-law distribution. (A) The same nodes as in Figure 12.9 but with different links and a different degree distribution. (B) The histogram of the degree distribution for the graph in panel A. In comparing this histogram with Figure 12.9B, note the occurrence of nodes with a higher-than-expected degree distribution, in this case nodes with eight and nine links. As the number of nodes and links increases, the degree distribution follows a power law, as shown in panel C. The degree distribution follows a straight line on a log–log plot.

12.10B, three nodes have eight or nine links and two have a single link. (To return to the degree exponent, hubs are increasingly important to the overall topology of the network as γ becomes smaller.) Furthermore, as shown in Figure 12.10C, the degree distribution for a network described by a power law is a straight line on a log–log plot, again reflecting that many nodes have few links and some nodes have many links.

The definition of a hub is familiar from other networks

Power-law distributions and networks with hubs are familiar in many everyday situations, which will be noted throughout the chapter. The term 'hub' is used in the same way for these biological networks as the airline industry uses it for airports with many connecting flights. A hub is a node whose degree is greater than would be predicted by random connections; more formally, a node whose degree has a random probability of less than 0.01 is defined as a hub. Looking at this another way, using a random distribution based on k, it is possible to calculate the probability that a node will have n or more links. If the probability is less than 0.01, the node is defined as a hub. In a network whose average degree is 3, as has been seen for many biological networks, a node with more than six links is a hub.

Other properties of networks will be introduced as we describe specific examples of biological interaction networks. In particular, the properties of hubs define many of the properties of biological networks. We now describe some particular examples of biological networks, beginning with the interactions between transcription factors and their DNA-binding sequences.

12.3 The interactions between transcription factors and DNA sequences

Transcriptional regulation occurs by the interaction of particular proteins—transcription factors—with specific DNA sequences—their binding sites. A simple example in which a conventional view of transcription is replotted as a network is shown in Figure 12.11. In considering transcriptional regulation in terms of a biological system or network, the components are the transcription factors together with their specific DNA target sequences. Figure 12.11A shows three genes with transcription factors binding to their upstream regulatory regions. Both the DNA-binding sites and the transcription factor-binding sites have been very thoroughly studied by assays with individual genes. For example, consensus binding sequences for many different transcription factors have been found using *in vitro* assays such as gel mobility shifts and DNA footprinting; enhancer sequences needed for specific expression have been identified using reporter gene constructs, as described in Chapter 8. Thus, it has been possible to reduce the complex system of transcriptional regulation in a cell or a tissue to its component parts and to study the interactions of those parts. But can the properties of the individual components be reassembled into a network of interactions, and what does that tell us about the logic of transcriptional regulation?

KEY ARTICLES

Blais, A. and Dynlacht, B.D. (2005). Constructing transcriptional regulatory networks. *Genes & Development*, **19**, 1499–1511.

Zhu, X., Gerstein, M., and Snyder, M. (2007). Getting connected: analysis and principles of biological networks. *Genes & Development*, **21**, 1010–24.

In order to analyze this network on a genome-wide scale, the specific DNA sequences bound by individual transcription factors have to be identified and the association between the regulatory sequence and the gene it regulates has to be established. Each of these interactions can be analyzed, although the process is not always simple. Some of the relevant questions when taking a network approach to studying the interactions between transcription factors and regulatory sequences are:

- Which and how many different transcription factors regulate the expression of each gene?
- How many different genes does one transcription factor regulate?

A. Transcription factors bound to regulatory regions

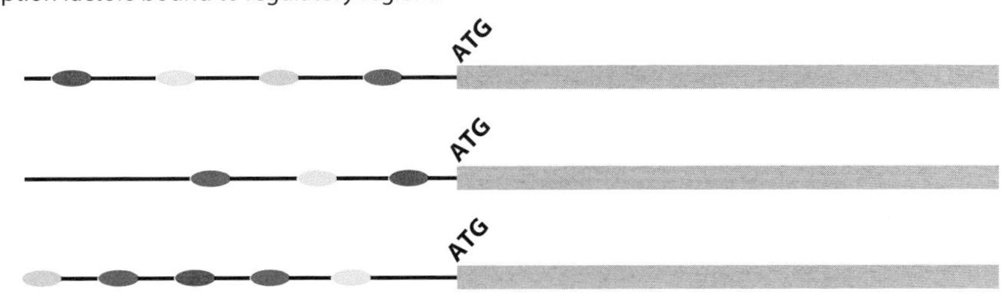

B. The binding sites for transcription factors

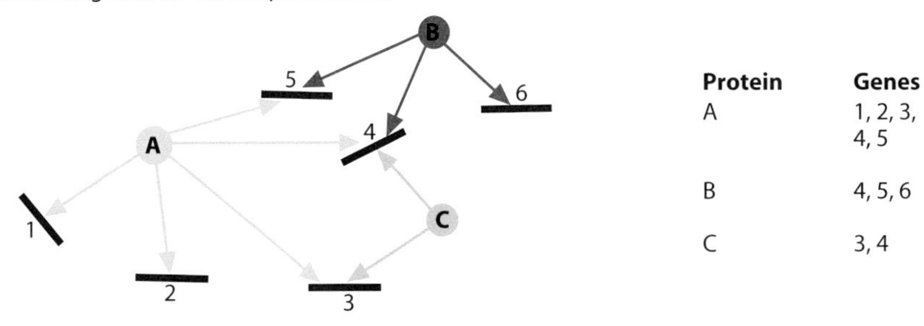

Protein	Genes
A	1, 2, 3, 4, 5
B	4, 5, 6
C	3, 4

C. Transcription factors bound to a regulatory region

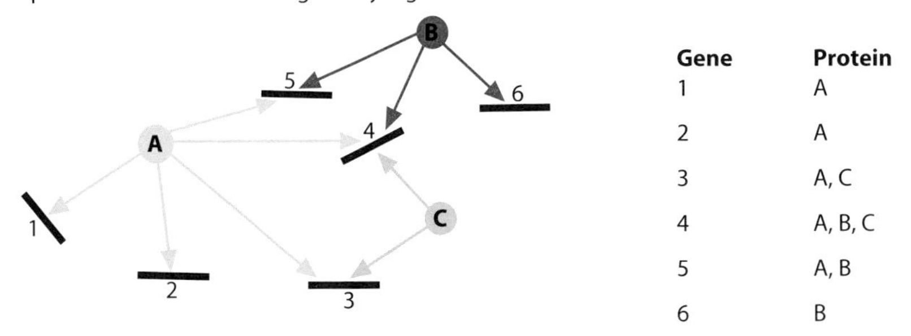

Gene	Protein
1	A
2	A
3	A, C
4	A, B, C
5	A, B
6	B

Figure 12.11 Depicting transcriptional regulatory networks. (A) Three different genes are shown with the transcription factor-binding sites in different colors in a conventional drawing of gene regulation. (B) The information is drawn from the perspective of transcription factor binding. The question for this experiment is: 'What genes are regulated by the same transcription factor?' For example, the red transcription factor shown as protein B binds to genes 4, 5, and 6. (C) The information is presented from the perspective of the genes being regulated. The question for this experiment is: 'What transcription factors regulate this gene?' For example, the gene shown as number 4 is regulated by transcription factors A, B, and C. The binding is shown with an arrow to indicate the direction of the interaction that binds that protein to the DNA sequence.

- Where are the regulatory sequences that serve as transcription factor-binding sites?
- How is the transcriptional response of different genes coordinated to change over time?

Identifying interactions between transcription factors and their binding sites

The network of interactions between transcription factors and their targets has been approached from different experimental directions. The more widely used methods have begun with a transcription factor and attempted to identify its regulatory target sequences wherever they occur in the genome. In effect, this approach asks: 'What regulatory sequences are bound by this protein?' The approach can be considered to be **protein based** as shown diagrammatically for a set of hypothetical genes and transcription factors in Figure 12.11B. Three different transcription factors are shown, with different numbers of regulatory binding sites for each transcription factor. One transcription factor has two binding sites, one has three, and one has five. To carry out such an experiment, the investigator began with a transcription factor and identified its binding sites throughout the genome.

In a second approach, less widely used so far, the investigator begins from the opposite perspective with the regulatory sequences themselves, and attempts to catalog all of the proteins that bind to the sequence. This perspective can be considered sequence or **gene centered**, and asks the question: 'What transcription factors bind to this regulatory region?' This approach is considered in Figure 12.11C using the same data as in Figure 12.11B, only plotted from the perspective of the genes being regulated. Thus, one gene has three transcription factors, two have two transcription factors, and three are regulated by a single transcription factor. A third strategy is **genome based**. This method uses computational approaches to identify potential regulatory sequences and transcription factors. We will describe strategies from all three approaches, beginning our discussion of transcriptional regulatory networks with a protein-based approach, i.e. with transcription factors.

Chromatin immunoprecipitation (ChIP) is a protein-based approach to surveying genome-wide binding sites

Transcription factors are readily classified into families based on their DNA-binding domain, such as C2H2 zinc-finger proteins or homeodomain proteins. Proteins of the same family usually bind to similar sequences. A quick search of the domain database **Pfam** on the NCBI website identifies more than 500 different DNA-binding domains used to classify proteins. It is estimated that 5–10 percent of proteins encoded in a eukaryotic genome are transcription factors. It is comparatively easy to identify a gene product as a potential transcription factor, but what are its binding sites and what genes does it regulate?

A conventional experiment to identify the binding site for a transcription factor relies on a purified transcription factor protein in one of several different *in vitro* binding assays. These binding assays have been expanded to a genome-wide scale using immunoprecipitation (IP) experiments. IP procedures for protein complexes were briefly described in Chapter 10. In typical IP experiments, an antibody raised against one protein—in this case, a transcription factor—is used to precipitate other proteins in a complex. The strategy has also been applied to chromatin to find not only the interacting proteins in the complex but, more significantly for this discussion, the DNA sequences associated with these protein complexes. The general procedure is referred to as **chromatin immunoprecipitation (ChIP)**, and is shown diagrammatically in Figure 12.12.

In ChIP, cells or tissues whose transcriptional patterns are being compared are grown under different conditions and then treated with an agent that produces reversible crosslinks between DNA-binding proteins and chromatin (Figure 12.12). An antibody against the transcription factor is then used to precipitate the chromatin complex, which includes both the transcription factor and the DNA sequences to which it is bound. In a standard ChIP assay, the presence of a specific DNA sequence is detected by PCR. When this experiment is done on a genome-wide scale, the DNA sequences isolated in the complex are hybridized to a DNA tiling microarray of the type described in Chapter 8 (Figure 12.12). These are called **ChIP on chip (ChIP–chip)** experiments, i.e. ChIP with microarray chips.

Hybridization to the microarray reveals which sequences and regions of the genome are bound by this specific transcription factor under these conditions. Because the positions of different sequences in the microaray chip are known, the locations of the transcription factor-binding sites are also known. As discussed in Chapter 8, tiling microarrays have short-sequence probes for both DNA strands in every region of the genome. The procedure is repeated, as a high-throughput technique, for different transcription factors and different experimental growth conditions. This reveals a transcriptional regulatory network of direct interactions between the transcription factor and its binding sites in the genome under different conditions.

More than 250 ChIP–chip experiments have been performed in yeast, and more than 10 000 interactions between transcription factors and binding sites have been detected by this method. The first analysis used epitope-tagged versions of 106 known transcription factors; the advantage of the epitope tag is that the same antibody can be used to precipitate different transcription factors. The strains were grown in three different media, and the interactions were tabulated. The table of interactions is used in the same way that the table of interactions was used in the networks plotted in Figures 12.9 and 12.10. Among the 6270 yeast genes in the genome, 37 percent (2343) were bound by one or more of these transcription factors, and a third were bound by

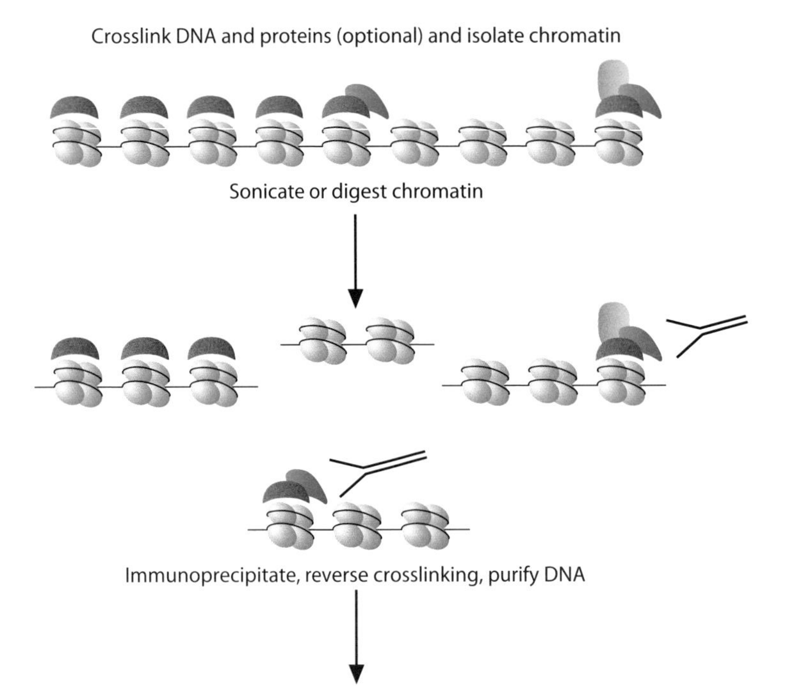

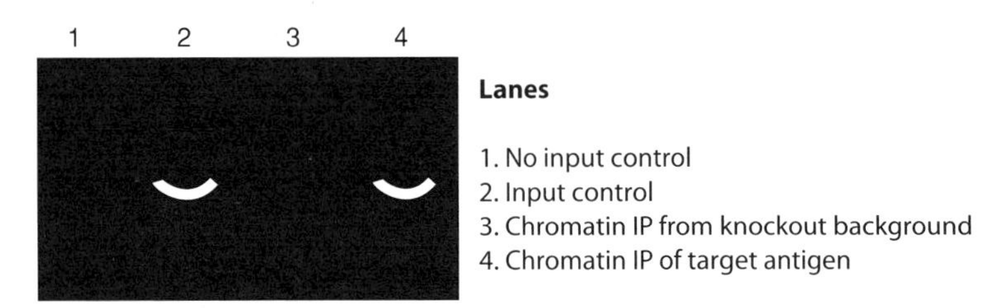

Figure 12.12 A ChiP–chip experiment. A transcription factor is bound to chromatin and crosslinked to form a stable complex. The chromatin is isolated and sonicated. An antibody to the transcription factor is used to immunoprecipitate the chromatin fragments, the crosslink is reversed to release the DNA, and the DNA is purified. In a ChIP–chip experiment, the DNA is used on a microarray to identify the binding sites for the transcription factor.

two or more of the proteins. However, these assays focus on the behavior of the transcription factor proteins themselves rather than on the genes being regulated. On average, a transcription factor bound to 38 different regulatory regions but the range is significant—from a few transcription factors that bound no regulatory regions under these growth conditions to one that bound 181 different regulatory regions.

KEY ARTICLES

Lee, T.I., Rinaldi, N.J., Robert, F., *et al.* (2002). Transcriptional regulatory networks in *Saccharomyces cerevisiae*. *Science*, **298**, 799–804.

Zhu, X., Gerstein, M., and Snyder, M. (2007). Getting connected: analysis and principles of biological networks. *Genes & Development*, **21**, 1010–24.

Genome-wide ChIP assays have given the most insights into transcriptional regulation in yeast so far

Genome-wide ChIP assays are being used to examine transcriptional regulatory networks in many different organisms, but some limitations of this approach should be recognized from the outset of our discussion. First, the technique depends on the availability of antibodies to the transcription factors of interest and on the specificity of the antibody to that one protein. A genome encodes several thousand transcription factors, so having antibodies to each of them is a formidable task. Secondly, the binding interactions can vary widely depending on the experimental conditions. Thus, the assay must be repeated for many different conditions to view all of the interactions regulating transcription. These limitations are worth noting, but they do not rule out the importance of the method; more transcription factors and more conditions can always be tested as the reagents become available. The most challenging limitation for ChIP–chip assays is that frequently transcription factors are present at very low levels in few cells and at specific times, particularly when considering the complex regulatory networks involved in metazoan development. Thus, many transient interactions that are critical to cellular differentiation could be missed. Nonetheless, the preliminary results in yeast are illuminating.

One potentially important distinction emerged between endogenous and exogenous network components in yeast (Figure 12.13). **Endogenous** network components (Figure 12.13A) are characteristic for genes whose expression is fundamental to the basic function of the cell and is restricted to particular times. Genes involved in the cell cycle are an example of endogenous components. As a rule, these genes tend to be regulated by a fairly high number of transcription factors, each with a few targets. Furthermore, these transcription factors are themselves regulated by other transcription factors so that the average path length in an endogenous network is slightly longer than the average path length for the total yeast cell. Therefore, the regulatory network for endogenous components is highly interconnected and is very specific and defined —a description of a set of genes expressed coordinately under specific conditions.

An **exogenous** network includes genes that respond to environmental stimuli. Figure 12.13B summarizes the properties of this network. These genes tend to be regulated by *few* transcription factors, each with *many* targets. The targets themselves are typically not other transcription factors, so that the average path length in an exogenous network is shorter. Such a network allows a more rapid but less highly coordinated response, involving genes with many different functions. One simplified inference is that the genes needed for the normal cellular bureaucracy and organization, when changes in transcription occur at predictable times, can be described as an endogenous network. Genes needed for responses to unpredictable situations are described by an exogenous network when a few genes respond quickly and directly contact many other targets.

A. Endogenous transcription network

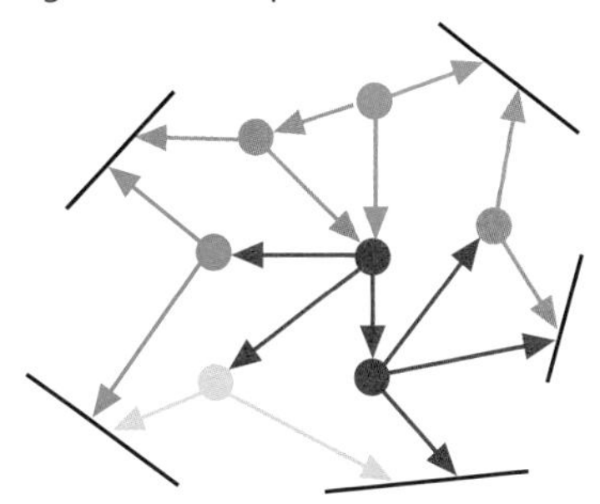

Complex interactions among transcription factors
Long path lengths
Few targets per transcription factor

B. Exogenous transcription network

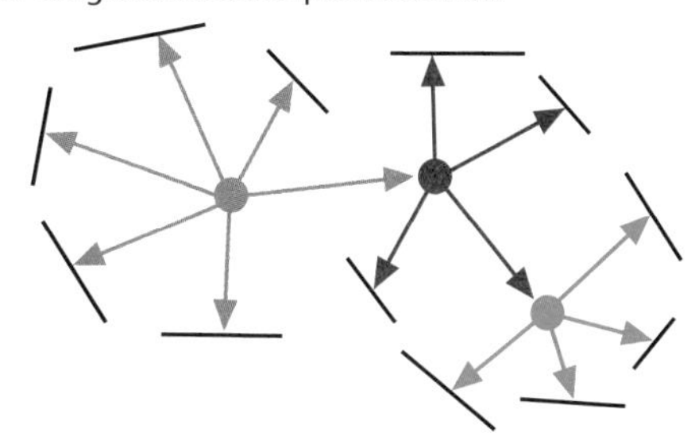

Simpler interactions among transcription factors
Short path lengths
Many targets per transcription factor

Figure 12.13 Endogenous and exogenous transcription networks. The networks show the transcription factors as colored circles with arrows representing the binding site to different regulatory regions. (A) In endogenous networks, the complexities of the interactions between the transcription factors and the longer path length between two different genes results in a highly integrated transcriptional response. (B) In exogenous networks, the short path length with many targets for each transcription factor allows a rapid transcriptional response.

KEY ARTICLES

Blais, A. and Dynlacht, B.D. (2005). Constructing transcriptional regulatory networks. *Genes & Development*, **19**, 1499–1511.

Luscombe, N.M., Babu, M.M., Yu, H., Snyder, M., Teichmann, S.A., and Gerstein, M. (2004). *Nature*, **431**, 308–12.

As additional ChIP–chip assays are performed under a wider range of conditions and with more organisms, further principles of transcriptional regulation will undoubtedly emerge, and it is likely that some of our original observations will prove to be naive. It is worth repeating that these networks are based on properties of the transcription factors, and rely on having reagents to identify the proteins.

A yeast one-hybrid assay is a gene-centered approach to identifying the transcription factors that bind to specific regulatory regions

The specific expression of a gene depends on the specificity of the sequences in its regulatory region, and thus on the transcription factors that bind to these sequences. An alternative approach to ChIP assays is to begin with the gene regions believed or known to contain regulatory sequences, and to find all of the binding proteins that interact with the regulatory region. The protocol has been termed a yeast one-hybrid (Y1H) approach, and is shown diagrammatically in Figure 12.14. Each of these methods has a parallel with the methods used to identify interactions among individual proteins described in Chapter 10. Just as ChIP–chip experiments parallel the co-immunoprecipitation procedures, the experiments to find all of the proteins binding to a particular regulatory region parallel the yeast two-hybrid (Y2H) procedures.

In a Y1H procedure, the bait vector has the predicted regulatory region fused as a transcriptional reporter upstream of the coding region of a selectable marker or reporter gene such as the *HIS3* or *lacZ* gene. The bait vector with the DNA regulatory sequences is integrated into a chromosomal location in the yeast genome. The prey library consists of plasmids, each with a cDNA for a known or suspected transcription factor fused to the GAL4 activation domain (AD), although any prey library could be used. (Recall from the Y2H assay that the GAL4 activation domain interacts with RNA polymerase II and the general transcription factors.) Thus, the prey are the transcription factors. The prey plasmids are transformed into the bait strains and, as with the Y2H screen, interactions are detected using reporter gene

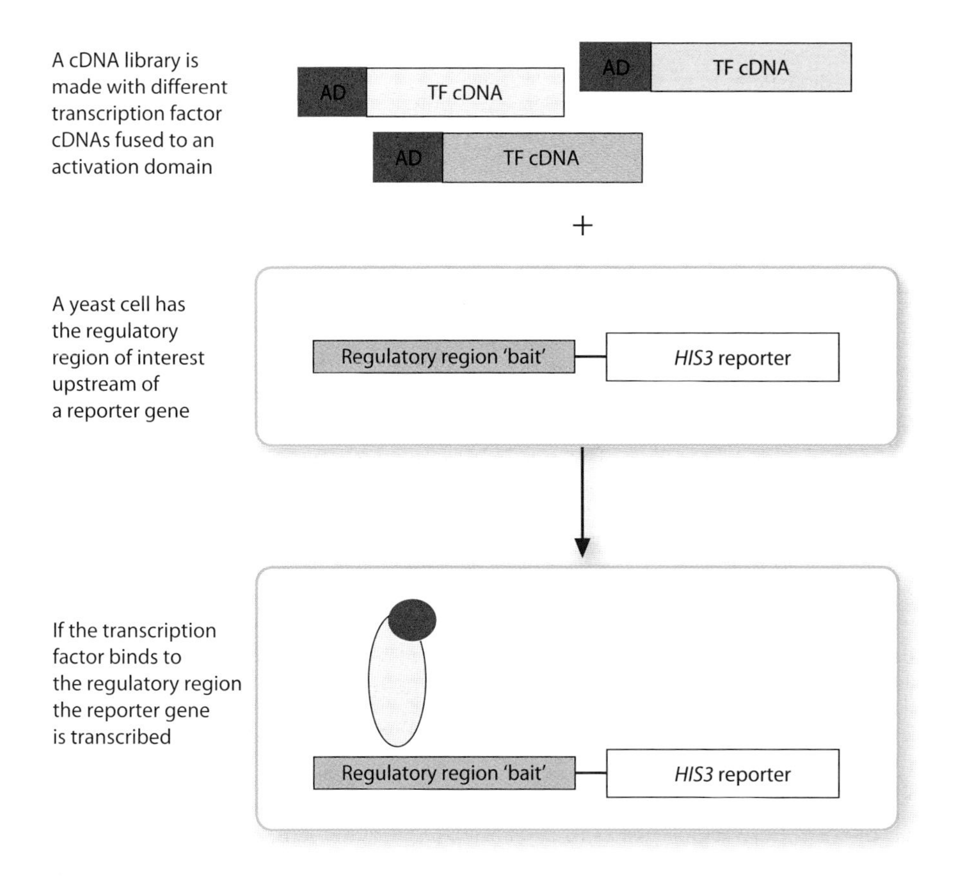

Figure 12.14 The yeast one-hybrid screen. A cDNA library is made with known or suspected transcription factors (TF cDNA) fused to the activation domain (AD) of the GAL4 protein (shown in pink). This library is transformed into yeast cells with the regulatory region of the gene of interest upstream of a reporter gene such as *HIS3*. If the transcription factor binds to the regulatory region, the activation domain can interact with RNA polymerase and the reporter gene is transcribed. In the example shown, the cell would grow in the absence of histidine.

expression. For example, in the diagram in Figure 12.14, an interaction between the bait sequence and a prey transcription factor results in the ability of the cell to grow in the absence of added histidine.

KEY ARTICLE

Deplancke, B., Mukhopadhyay, A., Ao, W., *et al.* (2006). A gene-centered *C. elegans* protein–DNA interaction network. *Cell*, **125**, 1193–1205.

Y1H assays are still being developed for genome-wide analysis, and the experiments so far have been for a rather limited subset of gene regulatory regions and transcription factors. One published screen examined the interactions between 72 different regulatory regions from genes expressed in the nematode endoderm used as baits and 117 transcription factors used as prey; a total of 283 interactions were detected and examined. The results were comparable to what was found for transcriptional regulatory networks using ChIP–chip assays. The number of transcription factors binding to a particular regulatory region ranged from one to 14, with a mode of four interacting proteins per regulatory region. On the other hand, a few transcription factors regulated many different targets—the maximum observed was 27 genes for one transcription factor. However, most transcription factors regulated very few genes. Seventy transcription factors regulated a single promoter and 28 regulated two promoters, so 98 of 117 (84 percent) transcription factors had just one or two gene targets.

This system of interactions in which many transcription factors control only one or two targets will result in a highly compartmentalized and fragmented network—imagine what communication would be like in a work environment where 84 percent of the people spoke to only one or two other people. The network is held together by the few highly connected transcription factors that interact with many genes, just as a few highly interactive and social coworkers can hold an office environment together. This situation is shown diagrammatically in Figure 12.15. In the set of interactions in the worm and in transcriptional regulatory networks in yeast, the transcriptional factors that interact with many targets are essential; elimination of these genes (by mutation or by RNAi) is approximately five times more likely to be lethal to the cell or the worm than the elimination of a gene with few target sequences.

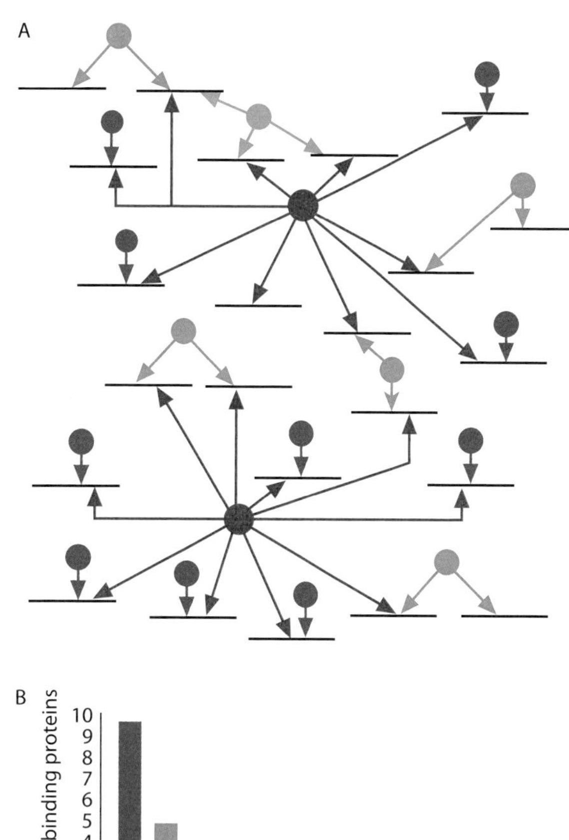

Figure 12.15 A hypothetical transcription factor regulatory network. The proteins are drawn as circles and the binding sites as lines, with the interactions depicted by an arrow. (A) Different transcription factors are shown with their binding sites indicated. (B) The distribution of the number of binding sites per transcription factor is plotted as a histogram. The colors in each part of the figure refer to the number of binding sites per transcription factor. Most transcription factors, shown in blue, regulate one or two genes. A few, shown in red, regulate many genes. The transcriptional network is held together by the transcription factors that regulate many genes.

The Y1H assay in worms tested the interaction of about 10 percent of the known or suspected transcription factors with a very small percentage of the regulatory regions, and in one highly defined differentiated tissue. This is clearly just a beginning, albeit a promising one. The number of different transcription factors and different potential targets is very large, so testing all of them by direct experimental means will be a daunting task. As a result, computational approaches to identify potential interactions are important in directing subsequent experimental confirmation.

Genome-based computational approaches to finding regulatory interactions are promising but need refinement

As more genomes are sequenced and annotated, it becomes very tempting to imagine that regulatory regions could be found computationally. After all, if one can compare an amino acid sequence with billions of other amino acid sequences to find the one that is most similar, there must be ways of comparing nucleic acid sequences to find similarity between the regulatory regions. Many different investigators are attempting to find regulatory regions by computational methods, using different approaches. Computational methods can also delineate transcriptional networks using a gene-centered or protein-centered approach. However, the problems remain formidable, and some of the problems are shown in Figure 12.16.

As with wet-laboratory approaches, computational methods to describe transcriptional regulatory networks can begin with the properties of either the transcription factor or the binding sites. For example, if the consensus binding sequence for a transcription factor is known, the genome can be scanned for these sequences close to the transcriptional start of the gene. We know from genes whose regulatory regions are defined by other assays that binding sequences are often found in clusters; it is uncommon to have a single binding site. The clustering of binding sites can also be used as a distinctive feature to recognize a genuine regulatory region.

Because the regulatory regions for most genes are located upstream of the gene, this region can be scanned for clusters of short sequences occurring more frequently than predicted by chance. Each of these approaches depends on the ability to compare a large number of sequences to compile a **position–weight matrix**, which records the frequency of particular bases at each site. Such a position-specific matrix helps to account for the fact that binding sites need not be an exact match to one another. Methods that examine only the sequence for similarities do not take into account the binding sequences of known transcription factors, which provide the opportunity to identify binding sites *de novo*.

Among the problems are the length of the regulatory regions and their position with respect to the coding

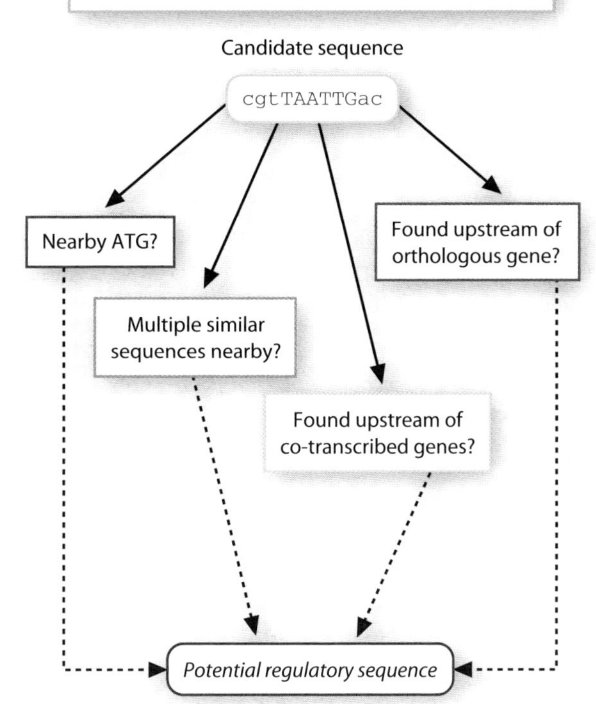

Figure 12.16 Computational approaches to finding transcription factor-binding sites. Some of the obstacles to finding transcriptional regulatory sites by computational methods are summarized at the top. A candidate sequence is found; this particular sequence could be a binding site for a transcription factor protein with a homeodomain. The most conserved bases are shown as capital letters, with other possible bases in the binding site shown as lower case letters. The candidacy of the possible sequence is appraised by several different computational methods, summarized here. For example, the nearby sequence can be scanned for an ATG start codon and for other copies of this sequence or related sequences. Genes identified by microarrays as co-expressed with this gene can be scanned for the candidate regulatory sequence. Orthologous genes from closely related species can be examined to determine if the candidate sequence is evolutionarily conserved. All of this information is integrated to determine if this is a potential regulatory sequence.

regions. Regulatory regions can be found in many different places with respect to the gene, from a few hundred base pairs to thousands of base pairs upstream of the start site, and even within the introns or the coding region. In addition, the binding site for regulatory interactions is often a short sequence, as little as 8 or 10 base pairs, and need not be an exact match. Thus, for a gene with a regulatory region of 3 kb, which is not atypical for a gene in worms, flies, or *Arabidopsis*, a binding site of 10 base pairs is 0.3 percent of the sequence. Therefore, the binding sites are small needles in fairly large haystacks. Furthermore, our experimental knowledge of transcriptional regulatory networks is probably not yet sufficiently advanced to allow very precise computational predictions of the binding sites. The problem becomes even more complex if the binding site is composed of, say, eight conserved bases among ten bases. Nonetheless, the potential power of computational approaches for this problem is appealing because it offers the possibility of a much more thorough description of the network. A computational approach could potentially predict that a particular sequence is a binding site even if that site is not being used under the particular experimental conditions tested, i.e. such an approach might provide a context-independent snapshot of the transcriptional network.

In practice, computational methods for predicting transcriptional binding sites can be integrated with microarray expression data, as shown diagrammatically in Figure 12.16. Such an integration arises from the recognition that genes being co-expressed are predicted to share some regulatory sequences and transcription factors. Thus, a computationally predicted binding sequence from one gene can be compared with the upstream regulatory region of a co-expressed gene to search for similarities. Computational predictions can also be integrated with evolutionary comparisons with closely related species, on the assumption that regulatory sequences are likely to be conserved, whereas other sequences will be subject to neutral substitutions and will not be so similar. This approach is becoming increasingly powerful as the genomes of closely related species are sequenced. For example, the regulatory region for a gene in *D. melanogaster* can be compared with the upstream regions of the orthologous genes in other *Drosophila* species to find the conserved sequences that comprise the best candidates for binding sites. There is no shortage of data for computational analysis, but the best methods so far have been able to identify only about 60 percent of the known regulatory sequences. This accuracy will certainly improve in the near future, as the predictive algorithms are refined and more genomes are sequenced and annotated.

Transcriptional regulatory networks are the first, and less studied, of the two types of interaction networks that we introduced early in the chapter. In the following sections, we discuss the application of network analysis to protein–protein interactions.

12.4 Interactions between proteins

Protein–protein interactions are the most completely studied interactions in the cell: a Y2H screen allows protein interactions to be studied on genome-wide scale. As described in Chapter 10, Y2H screens make it quite feasible to consider testing every protein for its ability to interact with every other protein encoded in the genome. In fact, this procedure has been done for protein interactions in budding yeast and is in progress in many other organisms including humans. In this assay, a protein is used as a bait to screen a library with every gene in the genome as potential prey. The interactions between the bait and the prey are found by a simple selection scheme, as described in Chapter 10. Despite the limitations of the Y2H screen (see Chapter 10), the method provides a rapid and powerful initial approach to compiling all of the protein–protein interactions. The complete set of protein–protein interactions that occur in an organism is referred to as an **interactome**. Complete sets of protein–protein interactionsfor eukaryotes other than yeast are not yet known, and so the term 'interactome' is also used to refer to the set of protein interactions detected to date.

KEY ARTICLES

Li, S., Armstrong, C.M., Bertin, N., *et al.* (2004). A map of the interactome network of the metazoan *C. elegans*. *Science*, **303**, 540–3.

Stelzl, U., Worm, U., Lalowski, M., *et al.* A human protein–protein interaction network: a resource for annotating the proteome. *Cell*, **122**, 957–68.

Uetz, P., Giot, L., Cagney, G., *et al.* (2000). A comprehensive analysis of protein–protein interactions in *Saccharomyces cerevisiae*. *Nature*, **403**, 623–7.

In compiling the data, each protein in the genome is plotted as a node on a graph. These nodes are connected

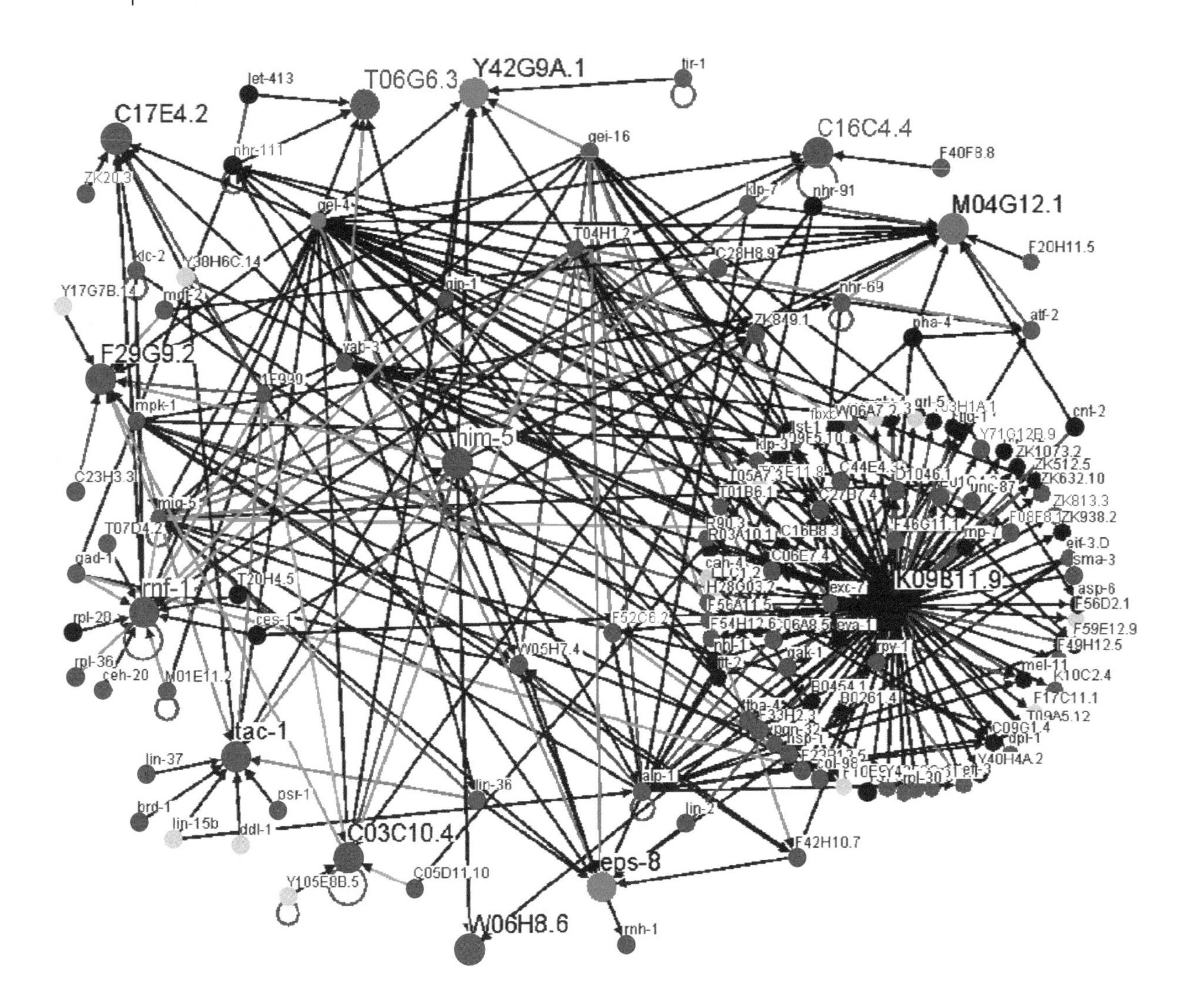

Figure 12.17 An example of a Y2H interactome. A small and incomplete interactome began with the gene *HIM-5* in *C. elegans*. The first- and second-degree interactions for *HIM-5* are plotted. Note that, even in this small network, hubs such as the gene represented as K09B11.9 (lower right) are evident.

by links, which represent the interactions detected by Y2H; we introduced these concepts in Figures 12.5 and 12.6. As described previously for other networks, the interactions among the first-degree interacting proteins are then plotted. The analysis results in a complex network, as shown for a worm gene in Figure 12.17. With such a network, it is possible to examine a single protein node and catalog all of its interacting proteins by following the links. This is one of the most generally useful results to arise from interactome experiments. Analysis of other properties of these networks is still preliminary but holds considerable interest and promise for revealing novel characteristics of cells and organisms.

Interactomes reveal possible functions for unknown genes

The interactome provides another extremely valuable way of inferring the function of a gene. Consider the gene F47G4.4 in *C. elegans*, shown in Figure 12.18. This gene was found or predicted from the genome sequence, as suggested by the gene name. The absence of a descriptive gene name tells us that no obvious function has been inferred from mutations of this gene; in all likelihood, none have been identified. Expression studies such as microarrays find that the gene is transcribed in many different cells at different stages. While all this is helpful information, it does not suggest a *function* for the gene. However, in a Y2H screen, the protein encoded by F47G4.4 interacts with a number of partner proteins that *have* been found in other organisms, particularly proteins associated with the mitotic spindle. Although

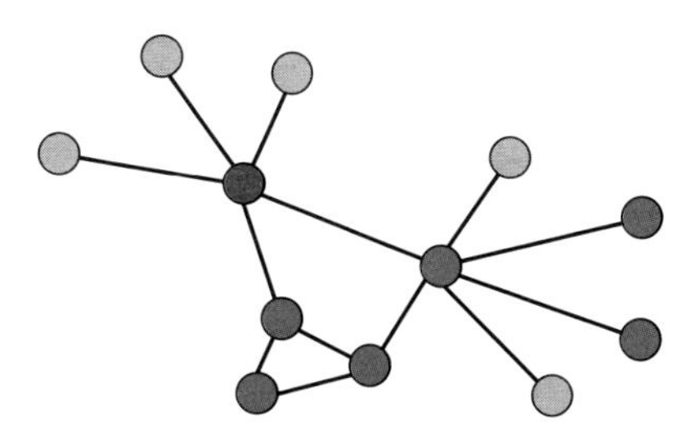

Figure 12.18 Inferring the function of a gene. The gene F47G4.4 from *C. elegans* is shown as the pink node. All five of its first-degree and some of its second-degree interactions are shown with the interactions indicated by the lines. The nodes in blue represent proteins known to be involved in the mitotic spindle. Those in gray have either an unknown function or a general function that is not specifically informative for the cellular function. Since so many of its first- and second-degree interactions are with spindle proteins, F47G4.4 is inferred to encode a spindle protein itself. In fact, RNAi and gene deletion experiments confirm this inference. (The data are replotted from information on Wormbase.)

we do not yet know the function of this gene, we can make a guess that it is something to do with spindle function. With this information, when investigators do RNAi experiments, prepare antibodies to the protein, or try to analyze mutations in the gene, they can pay attention to the structure and function of the spindle in dividing cells. (In fact, these experiments have shown that the protein encoded by F45G4.4 has a function in the mitotic spindle.)

An interactome can also suggest functions or interactions for a protein that may have been overlooked or not detected by other assays. For example, the original yeast interactome screen identified a protein involved in ribosome biogenesis that also interacted with proteins from the spliceosome. This suggested an unexpected connection between ribosome biogenesis and splicing. Furthermore, the same ribosomal biogenesis bait protein interacted with a protein involved in mRNA decapping (removing the 5′ cap from transcripts). Interestingly, the decapping protein also interacted with proteins from the spliceosome, suggesting that mRNA decapping, mRNA splicing, and ribosome biogenesis are physically and perhaps functionally linked to one another in a previously unknown manner.

The interactome network is built from both first- and second-degree interactions, which can be thought of as 'friends of friends'. An astute reader may realize that the 'friends of friends' aspect of biological networks has many analogous non-biological examples. For example, perhaps you routinely buy books, download music, or rent movies from an online vendor. After a few transactions, the vendor begins to make suggestions about other books, musicians, or movies that you may also like. In the language of websites, you have a purchasing profile. In the jargon of networks, your purchases are first-degree interactions. By comparing your network of first-degree interactions with the networks of other buyers, it is possible to find second-degree interactions and recommend additional purchases.

It is worth keeping this analogy in mind as we consider biological networks. For example, many consumers edit their purchasing profile so that gifts for others or a few poor choices are not used to make recommendations, i.e. these are not used to predict second-degree interactions that are of no interest. Similarly, because the Y2H screen yields some false positives, as described in Chapter 10, those working with an interactome try to recognize known false positives and exclude them from predictions of second-degree interactions. Predicting which second-degree interactions identify proteins that are functionally related to the original bait is one of the key goals of using the interactomes.

KEY ARTICLES

Barabási, A.L. and Oltvai, Z.N. (2004). Network biology: understanding the cell's functional organization. *Nature Reviews Genetics*, **5**, 101–14.

Keller, E.F. (2005). Revisiting 'scale-free' networks. *BioEssays*, **27**, 1060–8.

The protein network is represented by a graph

The type of analysis based on understanding individual proteins and genes is only the beginning of what the interactome can tell us. We are now going to return to some of the concepts about graphs and networks that were introduced early in the chapter and begin to apply them to protein interaction networks. Graph theory and the analysis of networks have been extensively applied to interactomes, with results that have already been illuminating and suggest that more can be learned.

We introduced the concept of degree distribution earlier in this chapter. From that section, we recall that the degree k is the number of direct links from a particular node, and that the degree distribution is a histogram representing the degrees for all of the nodes. We also saw that, in a random network, the degree distribution follows a Poisson distribution. A mean degree can be calculated from the degree distribution.

One property of a random network is that most nodes will have a degree that is scattered around the mean, with few nodes having many more or many fewer connections. Although most biological networks do not fit this degree distribution, a few do. For example, when the number of transcription factors that bind to a regulatory region in the gut of *C. elegans* is plotted as a histogram, the resulting distribution can be described by Poisson distribution as characteristic of a random network, ranging from one to 14 proteins per regulatory region with an average of four proteins bound. However, as noted above, a plot of the number of regulatory regions that are bound by different transcription factor does not follow a Poisson distribution.

Does the degree distribution of a protein–protein interaction network follow a Poisson distribution? Although the question can be answered with statistical rigor, we will use a more intuitive and observational approach to arrive at the same answer. The complete yeast interactome is shown in Figure 12.19. The mean *k* for the entire network is 3.8 so that, on average, a protein interacts with three or four other proteins in the Y2H screen. A quick look at Figure 12.19 suggests that this network does not arise from random interactions. Notice that many nodes have more than 10 links; in fact, some nodes have 20 links or even more. Nodes with as many as 10 links are expected to be exceptionally rare in a random network, but such nodes are relatively common in the known interactomes. The presence of such highly connected nodes is an indication that the network of protein interactions cannot be described as a random network. This observational conclusion is confirmed by more rigorous statistical tests that go beyond the scope of this book.

Figure 12.19 The yeast interactome. The complete or nearly complete set of protein–protein interactions found among the 6200 proteins in the yeast *S.cerevisiae*.

This result makes a certain biological sense. We know that the number of interactions varies widely for different proteins. Proteins often form highly ordered complexes that can have dozens of members. Other proteins work as dimers with highly specific partners. The assumption that protein interactions could be described by a random network that fits a Poisson distribution is at odds with our understanding of biological functions and the way that we understand how functional interactions have evolved. Thus, it is somewhat reassuring that the degree distribution of the yeast interactome does not follow a Poisson distribution.

Instead of being a random network, interactomes are characterized by a series of hubs. Recall that a hub is a node with many connections. What do these hubs tell us about biological interactions in the cell or organism? In some cases, as shown in Figure 12.20, the hubs correspond to known biological protein nanomachines, such as the polyadenylation complex or the proteasome. Other hubs correspond to other large protein complexes in the cell, such as the mitotic spindle or the secretory vesicle. In these cases, the interactome plot is re-creating what we already suspected from other types of analysis. Again, it is reassuring that a novel approach such as network analysis is identifying known proteins with many interactions. In effect, the identities of some of the hubs are the positive controls that tell us that the network approach is working as expected. Many other hubs do not correspond to protein complexes that we knew about previously. In these cases, the hub could be revealing some unsuspected property of cellular organization or evolution.

Scale-free networks

We noted earlier that the degree distribution for interactomes and many other biological networks follows a

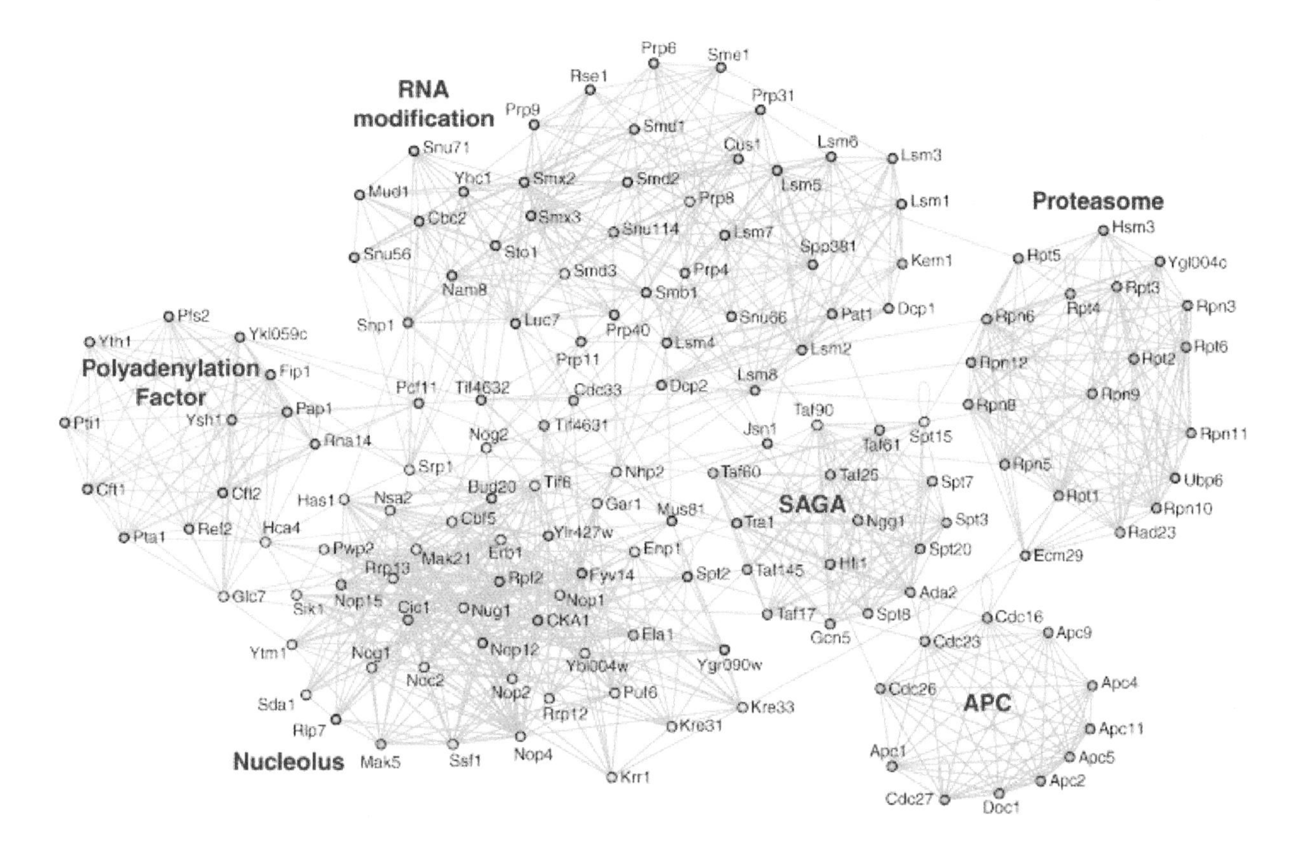

Figure 12.20 Some hubs from the yeast interactome. The yeast interactome shows that some proteins involved in common functions cluster together with extensive interactions among themselves. Among the identifiable clusters are proteins involved in RNA modification, the proteasome for protein degradation, the anaphase-promoting complex (APC), a histone acetyltransferase complex (SAGA), the nucleolus, and polyadenylation. (Reproduced from Bader *et al.* 2003, Figure 4.)

type of distribution described by a power law. In such a network, it is still possible to calculate the mean of the degree distribution, but the mean is not always helpful; there is no 'typical' node because the degree distribution is spread widely. Some of the early papers (which appeared in 1999) discussing degree distributions used the term **scale-free** to describe networks that obey a power-law distribution. This term was introduced from statistical physics and the study of phase transitions, and it has sometimes generated more confusion than clarity because not all of the properties of scale-free interactions in physics have yet been demonstrated in biological networks. We will use the term advisedly because it has become the most familiar one to those working on networks.

One property of a scale-free network is that the overall topology of the network is independent of its size. Both the degree distribution and the path length depend on the number of interactions in a network, but the overall topology is scale-free. This suggests that the properties observed with small and incomplete networks will also be found once interactomes have described the complete set of protein interactions that occur in an organism. However, this is not the property of scale-free networks that dominates the interpretation of biological networks.

Hubs have important biological properties

From the perspective of biological systems, the most important aspect of a scale-free network is the presence of hubs. Proteins identified as hubs have known biological properties, some of which may be a consequence of their large number of interactions. The analysis of hubs is only just beginning, but a few tentative conclusions seem appropriate, based in part on what is known about hubs from other types of scale-free network.

Deletion of a hub has profound consequences for the entire network

Recall that targeted gene knockouts have been made for every gene in yeast, as described in Chapter 7. We can ask if proteins that act as hubs in the network have a different behavior in the yeast deletion screens from proteins that are not hubs. The results are clear and somewhat predictable from the behavior of hubs in other networks. When the results from the gene deletion experiments are compared with the yeast interactome, we find that mutations that knock out a gene at a hub in yeast are about three times more likely to be lethal to the cell than ones that knock out a gene at random. Thus, elimination of a hub has profound consequences for the stability of the network or, in this case, the survival of the organism. This is the evidence that supports our statement above that hubs hold the network together.

The effect of deleting hubs is also a known property in other scale-free networks. Consider an example from the airline industry, when weather or other problems result in the closure of an airport. Targeted elimination of a hub tends to shut down the entire network. A snowstorm that closes O'Hare Airport in Chicago affects air traffic throughout the USA because O'Hare Airport is a hub. On the other hand, a similar storm that closes Grand Rapids airport an hour away in Michigan has only local effects. The Grand Rapids airport is not a hub, at least when considered from the perspective of the airline industry. (The perspective of the individual attempting to travel to or from Grand Rapids, Michigan, is probably different.) We will return to this property of scale-free networks in a later section, but the important lesson for now is that genes comprising hubs are more likely to be essential for survival than genes with few connections.

Hubs correspond to evolutionarily conserved genes

A second important property is that hub genes tend to correspond to more evolutionarily conserved genes, although there are numerous and important examples of non-hub nodes that are also highly conserved. This evolutionary conservation of hubs reveals another important feature of the interactome network: genes that are connected in one organism (such as yeast) are likely to interact in another organism (such as worms or humans). This increases both our understanding of the network in each organism and our insights into the organizational hierarchies of the cell.

The evolutionary conservation of a hub can be imagined as a consequence of the number of its connections. Every protein interaction that is important to the survival of the cell imposes a constraint on selection. In other words, components in any complex system that interact with many other components are necessarily resistant to change. Changes in a hub become less constrained if gene duplication occurs so that one hub is replaced by two nodes that are initially capable of the same interactions. The duplicate gene—a paralog of the original—may originally retain many of the same interactions. However, as the functions of the two paralogs diverge slightly, new interactions can be found.

12.5 Hubs and robustness

One of the most significant properties of scale-free networks such as the one described by the interactome is their **robustness**. Robustness refers to the resistance of the overall network to *random* changes in its individual components. The word 'robustness' has similar definitions in many other fields, including computer networks and evolutionary biology. The key word in the definition of robustness is 'random'. Although our attention is drawn to the hubs, most of the nodes in a scale-free network have very few links. The loss of a random node is far more likely to affect one with only a few links than it is a hub. Thus, most changes to nodes (or genes) in the network will have a minimal effect because they will only affect a node with few links.

The World Wide Web provides a familiar example of the importance of robustness. When doing internet searches, one frequently encounters dead links because a website has been changed or deleted. These lost websites correspond to the random loss of nodes. Despite the loss of this individual component, the overall Web continues to function, so the Web can be considered a robust network. In fact, one could make the case that the power of the World Wide Web comes from its robustness—its ability to withstand the random loss of a web page. Otherwise, every time a web page is changed or removed, the network (or at least the local network) would go down. Because individual web pages are constantly being deleted or changed, the World Wide Web would be useless if it were not robust.

Similarly, random changes in the yeast interactome network, and apparently in other biological networks, have little effect. The best genetic evidence for this conclusion comes from the targeted systematic deletions that are done in genome-wide screens. In these genetic screens, as described in Chapter 7, the functions of genes are knocked out without regard to whether the mutation confers a mutant phenotype. This is important to think about because a mutation that results in no mutant phenotype is not detected by a traditional genetic screen. In genome-wide screens, each gene was targeted for disruption and the effects (or lack of effects) were observed on the overall growth and phenotype of the yeast cell. These systematic deletions of open reading frames in yeast reveal that only about 20–25 percent of the genes are essential for growth under laboratory conditions, i.e. most changes have minimal or no effect on the growth of the yeast cell. This may seem like a surprising result based on our decades of genetics research: most mutations have little or no effect on the organism. However, knowing that the yeast interactome is a robust network, this result is more understandable. The result may also be applicable to thinking about other biological systems, as we speculate below.

It is important to compare these two conclusions from the genome-wide deletion screens in yeast. First, as noted above, deletion of a hub is likely to be lethal to the cell. Secondly, deletion of a random node is likely to have no phenotypic effect on the cell. Both of these are properties of scale-free networks; in fact, they are two manifestations of the same topology of the network. A scale-free network is robust regarding random changes but is vulnerable to attacks on hubs. Robustness is worth exploring in much more detail, particularly placing this property of organisms into the broader context of genetic analysis in the age of genomics.

Robustness has several possible origins

Attack vulnerability in a network is relatively easy to understand simply based on the definition of a hub. When a hub is lost, all of its many connections are also lost. But what underlies robustness? A number of possibilities suggest themselves, and all of them are likely to play a role. Two possible origins of robustness are shown in Figure 12.21: redundancy of a node and alternative paths between nodes. Both of these

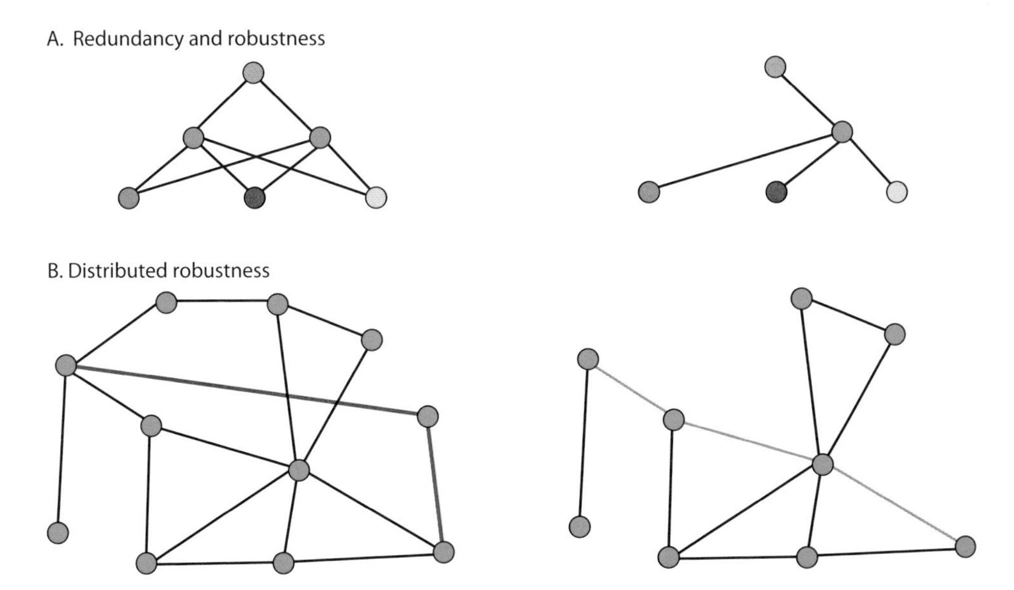

Figure 12.21 Two origins of robustness. Robustness arises from (A) redundancy or (B) distributed or alternative pathways. In panel A, the two green nodes are functionally redundant and share the same four interactions with other proteins. If one of the green nodes is eliminated (upper right), the same interactions occur with the remaining green node. A network with this set of interactions is expected to be robust. In panel B, the nodes are connected by the path shown by the heavy red line. The elimination of some of the nodes by mutation, as shown on the right, does not disrupt the network because alternative paths between the nodes exist.

possibilities were encountered in our discussion of genetic interactions in Chapter 10. They can be summarized as **duplicate genes** and **duplicate pathways**.

Robustness and duplicate genes

As was noted in Chapter 1, a key insight from genomic analysis is the prevalence of gene families. Thus, one reason for robustness is that many genes are members of gene families with functionally redundant members. This situation is shown diagrammatically in Figure 12.21A. If two nodes share the same links—if duplicate genes share the same interactions—loss of one node may have very little effect on the network. In fact, we encountered an example like this in Chapter 10 in the discussion of the acetyl cholinesterase (*ace*) genes in *C. elegans*. Recall from that discussion that there are three duplicate *ace* genes. Loss of one of the genes results in no mutant phenotype: the organism is robust because the genes are biochemically redundant. Mutant phenotypes are seen only when two or three of the genes are eliminated. Robustness of this type can also be termed **redundancy**—the genome has duplicate genes carrying out the same functions.

Robustness and duplicate pathways

A second explanation for robustness can be seen by examining the yeast interactome or any other biological network, and tracing the paths between two nodes that are not particularly close together. In particular, notice the paths as shown in the small imaginary network in Figure 12.21B. There are often many paths that lead between two nodes. This occurs even when the individual nodes themselves are not duplicates of one another and thus the proteins are not biochemically redundant. In other words, there is *functional overlap* in the network that arises from alternative paths between any two nodes. Elimination of one gene (node) with its interactions (links) simply means that a different path is followed. This property has been called **distributed robustness**. It arises, in part, from the inherent functional redundancy of the eukaryotic genome.

Again, we encountered examples of distributed robustness in *C. elegans* during the discussion of the pathways leading to vulval development (Chapter 10) and dauer larvae formation (Chapter 11). In each example, there are parallel genetic pathways leading to the same phenotype. The genes themselves are not paralogs of one another but the pathways have duplicate or overlapping effects. Elimination of a gene in one of the parallel pathways has no effect on the overall phenotype of the organism because the other pathway is still functional.

Mutant phenotypes are seen only when genes in both pathways are knocked out. In Chapters 10 and 11, we considered this from the point of view of synthetic enhancement, as a means of finding genes with related functions. However, synthetic enhancement also offers compelling evidence for the robustness of the network.

KEY ARTICLE

Lenski, R.E., Barick, J.E., and Ofria, C. (2006). Balancing robustness and evolvability. *PLoS Biology*, **4**, 2190–2.

Robustness is fundamental to evolution

The evolutionary advantages of robustness seem apparent. At least three possibilities come to mind. First, robustness might buffer the cell and organism against random fluctuations in the level of gene products, as illustrated in Figure 12.22. Although there is no change in the genes themselves, the cell or the organism experiences stochastic variation in the amount of a gene product. As a result of this variation, some links may be formed or lost. We currently have no accurate way of knowing the extent to which mRNA and protein levels vary in two cells making the same gene products 'at the same level'. This problem was discussed briefly in Chapter 8 when we considered biological noise and the accuracy of microarray data. Certainly a twofold variation in the amount of protein could occur and yet would be hard to detect by most current methods. The importance of buffering against small changes in gene expression was noted in Text Box 4.3 in our discussion of dominant and recessive alleles in Chapter 4; the same logic applies here in thinking about an interaction network. In the absence of some buffering mechanisms, variation in the levels of gene activity could have profound consequences for the organism. The ability of a robust network to buffer against stochastic fluctuations in gene product levels reduces the consequences of biological noise.

Secondly, as a consequence of robustness, most mutations are not lethal to the organism. This is

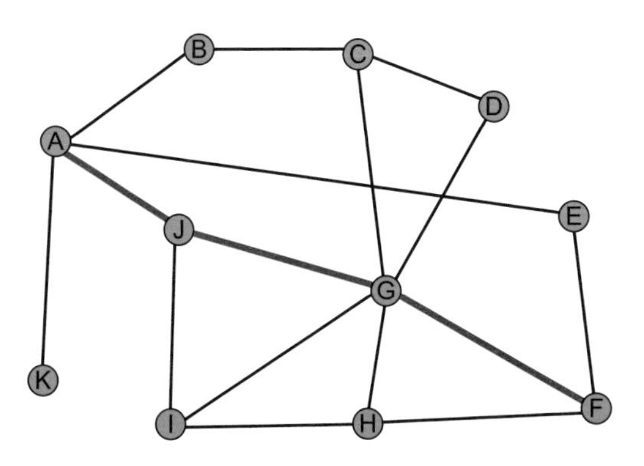

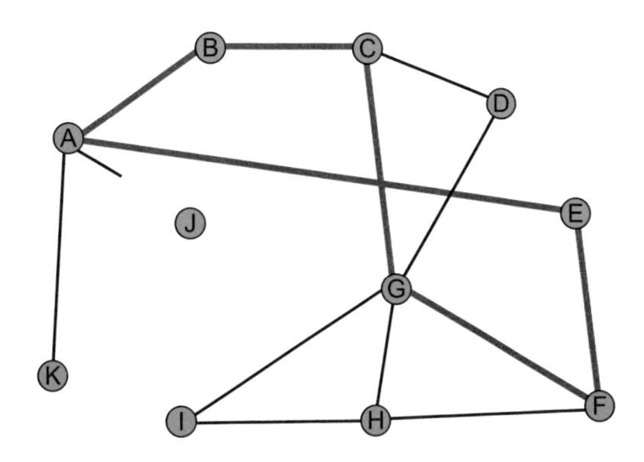

Figure 12.22 Robustness as a buffer against fluctuations in gene expression. A simple network is shown with the path between nodes A and F indicated in red. Because of a random fluctuation in gene expression at node J, the links between A and J, and J and I, are lost in the diagram on the right. However, other paths exist between A and F, as shown. As a result, the overall structure of the network is not disrupted.

A. Wild-type network

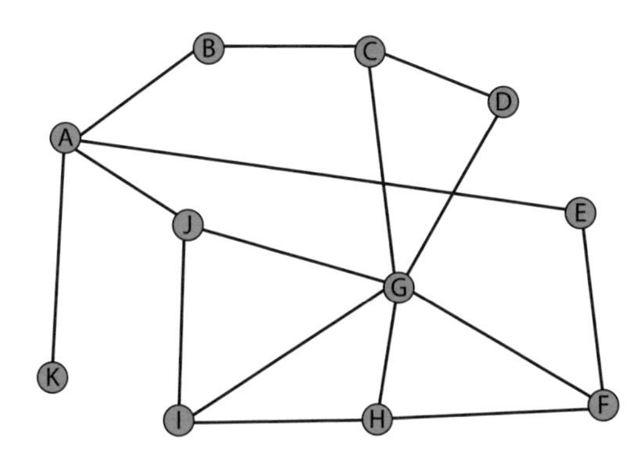

B. Networks with mutations

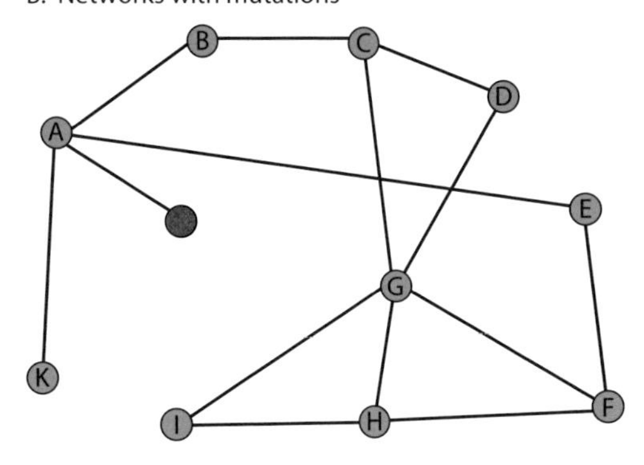

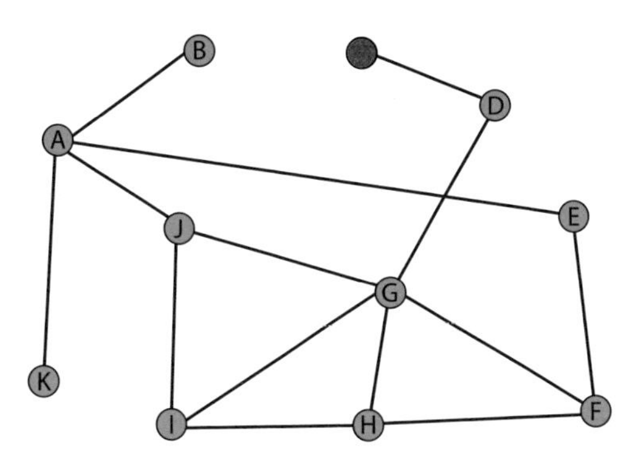

Figure 12.23 Robustness and the presence of mutations. The network in wild-type organisms is shown in panel A. Two different mutations are shown in red in panel B. Each mutation results in a node (or protein) that loses some of its interactions. However, the overall network structure is retained because other paths connecting all of the unaffected nodes are still present.

illustrated from a network perspective in Figure 12.23. In this diagram, two different mutations occurred in the network, each of which caused the node to lose some of its interactions. Situations like this are a fundamental fact of evolution since mutations occur constantly. If most mutations had phenotypic effects, a species simply could not survive. In Chapter 3, we saw the effects of diploidy on buffering against deleterious evolutionary change, but the same discussion holds for duplicate genes in a gene family and functional redundancy arising from parallel pathways.

A. Wild-type network

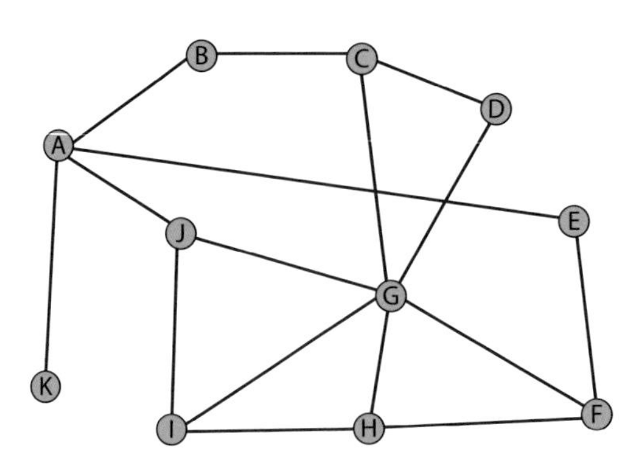

B. Mutations allow evolution of the network

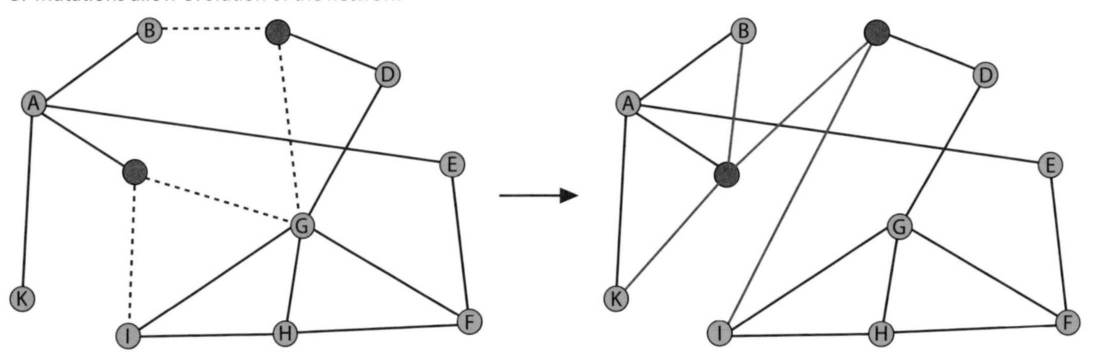

Figure 12.24 Robustness and evolutionary change. The same wild-type network used in Figure 12.23 is shown in panel A, and the same mutations are shown in panel B. In panel B, two mutations are shown in red, with the links lost by mutation indicated by dashed lines. Because the mutations will persist in the population, new links and interactions can occur, as shown in the red lines, and a new network can evolve.

Thirdly, because most mutations do not cause a major phenotypic change in the organism, different alleles (and thus genetic variability) will persist in a population. Some of the potential results of this are illustrated in Figure 12.24. As mutations persist, new connections are possible and a new network of interactions can evolve. In a general sense, most mutations are expected to be selectively neutral or nearly neutral because of robustness. This neutrality allows genetic variation to accumulate in the population.

The observations from the two preceding paragraphs go together. Genetic variability in a population is a necessary ingredient for evolutionary change. However, if a mutation results in the death or sterility of the organism, there is no possibility of evolutionary change and the mutation is eliminated immediately. Mutations that might otherwise be deleterious can be maintained because of the inherent robustness of the eukaryotic genome. The robustness of the gene interaction network provides both the flexibility and the stability needed for evolutionary change. It represents the balance between the genetic variation and phenotypic stability that natural selection requires.

12.6 The limitations of interactomes

The current versions of interactomes have some recognizable limitations. Some of these are largely technical: the known limitations of Y2H screens as discussed in Chapter 10, the quality of the cDNA library being used, the completeness with which the genome has been annotated and genes predicted, and so on. One can anticipate that some of these will become less significant as more complete interactomes are produced and new methods

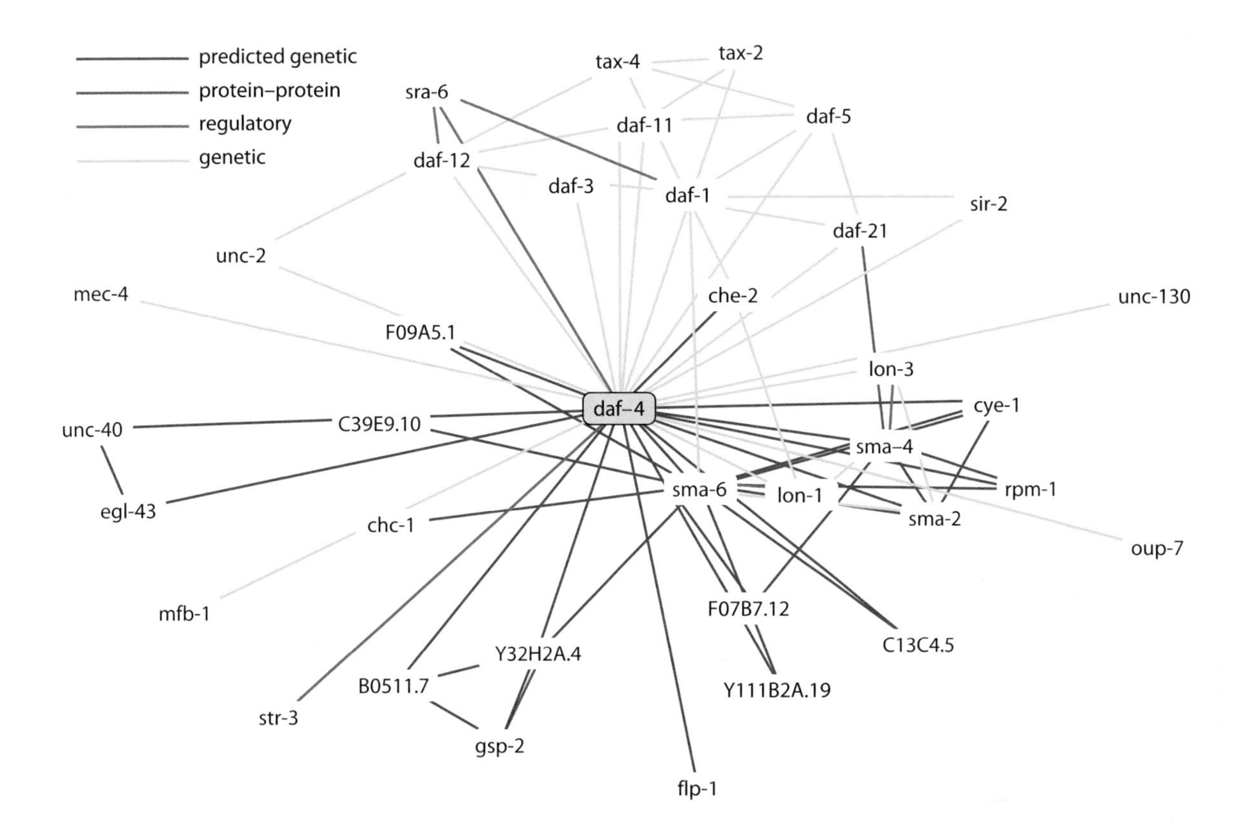

Figure 12.25 The interactions of *daf-4*. The *daf-4* gene is part of the TGF-β co-receptor in *C. elegans*, as discussed in Case Study 12.1. In this image from Wormbase, the protein and genetic interactions for *daf-4* and some of its interacting genes are shown. The colors of the connecting lines indicate the type of interaction, one style in which different interactions are integrated into one picture.

are developed. This does not mean that these limitations are unimportant; the Y2H screen will probably never be an ideal or definitive method to detect all of the physiologically significant protein interactions that occur in a cell or an organism. But the Y2H screen has the overriding virtue of being relatively simple to perform on a genome-wide scale, so some of its limitations need to be addressed in other ways.

A significant limitation is that the Y2H screen expresses all of the proteins at the same time and tests the ability of two proteins to interact. However, the expression pattern of a cell or an organism is highly dynamic. The interactome presents a static network in which all nodes and links exist at the same time. It can be compared to a map of an airport that shows all of the flights that land or take off from it, without including the time at which each flight occurs. Such a flight map would be cluttered and potentially misleading, when what is needed is an animation that shows arrivals and departures each hour. The dynamic aspect of networks is difficult to resolve, but this is the object of much current research in the field.

One promising approach is to integrate the Y2H data with other types of computationally available genomic data. We illustrate this in Figure 12.25. For example, microarrays can convey which genes (nodes) are transcribed at the same time. These microarray data can be used in principle to include a temporal component to the interactome. One problem is that microarray data reflect RNA expression, whereas the Y2H interactomes reflect protein expression. For a gene under translational control, transcription may occur at an earlier stage than the appearance of the protein. Thus, the microarray might give results that suggest that protein interactions could occur earlier than is actually possible. However, the converse is also true if the protein is particularly stable: a gene might be transcribed at one time but the resulting protein persists well after the time of expression measured by microarrays. In this case, the microarray might indicate that two genes are transcribed at different times and so their proteins could not interact, when in fact the proteins are present at the same time and could interact *in vivo*. A better approach would be to use protein expression from translational reporter gene constructs rather than microarray data. Attempts are being made to do this, but genome-wide reporter gene screens are not as easy to do as global RNA expression studies, so these results lie in the future.

There are also computational issues that arise from the very large size of the interaction and expression databases, as well as from the complexity of the data included in each. As noted in Chapter 7, biology in the genomics era is changing from a science focused on acquiring data to one focused on synthesizing data from complex sources. Nowhere is that transition more evident than in combining expression data with protein interaction data. Integrating such large and diverse datasets into one composite network is a challenge that lies just ahead.

12.7 Gene regulation networks

One of the principal limitations of the Y2H interactome networks is that they show the *presence* of an interaction without also showing its *importance*. For a geneticist, a description of the function of a gene will always include its mutant phenotype. However, this information is not included in the current versions of interactomes, although some of the combined networks are attempting to include genetic data. By genetic data, we mean the types of interactions describe in Chapters 10 and 11—suppressors, synthetic enhancers, and epistasis. In one sense, these genetic data are of a different type altogether than the interactome data. The genetic data do not measure a physical interaction and, as pointed out repeatedly during our discussion of genetic interactions, physical interactions between the gene products are not necessary to observe a functionally important genetic interaction. Furthermore, while a physical interaction between two proteins probably indicates a functional relationship between them, the nature of that functional relationship is not revealed by the interactome.

KEY ARTICLE

Bader, G.D., Heilbut, A., Andrews, B., Tyers, M., Hughes, T., and Boone, C. (2003). Functional genomics and proteomics: charting a multidimensional map of the yeast cell. *Trends in Cell Biology*, **13**, 344–56.

Let us imagine what information we would want to have to compile a complete network of interactions that comprise an organism. We can then compare that with our current state of knowledge. (Our current state of knowledge is different for each of the model organisms, so this is a general summary.) We summarize these and our progress in reaching them in Figure 12.26.

First, we require a complete list of all of the genes and gene products in the genome. This will address the question, 'What are the parts needed to make an organism?' As more genomes are sequenced and genome annotation improves, we are getting closer to having this information. We cannot yet predict the existence of all of the genes or gene products in the genome, but the situation is improving. The largest gaps in our knowledge are in predicting splice variants and their corresponding products, and in identifying regulatory RNA products. These are very active fields of study, so we can consider ourselves to be close to achieving this. The situation is not so clear if we think of the gene products as being mature and functional proteins because many proteins are post-translationally modified by cleavage, phosphorylation, glycosylation, and a host of other activities. We are far from this stage of knowledge, and most of what we will learn will come from proteomic approaches rather than genetics. We also still have a significant gap, of unknown size, in our knowledge of non-coding regulatory RNA molecules. This may prove to be the most significant gap of all.

Secondly, as shown in Figure 12.26, we would like to know the basic function of each of the gene products, i.e. which ones are transcription factors, which ones are kinases, and so on. This will address the question, 'What does each of these parts do?' The principal tool here is to identify sequence similarity to other proteins with known functions, to identify either orthologs in other

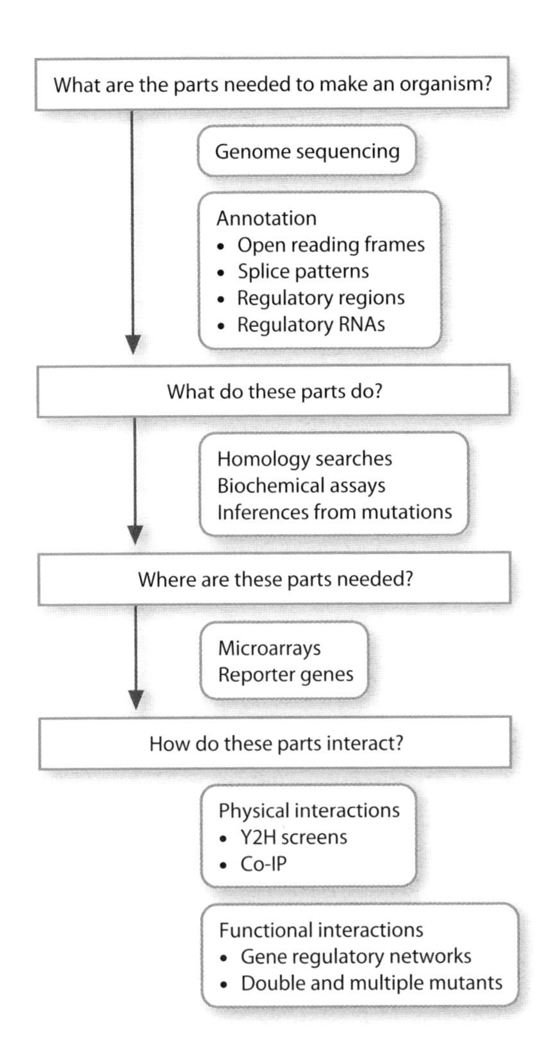

Figure 12.26 Requirements for an ideal systems approach. The flowchart represents some of the questions that are raised during a systems-based approach to genetic analysis. The circles represent some of the tools that are being applied to address these questions.

organisms or characteristic domains and amino acid motifs in the protein. Again, although we are very close to this in a computational sense, we are not so close in an experimental sense. We cannot quite yet infer the function of each gene product in the absence of direct biochemical tests. It seems unlikely that we will be able to do this in the near future, although rapid advances in proteomics could help make this easier. We may need to be satisfied for now with inferring the function of a large and increasing fraction of the gene products, and recognize that some parts may remain uncharacterized. (Perhaps these will be the parts left over when the system is assembled.)

The third piece of information that we need for our ideal complete network is the complete expression data of each gene and gene product. This would address the question, 'Where are these parts needed?', as summarized in Figure 12.26. Recognizing from our discussion in Chapters 8 and 9 that when a gene or gene product is expressed is not the same as when it is functional, we have still made considerable progess towards this goal. With microarray data, reporter genes, and proteomic approaches, the acquisition of these data may not be too far in the future. One of the greatest challenges at this stage is that the expression pattern and the functions are highly dynamic. Expression patterns change with the environment, the stage of the life cycle, and so on, but many of the tools appear to be in place.

The fourth piece of information for our complete and ideal network has been the main topic of this section of the book—the interactions between the gene products. This answers the question about how these parts can be assembled into a functional system. As described, this will have at least two aspects: the physical interactions that can be detected by screens such as the ChIP–chip assays and Y2H methods; and the functional interactions that are detected by genetic methods. This last seems like a formidable task. It requires having a mutation in each gene—ideally, more than one type of mutation in each gene—and making and analyzing double- and multiple-mutant strains. We will limit our ideal experiment to double-mutant strains because multiple mutants introduce a staggering number of additional interactions to consider and strains to construct.

The complete network of interactions has not been determined for any organism, although the information for *E. coli* probably comes the closest to being complete. Case Study 12.1 describes the integration of the protein interaction network and the genetic regulation network involving dauer larva formation in *C. elegans*. This is governed in part by a TGF-β signaling pathway, and the process has been used as an example in several previous chapters. The case study completes this picture (for now) by showing how a systems approach has been used. However, this is just one biological process, and is not a genome-wide analysis. Among eukaryotes, the most complete information has been determined for yeast, in experiments that we will consider next.

Case Study 12.1

A systems analysis of TGF-β signaling in *C. elegans*

The genetic pathway that leads to dauer larva formation in *C. elegans* was introduced in Case Study 9.1 and discussed in some detail in Chapter 11. As a brief summary, the dauer larva is an alternative to the third larval stage and forms under starvation or other stress conditions. Many genes involved in dauer formation have been described. Mutations in some genes result in constitutive dauer formation even when food is abundant; these are known as the *daf-c* genes. Mutations in other genes result in no dauer formation even when the population is starved; these are known as the *daf-d* (dauer defective) genes.

The pathway that leads to dauer formation is complicated, as described in Chapter 11 and shown again in Case Study Figure 12.1. Genes in the lower pathway in Case Study Figure 12.1A encode proteins in a TGF-β pathway, with the *daf-7* gene encoding the ligand. The correspondence between this branch of the dauer formation pathway and TGF-β signaling is shown in Case Study Figure 12.1B. TGF-β pathways are important in many

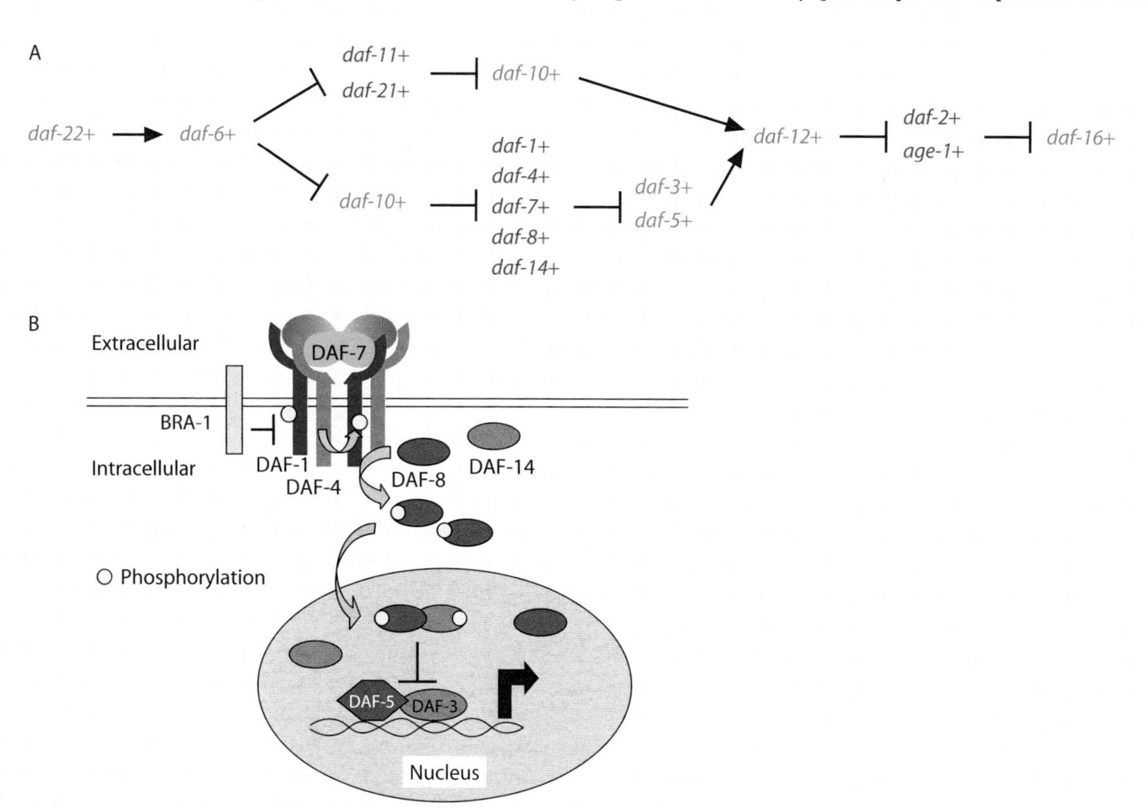

Figure C12.1 Panel A shows the pathway for dauer formation in *C. elegans*, as inferred from genetic interactions involving double mutants. Genes in blue are needed for dauer formation since mutations in these genes cannot form dauers. Genes in red are needed to prevent dauer formation, and mutations constitutively form dauers. Genes stacked at the same position in the pathway cannot be ordered using gene interactions alone. The pathway begins and ends with a single pathway, but has parallel pathways in the middle. For negative interactions, the downstream gene is epistatic. For positive interactions, the upstream gene is epistatic. The existence of parallel pathways was inferred by synthetic enhancement. Panel B shows the correspondence between the dauer formation pathway and the TGF-β pathway. The lower branch of the pathway in panel A is a TGF-β pathway, as shown here. DAF-7 is a ligand that binds to the receptor formed by DAF-1 and DAF-4. Downstream effectors are also shown.

processes in animal biology, and the proteins are evolutionarily conserved from worms to humans.

Because the TGF-β pathway has been studied in many different animals by a variety of genetic, biochemical, and molecular techniques, dauer formation is a good example of how network approaches can be integrated into our understanding of a fundamental biological process. In this case, a protein interaction network was constructed using six of the known proteins from the TGF-β pathway in *C. elegans* as bait in Y2H screens. The prey proteins were found from among a cDNA library with a comprehensive collection of worm genes. The six bait proteins are shown in Case Study Figure 12.2A. Two other proteins could not be used as bait: *daf-8* auto-activates when used as bait, so no interactions could be detected, and *daf-5* had not been molecularly identified when the study began.

Among the six bait proteins, no interactions were found for *daf-7* and *daf-12*. The lack of interactions for these two proteins probably illustrates two of the known limitations of Y2H screens discussed in Chapter 10—the reliance on bimolecular complexes and the inability to detect interactions that require protein modification. DAF-7 is the ligand that binds to a receptor encoded by a complex of DAF-1 and DAF-4; since Y2H only finds binary

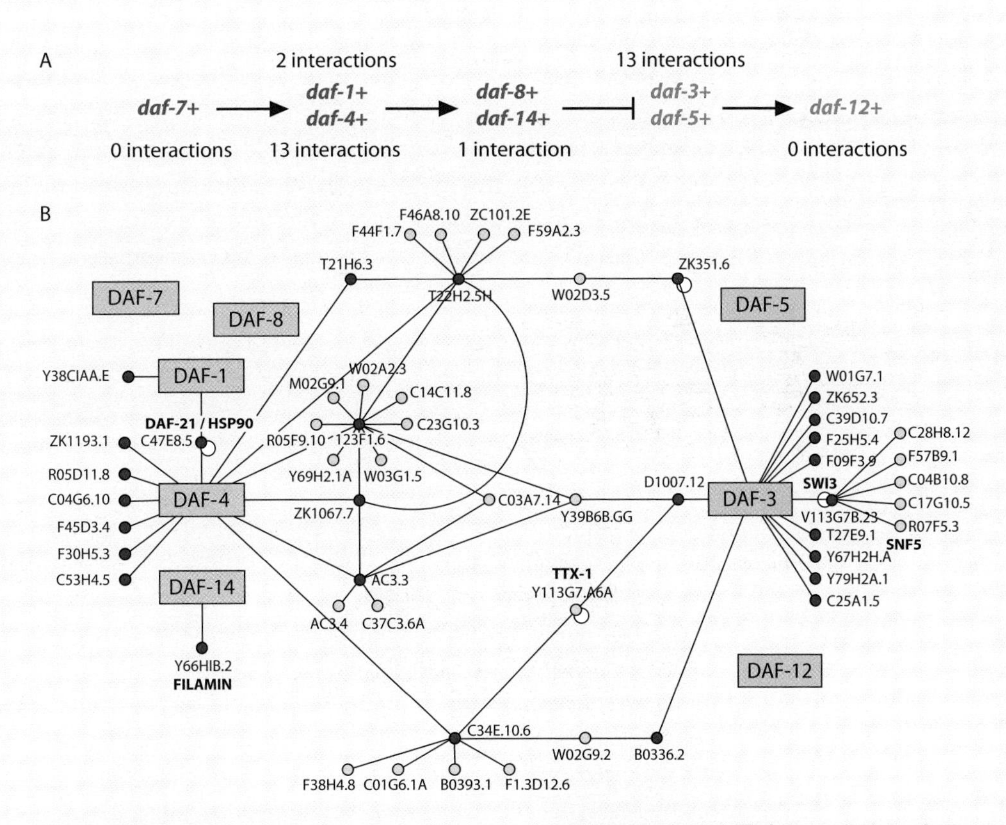

Figure C12.2 Panel A shows the genes and their protein products used as bait. Eight genes in the pathway are shown with their epistatic relationships. Six of these gene products were used as bait, with the number of interacting proteins shown adjacent to the gene name. DAF-8 could not be used as bait because it auto-activates in a Y2H screen. DAF-5 had not been cloned when these experiments began so it was not used as bait. DAF-7 and DAF-12 found no interacting proteins. Panel B shows the first- and second-degree interactions with these baits. The first-degree interactions are shown as blue balls, and the second-degree interactions are shown as yellow balls. Subsequent analysis showed that the gene known as W01G7.1, a first-degree interaction with DAF-3, corresponds to DAF-5. (Panel B reproduced from Tewari *et al.* 2004.)

interactions, a trimolecular complex formed from DAF-7, DAF-1, and DAF-14 would not be found. In addition, many proteins in the canonical TGF-β pathway are post-translationally modified, often by phosphorylation. Y2H screening misses interactions that require phosphorylation, which probably explains the failure to find interactions with DAF-12, a nuclear hormone receptor.

The network identified known members of the TGF-β pathway

Twenty-eight interacting proteins were identified with the other four bait proteins. Nineteen of these first-degree interactors were then used as bait in a further screen, and 27 second-degree interacting proteins were found. In total, 71 interactions involving 59 proteins were found. These are shown diagrammatically in Case Study Figure 12.2B. Given the known limitations of the Y2H screens, as noted above, these 71 interactions are probably just a subset of the actual number of proteins interacting with the bait proteins.

Of more concern for this analysis were any protein interactions that were detected by the Y2H screen but are not physiologically meaningful. For example, the proteins may not be expressed in the same cells at the same time in the worm but could still interact when expressed under the non-physiological conditions of the Y2H assay. In order to confirm some of the interactions by a different assay, epitope-tagged versions of a bait protein and a prey protein were co-expressed in mammalian cells; this type of co-immunoprecipitation is described in Chapter 10 as another means of detecting protein–protein interactions. Among the Y2H interactions tested, 89 percent were also found by the co-affinity precipitation assay using the epitope tags. This provides further evidence that most of these interactions are capable of occurring in the organism, although it does not conclusively prove that they do occur.

Interactions between proteins in the TGF-β pathway in other organisms had already been described before this set of experiments was done. One of the first tests of the approach was to ask if protein interactions previously detected by other means in other organisms are also found by the Y2H assays in worms. At least two examples of known interactions were also found by Y2H experiments in worms. For example, in mammalian cells, a protein known as filamen is found in a protein complex with SMAD2. The equivalent of the SMAD2 protein in the worm is DAF-12, and, as predicted from the results in mammals, one of its interacting proteins in worms is a nematode ortholog of filamen. Also in mammalian cells, the heat-shock protein HSP90 interacts with many proteins including the TGF-β receptor; in the Y2H assay, the nematode HSP90 protein DAF-21 interacts with the co-receptors DAF-1 and DAF-4. Case Study Figure 12.1A shows how *daf-21* had been tentatively placed in the other branch of the dauer formation pathway based on the behavior of an unusual dominant allele. However, the network shows that the gene belongs in *both* branches of the pathway. Other proteins in the network are also consistent with what is known or inferred from the TGF-β pathway in other cells, including the observation of the chromatin remodeling proteins SWI3 and SNF5 among the second-degree interactions with the DAF-3 protein.

Functional interactions were found by genetic assays

In the worm genome, many of the genes encode proteins whose functions are not known. Likewise, in the network of dauer formation genes, many of the gene products are

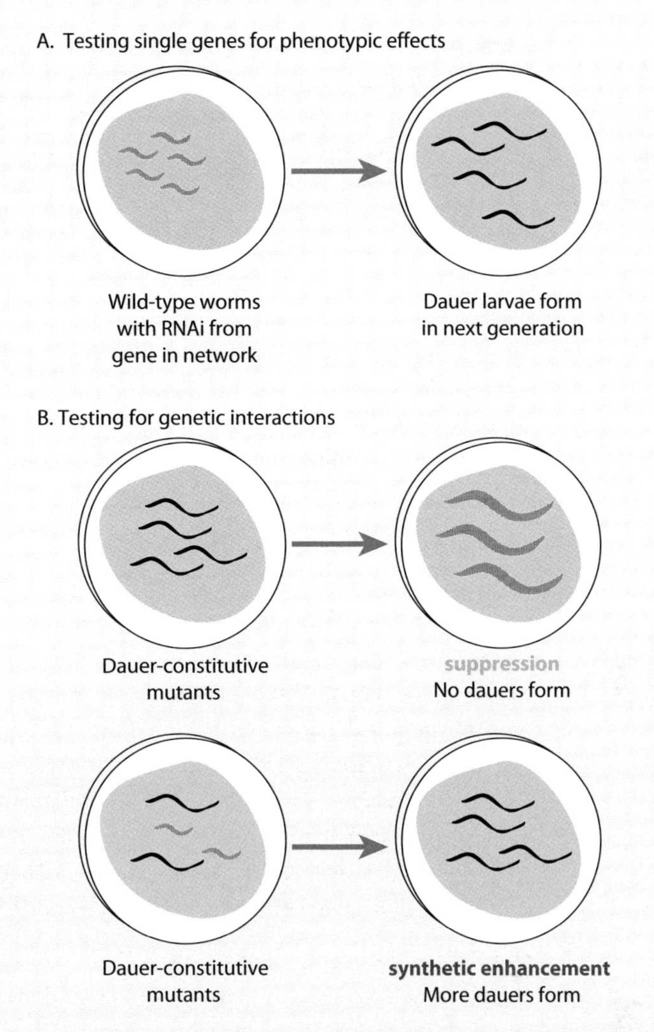

Figure C12.3 Testing for genetic effects. The genes identified in the network were tested for genetic effects using RNAi. (A) Each gene was tested for its effect on dauer formation. In the example shown, wild-type worms were grown on *E.coli* expressing one of the genes as double-stranded RNA. The offspring were examined for constitutive dauer formation. (B) The genes were tested for their ability to suppress or enhance other known dauer mutations. In the examples shown, a dauer-constitutive mutation could be suppressed so that no dauers form, or could be enhanced so that more dauers form.

functionally uncharacterized. Each of the genes for which no mutations were known were tested for effects on dauer formation by RNAi, as described in Chapter 7. If the gene is involved in dauer formation, a knockdown of its function by RNAi is expected to produce a Daf-c or Daf-d phenotype. The approach is shown in Case Study Figure 12.3A. Wild-type worms were placed onto bacteria expressing the double-stranded RNA corresponding to one of the genes from the network. The offspring were tested for their ability to form dauers. Three of the uncharacterized genes were found to produce dauer-constitutive Daf-c phenotypes, which identifies them as genes whose roles in dauer formation were previously unknown. Thus, the network approach found new genes that had been missed by other approaches. All three of these genes have mammalian orthologs, but none was thought to play any role in TGF-β signaling; an effect of these genes on TGF-β signaling remains to be tested in other organisms.

Comparing genetic interactions and physical interactions

The crucial test of the network approach is to test for *genetic* interactions among genes whose relationship was found by *physical* interactions. This test was done by performing RNAi assays on worms that also had a mutation in one of the known dauer genes. The assay is shown diagrammatically in Case Study Figure 12.3B. Known dauer mutant worms were placed on RNAi plates, as before, and suppression or synthetic enhancement of their phenotype was noted. Because the genetic pathway has been thoroughly analyzed using mutations and the mutant phenotypes for many of the genes have been clearly defined and quantified, as described previously, both suppression and synthetic enhancement could be tested. The results were a beautiful confirmation and extension of the previous research.

Thirteen different genetic interactions were found involving the known and unknown genes in the DAF pathway. For example, all three of the newly identified genes with RNAi phenotypes also showed suppression of *daf-12* in genetic assays. Following the same logic as described in Chapter 11 for epistatic pathways, this clearly places these three genes upstream of *daf-12*. Synthetic enhancement was also found, again by the methods described in Chapter 11. In all, five of the genes found *only* by the Y2H screen were functionally connected to the dauer pathway by these 13 genetic interactions. The combination of Y2H screening and genetic interaction networks provided a richer view of TGF-β signaling in worms than had been obtained previously relying solely on genetic interactions.

The pathway that integrates the genetic interactions with physical interactions is shown in Case Study Figure 12.4. The genes and proteins shown in the large circles have both genetic and physical interactions in the network. Many of these genes are evolutionarily conserved, suggesting that this network may also apply in other organisms in which similar tests are more difficult to perform. Given the importance of TGF-β signaling in human biology, including cancer, knowledge of the network and the interactions among its components offers insights that could not have been obtained easily in other ways.

The network-connected genetic and molecular identification of *daf-5*

The cloned gene known as W01G7.1 encoded a protein that interacted with DAF-3 in the Y2H assay. By RNAi assays, this gene also suppressed mutations in the *daf-c* genes *daf-1*, *daf-7*, *daf-8*, and *daf-14*, placing it well downstream in the genetic pathway. This is the same set of epistatic interactions shown by the dauer-defective gene *daf-5*. At the time of these experiments, *daf-5* had not yet been cloned. W01G7.1 and *daf-5* map to the same location in the genome, so it was natural to test whether W01G7.1 corresponds to *daf-5*. Confirmation of this was done by two of the methods described in Chapter 5, sequencing mutant alleles and complementing a *daf-5* mutant with the injected wild-type gene. When the W01G7.1 region of the genome was sequenced from a *daf-5* mutant allele, a missense mutation was identified in the gene, confirming that this clone is the *daf-5* gene. A transformation rescue experiment, as described in Chapter 5, also showed that W01G7.1 could complement a *daf-5* mutation.

An even more satisfying result was found when the wild-type *daf-5* gene was sequenced. The sequence of the wild-type gene indicates that the DAF-5 protein is a worm ortholog of the

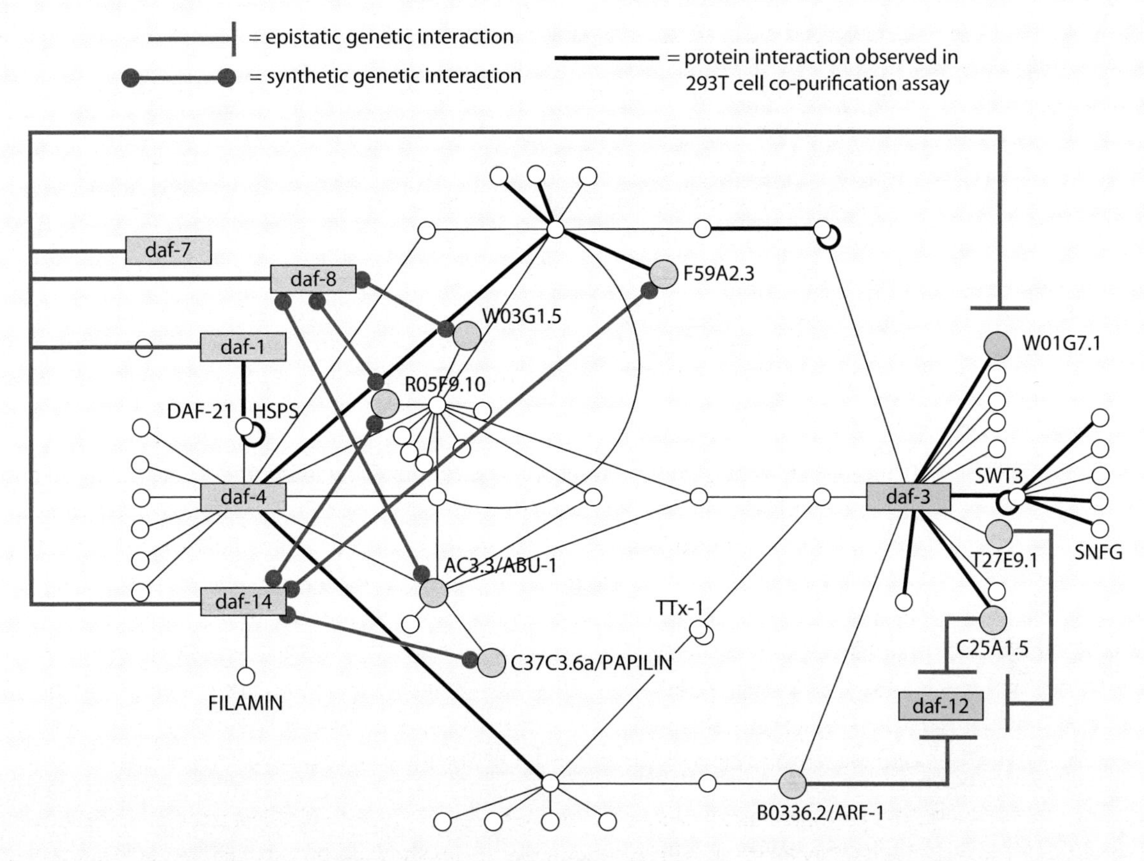

Figure C12.4 An integrated network of protein and genetic interactions. Protein interactions confirmed by immunoprecipitation are shown as black lines whereas those known only from Y2H assays are shown by gray lines. Genetic interactions are shown in red. Proteins and genes that exhibit both genetic and physical interactions, such as W03G1.5, are drawn as large circles. (Reproduced from Tewari *et al.* 2004.)

mammalian oncogene SNO/SKI. The SNO/SKI proteins in mammals are known to associate physically with the SMAD4 protein in the TGF-β pathway. SMAD4 in worms is DAF-3, and this protein was the bait used to find W01G7.1. This provides a striking confirmation of the parallels between the worm and human networks, and the use of one to illuminate the other.

Summary

We can summarize how the protein network and functional network analysis both confirmed and extended our knowledge of dauer formation and TGF-β signaling.

- Y2H in worms accurately reproduced known interactions among TGF-β components in mammals, including both the interaction of filamen with SMAD2 and the interaction of HSP90 with the TGF-β receptor.
- Extension of the network to second-degree interactions identified candidates for downstream genes in the TGF-β pathway.
- The protein network identified three previously unknown *daf-c* genes, all of them confirmed by RNAi analysis of the network.

- Functional interactions among the components could be detected by RNAi with mutant strains. These genetic interactions could be placed in the pathway of the known physical interactions among the proteins.
- *daf-5*, a gene known to be involved in dauer formation, was cloned based on its Y2H interactions. The results of functional analysis with RNAi confirmed the interactions that had been found by previous genetic analysis with mutant alleles of *daf-5*. Furthermore, the Y2H results confirmed a physical interaction that had been found in mammals. This indicates that the network found in *C. elegans* can probably be applied to the TGF-β network in mammals.

Further reading

Tewari, M., Hu, P.J., Ahn, J.S., *et al.* (2004). Systematic interactome mapping and genetic perturbation analysis of a *C. elegans* TGF-β signaling network. *Molecular Cell*, **13**, 469–82.

A yeast synthetic lethal network is the best progress towards a complete network of one type of genetic interaction

The targeted deletions for each of the yeast genes has shown us that only about 20–25 percent of genes are essential for growth under laboratory conditions. Overall, the genes that are essential are more likely to be evolutionarily conserved than the non-essential genes. For example, 38 percent of the essential genes have a clear-cut human ortholog, whereas about 20 percent of the non-essential genes have human orthologs.

KEY ARTICLE

Tong, A.H., Lesage, G., Bader, G.D., *et al.* (2004). Global mapping of the yeast genetic interaction network. *Science*, **303**, 808–13.

How can the prevalence of the non-essential genes be explained? We can think again about the two origins of robustness. One postulate is that some genes are not needed because they have paralogs in the yeast genome with a duplicate function, i.e. the function is biochemically redundant because there are duplicate genes. While this is certainly true in some cases, it is not the explanation for most of the non-essential genes. Only about 8.5 percent of the non-essential genes have a duplicate copy predicted to be carrying out the same function elsewhere in the genome. (Not surprisingly, fewer than 1 percent of the essential genes have an obvious paralog.) Thus, perhaps 70 percent of the genes in the yeast genome have functions that are not apparently needed for growth and division in the laboratory, at least as determined by analyzing single-mutant phenotypes.

However, as we considered earlier, there may be duplicate functions without sequence or biochemical similarity between the corresponding gene products. The genetic approach to studying functional redundancy is to use synthetic mutations, as discussed for vulval development or dauer formation in *C. elegans*. The presence of duplicate pathways is being examined on a genome-wide scale using synthetic lethality. In order to determine the extent of synthetic lethality, double-mutant strains are being constructed among many of the gene deletion strains. Different approaches are being used, since a complete test of all double-mutant strains would involve $(6500)^2$ strains. We are not nearly at this point yet. The initial analysis found about 4000 interaction phenotypes involving about 1000 genes. Thus, the synthetic lethal network is far from saturated. Not all of the genes have been tested in double-mutant strains, and not all the phenotypes of the double-mutant strains have been identified. In addition, for technical reasons that need not immediately concern us, a substantial fraction of synthetic lethal phenotypes (perhaps as many as 40 percent) could be missed in the genome-wide analysis.

If we begin with these estimates, that 1000 genes give about 4000 synthetic lethal phenotypes, then, because as many as 40 percent have been missed, the complete set of synthetic lethal phenotypes could be as many as 100 000—in a cell with about 6500 genes. (The human genome probably has about five times as many genes as yeast, so the number of potential gene interaction phenotypes in humans is enormous.) Nonetheless, despite the limitations of the data, the synthetic lethal network provides a significant addition to the interactome network by adding some genetic interaction data.

The results from genome-wide synthetic lethality screens are comparable to results from individual genes

The results of this genetic regulatory network analysis were similar to the results with the direct interactions from the interactome in that the number of synthetic lethal phenotypes per gene varies widely. A few genes interacted with only one other locus, whereas other genes had many more interactions. On average, a gene had 34 synthetic enhancers in these experiments. This is considerably more than the number of interactions that are observed in protein–protein interactions by the Y2H screen, which is less than four interactions per gene. The higher number of genetic interactions is expected since the interactome finds only direct protein contacts whereas, as was pointed out earlier, the synthetic lethal screen can find interactions between genes that act in the same process even if the gene proteins do not interact directly.

It is perhaps more instructive to determine *which* genes produce a synthetic lethal phenotype. Some of the results are shown diagrammatically in Figure 12.27. The synthetic enhancement phenotypes are most often observed when the two genes have similar single-mutant phenotypes. We encountered examples of these genes with dauer-constitutive mutations and with multivulva mutations in *C. elegans* (Chapter 10). Synthetic enhancement interactions are also frequently observed with genes whose products localize to the same cellular site and with genes encoding proteins known to be in a macromolecular complex. For example, the yeast gene *BNI1* encodes the protein formin, an evolutionarily conserved protein that is involved in the assembly of actin cables. BNI1 shows synthetic lethality with a number of other proteins involved in actin-based processes, including the other formin paralog in the genome, as well as proteins involved in cell polarity, bud emergence, and several other cell growth processes that require actin. We encountered synthetic enhancement interactions among genes that act at the same cellular site with the acetylcholinesterase genes in worms (Chapter 10) and will introduce an example involving protein secretion in yeast below. The yeast genes encoding the tubulin proteins (Case Study 10.1) provide examples of synthetic enhancement occurring with proteins in the same macromolecular complex, namely the mitotic spindle. Probably the most significant point is that synthetic enhancement interactions are extremely common, and as many as half of all non-essential genes appear to be essential in combination with at least one other non-essential gene.

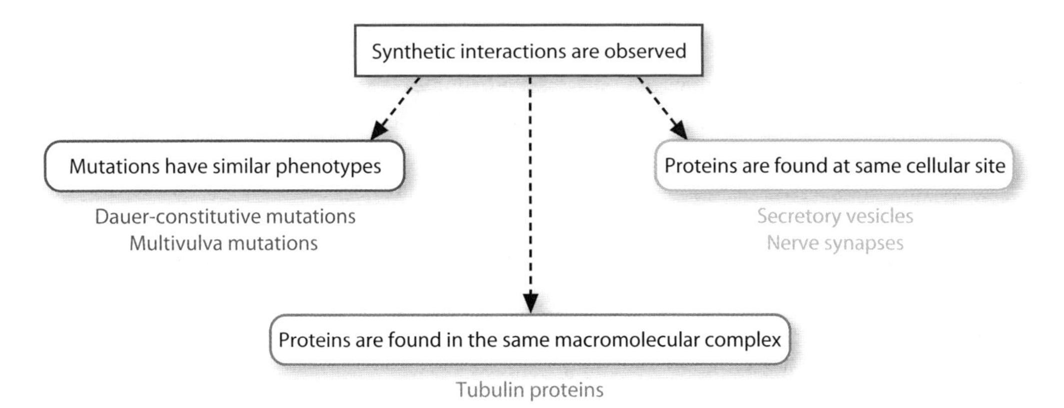

Figure 12.27 Predicting synthetic interactions. Synthetic enhancement is described in Chapters 10 and 11 as a genetic method of studying robustness. Based on mutations, more than half of the genes in yeast appear to have synthetic interactions with one or more other genes. The genes with synthetic interactions appear to share certain properties. The properties are shown in the colored ovals. A few examples of known synthetic interactions discussed in Chapters 10 and 11 are shown beneath the ovals.

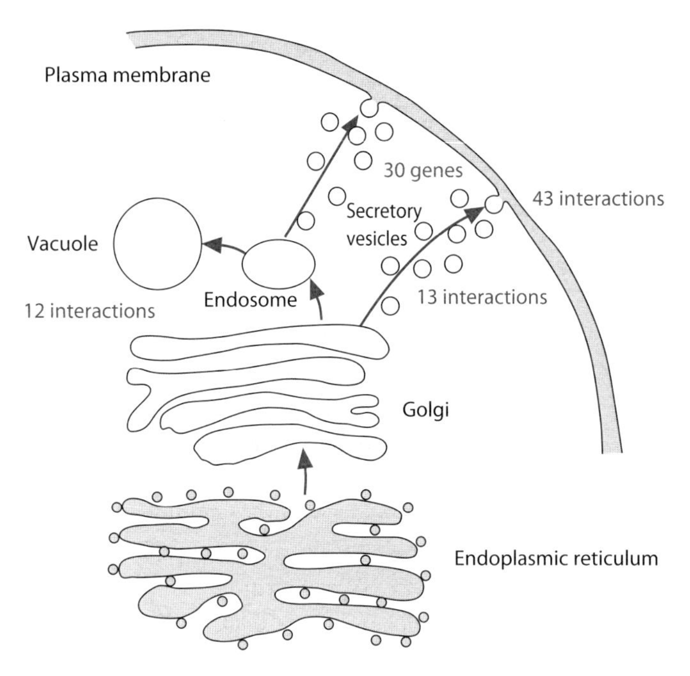

Figure 12.28 Genetic interactions in the yeast secretory pathway. Genetic analysis of the protein secretory pathway in yeast was introduced in Chapter 2. This figure summarizes an analysis of the synthetic interactions involving 30 genes involved in the fusion of the secretory vesicles with the plasma membrane. The number of interactions is shown in pink for different steps in the secretory pathway. Most of the synthetic interactions (43 in total) are among genes involved in membrane fusion, and many of the other interactions (13 in total) are with genes involved in other steps with the secretory vesicles. Only 12 interactions were with genes not directly involved with protein secretion.

KEY ARTICLE

Hartman, J.L., 4th, Garvik, B., and Hartwell, L. (2001). Principles for the buffering of genetic variation. *Science*, **291**, 1001–4.

A particularly extensive analysis of synthetic interactions has been performed with the secretory pathway in yeast, as shown in Figure 12.28. Recall that we introduced the secretory pathway in Chapter 2 as our example of a genetic screen for a fundamental cellular process. The secretory pathway is genetically complicated, involving 173 genes that act in 10 biochemically and morphologically distinguishable steps. Steps include translation of the vesicle to the Golgi apparatus, vesicle budding, fusion with the plasma membrane, and so on.

Many double mutants have been made involving genes with defects in some aspect of the secretory pathway, and a few general conclusions about synthetic enhancement have emerged from this analysis. Synthetic lethal interactions are commonly found among genes that affect the same step in the pathway and sometimes observed between genes that affect different steps. Synthetic lethal interactions are occasionally found between a secretory gene and a gene not involved in secretion, but these are less frequent than interactions between secretory genes. For example, 30 genes are involved in fusion of the secretory vesicle with the plasma membrane. There are 43 synthetic lethal interactions among these genes, and 13 additional synthetic lethal interactions with genes involved in other steps in the secretory process. There were also 12 synthetic lethal interactions with genes not involved with secretion. Thus, 63 percent (43 of 68) of the synthetic lethal interactions observed for these genes are found among genes that affect the same step, and an additional 19 percent (13 of 68) of the synthetic lethal interactions involved genes in the same pathway.

These results with synthetic enhancement in the yeast secretory pathway appear to be typical when expanded to other double-mutant combinations. At least half of the synthetic lethal interactions observed in yeast are between genes involved in the same biochemical or cellular step. About a quarter of the interactions are with genes involved in a different step in the same physiological process, and a quarter are with apparently unrelated genes. These results suggest that most, but certainly not all, synthetic lethal interactions will be observed among genes that have similar phenotypes or affect similar functions. Thus, as a first attempt to map the interactions that occur in a system, it may not be necessary to test mutations in every other gene in the genome for synthetic enhancement. A more strategic approach is to focus on genes with similar mutant phenotypes or that are involved in different steps in the same biological process.

12.8 Robustness, interactions, and risk factors: speculation

The synthetic lethal screens in yeast demonstrate that, whereas most genes have no mutant phenotype on their own, they can interact with mutations in other genes to produce a mutant phenotype. One of the most provocative aspects of these synthetic lethal screens arises when we think of them in a broader genetic context than the yeast cell in the laboratory, namely in human diseases and genetic variability. In laboratory organisms such as yeast, the only mutations that are present in the genome are those the geneticist knows about and wants to observe; at all other remaining loci, the cell or the organism has the wild-type allele.

As we noted at the beginning of the chapter, this genetic make-up is very different from what is found in a natural population such as humans. All of us have many mutations in our genome—we are homozygous for some and heterozygous for others. Some mutations were inherited from a parent; others have newly arisen in us. If the lessons from the gene disruptions in yeast and from the RNAi screens in worms and flies can be extended to humans, most of these mutations have little or no observable phenotypic effect on their own. This means that in a natural out-breeding population such as humans, mutations in most genes can accumulate without being selected against; since it has no phenotypic effect, the mutation is effectively neutral. In the population, there are many different alleles or mutations for each gene, not just the one allele defined to be wild-type. Most of these mutations by themselves will not affect the function of the gene.

But nearly all mutations in yeast that do not produce a mutant phenotype on their own will *interact* with mutations in another gene to produce a mutant phenotype. If synthetic enhancement interactions are this common among mutations in laboratory strains, we can speculate that the impact of synthetic enhancement on the human gene pool could be even more significant. After all, a natural population has far more variation than a laboratory strain, which is homozygous at most of its loci. Natural variation leads to many more possible allelic combinations than would ever be seen in the laboratory. Any human is a double or multiple mutant for many different loci, any of which could interact to produce a synthetic enhancement.

In addition, only one type of allele—the null allele—has been tested for synthetic lethal interactions in yeast. Hypomorphic, hypermorphic, and other types of allele have not yet been tested for synthetic interactions, so it is likely that additional interaction phenotypes will be seen. It is possible that the effects on the phenotype may be more subtle with mutations that do not knock out the locus, but even subtle effects on the phenotype are easily detected in humans. Furthermore, only pairwise combinations of mutations have been tested in yeast; multiple mutant strains have not been tested. A human is a multiple mutant. To add even more potential interactions to our thinking, the yeast experiments only examined synthetic lethality and did not consider other types of interaction such as suppression or epistasis, as discussed in Chapters 10 and 11. To sum up, human populations have more loci with genetic variation, more allelic variation at these loci, more possible types of interactions among these variable loci, and even more genes overall.

Genetic interactions could affect our risk for many genetic diseases

It is worth reflecting on the extent of possible interactions for a little longer. We can summarize the inferences from the yeast gene disruptions as follows. We carry in our genomes many mutations that do not by themselves result in a detectable phenotypic difference. However, each of these mutations is capable of interacting with other mutations in the genome that will produce a phenotype difference—probably dozens of other genes and possibly many more than that. The number of potential interactions among the estimated 30 000 genes in the human genome with variation at nearly every locus is enormous.

Now think about this variation in the context of multifactorial traits—the types of genetic disease that are the focus of much of human health. We know of genetic risk factors that predispose us to cancer, hypertension, heart disease, obesity, diabetes, and Alzheimer's disease, among many other traits. It is extremely likely that all of us carry in our genomes the genetic seeds of our own destruction. In fact, like the arms race during the Cold War, each of us probably carries enough destructive force in our genome to die many times over. The genetic traits that are most likely to affect our health and life span are multifactorial, showing complicated inheritance patterns with environmental effects as well.

It may be instructive to think of these multifactorial human disease traits in the same way that we think of synthetic enhancement in other organisms. We can use the hypothetical example of hypertension, for which many genetic risk factors and many contributing environmental factors are known. Having inherited a mutation (either homozygous or heterozygous) in one of the genes affecting blood pressure, we are susceptible to the disease. But the disease only develops when the second gene affecting blood pressure is also altered, possibly through an inherited mutation or a somatic mutation in the critical tissue during our own life span or even through an environmental effect on the gene product. Each of the hypothetical mutations alone does not result in hypertension, but a synthetic interaction between the two genes does produce the disease.

(As a quick aside, we are describing something different from the 'two-hit hypothesis' to explain the onset of many human cancers, although the intellectual connection is clear. According to the two-hit hypothesis, we inherit a mutant allele that predisposes us to cancer, but our cells are heterozygous for this mutation. When a second somatic mutation affects the other allele, cancer develops. With the type of synthetic interactions we are describing here, the mutant phenotype arises not from a second mutation in the same gene but instead through mutations in a *different* gene. As discussed in Chapter 10, the mutations in the interacting genes may be either homozygous or heterozygous.)

Is this a helpful way of thinking about the inheritance of complex traits? To put the question another way, should we be thinking about risk factors in terms of a network of genetic interactions in cells? The implications may be extremely important, but the relevant data are rather sparse. One study found as many as 30 different loci that contribute to lung cancer susceptibility in laboratory mice. This number is almost certainly an underestimate of the true number of susceptibility loci because the laboratory strains themselves showed much less genetic variation than would be found in natural strains. Thus, many potential 'susceptibility loci' would be homozygous for the non-susceptible allele, and would not be included in this estimate. However, even this estimate is sobering, since it considers only one complex trait under controlled laboratory conditions. In applying this to humans, we should recognize that many different complex traits have such risk factors and the risk factors are affected by environmental variation as well as other genes.

As discussed earlier (Chapter 5), even with the highly refined genetic map now available in humans, finding susceptibility loci for many human diseases has been challenging. Results have proved difficult to replicate, perhaps in part because different groups are mapping different loci or because of synthetic interactions among polymorphic loci in the population. It is a huge task to identify susceptibility loci for the multifactorial traits that will probably be the cause of our death.

The awesome power of yeast genetics that has made the task seem formidable may also help to provide some of the keys to its solution. Most of the interactions observed in yeast are between genes that function in the same subcellular location, the same complex, or the same physiological process. This may help us to know how to identify other genes involved in determining the risk. In addition, a remarkable number of genes in yeast, worms, or flies have clear-cut orthologs in humans. We might be able to assume that, since the sequence and the structure of the protein are evolutionarily conserved, its genetic interactions may also be conserved. It is safe to conclude that the days of model organisms are not yet passed.

As we unfolded a brief history of genetics in Chapter 1, we noted how each of its pioneers 'had it right', i.e. they laid an important conceptual foundation for those who came after. However, each one also had some of the foundational concepts wrong as well. With each genetic pioneer, some questions remained to be refined and re-answered by the next group. Most of the conceptual foundations were discovered to be too simple and some key information had been omitted—linkage or protein function or the DNA structure itself. We stand in a similar place. In this book, we have tried to show the conceptual foundation as it exists now. But even if this were done perfectly, we can be sure that our knowledge of genes and genomes is still rudimentary. As genomes are sequenced and annotated, we realize how much our view of genetics has been simplified.

When the number of predicted genes in the human genome was disclosed, many of us were discouraged to realize that we had only twice as many parts as animals like insects and worms. Surely a species that has produced specimens like Beethoven and Beckham, Shakespeare and Socrates, Mandela and Monet has more genetic capacity than that. We do. Our genes do not work as 30 000 individual pieces, each fitting into its own piece of the jigsaw puzzle that makes up a human. Rather, they enhance, suppress, or mask each other. They interact in vastly more ways than we can imagine or detect. As a result, we are distinctly more than the sum of our parts.

GLOSSARY

alignment a position-by-position comparison of two sequences, such as the amino acid sequences of two different proteins. A **global alignment** compares the entire length of the two sequences from one end to the other. A **local alignment** compares only the regions of highest similiarity. BLAST performs local alignments; the name is an acronym for Basic Local Alignment Sequence Tool.

allele one of the many alternative forms of a gene. Used very broadly to include both the phenotypic variation, such as round or wrinkled peas, and the molecular variation, such as a sequence alteration in a gene. See also **polymorphism**.

allelic series a collection of different mutant alleles for a gene that have different amounts of residual gene activity, and thus slightly different mutant phenotypes.

alternative splicing the processing of an RNA transcript to yield different mRNA sequences. In higher organisms, it is thought that most transcripts are alternatively spliced to yield related but different mRNA transcripts.

amorphic allele also known as a **null allele**. A mutant allele that eliminates the function of a gene. Most amorphic mutations are recessive.

annotation the process of identifying the functional components of a sequenced genome.

antimorphic mutation see **dominant-negative mutation**.

apoptosis the genetically programmed death of particular cells in multicellular animals.

ascus the sac structure that holds the haploid spores of yeast and other fungi.

association mapping a procedure to localize a gene for a mutant phenotype by examining the segregation of polymorphisms in a large population, attempting to identify polymorphisms that co-segregate with the mutant phenotype of interest.

attached X see **compound chromosome**.

auxotroph a mutant that cannot synthesize some essential nutrient for growth. Auxotrophic mutants can only grow if the nutrient is provided in the growth medium. The opposite of auxotroph is **prototroph**.

balanced heterozygote a heterozygous genotype in which a recessive mutation of interest is maintained with marker mutations on the other homologous chromosome. For example, *aB/Ab* is a balanced heterozygote.

balancer chromosome a genetically altered chromosome that cannot recombine with its normal homologous chromosome. Any recessive mutations on the normal homologous chromosome will be inherited together if they are maintained as a heterozygote with a balancer chromosome.

cell autonomous a gene whose phenotype corresponds to the genotype of the cell, regardless of the genotype or phenotype of adjacent cells. Most genes whose activity is cell autonomous encode proteins that work intracellularly.

chiasma (plural: chiasmata) the physical connections between homologous chromosomes that form as a result of recombination during meiosis I.

chimera an organism with cells of two or more different genotypes, produced by physically aggregating the cells. Compare with **mosaic**.

chromatin immunoprecipitation (ChIP) a procedure in which an antibody against a transcription factor protein is used to identify the binding sites for that protein in the genome.

co-immunoprecipitation (co-IP) a biochemical method to detect protein–protein interactions by using an antibody against one protein to also precipitate other proteins in a macromolecular complex.

cold-sensitive mutation a mutation that has a mutant phenotype at a slightly reduced growth temperature but a wild-type phenotype at a higher-than-normal growth temperature. This is the reverse of a **temperature-sensitive** mutation.

comparative genomics an experimental approach that examines the genomes of closely related species to identify sequences that could be functionally significant.

complementation test a genetic test to determine if two recessive mutations are alleles of the same gene. Mutations that are alleles of the same gene fail to complement.

compound chromosome a chromosome rearrangement in which two chromosomes have been fused together to make one large chromosome with a single centromere. A compound chromosome is the experimental equivalent of a Robertsonian fusion, which occurs during evolution to produce one metacentric chromosome from two telocentric chromosomes. One widely used compound chromosome is the attached X chromosome in *D. melanogaster*, in which two X chromosomes have been fused at their ends.

conditional mutation a mutation that exhibits a mutant phenotype only under certain environmental conditions, such as a slightly elevated growth temperature or particular nutritional conditions. **Temperature-sensitive** mutations are an example of conditional mutations.

core promoter also called the **minimal promoter**, or sometimes simply the **promoter**. The region upstream of a gene that is needed for transcription initiation by RNA polymerase. Many core promoters have a TATA box about 25 nucleotides (or –25) upstream of the start of transcription and other necessary regulation sequences at about –70 and –110.

counter-selection a genetic screen that allows selection for both a particular genotype and against the same genotype.

Cre a recombination enzyme from the bacteriophage P1 that catalyzes site-specific crossovers at the nucleotide sequences known as *loxP* sites. The Cre–*lox* recombination system is widely used to insert sequence or produce recombination at specific sites.

crossover suppressor a chromosome rearrangement such as an inversion that cannot recombine with the normal homologous chromosome. Since no recombinant gametes are produced, the chromosome is a crossover suppressor. Crossover suppressors are often used as **balancer chromosomes**.

dauer larva a stage in the life cycle of *Caenorhabditis elegans* that occurs under conditions of starvation, environmental stress, or crowding. A worm can enter the dauer larva stage as an alternative to the third larval stage and remain in the dauer stage indefinitely.

degree in a network diagram, the number of connections or links present for a particular node. The histogram displaying the degrees for all of the nodes in a network is called the **degree distribution**.

deletion also known as a **deficiency**. A chromosome rearrangement in which a contiguous piece of the chromosome has been removed and the remaining pieces fused together.

dipteran two-winged insects, including species of *Drosophila* and house flies.

distributed robustness robustness that arises from duplicate paths in a network even when the individual nodes do not have duplicate functions.

dominant-negative mutation also known as **antimorphic mutation**. A dominant mutation that eliminates or reduces the function of a gene, usually by interfering with the function of the normal product of the same gene.

duplication a chromosome rearrangement in which a contiguous piece of a chromosome has become duplicated or repeated. Duplications can be found adjacent to the normal site, in which case they are usually referred to as tandem duplications, or they may be found at another site.

effective lethal phase also known as the **terminal phenotype**. The stage of the life cycle at which a particular mutant arrests development.

electroporation a method that uses an electric current to introduce DNA or other molecules into a cell.

emergent property a characteristic of a biological system that arises from the interactions of its components as well as from the characteristics of those components themselves.

enhancer mutation see **synthetic enhancer mutation**.

enhancer sequence a sequence in the regulatory region of a gene that confers the specificity of transcription for that gene by activating transcription. Enhancer sequences are the binding sites of transcription factor proteins, and a gene may have many or few enhancer sequences.

enhancer trap a cloning and expression technique that integrates reporter gene constructs randomly into the genome and examining their expression. Those that land near an enhancer regulatory sequence will display a specific pattern of expression.

epistasis a genetic interaction between two genes in which the mutant phenotype of one gene masks the genotype and phenotype of the other gene.

epitope tag a short amino acid sequence, for which antibodies or easily used affinity columns are available, that is attached to another protein to allow its purification or analysis.

exon array a type of microarray in which the probes are from known or predicted exons.

expression profile the complete catalog of what genes are being transcribed in a particular tissue, stage, or environmental conditions. An expression profile is usually obtained from a microarray.

expressivity the phenomenon whereby individuals with the same mutant genotype have somewhat different mutant phenotypes. Expressivity is expressed qualitatively, such as 'variable expressivity' meaning that all of the individuals have a mutant phenotype but the mutant phenotype is not exactly the same.

F_1 screen also known as a **non-complementation screen**. A mutant screen that uses a pre-existing allele in a gene of interest to find additional alleles of the same gene. The screen looks for mutations in the F_1 generation that fail to complement the pre-existing allele.

floxed the nickname for a gene that is flanked by *loxP* sites. It is a specific target for deletion or alteration by the Cre recombinase.

Flp/FRT a system of site-specific recombination that occurs in the yeast 2 μm plasmid. Flp is the recombinase enzyme that catalyzes recombination at the FRT sequences.

gain-of-function mutation a mutation that over-expresses or produces a novel function for a gene. Many gain-of-function mutations are dominant.

gel mobility shift assay a method of identifying proteins that bind to a nucleotide sequence by testing their ability to retard the mobility of the sequence during gel electrophoresis.

gene family or a **protein family**. Genes that are highly related in DNA sequence and that encode proteins with highly similar amino acid sequences. The proteins have similar or even identical functions.

Gene Ontology a project to classify genes in many different organisms by the biological process that they affect, by their cellular or subcellular localization, and by their molecular or biochemical function. Such groupings allow the functional comparisons of genes from different species.

genome-wide screen a screen in which genome sequence information is used to design specific molecular tools for the analysis of all or nearly all of the identified genes.

global alignment see **alignment**.

gynandromorph an organism with both male and female cells and differentiation. Gynandromorphs arise in *D. melanogaster* from mitotic loss of an X chromosome.

haplo-insufficient gene a gene whose function is dose dependent, such that eliminating the function of one allele results in a mutant phenotype. A haplo-insufficient gene is recognized by dominant loss-of-function alleles.

high-copy suppressor a gene whose high copy number or over-expression of the wild-type gene suppresses a mutation in another gene. High-copy suppressors are functionally equivalent to over-producing dominant mutations because it is the wild-type function of the gene that produces the suppression.

holocentric chromosomes in which the spindle microtubules attach along the length rather than at one localized centromere. Chromosomes in *C. elegans* and other nematodes are holocentric, which has implications for methods for transformation and for producing mosaics.

homology the descent of two sequences or structures from a common ancestor. Homology is a qualitative term describing the evolutionary history and cannot be expressed quantitatively. Homology is often inferred from sequence similarity.

hub a node in a network with a very large number of links.

hybrid dysgenesis the syndrome of genetic effects arising from the movement of P-transposable elements in *D. melanogaster*.

hypermorphic allele also known as a **hypermorph**. A mutant allele that over-produces the normal function of the gene. Most hypermorphic mutations are dominant.

hypomorphic allele also known as a **hypomorph**. A mutant allele that reduces but does not completely eliminate the function of a gene. Most hypomorphic mutations are recessive. Compare with **null allele**.

imaginal disk a pocket of undifferentiated cells in *Drosophila* larvae and pupae that will give rise to structures such as the legs, wings, and genitalia in the adult. Different imaginal disks are the progenitors of different adult structures.

informational suppressor a suppressor mutation that affects a specific type of molecular lesion, such as a nonsense mutation, regardless of the function of the gene being suppressed. Informational suppressors affect many different genes, but will only affect certain alleles of the genes they suppress.

interactional suppressor a suppressor mutation that affects only specific alleles of a specific gene, typically because the suppressor gene and the target gene encode interacting gene products.

interactome the catalog of protein–protein interactions that occur in a tissue or an organism.

intragenic complementation a situation in which two mutations are known to lie in the same gene, but give a wild-type phenotype when heterozygous with each other, and so complement.

knock-in mutation a procedure in which the wild-type gene on the chromosome is replaced by a version that has been engineered *in vitro* with a specific mutation. The procedure is widely used in mice.

knockout mutation a mutant allele in which the wild-type allele has been severely disrupted or deleted. A knockout mutation is a null allele.

local alignment see **alignment**.

loss-of-function mutation a mutation that eliminates or reduces the function of a gene. Most loss-of-function mutations are recessive, and most recessive mutations arise from a loss or reduction in the function of a gene. See **null allele** and **hypomorphic allele**.

loxP the 34 base pair sequence that is the target for recombination by the Cre recombinase. The Cre–*lox* recombination system is widely used to insert sequence or produce recombination at specific sites.

microarray a hybridization-based procedure to examine the transcription of many genes simultaneously.

microRNA (miRNA) an RNA product of approximately 22 nucleotides that regulates the expression of another gene by making a double-stranded hybrid with a complementary sequence on the target gene's mRNA. The double-stranded RNA hybrid either targets the mRNA for degradation or blocks its translation.

missense mutation a mutation in which one amino acid in the protein sequence is replaced by another amino acid.

mitotic recombination also known as **somatic crossing-over**. Recombination that occurs in mitotic cells, usually as a result of experimentally induced double-stranded breaks. Mitotic recombination is used to make mosaics in *D. melanogaster*.

molecular bar code a nucleotide sequence not found in the genome that is used to specifically tag and identify a gene in genome-wide screens for mutations.

mosaic an organism with cells of two different genotypes produced by genetic methods such as somatic recombination or chromosome fragment loss. Compare with **chimera**.

mutagen an agent that induces mutations.

mutator a phenotype in which the organism has a high frequency of mutations. Often, mutator strains have very active transposable elements. Mutations that affect DNA repair also result in a mutator phenotype.

neomorphic allele a mutant allele that produces a novel phenotype, often by expressing the gene product ectopically, i.e. in a tissue or at a time that is different from wild-type.

noise the level of naturally occurring variation in the amount of a transcript or a protein.

non-allelic non-complementation a situation in which two recessive mutations that are not alleles of the same gene give a mutant phenotype when both are heterozygous.

nonsense mutation a mutation in which one codon in the mRNA is changed to become a stop codon, thus terminating the translation of the polypeptide.

null allele synonymous with **amorphic allele**. A mutant allele that eliminates the function of a gene. Most null mutations are recessive.

ortholog a functionally equivalent gene or protein in another species. Equivalence is usually recognized by highly similar nucleotide or amino acid sequences. The human β-globin gene and the mouse β-globin gene are orthologs of one another. Orthologs represent genes that were present in the common ancestor of the species being compared. Compare with **paralog**.

outcrossing a series of matings in which a mutagenized strain is crossed to a non-mutagenized strain so that regions of the mutagenized genome are replaced by the same region from an non-mutagenized genome. Outcrossing is done to removing extraneous mutations from a genetic strain.

P element a transposable element found in *D. melanogaster*, and widely used for as vector for transformation.

paralog a member of a gene or protein family within the same species. The human β-globin gene and the human γ-globin genes are paralogs of each other. Paralogs are derived from a common ancestral gene, often by a gene duplication event. Compare with **ortholog**.

penetrance the percentage of genotypic mutant individuals that exhibit a mutant phenotype. Penetrance ranges from zero to 100 percent.

permissive temperature the temperature at which a temperature-sensitive or cold-sensitive mutation exhibits a wild-type phenotype.

phene an individual component of a **pleiotropic** mutant phenotype, such as the effect on one tissue.

phenocopy a morphological mimic of a genetic phenotype that is produced by environmental means rather than a mutation.

phenocritical phase the stage in the life cycle when the first effects of a mutant are observed. In yeast, this is sometimes referred to as the **diagnostic landmark**.

pleiotropy multiple different phenotypic effects that occur together as a result of a single mutation. The separate phenotypic effects are called **phenes**.

polymorphism genetic variation at a locus. The terms 'allele', 'mutation', and 'polymorphism' have similar meanings but different connotations, with polymorphism being the most inclusive term. Polymorphisms use naturally occurring variations in the DNA sequence and may not have a detectable phenotypic variation. 'Mutation' usually implies a genetic change that has been experimentally induced, that occurs rarely, or that substantially alters the phenotype. 'Allele' often implies a detectable phenotype difference, which may arise from natural or experimentally induced variation.

polytene chromosome the large chromosome in certain insect tissues, particularly the salivary glands in *D. melanogaster*. The chromosomes in these cells undergo many rounds of DNA replication without strand separation, so that a polytene chromosome is large enough to have easily visible bands and other structural features. These are cytological landmarks for positioning genes on the *Drosophila* chromosomes.

positional cloning a cloning procedure based on the map location of the gene.

post-transcriptional gene silencing similar or identical to RNAi, the name of the process first identified in plants.

proteome the complete catalog of all of the proteins that are made by an organism, including the post-translational modifications.

prototroph a strain that can grow in the absence of an added nutrient because it has the wild-type function to synthesize that nutrient. The opposite of prototroph is **auxotroph**.

pseudo-gene a member of a gene family that is non-functional because of mutation.

redundancy robustness that arises from duplicate nodes with the same function in a network.

regulatory region the entire DNA sequence necessary for the gene to be transcribed in the proper tissue at the proper time and in the proper amount. The regulatory region includes all of the **enhancer** and **silencer sequences**. For most genes, the regulatory region is found upstream of the core promoter, but many exceptions are known in which the regulatory region includes sequences downstream of the gene or internal to the coding sequence, such as intronic sequences.

reporter gene a molecular construct that has the coding region of some protein that is easily assayed fused to the control region of the gene of interest. Among the proteins commonly used as reporters are β-galactoside (encoded by the *lacZ* gene of *E. coli*) and green fluorescence protein (GFP). Reporter genes can be either transcriptional or translational. A **transcriptional reporter** gene replaces most or all of the coding region of the gene of interest with the coding region of the reporter gene. A **translational reporter** gene places the coding region of the reporter gene in frame with the normal protein so that a fusion protein is made between the normal protein and the reporter protein.

restrictive temperature the temperature at which a temperature-sensitive or cold-sensitive mutation exhibits a mutant phenotype.

reverse genetics the process of beginning with a cloned gene and finding a mutant phenotype for it. The distinction between 'reverse genetics', which begins with a cloned gene, and 'genetic analysis', which begins with a mutant phenotype, is blurred, and the term is falling out of use.

RNA interference (RNAi) a widely used technique to reduce the expression of a gene using double-stranded RNA corresponding to part of the coding region of the gene. RNAi produces the equivalent of a mutant phenotype without altering the DNA sequence for the gene.

robustness the ability of a system or network to maintain its function and overall structure when random nodes are deleted.

saturation a situation in which all of the genes that could be identified by a particular mutant screen have been found.

segmental aneuploid a chromosome constitution that has extra or missing copies of portions of a chromosome as a result of a **duplication** or **deletion**. It is only that segment that is aneuploid rather than the entire chromosome.

selection a genetic screen in which only certain genotypes can grow. For example, a growth medium containing the antiobiotic kanamycin can be used to select for mutations that are kanamycin resistant.

sequence similarity a quantitative description of the relationship between two amino acid or nucleotide sequences. Sequence similarity is often used to infer **homology**, or descent from a common ancestor.

sex reversal a genetic mutation that causes the mutant to differentiate into a sexual phenotype that is different from its chromosomal phenotype. For example, the sex reversal mutation *tra* in *Drosophila* results in flies with two X chromosomes becoming males rather than females.

short interfering RNA (siRNA) the functional 22–25-nucleotide double-stranded RNA molecules involved in RNAi.

shuttle vector a plasmid that has genetic markers to allow it to be grown in both *E. coli* and yeast.

silencer sequence a sequence in the regulatory region of the gene that represses transcription. Silencer sequences are the functional opposite of **enhancer** sequences.

SNP (single-nucleotide polymorphism) genetic variation that arises from a change in a single nucleotide at a locus. SNPs (pronounced 'snips') are very frequent in natural populations and yield some of the most useful markers for **association mapping**.

somatic crossing-over see **mitotic recombination**.

spliceosome the complex of RNA and protein molecules responsible for RNA splicing.

suppressor mutation a mutation in a second gene that overcomes the phenotypic effects of another mutation. As a result of a suppressor mutation, the phenotype of the double mutant more closely resembles wild-type than does the phenotype of the single mutation. Compare with **synthetic enhancer mutation**.

synteny a general term for genes that are located on the same chromosome. Genes may be syntenic but sufficiently far apart on the chromosome that they segregate independently and thus are not linked in their inheritance. Synteny is important in comparing the genomes of closely related species because many linkage relationships have been preserved in evolution. A group of genes located together in two different species comprise a **syntenic block**.

synthetic enhancer mutation a mutation in a second gene that makes the phenotype of another mutation more severely mutant. In particular, each mutation alone has a wild-type phenotype, but the double mutant has a mutant phenotype. Compare with **suppressor mutation**.

temperature-sensitive mutation a mutation that has a mutant phenotype at a slightly elevated temperature but a wild-type phenotype at a lower growth temperature. The temperature that exhibits the wild-type phenotype is known as the **permissive temperature**. The temperature that exhibits the mutant phenotype is known as the **restrictive temperature**.

temperature-sensitive period (TSP) the time during which the activity of a gene is needed, as determined by temperature-shift experiments. The temperature-sensitive period is sometimes referred to as the execution point for the activity of a gene.

temperature shift an experiment in which a mutant with a temperature-sensitive mutation is shifted from one growth temperature to another to determine the time when the activity of the gene is needed. Temperature shifts from the permissive to the restrictive temperature are called shift-up experiments, whereas temperature shifts from the restrictive to the permissive temperature are called shift-down experiments.

terminal phenotype the mutant phenotype observed when a lethal mutation causes developmental arrest. Also known as the **effective lethal phase**.

tetrad the four haploid products of a single meiotic division. In yeast and other fungi, these four haploid spores are found together in the **ascus**.

thymidine dimer the covalent linkage of adjacent thymidine residues on the same DNA strand. Thymidine dimers are induced by UV light.

tiling array a type of microarray in which the probes are overlapping sequences from both DNA strands of the entire genome, without regard to the possible function of the sequences.

transcription factor a protein that binds to enhancer or silencer sequences to regulate transcriptional initiation of a gene.

transcriptome the complete catalog of all of the transcripts that are made by an organism, regardless of the tissue, stage, or environmental conditions.

transformation rescue introducing a wild-type copy of the gene to restore the normal gene function to a mutant.

transposable element a discrete DNA sequence element that can move from one location to another in the genome under the control of transposase, a class of enzyme encoded on the element itself.

transposase a class of enzyme encoded within a transposable element that catalyzes the excision and insertion of that transposable element. The transposase from one type of transposable element will not affect other types of transposable elements, but will affect other elements of some type regardless of their location in the genome.

transposon-tagging a procedure for finding mutations or cloning a gene using the random insertion of reporter gene constructs.

twin spots adjacent patches of somatic cells with different genotypes and phenotypes as a result of mitotic recombination.

yeast artificial chromosome (YAC) a linear cloning vector that can contain more than a megabase of sequence for propagation in yeast cells. In addition to the other DNA, the essential components of a YAC are the centromere, an origin of replication, and telomeres.

yeast one-hybrid (Y1H) assay a genetic method using transcription in yeast to identify all of the proteins that bind to the regulatory region of a gene.

yeast two-hybrid (Y2H) assay a genetic method using transcription in yeast to detect protein–protein interactions by co-expressing both genes on plasmids in the yeast cell.

INDEX

D

E

F

I

J

K

L

M

N

Q

R

S

T